CHEMISTRY IN CONTEXT

6th Edition

Graham Hill

John Holman

Nelson Thornes

Contents

Preface

Aims and intentions

Chemistry in Context has become the 'subject bible' to which students and teachers turn for a clear and authoritative explanation of post-16 and A Level Chemistry. Without sacrificing that authority or scholarship, this new edition is the most radical revision of the book since it was first published.

In a single concise textbook, *Chemistry in Context* presents chemistry as a unified subject using physical principles as a basis for the inorganic and organic sections which follow. Our continued aim has been to show chemistry in its wider context as a relevant and developing science which makes a vital contribution to society, industry and civilisation.

Our revisions for this Sixth Edition take account of syllabus (specification) changes, especially to those used by the very wide international readership. While maintaining the relevance of *Chemistry in Context* to students of the International Baccalaureate, we have paid particular attention to the syllabus provided by Cambridge International Examinations (CIE) and its influence in English-speaking countries throughout the world. At the same time, we have borne in mind the challenges now being faced by ambitious students in the UK with the introduction of A* grades.

Following the removal of material that is no longer required, we have:

▸ added new material to reflect modern developments in chemistry and the content of present A level courses with, for example, new chapters on Polymers, DNA and Proteins,
▸ included more in-text 'Quick questions' to encourage students to read more critically and maximise their understanding,
▸ introduced boxes with 'Definitions' and 'Key Points' to enable students to review and revise the content of each chapter more easily,
▸ added new contextual examples to reflect the increasingly international readership of the book.

In addition to Quick questions within the text there are Review questions suitable for tests, discussion, revision and homework at the end of each chapter and the accompanying website (www.chemistry-in-context.co.uk) contains answers to all these questions.

Our approach

The text draws on experimental evidence to develop key ideas and establish laws and theories. The book is intended to complement a course of practical work and consequently space is not devoted to instructions or suggestions for experiments. However, the companion volume '*Chemistry in Context Laboratory Manual*' would be an ideal source for such work.

Evidence from earlier editions indicates that the clear text, divided into short sections with informative figures, photos, diagrams, worked examples and graphs, makes *Chemistry in Context* suitable for students of all abilities.

In general, Chapters 1 to 19 (excluding Sections 5.12 – 5.16, 10.8 – 10.9, 13.7 – 13.9 and 19.4 – 19.5) cover AS courses where these apply. The exclusions just mentioned and Chapters 20 to 28 cover A2 requirements and complete the A Level course.

From the students and former students who contact us, we know that *Chemistry in Context* is appreciated by thousands of readers who have used it to bring them success in Advanced Level Chemistry. We hope you will find it equally valuable.

We would like to thank Dr Joanna Buckley for her help in producing some parts of this new edition and Sandy Marshall at Nelson Thornes for the efficient way in which she has steered the book through to production.

Graham Hill
John Holman

April 2011

How to get the most from this book

We know that *Chemistry in Context* is used both by independent learners and as a support for teachers. This new edition, with its accompanying website, is designed to provide complete support for students in whatever way they are learning.

We realise that getting ahead with post-16, AS and A Level courses can be a challenge. We have designed the book so that early chapters and initial sections in every chapter build on previous courses whether these are GCSE, IGCSE or AS Level. If you are using the book with your teacher, he or she may direct you to certain parts of the book or certain questions, and you should follow these instructions.

If you are using the book independently, we suggest you work through the text systematically in the following way.

- ▶ Use the Contents list at the front of the book and the Index to help you find the topic you are looking for.
- ▶ The text provides clear explanations and information divided into short sections. Important terms are in bold with concise 'Definitions' and 'Key Points' in the margin which you should learn and understand. The Key Points provide a summary for each chapter which will be very useful when you are revising.

- ▶ In many areas of the text, you will find 'Worked examples' to help you with calculations.
- ▶ At frequent intervals throughout the text there are 'Quick questions' to check your understanding of the ideas and information being covered. Get into the habit of answering all the Quick questions as you go along. You can often find the answers to these questions later in the chapter. And there are answers to the Quick questions on the website.
- ▶ At the end of each chapter there is a full set of 'Review questions' which you can use to check and reinforce your understanding of the work. Full answers to the Review questions can be found on the website – though we are sure you won't want to look at these until you have completed the questions!

In due course, we hope to produce further items to help you, so keep an eye on the website.

We wish you the very best with your studies.

1.1 Atoms and molecules

The first chemist to use the name 'atom' was John Dalton (1766–1844). Dalton used the word 'atom' to mean the smallest particle of an element. He then went on to explain how atoms could combine together to form molecules which he called 'compound atoms'.

For example, chlorine consists of particles of Cl_2 under ordinary conditions, but at very high temperatures these split up to form particles of Cl. So, molecules of chlorine are written as Cl_2 and atoms of chlorine are written as Cl.

⌃ Figure 1.1
John Dalton collecting 'marsh gas' (mainly methane) from rotting vegetation at the bottom of a pond. Dalton was born in 1766 in the village of Eaglesfield in Cumbria. He was the son of a handloom weaver. For most of his life, Dalton lived in Manchester and taught at what was then the Presbyterian College.

Quick Questions

1. Name one substance in Dalton's list of elements (Figure 1.2) which we now know is a compound and not an element.
2. Name an element whose molecules consist of two atoms under normal conditions, like chlorine.
3. Name an element that exists as separate single atoms under normal conditions.
4. How many atoms are there in one molecule of sugar (sucrose), $C_{12}H_{22}O_{11}$?
5. How many different types of atom are there in one molecule of sugar?

Most atoms have a radius of about 10^{-10} m, but the unit used in measuring atomic distances is usually the nanometre (nm).

$$1\,m = 10^9\,nm$$

$$10^{-9}\,m = 1\,nm$$

$$\therefore\ 10^{-10}\,m = 0.1\,nm$$

So, the radii of atoms are about 0.1 nm.

⌃ Figure 1.2
Dalton's symbols for the elements. In 1803, Dalton published his atomic theory. He suggested that all matter was composed of small particles which he called 'atoms'. Later, Dalton went on to suggest symbols for the atoms of different elements as shown above. 'Azote', the second element in Dalton's list, is now called nitrogen.

Atoms, of course, are far too small to be seen even with the most powerful light microscope. However, scientists have used electron microscopes to pick out individual atoms (Figure 1.3).

1.2 Comparing the masses of atoms

Individual atoms are far too small to be weighed, but, in 1919, F.W. Aston invented the mass spectrometer. This gave chemists an accurate method of comparing the relative masses of atoms and molecules.

A mass spectrometer separates atoms and molecules according to their mass and also shows the relative numbers of the different atoms and molecules present. Figure 1.4 shows a diagram of a simple mass spectrometer.

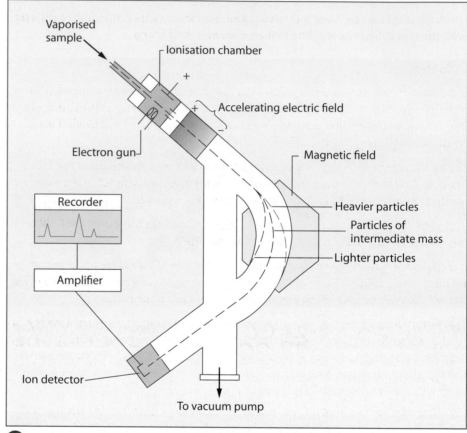

Figure 1.4
A diagram of a simple mass spectrometer

Before atoms can be detected in a mass spectrometer, they must be converted to positive ions in the vapour or gaseous state. Inside a mass spectrometer, there is a vacuum. This allows particles from the chemical under test to be studied without interference from particles in the air.

There are five stages in a mass spectrometer: vaporisation, ionisation, acceleration, deflection and detection.

Vaporisation

Gases, liquids and volatile solids vaporise when injected into the instrument just before the ionisation chamber. Less volatile solids are preheated to help them to vaporise.

Figure 1.3
In the 1990s, two scientists working for IBM introduced a little xenon into an evacuated container containing a small piece of nickel at −269 °C. Some xenon atoms stuck to the surface of the nickel. Then, using a special instrument called a 'scanning tunnelling microscope', they moved individual xenon atoms around the nickel surface to make the IBM logo. Each blue blob is the image of a single xenon atom.

Ionisation

Vaporised atoms and/or molecules pass into the ionisation chamber. Here they are bombarded with a beam of high-energy electrons. These knock electrons off the atoms or molecules in the sample forming positive ions:

$$e^- + X \longrightarrow X^+ + e^- + e^-$$

| high-energy electron | atom in sample | | positive ion | electron knocked out of X | high-energy electron retreating |

Acceleration

Positive ions, such as X^+, are now accelerated by an electric field.

Deflection

The accelerated ions pass into a magnetic field. As the ions pass through the magnetic field, they are deflected according to their mass and their charge.

Detection

If the accelerating electric field and the deflecting magnetic field stay constant, ions of only one particular mass-to-charge ratio will hit the ion detector at the end of the instrument. Ions of smaller mass-to-charge ratio will be deflected too much. Ions of greater mass-to-charge ratio will be deflected too little.

The ion detector is linked through an amplifier to a recorder. As the strength of the magnetic field is slowly increased, ions of increasing mass are detected and a mass spectrum similar to that shown in Figure 1.6 can be printed out.

By first using a reference compound with a known structure and relative molecular mass, the instrument can print a scale on the mass spectrum.

The relative heights of the peaks in the mass spectrum give a measure of the relative amounts of the different ions present. (Strictly speaking, it is the areas under the peaks and not the peak heights which give the relative amounts or abundances.)

Quick Questions

6 How many different ions are detected in the mass spectrum of naturally occurring magnesium in Figure 1.6?
7 What are the relative masses of these different ions?
8 The relative masses of the atoms which formed these ions are virtually the same as the relative masses of the ions. Explain why.
9 Estimate the relative proportions of the different ions.

1.3 Relative atomic masses – the ^{12}C scale

Chemists originally measured atomic masses relative to hydrogen. Hydrogen was chosen initially because it had the smallest atoms and these could be assigned a relative atomic mass of 1. At a later date, when scientists realised that one element could contain atoms of different mass (**isotopes**), it became necessary to choose a single isotope as the reference standard for relative atomic masses. Isotopes are studied more fully in Section 2.4.

In 1961, the isotope carbon-12 (^{12}C) was chosen as the new standard, because carbon is a solid which is much easier to store and transport than hydrogen (which is a gas).

On the ^{12}C scale, atoms of the isotope carbon-12 are assigned a relative atomic mass, or more correctly a **relative isotopic mass**, of exactly 12. So, the relative atomic mass

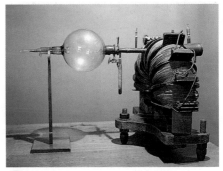

Figure 1.5
Aston's mass spectrometer. A vaporised sample of the element in the glass bulb on the left was bombarded by electrons. The ions produced were then accelerated by an electric field towards the magnetic field on the right (produced by the hundreds of coils in the electromagnet).

Note

Particles can only be attracted and accelerated into the spectrometer if they are **positively charged**.

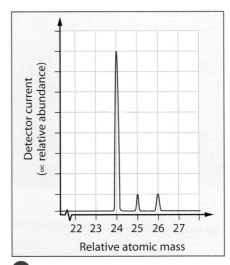

Figure 1.6
A mass spectrometer trace for naturally occurring magnesium

Definition

Isotopes are atoms of the same element with different masses.

Definitions

Chemists use the relative atomic mass scale to compare the masses of different atoms and isotopes. Atoms of the isotope carbon-12 are assigned a relative atomic mass of exactly 12.

The **relative atomic mass** of an element is the average mass of one atom of the element relative to one-twelfth the mass of one atom of carbon-12.

The **relative isotopic mass** of an isotope is the mass of one atom of the isotope relative to one-twelfth the mass of one atom of carbon-12.

The **relative molecular mass** of an element or compound is the sum of the relative atomic masses of all the atoms in its molecular formula.

The **relative formula mass** of a compound is the sum of the relative atomic masses of all the atoms in its formula.

of an element is the average mass of an atom of the element relative to one-twelfth the mass of an atom of the isotope carbon-12.

$$\textbf{Relative atomic mass} = \frac{\text{average mass of an atom of the element}}{\frac{1}{12} \times \text{the mass of one atom of carbon-12}}$$

Using the ^{12}C scale, the relative atomic mass of hydrogen is 1.008 and that of magnesium is 24.312. So, as the symbol for relative atomic mass is A_r, we can write:

$A_r(H) = 1.008$ and $A_r(Mg) = 24.312$ or simply H = 1.008 and Mg = 24.312 for short.

A few other relative atomic masses are listed in Table 1.1. The relative atomic masses of all elements are shown on page 474. In calculations using relative atomic masses, you will only be expected to work to one decimal place.

▼ Table 1.1
The relative atomic masses of some elements

Element	Symbol	Relative atomic mass
Carbon-12	^{12}C	12.000
Carbon	C	12.011
Chlorine	Cl	35.453
Copper	Cu	63.540
Hydrogen	H	1.008
Iron	Fe	55.847
Magnesium	Mg	24.312
Oxygen	O	15.999
Sulfur	S	32.064

Notice in Table 1.1 that the relative atomic mass of carbon is 12.011. This means that the average mass of a carbon atom is 12.011, not 12.000. This is because naturally occurring carbon contains a few atoms of carbon-13 and carbon-14 mixed in with those of carbon-12.

Quick Questions

Use Table 1.1 to answer the following questions:

10 Roughly, how many times heavier are:
   a carbon atoms than hydrogen atoms
   b magnesium atoms than carbon atoms?
11 Which element has atoms approximately twice as heavy as sulfur atoms?
12 Why do the values of relative atomic masses have no units?

Relative atomic masses can also be used to compare the masses of different molecules. These relative masses of molecules are called **relative molecular masses** (symbol M_r).

So, the relative molecular mass of water,

$$M_r(H_2O) = 2 \times A_r(H) + A_r(O) = (2 \times 1.0) + 16.0 = 18.0$$

and the relative molecular mass of chloromethane,

$$M_r(CH_3Cl) = A_r(C) + 3 \times A_r(H) + A_r(Cl) = 12.0 + (3 \times 1.0) + 35.5 = 50.5$$

Metal compounds, such as sodium chloride and copper sulfate, consist of giant structures containing ions, not molecules. So, it would be wrong to use the term 'relative molecular mass' for ionic compounds. Instead, chemists use the term '**relative formula mass**' for ionic compounds.

Quick Questions

13 What is the relative molecular mass of:
   a hydrogen, H_2
   b carbon disulfide, CS_2
   c sulfuric acid, H_2SO_4?
14 What is the relative formula mass of:
   a copper(II) chloride, $CuCl_2$
   b magnesium nitrate, $Mg(NO_3)_2$
   c hydrated iron(II) sulfate, $FeSO_4 \cdot 7H_2O$?

So, the relative formula mass of iron(III) sulfate,

$$M_r(Fe_2(SO_4)_3) = 2 \times A_r(Fe) + 3 \times A_r(S) + 12 \times A_r(O)$$
$$= (2 \times 55.8) + (3 \times 32.1) + (12 \times 16.0) = 399.9$$

1.4 Moles, molar masses and the Avogadro constant

Chemists measure the amount of a substance in moles. The word 'mole' comes from a Latin word meaning a heap or pile. One mole of an element or a compound has a mass equal to its relative atomic mass, its relative molecular mass or its relative formula mass in grams.

So, one mole of iron, Fe, is 55.8 g, one mole of magnesium, Mg, is 24.3 g. One mole of water, H_2O, is 18.0 g and one mole of chloromethane, CH_3Cl, is 50.5 g. Similarly, one mole of iron(III) sulfate, $Fe_2(SO_4)_3$, is 399.9 g.

These masses of 1 mole of different substances are usually called **molar masses** and given the symbol M.

So, the molar mass of iron, $M(Fe) = 55.8$ g per mole, usually written as $55.8 \, g \, mol^{-1}$. The molar mass of water, $M(H_2O) = 18.0 \, g \, mol^{-1}$ and the molar mass of iron(III) sulfate, $M(Fe_2(SO_4)_3) = 399.9 \, g \, mol^{-1}$.

Amounts in moles

Chemists often count in moles because one mole is the mass of 'one formula worth' of a substance. In fact, the mole is the SI unit for *amount* of substance whereas the kilogram is the SI unit for mass.

The *amount* of a substance is therefore measured in moles, which is usually abbreviated to 'mol'. So,

55.8 g of iron contain 1 mol of iron atoms,

111.6 g of iron contain 2 mol of iron atoms and

558.0 g of iron contain 10 mol of iron atoms.

These simple calculations show that:

$$\text{Amount of substance (mol)} = \frac{\text{mass of substance (g)}}{\text{molar mass (g mol}^{-1})}$$

The Avogadro constant

Since one atom of carbon is 12 times as heavy as one atom of hydrogen, it follows that 12 g of carbon will contain the same number of atoms as 1 g of hydrogen. In the same way, one atom of oxygen is 16 times as heavy as one atom of hydrogen, so 16 g of oxygen will also contain the same number of atoms as 1 g of hydrogen.

In fact, the relative atomic mass in grams (i.e. 1 mole) of every element (1 g hydrogen, 12 g carbon, 16 g oxygen, etc.) will contain the same number of atoms. Experiments show that this number is $6.02 \times 10^{23} \, mol^{-1}$. Written out in full, this is 602 000 000 000 000 000 000 000 per mole. The number is usually called the Avogadro constant in honour of the Italian scientist Amedeo Avogadro, and given the symbol L.

$$L = 6.02 \times 10^{23} \, mol^{-1}$$

The Avogadro constant is the number of atoms, molecules or formula units in one mole of any substance. Therefore, 1 mole of iron (55.8 g) contains 6.02×10^{23} Fe atoms, 1 mole of water (18.0 g) contains 6.02×10^{23} H_2O molecules and 1 mole of iron(III) sulfate (399.9 g) contains 6.02×10^{23} $Fe_2(SO_4)_3$ formula units.

△ Figure 1.7
One mole samples of (a) carbon, (b) sulfur, (c) magnesium, (d) copper and (e) iron

Key Point

The relative atomic mass in grams (molar mass) of any element contains 6.02×10^{23} atoms.

Definitions

The **Avogadro constant** ($6.02 \times 10^{23}\,mol^{-1}$) is the number of atoms in exactly 12 g of the isotope carbon-12.

A **mole** is the amount of substance which contains the same number of particles (atoms, molecules or formula units) as there are atoms in exactly 12 g of ^{12}C (i.e. 6.02×10^{23} particles).

The term 'mol' is the symbol for mole. It is *not* an abbreviation for 'molecule' or 'molecular'.

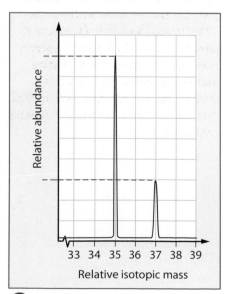

▲ Figure 1.8

One mole of sodium chloride (salt), NaCl (58.5 g), and one mole of water, H_2O (18.0 g)

1 mole of water contains 6.02×10^{23} H_2O molecules, 2 moles of water contain $2 \times 6.02 \times 10^{23}$ H_2O molecules and 10 moles of water contain $10 \times 6.02 \times 10^{23}$ H_2O molecules. Therefore,

$$\frac{\text{Number of atoms,}}{\text{molecules or formula units}} = \text{amount of substance in moles} \times \text{Avogadro constant}$$

Notice how important it is to specify exactly which particles you mean in discussing the number of moles of different substances. For example, the statement 'one mole of oxygen' is ambiguous. It could mean one mole of oxygen atoms (O), i.e. 16.0 g; or it could mean one mole of oxygen molecules (O_2), i.e. 32.0 g. To avoid this ambiguity, it is important to state the formula of the substance involved, for example, 'one mole of oxygen, O_2'.

Quick Questions

15 What is the amount in moles of:
   a 13.95 g of Fe atoms,
   b 7.1 g of Cl_2 molecules,
   c 15.19 g of $FeSO_4$?

16 What is the mass of:
   a one atom of carbon-12,
   b 6.02×10^{23} atoms of copper,
   c 0.5 mol of SCl_2,
   d 0.25 mol of SO_4^{2-} ions?

17 Using the Avogadro constant, calculate the number of:
   a S atoms in 4.125 g of sulfur,
   b N atoms in 8.0 g of NH_4NO_3,
   c SO_4^{2-} ions in 4 mol of $Fe_2(SO_4)_3$.

18 One cubic decimetre of Pellegrino natural mineral water contains 179 mg of calcium ions (Ca^{2+}) and 445 mg of sulfate ions (SO_4^{2-}). What are:
   a these masses of Ca^{2+} and SO_4^{2-} in grams,
   b the amount of Ca^{2+} and SO_4^{2-} in moles,
   c the numbers of Ca^{2+} and SO_4^{2-} ions?

1.5 Using mass spectra to calculate relative atomic masses

Look at the mass spectrum of naturally occurring chlorine in Figure 1.9. This shows that chlorine consists of a mixture of two isotopes with relative isotopic masses of 35 and 37. These isotopes can be labelled ^{35}Cl and ^{37}Cl respectively.

When chlorine is analysed in a mass spectrometer, the beam of ions separates into two paths producing two peaks in its mass spectrum corresponding to the two isotopes.

The heights of the two peaks in Figure 1.9 show that the relative proportions of ^{35}Cl to ^{37}Cl are 3:1. So, of every four chlorine atoms, three are chlorine-35 and one is chlorine-37.

On average, there are three atoms of ^{35}Cl and one atom of ^{37}Cl in every four atoms of chlorine. This is ¾ or 75% chlorine-35 and ¼ or 25% chlorine-37.

So the average mass of a chlorine atom on the ^{12}C scale, which is the relative atomic mass of chlorine is:

$$(¾ \times 35) + (¼ \times 37) = 26.25 + 9.25 = 35.5$$

Mass spectrometers can also be used to study compounds. When molecules of the vaporised compound are bombarded by electrons, some molecules lose just one electron while other molecules break into fragments. Because of the vacuum inside the mass spectrometer, chemists can study these molecular fragments which do not exist under normal conditions.

▲ Figure 1.9

A mass spectrum for naturally occurring chlorine

(graph axes: Relative abundance (vertical); Relative isotopic mass 33 34 35 36 37 38 39 (horizontal))

The peak with the highest mass in the spectrum is produced by ionising molecules without breaking them into fragments. So the mass of a 'molecular ion' gives the relative molecular mass of the compound.

$$e^- + M(g) \longrightarrow M^+(g) + \underbrace{e^- + e}$$

high-energy   molecule in   'molecular   electrons
electron   sample   ion'

By carefully studying the data from mass spectra, chemists can deduce:

► the isotopic composition (abundance) of elements,
► the relative isotopic masses of atoms,
► the relative atomic masses of elements,
► the relative molecular masses of compounds.

After identifying the fragments in mass spectra, chemists can also piece together the fragments and deduce the possible structures of compounds. This technique is particularly useful in determining the structures of newly synthesised compounds (Section 13.7).

Dating the polar ice-caps

The existence of naturally-occurring ^{18}O in addition to ^{16}O has been used to calculate the age of the polar ice-caps. Polar ice sheets have formed over thousands of years as water molecules evaporate from the oceans, move towards the poles and then fall as snow or hail.

However, $H_2{}^{16}O$ molecules with a lower mass than $H_2{}^{18}O$ molecules evaporate and diffuse more readily. So the ice collecting in the polar regions contains a higher proportion of $H_2{}^{16}O$ than water in the oceans. But, the process of evaporation and diffusion of the water molecules is also temperature-dependent and ice which collects at the Poles in winter has a higher proportion of $H_2{}^{16}O$ than ice collecting in the summer.

So, by measuring the ratio of $H_2{}^{16}O$ to $H_2{}^{18}O$ continuously through the depth of the polar ice-caps and counting the annual changes in the ratio $H_2{}^{16}O : H_2{}^{18}O$, the age of the ice-cap can be found. Between 1970 and 1983, a Russian expedition to the Antarctic collected an ice-core to a depth of 2000 metres. The bottom of this ice-core had formed about 150 000 years ago, but it was only about half-way down into the ice sheet.

1.6 Empirical and molecular formulae

It is possible to predict the formulae of most compounds, but the only certain way of knowing a formula is by experiment.

The word 'empirical' means 'from experiment' or 'from experience', so chemists use the term '**empirical formulae**' to describe formulae which have been calculated from the results of experiments.

An experiment to calculate an empirical formula involves three stages:

1 Measure the masses of elements which combine in the compound.
2 Calculate the number of moles of atoms which combine.
3 Calculate the simplest ratio for the atoms which combine.

This gives the empirical formula of the compound.

Worked example

When 10.00 g of ethene was analysed, it was found to contain 8.57 g of carbon and 1.43 g of hydrogen. What is its formula?

Quick Questions

19 Natural silicon in silicon-containing ores contains 92% silicon 28, 5% silicon-29 and 3% silicon-30.
   a What are the relative isotopic masses of the three silicon isotopes?
   b What is the relative atomic mass of silicon?
   c Samples of pure silicon obtained from ores, mined in different parts of the world, have slightly different relative atomic masses. Why is this?

Figure 1.10
After Ben Johnson had won the Men's 100 metres at the Olympic Games in 1988, he was required to give a urine sample for analysis. The test, involving mass spectrometry, showed that he had used illegal anabolic steroids to enhance his performance. Johnson was stripped of his title and had to return his gold medal.

An **empirical formula** shows the simplest whole number ratio for the atoms of each element in a compound.

A **molecular formula** shows the actual number of atoms of each element in one molecule of a compound.

Figure 1.11

This scientist is using a mass spectrometer to find the relative molecular mass of a substance in order to confirm its molecular formula.

Answer

	Carbon	Hydrogen
1 Masses of combined elements:	8.57 g	1.43 g
Molar mass of elements:	12.00 g mol^{-1}	1.00 g mol^{-1}
2 Moles of combined atoms:	$\dfrac{8.57\,\text{g}}{12.00\,\text{g mol}^{-1}}$	$\dfrac{1.43\,\text{g}}{1.00\,\text{g mol}^{-1}}$
=	0.714 mol	1.43 mol
3 Ratio of combined atoms:	$\dfrac{0.714}{0.714} = 1$	$\dfrac{1.43}{0.714} = 2$

Therefore, the empirical formula of ethene is CH_2.

This formula for ethene shows only the *simplest* ratio of carbon atoms to hydrogen atoms. The actual formula showing the correct number of carbon atoms and hydrogen atoms in one molecule of ethene could be CH_2, C_2H_4, C_3H_6, C_4H_8, etc. because all these formulae give CH_2 as the simplest ratio of atoms.

Experiments show that the relative molecular mass of ethene is 28, which corresponds to an actual formula of C_2H_4 and not CH_2. Formulae, such as C_2H_4 for ethene, which show the actual number of atoms of each element in one molecule of a compound are called **molecular formulae**.

The empirical formulae of some compounds like ethene can be calculated using combustion data in place of their composition by mass. The combustion data is obtained by burning the compound in oxygen to produce carbon dioxide and water. From the masses of carbon dioxide and water produced, it is possible to calculate the masses of carbon and hydrogen combined in the compound. These masses of carbon and hydrogen can then by used to calculate the empirical formula of the compound. Study the worked example below to see how this is done.

Worked example

A sample of a compound containing only carbon and hydrogen was burned completely in oxygen. All the carbon was converted to 3.38 g of carbon dioxide and all the hydrogen was converted to 0.692 g of water. What are the masses of carbon and hydrogen combined in the compound?

Answer

One mole of CO_2 contains one mole of C.

So, 44 g CO_2 contains 12 g of C.

$$\text{Fraction of C in } CO_2 = \frac{12}{44} = \frac{3}{11}$$

$$\therefore \textbf{ Mass of C in 3.38 g } CO_2 = 3.38 \times \frac{3}{11} = \textbf{0.92 g}$$

1 mole of H_2O contains 2 moles of H.

So, 18 g H_2O contains 2 g of H.

$$\text{Fraction of H in } H_2O = \frac{2}{18} = \frac{1}{9}$$

$$\therefore \textbf{ Mass of H in 0.692 g } H_2O = 0.692 \times \frac{1}{9} = \textbf{0.077 g}$$

20 Use the masses of carbon and hydrogen from the worked example on the right to calculate the empirical formula of the compound.

1.7 Predicting and writing chemical equations

When methane in natural gas burns on a hob, it reacts with oxygen in the air to form carbon dioxide and water. A word equation for the reaction is:

$$\text{methane} + \text{oxygen} \longrightarrow \text{carbon dioxide} + \text{water}$$

Word equations like this give the names of the reactants and products, but as chemists we should always aim to write **balanced chemical equations** using symbols and formulae.

There are three key stages in writing chemical equations:

1 Write a word equation:

$$\text{methane} + \text{oxygen} \longrightarrow \text{carbon dioxide} + \text{water}$$

2 Write symbols for elements and formulae for compounds in the word equation:

$$CH_4 + O_2 \longrightarrow CO_2 + H_2O$$

Remember that the elements oxygen, hydrogen, nitrogen and halogens exist as diatomic molecules containing two atoms, so they are written as O_2, H_2, N_2, Cl_2, Br_2 and I_2 in equations. All other elements are written as single atoms (e.g. Cu, C, Fe) in equations.

3 Balance the equation by writing numbers in front of the symbols and formulae in order to have the same number of each kind of atom on both sides of the equation.

In the equation above in Stage 2, there are 4 hydrogen atoms on the left and only 2 on the right. Therefore, H_2O on the right must be doubled ($2H_2O$). There are now 2 oxygen atoms in O_2 on the left, but 4 oxygen atoms in CO_2 and $2H_2O$ on the right. So, O_2 on the left must be doubled ($2O_2$). The balanced equation is:

$$CH_4 + 2O_2 \longrightarrow CO_2 + 2H_2O$$

Remember that *formulae must never be altered* in balancing an equation. The formula of methane is always CH_4 and never CH_2 or C_2H_4. Similarly, carbon dioxide is always CO_2 and water is always H_2O. Atoms in an equation can only be balanced by putting a number in front of a symbol or a formula, thus doubling or trebling, etc. the whole formula.

Normally state symbols are also included, so the final equation is:

$$CH_4(g) + 2O_2(g) \longrightarrow CO_2(g) + 2H_2O(l)$$

⬆ Figure 1.12
Methane (natural gas) burning on a hob

Key Point

Note the three stages in writing a balanced chemical equation:
1 writing a word equation
2 writing symbols for elements and formulae for compounds
3 balancing the equation .

Quick Questions

21 Write balanced chemical equations for the following word equations:
   a copper + oxygen $\longrightarrow$ copper(II) oxide
   b potassium + oxygen $\longrightarrow$ potassium oxide
   c sodium + water $\longrightarrow$ sodium hydroxide + hydrogen
   d magnesium + hydrochloric acid $\longrightarrow$ magnesium chloride + hydrogen
   e copper(II) oxide + sulfuric acid $\longrightarrow$ copper(II) sulfate + water
   f hydrogen + oxygen $\longrightarrow$ water

A balanced chemical equation tells us

▸ the reactants and products
▸ the formulae of reactants and products
▸ the relative amounts in moles of the reactants and products.

For example, the equation $CH_4(g) + 2O_2(g) \longrightarrow CO_2(g) + 2H_2O(l)$ tells us that:

▸ Methane reacts with oxygen to form carbon dioxide and water.
▸ The formula of methane is CH_4, oxygen is O_2, carbon dioxide is CO_2 and water is H_2O.
▸ 1 mole of methane (CH_4) reacts with 2 moles of oxygen molecules ($2O_2$) to produce 1 mole of carbon dioxide (CO_2) and 2 moles of water ($2H_2O$).

In addition, we can show the states of reactants and products by putting state symbols in the equation.

Using relative atomic masses it is also possible to calculate the relative masses of reactants and products involved in reactions (Section 1.8).

Chemical equations are very useful, but it is important to appreciate their limitations. An equation cannot tell us:

▶ the rate of the reaction – whether it is fast or slow, whether it will explode or take years to react
▶ how or why the reaction takes place.

For example, when natural gas (methane) burns, the balanced equation for the reaction is:

$$CH_4(g) + 2O_2(g) \longrightarrow CO_2(g) + 2H_2O(l)$$

The equation does not show that a spark or a flame is needed to start the reaction. Nor does it show that once the reaction has started it will continue as long as both methane and oxygen are available. The equation does not tell us why the methane reacts with oxygen or how the molecules and atoms behave in the reaction. Does the oxygen react with methane as O_2 molecules or split into separate oxygen atoms which then react? Also how and when do the carbon and hydrogen atoms in methane break apart? The equation cannot answer these questions.

Equations can only tell us about the overall chemical change – the amounts and states of the reactants and products. They cannot tell us anything about what happens on the pathway between the reactants and the products.

 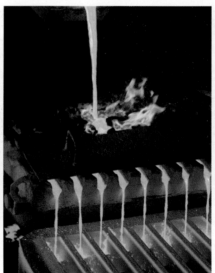 

△ Figure 1.13
The photographs show three stages in the manufacture of copper from copper ore (copper pyrites). The left-hand photo shows the huge open-cast copper mine at Brigham near Salt Lake City, USA. The ore is crushed and then reduced in furnaces to produce molten copper (centre). This can be made into solid bars or rolls of cable (right).

1.8 Calculations involving formulae and equations

In industry and research and everywhere that chemists carry out reactions, it is useful and sometimes essential to know how much product can be obtained from a particular amount of reactants. Chemists can calculate these quantities using formulae, equations and relative atomic masses.

Calculating the masses of reactants and products

There are four stages in calculating the masses of reactants and products in a chemical reaction:

1 Write a balanced equation for the reaction.
2 Note the amounts in moles of relevant reactants and products in the equation.
3 Calculate the masses of relevant reactants and products using relative atomic masses.
4 Scale the masses of relevant reactants and products to the required quantities.

Worked example

Calculate the mass of lime (calcium oxide, CaO) that can be obtained from 1 tonne (1000 kg) of pure limestone (calcium carbonate, $CaCO_3$).

Answer

1 $CaCO_3(s) \longrightarrow CaO(s) + CO_2(g)$

2 $1 \text{ mole } CaCO_3 \longrightarrow 1 \text{ mole } CaO$

3 $(40.1 + 12.0 + (3 \times 16.0)) \text{ g } CaCO_3 \longrightarrow (40.1 + 16.0) \text{ g } CaO$

$$100.1 \text{ g } CaCO_3 \longrightarrow 56.1 \text{ g } CaO$$

4 $\qquad 1 \text{ g } CaCO_3 \longrightarrow \dfrac{56.1}{100.1} \text{ g } CaO$

$\therefore 1000 \text{ kg } CaCO_3 \longrightarrow \dfrac{56.1}{100.1} \times 1000 \text{ kg } CaO$

$$= \textbf{560.4 kg of CaO (lime)}$$

▲ Figure 1.14
Lime kilns are used to decompose limestone to lime at high temperatures. This shows a traditional lime kiln in India.

Quick Questions

23 Iron reacts with chlorine to form iron(III) chloride:
   a How much chlorine reacts with 1.86 g of iron?
   b What mass of iron(III) chloride is produced?
24 A typical jumbo jet burns fuel at a rate of 200 kg per minute. Assuming jet fuel is $C_{11}H_{24}$, what mass of carbon dioxide does a jumbo jet produce per minute?
25 Aluminium is manufactured by electrolysis (electrolytic decomposition) of bauxite, Al_2O_3. How much aluminium can be obtained from 1 kg of pure bauxite?

Calculating the volumes of gases

The volume of any gas depends on its temperature and pressure.

Repeated measurements with different gases show that the volume of 1 mole of every gas occupies approximately 24 dm^3 ($24\,000 \text{ cm}^3$) under room conditions ($25\,°C$ and atmospheric pressure of 101 kPa). This volume of one mole of gas is described as the **molar volume** (symbol V_m). So, $V_m = 24 \text{ dm}^3 \text{mol}^{-1}$ under laboratory conditions at room temperature and pressure.

At standard temperature and pressure (s.t.p.), i.e. 273 K ($0\,°C$) and 101 kPa pressure, the molar volume is approximately 22.4 dm^3.

So under laboratory conditions, 1 mol of hydrogen, 1 mol of chlorine and 1 mol of methane all occupy 24 dm^3; 2 mol of each gas will occupy 48 dm^3 and 0.1 mol of each gas will occupy 2.4 dm^3, i.e.

$$\text{Amount of gas(mol)} = \frac{\text{volume of gas(dm}^3)}{\text{molar volume(dm}^3\,\text{mol}^{-1})}$$

Figure 1.15 shows the apparatus that can be used to measure the volume of hydrogen produced when magnesium reacts with dilute hydrochloric acid.

The **molar volume** of a gas is the volume of 1 mole. Under laboratory conditions, the molar volume of all gases is approximately $24\,dm^3\,mol^{-1}$ ($24\,000\,cm^3\,mol^{-1}$). At standard temperature and pressure (s.t.p), 273 K and 101 kPa pressure, the molar volume of all gases is approximately $22.4\,dm^3\,mol^{-1}$.

Quick Questions

26 Calculate the amount in moles at room temperature and pressure of:
   a $960\,cm^3$ of chlorine,
   b $144\,dm^3$ of methane,
   c $120\,cm^3$ of oxygen.

27 Calculate the volumes of the following gases at room temperature and pressure:
   a 5 mol of hydrogen,
   b 6.4 g of oxygen.

28 a Write an equation for the complete combustion of propane, C_3H_8 (Calor gas) with oxygen.
   b What volume of oxygen reacts with $2\,dm^3$ of propane and what volume of carbon dioxide is produced? (Assume measurements are made at room temperature and pressure.)

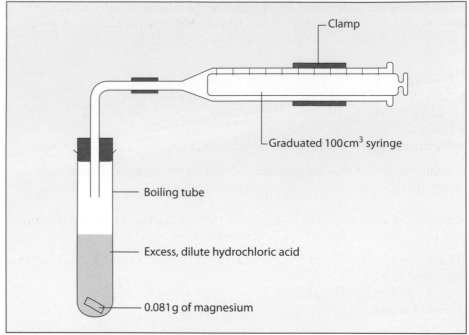

Figure 1.15
Measuring the volume of hydrogen produced when magnesium reacts with hydrochloric acid

When the 0.081 g of magnesium ribbon is added to the excess acid, a vigorous reaction occurs and hydrogen is evolved. When all the magnesium has reacted, effervescence stops and the syringe shows that $80\,cm^3$ of gas have been produced.

So, 0.081 g of Mg react with HCl(aq) to produce $80\,cm^3$ of H_2.

Now, $0.081\,g\ of\ Mg = \dfrac{0.081\,g}{24.3\,g\,mol^{-1}} = 0.0033\,mol\ Mg$

and $80\,cm^3\ of\ H_2 = \dfrac{80\,cm^3}{24\,000\,cm^3\,mol^{-1}} = 0.0033\,mol\ H_2$

$\therefore$ 0.0033 mol Mg produces 0.0033 mol H_2.

So, 1 mol Mg produces 1 mol H_2.

Assuming that the other product of the reaction is magnesium chloride, $MgCl_2$, the overall equation for the reaction is:

$$Mg(s) + 2HCl(aq) \longrightarrow MgCl_2(aq) + H_2(g)$$

Calculating the concentrations and volumes of solutions

As many reactions only take place in aqueous solution, it is important for chemists to know the concentrations of these solutions. Chemists usually measure the concentration of solutions in moles of solute per cubic decimetre ($mol\,dm^{-3}$) of solution.

$1\ decimetre\ (dm) = \frac{1}{10}\,metre = 10\,cm$

and $\quad 1\,dm^3 = (10\,cm)^3 = 1000\,cm^3$

So, a solution of sodium chloride, NaCl(aq), containing $1.0\,mol\,dm^{-3}$ has 1 mole of sodium chloride (58.5 g NaCl) in $1\,dm^3$ ($1000\,cm^3$) of solution. A solution of NaCl containing $0.2\,mol\,dm^{-3}$ contains 11.7 g ($0.2 \times 58.5\,g$) in $1\,dm^3$ of solution.

Notice that these concentrations are expressed as the number of moles in $1\,dm^3$ of *solution*, not $1\,dm^3$ of *solvent*. Here is the method which you should use to prepare $1\,dm^3$ of $1.0\,mol\,dm^{-3}$ sodium chloride solution (Figure 1.16).

▶ Weigh out 1 mol of NaCl (58.5 g).
▶ Dissolve this in about $500\,cm^3$ of distilled water in a beaker.
▶ Add this solution to the $1\,dm^3$ volumetric flask plus washings from the beaker.
▶ Add more distilled water up to the $1\,dm^3$ mark on the neck of the flask.
▶ Finally, mix the solution thoroughly.
▶ Can you see that $1\,dm^3$ of $1.0\,mol\,dm^{-3}$ NaCl contains less than $1\,dm^3$ of water?

A solution of sodium chloride containing $1.0\,mol\,dm^{-3}$ is usually written as:

$$[NaCl] = 1.0\,mol\,dm^{-3} \text{ or simply } [NaCl] = 1.0\,M$$

Notice that square brackets around a formula are used to indicate the concentration of a substance and that $mol\,dm^{-3}$ can be abbreviated to M.

The amounts of solute needed to make solutions of different concentration and volume can be worked out by simple proportion.

Worked example

What mass of solute is needed to make $250\,cm^3$ of a solution containing $0.1\,mol\,dm^{-3}$ sodium chloride, NaCl?

$1\,dm^3$ of 1.0 M NaCl(aq) contains	1 mol NaCl
$250\,cm^3$ of 1.0 M NaCl(aq) contains	$1 \times \dfrac{250}{1000}$ mol NaCl
$250\,cm^3$ of 0.1 M NaCl(aq) contains	$1 \times \dfrac{250}{1000} \times 0.1$ mol NaCl

$$= 0.025\,mol = 0.025\,mol \times 58.5\,g\,mol^{-1} = 14.625\,g\ NaCl$$

When ionic compounds like NaCl and H_2SO_4 dissolve, they are fully dissociated into **ions**.

For example: $\qquad H_2SO_4(aq) \longrightarrow 2H^+(aq) + SO_4^{2-}(aq)$

So, if the concentration of H_2SO_4 is $1\,mol\,dm^{-3}$, the concentration of H^+ ions is $2\,mol\,dm^{-3}$ and that of SO_4^{2-} is $1\,mol\,dm^{-3}$.

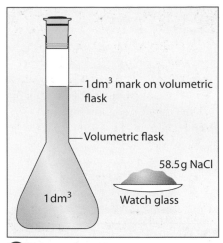

Figure 1.16
Preparing $1\,dm^3$ of $1.0\,mol\,dm^{-3}$ sodium chloride solution

1 dm³ mark on volumetric flask
Volumetric flask
58.5 g NaCl
1 dm³
Watch glass

Quick Questions

29 What is the concentration in $mol\,dm^{-3}$ of a solution containing:
  a $22.22\,g$ of calcium chloride, $CaCl_2$, in $500\,cm^3$ of solution,
  b $39.63\,g$ of ammonium sulfate, $(NH_4)_2SO_4$ in $400\,cm^3$ of solution,
  c $24.96\,g$ of hydrated copper sulfate, $CuSO_4 \cdot 5H_2O$ in $250\,cm^3$ of solution?
30 What mass of solute is required to make:
  a $3\,dm^3$ of $3.0\,mol\,dm^{-3}$ sodium hydroxide, NaOH,
  b $200\,cm^3$ of $0.2\,mol\,dm^{-3}$ sulfuric acid,
  c $50\,cm^3$ of $2.0\,mol\,dm^{-3}$ potassium carbonate, K_2CO_3?

1.9 Determining equations

There are three key stages in determining the equation for a particular reaction:

1 Find by experiment the masses or volumes of the reactants and products involved in the reaction.
2 Convert these masses or volumes to amounts in moles of the substances concerned.
3 Calculate the simplest whole number ratios for the amounts in moles of substances involved in the reaction.

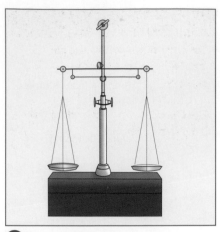

Figure 1.17
A drawing of the original balance which Dalton used to study the reacting quantities of different substances

Quick Questions

31 3.81 g of copper reacted with oxygen to produce 4.29 g of a copper oxide. Calculate a formula for the copper oxide and then write an equation for the reaction.

32 25 cm³ of a gas containing only nitrogen and oxygen decomposed to form 25 cm³ of nitrogen and 50 cm³ of oxygen. All the volumes were measured at the same temperature and pressure. Write an equation for the reaction.

33 18 cm³ of 1.0 M H_2SO_4 just reacted with 24 cm³ of 1.5 M NaOH to form sodium sulfate and water. Calculate the amounts in moles of sulfuric acid and sodium hydroxide reacting and write an equation for the reaction.

Worked example

0.27 g of aluminium reacts with 2.40 g of bromine to form aluminium bromide. What is the equation for the reaction?

Answer

1 Assuming the formula of aluminium bromide is $AlBr_3$, we know that:

$$0.27 \text{ g Al} + 2.40 \text{ g Br}_2 \longrightarrow 2.67 \text{ g AlBr}_3$$

2   Moles of Al reacting $= \dfrac{0.27}{27.0} \text{ mol} = 0.010 \text{ mol Al}$

   Moles of Br_2 reacting $= \dfrac{2.40}{2 \times 79.9} \text{ mol} = 0.015 \text{ mol Br}_2$

   Moles of $AlBr_3$ produced $= \dfrac{2.67}{27.0 + (3 \times 79.9)} = 0.010 \text{ mol AlBr}_3$

3 Ratio of moles of substances involved:

$Al : Br_2 : AlBr_3 = 0.010 : 0.015 : 0.010$

$$= \dfrac{0.010}{0.010} : \dfrac{0.015}{0.010} : \dfrac{0.010}{0.010}$$

$$= 1 \quad : \quad 1.5 \quad : \quad 1$$

Therefore, the simplest whole number ratio of $Al : Br_2 : AlBr_3 = 2 : 3 : 2$

So, the equation is: $2Al + 3Br_2 \longrightarrow 2AlBr_3$

1.10 Ionic equations

Many reactions involve ionic compounds. The part played in a reaction by the separate ions of these compounds can often be shown more clearly using an ionic equation. Here are five important types of reaction where ionic equations can be used.

(**Remember:** The only substances which contain ions are compounds of metals with non-metals (salts and bases) and aqueous acids.)

1 The reactions of metals with non-metals

When magnesium reacts with oxygen, the product is magnesium oxide – a solid ionic compound, $Mg^{2+}O^{2-}(s)$:

$$2Mg(s) + O_2(g) \longrightarrow 2Mg^{2+}O^{2-}(s)$$

2 The reactions of metals with acids

Earlier in this chapter we studied the reaction between magnesium and hydrochloric acid, HCl(aq), forming magnesium chloride solution, $MgCl_2(aq)$, and hydrogen.

The HCl(aq) and $MgCl_2(aq)$ are solutions of ionic compounds, fully dissociated into separated ions which are free to move apart. So, we can write an ionic equation as:

$$Mg(s) + 2H^+(aq) + 2Cl^-(aq) \longrightarrow Mg^{2+}(aq) + 2Cl^-(aq) + H_2(g)$$

By cancelling the Cl^- spectator ions which appear on both sides of the equation and take no part in the reaction, we get

$$Mg(s) + 2H^+(aq) \longrightarrow Mg^{2+}(aq) + H_2(g)$$

3 The reactions between acids and bases (neutralisation)

When hydrochloric acid reacts with sodium hydroxide solution, the equation for the reaction is:

$$HCl(aq) + NaOH(aq) \longrightarrow NaCl(aq) + H_2O(l)$$

However, HCl(aq), NaOH(aq) and NaCl(aq) consist of dissociated ions in aqueous solutions, so we can write an ionic equation as:

$$H^+(aq) + Cl^-(aq) + Na^+(aq) + OH^-(aq) \longrightarrow Na^+(aq) + Cl^-(aq) + H_2O(l)$$

Cancelling the Cl^- and Na^+ spectator ions on both sides of the equation gives:

$$H^+(aq) + OH^-(aq) \longrightarrow H_2O(l)$$

4 The precipitation of insoluble ionic solids

When solutions of silver nitrate and sodium chloride are mixed, an insoluble precipitate of white silver chloride forms:

$$AgNO_3(aq) + NaCl(aq) \longrightarrow AgCl(s) + NaNO_3(aq)$$

Ions in the aqueous solutions are fully dissociated, but those in the precipitate cling together as a solid and we write this in ionic equations as $Ag^+Cl^-(s)$. So, the ionic equation for the reaction is:

$$Ag^+(aq) + NO_3^-(aq) + Na^+(aq) + Cl^-(aq) \longrightarrow Ag^+Cl^-(s) + Na^+(aq) + NO_3^-(aq)$$

By cancelling spectator ions in this case, we get:

$$Ag^+(aq) + Cl^-(aq) \longrightarrow Ag^+Cl^-(s)$$

5 The reactions at electrodes during electrolysis

When liquids are electrolysed, the reactions which take place at the electrodes involve ions. These reactions which occur during electrolysis are discussed in Sections 6.4 to 6.5.

Quick Questions

34 Write ionic equations for the following reactions:
  a Iron + chlorine $\longrightarrow$ iron(III) chloride
  b Zinc + nitric acid $\longrightarrow$ zinc nitrate + hydrogen
  c Sodium + water $\longrightarrow$ sodium hydroxide + hydrogen
  d Potassium hydroxide + sulfuric acid $\longrightarrow$ potassium sulfate + water

Review questions

1 What amount in moles of:
  a Cl_2 are there in 7.1 g of chlorine,
  b $CaCO_3$ are there in 10.0 g of calcium carbonate,
  c Ag are there in 10.8 g of silver,
  d NH_3 are there in 3.4 g of ammonia,
  e S are there in 32 g of sulfur,
  f S_8 are there in 32 g of sulfur?

2 How many atoms are there in:
  a two moles of iron, Fe,
  b 0.1 moles of sulfur, S,
  c 18 g of water, H_2O,
  d 0.44 g of carbon dioxide, CO_2?

3 Figure 1.18 shows a mass spectrometer print-out for neon.

▶ Figure 1.18
A mass spectrometer print-out for neon

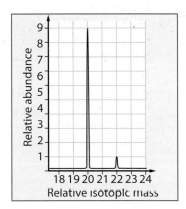

a How many isotopes are present in neon?
b What are the relative isotopic masses of the neon isotopes?
c What are the relative abundances of the neon isotopes?
d Calculate the relative atomic mass of neon.

4 A sample of bromine containing isotopes bromine-79 and bromine-81 was analysed in a mass spectrometer:

a How many peaks corresponding to Br^+ were recorded?

A 1   B 2   C 3   D 4

b How many peaks corresponding to Br_2^+ were recorded?

A 1   B 2   C 3   D 4

5 Figure 1.19 shows the mass spectrum of HCl. The peak at mass 36 corresponds to the molecular ion, $(H\,^{35}Cl)^+$. Assume that chlorine has only two isotopes, ^{35}Cl and ^{37}Cl.

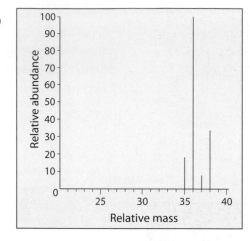

Figure 1.19

a What particle is responsible for the prominent peak at mass 38?

b What particles are responsible for the two peaks showing lower abundance?

c Explain the relative heights of the peaks at masses 36 and 38.

6 A garden fertiliser contains 15% by mass of nitrogen. The recommended usage is 30 g of fertiliser per 10 dm^3 of water. What is the concentration of nitrogen in this solution?

A 0.15 g dm^{-3} B 0.30 g dm^{-3}

C 0.45 g dm^{-3} D 3.00 g dm^{-3}

7 5.34 g of a salt of formula M_2SO_4 (where M is a metal) were dissolved in water. The sulfate ion was precipitated by adding excess barium chloride solution when 4.66 g of barium sulfate ($BaSO_4$) were obtained.

a How many moles of sulfate ion were precipitated as barium sulfate?

b How many moles of M_2SO_4 were in the solution?

c What is the formula mass of M_2SO_4?

d What is the relative atomic mass of M?

e Use a table of relative atomic masses to identify M.

8 2.4 g of a compound of carbon, hydrogen and oxygen gave on combustion, 3.52 g of CO_2 and 1.44 g of H_2O. The relative molecular mass of the compound was found to be 60.

a What are the masses of carbon, hydrogen and oxygen in 2.4 g of the compound?

b What are the empirical and molecular formulae of the compound?

9 A hydrate of potassium carbonate has the formula $K_2CO_3{\cdot}xH_2O$. 10.00 g of the hydrate leave 7.93 g of anhydrous salt on heating.

a What is the mass of anhydrous salt in 10 g of hydrate?

b What is the mass of water in 10 g of hydrate?

c How many moles of anhydrous salt are there in 10 g of hydrate?

d How many moles of water are there in 10 g of hydrate?

e How many moles of water are present in the hydrate for every 1 mole of K_2CO_3?

f What is the formula of the hydrate?

10 25 cm^3 of a solution of NaOH required 28 cm^3 of 1.0 mol dm^{-3} H_2SO_4 to neutralise it:

a Write an equation for the reaction.

b How many moles of H_2SO_4 were needed?

c How many moles of NaOH were thus neutralised?

d How many moles of NaOH are there in 25 cm^3 of solution?

e What is the concentration of the NaOH in mol dm^{-3}?

11 By calculating relative molecular masses, find whether it is cheaper to buy washing soda (sodium carbonate) as the decahydrate ($Na_2CO_3{\cdot}10H_2O$) at 4p per kg, or as the anhydrous salt (Na_2CO_3) at 8p per kg.

2 Atomic and electronic structure

2.1 Introduction

One hundred and fifty years ago in the 1860s, scientists still believed that atoms were tiny, solid spheres like snooker balls just as Dalton had predicted. Since then, they have obtained a great deal of evidence about the detailed structure of atoms.

Early in the nineteenth century, experiments involving electrolysis suggested that certain compounds contained **ions**. The formation of these ions from atoms could be explained by the loss or gain of negatively charged particles called electrons. This led to the idea that atoms consisted of a small positive nucleus surrounded by negative electrons.

Later, scientists discovered that the nuclei of atoms contained two kinds of particle – protons and neutrons. But, what is the evidence for these particles?

2.2 Evidence for atomic structure

1897 Thomson experiments with electrons

⌃ Figure 2.1
J. J. Thomson (1856–1940) investigating the conductivity of electricity by gases. Using discharge tubes (like that on the right of this photo), Thomson discovered electrons.

In 1897, J. J. Thomson was investigating the conduction of electricity by gases at very low pressure. When Thomson applied 15 000 volts across the electrodes of a tube containing a trace of gas, a bright green glow appeared on the fluorescent screen at the end of the tube. The green glow results from 'rays' travelling in straight lines from the negative terminal until they hit the fluorescent screen. Thomson called these 'rays' cathode rays (Figure 2.2). Experiments showed that a narrow beam of the 'rays' could be deflected by an electric field. When passed between charged plates, the 'rays' were always attracted towards the positive plate. This suggested that the 'rays' were negatively charged.

Key Point

Cathode rays consist of **electrons**: tiny negatively charged particles about 1840 times lighter than hydrogen atoms.

Quick Questions

1 Where do you think the electrons in the cathode rays have come from?
2 Why do the cathode rays cause the fluorescent screen to glow?
3 Which piece of Thomson's evidence showed that cathode rays could not be rays of light?
4 Assume that the Avogadro constant, $L = 6 \times 10^{23}\,\text{mol}^{-1}$. What is the mass of:
   a 6×10^{23} atoms of hydrogen,
   b 1 atom of hydrogen,
   c 1 electron?

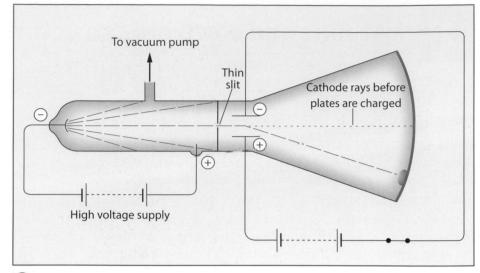

Figure 2.2
The effect of charged plates on cathode rays

Further study showed that the cathode rays were not electromagnetic radiation, but consisted of **electrons** – tiny negatively charged particles about 1840 times lighter than hydrogen atoms.

Thomson obtained electrons in further experiments with different gases in the tube and when the electrodes were made of different materials. This suggested that electrons were present in the atoms of *all* substances.

1899 Thomson's model of atomic structure

Thomson knew that atoms had no overall charge. So, as a result of his experiments, he suggested that atoms consisted of negative electrons embedded in a sphere of positive charge. Thomson believed, wrongly of course, that the mass of an atom was due only to its electrons. So, as one electron was $\frac{1}{1840}$ th of the mass of a hydrogen atom, Thomson concluded that each hydrogen atom must contain 1840 electrons.

Thomson's model of atomic structure was compared to a Christmas pudding. The electrons in a sphere of positive charge were like currants in a Christmas pudding.

1909 Rutherford and his colleagues explore the atom

Thomson's ideas of atomic structure led scientists to investigate how electrons were arranged inside atoms. Ernest Rutherford, who had been one of Thomson's research students at Cambridge University, had the idea of probing inside atoms using alpha-particles. Rutherford used alpha-particles from radioactive substances as 'nuclear bullets'. Alpha-particles are helium ions, He^{2+}, they are small, relatively heavy and positively charged. Two of Rutherford's colleagues at Manchester University, Hans Geiger and Ernest Marsden directed narrow beams of alpha-particles at very thin metal foil only a few atoms thick (Figure 2.3). Using Thomson's model of atomic structure, Rutherford expected most of the alpha-particles to pass straight through the foil or to be deflected slightly.

The results of Geiger and Marsden's experiment showed that:

▶ Most alpha-particles passed straight through the foil.
▶ Some of the alpha-particles were deflected by the foil.
▶ But, to everyone's surprise, one particle in every 10 000 appeared to rebound from the foil.

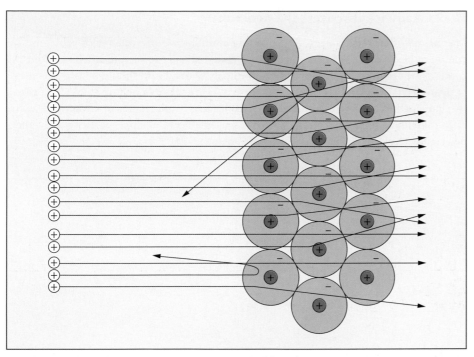

Figure 2.3
Investigating the deflection of alpha-particles by thin metal foil

Quick Questions

5 Why do most alpha-particles pass straight through the metal foil?
6 Why do some alpha-particles appear to rebound from the metal foil?
7 What factors do you think will affect the extent to which alpha-particles are deflected?

1911 Rutherford explains the structure of atoms

Alpha-particles are positively charged. This led Rutherford to suggest that some could rebound only if they came close to a concentrated region of positive charge.

Only a very small fraction of the alpha-particles rebounded, so Rutherford concluded that atoms in the metal foil consisted of a very small central nucleus composed of positively charged **protons**, where the mass of the atom was concentrated. This nucleus was then surrounded by a much larger region in which the electrons moved.

From the angles through which alpha-particles were deflected and rebounded, Rutherford calculated that the nucleus of an atom had a radius of about 10^{-14} m. This is about one ten-thousandth of the radius (10^{-10} m) of the whole atom.

Notice the difference between Rutherford's atomic model and that proposed by Thomson in 1899. Rutherford's atomic model has been compared to the Solar System. Rutherford pictured each atom as a miniature solar system with electrons orbiting the nucleus, like planets orbiting the Sun.

Rutherford also suggested that hydrogen, which has the smallest atoms, would have *one* proton in its nucleus, balanced by *one* orbiting electron. Atoms of helium which were next in size would have *two* protons in their nucleus balanced by *two* orbiting electrons and so on.

Although Rutherford's idea of planetary electrons has now been discarded, his idea of a small positive nucleus has been supported by many experiments.

Figure 2.4
These two might look like a pair of very ordinary old men from the 1930s, but nothing could be further from the truth. Thomson (left) and Rutherford are unquestionably two of the most brilliant scientists of the twentieth century.

Key Point

The nucleus is tiny compared to the size of a whole atom. The volume of the nucleus is only about one million millionth (i.e. 10^{-12}) of the total volume of an atom.

1932 Chadwick discovers the neutron

In spite of Rutherford's success in explaining atomic structure, one major problem could not be explained. If a hydrogen atom contains one proton and a helium atom contains two protons, then the relative atomic mass of helium should be two if that of hydrogen is one. Unfortunately, the relative atomic mass of helium is four and not two. James Chadwick, one of Rutherford's colleagues, was able to show where the extra mass in helium atoms came from. Chadwick used a beam of alpha-particles which he detected with a charged particle counter (Figure 2.5 (a)).

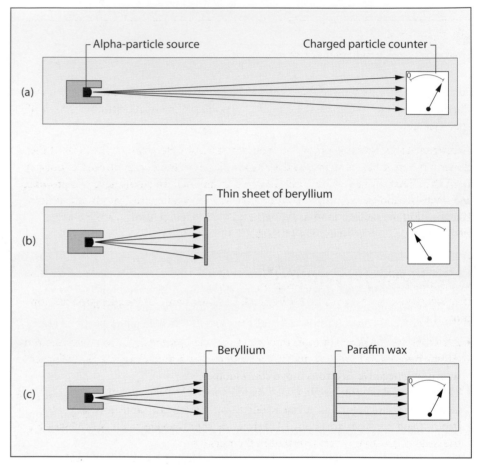

Figure 2.5
The experiment in which Chadwick found neutrons

When Chadwick placed a thin sheet of beryllium across the beam of alpha-particles, the counter registered nothing (Figure 2.5 (b)), showing that the alpha-particles are being stopped by the beryllium. However, when a piece of paraffin wax is placed between the beryllium and the counter, charged particles are detected again (Figure 2.5 (c)).

It seems that alpha-particles are stopped by the beryllium foil, yet charged particles are shooting out from the paraffin wax. How can this happen? What causes charged particles to be ejected from the paraffin wax? Chadwick provided an explanation.

Alpha particles striking the beryllium foil displace *uncharged* particles called **neutrons** from the nuclei of beryllium atoms. These uncharged neutrons cannot be detected by the charged particle counter, but they can displace positively charged protons from the paraffin wax which then affect the counter. A summary of Chadwick's explanation is shown in Figure 2.6.

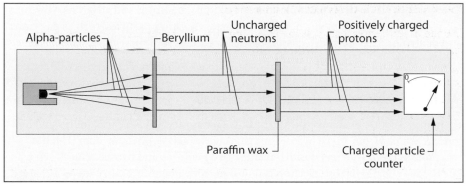

 Figure 2.6
Chadwick's explanation of his discovery of neutrons

Key Point

Nuclei contain two types of particle:

- protons with a relative mass of 1 and a relative charge of +1
- neutrons with a relative mass of 1 and a relative charge of 0.

Further experiments showed that neutrons had virtually the same mass as protons, so Chadwick was able to explain the difficulty concerning the relative atomic masses of hydrogen and helium.

Hydrogen atoms have one proton, no neutrons and one electron. As the mass of the electron is negligible compared to the masses of the proton and neutron, a hydrogen atom has a relative mass of one unit. In comparison, helium atoms have two protons, two neutrons and two electrons, so the relative mass of a helium atom is four units. This means that a helium atom is four times as heavy as a hydrogen atom and the relative atomic mass of helium is 4.0 (Table 2.1).

Table 2.1
The relative masses of hydrogen and helium atoms

	Hydrogen atoms	Helium atoms
Number of protons	1	2
Number of neutrons	0	2
Relative mass	1	4

2.3 Sub-atomic particles

Through the work of Thomson, Rutherford, Chadwick and their colleagues, we now know that:

- All atoms are composed of three important particles: protons, neutrons and electrons.
- Atoms have a small positive nucleus surrounded by a much larger region of space in which tiny negative electrons move continuously.
- The positive charge of the nucleus is due to positively charged protons. The nucleus also contains uncharged neutrons which have virtually the same mass as protons.
- Protons and neutrons are about 1840 times heavier than electrons, so virtually all the mass of an atom is concentrated in the nucleus.
- The positive charge on one proton is equal in size but opposite in sign to the negative charge on one electron.
- Atoms have equal numbers of protons and electrons, so the positive charges on the protons cancel out the negative charges on the electrons.

These properties of the key sub-atomic particles are summarised in Table 2.2.

Note

Protons and neutrons are collectively called **nucleons** because they occupy the nucleus.

Table 2.2
The relative mass, relative charge and position within atoms of protons, neutrons and electrons

Particle	Relative mass (atomic mass units)	Relative charge	Position within atoms
Proton	1	+1	Nucleus
Neutron	1	0	Nucleus
Electron	$\frac{1}{1840}$	−1	In space outside the nucleus

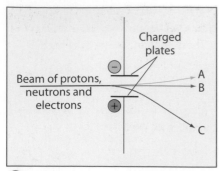

Figure 2.7
A narrow beam containing protons, neutrons and electrons entering an electric field

Quick Questions

8 Look closely at Figure 2.7. When the beam of protons, neutrons and electrons enters the electric field, it divides into three separate beams.
   a Which of the three separate beams contains:
   i protons, **ii** neutrons, **iii** electrons?
   b Why is the angle of deflection of beam C much greater than that of beam A?
   c How would the deflection of a beam of alpha-particles compare and contrast with the deflection of beam A?

The atoms of all elements are built up from protons, neutrons and electrons. Different atoms have different numbers of the three particles. Other particles, such as positrons and neutrinos, have been detected in particle-colliders, but these other particles are very unstable and only exist under extreme conditions.

Atoms and ions

From the structure of atoms which has just been described, it is easy to understand how ions are formed from atoms by losing or gaining electrons. A helium atom (He) has two protons (each with one positive charge), two neutrons and two electrons (each with one negative charge). If one electron is removed from a helium atom, it leaves a charged particle (an ion) with two protons and only one electron. So, the overall charge on the ion is one positive charge and its symbol can be written as He^+ (Figure 2.8).

If two electrons are removed from a helium atom, an ion with just two protons and two neutrons is left behind. The symbol for this ion is He^{2+} (Figure 2.8) which is an alpha-particle.

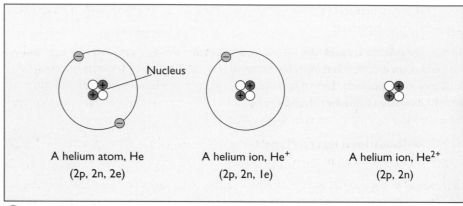

A helium atom, He
(2p, 2n, 2e)

A helium ion, He^+
(2p, 2n, 1e)

A helium ion, He^{2+}
(2p, 2n)

Figure 2.8
The atomic structure of a helium atom and helium ions

2.4 Proton number, nucleon number and isotopes

All the atoms of hydrogen have one proton, all the atoms of helium have two protons, all the atoms of lithium have three protons and so on. In fact, all the atoms of one particular element have the same number of protons and the atoms of different elements have different numbers of protons. This means that the number of protons in an atom tells us immediately which element it is. Because of this, scientists use the term '**proton number**' (symbol Z) for the number of protons in an atom.

So, hydrogen has a proton number of 1 ($Z = 1$), helium has a proton number of 2 ($Z = 2$) and lithium has a proton number of 3 ($Z = 3$). Notice that the proton number of an element is the same as the order of the element in the periodic table. So, sodium, the eleventh element in the periodic table, has a proton number of 11 and chlorine, the seventeenth element in the periodic table, has a proton number of 17.

As atoms have the same number of protons and electrons, their proton number provides information about their overall charge – the charge in the nucleus from protons and the charge outside the nucleus from electrons.

In addition to its charge, the other fundamental property of an atom is its mass and this is dictated by the number of protons plus neutrons (nucleons) in its nucleus. So, scientists use the term **nucleon number** (symbol A) for the number of protons plus neutrons in an atom.

So, hydrogen atoms with one proton and no neutrons have a nucleon number of one ($A = 1$). Helium atoms with two protons and two neutrons have a nucleon number of four ($A = 4$) and lithium atoms with three protons and four neutrons have a nucleon number of seven ($A = 7$).

After Aston had developed his mass spectrometer in 1919 (Figure 1.5, Section 1.2), it gradually became clear that most elements had atoms with different masses. These atoms of the same element with different masses are called **isotopes**. All the isotopes of one particular element have the *same proton number* because they have the same number of protons, but they have *different nucleon numbers* because they have different numbers of neutrons.

2.5 Isotopes, nucleon numbers and relative isotopic masses

Figure 2.9 shows a mass spectrum for naturally occurring copper. This shows that copper consists of two isotopes; each of these isotopes has a proton number of 29, but they have different relative isotopic masses of 63 and 65 respectively.

On the relative atomic mass scale, the mass of a proton (1.0074 units) is almost the same as that of a neutron (1.0089 units), while the mass of an electron (0.0005 units) is very small in comparison.

Now, as the relative masses of a proton and a neutron are both very close to one and the mass of an electron is negligible, it follows that all relative isotopic masses will be close to whole numbers. In fact, the relative isotopic mass of an isotope will be very close to its nucleon number (number of protons + neutrons) and the two are assumed to be identical in all but the most accurate work.

So, we can deduce from Figure 2.9 that the two isotopes of copper, copper-63 and copper-65, have nucleon numbers of 63 and 65 respectively.

Thus, atoms of copper-63, with a nucleon number of 63 and a proton number of 29, have 29 protons and 34 neutrons; whereas atoms of copper-65, with a nucleon number of 65 and a proton number of 29, have 29 protons and 36 neutrons.

We can write the symbol $^{63}_{29}Cu$ in order to specify the nucleon number and the proton number of the copper-63 atom.

Table 2.3 shows both the copper isotopes represented in this way. The nucleon number is written at the top left of the symbol and the proton number is written at the bottom left.

 Table 2.3
Specifying the nucleon number and proton number with the symbol of an isotope

Nucleon number → Proton number →	$^{63}_{29}Cu$	$^{65}_{29}Cu$
Number of protons + neutrons	63	65
Number of protons	29	29
Number of neutrons	34	36
Number of electrons	29	29

Isotopes have the same number of electrons and hence the same chemical properties because chemical properties depend on the arrangement and transfer of electrons. But isotopes have different numbers of neutrons and therefore different masses and different physical properties because physical properties often depend on the masses of particles. So, pure $^{63}_{29}Cu$ will have a lower density and lower melting point than $^{65}_{29}Cu$.

The **proton number** of an atom is the number of protons in its nucleus.

The term atomic number is sometimes used in place of proton number.

The **nucleon number** of an atom is the number of protons plus neutrons in its nucleus.

The term mass number is sometimes used in place of nucleon number.

Isotopes are atoms of the same element with the same proton number, but different nucleon numbers. Isotopes have the same number of protons, the same number of electrons, but different numbers of neutrons.

Note

Remember that the symbol for relative atomic mass is A_r, but the symbol for nucleon number is A.

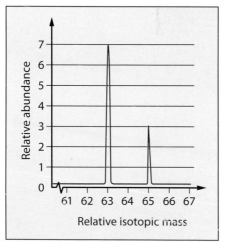

Relative abundance
Relative isotopic mass

Figure 2.9
A mass spectrum for naturally occurring copper

Figure 2.10

Chemical reactions involve the redistribution of electrons. The outcome can be explosive.

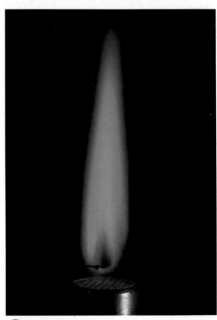

Figure 2.11

Copper(II) sulfate being heated strongly in a Bunsen flame

Quick Questions

9 Write symbols showing the nucleon number and proton number for:
   a a chlorine atom with 17 protons and 20 neutrons,
   b a sodium atom with 11 protons and 12 neutrons,
   c an aluminium ion with 13 protons, 14 neutrons and 10 electrons,
   d an oxide ion with 8 protons, 8 neutrons and 10 electrons.

10 Write down the number of protons, neutrons and electrons in the following atoms and ions:

   a $^{11}_{5}B$ b $^{81}_{35}Br$ c $^{209}_{83}Bi^{2+}$ d $^{31}_{15}P^{3-}$

11 Use Figure 2.9 to calculate the relative atomic mass of copper.

2.6 Evidence for the electronic structure of atoms

When chemical reactions take place, electrons in the outer parts of atoms are redistributed (Figure 2.10). The electrons may be transferred from one atom to another or they may be shared between the reacting atoms in a different way. Protons and neutrons, in the nuclei of atoms, take no part in chemical reactions.

It is possible to obtain information about the arrangement of electrons in atoms by studying the ease with which atoms lose electrons. The energy needed to remove one electron from each atom in a mole of gaseous atoms is known as the **first ionisation energy** (symbol ΔH_{i1}). So, the first ionisation energy of sodium is the energy required for the process:

$$Na(g) \longrightarrow Na^+(g) + e^- \qquad \Delta H_{i1} = +494\,kJ\,mol^{-1}$$

The most useful method of determining the ionisation energies of elements involves a study of their emission spectra. Using the data from spectra it is possible to determine the energy needed to remove electrons from atoms and then ions with increasing positive charges. This gives a series of successive ionisation energies represented by the symbols ΔH_{i1}, ΔH_{i2}, ΔH_{i3} and so on. For example:

$Be(g) \longrightarrow Be^+(g) + e^-$ First ionisation energy of beryllium, $\Delta H_{i1} = 900\,kJ\,mol^{-1}$

$Be^+(g) \longrightarrow Be^{2+}(g) + e^-$ Second ionisation energy of beryllium, $\Delta H_{i2} = 1760\,kJ\,mol^{-1}$

$Be^{2+}(g) \longrightarrow Be^{3+}(g) + e^-$ Third ionisation energy of beryllium, $\Delta H_{i3} = 14\,800\,kJ\,mol^{-1}$

$Be^{3+}(g) \longrightarrow Be^{4+}(g) + e^-$ Fourth ionisation energy of beryllium, $\Delta H_{i4} = 21\,000\,kJ\,mol^{-1}$

2.7 Obtaining ionisation energies from emission spectra

When sodium chloride is heated strongly in a Bunsen, the flame becomes a bright yellow. Other sodium compounds, such as sodium nitrate and sodium sulfate, also give a bright yellow light when heated in the same way. The compounds of some other elements also emit light when heated in this way. For example, the flame colour of calcium compounds is red (Figure 10.10, Section 10.7).

Quick Questions

12 What colour of light is emitted by the following when they are heated strongly on a nichrome wire in the hottest Bunsen flame?
   a copper compounds,
   b calcium compounds,
   c potassium compounds.

If the light emitted by these substances is examined using a spectroscope, it does *not* consist of a continuous range of colours like the spectrum of white light or the colours in a rainbow. Instead, the light emitted by these substances is composed of separate lines of different colours. In addition to these visible radiations, substances also give off radiations in the infrared and ultraviolet regions of the electromagnetic spectrum. These kinds of spectra are called **line emission spectra** (Figure 2.12).

In order to explain line emission spectra, scientists assume that the electrons in an atom can only exist at certain energy levels. Under normal conditions, the electrons in an atom or ion fill the lowest energy levels first. When sufficient energy is supplied to the atom, it is possible to promote (excite) an electron from a lower level to a higher one. This process is called **excitation.** The electron is unstable in the higher energy level, so it will emit the excess energy as radiation and drop back into a lower level.

Now, the energy difference between the higher and lower energy levels can only have certain fixed values because the energy levels themselves are fixed. This means that the radiation emitted when an electron falls from a higher to a lower energy level will only have certain fixed frequencies because the frequency of any radiation is determined by its energy. If these frequencies fall in the visible region of the electromagnetic spectrum, the radiation will have a specific colour.

The small amount of radiation emitted by an electron when it falls from a higher to a lower energy level is referred to as a **quantum** of radiation. The relationship between the energy, E, of a quantum of radiation and its frequency, f is:

$$E = h \times f$$

h is a constant called the Planck constant. The value of h is: 6.63×10^{-34} Js molecule $^{-1}$ or $6.63 \times 10^{-34} \times 6.02 \times 10^{23} = 3.99 \times 10^{-10}$ Js mol^{-1}

If sufficient energy is given to an atom, it is possible to excite an electron just beyond the highest energy level. In this case, the electron will escape and the atom becomes an ion. **Ionisation** has taken place.

The energy levels which electrons can occupy in atoms are numbered ($n = 1$, $n = 2$, $n = 3$, etc.). These numbers are sometimes referred to as the **principal quantum numbers** of the energy levels. They correspond to the shells of electrons. The level of lowest energy is given the principal quantum number 1, the next lowest 2, and so on. Figure 2.13 shows the energy levels in a hydrogen atom and the electron transitions to the $n = 1$ level. Lines corresponding to these transitions occur in the ultraviolet region of the hydrogen spectrum.

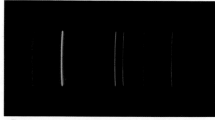

Figure 2.12
A photo of the line emission spectrum of helium in the visible region. The two most prominent lines have frequencies of 4.492×10^{14} Hz (red) and 5.106×10^{14} Hz (yellow).

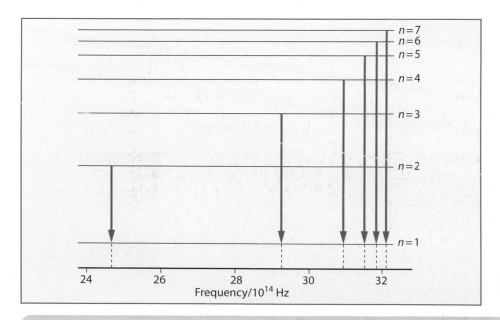

Figure 2.13
The energy levels in a hydrogen atom and electron transitions to the $n = 1$ level which produce lines in the ultraviolet region of the atomic hydrogen spectrum

Notice the following points from Figure 2.13.

1 As the energy levels get higher, they get closer. This means that the spectral lines get closer.
2 Eventually the energy levels come together. This means that the spectral lines also come together and the spectrum then becomes continuous.
3 In an atom, the highest possible energy level corresponds to the frequency at which lines in the spectrum come together. Just above the highest possible energy level of the atom, electrons will escape completely and ionisation occurs. So, by determining the frequency at which the converging spectral lines come together, we can find the ionisation energy of an element.

> **Figure 2.14**
Fireworks over the Sydney Harbour Bridge. The colours of fireworks are caused by electrons moving from higher to lower energy levels in atoms or ions. When electrons do this, radiation of a particular frequency is emitted which sometimes corresponds to the light of a particular colour.

> **Table 2.4**
Frequencies of electron transitions to the $n = 1$ level in hydrogen

Frequency $v/10^{14}$ Hz	Transition to which frequency corresponds		
24.66	$n = 2$	to	$n = 1$
29.23	$n = 3$	to	$n = 1$
30.83	$n = 4$	to	$n = 1$
31.57	$n = 5$	to	$n = 1$
31.97	$n = 6$	to	$n = 1$
32.21	$n = 7$	to	$n = 1$
32.37	$n = 8$	to	$n = 1$

Quick Questions

13 The frequencies in Table 2.4 correspond to the series of electron transitions to the $n = 1$ level in hydrogen.
   a Work out the difference in frequency, Δf, between successive lines in the series.
   b Plot a graph of Δf (vertically) against frequency, f. (Use the value of the lower frequency in plotting f.)
   c Use your graph to estimate the frequency when Δf becomes 0.

Worked example

An accurate value for the frequency when Δf becomes 0 for transitions to the $n = 1$ level in hydrogen is 32.7×10^{14} Hz. What is the ionisation energy of hydrogen?

Answer

When Δf becomes 0, the spectral lines come together and ionisation has occurred.

Therefore, the frequency corresponding to ionisation = 32.7×10^{14} Hz = 32.7×10^{14} s^{-1}
and the energy of radiation with this frequency,

$$E = h \times f = 3.99 \times 10^{-10} \, \text{Js mol}^{-1} \times 32.7 \times 10^{14} \, \text{s}^{-1}$$

$$= 130.5 \times 10^4 \, \text{J mol}^{-1} = 1305 \, \text{kJ mol}^{-1}$$

Therefore, the ionisation energy of hydrogen = $1305 \, \text{kJ mol}^{-1}$

2.8 Using ionisation energies to predict electronic structures – evidence for shells of electrons

Remember that beryllium has a proton number of 4 and therefore its atoms have only four electrons. Then, look carefully at the four successive ionisation energies of beryllium listed in Section 2.6. About twice as much energy is needed to remove the second electron from beryllium as to remove the first, and eight times as much energy is required to remove the third compared to the second. Then about one and a half times as much energy is needed to remove the fourth electron compared to the third. These values for the successive ionisation energies suggest that beryllium has two electrons which are relatively easy to remove and two electrons which are very difficult to remove.

Chemists have therefore deduced that a beryllium atom has two electrons in the lowest energy level ($n = 1$) and therefore very difficult to remove; and two other electrons in a higher energy level ($n = 2$) which are much easier to remove. This is represented on an energy level diagram in Figure 2.15.

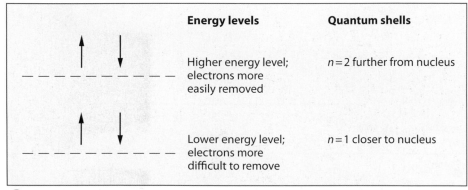

 Figure 2.15

Energy levels and quantum shells for electrons in a beryllium atom

The electron arrangement in beryllium has led chemists to write its electronic structure as 2, 2. Similarly, the electronic structure of lithium is written as 2, 1 because its emission spectrum suggests that it has two electrons in the $n = 1$ energy level (first quantum shell) and one electron in the $n = 2$ energy level (second quantum shell).

In Figure 2.15, the electrons have been represented by arrows. When an energy level is full, the electrons have paired up and in each of these pairs the electrons are spinning in opposite directions.

Chemists believe that paired electrons can only be stable when they spin in opposite directions. This is because the magnetic attraction which results from their opposite spins can counterbalance the electrical repulsion which results from their identical negative charges. In energy level diagrams like that in Figure 2.15, the opposite spins of the paired electrons are shown by drawing the arrows in opposite directions.

2.9 How are the electrons arranged in larger atoms?

The graph in Figure 2.16 shows a logarithmic plot of the successive ionisation energies of sodium. The proton number of sodium is 11. So, there are 11 electrons in a sodium atom and 11 successive ionisation energies for this element.

The successive ionisation energies of an element are all endothermic and their values become more and more endothermic. This is not surprising because each successive ionisation requires the removal of a negative electron from an ion of increasing positive charge.

The logarithmic plot in Figure 2.16 allows an extremely wide range of ionisation energies to be shown on the same graph, from $494\,kJ\,mol^{-1}$ to $159\,100\,kJ\,mol^{-1}$. Notice also in Figure 2.16 how there are two big jumps in value, between the first and second ionisation energies and the ninth and tenth ionisation energies. See if you can now deduce the electronic structure of sodium by answering Quick question 14.

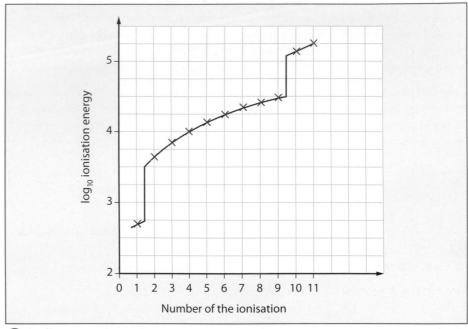

Figure 2.16
A graph of the logarithm to base 10 ($\log_{10}$) of the successive ionisation energies for sodium

Quick Questions

14 Look closely at Figure 2.16.
   a How many electrons does sodium have in the first shell close to the nucleus? These electrons will be the most difficult to remove.
   b How many electrons does sodium have in the second shell?
   c How many electrons does sodium have in the third shell far away from the nucleus and relatively easy to remove?
   d Write the electronic structure for sodium showing the number of electrons in each shell. (The electronic structure of beryllium is 2, 2.)

15 The first six ionisation energies of an element in $kJ\,mol^{-1}$ are 1060, 1900, 2920, 4960, 6270 and 21 270.
   a How many electrons are there in the outer shell of the atoms of this element?
   b Which group in the periodic table does the element belong to?

16 Sketch a graph of $\log_{10}$ ionisation energy against the number of each ionisation when all the electrons are successively removed from an aluminium atom.

Figure 2.17 (a) shows a sketch graph of $\log_{10}$ for the successive ionisation energies of potassium. This graph has been used to construct an energy level diagram (Figure 2.17 (b)) for the electrons in a potassium atom. This time there are three big jumps in the ionisation energies between the first and second, between the ninth and tenth and between the seventeenth and eighteenth ionisations.

These jumps suggest that the electronic structure of potassium is 2, 8, 8, 1 showing the number of electrons in each quantum shell as we move out from the nucleus.

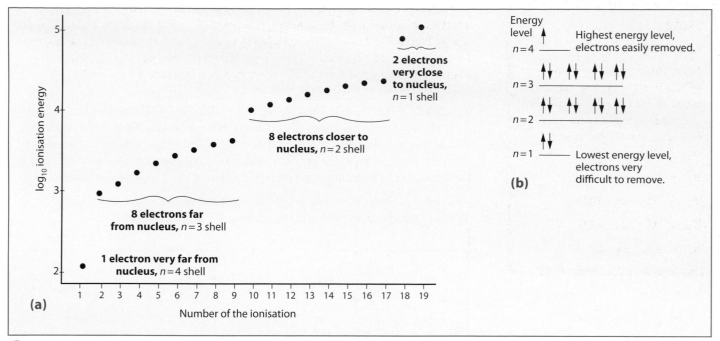

Figure 2.17

(a) A sketch graph of $\log_{10}$ of the successive ionisation energies for potassium

(b) An energy level diagram for the electrons in a potassium atom

Using ionisation energy data, it is possible to predict the electronic structures of other elements and Table 2.5 shows the electron structures of the first 20 elements in the periodic table.

Notice that the first shell is full when it contains two electrons and the second shell is full when it contains eight electrons. Although the electron structures in Table 2.5 suggest that the third shell can contain only eight electrons, it is possible for it to hold as many as 18 electrons.

Proton number	Element	Symbol	Electronic structure	Proton number	Element	Symbol	Electronic structure
1	Hydrogen	H	1	11	Sodium	Na	2, 8, 1
2	Helium	He	2	12	Magnesium	Mg	2, 8, 2
3	Lithium	Li	2, 1	13	Aluminium	Al	2, 8, 3
4	Beryllium	Be	2, 2	14	Silicon	Si	2, 8, 4
5	Boron	B	2, 3	15	Phosphorus	P	2, 8, 5
6	Carbon	C	2, 4	16	Sulfur	S	2, 8, 6
7	Nitrogen	N	2, 5	17	Chlorine	Cl	2, 8, 7
8	Oxygen	O	2, 6	18	Argon	Ar	2, 8, 8
9	Fluorine	F	2, 7	19	Potassium	K	2, 8, 8, 1
10	Neon	Ne	2, 8	20	Calcium	Ca	2, 8, 8, 2

Table 2.5
Electronic structures of the first 20 elements in the periodic table

As soon as the successive ionisation energies of a few elements had been measured, scientists realised that the quantum shells of elements correspond to the periods of elements in the periodic table. By noting where the first big jump comes in the successive ionisation energies of an element, it is possible to predict the group to which the element belongs.

For example, the first big jump in the successive ionisation energies for potassium comes after the first electron is removed. This suggests that potassium has just one electron in its outermost shell, so it must be in Group 1.

2.10 Evidence for sub-shells of electrons

Figure 2.18 shows the first ionisation energies of the first 40 elements in the periodic table plotted against proton number. The graph can be divided into sections ending in a noble (inert) gas. Each noble gas has a higher first ionisation energy than all the elements between it and the previous noble gas. These high first ionisation energies of noble gases are a clear indication of their stable electronic structures and unreactivity.

The points between one noble gas and the next in Figure 2.18 can be divided into sub-sections. These sub-sections provide evidence for sub-shells of electrons.

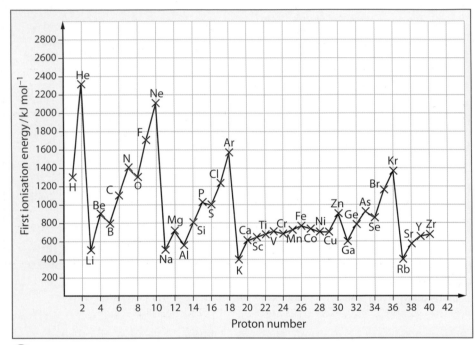

⌃ Figure 2.18
A graph of the first ionisation energies of the elements plotted against proton number

After both He and Ne in Figure 2.18, there are deep troughs followed by small intermediate peaks at Be and Mg. These are sub-sections with just two points. Immediately after Be and Mg there are similar sections of six points (B to Ne and Al to Ar). Further along in Figure 2.18, there is another sub-section of six points (Ga to Kr) and just before this is a sub-section of 10 points (Sc to Zn).

In fact, the points in Figure 2.18 between one noble gas and the next correspond to one shell of electrons and the sub-sections of points correspond to sub-shells of electrons. By studying ionisation energies and emission spectra in this way, scientists have come to the following conclusions.

▶ The first, $n = 1$ shell can hold 2 electrons in the same sub-shell.
▶ The second, $n = 2$ shell can hold 8 electrons; 2 in one sub-shell and 6 in a slightly higher sub-shell.
▶ The third, $n = 3$ shell can hold 18 electrons; 2 in one sub-shell, 6 in a slightly higher sub-shell and 10 in a still slightly higher sub-shell.
▶ The fourth, $n = 4$ shell can hold 32 electrons with sub-shells containing 2, 6, 10 and 14 electrons.

Electron structures

The shells of electrons with principal quantum numbers $n = 1$, $n = 2$, $n = 3$, etc. are sometimes referred to as the first quantum shell, second quantum shell, third quantum shell, etc.

The sub-shells that make up the main shells also have labels.

► Sub-shells that can hold up to 2 electrons are called s sub-shells.
► Sub-shells that can hold up to 6 electrons are called p sub-shells.
► Sub-shells that can hold up to 10 electrons are called d sub-shells.
► Sub-shells that can hold up to 14 electrons are called f sub-shells.

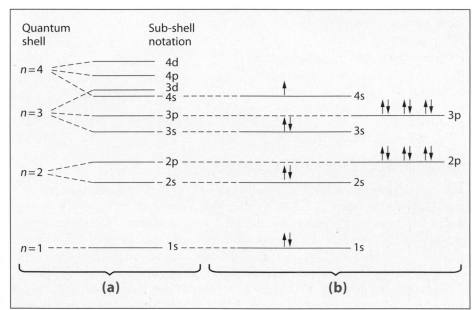

Figure 2.19
(a) The relative energy levels of sub-shells within the first four quantum shells
(b) The distribution of electrons in the sub-shells of a potassium atom

The relative energy levels of the various sub-shells within the first four quantum shells are shown on the left of Figure 2.19. On the right of Figure 2.19, we have shown the distribution of electrons in the sub-shells of a potassium atom. Compare Figure 2.19 with Figure 2.17 (b).

The electronic structure of an atom can be described simply in terms of the shells occupied by electrons (i.e. 2, 2 for beryllium) or more precisely in terms of the sub-shells occupied by electrons (i.e. $1s^2$, $2s^2$ for beryllium).

When sub-shells (energy sub-levels) are being filled, electrons always occupy the lowest available sub-shell first and the electrons do not begin to 'pair-up' until a sub-shell is half filled (Section 2.11).

Therefore from Figure 2.19, we can deduce that the order in which the sub-shells are filled is: 1s, 2s2p, 3s3p, 4s, 3d, 4p.

So, in terms of shells, the electronic structure of a potassium atom is 2, 8, 8, 1. But using a number (1, 2, 3, etc.) to denote the quantum shell, a letter (s, p, d or f) to denote the sub-shell and a superscript to indicate the number of electrons in the sub-shells, we can write the electronic structure of potassium more precisely as $1s^2$, $2s^2 2p^6$, $3s^2 3p^6$, $4s^1$ (Figure 2.20).

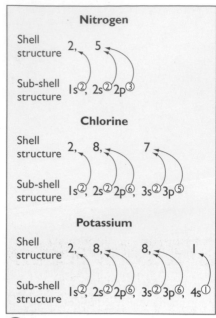

Nitrogen

Shell structure 2, 5

Sub-shell structure $1s^2, 2s^2 2p^3$

Chlorine

Shell structure 2, 8, 7

Sub-shell structure $1s^2, 2s^2 2p^6, 3s^2 3p^5$

Potassium

Shell structure 2, 8, 8, 1

Sub-shell structure $1s^2, 2s^2 2p^6, 3s^2 3p^6, 4s^1$

Figure 2.20

The relationship between the shell and sub-shell structure of nitrogen, chlorine and potassium

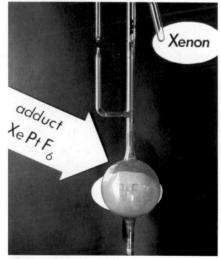

Xenon

adduct $XePtF_6$

Figure 2.21

The first noble gas compound was produced in 1962 by Neil Bartlett. While carrying out research, he allowed oxygen in the air to react with PtF_6 producing $O_2^+PtF_6^-$. He quickly realised that PtF_6 had removed an electron from O_2 forming O_2^+ and PtF_6^-. Then, as the first ionisation energy of xenon ($+1170\,kJ\,mol^{-1}$) was very similar to that of molecular oxygen ($+1175\,kJ\,mol^{-1}$), he thought it should be possible to make $Xe^+PtF_6^-$. Within days, he had produced $XePtF_6$ as a yellow/orange solid.

Notice in Figure 2.19 that the 3d sub-shell is just above the 4s sub-shell, but just below the 4p sub-shell. This means that once the 4s level is filled (at Ca in the periodic table), further electrons enter the 3d sub-shell, not the 4p sub-shell. Thus, scandium (the element after Ca) has the electronic structure $1s^2, 2s^2 2p^6, 3s^2 3p^6 3d^1, 4s^2$ and not $1s^2, 2s^2 2p^6, 3s^2 3p^6, 4s^2 4p^1$.

The filling of the 3d sub-shell is very important in the chemistry of elements from scandium, Sc, to zinc, Zn. These are known as d-block elements and we shall study them in some detail in Chapter 24.

Quick Questions

17 Write the electron structures for the following atoms and ions. (Hint: For Na this would be 2, 8, 1.)

    a N   **b** F^- **c** Mg^{2+}

    d P   **e** Ca^{2+} **f** V ($Z = 23$)

    g the atom with proton number 13

    h the ion with proton number 16 and charge 2–

18 Write the electronic sub-shell structures for the atoms and ions in question 17. (Hint: For Na this would be $1s^2, 2s^2 2p^6, 3s^1$.)

2.11 Trends in ionisation energies

Look again at the graph of the first ionisation energies of the elements against proton (atomic) number in Figure 2.18. Notice that, within a period (Li to Ne or Na to Ar), the first ionisation energy tends to rise as proton number increases. This increase in ionisation energy is associated with a decrease in metallic character from left to right across a period. When metals react, they readily lose electrons and form positive ions. So, not surprisingly, the first ionisation energies of metals are relatively low. But, the first ionisation energies gradually rise across a period as non-metal character increases and electrons are held more tightly by non-metals. This led chemists to wonder how ionisation energies were related to atomic structure.

Factors influencing the ionisation energies of elements

The ionisation energy of an atom is strongly influenced by three atomic parameters.

1 **The distance of the outermost electron from the nucleus**

    As this distance increases, the attraction of the positive nucleus for the negative electron decreases and consequently, the ionisation energy decreases.

2 **The size of the positive nuclear charge**

    As the nuclear charge becomes more positive with increasing proton number, its attraction for the outermost electron increases and consequently the ionisation energy increases.

3 **The screening (shielding) effect of inner shells of electrons**

    Electrons in inner shells exert a repelling effect on electrons in the outermost shell of an atom. Chemists say that the outermost electron is screened or shielded from the attraction of the positive nucleus by the repelling effect of inner electrons. This screening (shielding) effect means that the 'effective nuclear charge' is much less than the full positive charge in the nucleus.

In general, the screening effect by inner electrons is more effective the closer these inner electrons are to the nucleus. So,

▶ Electrons in shells of lower principal quantum number are more effective shields than those in shells of higher quantum number.

▶ Electrons in the same shell have a negligible shielding effect on each other.

This means that we need only consider inner shells of electrons in discussing the screening effect on an outermost electron.

The trend in ionisation energies across a period

Look at Figure 2.18 again.

Moving from left to right across any period there is a general increase in the first ionisation energy because:

▶ The nuclear charge is increasing.
▶ The distance of the outermost electron from the nucleus is decreasing because the increasing nuclear charge is pulling electrons closer to the nucleus, so the atomic radius is decreasing.
▶ The screening effect remains almost the same for elements in the same period.

Although there is a general increase in the first ionisation energies across any period, there are breaks in the overall pattern. For example, the first ionisation energy of beryllium is higher than that of boron the next element in Period 2.

The electron configuration of beryllium is $1s^2, 2s^2$, whereas that of boron is $1s^2, 2s^22p^1$. All the sub-shells in beryllium are filled, but the outer 2p sub-shell of boron contains only one electron.

From our studies earlier in this chapter, we know that filled electron shells, like those of helium and neon, are very stable. This means that their electron structures are difficult to disrupt and their first ionisation energies are high.

In a similar fashion, *there is also some extra stability associated with filled sub-shells*. This means that the electron structure of beryllium is rather more stable than we might have expected causing its first ionisation energy to be greater than that of boron.

In Period 2, there is another break in the general increase in first ionisation energies with nitrogen which has a higher first ionisation energy than oxygen. The electron structures of nitrogen and oxygen are $1s^2, 2s^22p^3$ and $1s^2, 2s^22p^4$ respectively. The half-filled 2p sub-shell in nitrogen, with one electron in each of the three 2p orbitals (Section 2.12), and its evenly distributed charge is more stable than the 2p sub-shell in oxygen which contains four electrons. This results in a higher first ionisation energy for nitrogen than oxygen.

The trend in ionisation energies down a group

Moving down any group in the periodic table, there is a general decrease in the first ionisation energy as the proton numbers increase. This is neatly illustrated by the first ionisation energies of the stable elements of Group 2 and Group 7 shown in Table 2.6.

 Table 2.6

The first ionisation energies of stable elements in Group 2 and Group 7

Group 2 element	Proton (atomic) number	First ionisation energy /kJ mol^{-1}	Group 7 element	Proton (atomic) number	First ionisation energy /kJ mol^{-1}
Beryllium	4	900	Fluorine	9	1680
Magnesium	12	736	Chlorine	17	1260
Calcium	20	590	Bromine	35	1140
Strontium	38	548	Iodine	53	1010
Barium	56	502			

Quick Questions

19 Look at the trend of first ionisation energies across Period 3 in Figure 2.18 and notice that magnesium has a first ionisation energy greater than that of aluminium.

a The electron structure of magnesium is $1s^2, 2s^22p^6, 3s^2$. Write the electron structure of aluminium in the same notation.

b Why is the first ionisation energy of magnesium higher than that of aluminium?

c Which other element in Period 3 has a higher first ionisation energy than the element immediately after it?

d Explain the relative values of the first ionisation energy for the element named in part c) and the element immediately after it.

Definition

An **orbital** is a region in space around the nucleus of an atom occupied by an electron or a pair of electrons with opposite spins.

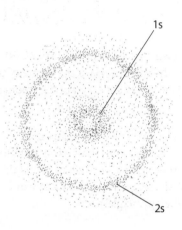

Figure 2.22
An electron density map for the charge clouds of electrons in a lithium atom

Before reading further, try to answer Quick question 20.

From the values in Table 2.6 and your answer to Quick question 20, you should realise that the distance and screening factors outweigh the nuclear charge factor and cause the first ionisation energies of elements to decrease down a group in the periodic table.

2.12 Electrons and orbitals

So far, we have described the electrons of atoms in shells and sub-shells at increasing distances from the nucleus.

Chemists have also used complex mathematics to calculate the probability of finding an electron at any point in an atom. Their calculations have led chemists to believe that there is a high probability that an electron or a pair of electrons will occupy certain regions, called **orbitals**, around the nucleus of an atom.

By calculating the probability of finding an electron in different regions of the orbital, it is possible to compose a density map showing how the electron is distributed throughout its orbital. Figure 2.22 shows one such electron density map for the 1s and 2s electrons in a lithium atom. The density maps are darkest where the electrons are most likely to be and lightest where the electrons are least likely. The charged cloud for the two 1s electrons and that for the single 2s electron are both spherical in shape.

The general shapes of orbitals are deduced from electron density plots by determining the boundary of the region in which there is a 95% chance of finding an electron or a pair of electrons. As a result of these studies, chemists believe that all s sub-shells contain one orbital, best described as a spherical annulus – like extra thick peel on an orange. On the other hand, p orbitals are not spherical, but approximately 'dumb-bell' shaped with the nucleus located between the two halves of the dumb-bell.

In fact, each p sub-shell has three separate p orbitals, each of which can hold a maximum of two electrons. This makes a total of six electrons in a filled p sub-shell.

Figure 2.23 shows the shapes of the three orbitals in a p sub-shell. They are identical except for their axes of symmetry which, like the axes of a three-dimensional co-ordinate system, are mutually at right angles. Thus, it is convenient to distinguish between the p orbitals by labelling them p_x, p_y and p_z. Electrons will, of course, occupy the three p orbitals singly at first because of their mutual repulsions. When a fourth electron is added to a p sub-shell, one of the three orbitals will contain a pair of electrons.

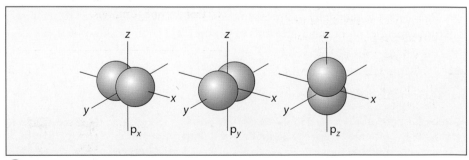

Figure 2.23
The shapes and relative positions of the three p orbitals in a p sub-shell

So, the first quantum shell contains just one sub-shell (1s) with one s orbital. The second quantum shell contains two sub-shells (2s and 2p) with a total of four orbitals, one 2s orbital and three 2p orbitals ($2p_x$, $2p_y$ and $2p_z$).

Figure 2.24

Suppose we magnified an atom one million million times to the size of the 2010 World Cup Stadium in Johannesburg. The nucleus would be the size of a pea at the centre of the pitch and the outermost electrons would be moving around at the edges of the stadium.

From their knowledge about sub-shells and orbitals, chemists have developed an 'electrons in boxes' notation for the electronic structures of atoms. Using this notation, each box represents an orbital. Using a set of three rules called the 'Aufbau principle', it is possible to work out the electronic structure of all atoms. In German, the word *Aufbau* means 'build up'. The three rules in the 'Aufbau principle' are:

1 Electrons enter the orbital of lowest energy first.
2 Orbitals can hold either one electron or two electrons with opposite spins.
3 Electrons occupy orbitals at the same sub-level singly before they pair up.

Figure 2.25 shows the electrons-in-boxes (orbitals) notation and also the sub-shell (s, p, d, f) notation for the electronic structures of beryllium, carbon, nitrogen and oxygen. These representations of the electronic structures of atoms are often called electronic configurations.

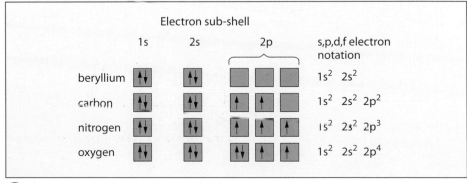

Figure 2.25

An electrons-in-boxes notation and a sub-shell notation for the electronic configurations of beryllium, carbon, nitrogen and oxygen

Review questions

1 a What are the proton (atomic) numbers of tellurium (Te) and iodine (I)?

   b What are the relative atomic masses of Te and I?

   c What are the numbers of protons, neutrons and electrons in the commonest isotopes of tellurium (^{128}Te) and iodine (^{127}I)?

   d Why does Te come before I in the periodic table?

   e Te comes before I in the periodic table, but Te has a larger relative atomic mass than I. Explain.

   f Look closely at the periodic table and write down the names of two other *pairs* of elements which are placed in the periodic table in the reverse order of their relative atomic masses.

2 The five main proposals in Dalton's Atomic Theory of matter were:

   • All matter is composed of tiny indestructible particles called atoms.

   • Atoms cannot be created or destroyed.

   • Atoms of the same element are alike in every way.

   • Atoms of different elements are different.

   • Atoms can combine together in small numbers to form molecules.

   a In the light of modern knowledge about atoms, isotopes, molecules and atomic structure, comment on the truth of each of these proposals.

   b Why is Dalton's Atomic Theory still useful in spite of these limitations?

3 Assume that the fluorine atom $^{19}_{9}$F is a sphere of diameter 10^{-10} m and that its nucleus is a sphere of diameter 10^{-14} m.

   a What is (i) the proton number, (ii) the nucleon number of $^{19}_{9}$F?

   b What is the *actual* mass of the nucleus in *one* $^{19}_{9}$F atom? (6×10^{23} ^{1}H atoms have a mass of 1 g.)

   c What is the density of the nucleus in a $^{19}_{9}$F atom?

   d What does the value in part c) suggest about the forces within the $^{19}_{9}$F nucleus?

   e What is the ratio of the volume of the atom to the volume of the nucleus in a $^{19}_{9}$F atom?

4 The accurate relative isotopic masses of five isotopes are shown below:

   $^{1}_{1}$H $^{2}_{1}$H (D) $^{12}_{6}$C $^{14}_{7}$N $^{16}_{8}$O
   1.0078   2.0141   12.0000   14.0031   15.9949

   a Calculate the accurate relative molecular masses for:

      i N_2 ii DCN   iii CO   iv C_2H_4 v C_2D_2

   b The relative molecular mass of a certain gas in a high-resolution mass spectrometer was 28.0171. What gas is probably under observation?

5 The following table shows the ionisation energies (in kJ mol^{-1}) of five elements lettered A, B, C, D and E.

Element	1st ionisation energy	2nd ionisation energy	3rd ionisation energy	4th ionisation energy
A	500	4600	6900	9500
B	740	1500	7700	10 500
C	630	1600	3000	4800
D	900	1800	14 800	21 000
E	580	1800	2700	11 600

   a Which of these elements is most likely to form an ion with a charge of 1+? Give reasons for your answer.

   b Which two of the elements are in the same group of the periodic table? Which group do they belong to?

   c In which group of the periodic table is element E likely to occur? Give reasons for your answer.

   d Which element would require the least energy to convert one mole of gaseous atoms into ions carrying two positive charges?

6 The electron energy levels of a certain element can be represented as $1s^2$, $2s^22p^6$, $3s^23p^4$.

   a Sketch a graph for the first seven ionisation energies of the element against the number of the ionisation.

   b What is the proton number of the element?

   c Draw an energy level diagram for the electrons in an atom of the element.

7 a List three factors which influence the size of the first ionisation energy of an element.

   b The first ionisation energies of the elements in Group I are shown below.

Element	First ionisation energy /kJ mol^{-1}
Li	520
Na	500
K	420
Rb	400
Cs	380

   Explain the change in the first ionisation energy as proton number increases.

c Explain, with examples, how ionisation energies can provide evidence for the arrangement of electrons in shells.

8 The graph in Figure 2.26 shows the first and second ionisation energies of the elements nitrogen to calcium.

a Why is there a large decrease in the first ionisation energy after neon and after argon?

b Why is the first ionisation energy of magnesium greater than that for aluminium?

c Why is the second ionisation energy of each element greater than the corresponding first ionisation energy?

d Why do the maxima for the two graphs occur at different proton numbers?

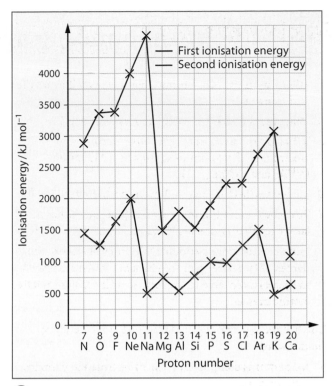

Figure 2.26
The first and second ionisation energies of the elements nitrogen to calcium

3 Chemical bonding and intermolecular forces

3.1 A theory for chemical bonding

Look at the information in Table 3.1, showing the electronic structures of the atoms and ions of the elements in Period 3.

Atoms of the first three elements in the period (Na, Mg and Al) *lose* electrons from their outermost shell to form positively charged ions (Na^+, Mg^{2+} and Al^{3+}). These ions have an electron structure like neon, the previous noble gas.

On the other hand, elements towards the end of the period in Groups VI and VII (S and Cl) *gain* electrons to form negatively charged ions (S^{2-} and Cl^-). These ions have the same electron structure as the next noble gas argon.

But, what happens to elements in the middle of the period such as silicon and phosphorus? These elements don't usually form ions in their compounds, but they do obtain an electron structure similar to that of a noble gas.

⌄ Table 3.1

Electronic structures of the atoms and ions of the elements in Period 3

Element	Na	Mg	Al	Si	P	S	Cl	Ar
Electron structure of the atom	2, 8, 1	2, 8, 2	2, 8, 3	2, 8, 4	2, 8, 5	2, 8, 6	2, 8, 7	2, 8, 8
No. of electrons in outermost shell of atom	1	2	3	4	5	6	7	8
Ion formed	Na^+	Mg^{2+}	Al^{3+}	–	–	S^{2-}	Cl^-	–
Electron structure of the ion	2, 8	2, 8	2, 8	–	–	2, 8, 8	2, 8, 8	–

Like other elements in Groups IV and V, silicon and phosphorus achieve electronic structures similar to noble gases by *sharing* electrons rather than by losing or gaining electrons.

Early in the 20th century, chemists realised that all the noble gases, except helium, had an outer shell containing eight electrons. This suggested that the electron structures of noble gases were responsible for their stability and inertness and led chemists to propose an **electronic theory of chemical bonding**.

Since the theory was first put forward in 1916, we now know of many compounds in which the elements do not have a noble gas structure. For example, most of the ions of transition metals (e.g. Fe^{2+}, Fe^{3+}, Cu^{2+}) do not have an electron structure like a noble gas, nor does the sulfur atom in SF_6. Nevertheless, the ideas first proposed still form the basis of theories of chemical bonding.

3.2 Transfer of electrons – ionic (electrovalent) bonding

Typical ionic compounds are formed when metals in Groups I and II, such as sodium and magnesium, react with non-metals in Groups VI and VII, such as oxygen and chlorine. When the reactions occur, electrons are *transferred* from the metal to the non-metal until the outer electron shells of the resulting ions are identical to those of noble gases.

Key Point

The electronic theory of chemical bonding

When elements form compounds, they either lose, gain or share electrons so as to achieve stable electron configurations similar to the next higher or lower noble gas in the periodic table.

Figure 3.1 shows what happens when this electron transfer from lithium to oxygen takes place with the formation of ions in lithium oxide. In Figure 3.1, the nucleus of each atom is represented by its symbol and the electrons in each shell are shown as circled dots or crosses around the symbol. Ions are drawn inside square brackets with the charge at the top right-hand corner.

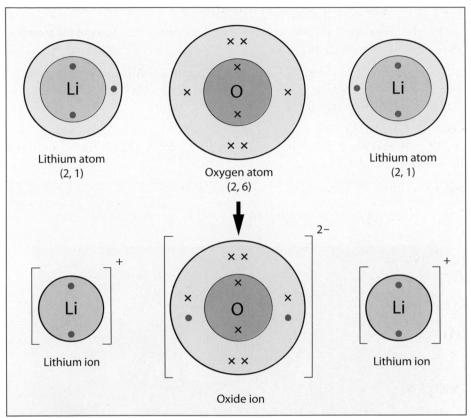

1 Look carefully at Figure 3.1.
   a What is the electronic structure of:
      i the lithium ion, ii the oxide ion?
   b Which noble gas has an electronic structure like:
      i Li^+, ii O^{2-} ?
   c Why do two lithium atoms react with only one oxygen atom?
2 Draw similar diagrams to Figure 3.1 to show the electron transfers which take place in the formation of:
   a magnesium oxide,
   b potassium sulfide,
   c calcium fluoride.

⌃ Figure 3.1
Transfer of electrons from lithium to oxygen in the formation of lithium oxide

Although the electrons of the different atoms in Figure 3.1 are shown by dots and crosses, you must not think that the electrons of lithium are any different from those of oxygen. All electrons are identical. They are shown differently in diagrams like that in Figure 3.1, so that you can follow their transfer more easily. Chemists describe diagrams like that in Figure 3.1 as 'dot-and-cross' diagrams.

Figure 3.2 shows the electron transfers which take place during the formation of sodium chloride and magnesium fluoride. In this figure, only those electrons in the outer shell of each atom are represented by dots or crosses around their symbol. The full electron structures are, however, shown in brackets below the symbols.

Remember when drawing 'dot-and-cross' diagrams, that the dots and crosses are simply a means of counting electrons. They cannot show the precise location of electrons within atoms because electrons are distributed in space as diffuse negative charge clouds.

The formation of ions in compounds such as sodium chloride, magnesium fluoride and lithium oxide involves a *complete transfer* of electrons. In these solid compounds, ions are held together by electrostatic attraction in **ionic (electrovalent) bonds**.

Definition

Ionic (electrovalent) bonding involves the complete transfer of electrons from one element to another forming oppositely charged ions. The ions are held together by strong electrostatic attractions between their opposite charges.

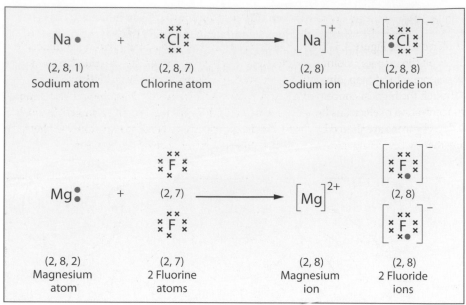

Figure 3.2
'Dot-and-cross' diagrams for the formation of sodium chloride and magnesium fluoride

Figure 3.3
The Taj Mahal in India is made of marble. This is mainly calcium carbonate ($CaCO_3$) which is held together by strong ionic bonds between calcium ions (Ca^{2+}) and carbonate ions (CO_3^{2-}).

3.3 The properties of ionic compounds

In ionic compounds, strong ionic bonds hold the oppositely charged ions firmly together (Figure 3.4). The arrangement of ions explains why ionic compounds:

▶ are hard crystalline substances,
▶ have high melting points and boiling points,
▶ are often soluble in water and other polar solvents which are attracted to the charged ions, but insoluble in non-polar solvents such as hexane (Section 3.10),
▶ do not conduct electricity as solids because the ions cannot move away from fixed positions in the solid structure,
▶ conduct electricity when they are molten or dissolved in water because the charged ions are then free to move towards the electrode of opposite charge (Section 6.4).

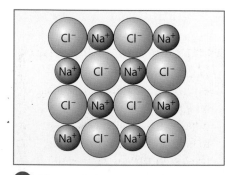

Figure 3.4
The arrangement of ions in one layer of a sodium chloride crystal

3.4 Sharing electrons – covalent bonding

Look closely at Figure 3.5. This shows an electron density map for a hydrogen molecule. Lines on the map join points with the same electron density in the way that contours on a geographical map join points at the same height above sea level. Notice that, although the highest concentration of electrons is near each nucleus, there is also a high concentration of electrons between the two nuclei. This suggests that in molecules such as H_2, electrons are shared by the two hydrogen atoms. In fact, the two atoms are held together by the strong attractions of their nuclei for the electrons in between.

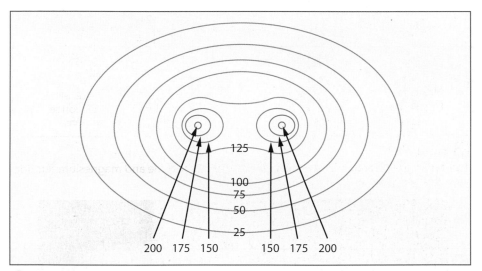

Figure 3.5
An electron density map for a hydrogen molecule. Units for the contours are electrons per nm^3.

Each hydrogen atom has only one electron. If, however, the two hydrogen atoms come close together, their 1s orbitals can overlap. The pair of electrons are then attracted to each nucleus and shared by each atom. This attraction to the shared electrons holds the nuclei together. A 'dot-and-cross' representation of this is shown in Figure 3.6. Each hydrogen atom now has two electrons, which is the same electronic structure as helium and the H_2 molecule is much more stable than an H atom. The shared pair of electrons has resulted in a **covalent bond**.

In most non-metals, atoms are joined together in small simple molecules such as hydrogen, H_2, chlorine, Cl_2, oxygen, O_2 and phosphorus, P_4. Most of the compounds of non-metals with other non-metals also have small simple molecules. These simple molecular compounds include hydrogen chloride, water, carbon dioxide and methane. Carbon and hydrogen form a vast range of organic compounds in which the atoms are held together by covalent bonds.

In all these simple molecular substances, each atom usually gains a noble gas electron structure as a result of sharing electrons. Figure 3.7 shows what happens in the case of chlorine, Cl_2. Each Cl atom has the electronic structure 2, 8, 7, with seven electrons in its outermost shell. By sharing one pair of electrons, both Cl atoms acquire an electron structure similar to argon (2, 8, 8). Notice in this case that the 'dot-and-cross' structures show only the electrons in the outermost shell of the atoms involved.

'Dot-and-cross' diagrams showing only the electrons in outer shells are a neat way of representing covalent bonds. Four more of these 'dot-and-cross' diagrams are shown in Figure 3.8. Note that each non-metal usually forms the same number of covalent bonds in all its compounds. This should help you to predict the structures of different molecules (Table 3.2).

Quick Questions

3 Use a periodic table to predict the charges on ions of the following elements:
   a barium,   b indium,
   c iodine,   d rubidium,
   e selenium.
4 a Write the formulae of sodium fluoride and magnesium oxide showing charges on the ions.
   b Use these formulae to explain why the melting point of magnesium oxide (2852 °C) is so much higher than that of sodium fluoride (993 °C).

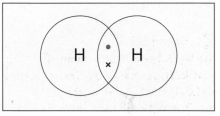

Figure 3.6
A 'dot-and-cross' diagram for the electron structure in a hydrogen molecule

Definition

A **covalent bond** involves the sharing of a pair of electrons between two atoms. In a normal covalent bond, each atom contributes one electron to the shared pair.

Note

Remember that all electrons are identical. They are only drawn differently as 'dots and crosses' to help you understand their transfer or sharing more easily.

Table 3.2
The number of covalent bonds usually formed by some common non-metals

Element	Symbol	Number of covalent bonds usually formed
Hydrogen	H	1
Chlorine	Cl	1
Oxygen	O	2
Sulfur	S	2
Nitrogen	N	3
Carbon	C	4

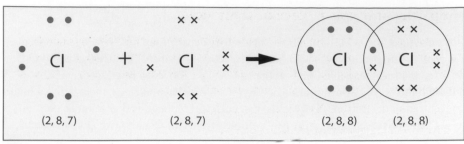

(2, 8, 7)   (2, 8, 7)   (2, 8, 8)   (2, 8, 8)

Figure 3.7
The formation of a covalent bond between two chlorine atoms

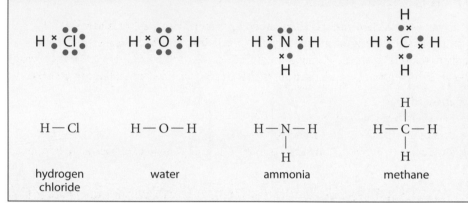

hydrogen chloride   water   ammonia   methane

Figure 3.8
'Dot-and-cross' diagrams for hydrogen chloride, water, ammonia and methane, plus a simpler way of representing these molecules by showing each covalent bond as a line

Double and triple bonds

A single covalent bond involves one shared pair of electrons. However, double and triple covalent bonds are also possible with two and three shared pairs respectively. The 'dot-and-cross' diagrams of oxygen (O_2) and carbon dioxide (CO_2) are shown in Figure 3.9, both of which contain double covalent bonds.

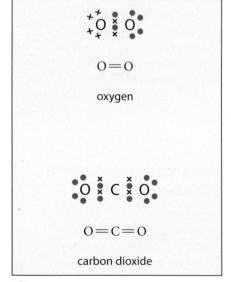

Figure 3.9
'Dot-and-cross diagrams for oxygen and carbon dioxide

Quick Questions

5 Draw electron 'dot-and-cross' diagrams for the following compounds.
   a ethane,   b ethene,   c nitrogen.

ethane   ethene   nitrogen

6 a Use Table 3.2 to draw structures for the following compounds using a single line to represent each covalent bond (e.g. for water this would be H—O—H).
   i H_2S
   ii NCl_3
   iii HCN
   iv NOCl
   b Construct 'dot-and-cross' diagrams for each of the structures you have drawn in part a.

Properties of simple molecular substances

The covalent bonds that hold atoms together within simple molecular substances are strong. So, the molecules do not readily break apart into atoms. But, the forces between separate molecules (intermolecular forces) are weak so it is easy to separate the molecules. This means that simple molecular substances:

▶ are usually gases or liquids at room temperature,
▶ have low melting points and boiling points,
▶ are usually more soluble in non-polar solvents such as hexane than in water,
▶ do not conduct electricity as solids, liquids or in solution because they have no ions or free electrons to carry the electric charge.

Giant molecular substances

A few non-metal elements, including carbon in diamond and graphite, and a few compounds of non-metals such as silicon(IV) oxide are composed of giant structures of billions upon billions of atoms linked together by covalent bonds (Section 4.13). The covalent bonds in these giant molecular structures are strong, which means that these substances:

▶ are hard solids,
▶ have very high melting points,
▶ do not usually conduct electricity,
▶ are insoluble in both polar solvents like water and non-polar solvents like hexane.

△ Figure 3.10
Quartz is silicon(IV) oxide. It is a very hard giant molecular substance in which vast numbers of silicon and oxygen atoms are held together by covalent bonds.

3.5 Describing covalent bonds in terms of overlapping orbitals

Chemists have developed the theory of atomic orbitals (Section 2.11) to explain the distribution of electrons in simple molecules. Molecular orbital theory extends the theory to molecules.

Molecular orbitals form when atomic orbitals overlap resulting in bonds between atoms. The shape of a molecular orbital, like an atomic orbital, shows the regions in space where there is a high probability of finding electrons.

A sigma (σ) bond is a single covalent bond formed by a pair of electrons when separate orbitals in two atoms in a molecule overlap. As a result of this overlap, there is a high electron density in the region between the two nuclei (Figure 3.11).

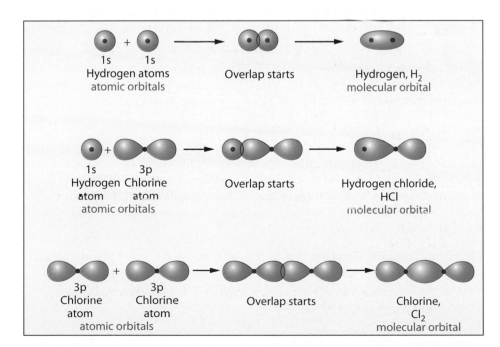

◁ Figure 3.11
The formation of sigma bonds in H_2, HCl and Cl_2

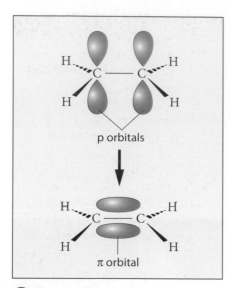

Figure 3.12
The formation of a π bond in ethene

Definition

A **co-ordinate bond (dative covalent bond)** involves the sharing of a pair of electrons between two atoms, both electrons in the bond being donated by one atom.

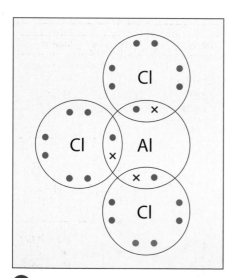

Figure 3.14
An electron 'dot-and-cross' diagram for AlCl₃

A sigma bond forms when:

- two s orbitals overlap as in H_2,
- an s orbital and a p orbital overlap as in HCl, or
- two p orbitals overlap as in Cl_2.

A pi (π) bond results when molecules have double and triple bonds between atoms. The first bond formed between a pair of atoms is always a sigma bond. When a second or third bond forms between a pair of atoms, atomic p orbitals on the two atoms *overlap sideways*. In the π bond which forms, the electron density is concentrated in two regions, one above and the other below the plane of the molecule on either side of the line joining the two nuclei (Figure 3.12).

3.6 Co-ordinate (dative covalent) bonding

In a normal covalent bond, each atom provides one electron towards the shared pair. However, in a few compounds, a bond is formed by the sharing of a pair of electrons both of which are provided by one atom. Chemists call this a **co-ordinate bond**, or alternatively a **dative covalent bond**. The word 'dative' meaning 'giving' is used because one atom gives both the electrons in forming the bond. Once a co-ordinate bond has formed, it is very much like a normal covalent bond.

In many molecules and ions, there are atoms with electron pairs in their outer shells which are not involved in any bonding between atoms. These unbonded electron pairs are called 'lone pairs'. Lone pairs are important in the formation of co-ordinate bonds and in determining the shapes of molecules. For example, water forms a co-ordinate bond when it reacts with a hydrogen ion, H^+ to produce an oxonium ion, H_3O^+. In a similar fashion, ammonia forms a co-ordinate bond when it reacts with a hydrogen ion to form an ammonium ion, NH_4^+ (Figure 3.13).

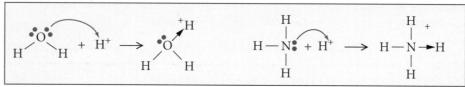

Figure 3.13
Co-ordinate bonds in the oxonium ion, H_3O^+ and the ammonium ion, NH_4^+

Notice in Figure 3.13 that the co-ordinate bonds are represented by an arrow in displayed formulae such as H_3O^+ and NH_4^+. The arrow points from the atom donating the pair of electrons to the atom accepting the electron pair.

Another compound in which co-ordinate bonding plays an important role is aluminium chloride. In the vapour phase at high temperatures, aluminium chloride consists of simple molecules of $AlCl_3$ with covalent bonds between the aluminium and chlorine atoms (Figure 3.14). This covalent bonding is unexpected because compounds of metals with non-metals like $AlCl_3$ are normally ionic. Notice also in Figure 3.14 that aluminium has only six electrons in its outer shell – two short of the noble gas structure for argon.

When gaseous aluminium chloride is cooled, $AlCl_3$ molecules join in pairs (dimerise) to form molecules of Al_2Cl_6. Monomers of $AlCl_3$ are held together in the Al_2Cl_6 dimer by co-ordinate bonds (Figure 3.15). Co-ordinate bonding is also present in the complex ions formed by metals (Section 24.7).

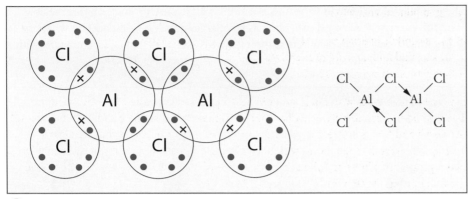

Figure 3.15
An electron 'dot-and-cross' diagram and a displayed structure to show the co-ordinate bonding in solid aluminium chloride, Al_2Cl_6

3.7 The shapes of simple molecules

Look closely at the 'dot-and-cross' diagrams and structures of the simple molecules in Figure 3.16.

Name	Beryllium chloride	Boron trichloride	Methane
'Dot-and-cross' diagram	:Cl × Be ×Cl:	(diagram of BCl_3)	(diagram of CH_4)
Structure	Cl — Be — Cl	(trigonal structure)	(tetrahedral structure)
Description of shape with respect to atoms	linear	trigonal	tetrahedral

Figure 3.16
Electron 'dot-and-cross' diagrams and structures for $BeCl_2$, BCl_3 and CH_4

Notice that the bonds in $BeCl_2$, BCl_3 and CH_4 spread out so as to be as far apart as possible. The three atoms in a $BeCl_2$ molecule are in a line and its shape is described as **linear**. The four atoms in BCl_3 are in the same plane with the chlorine atoms at the corners of a triangle. The shape is described as **trigonal** or trigonal planar. In CH_4, the four H atoms lie at the points of a tetrahedron with the C atom at its centre. This shape is described as **tetrahedral**.

Now look at the 'dot-and-cross' diagrams and structures of ammonia and water in Figure 3.17. In NH_3, the three H atoms form the base of a pyramid and the molecule is described as **pyramidal**. But why is ammonia pyramidal? Why is it not trigonal like BCl_3 which also has three atoms attached to a central atom?

The answer lies in the non-bonded lone pair of electrons on the nitrogen atom. This lone pair on the nitrogen atom occupies the fourth tetrahedral position around the N atom in the NH_3 molecule. Each of the N–H bonds in ammonia is composed of a region of negative

Quick Questions

7 **a** Draw 'dot-and-cross' diagrams to show outer shell electrons in:
        i NH_3 **ii** $AlCl_3$
    b Explain why NH_3 molecules readily react with $AlCl_3$ molecules to form $AlCl_3 \cdot NH_3$.

8 Use the displayed formulae of carbon monoxide and nitric acid below to draw their 'dot-and-cross' electronic structures.

C≡O   H—O—N (with two O)

Name	Ammonia	Water
'Dot-and cross' diagram	(diagram of NH_3)	(diagram of H_2O)
Structure	(pyramidal structure)	(bent structure)
Description of shape with respect to atoms	pyramidal	non-linear or V shaped

Figure 3.17
Electron 'dot-and-cross' diagrams and structures for ammonia and water

9 Why do the bonds in $BeCl_2$, BCl_3 and CH_4 get as far apart as possible?

10 Look carefully at the electron charge cloud model of H_2O in Figure 3.18.

  a How many lone pairs does the O atom in H_2O possess?

  b How many covalent bonds are there to the O atom in a H_2O molecule?

  c How many centres of negative charge are there around the O atom in H_2O?

  d What is the shape of a H_2O molecule with respect to negative centres around the central O atom?

  e Why is a H_2O molecule described as non-linear with respect to atoms and not linear?

11 What is unusual about the electron structures of Be and B in $BeCl_2$ and BCl_3 respectively?

Figure 3.19
The electronic structure and bonding in HCN and SO_2

Key Points

The **electron pair repulsion theory** says that the shapes of molecules and ions are dictated by the number of regions of negative charge in the outermost shell of their central atoms and that these regions of negative charge will repel each other and get as far apart as possible. The regions of negative charge may be lone pairs as well as single, double and triple bonds.

A **bond angle** is the angle between two covalent bonds in a molecule, ion or giant molecular structure.

charge similar to the lone pair. The nitrogen atom is therefore surrounded by four regions of negative charge which repel each other as far apart as possible. Consequently, the shape of ammonia, though pyramidal with respect to atoms, can be described as tetrahedral with respect to negative centres around the central nitrogen atom (Figure 3.18).

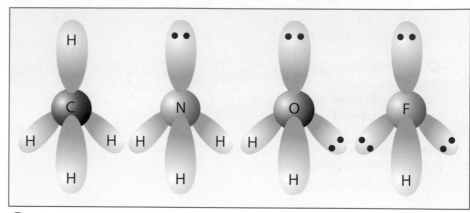

Figure 3.18
Electron charge cloud models of CH_4, NH_3, H_2O and HF

From the electron charge cloud models in Figure 3.18, you should appreciate that it is the number of regions of negative charge (not the number of covalent bonds) around the central atom which dictates the shape of a molecule. In predicting the shape of a molecule you must, therefore, count the number of negative centres around the central atom. Thus, CO_2 (Figure 3.9) and HCN (Figure 3.19) with two negative centres around their central C atoms are both linear. In CO_2, each double covalent bond counts as a single negative centre as does the triple covalent bond in HCN.

In contrast to CO_2, SO_2 has three centres of negative charge around the S atom and is therefore non-linear or V-shaped with respect to atoms (Figure 3.19).

Our discussions concerning the shapes of molecules are summarised in Table 3.3. Chemists have developed a very simple theory to explain and predict the shapes and **bond angles** of simple molecules and ions containing covalently bonded atoms. The simple theory is known as the **electron pair repulsion theory**.

Table 3.3
The number and angular separation of negative centres and the shapes of some simple molecules and ions

No. of negative centres around central atom	Angular separation of negative centres	Relative positions of negative centres	Example	Shape with respect to atoms	No. of lone pairs
2	180°	linear	$BeCl_2$, CO_2	linear	0
3	120°	trigonal	BCl_3, BF_3	trigonal	0
3	120°	trigonal	SO_2	non-linear or V-shaped	1
4	109°	tetrahedral	CH_4, NH_4^+, BH_4^-, CCl_4	tetrahedral	0
4	109°	tetrahedral	NH_3, H_3O^+	pyramidal	1
4	109°	tetrahedral	H_2O, OF_2	non-linear or V-shaped	2
6	90°	octahedral	SF_6	octahedral	0

The electron 'dot-and-cross' diagram and octahedral structure of sulfur hexafluoride are shown in Figure 3.20. There are six regions of negative charge around the central S atom in SF_6. These push each other as far apart as possible. This results in four F atoms arranged in a square around the S atom with the other two F atoms above and below the square. If lines are drawn from each F atom to its nearest neighbours, the result is an octahedron (a solid with eight sides). Each F atom is the same distance from the central S atom and the bond angles are all equal.

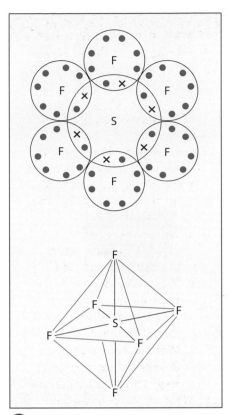

Quick Questions

12 Draw 'dot-and-cross' diagrams showing only electrons in the outer shell and predict the shapes of the following simple molecules:
   a BF_3 **b** SiH_4 **c** OF_2
13 Draw a 'dot-and-cross' diagram to explain why:
   a the H_3O^+ ion is pyramidal,
   b the NH_4^+ ion is tetrahedral.

The fine structure of methane, ammonia and water

Look carefully at the 'dot-and-cross' diagrams and the bond angles of methane, ammonia and water in Figure 3.21. Notice that all three substances have the same electron structure with four pairs of electrons in the outer shell of the central atom. However, in methane all four pairs of electrons are shared pairs, in ammonia three of the four pairs are shared pairs while the fourth pair is a lone pair. In water, there are also four pairs of electrons around the central oxygen atom – two shared pairs and two lone pairs.

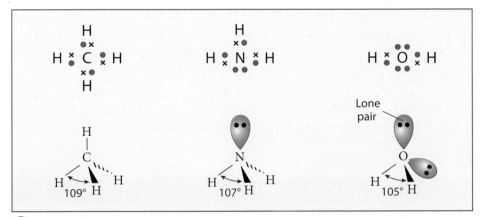

▲ Figure 3.20
The electron 'dot-and-cross' diagram and the octahedral structure of SF_6

▲ Figure 3.21
'Dot-and-cross' diagrams and bond angles in methane, ammonia and water

In CH_4, NH_3 and H_2O, the four pairs of electrons around the central atom take up tetrahedral positions, but lone pairs are closer to the central atom than shared pairs. This means that lone pairs have a stronger repelling effect than shared pairs. This explains why the bond angle in ammonia is less than that in methane and the bond angle in water is less than that in ammonia.

Key Point

A lone pair of electrons exerts a greater repelling effect than a bonded (shared) pair.

3.8 Metallic bonding

Metals are very important materials with an exceptional range of uses including bridges, cutlery, jewellery, ornaments, pipes, radiators and vehicles. X-ray analysis (Section 4.8) shows that the atoms in most metals are packed together as close as possible. This structure is usually described as **close packing**.

Figure 3.22 shows the arrangement of a few atoms in one layer of a metal crystal.

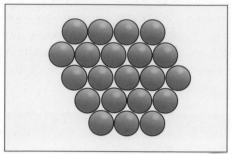
⬆ Figure 3.22
Close packing of atoms in one layer of a metal structure

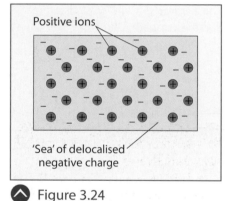
⬆ Figure 3.23
Close packing of atoms in adjacent layers of a metal structure

Notice that each atom in the middle of the layer 'touches' six other atoms in the same layer. When a second layer is placed on top of the first, atoms in the second layer sink into the dips between atoms in the first layer (Figure 3.23).

This close packing allows metal atoms in one layer to get as close as possible to those in adjacent layers above and below forming a giant structure of closely packed atoms. Each atom touches 12 other atoms altogether; 6 in the same layer, 3 in the layer above and 3 in the layer below. The regular arrangement of the metal atoms is usually described as a **giant lattice**.

Within the giant lattice, electrons in the outer shell of each atom can drift through the whole structure. These outer shell electrons, which are free to move and are shared by several atoms, are described as 'delocalised'. This has led chemists to picture metallic structures as giant lattices of positive ions with electrons moving around and between them in a 'sea' or 'cloud' of delocalised negative charge (Figure 3.24). The strong electrostatic attractions between the positive metal ions and the 'sea' of delocalised electrons bind the metal structure strongly together as a single unit. As a result of these strong forces between atoms, metals:

- have high melting points and boiling points,
- have high densities,
- are good conductors of heat and electricity,
- are malleable (Section 4.9).

Quick Questions

14 Explain the following terms:
   a lattice,   **b** delocalised electrons.
15 Use the ideas of close packing and delocalised electrons to explain why metals usually:
   a have high melting points,
   b have high densities,
   c are good conductors of electricity.

3.9 Polar bonds

When a covalent bond is between two identical atoms, such as Cl_2, the electrons are shared equally. But the electrons in a covalent bond will not be shared equally if the atoms joined by the bond are different. The nucleus of one atom will attract the electrons in the bond more strongly than that of the other atom. As a result of this,

Definitions

Metallic bonding involves a strong electrostatic attraction between a lattice of positive metal ions and a 'sea' of delocalised electrons.

Delocalised electrons are bonding electrons which are not fixed between two atoms in a bond. They are free to move throughout the structure.

one end of the bond will have a small excess of negative charge, δ– and the other end of the bond will have a small deficit of negative charge resulting in a partial positive charge, δ+. Covalent bonds in which this charge separation occurs are described as **polar** and the charge separation in the bond is known as **bond polarity**. The partial charges at the ends of covalent bonds result in a spectrum of polarity from zero, in molecules such as chlorine and oxygen, to high polarity, in molecules such as hydrogen chloride (Figure 3.25).

So, polar molecules, like HCl, have a tiny positive electric pole, labelled δ+ and a tiny negative electric pole labelled δ–. These two poles of opposite charge in a molecule are described as a **dipole**.

Electronegativity

Chemists use **electronegativity** values to determine the polarity of covalent bonds. The electronegativity of a particular element is a measure of its pull on electrons relative to other atoms.

Elements, such as chlorine and oxygen, which have a stronger pull on shared electrons in a covalent bond than elements such as hydrogen have higher values of electronegativity. So, in a polar O—H bond, there is a δ– charge on the oxygen atom and a δ+ charge on the hydrogen.

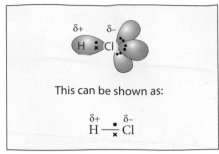

This can be shown as:

Figure 3.25
The polar covalent bond in a molecule of hydrogen chloride. The uneven distribution of electrons in the covalent bond leads to a partial charge of δ+ on the H atom and a partial charge of δ– on the Cl atom. The Greek letter, delta (δ) is used by scientists to indicate a small change in a physical quantity.

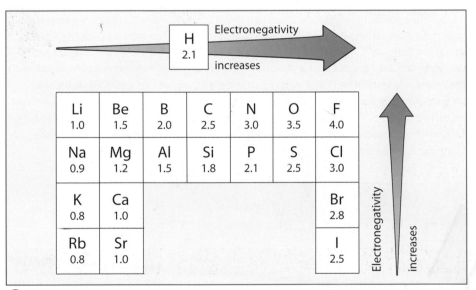

Figure 3.26
Electronegativity trends and values for elements in Periods 1, 2 and 3 and Groups I, II and VII in the periodic table

A number of electronegativity scales have been devised. The values in Figure 3.26 are from that prepared by Linus Pauling (1901–94).

Electronegativity values, such as those in Figure 3.26, compare one element with another, so it is useful to know the trends in electronegativity across the periods and down the groups in the periodic table. As expected, electronegativity increases across and up the periodic table towards the most reactive non-metals, oxygen, fluorine and chlorine. The greater the difference in electronegativity between the elements forming a covalent bond, the more polar the bond.

16 Why are there no values for helium and neon in Figure 3.26?
17 With which element in Figure 3.26 would:
   a chlorine form the least polar bonds,
   b phosphorus form the least polar bonds?
18 Put the following bonds in order of polarity from most to least polar:
   C—Br, C—Cl, C—H, C—I.
19 **a** Write the electronic shell structure for fluorine and chlorine.
   b Use your answers to part a and the concept of shielding to explain why
   fluorine is more electronegative than chlorine.

3.10 Polar and non-polar molecules

Figure 3.27 shows apparatus which can be used to check whether the molecules of a liquid are polar or non-polar. When a positive or negative charged rod is placed close to the jet of water, the water is deflected from its vertical path towards the charged rod. If, however, the water is replaced with tetrachloromethane or hexane, there is no deflection of the jet.

Now, look at the structures of the five simple molecules in Figure 3.28. Both the covalent bonds in water are polar and the water molecule is also polar. There are, however, molecules like carbon dioxide, tetrachloromethane and hexane that are non-polar even though they contain polar bonds. In carbon dioxide, the two C═O bonds are arranged symmetrically on the central carbon atom so that their polarities cancel each other out.

Similarly, the four polar C—Cl bonds in CCl$_4$ are also arranged symmetrically around the central carbon atom. Their polarities cancel each other out and CCl$_4$ molecules are non-polar. In hexane, the molecule is more complex than CO$_2$ and CCl$_4$, but the overall symmetry of similar bonds cancels out any overall polarity. In contrast to CO$_2$, CCl$_4$ and hexane, the polar bonds in water and trichloromethane are not symmetrical. The polarity does not cancel out and the molecules are polar.

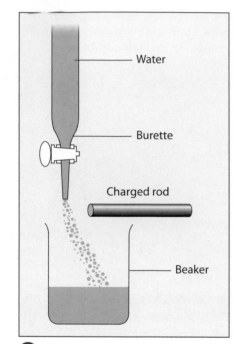

▲ Figure 3.27
The effect of a charged rod on a fine jet of water

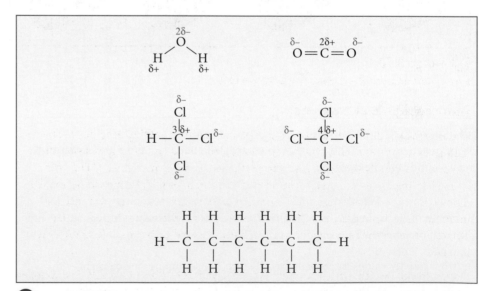

▲ Figure 3.28
Molecules with polar bonds can be polar or non-polar overall. Those which are polar overall have polar bonds in which the polarities do not cancel each other out. In those molecules which are non-polar overall, the bond polarities cancel out.

When a fine jet of a polar liquid runs past a charge rod as in Figure 3.27, the liquid is attracted to the rod whether it is positive or negative. This is because polar molecules in the jet will rotate and move until the charge on one part of the molecules is attracted to the opposite charge on the rod.

Definition

Polar molecules have polar bonds in which the polarities do not cancel each other out and the whole molecule is overall polar.

Quick Questions

20 Draw the shapes of the following molecules and show the polarity of their bonds. Then decide whether each molecule is overall polar or non-polar.

a $BeCl_2$ **b** HI **c** BF_3 **d** NH_3
e OF_2 **f** CH_4 **g** CH_3Cl **h** Br_2

3.11 Intermolecular forces

Earlier in this chapter, we studied the bonding between ions in ionic compounds, between atoms in covalently-bonded molecules and between atoms in metallic structures. We must now consider the forces between molecules.

Permanent dipole attractions

In the last section, we discovered that molecules with a non-symmetrical distribution of charge, such as H_2O and $CHCl_3$, were polar with permanent dipoles.

The interactions between these permanent dipoles explain the attractions between neighbouring molecules. The δ+ charge on one molecule tends to attract the δ– charge on other molecules and vice versa (Figure 3.29). These attractions between polar molecules are called **permanent dipole attractions**.

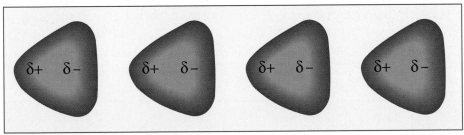

Figure 3.29
Permanent dipole attractions in polar molecules

Induced dipole attractions

The existence of dipole attractions will explain the forces holding together polar molecules in liquids such as water and trichloromethane, but what about non-polar molecules in liquids such as tetrachloromethane, bromine and hexane? How can we explain the forces between these non-polar molecules which have no permanent dipole? The fact that all these substances can exist as liquids is clear evidence for intermolecular forces between their molecules. If there were no intermolecular attractions between these non-polar molecules, it would be impossible to turn them into liquids or solids.

During the 1870s, the Dutch scientist Johannes van der Waals (1837–1923) developed a theory to explain the existence of intermolecular forces between non-polar molecules, such as bromine, tetrachloromethane and hexane.

The electrons in atoms and molecules are in continual motion. At any particular moment, the electron charge cloud around the molecule will not be perfectly symmetrical. There will be more negative charge on one side of the molecule than on the other. The molecule has an *instantaneous electric dipole* and this dipole will induce dipoles in neighbouring molecules. Positive dipoles will tend to induce negative dipoles and vice versa. In this way, **induced dipole attractions** exist between non-polar molecules (Figure 3.30). These induced dipoles will act first one way, then another way, continually forming and then disappearing due to electron movements.

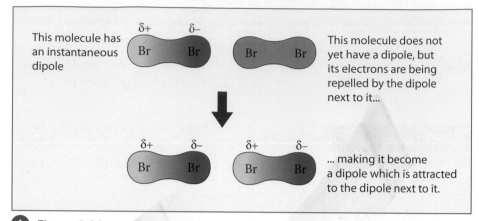

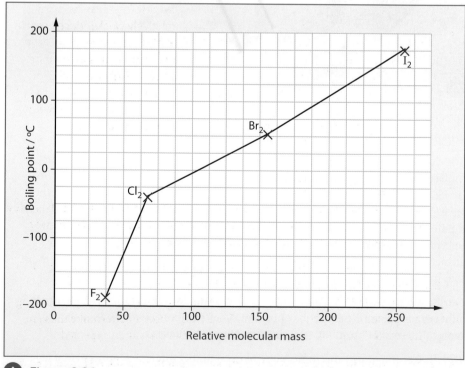

∧ Figure 3.30

Induced dipole attraction between non-polar bromine molecules

However, the force between induced dipoles is always an attraction: positives inducing negatives and vice versa. So, although the average dipole on every non-polar molecule over a period of time is zero, the resultant forces between molecules at any instant are not zero.

As the size of a molecule increases, the number of constituent electrons increases. As a result, the induced dipole attractions between molecules become stronger. This explains why the boiling points increase down Group 0 (the noble gases) and Group VII (the halogens), as in Figure 3.31.

∧ Figure 3.31

Boiling points for the elements in Group VII plotted against relative molecular mass

Induced dipole and permanent dipole attractions are sometimes classed together as van der Waals forces. They are, of course, much weaker than covalent, co-ordinate and ionic bonds. In general, their values are between one-hundredth and one-tenth the strength of covalent bonds.

For example, the energy of sublimation for solid chlorine (i.e. the energy required to overcome the induced dipole attractions between one mole of non-polar Cl_2 molecules) is only $25 \, kJ \, mol^{-1}$. In comparison, the bond energy of chlorine (i.e. the energy required to break one mole of Cl—Cl covalent bonds) is $244 \, kJ \, mol^{-1}$.

Finally, it is important to appreciate that induced dipole interactions will also occur in polar molecules in addition to the permanent dipole interactions we discussed earlier in this section.

Quick Questions

21 Why do the boiling points of noble gases increase as their relative atomic mass increases?
22 Which of the following molecules have a permanent dipole?
   a GeH_4 **b** ICl **c** SiF_4 **d** CH_2Cl_2 **e** CO_2

◀ Figure 3.32
Solid graphite in a soft pencil can be used as a lubricant for zips, because there are relatively weak induced dipole attractions between the layers of carbon atoms in graphite. The relatively weak induced dipole attractions allow the layers of graphite to move over each other smoothly.

3.12 Hydrogen bonding

Look closely at the graphs in Figure 3.33. These show the boiling points of the hydrides in Groups IV, V, VI and VII. Notice that the boiling points of the Group IV hydrides increase with increasing relative molecular mass from CH_4 to SnH_4.

However, the boiling points of the hydrides in Groups V, VI and VII do not follow a similar pattern. In Group VI for example, the boiling points rise with relative molecular mass from H_2S, through H_2Se to H_2Te, but H_2O has a much higher boiling point than expected. The hydrides in Group V and VII follow a pattern like those in Group VI. Here NH_3 and HF have much higher boiling points than expected.

How can we account for these unusually high boiling points of H_2O, NH_3 and HF? In water, liquid ammonia and liquid hydrogen fluoride, there must be unusually strong intermolecular forces. But why is this?

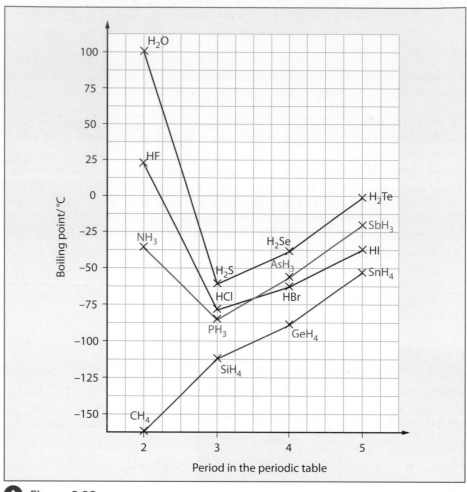

Figure 3.33
Patterns in the boiling points of the hydrides in Groups IV, V, VI and VII

H_2O, NH_3 and HF are all very polar because they contain the three most electronegative elements (oxygen, nitrogen and fluorine) bonded directly to hydrogen, which is weakly electronegative. This results in exceptionally polar molecules with stronger intermolecular forces than usual. These particularly strong intermolecular forces are known as **hydrogen bonds**.

What is a hydrogen bond?

Nitrogen, oxygen and fluorine are the three most electronegative elements. When they are bonded to a hydrogen atom, the electrons in the covalent bond are drawn towards the electronegative atom. Remember also that the H atom has no electrons other than its share of those in this covalent bond and these are being pulled away from it by the more electronegative N, O or F.

As the H atom has no inner shell of electrons, the single proton in its nucleus is exceptionally 'bare' and very attractive to any lone pair of electrons on another H_2O, NH_3 or HF molecule. Thus, H atoms attached to N, O or F can interpose themselves between two of these atoms. The H atoms are covalently bonded to one of the electronegative atoms and hydrogen bonded to the other (Figure 3.34).

The essential requirements for hydrogen bonding are:

▸ a hydrogen atom bonded to a highly electronegative atom and
▸ an unshared pair of electrons on a second electronegative atom.

In practice, this means that hydrogen bonding will usually occur in any substance containing a hydrogen atom attached to a nitrogen, oxygen or fluorine atom.

In the NH_3 molecule, there are three N—H bonds, but only one lone pair of electrons on the N atom. This means that there can be only one hydrogen bond per molecule. In water, however, there are two O—H bonds and two unshared electron pairs per molecule. This means that each H_2O molecule can form two hydrogen bonds per molecule.

Quick Questions

23 Which of the following solids contain(s) more than one type of chemical bond?
    A copper **B** graphite **C** ice
24 **a** How many H—F bonds are there in one HF molecule?
    b How many unshared electron pairs are there in one HF molecule?
    c What is the maximum number of hydrogen bonds formed by one HF molecule?
25 In which one of the following processes are hydrogen bonds broken?
    A $NH_3(g) \longrightarrow NH_2(g) + H(g)$
    B $NH_3(l) \longrightarrow N(g) + 3H(g)$
    C $NH_3(l) \longrightarrow NH_3(g)$

3.13 Estimating the strength of hydrogen bonds in water

Hydrogen bonds in water are intermolecular forces. They are part of the forces which hold the molecules together. When water is vaporised, energy is needed to separate the molecules. This energy which vaporises the water is needed to overcome the forces holding the water molecules together as a liquid.

Thus, the molar enthalpy change of vaporisation of water gives a measure of the intermolecular forces between one mole of water molecules. However, these intermolecular forces include van der Waals forces (permanent dipole and induced dipole attractions) as well as hydrogen bonds. So, how can we estimate the strength of hydrogen bonding alone?

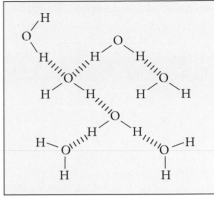

Figure 3.34
Hydrogen bonding in water

Definition

Hydrogen bonds are extra strong intermolecular attractions stronger than permanent dipole and induced dipole attractions, but weaker than covalent bonds.

Key Point

Hydrogen bonding occurs in any substance containing a hydrogen atom attached directly to a nitrogen, oxygen or fluorine atom.

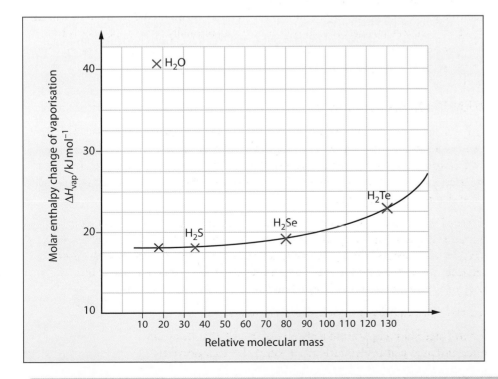

◀ Figure 3.35
A graph of the molar enthalpy changes of vaporisation for the hydrides of Group VI plotted against their relative molecular mass

26 In which of the following compounds will hydrogen bonding occur?
  a CH_3NH_2
  b CH_3Cl
  c HCN
  d CH_3OH
  e $HOCl$
  f CH_3OCH_3

27 The approximate strength of hydrogen bonding in water is $22\,kJ\,mol^{-1}$, whereas that in ammonia is $13\,kJ\,mol^{-1}$. Why do you think the strength of hydrogen bonds in water is about twice that in ammonia?

Figure 3.35 shows a graph of the molar enthalpy changes of vaporisation (symbol ΔH_{vap}) for the hydrides of elements in Group VI plotted against their relative molecular masses. Now, if we assume that H_2S, H_2Se and H_2Te have intermolecular forces due only to van der Waals forces (i.e. negligible hydrogen bonding), we can estimate a value for the strength of van der Waals forces in water.

We do this by extrapolating the curve in Figure 3.35 through the values of ΔH_{vap} for H_2S, H_2Se and H_2Te to the relative molecular mass of water. The value we want is marked by a red cross in Figure 3.35.

This gives a predicted molar enthalpy change of vaporisation for water of $+18.5\,kJ\,mol^{-1}$ assuming that water has only van der Waals forces.

Total strength of intermolecular forces
(H bonds + van der Waals forces) in water $= 40.7\,kJ\,mol^{-1}$

Estimated strength of van der Waals forces in water $= 18.5\,kJ\,mol^{-1}$

Therefore, approximate strength of H bonds in water $= 40.7 - 18.5 = 22.2\,kJ\,mol^{-1}$

Usually, the strength of H bonds are in the range $5-40\,kJ\,mol^{-1}$. So, hydrogen bonds are about one-tenth as strong as covalent bonds, but stronger than induced dipole and permanent dipole attractions. Remember, though, that molecules which are hydrogen bonded will also be attracted by permanent and induced dipoles. In general, we can say hydrogen bonds are weak forces, but strong enough to influence the physical properties of a substance.

3.14 The effect and importance of hydrogen bonding in water and ice

Water has unexpectedly high values for its melting point, boiling point, molar heat of fusion and molar heat of vaporisation. These high values result from the additional attractions between water molecules due to hydrogen bonding. This extra bonding between water molecules also causes high surface tension and high viscosity. The high surface tension of water provides a sort of 'skin effect' and it is possible for small but relatively dense articles, such as razor blades and beetles, to float on an undisturbed water surface (Figure 3.36).

The presence of two hydrogen atoms and two lone pairs in each water molecule results in a three-dimensional tetrahedral structure in ice. Each oxygen atom in ice is surrounded tetrahedrally by four others via two covalent bonds and two hydrogen bonds (Figure 3.38). This arrangement of water molecules in ice creates a very open structure which accounts for the fact that ice is less dense than water. When ice melts, the regular lattice breaks up and water molecules can then pack more closely. So water at $0\,°C$ has a greater density than ice at the same temperature.

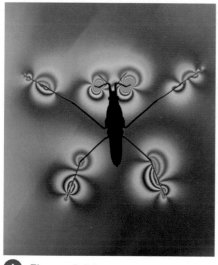

Figure 3.36

Walking on water! A pond skater resting on the surface of a pond – it's all done by hydrogen bonding and surface tension.

Figure 3.37

Ice and icebergs float on water. The arrangement of water molecules in ice creates a very open structure making it less dense than water.

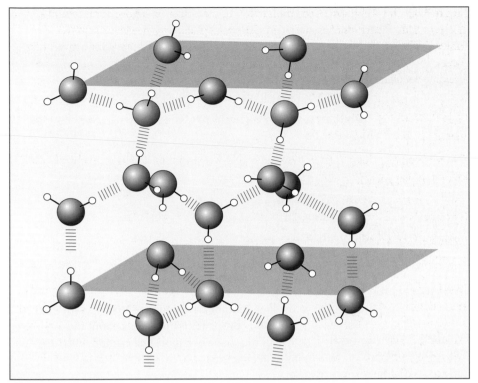

Figure 3.38
The structure of ice. Oxygen atoms are shown as red balls and hydrogen atoms as small white balls. Notice how oxygen atoms in the centre of the structure have 2 covalent bonds and 2 hydrogen bonds pointing towards other nearby oxygen atoms.

The anomalous physical properties of ice and water that result from hydrogen bonding have a great influence biologically and environmentally.

The fact that ice is less dense than water at 0 °C means that ponds and lakes freeze from the surface downwards. The layer of ice insulates the water below, preventing complete solidification. If ice were the more dense, water would freeze from the bottom upwards. In this case, ponds would freeze completely, killing fish, aquatic plants and other water-living creatures.

Water also has an unusually high boiling point owing to hydrogen bonds. Without these hydrogen bonds, water would be a gas under normal atmospheric conditions. Oceans, lakes and rivers would never exist and it would never rain!

The exceptionally high polarity and surface tension of water are important for both plants and animals. The high polarity of water means that polar and ionic substances are usually soluble in water. This allows plants to obtain the salts which they require for growth by absorption into the sap through their roots. The high surface tension of water then enables the sap to rise from the roots through the stems to the leaves and flowers.

In animals, essential ionic substances can dissolve in the water of food and drink and then get absorbed into their bloodstreams through the lining of the intestines.

There is more on hydrogen bonding related to proteins and DNA in Sections 28.3 and 28.5, respectively.

Review questions

1 Draw 'dot-and-cross' electron structures showing electrons in the outermost shell of each atom in the following compounds. Show the overall charge on each ion in the ionic compounds. (None of the compounds involve co-ordinate bonding except those in part c).)

  a   HCN, PCl_3, HNO_2, CH_3OH

  b $CaCl_2$, Na_3P, Al_2S_3

  c   CO, $BF_3 \cdot NH_3$, SO_2

2 How is the type of bonding in the chlorides of the elements Na, Mg, Al, Si, P and S related to

  a   their position in the periodic table,

  b   the number of electrons in the outermost shells of these elements?

3 Draw 'dot-and-cross' diagrams and predict the shapes with respect to atoms for molecules of the following compounds:

    SF_6, $POCl_3$, $SOCl_2$, H_3O^+, BF_3, C_2H_2

4 a   State the electronic configurations of the following atoms (e.g. Be would be 2, 2):

    C,   N,   O,   F

  b   Draw a series of 'dot-and-cross' diagrams to show the structures of the simplest hydrides formed by carbon, nitrogen, oxygen and fluorine.

  c   Sketch and describe the shapes of the molecules in part b and discuss the influence of any lone pairs of electrons on these shapes.

  d   What shape would you predict for the following?

    NH_4^+, NH_3, NH_2^-

5 X, Y and Z represent elements with proton numbers of 9, 19 and 34.

  a   Write the electronic structures for X, Y and Z (e.g. Be would be 2, 2).

  b   Predict the type of bonding which you would expect to occur between

    i   X and Y,   ii   X and Z,   iii   Y and Z.

  c   Draw 'dot-and-cross' diagrams for the compounds formed in part b, showing only the electrons in the outermost shell for each atom.

  d   Predict, giving reasons, the relative

    i   volatility,

    ii   electrical conductivity

    iii   solubility in water

    of the compound formed between X and Y compared with that formed between X and Z.

6 The structural formulae, boiling points and densities of the isomers pentane and 2,2-dimethylpropane are shown below.

	pentane	2,2-dimethyl-propane
Structural formula	$CH_3-CH_2-CH_2-CH_2-CH_3$	$CH_3-\overset{\displaystyle CH_3}{\underset{\displaystyle CH_3}{C}}-CH_3$
Boiling point/°C	36	9
Density /g cm^{-3}	0.626	0.591

  a   Why does pentane have a higher boiling point than 2,2-dimethylpropane? (Hint: Think about the way in which the shape of a molecule can influence its intermolecular forces.)

  b   Why does pentane have a higher density than 2,2-dimethylpropane?

  c   2-Methylbutane is an isomer of pentane and 2,2-dimethylpropane. How do you think its boiling point and density will compare with these two substances? Explain your answer.

7 The relative molecular masses and molar enthalpy changes of vaporisation for three of the hydrides of elements in Group V are given below.

Compound	Relative molecular mass	$\Delta H^{\ominus}_{vap}/$ kJ mol^{-1}
NH_3	17	+23.4
PH_3	34	+14.6
AsH_3	78	+17.5

  a   Plot a graph of $\Delta H^{\ominus}_{vap}$ against relative molecular mass for the three hydrides.

  b   Why is the value of $\Delta H^{\ominus}_{vap}$ for NH_3 unexpectedly high?

  c   Use your graph to estimate a value for the $\Delta H^{\ominus}_{vap}$ of NH_3 assuming that it has only van der Waals forces between molecules.

  d   Predict a value for the strength of hydrogen bonds in NH_3.

8 Suggest reasons for the following:

  a   The boiling points of water, ethanol and ethoxyethane ($CH_3CH_2OCH_2CH_3$, diethyl ether) are in the reverse order of their relative molecular masses, unlike those of their analogous sulfur compounds H_2S, C_2H_5SH and $C_2H_5SC_2H_5$.

  b BF_3 is non-polar, but NF_3 is polar.

4 States of matter

4.1 Evidence for moving particles

The Greeks were probably the first people to believe that matter was composed of particles. In 60 BCE, the Roman poet Lucretius suggested that matter existed in the form of invisible particles (Figure 4.1). He wrote about this idea in his poem, 'The Nature of the Universe.'

Unfortunately, the Greeks rarely performed experiments. Their theories remained nothing more than good ideas. They had, however, a vast range of everyday experience which supported their beliefs that matter was particulate. They knew, for instance, that a small amount of flavouring such as pepper or ginger could give a whole dish a strong, distinctive taste. This suggested that tiny particles in the pepper had spread throughout the whole dish.

The ancient Greeks also knew that when dyes such as Tyrian purple were dissolved in water, a tiny amount of the dye could colour an enormous volume of solution. This supported the idea that there must be many particles of the purple pigment in only a little solid and that the particles must be very small.

Quick Questions

1 Use the idea of particles to explain the following:
   a It is possible to smell the perfume a person is wearing from some distance.
   b Solid blocks of air freshener provide a pleasant smell in bathrooms. They disappear after a time without leaving any liquid.

Particles of vapour from the perfume and the air freshener which have a distinctive smell *mix* with air particles and *move* to other parts of the room. This mixing and movement of matter which results from the kinetic energy of moving particles is called **diffusion**.

The idea that particles in gases and liquids are moving has been confirmed by many other experiments involving diffusion.

Another phenomenon which provides strong evidence for moving particles is **Brownian Motion**.

During the 1820s, the botanist, Robert Brown, was carrying out a detailed study of pollen grains. At first, Brown believed that he would be able to observe the pollen grains more effectively through his microscope if they were suspended in water. Unfortunately, he found that the pollen continually jittered around in the water in a random manner. As a result of Brown's observation, the random movement of solid particles suspended in a liquid or in a gas is called Brownian Motion.

Observing the Brownian Motion of smoke particles in air

Brown's observation of pollen grains was the first recorded example of Brownian Motion. However, it is possible to watch Brownian Motion more conveniently using smoke particles in air. Figure 4.3 shows the apparatus you could use for this experiment.

Using a teat pipette inject smoke from a smouldering piece of string into the small glass smoke cell. Close the cell with a cover slip. When the illuminated cell is viewed through the microscope the smoke particles look like tiny jittering pin points of light.

▲ Figure 4.1
The Roman poet Lucretius. In 60 BCE Lucretius, in his poem 'The Nature of the Universe', suggested that matter was made up of invisible particles

▲ Figure 4.2
A male emperor moth can smell the pheromones (sex attractants) from a female moth as far as 8 km away

Quick Questions

2 a What particles are present in water? How are these particles moving?
   b What caused the pollen grains to jitter about in Brown's experiment? Why do they move so randomly and haphazardly?

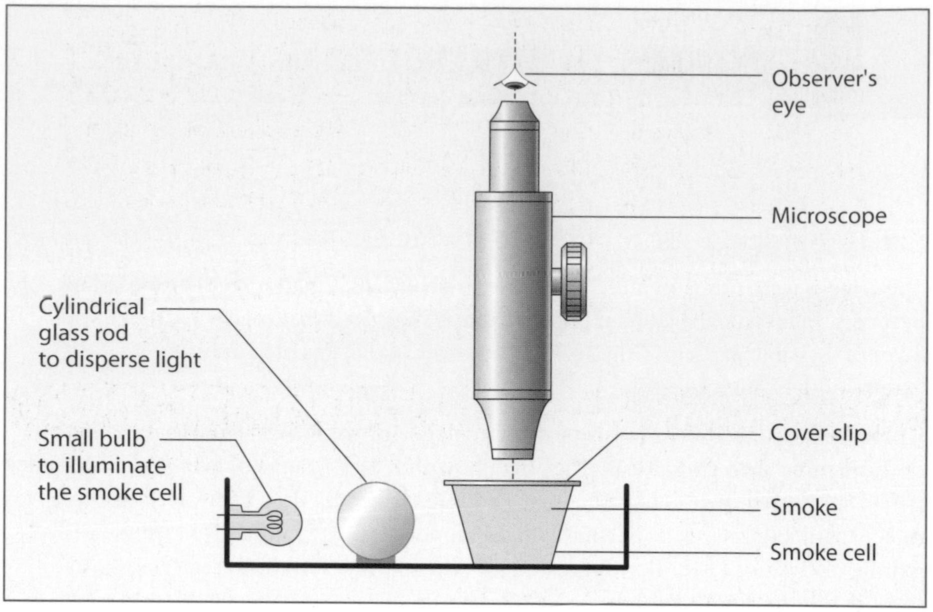

Figure 4.3
Observing the Brownian Motion of
smoke particles in air

Labels on the figure:
- Observer's eye
- Microscope
- Cylindrical glass rod to disperse light
- Cover slip
- Small bulb to illuminate the smoke cell
- Smoke
- Smoke cell

Quick Questions

3 Why do the smoke particles look like tiny pin points of light through the microscope?

4 What do you think will happen if the heat from the small lamp causes the temperature of the air and smoke in the cell to rise significantly?

The smoke particles which you see through the microscope are under constant bombardment from even smaller, invisible molecules in the air. Each individual smoke particle is bombarded haphazardly on one side or the other. So, it moves first this way and then that way in a random jittery motion. The movement of the smoke particles which you can see provides strong evidence for the movement of air molecules which, of course, you cannot see.

4.2 The kinetic–molecular theory

Our ideas about the movement of particles in solids, liquids and gases are summarised in the **kinetic–molecular theory** which is often simply called the kinetic theory. The word 'kinetic' is derived from the Greek word *kineo*, which means 'I move'. The main points in the kinetic–molecular theory can be summarised as follows:

1 All matter is composed of tiny, invisible particles. (Note that the term 'particles' is used rather loosely here to mean the basic units making up a substance which may be molecules, atoms or ions.)

2 Particles of different substances are different in size.

3 In solids, the particles are relatively close together. They have less energy than the same particles in the liquid and gaseous states at higher temperatures. Consequently, solid particles cannot overcome the strong forces of attraction holding them together. They can only vibrate about fixed positions in the solid crystal. So, particles in a solid have vibrational and rotational motion, but no translational motion (Figure 4.4).

4 In liquids, the particles are slightly further apart than in solids and have larger amounts of energy. Thus, they are able to overcome the forces between each other to some extent. They can move around each other even though they are close together. The liquid particles have vibrational, rotational and translational motion.

5 In gases, the particles are much more widely separated than those in solids and liquids and they have much more energy than the same particles in the solid and liquid states at lower temperatures. The gas particles have sufficient energy to overcome the forces of attraction between each other almost completely. They move rapidly, randomly and haphazardly into any space available.

6 An increase in temperature causes an increase in the average kinetic energy of particles. The kinetic energy is present as vibrational, rotational and translational energy in gases and liquids and as vibrational and rotational energy in solids.

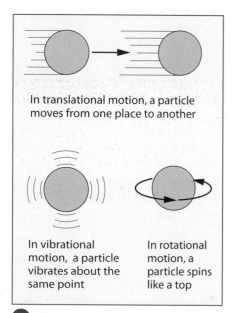

In translational motion, a particle moves from one place to another

In vibrational motion, a particle vibrates about the same point

In rotational motion, a particle spins like a top

Figure 4.4
The translational, vibrational and rotational motion of a particle

Property	Solids	Liquids	Gases
Rigidity/ fluidity	Rigid	Fluid	Fluid
Volume	Fixed	Fixed	Take volume of container
Shape	Fixed	Take shape of container below their surface	Take shape of whole container
Relative compressibility	Nil	Almost nil	Large
Relative density	Large	Large	Very small

◀ Table 4.1
Comparing the bulk (macroscopic) properties of solids, liquids and gases

Quick Questions

5 This question concerns some of the bulk (macroscopic) properties of solids, liquids and gases in Table 4.1. Use the kinetic–molecular model to answer the following questions:
  a Why are solids so rigid?
  b Why do liquids have a fixed volume, yet they take the shape of their container below their surface?
  c Why is the relative compressibility of gases so large?
  d Why are the relative densities of liquids similar to solids?

4.3 Changes of state

Solids, liquids and gases are called the **three states of matter**. A transition from one state to another is called a **change of state**. The kinetic–molecular model can be used to explain what happens when a substance changes from one state to another. A summary of the different changes of state is shown in Figure 4.6. These changes are usually caused by heating or cooling the substance.

▲ Figure 4.5
The particles in a gas are very far apart and the forces between such widely spaced particles are extremely weak. This means that gas particles will readily move away from each other and occupy the whole volume of their container, as in this massive hot air balloon.

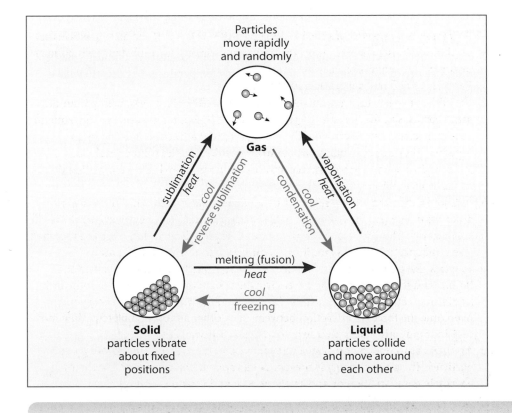

Key Point

The kinetic–molecular theory

Make sure you know and understand the main points in the kinetic–molecular theory summarised at the beginning of this section.

◀ Figure 4.6
Changes of state (red arrows represent endothermic changes and blue arrows exothermic changes)

Figure 4.7

The forces between metal atoms in steel are so strong that chains of steel can be used to lift very heavy loads.

Definitions

The **melting point** of a solid is the temperature at which it changes to a liquid at 1 atm pressure.

The **boiling point** of a liquid is the temperature at which it changes to a gas throughout the whole liquid at 1 atm pressure.

The **vapour pressure** of a liquid, at a particular temperature, is the pressure of the vapour in equilibrium with the liquid.

Figure 4.8

Hoar frost forms by reverse sublimation when water vapour in cold, dry air turns to ice.

Melting and freezing

Melting is an endothermic process. When a solid is heated, its particles gain energy. The particles vibrate faster and faster until they have sufficient energy to overcome the attractions of their neighbours and break away from their fixed positions. The particles begin to move around each other and the solid melts to form a liquid.

The temperature at which a solid melts tells us how strongly its particles are held together. Metals, like iron and steel, have high melting points. This suggests that there are strong forces between metal atoms and this is why metals can be used as girders and supports.

Freezing is an exothermic process – the reverse of melting. As a liquid cools, its particles slow down and eventually take up fixed positions about which they vibrate. The liquid has changed from a liquid to a solid.

Vaporisation and condensation

Vaporisation is an endothermic process. When a liquid is heated, its particles gain energy. The liquid particles are still very close together, but they can move around each other in a restricted way. On further heating, the particles move around faster and collide more frequently. Eventually, they have enough energy to overcome completely the attractions of other particles in the liquid and escape to form a gas.

Condensation is the reverse of vaporisation. When particles in a gas slow down, the gas condenses. Intermolecular attractions are strong enough to hold the less energetic particles close together and the liquid re-forms.

Evaporation and boiling both involve vaporisation. However, there are differences between the two processes. Evaporation takes place only at the surface of a liquid and can happen at any temperature. If the liquid is in a sealed container, particles which have already evaporated to form a vapour will collide with the liquid surface and condense. After a while, evaporation and condensation will take place at the same rate and the system reaches equilibrium (Section 7.2). The pressure of the vapour which is in equilibrium with the liquid is described as the **vapour pressure**.

When boiling occurs, particles evaporate in the body of the liquid as well as from the surface. The bubbles which form inside the liquid don't collapse because the vapour pressure inside the bubbles is equal to the atmospheric pressure above the liquid.

Boiling points tell us how strongly the particles are held together in liquids. Volatile liquids, like petrol, evaporate easily and boil at low temperatures because they have relatively weak forces between their particles.

Sublimation and reverse sublimation

A few substances, including iodine, change directly from solid to gas and vice versa on heating and cooling. When crystals of iodine are heated gently, purple iodine vapour is given off from the solid without any liquid being formed. The solid has *sublimed* changing directly from solid to gas.

As the purple iodine vapour reaches the cooler parts of the tube, solid iodine is deposited. This is reverse sublimation.

4.4 The gaseous state

During the seventeenth, eighteenth and nineteenth centuries, a great deal of scientific research involved investigations of the physical and chemical properties of gases. Careful investigations were carried out in order to discover how the volume of a gas varied with changes in temperature and pressure.

Boyle's law

As early as 1662, Robert Boyle had discovered that the volume of a fixed amount of gas increased if the pressure on it was reduced and vice versa. He also found that the volume changed in inverse proportion to the pressure. For example, if he reduced the pressure to half, then the volume doubled. If he reduced the pressure to a quarter, then the volume quadrupled. This type of relationship, where one quantity goes up as the other goes down and vice versa, is described as 'inversely proportional'. The relationship between volume and pressure is known as **Boyle's law** which can be expressed mathematically by the equation:

$$V \propto \frac{1}{p}$$

Alternative ways of writing the equation are:

$$V = \text{constant} \times \frac{1}{p} \quad \text{or}$$

$$pV = \text{constant}$$

A sketch graph showing the relationship between V and p is shown in Figure 4.9.

> ### Note
>
> The symbol $\propto$ means 'proportional to'. This means that if the term on the right of the equation increases, then the term on the left increases in the same ratio. So, we can say, either V is proportional to one over p or V is inversely proportional to p.

Definition

Boyle's law says that the volume of a fixed amount of gas is inversely proportional to the pressure, provided the temperature is constant.

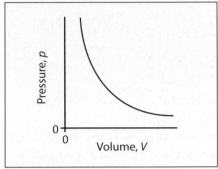

Figure 4.9
A sketch graph showing the relationship between V and p for a fixed amount of gas at constant temperature

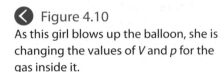

Figure 4.10
As this girl blows up the balloon, she is changing the values of V and p for the gas inside it.

Charles' law

About a century after Boyle's experiments, the Frenchman Jacques Charles studied the effects of temperature on the volume of a fixed amount of gas at constant pressure. In 1787, Charles discovered that for a fixed amount of gas at constant pressure, the volume increased exactly in line with rising temperature (Figure 4.11). What is more, if the temperature is measured in kelvin, K, the volume increases by the same proportion as the temperature increases. This linear relationship between volume and temperature can be expressed mathematically as:

$$V \propto T$$

where T is the temperature in kelvin. The relationship is usually called **Charles' law**.

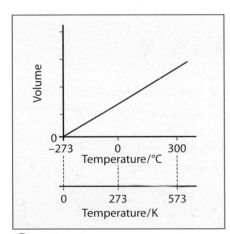

Figure 4.11
A sketch graph showing the relationship between the volume and temperature for a fixed amount of gas at constant pressure

Combining Boyle's law and Charles' law

By combining Boyle's law, $V \propto \dfrac{1}{p}$ and Charles's Law, $V \propto T$ we get

$$V \propto \frac{T}{p} \text{ for a fixed amount of gas}$$

$$\text{or } pV \propto T \text{ or } \frac{pV}{T} = \text{constant}$$

So, for a fixed amount of gas under two different sets of conditions denoted by the subscripts 1 and 2,

$$\frac{p_1 V_1}{T_1} = \frac{p_2 V_2}{T_2}$$

This last equation enables the volume of a gas (V_2) to be calculated under any conditions of temperature and pressure (say T_2 and p_2), provided its volume (V_1) is known under some other conditions of temperature (T_1) and pressure (p_1).

Worked example

A large balloon contains $10\,dm^3$ of air at 1 atm pressure and 30 °C. What volume will the air occupy if the conditions change to –10 °C and 0.9 atm pressure?

Quick Questions

6   A balloon contains $1000\,cm^3$ of air. What volume will the air occupy if:
   a the pressure of air is reduced to half its initial value?
   b the temperature of the air changes from 273 K to 364 K?

Answer

Remember that temperatures must be measured in kelvin, K.

p_1 = initial pressure = 1 atm $\qquad$ p_2 = final pressure = 0.9 atm

V_1 = initial volume = $10\,dm^3$ $\qquad$ V_2 = final volume = ?

T_1 = initial temperature = 30 °C $\qquad$ T_2 = final temperature = –10 °C

$\qquad\qquad$ = 303 K $\qquad\qquad\qquad\qquad\qquad\qquad$ = 263 K

Substituting these values in $\quad \dfrac{p_1 V_1}{T_1} = \dfrac{p_2 V_2}{T_2}$, we get

$$\frac{1 \times 10}{303} = \frac{0.9 \times V_2}{263}$$

So, final volume of air, $\qquad V_2 = \dfrac{1 \times 10}{303} \times \dfrac{263}{0.9}\,dm^3 = 9.64\,dm^3$

The Ideal gas equation

By combining the mathematical expressions for Boyle's law and Charles' law, we obtained the equation:

$$\frac{pV}{T} = \text{constant}$$

Note

Representing temperature

The symbol T represents the temperature in kelvin, K. 0 K is the lowest possible temperature, sometimes called absolute zero.

0 K is –273 °C

273 K is 0 °C

373 K is 100 °C

t K is (t – 273) °C

But what is the value of the constant? Its value will, of course, depend on the amount of gas involved. So, let's calculate the value of the constant for one mole of gas. For one mole of gas, the constant is called the **Gas constant** and given the symbol **R**.

We know that one mole of gas occupies $22.4\,dm^3$ at one atmosphere pressure and 273 K. However, we must calculate R in SI units.

Converting to SI units: $\quad 22.4\,dm^3 = 0.0224\,m^3$

$$1\,\text{atm} = 101\,325\,N\,m^{-2} = 101\,325\,Pa$$

So, in SI units at 273 K,

$$R = \frac{pV}{T} = \frac{101\,325\,\text{N}\,\text{m}^{-2} \times 0.0224\,\text{m}^3\,\text{mol}^{-1}}{273\,\text{K}}$$

$$R = 8.31\,\text{N}\,\text{m}\,\text{K}^{-1}\,\text{mol}^{-1} = 8.31\,\text{J}\,\text{K}^{-1}\,\text{mol}^{-1}$$

So, for 1 mole of gas, $pV = RT$ where $R = 8.31\,\text{J}\,\text{K}^{-1}\,\text{mol}^{-1}$

And for n moles of gas, $pV = nRT$ which is called the **Ideal gas equation** or the **General gas equation**.

A gas which obeys this equation is called an 'ideal' gas. In practice, real gases obey the equation most closely at low pressure and high temperature. Under these conditions, a gas behaves most like a gas and is least likely to form a liquid.

Quick Questions

For each part of Quick questions 7 and 8, remember to label the axes and show the zero for each axis.

7 Draw separate sketch graphs of:
    a p against V,
    b p against $\dfrac{1}{V}$,
    c pV against p,
    for a fixed amount of an ideal gas at constant temperature.
8 Plot separate sketch graphs for:
    a pV against $T(\text{K})$,
    b $\dfrac{pV}{T}$ against $T(\text{K})$,
    for a fixed amount of an ideal gas.

Mixtures of gases

In a gas, the molecules are well separated with plenty of space in between. All this 'empty space' means that gases can easily mix together exerting their own separate pressures on any container. Figure 4.12 shows two identical cylinders. The first contains gas A at 1 atm pressure ($p_A = 1$ atm). The second cylinder contains gas B at 2 atm pressure ($p_B = 2$ atm). Now, suppose all the gas from the first cylinder is forced into the second cylinder, what is the final pressure?

With all the space in which to move, molecules of each gas exert the same pressure as if they are the only gas in the cylinder (Figure 4.13).

So, the total pressure, $p_{\text{total}} = p_A + p_B = 1\,\text{atm} + 2\,\text{atm} = 3\,\text{atm}$

The conclusion we have just reached is an example of **Dalton's law of partial pressures**. More generally, this can be expressed mathematically as:

$$p_{\text{total}} = p_1 + p_2 + p_3 + \ldots + p_n$$

where p_{total} is the total pressure of the mixture, p_1 is the pressure caused by gas 1, usually described as the partial pressure of gas 1, p_2 is the partial pressure of gas 2, etc.

4.5 The kinetic theory and ideal gases

The ideal gas equation arose from the experiments of scientists as long ago as the 17th century. Although it allows us to understand and follow what happens when conditions such as temperature and pressure change, it does not allow us to understand and make predictions at a molecular level. In the 19th century, scientists were able to overcome this situation by developing the kinetic theory (the kinetic–molecular theory). One of the key

Note

The international accepted unit of pressure is the pascal, Pa. One pascal is defined as a pressure of one newton per square metre, (i.e. $1.0\,\text{Pa} = 1.0\,\text{N}\,\text{m}^{-2}$).

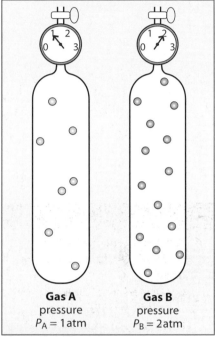

▲ Figure 4.12
Gases A and B in separate identical cylinders

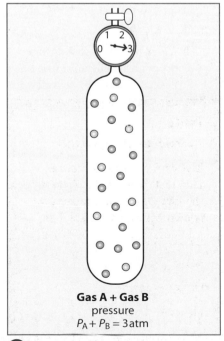

▲ Figure 4.13
Gases A and B in the same cylinder

points of the kinetic theory is that gases consist of molecules in constant random motion colliding with each other and with the walls of their container. The pressure of the gas results from molecules of the gas continually bombarding the walls of their container.

Assumptions of the kinetic theory for ideal gases

There are two fundamental assumptions in the kinetic theory for ideal gases:

1 Molecules of an ideal gas *occupy negligible volume* and can be treated as point masses.
2 Molecules of an ideal gas *exert no forces on each other* and collide with each other and the walls of their container without loss or gain of energy.

The behaviour of real gases

Look carefully at Figure 4.14 which shows the values of $\frac{pV}{RT}$

with increasing pressure for nitrogen at three different temperatures.

For n moles of an ideal gas $\quad pV = nRT$,

So, for 1 mole of an ideal gas, $\quad pV = RT \quad$ and $\quad \frac{pV}{RT} = 1$

Therefore, if nitrogen behaved as an ideal gas under all conditions, its value of $\frac{pV}{RT}$ would follow the black line in Figure 4.14.

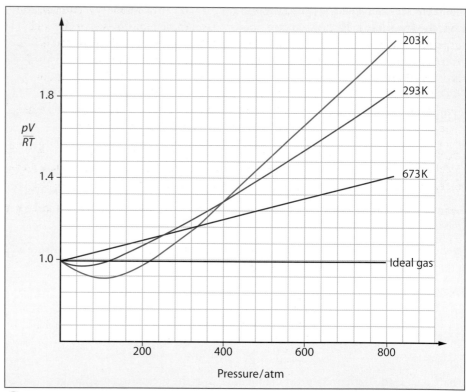

Figure 4.14

Variation of $\frac{pV}{RT}$ with increasing pressure for N_2 at 673 K, 293 K and 203 K

From Quick question 9, you should appreciate that nitrogen deviates more and more from ideal behaviour as the pressure increases and the temperature decreases. This tallies with what we might expect because as the pressure increases and the temperature falls, a gas is more likely to condense becoming less like a gas and more like a liquid.

All this suggests that nitrogen is more likely to approach ideal behaviour at low pressure and high temperature.

Other real gases show similar deviations from ideal behaviour to nitrogen. These deviations can be understood by reconsidering the two fundamental assumptions of the kinetic theory of gases stated earlier in this section.

1 Molecules occupy a negligible volume

Strictly speaking, the molecules in a gas do not have a negligible volume. Each one has a finite size which excludes a certain volume of the container from all the others. If we call this 'excluded' volume $b\,dm^3\,mol^{-1}$, then the 'true' volume in which the molecules move is $(V - nb)$.

Therefore, the equation $pV = nRT$ should be amended to $p(V - nb) = nRT$.

All this tallies with the fact that real gases behave more like ideal gases and obey $pV = nRT$ at low pressure when the volume of the molecules, nb, is negligible compared with the total volume, V.

2 Molecules exert no forces on each other

Unfortunately, intermolecular forces cannot be neglected. This is particularly so at very high pressures when molecules are relatively close.

The pressure of a gas results from molecules of the gas bombarding the walls of their container. Within the bulk of a gas, intermolecular forces from one direction will be cancelled by those from the opposite direction. But those molecules near the wall experience an overall force tending to pull them back into the bulk. This results in a measured pressure less than the 'true' pressure. The size of the 'pressure reduction' is proportional to both the concentration of molecules near the wall ($\propto n/V$) and the concentration of molecules within the bulk (also $\propto n/V$). Thus, the 'pressure reduction' can be written as $a(n/V)^2$ where a is a constant. By adding $a(n/V)^2$ to the measured pressure, p, we obtain the corrected pressure term $[p + a(n/V)^2]$.

Again this tallies with our predictions. We would expect real gases to behave more like ideal gases at low pressures when their molecules are further apart exerting less force on each other, and at high temperatures when their molecules are moving faster.

Finally, if we replace p and V in the Ideal gas equation with their corrected terms, we get:

$$\left(p + a\left(\frac{n}{V}\right)^2\right)\left(V - nb\right) = nRT$$

- p: Measured pressure
- $a\left(\frac{n}{V}\right)^2$: Correction for forces between molecules
- V: Measured volume
- nb: Correction for volume of molecules

Key Points

An **ideal gas** is one which obeys the **Ideal gas equation**, $pV = nRT$.

Ideal gases are sometimes called perfect gases.

Real gases obey the Ideal gas equation most closely at low pressure and high temperature.

Note

The constant, b, has a value of about $0.03\,dm^3\,mol^{-1}$ for many gases. As one mole of many gases occupies approx. $0.03\,dm^3$ on liquefaction, this provides some evidence that nb is a good correction for the volume of molecules.

Note

Real gases approach ideal behaviour and obey the Ideal gas equation closely at

- low pressure,
- high temperature,

when they are most like gases and least likely to condense forming a liquid.

Real gases do not obey the Ideal gas equation at

- very high pressure,
- very low temperature,

when they are most likely to condense forming a liquid.

4.6 Determining the relative molecular masses of gases and volatile liquids

The most accurate and convenient method of measuring relative molecular masses is by mass spectrometry. Using this method, relative molecular masses can be obtained in a few minutes with an accuracy of one part in a million.

Originally, the relative molecular masses of gases and volatile substances were obtained by methods dependent on the Ideal or General gas equation.

Finding the relative molecular mass of a gas by direct weighing

Using the Ideal (General) gas equation, $pV = nRT$, and $n = \dfrac{m}{M}$ where m is the mass of gas whose molar mass is M,

we can write

$$pV = \frac{m}{M}RT$$

Therefore

$$M = \frac{m}{pV}RT$$

By measuring m, p, V and T for a sample of gas and assuming R, the Gas constant equals $8.31\,\mathrm{J\,K^{-1}\,mol^{-1}}$, it is possible to calculate M. The relative molecular mass of the gas, M_r, is then equal to the numerical value of M.

A container of known volume (V) is weighed full of gas at pressure p and temperature T. The same container is then weighed again after complete evacuation in order to obtain the mass of gas inside (m).

This method is very accurate provided the following precautions are taken.

1 The container must be large, so as to give an appreciable mass of gas. Gases are so light that any slight error in weighing produces a large percentage error if the mass of the substance weighed is very small.
2 All weighings must be carried out at the same temperature and pressure so that the upthrust on the container remains constant.

Finding the relative molecular mass of a volatile liquid

Using a steam jacket, it is possible to adapt the method just described for use with volatile liquids. Although the method gives only an approximate value for the relative molecular mass, it enables an accurate value to be obtained if the empirical formula of the substance can be determined.

The principle of the experiment is the same as that just described for gases using the equation:

$$M = \frac{m}{pV}RT$$

Thus, the molar mass (M) of the liquid can be calculated by finding the volume of vapour (V) formed from a known mass of liquid (m) at a measured temperature (T) and pressure (p), provided we assume the value of the gas constant (R).

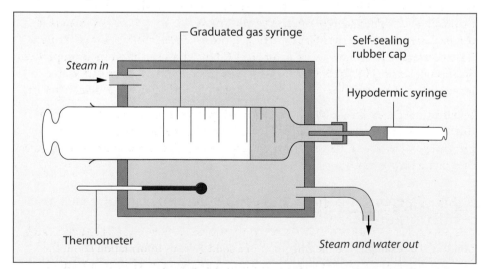

> Figure 4.15
Determination of the relative molecular mass of a volatile liquid

Figure 4.15 shows a suitable apparatus to use. Draw a few cubic centimetres of air into the graduated gas syringe and fit the self-sealing rubber cap over the nozzle. Pass steam through the outer jacket until the thermometer reading and the volume of air in the syringe become steady. Continue to pass steam through the jacket and record the temperature (T) and the volume of air in the syringe.

Now fill the hypodermic syringe with about $1 \, cm^3$ of the liquid under investigation, ensuring there is no air in the needle. Weigh the hypodermic syringe and its contents and then push the needle through the self-sealing cap of the graduated gas syringe. Inject about $0.2 \, cm^3$ of liquid into the large syringe and withdraw the hypodermic syringe. Immediately, re-weigh the hypodermic syringe and its contents.

The liquid injected into the graduated gas syringe will evaporate. The final volume of air plus vapour in the graduated syringe should be recorded when the volume becomes steady. Finally, record the atmospheric pressure.

Here are some typical results:

$$\text{Mass of liquid, X, vaporised, } m = 0.16 \, g$$
$$\text{Initial volume of air in graduated gas syringe} = 10 \, cm^3$$
$$\text{Final volume of air + vapour in gas syringe} = 56 \, cm^3$$
$$\therefore \text{ Volume of X vaporised} = 46 \, cm^3$$
$$= 46 \times 10^{-6} \, m^3$$
$$\text{Temperature of X} = 100\,°C = 373 \, K$$
$$\text{Pressure of X} = 1 \, atm = 101\,325 \, N\,m^{-2}$$
$$\text{Gas constant, } R = 8.31 \, J\,K^{-1}\,mol^{-1}$$

Calculation

$$\text{Using } M = \frac{m}{pV}RT$$

$$M = \frac{0.16}{101\,325 \times 46 \times 10^{-6}} \times 8.31 \times 373 = 106.4 \, g\,mol^{-1}$$

$\therefore$ relative molecular mass of X, $M_r \approx 106$

Quick Questions

10 **a** Why is the apparatus in Figure 4.15 unsuitable for liquids which boil above $100\,°C$?

b Why should the hypodermic syringe be handled as little as possible between weighings?

c Suggest two sources of error in this experiment.

d The substance X, for which the typical results are given, has the percentage composition by mass of 22.0% C, 4.6% H and 73.4% Br.

  i Calculate the empirical formula of X.

  ii What is the molecular formula of X?

  iii What is the accurate relative molecular mass of X?

4.7 The solid state

One of the major achievements of chemistry has been the synthesis of new solid materials, such as ceramics, plastics, fibres and alloys. Many of these new materials can be designed to have specific properties. In order to design these new substances, it is necessary to know how the structure of solid materials can affect their properties. So, it is not surprising that one of the most important aspects of chemistry is the investigation of the structure of solids and the bonds holding their particles together.

In this part of the chapter, we shall begin by looking at the methods used to investigate the structures of solid materials. Then we shall consider the lattice structures and properties of different types of solid and the way in which their properties are dictated by structure and bonding.

4.8 Evidence for the structure of solids

The physical properties of a material can often provide evidence for its structure and bonding. For example, the melting point of a solid gives us information about the forces of attraction between its particles, whilst the effect of an electric current on the molten solid can tell us something about the nature of the constituent particles. However, the most reliable evidence for the structure of solids comes from X-ray diffraction.

⌃ Figure 4.16
A small cylindrical magnet floating freely above a cylindrical superconducting ceramic made from yttrium, barium and copper oxide. The visible vapour is from liquid nitrogen which keeps the ceramic in its superconducting temperature range. Superconducting ceramics are leading to new advances in technology.

Figure 4.17

This photograph shows strands of glass fibres. Glass fibres have been developed for use in fibre optics, medical research and for strengthening other materials.

Figure 4.18

Professor Dorothy Hodgkin, who won a Nobel Prize in 1964 for her work in determining the structure of various biological molecules including vitamin B_{12}, using X-ray crystallography

Key Point

The lattice structures of crystalline solids can be determined by studying the diffraction patterns obtained when X-rays are scattered by particles in the crystals.

Quick Questions

11 Look closely at the electron density map of naphthalene in Figure 4.20.
   a How many atoms can be located from the electron density map?
   b Which atoms are they?
   c Why are the other atoms not evident from the electron density map?

Evidence from X-ray diffraction

Look through a piece of thin stretched cloth (possibly your handkerchief) at a bright light. The pattern you see is caused by the deflection and interference of light as it passes through the regularly spaced threads of the fabric. This deflection of light is called **diffraction** and the patterns produced are **diffraction patterns**.

When the cloth is rotated in front of the light, the diffraction pattern also rotates. If the cloth is stretched so that the strands of the fabric get closer, then the pattern spreads out further. If the strands are arranged differently the pattern changes. From the diffraction pattern which we can observe, it is possible to deduce how the strands are arranged in the fabric. The same idea is used to determine the arrangement of particles in the lattice structures of crystalline solids.

The regular shapes of crystals suggest that their constituent particles (atoms, ions or molecules) are arranged in a regular pattern. Until the early years of the twentieth century, scientists could only guess at the arrangement of invisible atoms or ions in a crystal. Then, Sir Lawrence Bragg (1890–1971) realised that X-rays could be used to investigate crystal structures because their wavelengths are about the same as the distances between atoms in a crystal.

When a narrow beam of X-rays falls on a crystal composed of regularly spaced atoms or ions the X-rays are scattered (reflected) in different directions. In most instances, waves from the reflected X-rays interfere with and destroy each other. However, because of the regular arrangement of particles, it is possible for the X-rays to be scattered by the particles so that their waves coincide and reinforce each other (Figure 4.19). This means that the scattering patterns from individual particles combine to produce an overall diffraction pattern which can be recorded as a series of spots or darkened areas on X-ray film or plate.

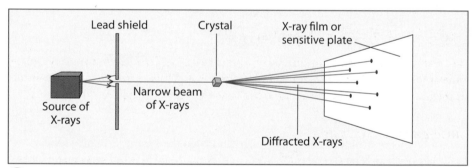

Figure 4.19

Using X-rays to investigate the arrangement of atoms or ions in a crystalline solids

Electron density maps

When X-rays strike a crystal, they are diffracted by the electrons in the atoms or ions. Consequently, the larger the atom and the more electrons it possesses, the darker the spot will be on the diffraction pattern. By analysing both the positions and the intensities of the spots on a diffraction pattern, it is possible, using computers, to determine the charge density of electrons within the crystal. The charge density is measured in terms of electrons per cubic nanometre. Points of equal density in the crystal are joined by contours giving an electron density 'map' similar to the one in Figure 4.20.

By mounting electron density maps on clear plastic or Perspex it has been possible to assemble accurate three-dimensional structures of complex substances such as proteins and nucleic acids. Three-dimensional visualisations can also be produced by computers. It is then possible to see the positions of different atoms within the molecules. There are many different arrangements in which atoms or ions can be packed in repeating units to form crystals. A few of these arrangements that occur in natural crystals are considered in the following sections.

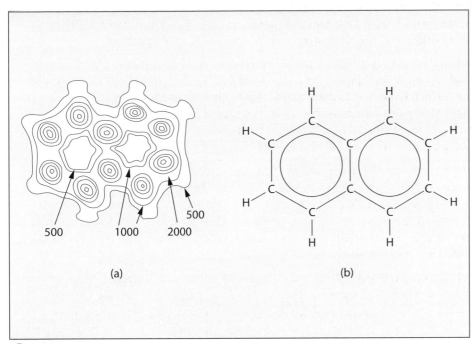

Figure 4.20
(a) An electron density map of naphthalene (contours in electrons per nm³)
(b) The structural formula of naphthalene

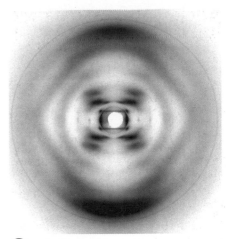

Figure 4.21
An X-ray diffraction photograph of DNA

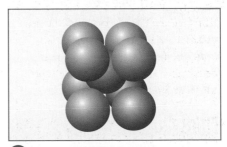

Figure 4.22
Some fruits and vegetables can be stacked in the same way that metal atoms pack in close-packed structures.

4.9 Giant metallic structures

Metals are exceptionally useful and versatile materials. Just look around you and notice some of their uses. Detailed X-ray analysis shows that there are two possible structures for metals: close-packed structures and body-centred cubic structures.

Close-packed structures

As we found in Section 3.8, most metals have a **close-packed structure** in which each atom within the metal touches six others in its own layer (Figure 3.22) plus three atoms in the layer above and three atoms in the layer below (Figure 3.23). This means that one atom in the middle of a close-packed metal structure touches 12 others. Chemists summarise this by saying that the **co-ordination number** is 12.

Metals with a close-packed structure and therefore a co-ordination number of 12 include calcium, magnesium, aluminium, zinc, lead, copper, silver, gold and platinum.

Body-centred cubic structures

Some metals do not have their atoms in a close-packed structure. Instead, these metals have giant lattices with a **body-centred cubic structure** (Figure 4.23).

As its name suggests, this structure is basically cubic with an atom at the centre of each cube. In this case, each atom is surrounded by eight others. Thus, the co-ordination number is 8 and the giant lattice is a little more open than in close-packed structures.

All the alkali metals and iron have a body-centred cubic structure, but there is no clear pattern between the structure of a metal and its position in the periodic table.

Figure 4.23
A few metal atoms in the body-centred cubic structure

Definition

The **co-ordination number** of an atom or ion is the number of its nearest neighbours.

Key Point

Metals form either close-packed structures (co-ordination number = 12) or body-centred cubic structures (co-ordination number = 8).

4.10 The properties and uses of metals

The properties of metals depend on their giant structure and the strong bonding between positive ions and a 'sea' of delocalised electrons (Section 3.8). Metals have high melting points, high densities and good conductivity. They are also malleable.

There are no rigid, directed bonds in a metal, so layers of atoms can slide over each other when a strong enough force is applied. This relative movement of layers in the metal lattice is called **slip**. After slipping, the atoms settle into new positions and the crystal structure is restored. So, a metal can be hammered into different shapes (malleable) or drawn out into a wire (ductile). Figure 4.24 shows what happens when slip occurs.

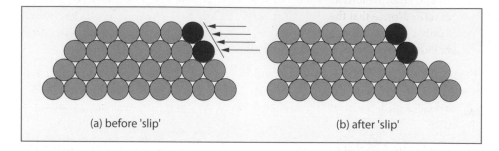

(a) before 'slip' (b) after 'slip'

> Figure 4.24
The arrangement of metal atoms
a) before, and b) after slip

Metallurgists can modify the properties of metals including increasing their strength, by alloying. Metals will readily form alloys since the metallic bond in which positive ions attract delocalised electrons is common to all metals. The presence of small quantities of a second element in the metal frequently increases its strength.

Both copper and aluminium have important uses both as pure metals and as alloys. Copper is a very useful metal because of its malleability, ductility, high conductivity and its ability to form alloys. The pure metal is used extensively as an electrical conductor in wires and cables. Copper is also used in central heating systems as pipes and radiators. All these uses of pure copper depend on the fact that it is malleable and ductile.

Copper also provides a range of alloys with other metals. Alloying with zinc produces various types of brass which are harder and stronger than pure copper. Atoms of zinc are slightly larger than atoms of copper. These larger zinc atoms interrupt the orderly arrangement in the lattice and prevent slip. Copper also alloys readily with nickel and gold. In many countries, 'silver' coins are 75% copper and 25% nickel, whilst 9 carat gold is about two-thirds copper and one-third gold.

Aluminium, like copper, has extensive properties both as the pure metal and in alloys. Most of these uses are only possible because of the thin, non-porous, protective oxide layer which forms on aluminium and its alloys. During the last 40 years, aluminium alloys have been used more and more, particularly in the aircraft industry. These include duraluminium containing 4% copper which is light and also very strong.

Quick Questions

12 What properties of aluminium (besides corrosion resistance) make it suitable for:
 a milk bottle tops,
 b baking foil,
 c aircraft bodywork,
 d electricity cables,
 e pans,
 f tent frames?
13 a Why does aluminium not corrode away like iron?
 b Aluminium and iron are the two most widely used metals. What advantages does aluminium have for some uses in addition to its corrosion resistance?

Key Point

From bonding and structure to properties and uses

Notice how the bonding and structure of a substance determines its properties and how its properties then determine its uses.

> Figure 4.25
Slip occurs when metals are twisted or beaten into different shapes.

4.11 Giant ionic structures

Ionic structures are formed when elements with large differences in electronegativity form compounds. Electrons are transferred from elements of low electronegativity (metals) to those of high electronegativity (non-metals). The oppositely charged ions which result are held together by strong electrostatic forces of attraction in ionic bonds (Section 3.2).

X-ray analysis shows that the particles in different ionic structures can be arranged in different patterns. One of the simplest and most common structures for ionic compounds is the cubic arrangement of ions as in sodium chloride and magnesium oxide. Figure 4.27 shows a space-filling model of the structure of sodium chloride. The sodium and chloride ions are shown as solid spheres. The sizes of the ions are in the correct ratio. Notice that the ions are arranged in a cubic pattern. Although Figure 4.27 shows only a few Na^+ and Cl^- ions, there will be billions upon billions of ions in even the smallest visible crystal of sodium chloride.

The positions of Na^+ and Cl^- ions in the cubic lattice of sodium chloride are emphasised in Figure 4.28. The small blue and orange circles in Figure 4.28 represent the centres of Na^+ and Cl^- ions, respectively. The solid lines in the diagram show the cubic geometry of the lattice.

Notice in Figure 4.28 that each positive sodium ion is surrounded by six Cl^- ions and each negative chloride ion is surrounded by six Na^+ ions. The six Cl^- ions around the central Na^+ ion in Figure 4.28 have been numbered.

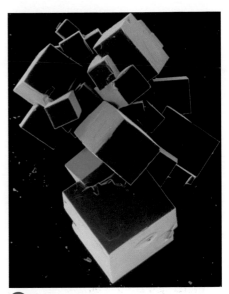

Figure 4.26
These crystals of sodium chloride show their cubic shape quite clearly.

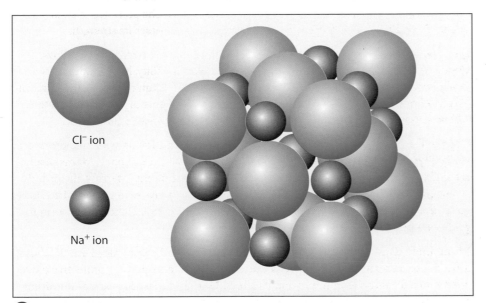

Figure 4.27
A space-filling model of the structure of sodium chloride

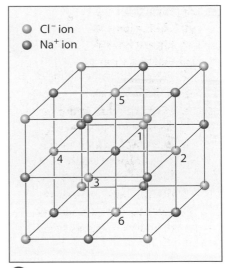

Figure 4.28
The structure of sodium chloride

Four of the Cl^- ions (numbered 1, 2, 3 and 4) are in the same horizontal layer of the crystal as the central Na^+ ion. One Cl^- ion (number 5) is in the layer above. The final Cl^- ion (number 6) is in the layer below.

The structure of sodium chloride is said to have 6:6 co-ordination because the Na^+ ions have a co-ordination number of 6 and the Cl^- ions also have a co-ordination number of 6.

Measuring the size of ions

X-ray measurements on ionic solids can be presented in the form of electron density maps in order to determine the size of different ions. Figure 4.29 shows an electron density map for sodium chloride. The circular contours suggest that the electron distribution in these ions is spherical. The spacing of the contour lines enables us to distinguish particles with different numbers of electrons.

Key Point

Sodium chloride and magnesium oxide form cubic structures in which the co-ordination numbers of their ions are described as 6:6.

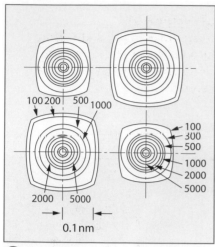

Figure 4.29
An electron density map for sodium chloride. (Electron densities are expressed as electrons per nm^3.)

Quick Questions

14 Look closely at Figure 4.29.
  a Which are the sodium ions?
  b What is the interionic distance between neighbouring Na$^+$ and Cl$^-$ ions?
  c What is the ionic radius of: (i) Na$^+$, (ii) Cl$^-$?
15 a Why do molten ionic compounds conduct electricity?
  b Why are solid ionic compounds non-conductors?
16 Why are ionic compounds such as sodium chloride soluble in water? (Hint: Why might water molecules be attracted to ions, such as Na$^+$ and Cl$^-$, in sodium chloride?)

Using electron density maps like that in Figure 4.29, it is possible to compile tables of ionic radii. However, the size of a particular ion can vary slightly depending on the size and charge of other ions in the crystal. So, values of the ionic radius for one particular ion do not always agree.

4.12 The properties of ionic solids

Ionic compounds, like sodium chloride, magnesium oxide and copper sulfate are usually hard solids with high melting points and boiling points. They are also good conductors of electricity when molten, but non-conductors when solid. Very often, they are soluble in polar solvents such as water, but insoluble in non-polar solvents such as hexane.

What happens when ionic solids dissolve in water?

When sodium chloride dissolves in water, the crystal lattice is broken up forming separate Na$^+$ and Cl$^-$ ions in aqueous solution. Where does the energy required to separate the oppositely charged ions come from? We have seen already that water contains highly polar molecules. The positive ends of polar water molecules are attracted to negative ions in the crystal, and negative ends of the water molecules are attracted to positive ions. The formation of ion–solvent bonds results in a release of energy. This is sufficient to cause the detachment of ions from the lattice (Figure 4.30).

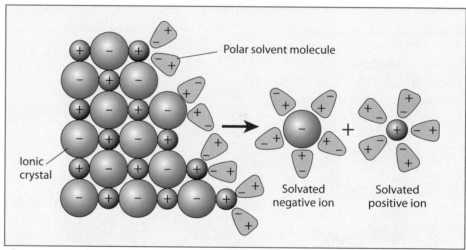

Figure 4.30
Polar solvent molecules dissolving an ionic solid

Thus, ionic crystals will often dissolve in polar solvents such as water, ethanol and propanone (acetone).

The attachment of polar solvent molecules to ions is known generally as **solvation**. Chemists say that the ions are **solvated**. Very often the solvent is water and in this specific case the ions are said to be **hydrated**. Figure 4.31 shows the structures of hydrated positive and negative ions.

Why are ionic solids insoluble in non-polar solvents?

Non-polar liquids such as hexane cannot solvate ionic solids. The reasons for this are not difficult to see. Non-polar molecules are held together by weak intermolecular forces from induced dipole attractions. These forces are much smaller in magnitude than the forces between ions in an ionic crystal. The ion–ion attractions are in fact much stronger than either the solvent–solvent interactions or the ion–solvent interactions. So the non-polar solvent molecules cannot penetrate the ionic lattice. Thus, sodium chloride and other ionic compounds are virtually insoluble in hexane.

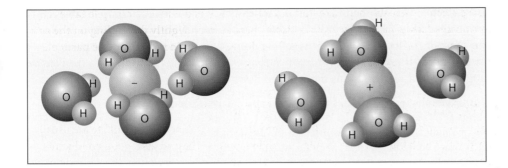

 Figure 4.31
The structures of hydrated negative and positive ions. Only those water molecules in the horizontal plane are shown.

4.13 Giant molecular (giant covalent) structures

A few non-metals, including carbon and silicon, and some compounds of non-metals, such as silicon(IV) oxide (Section 4.14), silicon carbide and boron nitride, consist of giant structures of atoms held together by covalent bonds.

The structure of diamond

In diamond, every carbon atom can be imagined to be at the centre of a regular tetrahedron surrounded by four other carbon atoms whose centres are at the corners of the tetrahedron (Figure 4.32). Within the structure, every carbon atom forms four covalent bonds by sharing electrons with each of its four nearest neighbours.

Silicon (Section 23.3) and silicon carbide (SiC, used in the abrasive 'carborundum') exist in a similar crystal structure to diamond. In each of these structures, atoms are linked by localised electrons in strong covalent bonds throughout the whole three-dimensional lattice.

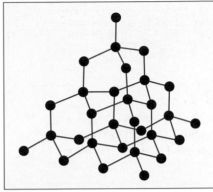

Figure 4.32
Part of the giant molecular structure of diamond

Figure 4.33
A glass engraver using a diamond-tipped wheel to produce a design on a glass goblet

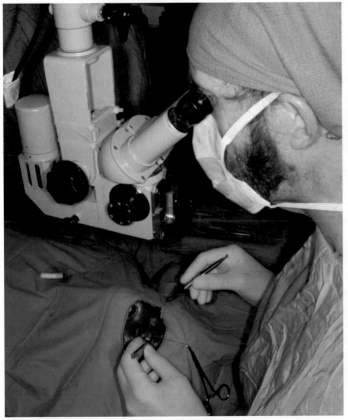

Figure 4.34
A special diamond-tipped cutting tool being used for eye surgery

Quick Questions

17 a Why is the structure of diamond so stable?
    b What is the co-ordination number of carbon atoms in diamond?
    c Will carbon atoms on the outside of a diamond form four covalent bonds?

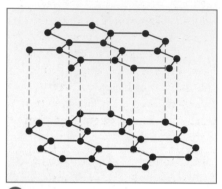

Figure 4.35
The giant molecular structure of graphite. The layers are vast sheets of carbon atoms and the sheets are piled one above the other. Within each layer there are strong covalent bonds, but the bonding between layers is relatively weak.

It is therefore very difficult to distort a covalently bonded crystal, because this would involve breaking covalent bonds. Consequently, diamond, silicon and silicon carbide are hard and brittle with very high melting points and very high boiling points. Furthermore, the localised electrons in strong covalent bonds cannot move freely in an applied electric field and thus these materials do not conduct electricity.

Almost all the industrial uses of diamond and silicon carbide depend on their hardness. Diamond is one of the hardest natural substances. Diamonds unsuitable for gemstones are used in glass cutters and in diamond-studded saws. Powdered diamond and carborundum (SiC) are used as abrasives for smoothing very hard materials.

The structure of graphite

In graphite, the carbon atoms are arranged in flat, parallel layers. Each layer contains trillions of hexagonally arranged carbon atoms (Figure 4.35). Each carbon atom is covalently bonded to three other atoms in its layer. Each layer can be viewed as a two-dimensional giant layer lattice. The carbon–carbon bond length within a layer is 0.142 nm. This suggests some partly double-bond character, because the value is intermediate in length between a single and a double carbon–carbon bond. The strong covalent bonds within the layers account for the very high melting point (3730 °C) of graphite. Owing to its high melting point, graphite is used to make crucibles for molten metals. The covalent bonds between carbon atoms within the layers of graphite are so strong, that many modern **composites** include graphite fibres for greater tensile strength.

Figure 4.36
Graphite fibres are used to reinforce the shafts of broken bones, golfclubs and tennis rackets, like the one being used by Venus Williams.

The distance between the layers in graphite is 0.335 nm, which is much longer than a single carbon–carbon bond. As a result of this, the bonding between the layers is relatively weak and the layers can slide over each other easily. This accounts for the softness of graphite and its use as a lubricant.

The bonding in graphite can be pictured as three trigonally arranged covalent bonds. These are formed by three of the four outer shell electrons of carbon, whilst the fourth electron is delocalised over the whole layer. This delocalisation of electrons, similar to that in metals, results in graphite conducting electricity and appearing shiny. The electrical conductivity of graphite enables it to be used as electrodes in industry and as the positive terminal in dry cells.

Diamond and graphite are examples of **allotropes** – different forms of the same element in the same state. Another example of allotropes are oxygen (O_2) and ozone (O_3) which are both gases.

Until 1985, chemists thought there were just two forms of solid carbon – diamond and graphite. Then, in 1985, Harry Kroto and his research team at the University of Sussex prepared buckminsterfullerene, C_{60}. This was a black solid with a simple molecular structure. Since 1985, a large group of similar compounds have been produced. The group are collectively known as fullerenes and are discussed further in Section 4.16.

4.14 Silicon(IV) oxide, silicates and ceramics – more giant structures

After water, sand is probably the most common material on Earth. Sand is mainly silicon(IV) oxide, SiO_2, sometimes known as silica. There are several forms of SiO_2 with different crystal structures. These different solid forms of the same compound are called **polymorphs**. Polymorphism in compounds can be compared to **allotropy** in elements (Section 4.13).

The most common form of silicon(IV) oxide is quartz. Sand is an impure form of quartz. Its brown colour is due to impurities of iron(III) compounds.

The structure of sand is based on tetrahedra of silicon atoms covalently bonded to four oxygen atoms (Figure 4.37).

In silicon(IV) oxide, each SiO_4 tetrahedron shares the oxygen atoms at its corners with four other SiO_4 tetrahedra. Therefore, each silicon atom has a half-share in four oxygen atoms. So, its formula is SiO_2 and silicon(IV) oxide is a pure giant molecular (giant covalent) structure. The strong covalent bonds linking the three-dimensional network of atoms in silicon(IV) oxide (bond energy, $E(Si-O) = 464\,kJ\,mol^{-1}$, compared with $E(C-C) = 346\,kJ\,mol^{-1}$) account for its hardness, very high melting point, electrical and thermal insulating properties. Its structure and physical properties are similar to those of diamond.

Anions derived from silica are called **silicates**. Some of them are very important compounds. All silicates, like SiO_2, have structures based on SiO_4 tetrahedra.

In silica itself, each SiO_4 tetrahedron shares its corners with four others. If *none* of the oxygen atoms are shared, then the silicate ion, SiO_4^{4-}, results. Between these extremes, various ring, chain and sheet structures are possible for silicates (Figure 4.38).

In each of these structures, negative charges on the silicate chains and sheets are balanced by cations, such as Na^+, Ca^{2+}, Mg^{2+} and Al^{3+} held within the giant covalent structure. Just like SiO_2, these silicate structures are hard yet brittle materials with high melting points and electrical insulating properties.

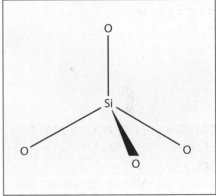

Figure 4.37
The SiO_4 tetrahedron is the basic unit for the structure of silicon(IV) oxide, sand and all silicates.

Figure 4.38
Chain and sheet silicates are based on SiO_4 tetrahedra.

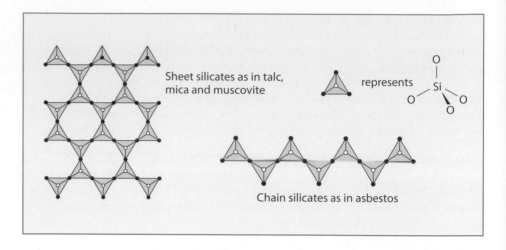

In chain silicates, the charge on each tetrahedron is 2–, because two of the four oxygen atoms are not bonded to other tetrahedra. Notice also that silicates have bonding characteristics of both giant covalent and giant ionic structures.

Probably the most important products from silicon(IV) oxide and silicates are the various kinds of glass. The glass formed by cooling pure quartz has excellent optical properties. It is used to make lenses, but is very brittle and easily breaks. For most uses, sodium carbonate is added to the molten silicon(IV) oxide before it cools. This produces tougher soda-glass, containing silicon(IV) oxide and sodium silicate, used for bottles, jars and windows. Addition of boron(III) oxide (B_2O_3) to this glass leads to borosilicate glass, better known as Pyrex. Borosilicate glass expands very little on heating and is less likely to crack. This is the glass used for laboratory and oven glassware.

Quick Questions

19 Look carefully at Figure 4.38.
  a How many oxygen atoms in each SiO_4 tetrahedron are bonded to other tetrahedra in sheet silicates?
  b What is the charge on each tetrahedron in sheet silicates?
  c Talcum powder is made from talc, a sheet silicate mineral. Why do you think talcum powder feels smooth when rubbed between your fingers?
  d Thin sheets can be stripped off flat pieces of mica. Why is this?

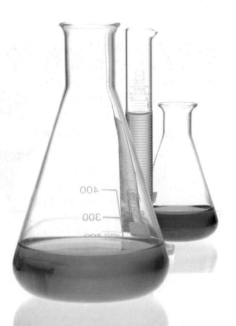

Figure 4.39
Laboratory and oven glassware is usually made from borosilicate glass (Pyrex).

Another set of compounds closely related to the silicates is the **aluminosilicates**. In these compounds, some silicon atoms are replaced by aluminium atoms. Impure aluminosilicates contain other metals, such as magnesium and calcium. Clay is probably the most important aluminosilicate. It has a sheet structure which allows water molecules to get between the sheets when the two are mixed. So, when clay is wet the sheets move over each other and the material can be moulded into different shapes.

When the clay is fired, water is driven out of the structure and a three-dimensional network incorporating both covalent and ionic bonds is formed.

The products from fired clay include china, pottery, bricks and concrete plus materials used for furnace linings and electrical insulators. They are hard and strong (although brittle), heat resistant and chemically unreactive. In general, the best quality aluminosilicates, known as china clay or kaolin, are used for crockery. Traditionally, these products moulded from clay at room temperature and then hardened by heat were known as **ceramics**. Today, the term 'ceramics' is used more generally to include giant molecular (giant covalent) structures with ions enclosed in the network. So, ceramics include glasses and other silicate structures as well as the products from the fired clay.

4.15 Simple molecular structures

Non-metal elements and compounds of non-metals such as iodine, carbon dioxide, water, naphthalene ($C_{10}H_8$) and sugar ($C_{12}H_{22}O_{11}$)are usually composed of simple molecules. In these molecular substances, the atoms are joined together *within* the molecule by strong covalent bonds. But the separate molecules are attracted to each other by much weaker intermolecular forces.

X-ray diffraction measurements on crystals of molecular substances have been used to determine their lattice structures. Figure 4.40 shows the arrangement of I_2 molecules in solid iodine. The arrangement of molecules in the crystal lattice is described as **face-centred cubic**. The molecules are arranged in a *cube* with a molecule at each corner and a molecule at the *centre* of each *face*.

The packing of molecules in simple molecular crystals is often more complicated than the packing of simple atoms or ions. Even so, each substance is found to have a characteristic and uniform lattice arrangement.

Now, although the atoms within these simple molecules are joined by strong covalent bonds, the separate molecules are only held together by weak intermolecular forces. This means that the molecules can be separated easily. Hence the crystals of simple molecular substances are usually soft with low melting points and low boiling points.

In addition, molecular compounds contain neither delocalised electrons (like metals) nor ions (like ionic compounds). So, they cannot conduct electricity.

Non-polar molecular compounds such as iodine and naphthalene are almost insoluble in polar solvents such as water, though they are usually very soluble in non-polar solvents such as hexane.

Why are non-polar substances such as iodine insoluble in water, but soluble in hexane?

In liquids of high polarity like water, there are strong water–water attractions. These are considerably stronger than either iodine–iodine attractions or iodine–water attractions. Consequently, iodine molecules cannot penetrate the water structure and there is little tendency for water molecules to solvate non-polar iodine molecules. Iodine is therefore almost insoluble in water.

In non-polar liquids, such as hexane, there are weak induced dipole attractions. These hexane–hexane attractions are similar in strength to iodine–hexane and iodine–iodine induced dipole attractions. Thus, it is easy for hexane molecules to penetrate into the iodine crystal and solvate the iodine molecules. Consequently, iodine dissolves easily in hexane.

4.16 Fullerenes

Fullerenes are allotropes of diamond and graphite but they are fundamentally different because they have simple molecular structures. The first fullerene to be prepared was named 'buckminsterfullerene' because the structure of its C_{60} molecule resembled the football-like (geodesic) dome invented by the American engineer Robert Buckminster Fuller (Figure 4.41). Not surprisingly, it was quickly nicknamed 'bucky ball' and 'footballene'.

Since the discovery of C_{60}, other fullerenes have been prepared including C_{32}, C_{50} and C_{240}. These fullerenes are black solids soluble in various non-polar solvents. This property has led to the use of C_{60} in printing ink and mascara.

Once chemists had understood the structure of fullerenes, they were able to produce them in the form of cylinders as well as spheres. The diameter of the cylinders range from 5 nm to 15 nm which is why they are called 'nanotubes'. Nanotubes can be open at both ends or capped with carbon atoms from half a C_{60} molecule.

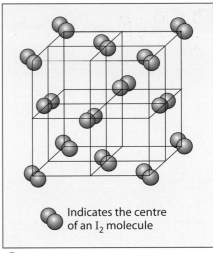

Indicates the centre of an I_2 molecule

⌃ Figure 4.40
The crystal structure of iodine

Quick Questions

20 Look closely at Figure 4.40.
  a How many nearest neighbours does an I_2 molecule have in the same layer?
  b How many nearest neighbours does an I_2 molecule have in either the layer above or below?
  c Is the co-ordination number of iodine molecules in the crystal 4, 6, 8 or 12?

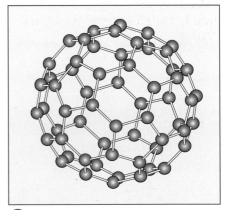

⌃ Figure 4.41
The structure of buckminsterfullerene, C_{60}. With care, you can count all 60 carbon atoms!

21 a In what way does the bonding in C_{60} resemble that in graphite?

b What is the co-ordination number of carbon atoms in C_{60}?

c C_{60} does not conduct electricity like graphite. Why is this?

Definition

Nanotechnology is the term used to describe technological developments and applications using materials and particles on the nanometre scale.

Nanotubes have fascinating properties. Lengthways, they can conduct electricity and this has opened up the possibility of their use as microminiature molecular wires. There are reports of integrated circuits using nanotubes to connect components in place of copper wire. Another potential application of nanotubes is as molecular-sized needles. Scientists have moved biological molecules along the inside of nanotubes leading to the possibility of using them as ultra-small needles for injecting the molecules of drugs into individual cells in the treatment of cancer.

These potential uses of nanotubes as microminiature electrical wires and needles are good examples of **nanotechnology**.

4.17 Materials as finite resources – recycling

Manufactured goods, whether metals from mineral ores, plastics from crude oil or ceramics from clay are produced at a heavy cost to society. Their production uses up valuable materials, it consumes fuels and energy and it disrupts wildlife and the environment. The materials that we take from the Earth like metal ores, crude oil and clay are, of course, **finite resources**. They cannot last forever. Geologists can estimate how long the known **reserves** of these materials will last and their estimations are alarming (Figure 4.42).

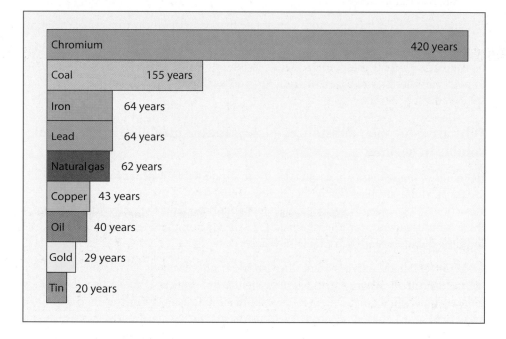

> Figure 4.42
How long will our reserves of different materials last?

The time-scales for the different materials in Figure 4.42 emphasise the crucial importance of **recycling** manufactured goods.

Recycling is important because:

▶ **It saves reserves** for the future.
▶ **It saves energy and fuels**. Think of the enormous amounts of electricity needed to extract aluminium and the amount of iron ore and coke used to extract iron.
▶ **It saves money**. The high cost of extracting and concentrating ores makes the recycling of metals particularly economical and worthwhile. For example, recycling aluminium requires only 5% of the energy needed to produce aluminium from bauxite.
▶ **It saves the environment**. Land that would have been used for mining metal ores, clay and sand or drilling for oil is saved.
▶ **It reduces waste** and avoids the problems of waste disposal.

Definitions

Finite resources are materials which cannot last forever.

Reserves are the present known amounts of different materials.

Recycling is the reclamation and re-use of materials.

Sometimes, recycling is not economical owing to the cost of collecting, sorting and processing the waste material. The higher the value of a material, the more economical it is to recycle. As a result of this, virtually all gold is recycled, but only 40% of aluminium is recycled.

Recycling plastics presents a particular problem because of the difficulty in identifying the type of polymer used. It is easy to separate iron from copper in recycling, but PVC and polythene are much more difficult to distinguish and separate. One way of overcoming this problem is to label polymer goods with an appropriate symbol during the production process (see Figure 27.19, Section 27.5).

4.18 Comparing typical solid structures

The structure, bonding and properties of the four common solid structures (giant metallic, giant molecular, giant ionic and simple molecular) are summarised and compared in Table 4.2.

Quick Questions

22 Use Table 4.2 to consider each of the following in the solid state:
Cu, Si, NH_3, NaI, Xe.
Which solid:
a is a good electrical conductor,
b is a poor electrical conductor, but conducts on melting,
c is hard and brittle and insoluble in water,
d has hydrogen bonds between molecules,
e has the lowest melting point,
f is most soluble in water?

Key Point

Use Table 4.2 to help you understand and recall the structure, bonding and properties of the four typical solid structures.

 Table 4.2
Comparing typical solid structures

	Giant metallic	Giant molecular (Giant covalent)	Giant ionic	Simple molecular
1 Structure				
i Examples	Na, Fe, Cu	Diamond, SiC, SiO_2	Na^+Cl^-, $Ca^{2+}O^{2-}$, $(K^+)_2SO_4^{2-}$	I_2, S_8, $C_{10}H_8$, CH_4
ii Constituent particles	Atoms	Atoms	Ions	Molecules
2 Bonding in the solid				
	Attraction of outer mobile electrons for positive metal ions creates strong metallic bonds	Atoms are linked through the whole structure by very strong covalent bonds	Attraction of positive ions for negative ions results in strong ionic bonds	Molecules are held together by relatively weak intermolecular forces
3 Properties				
i m.pt. and b.pt. State at room temp.	High Usually solid	Very high Solid	High Solid	Low Usually gases or volatile liquids
ii Hardness/ malleability	Hard, yet malleable	Very hard and brittle	Hard and brittle	Soft if solid
iii Conductivity	Good conductors when solid or liquid	Non-conductors (graphite is an exception)	Non-conductors when solid. Conductors when molten or in aqueous solution	Non-conductors when solid, liquid and in aqueous solution
iv Solubility	Insoluble in polar and non-polar solvents. Some react with water	Insoluble in all solvents	Some are soluble in polar solvents, insoluble in non-polar solvents	Some are soluble in polar solvents. Most are soluble in non-polar solvents

Review questions

1 a What do you understand by the terms:

   i atom, ii molecule, iii ion?

   b Use one or more of the terms in part **a** to describe the following structures:

   i copper, ii solid carbon dioxide, iii graphite.

   c How are the properties of copper and graphite related to their structure and bonding?

2 Consider the following five types of crystalline solids:

   A Metallic,

   B Ionic,

   C Giant molecular (macromolecular),

   D Composed of monatomic molecules, and

   E Composed of molecules containing a small number of atoms.

   Select the letter (A–E) for the structure most likely to show the following properties.

   a An element which conducts electricity and boils at 1600 °C to form gaseous monatomic atoms.

   b A solid which melts at −250 °C.

   c A solid with a very high molar enthalpy change of vaporisation which does not conduct when liquid.

   d A hard, brittle solid which easily cleaves.

   e A substance which boils at −50 °C and decomposes at high temperatures.

3 Diamond is one of the hardest natural substances. Before the production of carborundum, powdered diamond was the most widely used abrasive. Carborundum (silicon carbide) is made by heating coke with sand (silicon(IV) oxide) at 2500 °C.

   a What is an abrasive?

   b Write an equation for the manufacture of carborundum from coke and sand.

   c Why do you think carborundum has superseded diamond as the most widely used abrasive?

   d 'Diamonds are a girl's best friend'! Why? Carborundum has not superseded diamonds in this context. Why not?

   e It has been said that the discovery of carborundum enabled the industrial revolution to occur swiftly during the late nineteenth and early twentieth centuries. Why was this?

4 Discuss and explain the following:

   a Silicon(IV) oxide is a solid at room temperature which does not melt until 1973 K, whereas carbon dioxide (m.pt. 217 K) is a gas at room temperature.

   b Both calcium oxide and sodium chloride have a simple cubic structure and similar interionic distances, yet calcium oxide melts at 2973 K whereas sodium chloride melts at 1074 K.

   c Glucose ($C_6H_{12}O_6$) is much more soluble in water than in hexane, but cyclohexane (C_6H_{12}) is much more soluble in hexane than in water.

5 a What is meant by the term co-ordination number?

   b Name two metals with a close-packed structure and two metals with a body-centred cubic structure.

   c What is the co-ordination number of atoms

   i in a close-packed structure,

   ii in a body-centred cubic structure?

   d Suppose one particular metal can have either a close-packed or a body-centred cubic structure. In which of these two forms would you expect it to have:

   i the higher density,

   ii the higher melting point,

   iii the greater malleability?

   Explain your choice in each case.

6 The following data apply to the compounds XCl_x and YCl_y

	Melting point /°C	Boiling point /°C	Solubility in water /g per 100 g	Solubility in hexane /g per 100 g
XCl_x	801	1443	37	0.063
YCl_y	−22.6	76.8	0.08	miscible with hexane

   a What types of bond(s) are present in these two chlorides?

   b Explain clearly how the bonding in each chloride leads to such great differences in volatility and solubility.

7 a Plot pV against T and pV/T against p for a constant number of moles of ideal gas.

   b Draw a sketch graph to show how the pressure of a constant mass of ideal gas will vary as the temperature rises from absolute zero in a container of constant volume. How will the graph change if the gas tends to dissociate ($X_2 \longrightarrow 2X$) as temperature increases?

   (For all graphs label the axes and show the zero for each axis.)

8 A balloon can hold 1000 cm^3 of air before bursting. The balloon contains 975 cm^3 of air at 5 °C. Will it burst when it is taken into a house at 25 °C? Assume that the pressure of the gas in the balloon remains constant.

9 a A mixture of two gases in a container exerts a pressure of 800 mm Hg and occupies a volume of 400 cm³. If one of these gases (A) occupies a volume of 300 cm³ at the same temperature and pressure, what pressure does the other gas (B) exert in the mixture?

b Consider the following gases at room temperature, helium, hydrogen, hydrogen chloride, oxygen. Which one of the gases would behave:

   i   most like an ideal gas at room temperature,

   ii least like an ideal gas at room temperature?

10 0.50 g of a volatile liquid was introduced into a globe of 1000 cm³ capacity. The globe was heated to 91 °C so that all the liquid vaporised. Under these conditions the vapour exerted a pressure of 0.25 atm. What is the relative molecular mass of the liquid?

11 Consider the following five substances in the solid state:

  A sodium   **D** argon

  B silicon   **E** potassium bromide

  C tetrachloromethane

Select the letter (A–E) for the substance most likely to show the structure or property described in each of the following:

a A monatomic substance held together by van der Waals forces.

b A compound of low melting point composed of small molecules.

c A network solid of covalently bonded atoms.

d A non-conducting solid which melts to form a liquid which conducts electricity.

e A substance which exists as a liquid over a temperature range of only 3 K.

f A substance which is decomposed by an electric current in the liquid state.

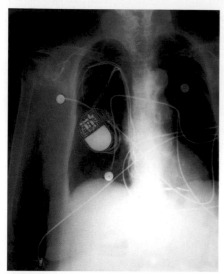

Figure 5.1

This heart pacemaker, fitted over the ribs, provides energy from a miniature lithium battery. Wires from the pacemaker battery supply energy as electrical impulses to maintain a regular heartbeat.

Key Point

The transfer of energy to or from chemicals plays a crucial part in chemical processes in industry and in living things.

5.1 Energy and energy changes

Energy is the most precious commodity we have. Without it there could be no life, no warmth, no movement. Energy gives us the power to do work. In every country, people's living standards are closely related to the availability of energy.

From the earliest times, people worshipped the Sun. This is not surprising. Through the process of photosynthesis, the Sun provides us with most of our food and, over millions of years, it has created our supplies of fossil fuels – coal, oil and natural gas.

The transfer of energy to or from chemicals plays a crucial part in the chemical processes in industry and in living things. Consequently, the study of these energy changes are as important as the study of the changes in the materials themselves.

Our present-day living conditions rely heavily on the availability of energy in its various forms. Chemical energy is converted to thermal energy when fuels such as wood, oil and coal are burnt in our homes and in industry, and vast quantities of chemical energy are converted to mechanical energy each day from the petrol and diesel burnt in our vehicles!

Within our own bodies, energy changes are also vital. Foods such as fats and carbohydrates are important biological fuels. During metabolism, the chemical energy in these foods is converted into heat (thermal energy) to keep us warm, into mechanical energy in our muscles and into electrical energy in the signals within our nerve fibres.

5.2 The ideas and language of thermochemistry

Most chemical reactions are accompanied by energy changes, usually in the form of heat or more precisely the **heat content** of the materials which are reacting. The correct term for heat content is **enthalpy**, symbol H. Normally, this change in the heat content or enthalpy shows up as a change in temperature. Indeed, the change in temperature when substances react often provides evidence that a chemical change has taken place.

When an **exothermic reaction** occurs, heat is given out and the temperature of the products rises. Eventually, the temperature of the products falls to room temperature as the heat produced is lost to the surroundings (Figure 5.2(a)). Thus, the heat content (enthalpy) of the products (H_2) is less than that of the reactants (H_1). Since the materials have *lost* heat, we can see that the **enthalpy change of the reaction**, ΔH (sometimes called the **heat of reaction**) is negative (Figure 5.2(b)).

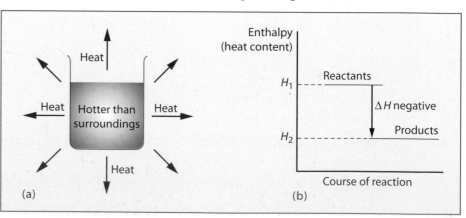

Figure 5.2

In an exothermic reaction, heat is lost to the surroundings and ΔH is negative.

For example, when magnesium reacts with oxygen, heat is evolved:

$$2Mg(s) + O_2(g) \longrightarrow 2MgO(s) \qquad \Delta H = -1204 \, kJ \, mol^{-1}$$

Chemical energy in the magnesium and in the oxygen is partly transferred to chemical energy in the magnesium oxide and partly evolved as heat. Thus, the magnesium oxide has less energy than the starting materials, magnesium and oxygen (Figure 5.3). The value of ΔH, the enthalpy change of the reaction, relates to the amounts shown in the equation, i.e. 2 moles of Mg atoms, 1 mole of O_2 molecules and 2 moles of MgO.

When an **endothermic reaction** occurs, the heat required for the reaction is taken from the reacting materials themselves. At first, the temperature of the products falls below the initial temperature, but eventually the temperature of the products rises to room temperature again as heat is *absorbed* from the surroundings. In this case, the enthalpy of the products is greater than that of the reactants and the enthalpy change, ΔH, is positive (Figure 5.4).

We can summarise these ideas as: Enthalpy change = $\Delta H = H_2 - H_1$

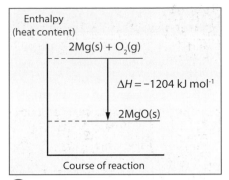

Figure 5.3
An enthalpy level diagram for the exothermic reaction of magnesium with oxygen

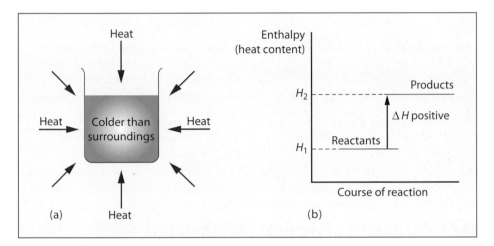

Figure 5.4
In an endothermic reaction, heat is gained from the surroundings and ΔH is positive.

Remember that ΔH refers only to the energy change *for the reacting materials*. The surroundings will obviously gain whatever heat the reacting materials lose, and vice versa. Thus, the *total* energy is unchanged during a chemical reaction. This important concept is known as the **Law of Conservation of Energy** and also as the **First Law of Thermodynamics**.

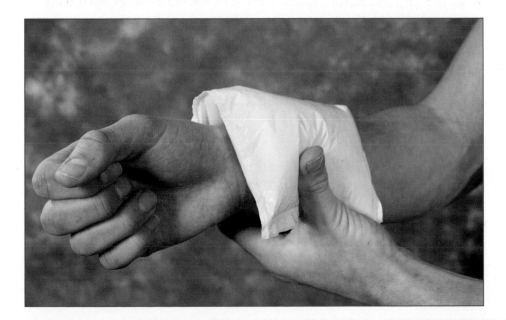

Key Points

In an **exothermic reaction**, heat is lost from the reacting materials and ΔH is negative.

In an **endothermic reaction**, heat is gained by the reacting materials and ΔH is positive.

Figure 5.5
Squeeze a cold pack and it gets cold. Chemicals in the cold pack react endothermically, taking in heat from the pack which gets cold. This cools the injury and helps to reduce painful swelling.

The law of conservation of energy
Energy may be exchanged between materials and their surroundings, but the total energy of the materials and surroundings remains constant.

1 Which of the following changes are exothermic and which are endothermic?
   a freezing water,
   b metabolising chocolate,
   c evaporating water,
   d burning sulfur.
2 When 0.1 mol of solid calcium oxide reacts with excess water to form calcium hydroxide solution 107 kJ of heat are produced.
   a Write an equation for the reaction including state symbols.
   b How much heat is produced when 1 mole of calcium oxide reacts with excess water?
   c Draw an enthalpy level diagram for the reaction like that in Figure 5.3 showing the enthalpy change.

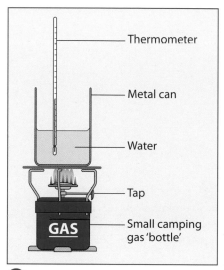

- Thermometer
- Metal can
- Water
- Tap
- Small camping gas 'bottle'

GAS

Figure 5.6
Measuring the enthalpy change when butane burns

5.3 Measuring enthalpy changes

Enthalpy changes during combustion

The heat given out when fuels burn can be measured by allowing the burning fuels to heat water. From the heat given to the water we can determine the enthalpy change per gram or per mole of fuel. Figure 5.6 shows the apparatus that was used to measure the heat given out when butane burns from a small camping gas bottle. **Do not attempt this experiment unless your chemistry teacher is present.**

We will assume that all the heat produced from the burning butane heats up the water. The results of the experiment are shown in Table 5.1.

 Table 5.1

Mass of butane 'bottle' at start of experiment	= 955.79 g
Mass of butane 'bottle' at end of experiment	= 954.34 g
∴ mass of butane burnt	= 1.45 g
Volume of water in can	= 250 cm³
∴ mass of water in can	= 250 g
Initial temperature of water	= 20 °C
Final temperature of water	= 60 °C
∴ rise in temp. of water	= 40 °C

▶ From the mass of water in the can and its temperature rise, we can work out the heat produced.
▶ From the loss in mass of the butane 'bottle', we can find the amount of butane burnt.
▶ We can then calculate the heat produced when 1 mole of butane burns.

The **specific heat capacity** of water is $4.2 \, \text{J g}^{-1} \text{K}^{-1}$. This means that:

4.2 J will raise the temp. of 1 g of water by   1 K (1 °C)

So, $m \times 4.2$ J will raise the temp. of m g of water by   1 K (1 °C)

And $m \times 4.2 \times \Delta T$ J will raise the temp. of m g of water by ΔT K (ΔT°C)

In general, if m g of a substance with a specific heat capacity of c J g^{-1} K^{-1} changes in temperature by ΔT K, then

The **enthalpy change** (energy change) $= m / \text{g} \times c / \text{J g}^{-1} \text{K}^{-1} \times \Delta T / \text{K} = mc\Delta T / \text{J}$

So, from the results of the experiment we can write:

Enthalpy change from burning 1.45 g butane $= 250\,g \times 4.2\,J\,g^{-1}\,K^{-1} \times 40\,K$

$$= 42\,000\,J$$

Therefore, enthalpy change from 1 mole (58 g) of burning butane $= \dfrac{58\,g}{1.45\,g} \times 42\,000\,J$

$$= 1\,680\,000\,J$$

$$= 1680\,kJ$$

As the reaction is exothermic, we can write:

$$C_4H_{10}(g) + 6\tfrac{1}{2}O_2(g) \longrightarrow 4CO_2(g) + 5H_2O(l) \qquad \Delta H = -1680\,kJ\,mol^{-1}$$

The enthalpy change when one mole of a fuel burns completely is generally called the **enthalpy change of combustion** of the fuel and the symbol for this is ΔH_c.

So, $\Delta H_c(C_4H_{10}(g)) = -1680\,kJ\,mol^{-1}$

Errors in our experiment

The accepted value for the enthalpy change of combustion of butane, $\Delta H_c(C_4H_{10}(g))$, is $-2877\,kJ\,mol^{-1}$. The result we have obtained is far less than this, so what are the errors in our experiment?

1 We have assumed that all the heat from the burning butane heats the water. In practice, a large proportion of the heat transfers to the metal can and to the surrounding air.
2 During heating, the flame can be affected by draughts and other air movements, which also reduce the heat that reaches the water.
3 Finally, the butane may burn incompletely, leaving soot on the base of the can.

All three of these sources of error reduce the heat transferred to the water, leading to a result well below the accepted value.

Accurate values of enthalpy changes of combustion can be obtained using a bomb calorimeter (Figure 5.7). The apparatus is specially designed to avoid heat losses by surrounding the 'bomb' first with water and then an insulating air jacket. Problems of incomplete combustion are overcome by heating the test sample in excess oxygen at high pressure.

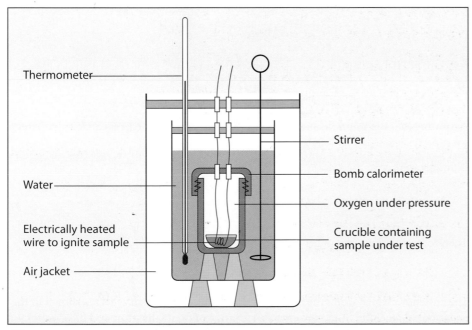

Figure 5.7
A bomb calorimeter

Thermometer

Water

Electrically heated wire to ignite sample

Air jacket

Stirrer

Bomb calorimeter

Oxygen under pressure

Crucible containing sample under test

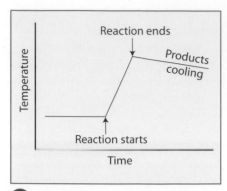

Figure 5.8
A graph of temperature against time for a reaction in a bomb calorimeter

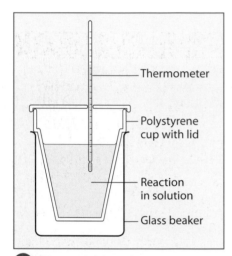

Figure 5.9
Measuring the enthalpy change for a reaction in solution

Heat losses can be eliminated completely if the experiment is repeated with an electrical heating coil in place of the reactants. The current in the coil is carefully adjusted to give a temperature / time curve identical to that with the reactants (Figure 5.8). By recording the current, voltage and time during this electrical calibration, it is possible to calculate the electrical energy supplied with great accuracy. This electrical energy is exactly the same as the energy change in the chemical reaction.

Enthalpy changes of combustion are very important in the food and fuel industries. Chemists use bomb calorimeters similar to that in Figure 5.7 (on the previous page) to measure the energy values of foods and fuels. Nutritionists can then give more accurate advice about different foods and public utilities, providing gas and electricity, can relate the prices of fuels to their energy values.

Enthalpy changes in solution

Many reactions take place in solution and the enthalpy changes for these reactions can be measured fairly accurately using the apparatus in Figure 5.9.

The polystyrene cup with lid is an excellent insulator. It also has a very small heat capacity so any heat lost to it or gained from it can be ignored relative to the heat given to or taken from the solution.

If the solutions are dilute, we can assume they have the same density and specific heat capacity as water.

Worked example

When $50\,cm^3$ of $1.0\,mol\,dm^{-3}$ sodium hydroxide was added to $50\,cm^3$ of $1.0\,mol\,dm^{-3}$ hydrochloric acid, the temperature rose by $6.2\,°C$. What is the enthalpy change for the neutralisation reaction which occurs?

Answer

The equation for the reaction is:

$$NaOH(aq) + HCl(aq) \longrightarrow NaCl(aq) + H_2O(l)$$

Amount of NaOH reacting $= \dfrac{50}{1000}\,dm^3 \times 1.0\,mol\,dm^{-3}$

$= 0.05\,mol$

Amount of HCl reacting $= \dfrac{50}{1000}\,dm^3 \times 1.0\,mol\,dm^{-3}$

$= 0.05\,mol$

Assuming that the solutions have the same density and specific heat capacity as water, we can say:

Enthalpy change in the solution $= m \times c \times \Delta T$

$= 100\,g \times 4.2\,J\,g^{-1}K^{-1} \times 6.2\,K$

$= 2604\,J$

Therefore, enthalpy change per mole $= \dfrac{2604\,J}{0.05\,mol}$

$= 52\,080\,J\,mol^{-1}$

The enthalpy change when the molar amounts of acid and alkali react, as shown in the equation, is generally called the **enthalpy change of neutralisation** and the symbol for this is $\Delta H_{neutralisation}$.

So, $NaOH(aq) + HCl(aq) \longrightarrow NaCl(aq) + H_2O(l)$ $\quad \Delta H_{neutralisation} = -52\,kJ\,mol^{-1}$

3 When excess powdered zinc is added to $50 \, cm^3$ of $0.2 \, mol \, dm^{-3}$ copper(II) sulfate solution, the temperature rises by $9 \,°C$.

   a Write an equation for the reaction.

   b Calculate the enthalpy change for the molar amounts in the equation.

4 The enthalpy change of vaporisation of water is $+ 44 \, kJ \, mol^{-1}$. An electric kettle rated at $3 \, kW$ ($1 \, kW = 1 \, kJ \, s^{-1}$) boils water at $100 \,°C$. How long will it take to evaporate 1 mole of water if the kettle is not switched off?

5 On adding $8.0 \, g$ of ammonium nitrate, NH_4NO_3, to $100 \, cm^3$ of water, the temperature falls by $6 \,°C$.

   a Is the enthalpy change exothermic or endothermic?

   b What is the enthalpy change for the above process?

   c What is the enthalpy change when 1 mole of ammonium nitrate dissolves under the same conditions?

5.4 The standard conditions for thermochemical measurements

In order to compare energy changes fairly, it is important that the conditions under which measurements and reactions are carried out are the same in all cases. These conditions for the fair comparison of thermochemical changes, including enthalpy changes, are called **standard conditions**.

These standard conditions are:

▶ a temperature of 298 K (25 °C),

▶ a pressure of 1 atmosphere ($\approx 100 \, kPa$),

▶ substances in their most stable state at 298 K and 1 atmosphere pressure (the standard state),

▶ solutions at a concentration of $1.0 \, mol \, dm^{-3}$.

Any enthalpy change measured under these conditions is described as a **standard enthalpy change** and given the symbol $\Delta H^{\ominus}_{298}$ or simply $\Delta H^{\ominus}$, pronounced 'delta H standard'.

Thus, $\Delta H^{\ominus}_{298}$ for the reaction

$$2H_2(g) + O_2(g) \longrightarrow 2H_2O(l)$$

must relate to gaseous hydrogen, gaseous oxygen and liquid water (not steam).

In the case of elements which exist as different allotropes, and compounds which exist as different polymorphs, the most stable form at 298 K and 1 atm is chosen as the standard. Consequently, $\Delta H^{\ominus}_{298}$ values for reactions involving carbon should relate to the allotrope graphite rather than diamond.

5.5 Standard enthalpy changes

Enthalpy changes of formation and atomisation

As you will learn later, one of the most important measurements for a substance is its **standard enthalpy change of formation**.

The standard enthalpy change of formation of a substance, $\Delta H^{\ominus}_f$ is the heat evolved or absorbed when one mole of the substance is formed from its elements.

The standard enthalpy change of formation of a substance is given the symbol $\Delta H^{\ominus}_f$. The superscript $^{\ominus}$ indicates standard conditions and the subscript $_f$ refers to the formation reaction.

Quick Questions

6 Consider the reaction
$$2H_2(g) + O_2(g) \rightarrow 2H_2O(l)$$
$$\Delta H^{\ominus}_{298} = -575 \, kJ \, mol^{-1}$$

   a Is the reaction endothermic or exothermic?

   b Which have the greater enthalpy, the products or the reactants?

   c What is the value of $\Delta H^{\ominus}_{298}$ for:

     i $2H_2O(l) \rightarrow 2H_2(g) + O_2(g)$

     ii $H_2(g) + \frac{1}{2}O_2(g) \rightarrow H_2O(l)$

   d Given also that
$$H_2O(l) \rightarrow H_2O(g)$$
$$\Delta H^{\ominus}_{298} = +44 \, kJ \, mol^{-1},$$
calculate the value of $\Delta H^{\ominus}_{298}$ for:
$$2H_2(g) + O_2(g) \rightarrow 2H_2O(g)$$

Key Point

It is important to know the standard conditions for thermochemical measurements.

Definition

The **standard enthalpy change of a reaction**, $\Delta H^{\ominus}_r$, is the heat evolved or absorbed when the molar amounts of reactants as stated in the equation react under standard conditions.

Thus, the statement $\Delta H_f^\ominus(MgO(s)) = -602\,kJ\,mol^{-1}$ relates to the formation of 1 mole of magnesium oxide from 1 mole of Mg atoms and 0.5 mole of O_2 molecules, i.e.

$$Mg(s) + \tfrac{1}{2}O_2(g) \longrightarrow MgO(s)$$

An energy level diagram for the standard enthalpy change of formation of water $(\Delta H_f^\ominus(H_2O(l)) = -286\,kJ\,mol^{-1})$ is shown in Figure 5.10. This shows that, under standard conditions, water has a lower energy content than the hydrogen and oxygen from which it is formed.

One important consequence of the definition of standard enthalpy change of formation is that the enthalpy change of formation of an element in its standard state under standard conditions is zero, since no heat change is involved when an element is formed from itself.

$$Cu(s) \longrightarrow Cu(s) \qquad \Delta H_f^\ominus(Cu(s)) = 0\,kJ\,mol^{-1}$$
$$H_2(g) \longrightarrow H_2(g) \qquad \Delta H_f^\ominus(H_2(g)) = 0\,kJ\,mol^{-1}$$

Obviously, $\Delta H_{298}^\ominus$ for the process $H_2(g) \longrightarrow H_2(g)$ is zero, but this is not so for the process $H_2(g) \longrightarrow 2H(g)$. This involves atomisation, i.e. the conversion of H_2 molecules to single H atoms. Thus, the **standard enthalpy change of atomisation of an element** is the enthalpy change when one mole of gaseous atoms are formed from the element.

Therefore, the standard enthalpy change of atomisation of hydrogen $(\Delta H_{at}^\ominus(H(g))$ refers to the process:

$$\tfrac{1}{2}H_2(g) \longrightarrow H(g) \qquad \Delta H_{at}^\ominus(H(g)) = +218\,kJ\,mol^{-1}$$

The enthalpy changes of atomisation of many solid and liquid elements can be obtained using specific heat capacities and molar enthalpy changes of fusion and vaporisation.

In discussing the heat changes in chemical reactions, it is useful to give each compound a definite heat content or enthalpy. For convenience, all elements are assigned a heat content of zero under standard conditions (Figure 5.10). The standard enthalpy change of formation of a compound then provides a measure of the heat content of the compound relative to its constituent elements. Remember, however, that these enthalpy values are *only* relative values. They give no information about the absolute energy content of a substance. This will depend on the potential energy in the electrical and nuclear interactions of the constituent particles plus the kinetic energy possessed by the atoms and sub-atomic particles.

Enthalpy changes of combustion and neutralisation

Two other enthalpy changes of some importance are the standard enthalpy changes of combustion and neutralisation. We have already considered each of these enthalpy changes in Section 5.3, and their definitions are now shown in the margin.

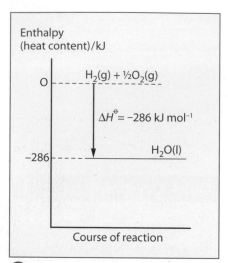

Figure 5.10
An energy level diagram for the standard enthalpy change of formation of water

Definitions

The **standard enthalpy change of formation of a substance**, $\Delta H_f^\ominus$, is the enthalpy change when 1 mole of the substance forms from its elements under standard conditions.

The **standard enthalpy change of atomisation of an element**, $\Delta H_{at}^\ominus$, is the enthalpy change when one mole of gaseous atoms form from the element under standard conditions.

The **standard enthalpy change of combustion of a substance**, $\Delta H_c^\ominus$, is the enthalpy change when 1 mole of the substance is completely burnt in oxygen under standard conditions.

The **standard enthalpy change of neutralisation**, $\Delta H_{neutralisation}^\ominus$, is the enthalpy change when the molar amounts of acid and alkali as shown in the equation react under standard conditions.

Quick Questions

7 Why is it necessary to say 'completely burnt in oxygen' in the definition of the standard enthalpy change of combustion, $\Delta H_c^\ominus$?

8 a Draw the enthalpy changes of combustion of both graphite and diamond on the same energy level diagram.
$\Delta H_c^\ominus(C(graphite)) = -393\,kJ\,mol^{-1}$
$\Delta H_c^\ominus(C(diamond)) = -395\,kJ\,mol^{-1}$
   b Which allotrope has the larger enthalpy (energy content)?
   c Which allotrope is the more stable?
   d What is the enthalpy change for the process C (graphite) → C (diamond)?

5.6 Measuring standard enthalpy changes of formation

The standard enthalpy changes of formation of carbon dioxide, magnesium oxide and many other oxides can be measured directly using a bomb calorimeter similar to that discussed in Section 5.3.

However, there are many compounds for which $\Delta H_f^{\ominus}$ cannot be measured directly. Imagine the difficulty in trying to measure the standard enthalpy change of formation of carbon monoxide, $\Delta H_f^{\ominus}(g)(CO(g))$. The equation for this is:

$$C_{(graphite)} + \tfrac{1}{2}O_2(g) \longrightarrow CO(g)$$

How could you possibly react graphite with oxygen so that it produces only carbon monoxide and no carbon dioxide?

Aluminium oxide (Al_2O_3) provides a different problem in attempts to measure its standard enthalpy changes of formation because a protective layer of oxide coats the aluminium.

As a result of these problems chemists have had to obtain standard enthalpy changes of formation indirectly. One indirect method involves using the enthalpy changes of combustion of the compound and its constituent elements.

Enthalpy changes of formation from enthalpy changes of combustion

Figure 5.11 shows how the enthalpy change of formation of carbon monoxide can be obtained using the enthalpy changes of combustion of carbon monoxide and graphite. Notice how the three processes have been linked in an energy cycle in Figure 5.11. This shows two routes for converting graphite and oxygen to CO_2. One of these is the direct route straight from graphite and oxygen to CO_2. The alternative route goes via CO. It would seem reasonable that the overall enthalpy change for the conversion of graphite to carbon dioxide is independent of the route taken, so we can write

$$\Delta H_1 = \Delta H_2 + \Delta H_3 \qquad \text{Equation (1)}$$

Now
$$\Delta H_1 = \Delta H_c^{\ominus}(C(graphite)) = -393 \text{ kJ mol}^{-1},$$

$$\Delta H_3 = \Delta H_c^{\ominus}(CO(g)) = -283 \text{ kJ mol}^{-1}$$

and
$$\Delta H_2 = \Delta H_f^{\ominus}(CO(g))$$

But from Equation (1) above, $\Delta H_2 = \Delta H_1 - \Delta H_3$

So,
$$\Delta H_f^{\ominus}(CO(g)) = -393 - (-283)$$
$$= -110 \text{ kJ mol}^{-1}$$

The argument we used to obtain Equation (1) above is a specific example of **Hess' law**. This states that the enthalpy change of a reaction is the same whether the reaction occurs in one step or in a series of steps.

Of course, Hess' law is simply an application of the more fundamental law of conservation of energy. If Hess' law were not true, then we could create or destroy energy by going one way or the other round an energy cycle, such as that in Figure 5.11.

Figure 5.12 shows a diagram to illustrate Hess' law more generally.

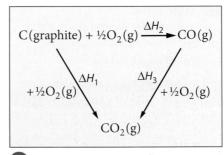

Figure 5.11
An energy cycle incorporating the enthalpy change of formation of carbon monoxide

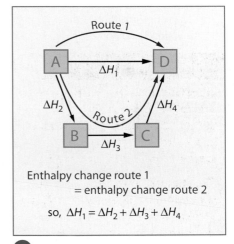

Figure 5.12
A diagram to illustrate Hess' law

Definition

Hess' law states that the enthalpy change in converting reactants to products is the same regardless of the route taken, provided the initial and final conditions are the same.

Quick Questions

9 The equation for the standard enthalpy change of formation of methane, $\Delta H_f^{\ominus}(CH_4(g))$ is: C (graphite) + 2H$_2$(g) $\longrightarrow$ CH$_4$(g)
   Use the following values to calculate, $\Delta H_f^{\ominus}(CH_4(g))$:
   $\Delta H_c^{\ominus}(C(graphite)) = -393 \text{ kJ mol}^{-1}$
   $\Delta H_c^{\ominus}(H_2(g)) = -286 \text{ kJ mol}^{-1}$
   $\Delta H_c^{\ominus}(CH_4(g)) = -890 \text{ kJ mol}^{-1}$

Figure 5.13
Edwin Aldrin, the second man on the moon, steps from the ladder of the Apollo 11 Lunar Module.

Quick Questions

10 Write equations for the standard enthalpy change of formation of:
  a sodium chloride, NaCl,
  b copper sulfate, $CuSO_4$,
  c ethanol, C_2H_5OH.

11 Tin is manufactured by heating tinstone, SnO_2 at high temperature with coke (carbon). The other product of the reaction is carbon dioxide.
  a Write an equation for the reaction involved.
  b Calculate the standard enthalpy change for the reaction involved using the data below.
  $\Delta H_f^\ominus(SnO_2(s)) = -581 \, kJ \, mol^{-1}$
  $\Delta H_f^\ominus(CO_2(g)) = -394 \, kJ \, mol^{-1}$

Figure 5.14
An energy cycle incorporating the reaction of methylhydrazine with dinitrogen tetraoxide

5.7 Using Hess' law and enthalpy changes of formation to calculate the energy changes in reactions

In the last section, we used Hess' law to calculate the standard enthalpy change for the formation of carbon monoxide which cannot be measured directly.

In fact, using Hess' law and enthalpy changes of formation it is possible to calculate the enthalpy change of any reaction indirectly. Let's see how this can be done.

Worked example

On 21 July 1969, the Apollo 11 project landed the first man on the Moon. During this project, engines of the lunar module used methylhydrazine (CH_3NHNH_2) and dinitrogen tetraoxide (N_2O_4). These liquids were carefully chosen since they ignite spontaneously and very exothermically on contact. How can we calculate the enthalpy change for the reaction?

1 Write the equation for the reaction:

$$4CH_3NHNH_2(l) + 5N_2O_4(l) \longrightarrow 4CO_2(g) + 12H_2O(l) + 9N_2(g)$$

2 Draw an energy cycle by adding the formation equations from the same elements to both sides of the equation as in Figure 5.14, adding $\Delta H_f^\ominus$ terms on the upward arrows.
  Remember that, for any element, $\Delta H_f^\ominus$(element) $= 0 \, kJ \, mol^{-1}$.

3 Apply Hess' law to the cycle.
  By Hess' law, the total enthalpy change for the formation of carbon dioxide, water and nitrogen will be the same whether they are formed directly from their elements or via the intermediates, CH_3NHNH_2 and N_2O_4.

$$\therefore \; 4\Delta H_f^\ominus(CH_3NHNH_2(l)) + 5\Delta H_f^\ominus(N_2O_4(l)) + \Delta H_{reaction}^\ominus$$
$$= 4\Delta H_f^\ominus(CO_2(g)) + 12\Delta H_f^\ominus(H_2O(l))$$

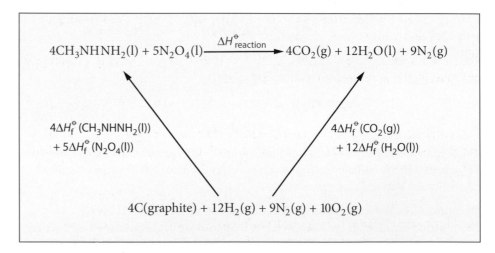

4 Insert the $\Delta H_f^\ominus$ values in the equation obtained by applying Hess' law and calculate the enthalpy change for the reaction.
  Be careful with the signs of $\Delta H^\ominus$ values by putting the value and sign for a quantity in brackets when adding and subtracting.

$(\Delta H_f^\ominus(CH_3NHNH_2(l)) = +53 \, kJ \, mol^{-1}, \qquad \Delta H_f^\ominus(N_2O_4(l)) = -20 \, kJ \, mol^{-1},$

$\Delta H_f^\ominus(CO_2(g)) = -393 \, kJ \, mol^{-1}, \qquad \Delta H_f^\ominus(H_2O(l)) = -286 \, kJ \, mol^{-1}.)$

$$\therefore \; [4\,(+53) + 5(-20)] \, kJ \, mol^{-1} + \Delta H_{reaction}^\ominus$$
$$= [4(-393) + 12(-286)] \, kJ \, mol^{-1}$$

so, $\Delta H^{\ominus}_{reaction} = -5116\,kJ\,mol^{-1}$

Therefore

$$4CH_3NHNH_2(l) + 5N_2O_4(l) \longrightarrow 4CO_2(g) + 12H_2O(l) + 9N_2(g)\quad \Delta H^{\ominus} = -5116\,kJ\,mol^{-1}$$

The enthalpy changes in other reactions can be calculated in a similar fashion to this by drawing an energy cycle to include the reaction involved and the reactions of those elements which form both the reactants and products (Figure 5.15).

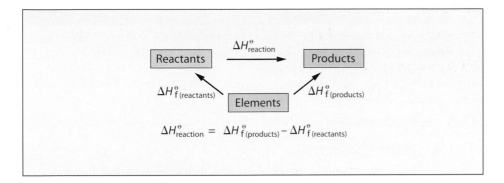

$$\Delta H^{\ominus}_{reaction} = \Delta H^{\ominus}_{f(products)} - \Delta H^{\ominus}_{f(reactants)}$$

◁ Figure 5.15
A general energy cycle to calculate the enthalpy change in any reaction

5.8 Using standard enthalpy changes of formation to predict the relative stabilities of compounds

Most compounds are formed exothermically from their elements. Thus, the standard enthalpy changes of formation of water, carbon dioxide, aluminium oxide and many other compounds are negative. These compounds are therefore at a lower energy level than their constituent elements. This means that the compounds are energetically more stable than the elements from which they are formed. But, consider the following problem.

The standard enthalpy change of formation of hydrogen peroxide is $-188\,kJ\,mol^{-1}$. From this, we would expect H_2O_2 to be stable. But, H_2O_2 decomposes fairly readily into water and oxygen. How can this be explained?

The answer lies in the fact that $\Delta H^{\ominus}_f(H_2O_2(l))$ only describes the stability of hydrogen peroxide *relative to its elements*:

$$H_2(g) + O_2(g) \longrightarrow H_2O_2(l) \qquad \Delta H = -188\,kJ\,mol^{-1}$$

H_2O_2 is obviously more stable than its elements, but on decomposition it produces not $H_2(g) + O_2(g)$ for which $\Delta H^{\ominus}$ is $+188\,kJ\,mol^{-1}$, but $H_2O(l) + \frac{1}{2}O_2(g)$ for which $\Delta H^{\ominus}$ is $-98\,kJ\,mol^{-1}$, i.e.

$$H_2O_2(l) \longrightarrow H_2O(l) + \frac{1}{2}O_2(g) \qquad \Delta H = -98\,kJ\,mol^{-1}$$

Thus, hydrogen peroxide is energetically stable with respect to its elements, but unstable with respect to water and oxygen. *This example shows how important it is to specify with respect to what substances a compound is stable or unstable.*

A few compounds, such as carbon disulfide (CS_2) and nitrogen oxide (NO) are formed endothermically from their elements:

$$\frac{1}{2}N_2(g) + \frac{1}{2}O_2(g) \longrightarrow NO(g) \qquad \Delta H^{\ominus} = +90\,kJ\,mol^{-1}$$

These compounds have positive standard enthalpy changes of formation. They are therefore energetically unstable with respect to their elements. So, why don't these compounds decompose instantaneously into their constituent elements?

⌃ Figure 5.16
The launch of Apollo 11 from the Kennedy Space Centre, Florida, on 16 July, 1969. Four days later, two of the crew members, Neil Armstrong and Edwin Aldrin, were the first people to step onto the Moon. Apollo 11 was launched using a Saturn V rocket which burns 15 tonnes of kerosine/oxygen mixture every second.

Carbon disulfide and nitrogen oxide can be stored for long periods at room temperature and pressure. They do, however, begin to decompose at high temperatures. In order to explain the unexpected stability of these compounds, we must distinguish between *energetic* stability and *kinetic* stability.

Carbon disulfide and nitrogen oxide are energetically unstable with respect to their elements. But at low temperatures and pressures the decomposition reactions are so slow that they are kinetically stable.

Quick Questions

12 Assuming that kerosine is $C_{11}H_{24}$, calculate the standard enthalpy change when 1 mole of kerosine burns completely in oxygen, using the data below:
$$\Delta H_f^{\ominus} (C_{11}H_{24}(l)) = -327\,kJ\,mol^{-1}$$
$$\Delta H_f^{\ominus} (CO_2(g)) = -394\,kJ\,mol^{-1}$$
$$\Delta H_f^{\ominus} (H_2O(l)) = -286\,kJ\,mol^{-1}$$

13 The following data shows that diamond is unstable with respect to graphite:
$$C\,(diamond) \longrightarrow C\,(graphite) \quad \Delta H^{\ominus} = -2\,kJ\,mol^{-1}$$
Why then do diamonds not turn into graphite?

14 Colourless nitrogen monoxide remains unchanged on its own, but it reacts rapidly with oxygen to form brown fumes of nitrogen dioxide:
$$\Delta H_f^{\ominus} (NO(g)) = +90\,kJ\,mol^{-1}$$
$$\Delta H_f^{\ominus} (NO_2(g)) = +33\,kJ\,mol^{-1}$$
   a Why is nitrogen monoxide described as energetically unstable, but kinetically stable with respect to its elements?
   b Use the data above to describe the stability of nitrogen monoxide relative to nitrogen dioxide:
      i   in the presence of oxygen,
      ii in the absence of oxygen.

15 a Why are petrol/oxygen mixtures stable at room temperature before a spark is applied?
   b How does a spark initiate a reaction between petrol and oxygen?

5.9 Predicting whether reactions will occur

Strike a match. It catches fire and burns quickly. Light a candle and it burns quietly. Both of these are exothermic reactions in which energy is lost when the reaction occurs. ΔH for each reaction is negative and the products of the reaction are more stable than the reactants.

In fact, many reactions which are exothermic just keep going once started. As a general rule, exothermic reactions are more likely to occur than endothermic reactions. Because of this, the sign of ΔH can be used to predict whether a reaction will occur. However, this rule does not always give reliable predictions for three main reasons.

1 Predictions from $\Delta H^{\ominus}$ values relate to standard conditions (i.e. 298 K and 1 atm pressure) and the likelihood of a reaction may be very different under different conditions or in the presence of a catalyst.

2 Values of $\Delta H^{\ominus}$ show the relative energetic stabilities of the reactants and products of a reaction. They say nothing about the kinetic stability of the products relative to the reactants. In other words, $\Delta H^{\ominus}$ values are no guide to the *rate* of a reaction. In fact, some highly exothermic reactions never occur because the reaction rate is so slow and the reactants are kinetically stable with respect to the products (Figure 5.17).

3 In order to make accurate predictions about the relative energy levels of reactants and products, it is necessary to consider not only the energy lost or gained by the reacting system but also energy changes inside that system. For example, when a gas

Figure 5.17
The reaction of paper (cellulose) with oxygen in the air is very exothermic. Paper and oxygen are energetically very unstable relative to their products (CO_2 and water). Fortunately, however, the reactants are kinetically stable and it is safe to read the paper at leisure.

is produced in a reaction or a solid dissolves in a liquid, there is a marked increase in disorder and an increase in the number of ways in which energy is distributed within the system. This disorder of energy distribution among the particles within a system can be compared to the disorder of thermal energy (heat) when it spreads out and gets dissipated to the surroundings in an exothermic reaction. The disorder of a system is measured as entropy.

At normal temperatures, the additional entropy within a system, following a reaction, is usually unimportant. Even so, it explains why certain endothermic reactions occur spontaneously. For example, many solids dissolve easily in water in spite of the fact that the process is endothermic. The reason for this is that, although the enthalpy change is positive, there is an enormous increase in entropy as the solid dissolves.

5.10 Enthalpy changes and bond energies

When chemical reactions take place, the bonds in reactant molecules must break before new bonds can form in the products. For example, when hydrogen reacts with chlorine:

$$H_2(g) + Cl_2(g) \longrightarrow 2HCl(g)$$

bonds in the H_2 and Cl_2 molecules first break to form H and Cl atoms (Figure 5.18). Then, new bonds form between the H and Cl atoms to produce HCl (hydrogen chloride).

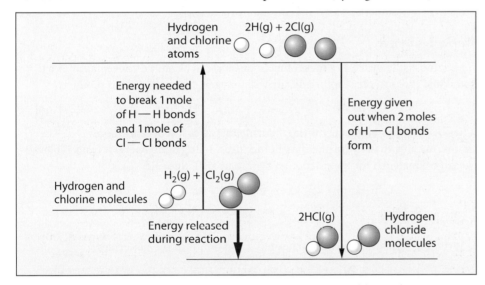

Obviously, energy must be supplied to break the bonds between atoms, so energy must be released when the reverse occurs and bonds form.

Chemical reactions involve bond breaking followed by bond making. This means that the enthalpy change of a reaction is the energy difference between the bond breaking and bond making processes.

When hydrogen and chlorine react, more energy is released in making two new H—Cl bonds in the two HCl molecules than in breaking the bonds in one H_2 molecule and one Cl_2 molecule. So, the overall reaction is exothermic.

A definite quantity of energy, known as the **bond energy** or **bond dissociation energy** can be associated with each type of bond. This energy is taken in when the bond is broken and given out when the bond is formed.

Use of the term bond *dissociation* energy has led to the symbol D for bond energies. So the H—Cl bond energy is written as $D(\text{H—Cl}) = 431 \, \text{kJ mol}^{-1}$. In measuring and using bond energies, chemists distinguish between the terms 'bond energy' and 'average bond energy'.

Figure 5.18
An enthalpy level diagram for the reaction between hydrogen and chlorine

Table 5.2

The average bond energies of some common bonds

Bond	Average bond energy $\bar{D}(A-B)$ /kJ mol^{-1}	Bond	Average bond energy $\bar{D}(A-B)$ /kJ mol^{-1}
C—H	410	O—H	460
C—C	350	O—O	150
C=C	610	O=O	496
C—Cl	340	N—N	160
C—F	495	N≡N	994
H—H	436	N—H	390
H—Cl	431	Cl—Cl	244
C—O	360	F—F	158
C=O	740	H—F	562

Bond energies are precise values for specific bonds in particular compounds (e.g. for the C—Cl bond in CH_3Cl). On the other hand, average bond energies, symbol $\bar{D}$ are average values for one kind of bond in different compounds (e.g. an average value for the C—Cl bond in all compounds).

Bond energies are sometimes called **bond enthalpies**.

Results show that the bond energy of one particular kind of bond can vary slightly in different compounds. For example, the bond energy for the single carbon–carbon bond, C—C, has been determined using several different compounds. The results range from 330 to 360 kJ mol^{-1}. This shows that a bond energy value depends to some extent on the environment of the bond within a particular molecule. Thus the strength of a particular C—C bond will vary slightly according to which atoms are attached to the two carbon atoms.

Table 5.2 shows the average bond energies of some common bonds.

5.11 Using bond energies

Bond energies are useful in comparing the strengths of different bonds and in understanding the bonding in different compounds. However, the most important use of bond energies is in estimating the enthalpy change in chemical reactions involving molecular substances with covalent bonds. These estimates are particularly useful when experimental measurements cannot be made, as in the following example.

Worked example

Hydrazine is often used as a rocket fuel because it can be stored conveniently as a liquid and it reacts very exothermically with oxygen, forming gaseous products.

$$N_2H_4(g) + O_2(g) \longrightarrow N_2(g) + 2H_2O(g) \qquad \Delta H^\ominus = -622 \text{ kJ mol}^{-1}$$

It has been suggested that hydrazine/fluorine mixtures might react more exothermically than hydrazine/oxygen mixtures. Using bond energies from Table 5.2 we can estimate ΔH for the reaction of hydrazine with fluorine:

$$N_2H_4(g) + 2F_2(g) \longrightarrow N_2(g) + 4HF(g)$$

Answer

1 Write the equation showing all the atoms and bonds in the molecules.

$$\begin{matrix} H & & H \\ \backslash & & / \\ N & — & N \\ / & & \backslash \\ H & & H \end{matrix} \quad + 2F\!-\!F \longrightarrow N\!\equiv\!N + 4H\!-\!F$$

2 Count the number bonds broken and bonds formed.

Bonds broken	kJ mol^{-1}		Bonds formed	kJ mol^{-1}
one N—N	+160		one N≡N	−994
four N—H	+(4 × 390)		four H—F	−(4 × 562)
two F—F	+(2 × 158)			

3 Calculate the enthalpy change knowing that this is the difference between the energy needed to break bonds and the energy released as bonds form.

So $\qquad \Delta H_r = +160 + (4 \times 390) + (2 \times 158) - 994 - (4 \times 562)$

$\qquad\qquad\qquad = +2036 - 3242 = -1206 \text{ kJ mol}^{-1}$

For most reactions, the values of ΔH estimated from average bond energies agree quite closely with experimental values. This has further established the usefulness of bond energies.

Quick Questions

16 Using bond energies from Table 5.2, calculate the enthalpy change for the reaction shown in Figure 5.18.

17 Write equations to explain that $D(H—H) = 2 \times \Delta H^\ominus_{at}(H(g))$

18 a Calculate the bond energy for the O—H bond in water from the data below.

$H_2O(g) \longrightarrow HO(g) + H(g)$
$\qquad \Delta H^\ominus = 498 \text{ kJ mol}^{-1}$
$HO(g) \longrightarrow H(g) + O(g)$
$\qquad \Delta H^\ominus = 428 \text{ kJ mol}^{-1}$

b Why is your answer to part a not the same as the average O—H bond energy in Table 5.2.

19 Use the data in Table 5.2 to estimate the enthalpy change for the following reactions.

a $N_2(g) + 3H_2(g) \longrightarrow 2NH_3(g)$

b $H_2C\!=\!CH_2(g) + H_2(g) \longrightarrow CH_3CH_3(g)$

$\Delta H_{at}(Na(g))$ and $\Delta H_{il}(Na(g))$, are written upwards followed by those for chlorine, $\Delta H_{at}(Cl(g))$ and $E_a(Cl(g))$. Finally, the cycle is completed with the lattice energy of sodium chloride, $\Delta H_{latt}(Na^+Cl^-(s))$.

▶ Exothermic processes are represented by downward arrows and endothermic processes by upward arrows.

All the processes in the cycle can be measured experimentally, except the lattice energy. So, by applying Hess' law, it is possible to calculate the lattice energy.

Worked example

Calculate the lattice energy of sodium chloride using the information in Figure 5.21.

Answer

Apply Hess' law to the cycle in Figure 5.21. Remember that an exothermic change in one direction will become an endothermic change with the opposite sign if it is reversed.

Therefore $\Delta H_{latt}(Na^+Cl^-(s)) = [+349 -122 -494 -107 -411]\,kJ\,mol^{-1}$

$$= -785\,kJ\,mol^{-1}$$

5.14 Testing the model of ionic bonding

Essentially, the lattice energy of an ionic compound is the energy change which occurs when well-separated ions are brought together in forming the crystal. So, it is possible to calculate a theoretical value for the lattice energy of a compound by considering the interionic attractions and repulsions within the lattice (Figure 5.22).

Theoretical lattice energies are calculated by assuming that the only bonding in the compound is ionic in which discrete spherical ions carry an evenly distributed charge. Once the theoretical lattice energy has been calculated, it can be compared with the experimental lattice energy obtained via a Born–Haber cycle. By comparing the two values, it is possible to determine the extent to which the bonding in a compound is truly ionic.

The theoretical lattice energies of some ionic substances are compared with their corresponding experimental values in Table 5.4. Notice the similarity between the theoretical and experimental lattice energies for the three sodium halides. The difference between the two values is less than 3%. This close agreement provides strong evidence that the simple model of an ionic lattice is very satisfactory for sodium halides.

Now look at the theoretical and experimental lattice energies for the silver halides in Table 5.4. For these compounds the theoretical values are at least 8% less than the experimental values. This suggests that the bonding in silver halides, and particularly silver iodide, is clearly stronger than the ionic model predicts.

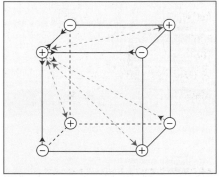

⌃ Figure 5.22
A few of the attractions (red) and repulsions (blue) which must be considered in calculating a theoretical lattice energy

Note

In comparing lattice energies which are always negative, it is ambiguous to use the words 'larger' or 'smaller'. It is better to describe one lattice energy as 'more exothermic' or 'less exothermic' than another.

An exothermic change in one direction becomes endothermic with the opposite sign if it is reversed.

Key Point

The close agreement between theoretical and experimental lattice energies for many ionic compounds provides good evidence for the ionic model of discrete spherical ions with evenly distributed charge.

⌄ Table 5.4
Theoretical and experimental lattice energies

Compound	Theoretical (calculated) lattice energy / kJ mol^{-1}	Experimental lattice energy (obtained via Born–Haber cycle) / kJ mol^{-1}	Percentage difference between theoretical and experimental lattice energies
NaCl	−770	−776	0.8
NaBr	−735	−742	1.0
NaI	−687	−705	2.6
AgCl	−833	−905	8.6
AgBr	−816	−891	9.2
AgI	−778	−889	14.3

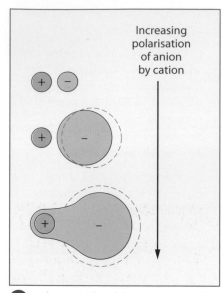

Figure 5.23
Increasing polarisation of the negative charge clouds around anions as their radius increases. (The dashed circles show the shapes of the isolated anions before polarisation.)

Increasing polarisation of anion by cation

Definition

Polarisation is the distortion of electron clouds in molecules and ions by nearby positive charges.

The values in Table 5.4 show that silver halides, although mainly ionic, have a significant degree of covalent bonding. The partly covalent nature of the bonds can be interpreted by assuming that positive metal ions can attract the outermost electrons of negative ions, pulling these electrons into the spaces between the ions. This distortion of the electron clouds around anions by positively charged cations is an example of **polarisation**. As a result of this polarisation, there is a significant degree of covalent bonding in some ionic compounds. Figure 5.23 shows three examples of ionic bonding with increasing degrees of polarisation, leading to electron sharing and therefore increasing covalent character.

In general,

▸ a cation has more polarising power as its charge increases and its radius decreases,
▸ an anion is more polarisable as its radius increases.

This means that iodide ions are more polarisable than bromide ions and bromide ions are more polarisable than chloride ions.

Quick Questions

25 Four values for lattice energy (in kJ mol^{-1}) are -2440, -2327, -1985 and -1877. The four compounds to which these values relate are $BaBr_2$, BaI_2, $MgBr_2$ and MgI_2. Match the formulae with the values and explain your choices.
26 The experimental value of the lattice energy of lithium iodide is greater than the calculated value by 21 kJ mol^{-1}, but the experimental value for rubidium iodide is greater than the calculated value by only 11 kJ mol^{-1}. Why is this?
27 Which one of the following chlorides has bonding that can be described as ionic with some covalent character?
    A KCl   B $CaCl_2$ C $AlCl_3$ D $SiCl_4$

5.15 Enthalpy changes when ionic substances dissolve

When ionic solids dissolve in water, heat is usually evolved or absorbed. Why is this? In order to answer the question, we need to consider the enthalpy changes which take place when ionic compounds dissolve.

When one mole of sodium chloride dissolves in water to produce a solution containing $1 \, mol \, dm^{-3}$ under standard conditions, the enthalpy of the system increases by 4 kJ. This enthalpy change is called the **standard enthalpy change of solution** of sodium chloride, $\Delta H^{\ominus}_{soln}(NaCl(s))$. The process can be summarised as:

$$Na^+Cl^-(s) + aq \longrightarrow Na^+(aq) + Cl^-(aq) \qquad \Delta H^{\ominus}_{soln} = +4 \, kJ \, mol^{-1}$$

In the equation 'aq' means 'addition of water'.

Sodium chloride and many other substances dissolve easily in water even though the processes are endothermic. This is further evidence that ΔH values are not a reliable guide to the likelihood of a process or reaction taking place. In these cases, the increase in disorder (entropy) of the system drives the process forward as regularly spaced ions in the crystal lattice dissolve to form freely moving ions in solution.

In order to understand the enthalpy change when ionic solids like sodium chloride dissolve, the overall process can be divided into two stages as shown in Figure 5.24. In fact, the stages overlap with one another because some ions will be completing the second stage while others are only starting the first.

▸ In the first stage, ions in the solid ionic lattice, $Na^+Cl^-(s)$, are separated into free, well-spaced gaseous ions, $Na^+(g)$ and $Cl^-(g)$. This is the reverse of the lattice energy process.
$$Na^+Cl^-(s) \longrightarrow Na^+(g) + Cl^-(g) \qquad -\Delta H_{latt} = +776 \, kJ \, mol^{-1}$$

► In the second stage, the gaseous $Na^+(g)$ and $Cl^-(g)$ ions are hydrated by polar water molecules to form a solution of sodium chloride, $Na^+(aq) + Cl^-(aq)$. Under standard conditions, this second stage involves the **standard enthalpy change of hydration** of both $Na^+(g)$ and $Cl^-(g)$, $\Delta H^{\ominus}_{hyd}(Na^+) + \Delta H^{\ominus}_{hyd}(Cl^-)$.

$$Na^+(g) + Cl^-(g) + (aq) \longrightarrow Na^+(aq) + Cl^-(aq)$$
$$\Delta H^{\ominus}_{hyd}(Na^+) + \Delta H^{\ominus}_{hyd}(Cl^-) = -772\,kJ\,mol^{-1}$$

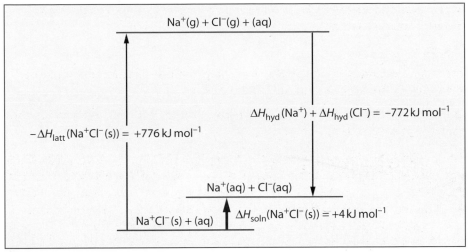

Figure 5.24
A Born–Haber cycle connecting the enthalpy change of solution, the enthalpy changes of hydration and lattice energy

Enthalpy changes of hydration are sometimes simply called **hydration energies**. From Figure 5.24 it is possible to appreciate why sodium chloride and other ionic compounds dissolve in water. Although the energy required to separate the ions (the reverse lattice energy) is so large, the energy released when the ions are hydrated is almost as large.

In the case of ionic solids, the reverse lattice energy is always endothermic as it involves separating ions in the solid lattice. On the other hand, the hydration energies are always exothermic since this involves the attraction of ions for polar water molecules. This means that the overall enthalpy change of solution will depend on whether the endothermic or the exothermic stage is greater.

Notice from Figure 5.24 that the relationship between the enthalpy change of solution, the reverse lattice energy and the hydration energies is:

$$\Delta H_{soln} = \underset{\text{endothermic stage}}{-\Delta H_{latt}} + \underset{\text{exothermic stage}}{\Sigma\Delta H_{hyd}}$$

In the case of sodium chloride, the endothermic stage is marginally greater than the exothermic stage. So, the enthalpy change of solution has a small positive value. The lattice energy and the hydration energies for ionic compounds are nearly always large values. As a result, the enthalpy change of solution, which is the difference between these two values, is positive in some cases and negative in others.

The overall hydration energy in Figure 5.24 is, of course, the sum of the separate hydration energies of Na^+ and Cl^-. Clearly, the individual hydration energy for Na^+ cannot be measured directly because sodium ions always exist in combination with anions. However, it is often useful to have some measure, however uncertain, of the individual hydration energies of particular ions. Various attempts have been made to estimate these values from the overall hydration energies of ionic compounds. The

Definitions

The **standard enthalpy change of solution** is the enthalpy change when one mole of a compound dissolves in water to produce a solution containing $1\,mol\,dm^{-3}$ under standard conditions.

The **standard enthalpy change of hydration (hydration energy)** is the enthalpy change when one mole of gaseous ions dissolves in water to form a solution containing $1\,mol\,dm^{-3}$ of the ions under standard conditions, e.g. for sodium ions:

$$Na^+(g) + aq \longrightarrow Na^+(aq)$$
$$\Delta^{\ominus}_{hyd}(Na^+) = -391\,kJ\,mol^{-1}$$

Note

The symbol Σ in the equation on the left means 'sum of' so $\Sigma\Delta H_{hyd}$ means the sum of the relevant enthalpy changes of hydration.

Table 5.5

Enthalpy changes of hydration (hydration energies) of some ions

Ion	ΔH_{hyd} /kJ mol^{-1}	Ion	ΔH_{hyd} /kJ mol^{-1}
H$^+$	−1075	F$^-$	−457
Li$^+$	−499	Cl$^-$	−381
Na$^+$	−391	Br$^-$	−351
K$^+$	−305	I$^-$	−307
Mg^{2+}	−1891		
Ca^{2+}	−1562		
Al^{3+}	−4613		

Quick Questions

28 Why are the hydration energies of anions and cations both negative?

29 Calculate the hydration energy of MgCl$_2$(s).

30 Why does the hydration energy get more exothermic along the series Na$^+$, Mg^{2+}, Al^{3+}?

31 Why does the hydration energy get less exothermic along the series F$^-$, Cl$^-$, Br$^-$, I$^-$?

Quick Questions

32 Choose examples from Table 5.3, Section 5.12, to illustrate that lattice energies become:
   a more exothermic as the charges on ions increase,
   b less exothermic as the radii of ions increase.

convention we shall use in this book is to accept −1075 kJ mol^{-1} as the hydration energy for H$^+$. Using this standard value, chemists have been able to work out the hydration energies of other ions. A few individual hydration energies are listed in Table 5.5.

Figure 5.25

Harvesting salt from lagoons in Brittany, France. The salt (sodium chloride) crystallises from solution in an exothermic process in which solid Na$^+$Cl$^-$(s) is more energetically stable than its aqueous ions, Na$^+$(aq) + Cl$^-$(aq).

The effect of ionic charge on lattice energies and enthalpy changes of hydration

Both lattice energies and enthalpy changes of hydration are exothermic processes because they involve attractions between opposite charges. Lattice energies involve attractions between oppositely charged ions whilst enthalpy changes of hydration involve the attractions of ions for δ+ and δ− charges on polar water molecules.

In addition, both processes become more exothermic as the charges on the ions increase and become more strongly attracted to any opposite charges.

The effect of ionic radius on lattice energies and enthalpy changes of hydration

As the radius of an ion increases, its charge density decreases. This reduces its attraction for oppositely charged ions and polar water molecules. So, both lattice energies and enthalpy changes of hydration become less exothermic as ionic radii increase.

5.16 Energy for life

Every living organism, whether animal or plant, needs energy and this energy is closely connected with the formation and hydrolysis of the chemical known as ATP.

The reactions that maintain life involve sequences of reactions linked in various pathways. These reactions are needed to maintain the functions of cells such as muscle contractions, movement of electric charge in nerve cells and the synthesis of essential chemicals for life. Each reaction in these pathways is catalysed by its own particular enzyme, with the product of one reaction becoming the **substrate** (reactant) for the next reaction. Many of the steps in these pathways need small amounts of readily-accessible energy.

The key chemical in providing readily accessible energy for the metabolism of all cells is ATP – **ad**enosine **trip**hosphate (Figure 5.26). ATP consists of three distinct parts – adenine, an organic base, bonded to a 5-carbon sugar, ribose, which is then linked to a short triphosphate chain with three phosphate groups.

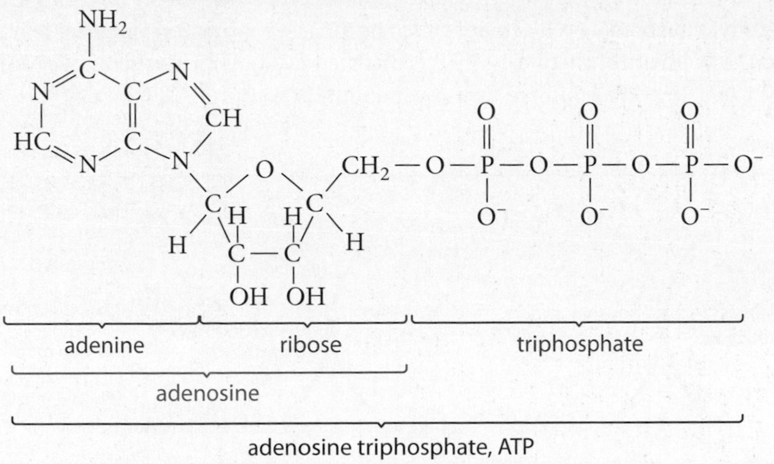

adenosine triphosphate, ATP

◀ Figure 5.26
The structure of ATP

The breakdown of ATP, which involves **hydrolysis**, is an exothermic reaction. And, it is the energy released in this reaction that is used to drive the enzyme catalysed reactions in the pathways described above.

ATP is hydrolysed to ADP (adenosine diphosphate) plus an H^+ ion and a hydrogenphosphate ion, HPO_4^{2-}.

$$ATP + H_2O \longrightarrow ADP + H^+ + HPO_4^{2-} \quad \Delta H = -30\,kJ\,mol^{-1}$$

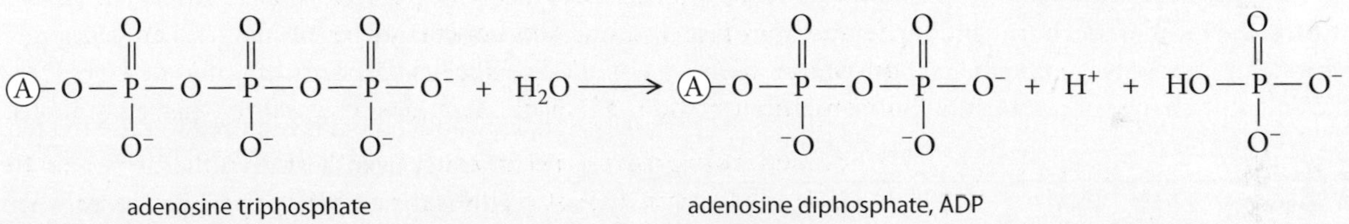

adenosine triphosphate

adenosine diphosphate, ADP

When hydrolysis occurs, energy is required to break bonds in water and between the phosphate groups, but this is more than balanced by the energy released as the products form. So the overall reaction is exothermic.

Although this hydrolysis of ATP to ADP is energetically favourable, the activation energy (Section 8.7) for the reaction is high. This allows control of the hydrolysis by enzymes, because the reaction does not happen until it is triggered by an enzyme, ATPase.

In order to sustain life, all living things require a continuous, steady supply of ATP. Fortunately, the hydrolysis of ATP to ADP is reversible but this regeneration of ATP is endothermic and therefore energetically unfavourable.

However, the energy needed to regenerate ATP can be provided in both animals and plants by the **respiration** of glucose. Animals get glucose from the digestion of carbohydrates in food. Plants get glucose from photosynthesis. During respiration, glucose is ultimately oxidised to carbon dioxide and water through a sequence of complex enzyme catalysed reactions, most of which are energetically favourable. The complete oxidation of one mole of glucose can produce 38 moles of ATP.

Definitions

A **substrate** is a reactant in an enzyme-catalysed reaction.

Hydrolysis is a reaction in which water molecules are split and then added to the fragments of other reactants. The word 'hydrolysis' is derived from two separate words – 'hydro' meaning 'water' and 'lysis' which means 'splitting'.

Phosphorylation is a reaction in which a phosphate group is added to another reactant.

Respiration is the metabolic process in cells that oxidises glucose and produces ATP.

Metabolism is the chemical reactions that take place in a living organism to sustain all life processes.

So overall, the **metabolism** and activity of cells is controlled by two pairs of coupled processes involving ATP (Figure 5.27). In the first coupling, food (mainly glucose) is oxidised to carbon dioxide and water while ADP, H⁺ and HPO₄²⁻ are converted to ATP. This energetically unfavourable **phosphorylation** process is 'driven' by the energetically favourable oxidation of food.

In the second coupling, the energetically favourable hydrolysis of ATP 'drives' various energetically unfavourable cell activities such as muscle contraction, 'messages' in nerve cells and the synthesis of chemicals. ATP is involved in so many biochemical processes in your body that an average human turns over about 0.1 tonne of ATP per day.

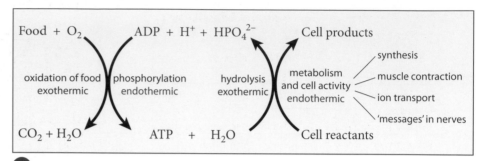

Figure 5.27
The central role of ATP in metabolism and cell activity

Review questions

1 Figure 5.28 shows the apparatus used to determine the enthalpy change of combustion of ethanol. Heat produced by the burning fuel warms a known mass of water. By measuring the mass of fuel burnt and the temperature rise of the water, it is possible to obtain an approximate value for the enthalpy change of combustion of the fuel.

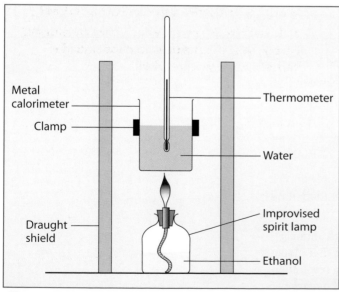

Figure 5.28

Volume of water in calorimeter $= 400\,\mathrm{cm}^3$
Initial temperature of water $= 12\,°\mathrm{C}$
Final temperature of water $= 22\,°\mathrm{C}$
Mass of ethanol burnt $= 0.92\,\mathrm{g}$
(Specific heat capacity of water $= 4.2\,\mathrm{J\,g^{-1}\,K^{-1}}$)

a How much heat is required to raise the temperature of the water from 12 °C to 22 °C?
(This is the amount of heat produced when 0.92 g of ethanol burn.)

b How much heat would be produced when 1 mole of ethanol burns?

c Why is the answer to **b** not described as the *standard* enthalpy change of combustion of ethanol?

d An accurate value for $\Delta H_c^\ominus(C_2H_5OH(l))$ is $-1368\,\mathrm{kJ\,mol^{-1}}$. Mention three serious errors in the simple experiment which could be responsible for the poor result.

2 Titanium can be extracted from the mineral rutile, TiO_2 by heating with carbon. The other product of the reaction, besides titanium, is carbon monoxide.

a Write an equation for the reaction including state symbols.

b Calculate the standard enthalpy change for the reaction using the following data.
$\Delta H_f^\ominus(TiO_2(s)) = -940\,\mathrm{kJ\,mol^{-1}}$,
$\Delta H_f^\ominus(CO(g)) = -110\,\mathrm{kJ\,mol^{-1}}$

3 Two campers are desperately short of camping gas, yet they badly need a hot drink. They estimate that they have $1.12\,dm^3$ of camping gas (measured at 0 °C and 1 atm) in their 'gas bottle'.

a What is the maximum volume of water (at 20 °C) which they could boil in order to make some hot coffee? (Assume that camping gas is pure butane, $\Delta H_c(C_4H_{10}(g)) = -3000\,kJ\,mol^{-1}$ and that 75% of the heat evolved in burning the gas is absorbed by the water.)

b State any other assumptions you make.

4 A possible mechanism for the reaction of fluorine with methane is

$$CH_4 + F_2 \xrightarrow{\text{slow}} CH_3{\cdot} + HF + F{\cdot}$$

$$CH_3{\cdot} + F{\cdot} \xrightarrow{\text{fast}} CH_3F$$

a Use the bond energies in Table 5.2, Section 5.10, to calculate the enthalpy change in each step of the reaction mechanism.

b Is the suggested mechanism viable? Explain your answer.

c Write the equations for a reaction between CH_4 and Cl_2, assuming a similar mechanism to that for CH_4 and F_2 and calculate the enthalpy change in each step.

d Is such a mechanism viable for the CH_4/Cl_2 reaction? Explain your answer.

e Why is it that fluorine will react with methane in the dark, whereas chlorine only reacts appreciably with methane in sunlight?

5 a Explain the terms lattice energy, enthalpy change of hydration and enthalpy change of solution with reference to the hypothetical substance, X^+Y^-.

b Draw an energy cycle relating the three terms in part **a**.

c Calculate the enthalpy change of hydration of potassium iodide, assuming that its enthalpy change of solution is $+21\,kJ\,mol^{-1}$ and its lattice energy is $-642\,kJ\,mol^{-1}$.

6 a Explain what is meant by the terms:

i ionisation energy,

ii atomisation energy.

b Draw a complete, fully labelled Born–Haber Cycle for the formation of potassium bromide.

c Using the information in the table below, calculate the lattice energy of potassium bromide.

Reaction	$\Delta H/kJ\,mol^{-1}$
$K(s) + \frac{1}{2}Br_2(l) \longrightarrow K^+Br^-(s)$	−392
$K(s) \longrightarrow K(g)$	+90
$K(g) \longrightarrow K^+(g) + e^-$	+420
$\frac{1}{2}Br_2(l) \longrightarrow Br(g)$	+112
$Br(g) + e^- \longrightarrow Br^-(g)$	−342

d The values of the lattice energies of the other potassium halides are:

Compound	KF	KCl	KI
Lattice energy/kJ mol^{-1}	−813	−710	−643

What explanation can you give for the trend in these values?

7 The enthalpy changes involved in the synthesis of calcium oxide are represented in a Born–Haber cycle below. (The numerical values printed beside the cycle are in $kJ\,mol^{-1}$.)

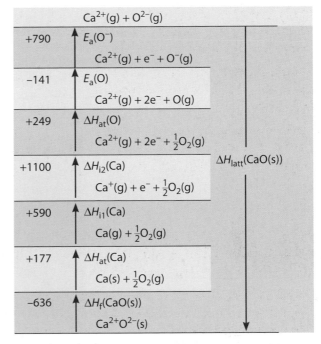

a Calculate the lattice energy for calcium oxide, $\Delta H_{latt}(CaO(s))$.

b Why is $\Delta H_{i2}(Ca)$ greater than $\Delta H_{i1}(Ca)$?

c Why is $E_a(O)$ negative whereas $E_a(O^-)$ is positive?

d State and explain how the value of the first ionisation energy of magnesium would compare with the corresponding value for calcium.

6 Redox

Note

The term 'redox' is used as an abbreviation for reduction and oxidation.

6.1 Introduction

The term **redox** is used by chemists as an abbreviation for the processes of **red**uction and **oxi**dation. These two processes usually occur simultaneously. Redox reactions include processes such as burning, rusting and respiration. Originally, chemists had a very limited view of redox, using it to account for only the reactions of oxygen and hydrogen.

Nowadays, our ideas of redox have been extended to include all electron-transfer processes. An important feature of some electron-transfer redox reactions is that the energy of the chemical reaction may be released in the form of electrical energy and harnessed to provide electricity. This is what happens in the dry cell of a small torch, in the button cell of a hearing aid and in the battery of a motor car. Electron-transfer also occurs during electrolysis when electricity is used to decompose compounds.

> Figure 6.1
Antoine Lavoisier (1743–1794) was the first chemist to explain the redox reactions which occur during burning. In 1775, Lavoisier was appointed to a post at the French government munitions factory. Here he carried out experiments in combustion (burning) and respiration.

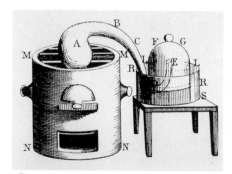

Figure 6.2
Lavoisier's apparatus for preparing oxygen. Red mercury(II) oxide was heated in the retort (A) on the left. The oxygen which formed was collected in the bell jar (E) above mercury.

6.2 Redox processes in terms of electron transfer

When metals react with oxygen they form oxides. For example magnesium and sodium burn brightly in air to form white ionic oxides, $Mg^{2+}O^{2-}$ and $(Na^+)_2O^{2-}$:

$$2Mg(s) + O_2(g) \longrightarrow 2Mg^{2+}O^{2-}(s)$$

$$4Na(g) + O_2(g) \longrightarrow 2(Na^+)_2O^{2-}(s)$$

The metal is oxidised and the oxygen is reduced. During this process the metal atoms lose electrons to form positive ions and oxygen atoms gain electrons to form negative oxide ions, O^{2-}. The oxygen takes the electrons given up by the metal.

$$2Mg \longrightarrow 2Mg^{2+} + 4e^-$$

$$O_2 + 4e^- \longrightarrow 2O^{2-}$$

Electron-transfer reactions such as this are called **redox reactions**.

1 a Copy and balance the following equation for the decomposition of red mercury(II) oxide:
$$HgO(s) \longrightarrow Hg(l) + O_2(g)$$
b Show the charges on HgO in the equation and explain which ion is oxidised and which is reduced.

2 Do the following processes involve oxidation, reduction, both or none of these?
a $2H^+ + 2e^- \longrightarrow H_2$
b $Cu^+ \longrightarrow Cu^{2+} + e^-$
c $Ag^+ + Cl^- \longrightarrow AgCl$
d $NH_3 + H^+ \longrightarrow NH_4^+$
e $2Cu^+ \longrightarrow Cu^{2+} + Cu$

▲ Figure 6.3
Copper wire suspended in silver nitrate solution

▲ Figure 6.4
Tutankhamun's gold funerary mask: gold is such an unreactive metal that it remains unoxidised and untarnished after centuries.

In the equations above, Mg loses electrons and is oxidised to Mg^{2+}; O_2 gains electrons and is reduced to O^{2-}.

This leads to the following definitions:

Oxidation is the loss of electrons.

Reduction is the gain of electrons.

Oxidising agents, such as oxygen, accept electrons.

Reducing agents donate electrons.

When metals react they lose electrons and form their ions. Thus, metals are *oxidised* and act as *reducing agents* in their reactions. Notice that the oxidised substance (magnesium in the above example) acts as the reducing agent and the reduced substance (oxygen in the above example) acts as the oxidising agent.

When redox is viewed in terms of electron transfer it is easy to see why oxidation and reduction always take place together. One substance cannot lose electrons and be oxidised unless another substance gains electrons and is reduced.

Half-equations

The separate equations showing which substance gains electrons and which substance loses electrons in a redox reaction are called **half-equations**. Half-equations help to show what is happening in terms of electrons during a reaction. Two half-equations can be combined to give the overall balanced equation. It is important to remember that half-equations cannot occur on their own.

When copper wire is suspended in silver nitrate solution, solid silver deposits on the copper wire and a blue solution of copper(II) nitrate starts to form (Figure 6.3).

During the reaction, copper atoms in the wire, $Cu(s)$, give up electrons and go into solution as blue copper(II) ions, $Cu^{2+}(aq)$. The half-equation for this is:

$$Cu(s) \longrightarrow Cu^{2+}(aq) + 2e^-$$

At the same time, silver ions in the solution, $Ag^+(aq)$, take the electrons given up by the copper atoms and form the solid deposit of silver, $Ag(s)$.

$$2Ag^+(aq) + 2e^- \longrightarrow 2Ag(s)$$

Adding the two half-equations together, we get the overall ionic equation:

$$Cu(s) + 2Ag^+(aq) \longrightarrow Cu^{2+}(aq) + 2Ag(s)$$

6.3 Important types of redox reaction

There are four important and relatively common types of redox reaction.

1 The reactions of metals with non-metals

In this case, metals give up electrons to form positive ions and non-metals (O_2, Cl_2, S) take these electrons to form negative ions (O^{2-}, Cl^-, S^{2-}). For example:

$$Fe(s) + S(s) \longrightarrow Fe^{2+}S^{2-}(s) \quad \begin{cases} Fe \longrightarrow Fe^{2+} + 2e^- \\ S + 2e^- \longrightarrow S^{2-} \end{cases}$$

The metal loses electrons and is oxidised. The non-metal gains electrons and is reduced. Metals higher in the activity (electrochemical) series lose electrons more easily than the less reactive metals lower down. Thus, moving down the activity series, metals become weaker reducing agents.

2 The reactions of metals with water

Metals at the top of the activity series (K, Na, Ca and Mg) are sufficiently reactive to form hydrogen with water.

$$Ca(s) + 2H_2O(l) \longrightarrow Ca^{2+}(OH^-)_2(s) + H_2(g)$$

The next few metals in the activity series (e.g. Al, Zn and Fe) do not react noticeably with water, but they will react with steam to form hydrogen.

In these reactions the metal atoms are oxidised to form positive ions. The electrons which they release are accepted by water molecules, which are reduced to hydroxide ions (OH^-) and hydrogen (H_2).

$$Ca \longrightarrow Ca^{2+} + 2e^-$$
$$2H_2O + 2e^- \longrightarrow 2OH^- + H_2$$

3 The reactions of metals with acids

In this case, the metal atoms lose electrons which are taken by aqueous H^+ ions in the acid to form H_2. For example:

$$Zn(s) + 2H^+(aq) \longrightarrow Zn^{2+}(aq) + H_2(g)$$

The half-equations are:

$$Zn \longrightarrow Zn^{2+} + 2e^-$$
$$2H^+ + 2e^- \longrightarrow H_2$$

The reactivity of metals with acids depends on the ease with which the metal loses electrons to form its aqueous ions. Metals which form ions more easily than hydrogen therefore react with acids releasing electrons, which are taken by H^+ ions in the acid.

Metals below hydrogen in the activity series, such as copper and silver, form ions less readily than hydrogen. So, these metals do not react with acids to form hydrogen.

4 Reactions at the electrodes during electrolysis

During electrolysis, cations are attracted to the negatively charged cathode where they gain electrons and are reduced. For example, during the electrolysis of molten lead(II) bromide:

at the cathode (–)

$$Pb^{2+} + 2e^- \longrightarrow Pb$$

At the same time, Br^- anions are attracted to the positively charged anode and oxidised by the loss of electrons.

at the anode (+)

$$2Br^- \longrightarrow Br_2 + 2e^-$$

In the next section, electrolysis is studied in more detail.

Figure 6.5
Photochromic spectacles in low light intensity (above) and in strong light (below). Photochromic glass darkens on exposure to strong light and becomes clearer when the light is poor. The glass contains small particles of silver chloride and copper(I) chloride. When the light is strong, Ag^+ ions oxidise Cu^+ to Cu^{2+}.

$$Ag^+(s) + Cu^+(s) \longrightarrow Ag(s) + Cu^{2+}(s)$$

The silver produced reflects light from the glass and cuts out transmission (glare) to the eyes. In low light intensity, copper(II) ions oxidise Ag back to Ag^+ ions and the glass becomes clear again.

Quick Questions

3 Write balanced equations for the following reactions and then deduce the half-equations which show electron loss and gain more clearly.
   a sodium with oxygen,
   b potassium with water,
   c magnesium with dilute sulfuric acid,
   d electrolysis of molten sodium chloride.

6.4 Electrolysis

Electrolysis involves the decomposition of a molten or aqueous compound by electricity. The process takes place in an electrolytic cell and the compound decomposed is called an **electrolyte**.

Electrolysis is important in industry. It is used:

- to extract sodium, aluminium (Section 6.5) and other reactive metals from their purified ores,
- to manufacture chlorine and sodium hydroxide (Section 11.3),
- to purify 'blister copper' (Section 6.5).

The energy that causes the chemical changes during electrolysis is provided by an electric current. An electric current is simply a flow of electrons. The electrons flow from the negative terminal of the cell or battery (power supply) to the negative electrode (**cathode**) in the electrolyte and then from the positive electrode (**anode**) back to the positive terminal of the battery (Figure 6.6).

Explaining electrolysis

When an electric current passes through molten sodium chloride, sodium is produced at the cathode and chlorine forms at the anode (Figure 6.6).

During electrolysis, positive Na^+ ions in the electrolyte are attracted to the negative cathode and negative Cl^- ions in the electrolyte are attracted to the positive anode.

When Na^+ ions reach the cathode, they combine with negative electrons from the battery forming neutral sodium atoms.

$$Na^+ \quad + \quad e^- \quad \longrightarrow \quad Na$$

| sodium ion in sodium chloride electrolyte | electron on cathode from the battery | sodium atom in the molten metal |

At the anode, Cl^- ions lose electrons to the positive anode forming neutral chlorine atoms.

$$Cl^- \quad \longrightarrow \quad e^- \quad + \quad Cl$$

| chloride ion in electrolyte | electron given to anode | chlorine atom |

The Cl atoms immediately join up in pairs to form molecules of chlorine gas, Cl_2.

$$Cl + Cl \longrightarrow Cl_2$$

During electrolysis:

- A metal or hydrogen forms at the negative cathode. This confirms that metals and hydrogen have positive ions. These ions are called **cations** because they are attracted to the cathode.
- A non-metal (except hydrogen) forms at the positive anode. This confirms that non-metals (except hydrogen) have negative ions. These ions are called **anions** because they are attracted to the anode.

The equations for the electrolysis of molten sodium chloride show that Na^+ ions remove electrons from the cathode and Cl^- ions give up electrons to the anode. In this way, the electric current is carried through the electrolyte by ions moving to the electrode of opposite charge.

The electrolysis of other molten and aqueous substances can also be explained in terms of ions.

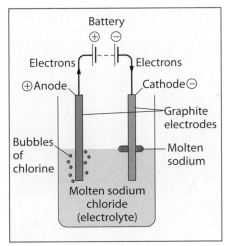

Figure 6.6
Electrolysis of molten sodium chloride

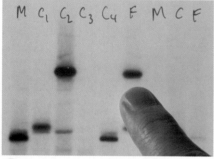

K	K⁺
Na	Na⁺
Mg	Mg²⁺
Al	Al³⁺
Zn	Zn²⁺
Fe	Fe²⁺
Pb	Pb²⁺
H	H⁺
Cu	Cu²⁺
Ag	Ag⁺

metals become less reactive (i.e. less likely to form ions) →

ions become more likely to form atoms →

Figure 6.7
The electrochemical series, showing the reactivity of atoms and ions.

Figure 6.8
DNA fingerprinting uses a process called electrophoresis. Electrophoresis of DNA involves the migration of charged particles from a person's DNA. The charged particles move over filter paper or through a gel in an electric field of several thousand volts. The 'fingerprinting' of DNA in body fluids is used in crime detection and to prove paternity. This photo shows bands of DNA from different people after electrophoresis. The pattern of DNA bands is unique to each individual, but some bands are shared by related people. The bands in this photo suggest that F is the father of child C_2.

Quick Questions

4 Write half-equations for the products at graphite electrodes when the following are electrolysed:
 a dilute sulfuric acid,
 b concentrated sodium chloride solution,
 c silver nitrate solution.

6.5 Explaining the electrolysis of mixtures

Electrolysis is more complicated if there is more than one cation or more than one anion present in the electrolyte. The product may also be different if the electrolyte is molten rather than in aqueous solution. In most cases, however, there is only one product at each electrode.

▶ **When there is more than one cation in the electrolyte**, the ion discharged at the cathode can be predicted from the **electrochemical (reactivity) series** (Figure 6.7).

Metals like potassium and sodium which are keen to form ions do not readily re-form the metal during electrolysis. *So, the ions of metals lower down in the electrochemical series are discharged in preference to those higher up.*

This simple rule is, however, complicated by two other factors.

1 If a cation is present in very high concentration, it may be discharged in preference to one below it in the electrochemical series at much lower concentration.
2 Hydrogen is discharged from aqueous solutions of the salts of metals above it in the electrochemical series. So, hydrogen is produced at the cathode when aqueous sodium chloride is electrolysed.

In pure water and in aqueous solutions like NaCl(aq), water is slightly ionised.

$$H_2O(l) \rightleftharpoons H^+(aq) + OH^-(aq)$$

Even though the concentration of H^+ ions in aqueous solution is only about $10^{-7}\,mol\,dm^{-3}$, H^+ ions are still discharged in preference to cations, such as K^+, Na^+ and Mg^{2+} at much higher concentrations.

Cathode (–) $2H^+(aq) + 2e^- \longrightarrow H_2(g)$

▶ **When there is more than one anion present in the electrolyte**, experiments show that the order of discharge is

$$SO_4^{2-}, NO_3^-, Cl^-, OH^-, Br^-, I^-$$
ions discharge more easily →

This simple rule, like that related to cations, is complicated by two other factors.

1 If an anion is present in very high concentration, it may discharge in preference to one that would be discharged more readily if their concentrations were equal.
2 Oxygen from OH^- ions is usually discharged from the aqueous solutions of nitrates and sulfates, even though the concentration of OH^- ions from the water is as low as $10^{-7}\,mol\,dm^{-3}$.

Anode (+) $4OH^-(aq) \longrightarrow 2H_2O(l) + O_2 + 4e^-$

Worked example

What are the products at the electrodes when aqueous sodium sulfate is electrolysed?

Answer

	Cations	Anions
Ions present from sodium sulfate	Na⁺(aq)	SO₄²⁻(aq)
Ions present from water	H⁺(aq)	OH⁻(aq)

Cathode $H^+(aq)$ ions are discharged in preference to $Na^+(aq)$
$2H^+(aq) + 2e^- \longrightarrow H_2(g)$

Anode $OH^-(aq)$ ions are discharged in preference to $SO_4^{2-}(aq)$
$4OH^-(aq) \longrightarrow 2H_2O(l) + O_2(g) + 4e^-$

∴ Hydrogen is produced at the cathode and oxygen at the anode.

6.6 Electrolysis in industry

Reactive metals, like sodium, magnesium and aluminium, cannot be obtained by reducing their ores to the metal with carbon (coke) like iron and lead. These metals can only be obtained by electrolysis of their molten compounds. Electrolysis of their aqueous compounds cannot be used either because hydrogen (from the water) rather than the metal is produced at the cathode.

Metals low in the reactivity series, such as copper and silver, can be obtained by reduction of their compounds or by electrolysis of their aqueous solutions. When their aqueous compounds are electrolysed, the metal is produced at the cathode rather than hydrogen (from the water).

Figure 6.9
Rows of electrolytic cells at Alcan's aluminium smelter at Kitimat in British Columbia, Canada

Extracting aluminium by electrolysis

Aluminium is extracted by the electrolysis of pure aluminium oxide dissolved in molten cryolite (sodium hexafluoroaluminate(III), Na_3AlF_6). The pure aluminium oxide is obtained from bauxite, impure hydrated $Al_2O_3 \cdot 2H_2O$. Although clay is the most abundant source of aluminium in aluminosilicates (Section 4.14), the metal cannot be extracted economically from clay.

Pure aluminium oxide does not melt until 2072 °C and this makes its electrolysis uneconomic. Fortunately, the electrolytic process can be carried out economically by dissolving the oxide in molten cryolite (Figure 6.10). This has a much lower melting point than aluminium oxide. Even so, the temperature of the molten electrolyte must be raised to 850 °C and maintained at this temperature. The energy required for this is enormous and the process is usually carried out near sources of cheap hydroelectricity.

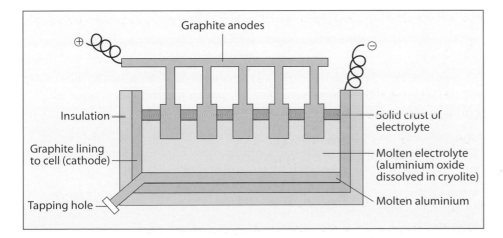

Figure 6.10
The electrolytic cell for extraction of aluminium

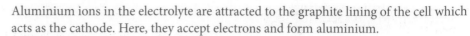

Quick Questions

5 Look closely at the photo in Figure 6.9. What is the white powder on the floor?

6 During the electrolytic extraction of aluminium, why do you think:

 a sodium ions from cryolite are not discharged at the cathode,

 b fluoride ions from cryolite are not discharged at the anode?

Aluminium ions in the electrolyte are attracted to the graphite lining of the cell which acts as the cathode. Here, they accept electrons and form aluminium.

Cathode (–) $Al^{3+}(l) + 3e^- \longrightarrow Al(l)$

Molten aluminium collects at the bottom of the cell and is tapped off at intervals. It takes about 16 kilowatt-hours of electricity to produce 1 kg of aluminium.

Oxide ions, O^{2-}, in the electrolyte are attracted to the graphite anodes to which they give up their electrons and form oxygen gas.

Anode (+) $2O^{2-}(l) \longrightarrow O_2(g) + 4e^-$

The oxygen reacts with the graphite anodes to form oxides of carbon at the high temperatures involved. As a result, the anodes are slowly burnt away and must be replaced from time to time.

Purifying copper by electrolysis

> **Figure 6.11**
Sparks fly during the smelting of copper ore. Initially, the ore is heated in air to a high temperature resulting in a very exothermic reaction with oxygen.

The first stage in the extraction of copper is the reduction of copper ores (usually copper pyrites). This involves heating with the correct amount of air.

$$2CuFeS_2(s) + 5O_2(g) \longrightarrow 2Cu(s) + 2FeO(s) + 4SO_2(g)$$

Most of the iron(II) oxide, FeO, can be removed from the solid products by heating the mixture with silica (silicon(IV) oxide, SiO_2)after the furnace has been closed. The iron(II) oxide reacts with the silica to form molten iron(II) silicate (slag) which floats on the molten mixture and can be tapped off separately.

$$FeO(s) + SiO_2(s) \longrightarrow FeSiO_3(l)$$
iron(II) silicate (slag)

The copper extracted at this stage is known as 'blister copper'. It releases bubbles of SO_2 as it solidifies and therefore gets a blistered appearance. The blister copper still contains 2 or 3% of impurities, mainly iron and sulfur.

> **Figure 6.12**
Electrolytic purification of copper. Pure copper has deposited on these sheets of the metal.

The final stage in the extraction of copper involves purification of the blister copper by electrolysis of copper(II) sulfate solution. Thin sheets of pure copper are used as cathodes and sheets of impure blister copper are the anodes.

During electrolysis, Cu^{2+} ions are attracted to the cathodes where they gain electrons and deposit on the cathode as pure copper.

| Cathode (−) | $Cu^{2+}(aq) + 2e^- \longrightarrow Cu(s)$ |

Sulfate ions, SO_4^{2-}, in the electrolyte are attracted to the anode, but they are not discharged. Instead, copper atoms which make up the anode, give up two electrons each and go into solution as Cu^{2+} ions.

| Anode (+) | $Cu(s) \longrightarrow Cu^{2+}(aq) + 2e^-$ |

So, during the electrolysis, copper 'dissolves away' from the impure 'blister' anode, whilst a thickening deposit of pure copper appears on the cathode. The overall result is that copper metal is transferred from the anode to the cathode.

Quick Questions

7   **a**   Impure blister copper usually contains iron as an impurity. The iron dissolves from the anode, during electrolysis, in preference to copper, but does not deposit on the pure copper cathode. Why is this?

     b   Some samples of blister copper contain silver as an impurity which can be recovered as a valuable by-product. What do you think happens to the silver during the electrolytic purification of the blister copper?

6.7 Oxidation numbers

Many reactions involve a *complete* transfer of electrons from one substance to another and it is easy to see that they involve redox. These processes usually have ions as either the reactants or the products. In some cases, both the reactants and the products are ions.

There are, however, some reactions in which it is difficult to appreciate the redox processes involved because there is no obvious transfer of electrons from one substance to another. For example, the reactions

$$2H_2(g) + O_2(g) \longrightarrow 2H_2O(l) \quad \text{and} \quad C(s) + O_2(g) \longrightarrow CO_2(g)$$

both involve redox, but it is difficult to pick out the separate oxidation and reduction processes because there is no clear electron loss and gain.

In order to overcome this problem, the concept of **oxidation number** (or oxidation state) was introduced. This provided a similar, but alternative, definition of redox to that involving electron transfer. An oxidation number is a number assigned to an atom or an ion to describe its relative state of oxidation or reduction.

Using oxidation numbers it is possible to decide whether redox has occurred in any chemical reaction.

Assigning oxidation numbers

▶ Atoms in uncombined elements are given an oxidation number of zero. Thus, the oxidation number of Mg is 0, as is the oxidation number of chlorine atoms in Cl_2.
▶ For simple ions, the oxidation number is simply the charge on the ion. Thus, the oxidation numbers of Cl^-, Fe^{2+} and Fe^{3+} are −1, +2 and +3, respectively.
▶ For compounds, the oxidation numbers of the atoms within them are obtained by assuming the compounds are *wholly ionic* and then working out the charge associated with each atom. For example:

H_2O $((H^+)_2O^{2-})$	Ox. No. of H in H_2O	= +1
	Ox. No. of O in H_2O	= −2
HCl (H^+Cl^-)	Ox. No. of H in HCl	= +1
	Ox. No. of Cl in HCl	= −1

In assigning oxidation numbers in this way it is necessary to assume that the electrons in each bond of the molecule or ion belong to the *more electronegative atom* (i.e. the atom with the greater attraction for electrons).

Figure 6.13
Electrical gear under the bonnet of a car powered by fuel cells. In a fuel cell, electrical energy is produced by the controlled oxidation of a fuel. The fuel cells store enough charge for about 75 km of driving before they need to be recharged.

Key Point

There are two related definitions of redox, one involving electron transfer, the other involving oxidation numbers.

Definition

An **oxidation number** is a number assigned to an atom or ion to describe its relative state of oxidation or reduction.

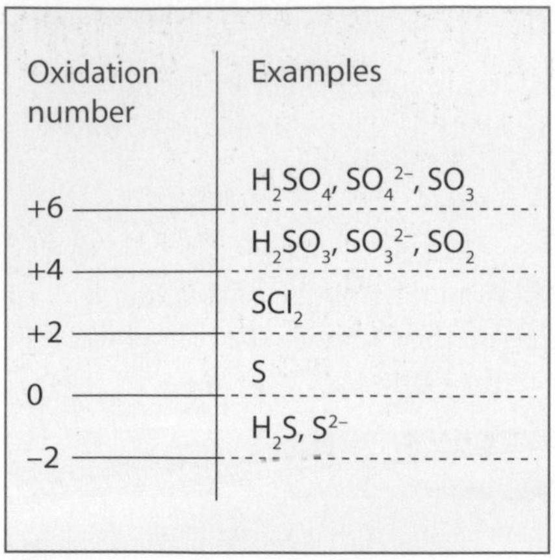

Figure 6.14
An oxidation number chart for sulfur

Since hydrogen is the least electronegative non-metal, the oxidation number of H in its compounds is usually +1. For example:

$$PH_3(P^{3-}(H^+)_3)$$

Ox. No. of H in PH_3	= +1
Ox. No. of P in PH_3	= −3

In metal hydrides, however, the metal is more electropositive than hydrogen and in this case the oxidation number of H is −1. For example:

$$Na^+H^-$$

Ox. No. of H in NaH	= −1
Ox. No. of Na in NaH	= +1

Apart from fluorine, oxygen is the most electronegative element. This means that the oxidation number of oxygen in its compounds is usually −2.

$$Na_2O((Na^+)_2O^{2-})$$
$$CO_2(C^{4+}(O^{2-})_2)$$

Ox. No. of O in Na_2O	= −2
Ox. No. of Na in Na_2O	= +1
Ox. No. of O in CO_2	= −2
Ox. No. of C in CO_2	= +4

However, in OF_2 the oxidation number of oxygen is +2 and in peroxides the oxidation number of oxygen is −1.

$$OF_2(O^{2+}(F^-)_2)$$
$$Na_2O_2((Na^+)_2(O_2)^{2-})$$

Ox. No. of O in OF_2	= +2
Ox. No. of F in OF_2	= −1
Ox. No. of O in Na_2O_2	= −1
Ox. No. of Na in Na_2O_2	= +1

The points we have just made about oxidation numbers can be summarised in six simple rules.

Quick Questions

8 What are the oxidation numbers of each element in the following?
$MgCl_2$, SO_2, CO, NaOH, PCl_3, SO_4^{2-}, MnO_4^-

9 What are the oxidation numbers of sulfur in the following compounds?
$NaHSO_4$, CS_2, SO_2Cl_2, Na_2S, S_2Cl_2

Rules for oxidation numbers

1 The oxidation number of atoms in uncombined elements is 0.

2 In neutral molecules, the sum of the oxidation numbers is 0.

3 In ions, the sum of the oxidation numbers equals the charge on the ion.

4 In any substance, the more electronegative atom has the negative oxidation number, and the less electronegative atom has the positive oxidation number.

5 The oxidation number of hydrogen in all its compounds, except metal hydrides, is +1.

6 The oxidation number of oxygen in all its compounds, except in peroxides and in OF_2, is −2.

Some elements have five or more possible oxidation states. The principal oxidation states of sulfur are shown in an oxidation number chart in Figure 6.14.

Oxidation numbers and nomenclature

Oxidation numbers are used in the systematic naming of compounds even though the names are often cumbersome and complex. There are, however, a few basic rules which can be followed in naming compounds.

▶ Compounds containing only two elements have names ending in '-ide' with the more electronegative element coming second. For example: sodium oxide (Na_2O), magnesium chloride ($MgCl_2$) and silicon tetrachloride ($SiCl_4$).

▶ Simple molecular compounds containing only two elements often include reference to the number of different atoms without stating oxidation numbers. Hence, CO_2 – carbon dioxide, $SiCl_4$ – silicon tetrachloride and S_2Cl_2 – disulfur dichloride.

▶ Ionic compounds containing metals with variable oxidation states do, however, include the oxidation number of these elements. For example, $FeSO_4$ – iron(II) sulfate, $FeCl_3$ – iron(III) chloride.

- Remember, though, that Roman numerals are always used to indicate the oxidation number of an element within a compound to prevent any confusion with the electrical charge on an ion. Thus, CuO is named copper(II) oxide and *not* copper(2) oxide.
- The traditional names for oxoacids end in '-ic' or '-ous', the '-ic' ending being used for the acid in which the central atom has the higher oxidation number. For example, H_2SO_4 – sulfuric acid, H_2SO_3 – sulfurous acid and HNO_3 – nitric acid, HNO_2 – nitrous acid.
- The traditional names for the salts of these oxoacids end in '-ate' and '-ite'. For example, Na_2SO_4 – sodium sulfate, Na_2SO_3 – sodium sulfite and $Cu(NO_3)_2$ – copper(II) nitrate, $Cu(NO_2)_2$ – copper(II) nitrite.
- The full systematic names for these oxoacids and oxosalts include the number of oxygen atoms and the oxidation numbers of the central atoms. For example, tetraoxosulfuric(VI) acid for H_2SO_4 and trioxosulfate(IV) for SO_3^{2-}. In some instances, the semi-systematic names are also used such as sulfuric(VI) acid for H_2SO_4 and sulfate(IV) for SO_3^{2-}.

In this book we will use the names sulfuric, sulfurous, sulfate, sulfite, nitric, nitrous, nitrate and nitrite as recommended by IUPAC (The International Union of Pure and Applied Chemistry) in preference to sulfuric(VI), sulfuric(IV), sulfate(VI), sulfate(IV), nitric(V), nitric(III), nitrate(V) and nitrate(III).

6.8 Explaining redox in terms of oxidation numbers

We are now in a position to define oxidation and reduction in terms of oxidation number. This definition of redox is an alternative to that involving electron transfer, although the two definitions are quite closely related. An atom is said to be oxidised when its oxidation number increases (becomes more positive) and reduced when its oxidation number decreases (becomes more negative).

Consider the following reactions.

a
$$\overset{0}{2Na}(s) + \overset{0}{Cl_2}(g) \longrightarrow \overset{+1\ -1}{2NaCl}(s)$$

The oxidation number of sodium has increased from 0 to +1: it has been oxidised. The oxidation number of chlorine has decreased from 0 to –1: it has been reduced.

b
$$\overset{+1}{H^+}(aq) + \overset{-2\ +1}{OH^-}(aq) \longrightarrow \overset{+1\ -2}{H_2O}(l)$$

This ionic equation summarises the neutralisation of an acid with an alkali. Notice that the oxidation number of each element remains unchanged and so the reaction does not involve redox.

c Another important ionic reaction which does not involve redox is precipitation. Here again the oxidation number of each element remains unaltered during the reaction.

$$\overset{+1}{Ag^+}(aq) + \overset{-1}{Cl^-}(aq) \longrightarrow \overset{+1\ -1}{AgCl}(s)$$

$$\overset{+2}{Ba^{2+}}(aq) + \overset{+6\ -2}{SO_4^{2-}}(aq) \longrightarrow \overset{+2\ +6\ -2}{BaSO_4}(s)$$

d
$$\overset{+2\ -2}{CO}(g) + \overset{0}{\tfrac{1}{2}O_2}(g) \longrightarrow \overset{+4\ -2}{CO_2}(g)$$

In this case, the oxidation number of carbon rises from +2 to +4: it has been oxidised. The oxidation number of the oxygen atom in CO remains unchanged, but the elemental oxygen (O_2) is reduced: its oxidation number falls from 0 to –2.

Quick Questions

10 Write the formulae of the following compounds:
  a dinitrogen trioxide
  b phosphorus pentachloride
  c aluminium nitride
  d iron(III) nitrate
  e silver sulfite

Definitions

Oxidation is an increase in oxidation number.

Reduction is a decrease in oxidation number.

Figure 6.15
An unusual redox reaction!

Figure 6.16
This premature baby is being supplied with oxygen in an incubator. The oxygen is needed for the baby's respiration, which is an important redox reaction.

Quick Questions

11 Which of the following reactions involve redox?
 a $Cl_2 + 2OH^- \longrightarrow Cl^- + ClO^- + H_2O$
 b $Cu^{2+} + 2OH^- \longrightarrow Cu(OH)_2$
 c $H_2O + SO_3 \longrightarrow H_2SO_4$
 d $2CrO_4^{2-} + 2H^+ \longrightarrow Cr_2O_7^{2-} + H_2O$

12 Write definitions for oxidising agent and reducing agent in terms of oxidation numbers.

13 Look at Figure 6.15. Suggest possible substances which might get oxidised and reduced when the fire eater performs.

6.9 The advantages and disadvantages of oxidation numbers

Oxidation numbers can help us decide whether or not redox is involved in a particular process. Oxidation numbers, for example, show that neutralisation and precipitation are not redox reactions, even though they involve ions.

The second advantage in using oxidation numbers is that they allow us to see exactly which part of a molecule or an ion is reduced or oxidised. For example, the half-equation

$$MnO_4^-(aq) + 8H^+(aq) + 5e^- \longrightarrow Mn^{2+}(aq) + 4H_2O(l)$$

shows that MnO_4^- and H^+ ions are reduced to Mn^{2+} and $4H_2O$. But, which element or elements in MnO_4^- and H^+ are reduced? If oxidation numbers are assigned to the atoms in the half-equation

$$\overset{+7\,-2}{MnO_4^-}(aq) + \overset{+1}{8H^+}(aq) + 5e^- \longrightarrow \overset{+2}{Mn^{2+}}(aq) + \overset{+1\,-2}{4H_2O}(l)$$

it is clear that manganese is the reduced element because its oxidation number changes from +7 to +2.

The main disadvantage of oxidation numbers is that they can lead to a misunderstanding about the structure of molecular substances. No physical or structural significance can be attached to oxidation numbers of atoms in molecular substances. The oxidation number of carbon in CO_2 is +4, but it must *not* be supposed that there is a charge of +4 on the carbon atom.

In a few cases ambiguities can arise with oxidation numbers. For example, the rules for assigning oxidation numbers suggest that each sulfur atom in the thiosulfate ion, $S_2O_3^{2-}$, has an oxidation number of +2. However, the structure of the $S_2O_3^{2-}$ ion shows that the two sulfur atoms in it are quite different. One sulfur atom is at the centre of a tetrahedron bonded to the other four atoms (one S and three O atoms). This is similar to the S atom in the SO_4^{2-} ion. With this in mind, we could assign an oxidation number of +6 to the central S atom in $S_2O_3^{2-}$ (similar to the central S atom in SO_4^{2-}) and an oxidation number of –2 to each of the surrounding atoms, including the second sulfur atom.

Two further problems with oxidation numbers concern their use with organic compounds.

The carbon atoms in CH_4, C_2H_6 and C_3H_8 all have four covalent bonds. In spite of this similarity, they have different oxidation numbers i.e -4, -3 and $-2\frac{2}{3}$ respectively.

The other problem (which C_3H_8 highlights) is that in some compounds, atoms can have oxidation numbers which are not whole numbers.

In spite of these disadvantages the concept of oxidation number is still very useful.

Review questions

1   What are the oxidation numbers of
   a   chlorine in HCl, $HClO$, ClO_3^-, PCl_3, Na_3AlCl_6, $POCl_3$,
   b   nitrogen in N_2O, NO, NO_2, NO_3^-, N_2H_4, HCN?

2   a   Write the formulae for substances containing sulfur in which it shows the following oxidation states: -2, -1, 0, $+1$, $+2$, $+4$, $+6$.
   b   The following equations summarise redox reactions involving sulfur compounds. Deduce the oxidation number of all the atoms and ions in these equations and hence determine precisely which species is oxidised and which is reduced.
       i $2MnO_4^- + 6H^+ + 5SO_3^{2-} \longrightarrow 2Mn^{2+} + 3H_2O + 5SO_4^{2-}$
       ii $2NaI + 3H_2SO_4 \longrightarrow 2NaHSO_4 + 2H_2O + I_2 + SO_2$
       iii $S_2O_3^{2-} + 2H^+ \longrightarrow S + SO_2 + H_2O$
       iv $SO_3^{2-} + H_2O + 2Ce^{4+} \longrightarrow SO_4^{2-} + 2H^+ + 2Ce^{3+}$
       v $2S_2O_3^{2-} + I_2 \longrightarrow S_4O_6^{2-} + 2I^-$

3   Which of the following may be regarded as redox reactions? Explain your answers.
   a $Cu^{2+} + 4NH_3 \longrightarrow Cu(NH_3)_4^{2+}$
   b $Ca^{2+} + 2F^- \longrightarrow CaF_2$
   c $Ca + F_2 \longrightarrow CaF_2$
   d $2CCl_4 + CrO_4^{2-} \longrightarrow 2COCl_2 + CrO_2Cl_2 + 2Cl^-$

4   Write redox half-equations for the following reactions:
   a   When copper is added to concentrated nitric acid the solution becomes pale blue, and brown fumes of nitrogen dioxide are produced.
   b   When potassium iodide is added to acidified hydrogen peroxide a brown colour appears.
   c   Manganese(IV) oxide oxidises concentrated hydrochloric acid to chlorine.
   d   When zinc is added to silver nitrate solution, crystals of silver form on the zinc surface.

5   a   What do you understand by the terms 'oxidation' and 'reduction'?
   b   In each of the following reactions say what (if anything) has been reduced and what has been oxidised.
       Write electron-transfer equations to explain your answers:
       i $2FeCl_2 + Cl_2 \longrightarrow 2FeCl_3$
       ii $CuO + H_2 \longrightarrow Cu + H_2O$
       iii $3Cu + 8HNO_3 \longrightarrow 3Cu(NO_3)_2 + 4H_2O + 2NO$
       iv $2Na + H_2 \longrightarrow 2NaH$

6   Certain features of the element vanadium, V, are presented in Figure 6.17. Consider the diagram carefully.
   a   What is the oxidation number of vanadium in the compounds A–I?
   b   What can you deduce about the oxidising power of the halogens towards vanadium?
   c   How does the action of chlorine on vanadium compare with the action of hydrogen chloride?

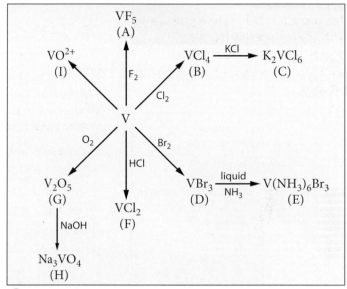

▲ Figure 6.17

7   10 g of an impure iron(II) salt were dissolved in water and made up to 200 cm^3 of solution. 20 cm^3 of this solution, acidified with dilute sulfuric acid, required 25 cm^3 of 0.04 $mol\,dm^{-3}$ $KMnO_4(aq)$ before a faint pink colour appeared.
   a   Write a balanced ionic equation (or half-equations) for the reaction of acidified manganate(VII) ions, $MnO_4^-(aq)$ with iron(II) ions.
   b   How many moles of iron(II) ions react with one mole of MnO_4^- ions?
   c   How many moles of Fe^{2+} react with 25 cm^3 of 0.04 $mol\,dm^{-3}$ $KMnO_4(aq)$?
   d   How many grams of Fe^{2+} are there in the 200 cm^3 of original solution? ($Fe = 56$)
   e   Calculate the percentage by mass of iron in the impure iron(II) salt.

8   In an experiment, 100 cm^3 of a 0.20 $mol\,dm^{-3}$ solution of metal ion, X^{3+}, reacted exactly with 50 cm^3 of 0.2 $mol\,dm^{-3}$ aqueous sodium sulfite. The half-equation for oxidation of sulfite ion is:
   $SO_3^{2-}(aq) + H_2O(l) \longrightarrow SO_4^{2-}(aq) + 2H^+ + 2e^-$
   What is the final oxidation number of X after the reaction?

7 Equilibria

7.1 Reversible reactions

Baking a cake, boiling an egg and burning natural gas (methane) all involve one-way reactions. When a cake is baked or an egg is boiled, chemical reactions take place in the cake and the egg. It is impossible to take the cake and turn it back into flour, sugar, water and fat. This also applies to the boiled egg and to the carbon dioxide and water produced when natural gas burns:

$$CH_4(g) + 2O_2(g) \longrightarrow CO_2(g) + 2H_2O(g)$$
methane in
natural gas

No matter what you do, the egg cannot be 'unboiled'. Neither can the carbon dioxide and water be turned back into methane and oxygen. Reactions like this which cannot be reversed are called **irreversible reactions**. Most of the chemical reactions that you have studied so far are also irreversible, but there are some reactions which can be reversed.

For example, when blue (hydrated) copper(II) sulfate is heated, it decomposes to white anhydrous copper(II) sulfate and water vapour:

$$CuSO_4 \cdot 5H_2O(s) \longrightarrow CuSO_4(s) + 5H_2O(g)$$
blue   white

If water is now added the change can be reversed and blue hydrated copper(II) sulfate re-forms:

$$CuSO_4(s) + 5H_2O(l) \longrightarrow CuSO_4 \cdot 5H_2O(s)$$
white   blue

These two processes can be combined in one equation as:

$$CuSO_4 \cdot 5H_2O(s) \underset{\text{mix reactants}}{\overset{\text{heat}}{\rightleftharpoons}} CuSO_4(s) + 5H_2O(l)$$

Reactions like this, which can be reversed by changing the conditions or adding and removing reagents, are called **reversible reactions**.

Another reaction which can be reversed by changing the conditions involves ammonia, hydrogen chloride and ammonium chloride. Ammonia and hydrogen chloride will react at room temperature to form a white smoke which is a suspension of solid ammonium chloride (Figure 7.2).

$$NH_3(g) + HCl(g) \longrightarrow NH_4Cl(s)$$

If ammonium chloride is heated, this reaction is reversed. The ammonium chloride decomposes forming ammonia and hydrogen chloride:

$$NH_4Cl(s) \longrightarrow NH_3(g) + HCl(g)$$

During a reversible reaction, the reactants are sometimes completely changed to the products. But, in other cases, the reactants are not *completely* converted to the products. For example, if ammonium chloride is heated in a sealed container, only part of the solid will decompose (Figure 7.3). Gaseous ammonia and hydrogen chloride occupy the space above some of the unchanged ammonium chloride. If the temperature stays constant, the materials inside the container reach a stage at which neither the gases nor the solid seem to change. When reactants and products are both present in a closed system and no further changes appear to be taking place, chemists

Figure 7.1
Baking a cake involves irreversible reactions.

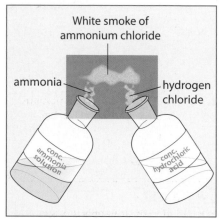

Figure 7.2
Fumes of ammonia and hydrogen chloride reacting to form ammonium chloride

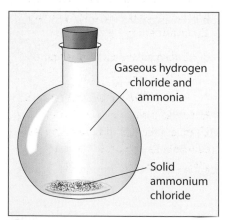

Figure 7.3
In a closed container at 100 °C, solid ammonium chloride comes to equilibrium with gaseous ammonia and hydrogen chloride.

say that the substances are in **equilibrium**. The system must be closed so no reactants or products can escape. When substances are in equilibrium like this, the reversible arrows sign ($\rightleftharpoons$) in the equation is replaced with the equilibrium arrows sign ($\rightleftharpoons$).

$$NH_4Cl(s) \rightleftharpoons NH_3(g) + HCl(g)$$

At equilibrium, this reaction may be well to the left with most of the ammonium chloride unchanged and very little ammonia and hydrogen chloride present. Alternatively, it may be well be to the right or at some point in between. The actual position will depend on the conditions of temperature and pressure in the container.

Quick Questions

1 Figure 7.4 shows what happens when ammonium chloride is heated forming ammonia (which is alkaline) and hydrogen chloride (which is acidic). The red litmus paper first turns blue, then both pieces of litmus paper turn red.
   a Why do the gases produced separate as they pass up the tube through the glass wool?
   b Which gas is detected first and why?
   c Why does a white smoke form above the tube?
2 Clothes that have been washed and are wet stay wet if kept in a plastic laundry bag, but they dry out if hung on a line. Why is this?
3 When purple hydrated cobalt chloride, $CoCl_2 \cdot 6H_2O(s)$, is heated, it changes to blue anhydrous cobalt chloride.
   a Write an equation for this reaction.
   b How is the reaction reversed?
   c How is this reaction used as a test for water?

7.2 Equilibria in physical processes

Liquid–vapour equilibrium: vapour pressure

When liquid bromine is shaken in a stoppered flask, some of it evaporates, forming an orange gas. Gradually, the gas becomes thicker. Eventually, the colour of the gas does not change any more. However much we shake the flask, its colour remains constant. The constant colour of the gas in the flask suggests that a position of equilibrium has been reached. Some of the bromine has formed a vapour and some of it remains as a liquid. *At equilibrium, macroscopic (large scale) properties (such as the intensity of the orange vapour) are constant.* This equilibrium can be summarised as:

$$Br_2(l) \rightleftharpoons Br_2(g)$$

The equilibrium sign ($\rightleftharpoons$) shows that both bromine liquid and bromine gas are present in the flask. But, what is happening to the molecules in the flask at equilibrium? Do all the gas molecules remain as gas while all the liquid molecules remain as liquid (a **static equilibrium**)? Or, are some gas molecules becoming liquid while an equal number of liquid molecules become gas (a **dynamic equilibrium**)?

Experiments show that liquid and gas molecules move around randomly. So, it is likely that molecules in the flask are in a dynamic rather than a static equilibrium. If this is so, the rate at which molecules leave the liquid surface and enter the vapour must equal the rate at which other molecules in the vapour return to the liquid. Random molecular movement may occur even after all the external signs of change have disappeared, but we cannot see the molecules moving. Nor can we measure the rate at which they enter or leave the vapour.

The differences between a static and a dynamic equilibrium are illustrated very neatly in Figure 7.5.

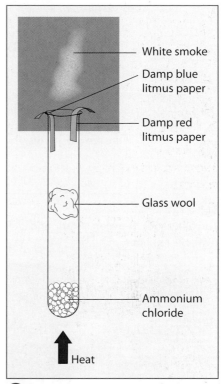

Figure 7.4
Investigating what happens when ammonium chloride decomposes

Labels: White smoke; Damp blue litmus paper; Damp red litmus paper; Glass wool; Ammonium chloride; Heat

Definitions

A **reversible reaction** is a reaction which can be reversed by changing the conditions.

An **irreversible reaction** is one that cannot be reversed.

Equilibrium in a chemical reaction is a state of balance in which no change takes place in the amounts of reactants and products in a closed system.

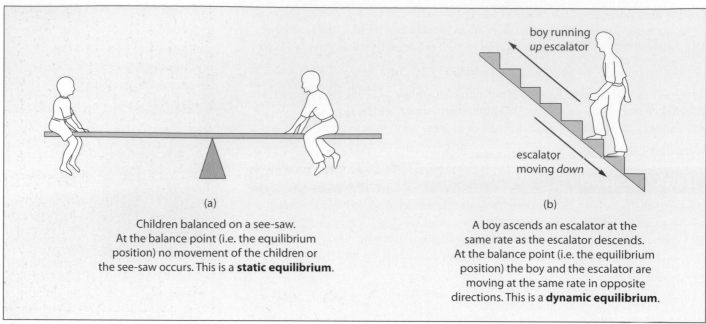

(a)

Children balanced on a see-saw.
At the balance point (i.e. the equilibrium
position) no movement of the children or
the see-saw occurs. This is a **static equilibrium**.

(b)

A boy ascends an escalator at the
same rate as the escalator descends.
At the balance point (i.e. the equilibrium
position) the boy and the escalator are
moving at the same rate in opposite
directions. This is a **dynamic equilibrium**.

Figure 7.5
The difference between a static and a
dynamic equilibrium

Figure 7.7
One of the key factors which dictates
the position of equilibrium in any
physical or chemical process is the
tendency for materials to exist in the
lowest energy state. Will equilibrium
ever be achieved in the waterfall?

Quick Questions

4 a Will the rate of condensation
of gas molecules still equal the
rate of evaporation of liquid
molecules in Figure 7.6 (b)?
Explain.

b What happens to the
concentration of molecules
in the gas phase as time goes
on? Explain.

c How and when is equilibrium
attained once more?

Consider the equilibrium between bromine liquid and bromine vapour once again.
Figure 7.6 (a) shows perfect balance between the rate of evaporation and the rate of
condensation at equilibrium.

Now, suppose some of the vapour is suddenly removed without affecting the system in
any other way (Figure 7.6 (b)) and answer Quick question 4.

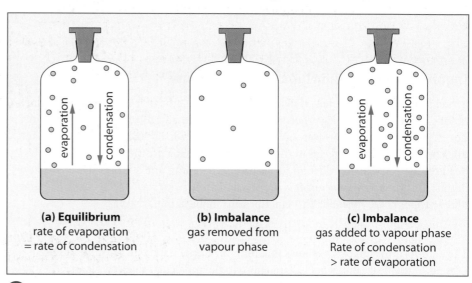

(a) Equilibrium
rate of evaporation
= rate of condensation

(b) Imbalance
gas removed from
vapour phase

(c) Imbalance
gas added to vapour phase
Rate of condensation
> rate of evaporation

Figure 7.6
Liquid–vapour equilibria

Suppose, now, the equilibrium is disturbed by injecting more bromine vapour into
the flask (Figure 7.6 (c)). The concentration of gas molecules suddenly rises and the
rate of condensation increases. Condensation occurs faster than evaporation. Thus the
concentration of gas molecules in the vapour phase decreases. Gradually, the rate of
condensation falls until finally the rate of condensation equals the rate of evaporation.
The system is in equilibrium once more and the vapour pressure of the bromine is the
same as it was before the equilibrium was disturbed.

Notice that *equilibrium can be reached from either direction*. We can start with either
very little bromine vapour or excess bromine vapour. *Provided we have a closed
container (system)*, equilibrium will eventually be reached.

Solute–solution equilibrium: solubility

When one teaspoon of sugar is added to a cup of tea, all the sugar dissolves forming a solution. As more sugar is added, this also dissolves at first, but a stage is eventually reached when no more will dissolve. The solution (tea) is now saturated with solute (sugar) at the temperature involved. Solute particles in the undissolved sugar are in equilibrium with solute particles in the solution.

$$sugar(s) \rightleftharpoons sugar(aq)$$

Provided the system is closed, no solvent can escape and the amounts of dissolved and undissolved solute remain constant. As in the liquid–vapour equilibrium, macroscopic properties have become constant. In this case, the constant macroscopic properties are the concentration of the solution and the amount of undissolved solute.

Investigating the nature of equilibrium

How are the molecules behaving when equilibrium is established between a solute and its saturated solution? No changes are apparent at a macroscopic level, but what is happening at a molecular (microscopic) level?

It would seem reasonable to predict that the equilibrium in this case is also dynamic. This would involve particles leaving the undissolved solute and going into solution at the same rate as dissolved particles rejoin the solid.

In order to check these ideas about a dynamic equilibrium, we can 'label' some of the particles in either the saturated solution or in the undissolved solute. This is done by using a radioactive isotope.

If solid radioactive $^{212}PbCl_2$ is added to a saturated solution of non-radioactive $PbCl_2(aq)$ at a constant temperature, no increase in the amount of dissolved lead chloride can occur. But, if our ideas about a dynamic equilibrium are correct, there will be an interchange of Pb^{2+} and Cl^- ions between the undissolved solid and the saturated solution. As a result, radioactive $^{212}Pb^{2+}$ ions will appear in the solution.

The background radiation of the saturated aqueous lead chloride is first measured. It is low because this $PbCl_2$ is non-radioactive. The saturated $PbCl_2(aq)$ is then shaken with radioactive solid $^{212}PbCl_2$ for 10 minutes. After this, the saturated solution and undissolved solute are separated by centrifuging and the radioactivity of the saturated solution is measured a second time. The mixing and centrifuging are repeated at intervals and the radioactivity of the saturated solution is measured at each stage. Figure 7.9 shows how the radioactivity of the solution changes with time.

The increasing radioactivity of the saturated solution suggests that solid radioactive $^{212}PbCl_2$ must have dissolved. But, the solution was already saturated before any $^{212}PbCl_2$ dissolved. So, the radioactive $^{212}Pb^{2+}$ ions dissolving from the solid must have replaced non-radioactive particles which have crystallised onto the solid from the solution. This means that a dynamic equilibrium exists between the undissolved solute and the saturated solution:

$$PbCl_2(s) \rightleftharpoons PbCl_2(aq)$$

At equilibrium, microscopic (molecular-scale) processes continue, but these are in balance. The rate of the forward process in which $PbCl_2(s)$ dissolves is exactly the same as the rate of the reverse process in which $Pb^{2+}(aq) + Cl^-(aq)$ ions recrystallise on the solid. The system is in dynamic equilibrium.

7.3 Characteristic features of a dynamic equilibrium

The last section has highlighted four important features of dynamic equilibria.

1 *At equilibrium, macroscopic (large scale) properties are constant* under the given conditions of temperature, pressure and initial amounts of substances.

Figure 7.8
When the atmosphere is saturated with water vapour, water is in equilibrium with its vapour.

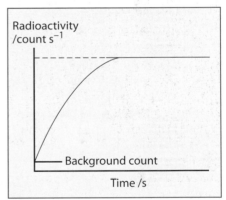

Figure 7.9
Following the radioactivity of a saturated solution in contact with its radioactive solute

Quick Questions

5 Figure 7.9 shows that the radioactivity of the saturated solution eventually reaches a constant value. Why is this?

Dynamic equilibria:

▶ only exist in closed systems,
▶ have constant macroscopic properties,
▶ have continuing microscopic properties,
   Rate of forward process
   = rate of back process
▶ can be attained from either direction.

Quick Questions

6 a Why does the decomposition of limestone in a rotary lime kiln never come to equilibrium?

 b Suggest an important use for lime.

7 A lime kiln produces 0.5 tonnes of lime (calcium oxide, CaO) from 1.0 tonne of limestone (calcium carbonate, $CaCO_3$).

 a What is the theoretical (maximum) yield of calcium oxide from 1.0 tonne of calcium carbonate?

 b What is the actual percentage yield of calcium oxide from the lime kiln?

2 *The equilibrium can be attained from either direction.* Changes of this kind are reversible.

3 *Equilibrium can only be achieved in a closed system.* A closed system is one in which there is no loss or gain of materials or energy to or from the surroundings. An open system may allow matter or energy to escape or to enter.

4 *At equilibrium, microscopic (molecular-scale) processes continue but these are in balance. This means that no overall macroscopic changes occur.* The particles participate in both forward and reverse processes. The rate of the forward process is equal to the rate of the reverse process so that no net change results.

7.4 Equilibria in chemical reactions

The decomposition of calcium carbonate

When calcium carbonate is heated strongly, it decomposes forming calcium oxide and carbon dioxide:

$$CaCO_3(s) \xrightarrow{\text{heat}} CaO(s) + CO_2(g)$$

Normally in an open container, CO_2 escapes well away from the solid CaO, so their recombination to form $CaCO_3$ never occurs. The system can never reach equilibrium.

If, however, a few grams of $CaCO_3(s)$ are heated at 800 °C in a *closed* evacuated container, only part of the solid is decomposed no matter how long it is heated. The pressure of CO_2 inside the container rises, and then remains steady at a constant value. As long as the temperature stays at 800 °C, the pressure remains constant at 25 kPa (0.25 atm). The reaction has reached an equilibrium. Constant macroscopic properties have been achieved in a closed system.

The system also comes to equilibrium with the same macroscopic composition when approached from the opposite direction.

Solid CaO is first placed in the reaction vessel. This is then evacuated and refilled with CO_2 at a pressure well above 25 kPa. Finally, the container is closed and maintained at 800 °C. The pressure begins to fall and becomes steady at 25 kPa (0.25 atm). The same equilibrium pressure of CO_2 results whether we start with $CaCO_3$ or with CaO and CO_2. The equilibrium can be summarised as

$$CaCO_3(s) \rightleftharpoons CaO(s) + CO_2(g)$$

Figure 7.10

A rotary lime kiln. The kiln is viewed from the lower heated end. Raw limestone slurry is pumped in at the top end. As the slurry flows down the kiln, it is first dried and then decomposed to lime at the lower heated end. The kiln is inclined at a gradient of 1 in 30. It makes one revolution per minute. Millions of tonnes of lime (calcium oxide) are produced annually by heating limestone (calcium carbonate)in large industrial kilns.

In this case, $CaCO_3$ is decomposing to form $CaO + CO_2$ at the same rate as $CaO + CO_2$ are reforming $CaCO_3$. The rate of the forward reaction equals the rate of the back reaction.

The reaction of hydrogen with iodine

Hydrogen does not react with iodine at room temperature. Even at 500 K the reaction which produces hydrogen iodide is slow and only partial. If hydrogen and iodine are heated at 500 K in a closed system (i.e. stoppered container), the intensity of the purple iodine vapour slowly falls as hydrogen iodide is produced. Eventually, the colour intensity of the iodine vapour does not change any more and the composition of the mixture becomes constant. The reaction seems to have reached an equilibrium because macroscopic properties, such as the intensity of purple iodine, in the closed system have become constant:

$$H_2(g) + I_2(g) \rightleftharpoons 2HI(g)$$

If 1 mol of hydrogen and 1 mol of iodine are heated at a constant temperature of 500 K in a 1 dm^3 stoppered container, the system eventually reaches an equilibrium with 0.14 mol H_2, 0.14 mol I_2 and 1.72 mol HI (Figure 7.11 (a)).

The equilibrium can also be studied in the opposite direction starting from hydrogen iodide. If 2 mol of hydrogen iodide are heated at a constant temperature of 500 K in a 1 dm^3 stoppered container, the system comes to exactly the same equilibrium as before with 0.14 mol H_2, 0.14 mol I_2 and 1.72 mol HI (Figure 7.11 (b)).

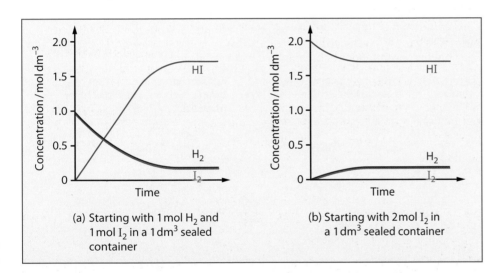

(a) Starting with 1 mol H_2 and 1 mol I_2 in a 1 dm^3 sealed container

(b) Starting with 2 mol I_2 in a 1 dm^3 sealed container

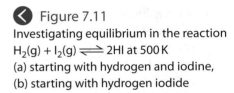

Figure 7.11
Investigating equilibrium in the reaction $H_2(g) + I_2(g) \rightleftharpoons 2HI$ at 500 K
(a) starting with hydrogen and iodine,
(b) starting with hydrogen iodide

Definition

A **dynamic equilibrium** is one in which changes continue at a molecular level with forward and reverse reactions taking place at the same rate in a closed system and, therefore, resulting in no overall change at a macroscopic level.

Quick Questions

8 a What is happening to the molecules of H_2, I_2 and HI in the container at equilibrium?

   b At equilibrium, how is the rate of formation of HI related to the rate of decomposition of HI?

   c Why are the curves for H_2 and I_2 always together in Figure 7.11?

   d What is the relationship between the decrease in amounts of H_2 and I_2 and the increase in amount of HI in Figure 7.11 (a)?

   e Why do the curves in Figure 7.11 (a) become horizontal at the same time?

7.5 How far? Studying the equilibrium of a solute between two immiscible solvents

When chemists ask the question 'How far?', they want to know what proportion of the reactants have been converted to products when a system reaches equilibrium. This is called the *position* of equilibrium.

In order to investigate equilibrium in detail, it is useful to study the equilibrium which occurs when a solute distributes itself between two immiscible solvents. For example, if a small crystal of iodine is shaken with 5 cm^3 of aqueous potassium iodide, a brown solution forms. When 5 cm^3 of hexane is carefully added to the mixture, it simply floats on top of the aqueous brown solution as a clear liquid. Water and hexane do not mix with each other; we say they are immiscible. But when the mixture is shaken, iodine from the KI solution begins to dissolve in the hexane forming a purple solution. If the shaking continues, the intensity of the purple colour in the hexane increases whilst the brown colour of the KI solution becomes paler.

Eventually, the colours of the two solutions remain constant. The iodine has distributed (partitioned) itself between the two solvents and an equilibrium has been reached. No matter how much the mixture is shaken, no further changes in colour intensity occur. The concentrations of iodine in the two solutions remain constant.

Other solutes which are soluble in two immiscible solvents will also partition themselves between both solvents when the three substances are shaken together. These are further examples of equilibrium systems. But, is there a relationship between the equilibrium concentrations of the solute in the two solvents? We can investigate this in the next experiment.

Investigating the partition of butanedioic acid ($HOOCCH_2CH_2COOH$) between water and ethoxyethane (ether, $CH_3CH_2OCH_2CH_3$)

1 g of butanedioic acid was added to a separating funnel containing 25 cm^3 of water and 25 cm^3 of ethoxyethane (ether). This was shaken until all the solid had dissolved and equilibrium was reached. The mixture was then left to stand for some time. This allowed the organic and aqueous layers to separate as fully as possible.

The concentration of butanedioic acid in each layer was then determined by titration against sodium hydroxide solution of known concentration using phenolphthalein indicator.

Further experiments were carried out using different initial amounts of butanedioic acid between 0.5 and 1.5 g. The results obtained are shown in Table 7.1.

 Table 7.1

Equilibrium concentrations of butanedioic acid in ether and in water

Experiment number	Equilibrium concentration of butanedioic acid in ether layer /mol dm^{-3}	Equilibrium concentration of butanedioic acid in water layer /mol dm^{-3}
1	0.023	0.152
2	0.028	0.182
3	0.036	0.242
4	0.044	0.300
5	0.052	0.358
6	0.055	0.381

When these equilibrium concentrations in Table 7.1 are plotted graphically, a straight line is obtained (Figure 7.12).

The results in Figure 7.12 show that the ratio:

$$\frac{\text{concentration of butanedioic acid in ether}}{\text{concentration of butanedioic acid in water}} \text{ is constant.}$$

Using square brackets, $[\ldots]_{eqm}$, to denote the concentrations at equilibrium, we can deduce from the graph in Figure 7.12 that

$$\frac{[\text{butanedioic acid (ether)}]_{eqm}}{[\text{butanedioic acid (water)}]_{eqm}} = 0.15$$

This ratio is known as the **partition coefficient** for butanedioic acid between ether and water. *The partition coefficient is independent of the amount of solute taken and also independent of the volumes of the solvents used.* Notice that it is *the ratio of concentrations and not the ratio of masses* of solute which matter. Other investigations of the distribution of solutes between immiscible liquids at equilibrium give similar results. In each case, the partition coefficient is constant provided:

Note

Square brackets […] are used to denote the concentration of the bracketed substance in mol dm^{-3}. Don't confuse these square brackets with those used for complex ions such as $[Cu(NH_3)_4]^{2+}$.

Definition

The **partition coefficient** of solute X distributed between two immiscible solvents A and B is defined as:

$$\frac{[\text{X(solvent A)}]_{eqm}}{[\text{X(solvent B)}]_{eqm}} = \text{a constant,}$$

at constant temperature

1 The temperature is constant.
2 The solvents are immiscible and do not react with each other.
3 The solute does not react, or change in the solvents.

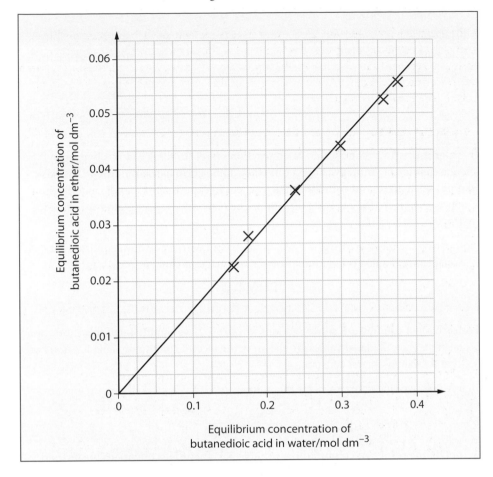

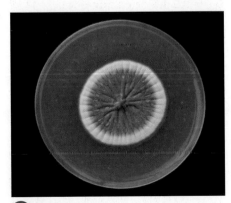

Figure 7.12
A graph of the equilibrium
concentration of butanedioic acid in
ether against that in water

7.6 Solvent extraction

The most important application of partition is solvent extraction in the laboratory
and in industry. Organic compounds are generally more soluble in non-polar
solvents such as, hexane and ethoxyethane (ether) than in water. These solvents are
themselves immiscible with water. Thus, organic compounds can be extracted from
aqueous solutions or suspensions by shaking with a non-polar organic solvent and
then separating the two layers. The pure organic compound can then be obtained by
distilling off the solvent. For example, penicillin is extracted from a dilute aqueous
solution using trichloromethane (Figure 7.13). Phenylamine (aniline) can be reclaimed
from a mixture with water by using ether. Iodine can be extracted from an aqueous
mixture using a hydrocarbon solvent such as cyclohexane.

7.7 The equilibrium constant

In Section 7.5, we obtained a constant value for the ratio of concentrations of a solute
between two immiscible solvents at equilibrium. Thus, it would seem sensible to
investigate whether there is a similar relationship between the concentrations of
reactants and products at equilibrium in chemical reactions.

Table 7.2 shows information concerning the equilibrium between hydrogen, iodine
and hydrogen iodide at 731 K:

$$H_2(g) + I_2(g) \rightleftharpoons 2HI(g)$$

Figure 7.13
Penicillium mould growing in a Petri
dish on a nutrient jelly. Penicillin can be
washed from the mould into aqueous
solution and then extracted using
trichloromethane.

Experiment number	Initial concentrations/mol dm^{-3}			Equilibrium concentrations/mol dm^{-3}		
	$[H_2(g)]$	$[I_2(g)]$	$[HI(g)]$	$[H_2(g)]_{eqm}$	$[I_2(g)]_{eqm}$	$[HI(g)]_{eqm}$
1	2.40×10^{-2}	1.38×10^{-2}	0	1.14×10^{-2}	0.12×10^{-2}	2.52×10^{-2}
2	2.40×10^{-2}	1.68×10^{-2}	0	0.92×10^{-2}	0.20×10^{-2}	2.96×10^{-2}
3	2.44×10^{-2}	1.98×10^{-2}	0	0.77×10^{-2}	0.31×10^{-2}	3.34×10^{-2}
4	2.46×10^{-2}	1.76×10^{-2}	0	0.92×10^{-2}	0.22×10^{-2}	3.08×10^{-2}
5	0	0	3.04×10^{-2}	0.345×10^{-2}	0.345×10^{-2}	2.35×10^{-2}
6	0	0	7.58×10^{-2}	0.86×10^{-2}	0.86×10^{-2}	5.86×10^{-2}

> **Table 7.2**
Initial and equilibrium concentrations of H_2, I_2 and HI at 731K

In experiments 1, 2, 3 and 4 the sealed reaction vessel contains gaseous hydrogen and gaseous iodine initially. After a time the composition of the mixture in the flask remains unchanged and equilibrium is reached. In experiments 5 and 6, equilibrium is approached from the opposite direction. The reaction vessel contains only gaseous hydrogen iodide initially.

Comparing the reaction $H_2(g) + I_2(g) \rightleftharpoons 2HI(g)$ with a partition equilibrium, it is reasonable to calculate the ratio of product and reactant concentrations as

$$\frac{[HI(g)]_{eqm}}{[H_2(g)]_{eqm} \times [I_2(g)]_{eqm}}$$

For each set of equilibrium concentrations in Table 7.2, we want to see if this expression is constant. The calculated values are shown in the second column of Table 7.3. You can see that the values are *not* constant.

If, however, we calculate values for the ratio

$$\frac{[HI(g)]^2_{eqm}}{[H_2(g)]_{eqm}[I_2(g)]_{eqm}}$$

we obtain the results in column three of Table 7.3. The values obtained this time are constant, within the limits of experimental error. Therefore, we can write

$$\frac{[HI(g)]^2_{eqm}}{[H_2(g)]_{eqm}[I_2(g)]_{eqm}} = \text{a constant} = 46.8 \text{ at } 731K$$

Notice that the ratio of concentrations which gives a constant value is related to the number of moles of each substance in the balanced equation.

$$H_2(g) + I_2(g) \rightleftharpoons 2HI(g)$$

The ratio uses the *second* power of [HI] ($[HI]^2$) in the equilibrium expression and the first power for both [H_2] and [I_2]. Thus, *the power to which we have raised the concentration of a substance in the equilibrium expression is the same as its coefficient in the balanced equation.*

Quick Questions

10 a According to the equation
$H_2(g) + I_2(g) \rightleftharpoons 2HI(g)$
one mole of H_2 reacts with one mole of I_2 forming two moles of HI. Check the data from experiments 1 and 2 in Table 7.2 to see whether:
no. of moles H_2 reacted
= no. of moles I_2 reacted
= $\frac{1}{2}$ no. of moles HI formed.
 b Why is $[H_2(g)]_{eqm} = [I_2(g)]_{eqm}$ in experiments 5 and 6?
 c How could you show that the system involving H_2, I_2 and HI is in dynamic equilibrium?

Experiment number	$\dfrac{[HI(g)]_{eqm}}{[H_2(g)]_{eqm}\,[I_2(g)]_{eqm}}$	$\dfrac{[HI(g)]^2_{eqm}}{[H_2(g)]_{eqm}\,[I_2(g)]_{eqm}}$
1	1840	46.4
2	1610	47.6
3	1400	46.7
4	1520	46.9
5	1970	46.4
6	790	46.4

> **Table 7.3**
Possible equilibrium constant expressions for the reaction
$H_2 + I_2 \rightleftharpoons 2HI$ at 731K

The value of this ratio of concentrations at equilibrium is known as the **equilibrium constant**. The equilibrium constant is represented by the symbol K_c. Thus, for the reaction $H_2(g) + I_2(g) \rightleftharpoons 2HI(g)$

$$K_c = \frac{[HI(g)]^2_{eqm}}{[H_2(g)]_{eqm}[I_2(g)]_{eqm}}$$

Very often, the 'eqm' subscripts are omitted from the concentration terms because it is assumed that the concentrations in the expression for K_c are the equilibrium ones. Hence, the expression for K_c is shortened to

$$K_c = \frac{[HI(g)]^2}{[H_2(g)][I_2(g)]}$$

The subscript 'c' indicates that K_c is expressed in concentrations.

The ratio, $\dfrac{[HI(g)]^2}{[H_2(g)][I_2(g)]}$ is known as the reaction quotient, Q.

The value of Q when the reaction reaches equilibrium is the equilibrium constant. You can calculate Q at any stage in the reaction, but it is only at equilibrium that Q equals K_c.

> **Note**
>
> Strictly speaking, equilibrium concentrations should be indicated by the subscript 'eqm' after the square brackets but this is often omitted.

7.8 The equilibrium law

The equilibria in many other chemical reactions have also been studied. In each case the equilibrium constant relates to a balanced equation in a similar manner to that for the hydrogen, iodine and hydrogen iodide system.

For example, the reaction

$$2SO_2(g) + O_2(g) \rightleftharpoons 2SO_3(g)$$

has an equilibrium constant in which

$$K_c = \frac{[SO_3(g)]^2}{[SO_2(g)]^2[O_2(g)]}$$

These observations lead to the general statement known as the **equilibrium law** or the **law of chemical equilibrium**.

If an equilibrium mixture contains substances A, B, C and D related by the equation

$$aA + bB \rightleftharpoons cC + dD$$

it is found experimentally that $\dfrac{[C]^c[D]^d}{[A]^a[B]^b} = K_c$

K_c, *the equilibrium constant, is constant at a given temperature.*

In writing expressions for the equilibrium constants of reactions, there are important conventions. Concentrations of substances on the right-hand side of the equation are written in the numerator, on the top. Concentrations of substances on the left-hand side are written in the denominator, on the bottom.

Thus, it is essential to relate any numerical value for an equilibrium constant to the particular equation concerned. Suppose the equilibrium constant for the reaction

$$H_2(g) + I_2(g) \rightleftharpoons 2HI(g) \text{ equals } x$$

$$\text{Then, } K_c = \frac{[HI]^2}{[H_2][I_2]} = x$$

The equilibrium constant for the reverse reaction $2HI \rightleftharpoons H_2 + I_2$ at the same temperature is

$$K'_c = \frac{[H_2][I_2]}{[HI]^2} = \frac{1}{x}$$

> **Definition**
>
> **The equilibrium law**
>
> If an equilibrium mixture contains substances A, B, C and D related by the equation
>
> $$aA + bB \rightleftharpoons cC + dD$$
>
> it is found experimentally that $\dfrac{[C]^c[D]^d}{[A]^a[B]^b} = K_c$
>
> K_c is called the equilibrium constant which is constant at a given temperature.

11 Write K_c expressions for the following equations.

   a $N_2(g) + 3H_2(g) \rightleftharpoons 2NH_3(g)$
   b $2NH_3(g) \rightleftharpoons N_2(g) + 3H_2(g)$
   c $2NO(g) + O_2(g) \rightleftharpoons 2NO_2(g)$
   d $NO(g) + \frac{1}{2}O_2(g) \rightleftharpoons NO_2(g)$

12 a Assuming that the value of K_c for the reaction in question 11a is x, calculate the value of K_c for the reaction in 11b.

   b Assuming that the value of K_c for the reaction in question 11c is y, calculate the value of K_c for the reaction in 11d.

13 What are the units of K_c for the reactions in each of questions 11a to 11d?

And, as you would expect, $K'_c = \dfrac{1}{K_c}$

On the other hand, the equilibrium constant for the reaction

$$\tfrac{1}{2}H_2(g) + \tfrac{1}{2}I_2(g) \rightleftharpoons HI(g)$$

is

$$K''_c = \frac{[HI]}{[H_2]^{\frac{1}{2}}[I_2]^{\frac{1}{2}}} = \sqrt{x}$$

K_c has no units in reactions with equal numbers of particles on both sides of the equation. This is because the concentration units cancel out in the expression for K_c. For example, for the reaction

$$H_2(g) + I_2(g) \rightleftharpoons 2HI(g)$$

$$K_c = \frac{[HI(g)]^2}{[H_2(g)][I_2(g)]} \qquad \text{Units} = \frac{(\cancel{\text{mol dm}^{-3}})^2}{(\cancel{\text{mol dm}^{-3}})(\cancel{\text{mol dm}^{-3}})}$$

For reactions in which the numbers of reactant and product particles are not equal, K_c will, of course, have units.

It is particularly important to remember that equilibrium constants vary with temperature. Therefore, when you state an equilibrium constant value, you must always give the temperature.

Determination of the value of equilibrium constants by experiment

The essential stages in determining an equilibrium constant are listed below.

1 Write the balanced equation.
2 Mix known molar amounts of either the reactants or the products or both.
3 Allow the mixture to reach equilibrium in a closed system.
4 Determine the equilibrium concentration of at least one substance in the equilibrium mixture.
5 Deduce the equilibrium concentrations of the other substances in the mixture.
6 Substitute the calculated equilibrium concentrations in the expression for K_c.
7 Repeat the determination of K_c using different initial concentrations at the same temperature.

In determining and using equilibrium constants it is important to appreciate the following points:

Key Points

▶ The equilibrium law only applies to systems in equilibrium.
▶ K_c is constant at one particular temperature. If the temperature changes, the value of K_c changes.
▶ The only factor which affects the value of an equilibrium constant is temperature. The values of K_c and K_p (Section 7.9) are unaffected by changes in concentration or pressure provided the temperature remains constant.

▶ *The numerical value of K_c is unaffected by changes in concentration of either reactants or products*. Obviously, when more reactant is suddenly added to a system in equilibrium, more of the products will tend to form. Eventually the system will adjust itself to a new equilibrium position in which the concentrations of reactants and products give the same numerical value for K_c.
▶ *The magnitude of K_c provides a useful indication of the extent of a chemical reaction.* A large value for K_c indicates a high proportion of products to reactants (i.e. an almost complete reaction). A low value for K_c indicates that only a small fraction of reactants have been converted to products.
▶ *Although the equilibrium constant for a reaction indicates the extent of the reaction, it gives no information about the rate of reaction.* K_c tells us *how far*, but *not how fast* the reaction goes. In fact, the extent and the rate of a reaction are quite independent. For example, the conversion of sulfur dioxide and oxygen to sulfur trioxide at 450 °C occurs very slowly but almost completely. In contrast, the conversion of nitrogen oxide and oxygen to nitrogen dioxide at the same temperature occurs rapidly but only partially.

The following worked examples will help you to understand the ideas we have covered so far. It will show you how equilibrium constants and equilibrium concentrations can be determined.

Worked example – determining an equilibrium constant

When 1 mol of hydrogen iodide reacts to form H_2 and I_2 in a 1.0 dm^3 vessel at 440 °C, only 0.78 mol of HI are present at equilibrium. What is the equilibrium constant at this temperature for the reaction, $2HI(g) \rightleftharpoons H_2(g) + I_2(g)$?

	2HI $\rightleftharpoons$	H_2	+	I_2
Initial amounts/mol	1	0		0
Amounts at equilibrium/mol	0.78	0.11		0.11

Since 0.22 mol of HI have decomposed, 0.11 mol of H_2 and 0.11 mol of I_2 must have formed at equilibrium.

Concentration at equilibrium/mol dm^{-3} $\dfrac{0.78\,\text{mol}}{1.0\,\text{dm}^3}$ $\dfrac{0.11\,\text{mol}}{1.0\,\text{dm}^3}$ $\dfrac{0.11\,\text{mol}}{1.0\,\text{dm}^3}$

$$K_c = \frac{[H_2(g)][I_2(g)]}{[HI(g)]^2} = \frac{0.11 \times 0.11}{(0.78)^2} = 0.02$$

$\Rightarrow K_c$ for the reaction = 0.02 at 440 °C

In this case, the units cancel and K_c has no units.

Calculating concentrations at equilibrium

Suppose 1 mol of hydrogen and 1 mol of iodine are mixed together in a 1.0 dm^3 vessel at 440 °C. What are the concentrations of HI, H_2 and I_2 at equilibrium?

	2HI $\rightleftharpoons$	H_2	+	I_2
Initial amounts/mol	0	1		1
Amounts at equilibrium/mol	$2x$	$1-x$		$1-x$

If x mol of H_2 and x mol of I_2 react ('disappear') then $2x$ mol of HI will form.

Concentration at equilibrium/mol dm^{-3} $\dfrac{2x}{1}$ $\dfrac{1-x}{1}$ $\dfrac{1-x}{1}$

$$\Rightarrow K_c = \frac{(1-x)(1-x)}{(2x)^2} = 0.02$$

Taking the square roots of both sides of the equation, we get:

$$\frac{(1-x)}{2x} = 0.14$$

So $\quad 1-x = 2x \times 0.14 = 0.28x$

i.e. $\quad 1 = 1.28x$ and $x = \dfrac{1}{1.28}$

$\therefore \quad x = 0.78\,\text{mol dm}^{-3}$

Hence at equilibrium, the concentration of HI = 1.56 mol dm^{-3}, the concentration of H_2 = 0.22 mol dm^{-3} and the concentration of I_2 = 0.22 mol dm^{-3}.

Since the volume of the vessel is 1.0 dm^3, these are also the numbers of moles.

> **Note** the key stages in equilibrium calculations:
>
> 1 Write the equation.
>
> 2 Write down the initial amounts present.
>
> 3 Write down the amounts at equilibrium using terms involving x for equilibrium concentrations if necessary.
>
> 4 Calculate the equilibrium concentrations.
>
> 5 Substitute the values in K_c to calculate K_c or x as appropriate.

14 The equilibrium constant for the reaction

$$2NO_2(g) \rightleftharpoons N_2O_4(g)$$

at 298 K is $200 \, mol^{-1} \, dm^3$.

a Write an expression for the equilibrium constant for the reaction.

b If the concentration of $N_2O_4(g)$ in the equilibrium mixture at 298 K is $0.02 \, mol \, dm^{-3}$, what is the concentration of $NO_2(g)$?

c Calculate the equilibrium constant at 298 K for the reaction:

$$\tfrac{1}{2}N_2O_4(g) \rightleftharpoons NO_2(g)$$

15 The equilibrium constants for the synthesis of hydrogen chloride, hydrogen bromide and hydrogen iodide at a particular temperature are given below.

$$K_c$$

$$H_2(g) + Cl_2(g) \rightleftharpoons 2HCl(g) \qquad 10^{17}$$

$$H_2(g) + Br_2(g) \rightleftharpoons 2HBr(g) \qquad 10^{9}$$

$$H_2(g) + I_2(g) \rightleftharpoons 2HI(g) \qquad 10$$

a What do the values of K_c tell you about the extent of each reaction?

b Which of these reactions would you regard as virtually complete?

7.9 Equilibrium constants in gaseous systems

The partial pressure of a gas is proportional to its concentration. So, the equilibrium constant of a gaseous reaction can be expressed either in terms of partial pressures or in terms of concentrations.

Equilibrium constants are normally expressed in terms of concentrations using the symbol, K_c. For reactions involving gases, however, it is usually more convenient to use the partial pressures of the gases rather than their molar concentrations.

Using the ideal gas equation,

$$pV = nRT$$

$$\Rightarrow \quad p = \frac{n}{V}RT$$

In this equation, p is the pressure of the gas, n is the amount of gas, V is the volume of gas, T is the temperature and R is the gas constant.

So, as $\dfrac{n}{V}$ represents the concentration of gas in amount per unit volume, at constant temperature we can say:

$$p \propto [\text{gas}]$$

where [gas] is the concentration of the gas in $mol \, dm^{-3}$. Thus, at a constant temperature, the pressure of a gas is proportional to its concentration.

This means that for the equilibrium

$$H_2(g) + I_2(g) \rightleftharpoons 2HI(g)$$

we can write either

$$K_c = \frac{[HI(g)]^2}{[H_2(g)][I_2(g)]}$$

$$\text{or } K_p = \frac{(p_{HI})^2}{(p_{H_2})(p_{I_2})}$$

The symbol 'p_{HI}' means 'the partial pressure of HI in the mixture' (Section 4.4). When equilibrium constants are expressed and calculated in terms of partial pressures, the symbol used for the equilibrium constant is K_p. In calculating values of K_p, pressures can be expressed in pascals or in atmospheres.

$$1 \text{ pascal(Pa)} = 1 \, N \, m^{-2}; \ 1 \text{ atm} \approx 10^5 \, Pa.$$

16 a Write K_p expressions for the following equations:

i $N_2(g) + 3H_2(g)$
$\rightleftharpoons 2NH_3(g)$

ii $N_2O_4(g) \rightleftharpoons 2NO_2(g)$

b What are the units of K_p for the equations in part a?

c At 60 °C and a total pressure of 1.0 atm, N_2O_4 is 50% dissociated into NO_2. What is the value of K_p for this dissociation at 60 °C?

7.10 The effect of concentration changes on equilibria

When a system in equilibrium is suddenly disturbed, it will respond until the equilibrium is eventually restored.

Consider the equilibrium

$$Fe^{3+}(aq) + NCS^-(aq) \rightleftharpoons [Fe(NCS)]^{2+}(aq)$$
pale yellow   colourless   deep red

When dilute iron(III) nitrate solution is added to an equal volume of dilute potassium thiocyanate, a red solution is produced. The red colour is due to the formation of thiocyanatoiron(III) ions, $[Fe(NCS)]^{2+}(aq)$. The system forms an equilibrium mixture containing unreacted Fe^{3+}, unreacted NCS^- and the product $[Fe(NCS)]^{2+}$. But what happens to the equilibrium when one of the concentrations is suddenly changed?

If a soluble iron(III) salt is added to the equilibrium solution, the colour of the solution becomes deeper red (Figure 7.14).

A new state of equilibrium is quickly attained in which the concentration of $[Fe(NCS)]^{2+}(aq)$ is obviously greater than before the addition of Fe^{3+}. Increasing the concentration of Fe^{3+} has increased the concentration of $[Fe(NCS)]^{2+}(aq)$. In the same way, the concentration of $[Fe(NCS)]^{2+}(aq)$ also rises when a soluble thiocyanate is added to the system. On the other hand, removal of Fe^{3+} or NCS^- from the equilibrium mixture causes the solution to become paler. A decrease in the concentration of Fe^{3+} or NCS^- results in the conversion of some $[Fe(NCS)]^{2+}$ into Fe^{3+} and NCS^- which has the effect of replacing the substance removed.

The results of these experiments can be summarised by the following statement:

> If the concentration of one of the substances in a reversible equilibrium is altered, the *equilibrium* will shift to oppose the change in concentration.

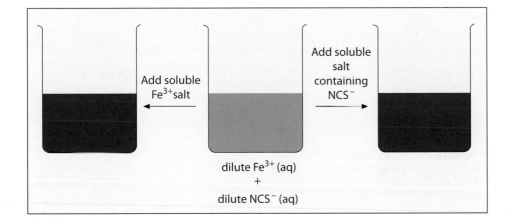

dilute Fe^{3+} (aq)
+
dilute NCS^- (aq)

Thus, if a reactant is added to a system in equilibrium, a reaction will occur which uses up the added reactant. Conversely, if a reactant is removed, a reaction will occur which replenishes the removed reactant. The boxed statement above is a specific application of an important generalisation known as **Le Chatelier's principle**.

During the 1880s, the Frenchman, Henri Louis Le Chatelier (1850–1936) studied the effects of different factors such as temperature, pressure and concentration on equilibria. After studying various equilibria, Le Chatelier proposed the following generalisation now known as Le Chatelier's principle:

If a system in equilibrium is subjected to a change, processes will occur which tend to counteract the change imposed.

Key Point

Although temperature is the only factor which changes the *value* of an equilibrium constant, the *position* of equilibrium can be altered by changing concentration, pressure or temperature. These effects are studied in the next few sections.

◀ Figure 7.14
The effect of adding Fe^{3+} and NCS^- on the equilibrium:

$$Fe^{3+}(aq) + NCS^-(aq) \rightleftharpoons [Fe(NCS)]^{2+}(aq)$$

Definition

Le Chatelier's principle says that if a system in equilibrium is subjected to a change, processes occur which tend to remove the change imposed.

Although the concentration of individual substances in an equilibrium may vary, the equilibrium constant is always the same at one particular temperature. This, of course, is the crucial point of the equilibrium law. We can emphasise this further by considering the effect of suddenly increasing the concentration of hydrogen in an equilibrium mixture of $H_2(g)$, $I_2(g)$ and $HI(g)$. In the initial equilibrium mixture (Figure 7.15),

$$[HI(g)] = 0.07 \text{ mol dm}^{-3}, \; [H_2(g)] = 0.01 \text{ mol dm}^{-3} \text{ and } [I_2(g)] = 0.01 \text{ mol dm}^{-3}$$

$$\therefore \quad K_c = \frac{[HI(g)]^2}{[H_2(g)][I_2(g)]} = \frac{(0.07 \text{ mol dm}^{-3})^2}{(0.01 \text{ mol dm}^{-3}) \times (0.01 \text{ mol dm}^{-3})} = 49$$

When the concentration of $H_2(g)$ is suddenly doubled,

$$\Rightarrow \quad \frac{[HI(g)]^2}{[H_2(g)][I_2(g)]} = \frac{(0.07 \text{ mol dm}^{-3})^2}{(0.02 \text{ mol dm}^{-3}) \times (0.01 \text{ mol dm}^{-3})} = 24.5 < K_c$$

The system is no longer in equilibrium. In order to restore the equilibrium, the concentration of $HI(g)$ must rise, whilst those of $H_2(g)$ and $I_2(g)$ must fall. This is achieved by conversion of *some* of the hydrogen and iodine in the mixture to hydrogen iodide:

$$H_2(g) + I_2(g) \longrightarrow 2HI(g)$$

When equilibrium is restored once more (Figure 7.15), we find that

$$[HI(g)] = 0.076 \text{ mol dm}^{-3}, \; [H_2(g)] = 0.017 \text{ mol dm}^{-3} \text{ and } [I_2(g)] = 0.007 \text{ mol dm}^{-3}$$

$$\Rightarrow \quad \frac{[HI(g)]^2}{[H_2(g)][I_2(g)]} = \frac{(0.076 \text{ mol dm}^{-3})^2}{(0.017 \text{ mol dm}^{-3}) \times (0.007 \text{ mol dm}^{-3})} = 49 = K_c$$

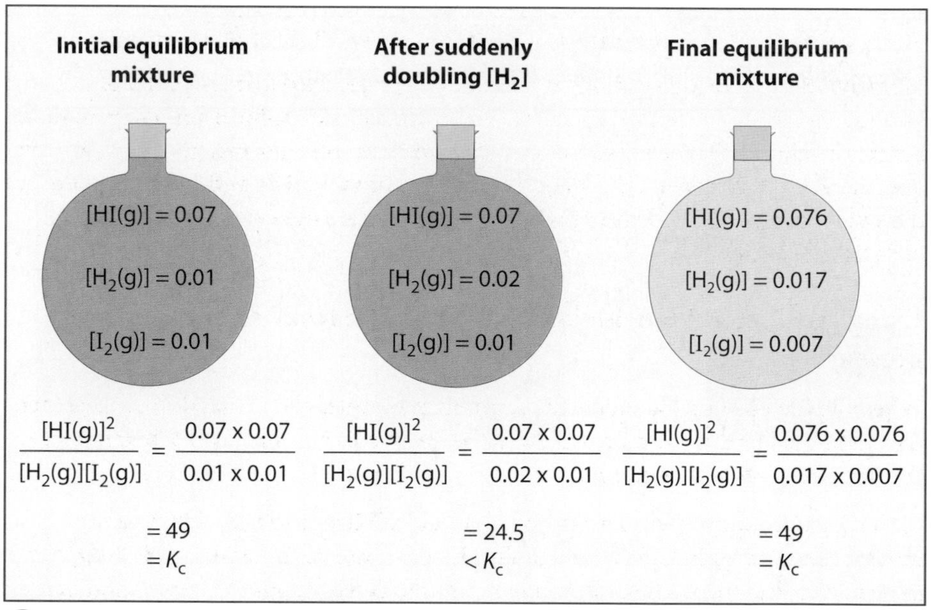

Initial equilibrium mixture	After suddenly doubling $[H_2]$	Final equilibrium mixture
$[HI(g)] = 0.07$	$[HI(g)] = 0.07$	$[HI(g)] = 0.076$
$[H_2(g)] = 0.01$	$[H_2(g)] = 0.02$	$[H_2(g)] = 0.017$
$[I_2(g)] = 0.01$	$[I_2(g)] = 0.01$	$[I_2(g)] = 0.007$
$\dfrac{[HI(g)]^2}{[H_2(g)][I_2(g)]} = \dfrac{0.07 \times 0.07}{0.01 \times 0.01}$	$\dfrac{[HI(g)]^2}{[H_2(g)][I_2(g)]} = \dfrac{0.07 \times 0.07}{0.02 \times 0.01}$	$\dfrac{[HI(g)]^2}{[H_2(g)][I_2(g)]} = \dfrac{0.076 \times 0.076}{0.017 \times 0.007}$
$= 49$ $= K_c$	$= 24.5$ $< K_c$	$= 49$ $= K_c$

⬆ Figure 7.15
The effect of suddenly increasing the concentration of one species in a mixture at equilibrium. (All concentrations are in mol dm^{-3}.)

Notice that only part of the added hydrogen is used up in restoring equilibrium. The concentration of $H_2(g)$ was suddenly doubled from 0.01 mol dm^{-3} in the initial equilibrium to 0.02 mol dm^{-3}. When equilibrium is achieved once more the final concentration of hydrogen is not 0.01 mol dm^{-3} but $0.017 \text{ mol dm}^{-3}$. Obviously, $[HI(g)]$ in the final equilibrium is greater than that in the initial equilibrium, whilst $[I_2(g)]$ in the final equilibrium is less than that initially.

Quick Questions

17 **a** Use the equilibrium law to predict and explain the effect of adding pure oxygen to the following equilibrium:

$$2SO_2(g) + O_2(g) \rightleftharpoons 2SO_3(g)$$

b Show how your prediction agrees with Le Chatelier's principle.

c Predict the effect of compressing the mixture to half its volume, doubling the pressure.

7.11 The effect of pressure changes on equilibria

If the partial pressure of *only one* of the gases in an equilibrium mixture is changed, the overall effect can be predicted like those involving changes in concentration in the last section. But what happens when the *total* pressure of gases in equilibrium is suddenly increased or decreased? In this case, the partial pressures of *all* the gases increase or decrease. Consider, first, the reaction

$$N_2(g) + 3H_2(g) \rightleftharpoons 2NH_3(g)$$

for which

$$K_p = \frac{(p_{NH_3})^2}{p_{N_2}(p_{H_2})^3}$$

Now, suppose the equilibrium partial pressures of nitrogen, hydrogen and ammonia are a, b and c Pa, respectively. Thus:

$$K_p = \frac{(p_{NH_3})^2}{p_{N_2}(p_{H_2})^3} = \frac{c^2}{ab^3}$$

What happens when the total pressure is suddenly doubled? What do the equilibrium law and Le Chatelier's principle predict will happen after this change in pressure?

Using the equilibrium law to predict the results of a change in pressure

When the total pressure is suddenly doubled, all of the partial pressures are doubled. Hence

$$p_{N_2} = 2a, \ p_{H_2} = 2b \text{ and } p_{NH_3} = 2c$$

$$\Rightarrow \qquad \frac{(p_{NH_3})^2}{p_{N_2} \cdot (p_{H_2})^3} = \frac{(2c)^2}{2a.(2b)^3} = \frac{4c^2}{2a.8b^3} = \frac{1}{4} \frac{c^2}{ab^3}$$

Momentarily, the reaction quotient, Q (Section 7.7) is reduced to one-quarter of its value at equilibrium. Therefore, nitrogen and hydrogen react to form ammonia, so that c increases and a and b decrease, until equilibrium is restored once more. Table 7.4 shows how the percentage of ammonia in the equilibrium mixture rises as the total pressure on the system increases. This explains why high pressures are used in the manufacture of NH_3 from N_2 and H_2.

Using Le Chatelier's principle to predict the results of a change in pressure

When the total pressure is suddenly increased, molecules are crowded closer together. The additional pressure can be relieved if the molecules are able to react and reduce the number of molecules present.

In the reaction we are considering, one molecule of nitrogen reacts with three molecules of hydrogen to form two molecules of ammonia. In other words, four molecules of gas are reacting to form only two molecules of gas. This reduction in the total number of gas molecules results in a reduction in the total pressure. Hence, any increase in pressure on the $N_2/H_2/NH_3$ system at equilibrium can be relieved by a conversion of nitrogen and hydrogen to ammonia. Conversely, a decrease in pressure will favour the formation of nitrogen and hydrogen. This results in an increase in the number of gas molecules present, thereby counteracting the pressure reduction.

In general, for gaseous reactions an increase in pressure favours the reaction which produces fewer molecules of gas, and a decrease in pressure favours the gaseous reaction which produces more molecules of gas.

On the other hand, pressure has no effect on a gaseous reaction if there is no change in the number of molecules (Quick question 19).

▼ Table 7.4
The effect of pressure on the equilibrium percentage of ammonia in the system: $N_2 + 3H_2 \rightleftharpoons 2NH_3$

Total pressure /atm	Equilibrium percentage of NH_3 at 723 K
1	0.24
50	9.5
100	16.2
200	25.3

▲ Figure 7.16
The following equilibria are established in this bottle of sparkling water

$$CO_2(g) \rightleftharpoons CO_2(aq)$$

$$CO_2(aq) + H_2O(l) \rightleftharpoons HCO_3^-(aq) + H^+(aq)$$

Quick Questions

18 a Why is the screw cap important for equilibria inside the bottle?

b Explain what happens to the acidity of the sparkling water after it is opened.

19 Suppose the partial pressures of H_2, I_2 and HI in the equilibrium:

$$H_2(g) + I_2(g) \rightleftharpoons 2HI(g)$$

are x, y and z respectively.

a Write the equilibrium constant, K_p, in terms of x, y and z.

b Suppose the overall pressure is halved. What are the partial pressures of H_2, I_2 and HI now?

c What is the value of the reaction quotient

$$\frac{(p_{HI})^2}{(p_{H_2})(p_{I_2})}$$

when the overall pressure is halved?

d Explain, using Le Chatelier's principle, why pressure changes have no effect on this system?

7.12 The effect of catalysts on equilibria

The equilibrium constant expression includes *only* those substances shown in the overall balanced equation. Catalysts do not appear in the overall equation for a reaction. Therefore, it is not surprising that they have no effect on the equilibrium position.

Experiments show that catalysts increase the *rates* of both forward and backward reactions in an equilibrium. So they enable equilibrium to be achieved much more rapidly, but they do not alter the concentrations of reacting substances at equilibrium.

7.13 The effect of temperature changes on equilibria

Although the equilibrium concentrations of reactants and products can vary over a wide range, the value of the equilibrium constant remains constant at one particular temperature. This means that K_c and K_p are unaffected by catalysts or by changes in pressure and concentration.

The equilibrium constant does, however, vary with temperature. Table 7.5 shows the values of K_p at different temperatures for three important reactions. The corresponding enthalpy changes for the complete conversion of reactants to products are also shown.

Notice that the two exothermic reactions have K_p values which *decrease* with an increase in temperature. In contrast, the endothermic reaction has K_p values which *increase* as the temperature rises. Evidence from other investigations confirms this pattern of results. In general, it is found that:

1 equilibria in which the **forward reaction is exothermic (i.e. $\Delta H^\ominus$ negative)** have equilibrium constants that decrease as temperature rises,

2 equilibria in which the **forward reaction is endothermic (i.e. $\Delta H^\ominus$ positive)** have equilibrium constants that increase as temperature rises.

$N_2(g) + 3H_2(g) \rightleftharpoons 2NH_3(g)$ $\Delta H^\ominus = -92\,kJ\,mol^{-1}$ $K_p = \dfrac{(p_{NH_3})^2}{p_{N_2}(p_{H_2})^3}$		$N_2O_4(g) \rightleftharpoons 2NO_2(g)$ $\Delta H^\ominus = +57\,kJ\,mol^{-1}$ $K_p = \dfrac{(p_{NO_2})^2}{p_{N_2O_4}}$		$2SO_2(g) + O_2(g) \rightleftharpoons 2SO_3(g)$ $\Delta H^\ominus = -197\,kJ\,mol^{-1}$ $K_p = \dfrac{(p_{SO_3})^2}{(p_{SO_2})^2 p_{O_2}}$	
T/K	K_p/atm^{-2}	T/K	K_p/atm	T/K	K_p/atm^{-1}
400	1.0×10^2	200	1.9×10^{-6}	600	3.2×10^3
500	1.6×10^{-1}	300	1.7×10^{-1}	700	2.0×10^2
600	3.1×10^{-3}	400	5.1×10	800	3.2×10
700	6.3×10^{-5}	500	1.5×10^3	900	6.3
800	7.9×10^{-6}	600	1.4×10^4	1000	2.0

> Table 7.5
Values of K_p for three different reactions at various temperatures

Key Point

Catalysts affect neither the position of equilibrium nor the values of K_c and K_p.

Le Chatelier's principle can be used once again to predict the effect of temperature on chemical systems in equilibrium. Changing the temperature of a system in equilibrium provides a constraint. The system will try to remove the constraint. Hence, increase in temperature favours the endothermic process which will absorb the additional heat. Alternatively, decrease in temperature favours the exothermic process. This may be summarised as:

$$A + B \xrightarrow[\text{endothermic process favoured by increase in temperature}]{\text{exothermic process favoured by decrease in temperature}} C + D \qquad \Delta H = -x\,kJ\,mol^{-1}$$

The following demonstration illustrates the effect of temperature on the position of an equilibrium. Three identical sealed tubes are prepared containing dark brown nitrogen dioxide (NO_2) in equilibrium with pale yellow dinitrogen tetraoxide (N_2O_4).

$$N_2O_4(g) \rightleftharpoons 2NO_2(g)$$

pale yellow   dark brown

Initially, all three tubes contain the same amounts of NO_2 and N_2O_4 and they have the same pale brown appearance.

One tube is now placed in iced water. A second tube is left at room temperature and the third tube is placed in hot water (Figure 7.17).

The tube in cold water becomes much paler whilst that in hot water turns dark brown. This shows that the equilibrium in the reaction is displaced towards the formation of darker NO_2 at higher temperatures. This is the endothermic direction. The equilibrium moves in the exothermic direction towards the paler N_2O_4 at lower temperatures.

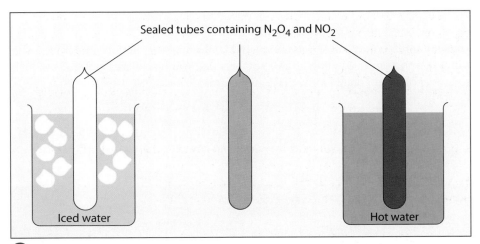

Sealed tubes containing N_2O_4 and NO_2

Iced water   Hot water

Figure 7.17
The effect of temperature change on the equilibrium: $N_2O_4 \rightleftharpoons 2NO_2$

7.14 Acids, bases and equilibria

During the nineteenth century, chemists began to think of acids as substances which dissociate (split up) in water to produce hydrogen ions, H^+.

For example: $HCl(aq) \longrightarrow H^+(aq) + Cl^-(aq)$

In time, and with the development of knowledge concerning atomic structure, chemists realised that the H^+ ion was simply a proton and it was very unlikely that such a small ion could exist independently in aqueous solution.

Therefore, H^+ ions with their high charge density were believed to associate with polar water molecules in aqueous solutions as H_3O^+ ions.

Consequently, the dissociation of acids in water were represented more accurately as:

$$HCl(aq) + H_2O(l) \longrightarrow H_3O^+(aq) + Cl^-(aq)$$

However, when the same indicator is added to different acids with the *same* concentration, different pHs are recorded. Figure 7.18 shows the results when $0.1 \, mol \, dm^{-3}$ solutions of various acids were tested with universal indicator. The different pH values mean that some acids produce H_3O^+ ions more readily than others. The results in Figure 7.18 show that hydrochloric, nitric and sulfuric acids produce more H_3O^+ ions than the other acids tested. Because of this, these three acids are described as **strong acids** and the others as **weak acids**.

Quick Questions

20 Look at the information in Table 7.5.
 a How does the proportion of ammonia in the $N_2/H_2/NH_3$ system change as temperature increases?
 b What is the value of $\Delta H^\ominus$ for this reaction
 $$2NH_3(g) \rightleftharpoons N_2(g) + 3H_2(g)?$$
 What is the value of K_p for the reaction at 400 K?
 c Use Le Chatelier's principle to predict the effect of increasing temperature on K_p for the reaction in part b.

Key Points

Temperature is the only factor which influences the values of K_c and K_p.

Equilibria in which the forward reaction is exothermic have equilibrium constants that decrease as temperature rises.

Equilibria in which the forward reaction is endothermic have equilibrium constants that increase as temperature rises.

Note

The correct name for H_3O^+ is the oxonium ion. Hydrogen ions in aqueous solution are correctly represented as $H_3O^+(aq)$, but it is often acceptable to use the symbol $H^+(aq)$.

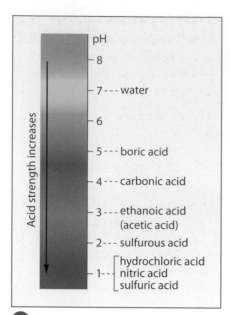

Figure 7.18

The pH values of 0.1 mol dm^{-3} solutions of various acids

Definitions

Strong acids and strong bases are completely dissociated into ions in aqueous solution.

Weak acids and weak bases are only partially dissociated into ions in aqueous solution.

Note

Bases which are soluble in water are often called alkalis.

Key Points

Notice the difference between the terms 'concentration' and 'strength' in relation to acids and bases.

Concentration tells you *how much acid or base is dissolved* in a solution and we use the words 'concentrated' and 'dilute'.

Strength tells you *how much of the acid or base is dissociated into ions* and we use the words 'strong' and 'weak'.

Strong acids produce more H_3O^+ ions than weak acids of the same concentration because they are completely dissociated into ions in aqueous solution, whereas weak acids are only partially dissociated. For example, in a solution of ethanoic (acetic) acid, the dissociated ions, H_3O^+ and CH_3COO^- are in equilibrium with undissociated molecules of CH_3COOH and water. So we can represent the dissociation of ethanoic acid as:

$$CH_3COOH(aq) + H_2O(l) \rightleftharpoons H_3O^+(aq) + CH_3COO^-(aq)$$

The only common strong acids are hydrochloric, nitric and sulfuric acids. Common weak acids include sulfurous acid (H_2SO_3), nitrous acid (HNO_2) and carbonic acid (H_2CO_3). Ethanoic acid (CH_3COOH) and other organic acids are also weak acids.

Bases can also be regarded as strong and weak in a similar way to acids. Whereas acids produce H_3O^+ ions, bases produce OH^- ions. The commonest strong bases are sodium hydroxide (NaOH), potassium hydroxide (KOH) and calcium hydroxide ($Ca(OH)_2$). In aqueous solution, these strong bases are completely dissociated into ions.

$$NaOH(aq) \longrightarrow Na^+(aq) + OH^-(aq)$$

$$Ca(OH)_2(aq) \longrightarrow Ca^{2+}(aq) + 2OH^-(aq)$$

The only common weak base is ammonia (NH_3). In aqueous solution, ammonia reacts with water to form an equilibrium mixture containing undissociated molecules of NH_3 and water plus ammonium ions, NH_4^+ and hydroxide ions, OH^-.

$$NH_3(aq) + H_2O(l) \rightleftharpoons NH_4^+(aq) + OH^-(aq)$$

As ammonia is only partially dissociated, its 0.1 mol dm^{-3} solution is less alkaline (pH = 11) than a solution of sodium hydroxide of the same concentration (pH = 13).

Figure 7.19

Citrus fruits, such as oranges, lemons and limes contain citric acid. This is a relatively weak acid.

Quick Questions

21 Use the words 'concentrated' or 'dilute' and 'strong' or 'weak' to describe:
   a white wine vinegar containing 5% ethanoic acid,
   b drain unblocker liquid containing 95% sulfuric acid.
22 **a** Write equations for the following reactions:
      i sodium oxide with dilute hydrochloric acid,
      ii sodium sulfide with dilute hydrochloric acid.
   b Compare the equations in part a. Is reaction ii an acid–base reaction? Explain your answer.

7.15 The Brønsted–Lowry theory of acids and bases

Until the early part of the 20th century, chemists defined acids as substances which dissociate in water to produce hydrogen ions, H^+ and bases as substances which react with H^+ ions to form water.

For example, Acid:

$$HCl(aq) \longrightarrow H^+(aq) + Cl^-(aq)$$

Base:

$$Cu(OH)_2(s) + 2H^+(aq) \longrightarrow Cu^{2+}(aq) + 2H_2O(l)$$

Soluble bases, like sodium hydroxide, were called alkalis and these were regarded as substances which dissolve in water producing OH^- ions:

$$NaOH(s) + aq \longrightarrow Na^+(aq) + OH^-(aq)$$

During the early 20th century, it became clear that these definitions were too limited. They only applied to aqueous solutions and bases were restricted to substances which reacted with H^+ ions to form water.

In order to widen the scope of acid–base reactions and include non-aqueous systems, the Danish scientist, Johannes Brønsted (1879–1947) and the Englishman, Thomas Lowry (1874–1936) independently suggested broader definitions for acids and bases. According to their definitions, **acids** are proton (H^+ ion) donors and **bases** are proton (H^+ ion) acceptors.

Thus, the relationship between an acid and its corresponding base is:

$$\begin{array}{ccccc} HB & \rightleftharpoons & H^+ & + & B^- \\ \text{acid} & & \text{proton} & & \text{base} \\ \text{(proton donor)} & & & & \text{(proton acceptor)} \end{array}$$

HB and B^- are said to be *conjugate* and to form a *conjugate acid–base pair*. HB is the conjugate acid of B^- and B^- is the conjugate base of HB.

According to the Brønsted–Lowry theory, acid salts (such as $NaHSO_4$) and ammonium ions are recognised as acids. Bases now include all anions, water and ammonia as well as oxide and hydroxide ions. These points are illustrated in the following equations.

Acid		**Base**		**Base**		**Acid**
HSO_4^-	$+$	OH^-	$\rightleftharpoons$	SO_4^{2-}	$+$	H_2O
NH_4^+	$+$	OH^-	$\rightleftharpoons$	NH_3	$+$	H_2O
$2H_3O^+$	$+$	S^{2-}	$\rightleftharpoons$	$2H_2O$	$+$	H_2S
H_3O^+	$+$	NH_3	$\rightleftharpoons$	H_2O	$+$	NH_4^+

Notice in the equations above that *water can act as a base and as an acid*. What is more, it does this simultaneously during its dissociation:

$$\begin{array}{cccc} H_2O + H_2O & \rightleftharpoons & OH^- + H_3O^+ \\ \text{acid} \quad \text{base} & & \text{base} \quad \text{acid} \end{array}$$

Relative strengths of acids and bases

Using the definition of acids as proton donors and bases as proton acceptors, we can arrange all acids and bases in a 'league table'. This shows the order of their relative strengths (Table 7.6).

In the **Brønsted–Lowry theory**, an **acid** is a proton (H^+ ion) donor, a **base** is a proton (H^+ ion) acceptor.

Monobasic acids donate one proton (H^+ ion) per molecule, e.g. nitric acid, HNO_3 and ethanoic acid, CH_3COOH. Monobasic acids are sometimes called monoprotic acids.

Dibasic acids donate two protons (H^+ ions) per molecule, e.g. sulfuric acid, H_2SO_4 and carbonic acid, H_2CO_3. Dibasic acids are sometimes called diprotic acids

Figure 7.20

A life-saving acid–base reaction. During the Apollo 13 space project, the astronauts discovered that carbon dioxide was building up in the spacecraft as they travelled back towards the Earth. By an ingenious use of lithium hydroxide, they were able to repair the air conditioning equipment. This photo shows test pilot Scott Macleod holding one of the lithium hydroxide containers. Lithium hydroxide is a base which neutralised the carbon dioxide, which is an acidic oxide.

Key Point

Acid–base reactions involve competition between bases for protons (H^+ ions).

 Table 7.6
The relative strengths of various acids and bases

Acid strength increases →

Base strength decreases →

Name of acid	Acid $\rightleftharpoons$ H^+ + Base		Name of base
ethanol	C_2H_5OH $\rightleftharpoons$	$H^+ + C_2H_5O^-$	ethoxide
water	H_2O $\rightleftharpoons$	$H^+ + OH^-$	hydroxide
ammonium	NH_4^+ $\rightleftharpoons$	$H^+ + NH_3$	ammonia
hydrogen sulfide	H_2S $\rightleftharpoons$	$2H^+ + S^{2-}$	sulfide
ethanoic acid	CH_3COOH $\rightleftharpoons$	$H^+ + CH_3COO^-$	ethanoate
sulfurous acid	H_2SO_3 $\rightleftharpoons$	$2H^+ + SO_3^{2-}$	sulfite
oxonium	H_3O^+ $\rightleftharpoons$	$H^+ + H_2O$	water
sulfuric acid	H_2SO_4 $\rightleftharpoons$	$2H^+ + SO_4^{2-}$	sulfate
hydrochloric acid	HCl $\rightleftharpoons$	$H^+ + Cl^-$	chloride

If an acid is weak (i.e. has little tendency to donate protons), it follows that its conjugate base is strong. The base will have a strong affinity for protons. For example, hydrogen sulfide is a weak acid, but the sulfide ion is a strong base. Thus, in Table 7.6 the acids increase in strength down the page, but their conjugate bases gradually become weaker. The dissociation of acids is discussed further in Sections 21.6–21.12.

7.16 Acid–base reactions: competition for protons

When dilute hydrochloric acid is added to a solution of sodium sulfide, hydrogen sulfide is produced:

$$S^{2-}(aq) + \underbrace{2H^+(aq) + 2Cl^-(aq)}_{\text{hydrochloric acid}} \longrightarrow H_2S(g) + 2Cl^-(aq)$$

This reaction shows that hydrochloric acid is a stronger acid than hydrogen sulfide. It donates protons (H^+ ions) to sulfide ions in forming H_2S. Alternatively, we could say that sulfide is a stronger base than chloride. In competition for protons, sulfide wins convincingly.

A similar reaction occurs when dilute hydrochloric acid is added to carbonates. In this case, the $HCl(aq)$ donates H^+ ions to carbonate ions forming carbonic acid (H_2CO_3), because hydrochloric acid is a stronger acid than carbonic acid. The carbonic acid is unstable and decomposes to form carbon dioxide and water:

$$2H^+(aq) + 2Cl^-(aq) + CO_3^{2-}(s) \longrightarrow 2Cl^-(aq) + H_2CO_3(aq)$$

$$\text{Then } H_2CO_3(aq) \longrightarrow H_2O(l) + CO_2(g)$$

Thus, *acid–base reactions involve competition between bases for protons*. In this respect, they can be compared to redox reactions which involve competition between oxidising agents for electrons (Section 20.1).

When dilute hydrochloric acid or sulfurous acid is added to a solution of sodium benzoate, a white precipitate of benzoic acid appears:

$$\underset{\text{benzoate ion}}{C_6H_5COO^-(aq)} + \underset{\text{hydrochloric acid}}{H^+(aq) + Cl^-(aq)} \longrightarrow \underset{\text{benzoic acid}}{C_6H_5COOH(s)} + Cl^-(aq)$$

$$\underset{\text{benzoate ion}}{C_6H_5COO^-(aq)} + \underset{\text{sulfurous acid}}{H_2SO_3(aq)} \longrightarrow \underset{\text{benzoic acid}}{C_6H_5COOH(s)} + HSO_3^-(aq)$$

Hence, both hydrochloric acid and sulfurous acid are stronger than benzoic acid. However, when ethanoic acid is added to a solution of sodium benzoate, no apparent reaction occurs. Ethanoic acid does not protonate benzoate ions because ethanoic acid is a weaker acid than benzoic acid. Thus, we could place benzoic acid between ethanoic acid and sulfurous acid in Table 7.6 showing the relative strengths of acids.

Quick Questions

23 a What is the conjugate acid of:
       i CH_3COO^- ii NH_3 iii HCO_3^-?
    b What is the conjugate base of:
       i H_3O^+ ii HSO_4^- iii NH_3?

24 When universal indicator solution is added to separate portions of
    i ammonium chloride solution, **ii** dilute phenol (C_6H_5OH) in water and
    iii water, solutions **i** and **ii** produce an orange colour, but **iii** is yellow green.
    a What can you conclude from this about the relative acidities of the liquids, **i**, **ii** and **iii**?
    b When magnesium turnings are added to the three liquids above, **i** evolves hydrogen most vigorously and **iii** least vigorously. What can you conclude from this about the relative acidities of **i**, **ii** and **iii**?
    c Where would you place phenol in Table 7.6?

Review questions

1 0.6 g of a solute, X, was shaken with an immiscible mixture of 20 cm^3 trichloromethane and 100 cm^3 water at 298 K.

   At equilibrium, analysis showed that the trichloromethane contained 0.5 g of X and the water contained 0.1 g of X.

   a Calculate the partition coefficient of X between trichloromethane and water at 298 K.

   b Partition coefficients have no units. Why is this?

   c Why should the temperature always be quoted with a measured partition coefficient?

2 The Mogul Oil Company is worried about the impurity, M, in its four-star petrol. 1 dm^3 of petrol contains 5 g of M. In an effort to reduce the concentration of M in the petrol, Mogul have discovered the secret solvent, S. The partition coefficient of M between petrol and S is 0.01 at 298 K.

   a What is meant by the term 'partition coefficient'?

   b Explain the principles of solvent extraction.

   c Calculate the total mass of M removed from 1 dm^3 of petrol by shaking it with 100 cm^3 of solvent, S at 298 K.

3 5 mol of ethanol, 6 mol of ethanoic acid, 6 mol of ethyl ethanoate and 4 mol of water were mixed together in a stoppered bottle at 15 °C. After equilibrium had been attained the bottle was found to contain only 4 mol of ethanoic acid.

   a Write an equation for the reaction between ethanol and ethanoic acid to form ethyl ethanoate and water.

   b Write an expression for the equilibrium constant, K_c, for this reaction at 15 °C.

   c How many moles of ethanol, ethyl ethanoate and water are present in the equilibrium mixture?

   d What is the value of K_c for this reaction?

   e Suppose 1 mol of ethanol, 1 mol of ethanoic acid, 3 mol of ethyl ethanoate and 3 mol of water are mixed together in a stoppered flask at 15 °C. How many moles of:

      i ethanol,   ii ethyl ethanoate,

      are present at equilibrium?

4 At a certain temperature and a total pressure of 10^5 Pa, iodine vapour contains 40% by volume of I atoms:
   $$I_2(g) \rightleftharpoons 2I(g)$$
   a Calculate K_p for the equilibrium.

   b At what total pressure (without temperature change) would the percentage of I atoms be reduced to 20%?

5 a Write expressions for both K_p and K_c for the gaseous equilibria:

      i $2NO(g) + O_2(g) \rightleftharpoons 2NO_2(g)$

      ii $NO(g) + \frac{1}{2}O_2(g) \rightleftharpoons NO_2(g)$

   b What are the units of K_p and K_c for the two equilibria in part a?

6 Consider the following reaction:

$$H_2(g) + I_2(g) \rightleftharpoons 2HI(g)$$

a Write an expression for the equilibrium constant in terms of partial pressures.

At a certain temperature, analysis of an equilibrium mixture of the gases yielded the following results:

$p_{H_2} = 2.5 \times 10^4\,Pa$, $\quad p_{I_2} = 1.6 \times 10^4\,Pa$,

$p_{HI} = 4.0 \times 10^4\,Pa$.

b Calculate the equilibrium constant for the reaction. What are its units?

c In a second experiment at the same temperature, pure hydrogen iodide was injected into the flask at a pressure of $6 \times 10^4\,Pa$. What are the partial pressures of hydrogen, iodine and hydrogen iodide at equilibrium?

7 At 25 °C, the value of K_c for the following system is 10^{10}:

$$Sn^{2+}(aq) + 2Fe^{3+}(aq) \rightleftharpoons Sn^{4+}(aq) + 2Fe^{2+}(aq)$$

a Write an expression for K_c for this reaction.

b Explain why K_c has no units.

c What is the value of K_c for:

i $Sn^{4+}(aq) + 2Fe^{2+}(aq) \rightleftharpoons Sn^{2+}(aq) + 2Fe^{3+}(aq)$

ii $2Sn^{2+}(aq) + 4Fe^{3+}(aq) \rightleftharpoons 2Sn^{4+}(aq) + 4Fe^{2+}(aq)$?

8 At 200 °C, K_c for the reaction:

$$PCl_5(g) \rightleftharpoons PCl_3(g) + Cl_2(g) \qquad \Delta H^\ominus = +124\,kJ\,mol^{-1}$$

has a numerical value of 8×10^{-3}.

a Write an expression for K_c for the reaction.

b What are the units of K_c?

c What is the value of K_c for the reverse reaction at 200 °C and what are its units?

d How will the amounts of PCl_5, PCl_3 and Cl_2 in the equilibrium mixture change (increase, decrease or stay the same) if: **i** more PCl_5 is added, **ii** the pressure is increased, **iii** the temperature is increased?

e What would be the effect on K_c if: **i** more PCl_5 is added, **ii** the pressure is increased, **iii** the temperature is increased?

f A sample of pure PCl_5 was introduced into an evacuated vessel at 200 °C. When equilibrium was reached, the concentration of PCl_5 was $0.5 \times 10^{-1}\,mol\,dm^{-3}$. What are the concentrations of PCl_3 and Cl_2 at equilibrium?

9 The densities of diamond and graphite are 3.5 and $2.3\,g\,cm^{-3}$ respectively. The change from graphite to diamond is represented by the equation:

$$C(graphite) \rightleftharpoons C(diamond) \qquad \Delta H = +2\,kJ\,mol^{-1}$$

Is the formation of diamond from graphite favoured by:

a high or low temperature,

b high or low pressure?

Explain your answers.

Rates of reaction

8.1 Introduction

The rates of chemical reactions are just as important to you as they are to industrialists and chemical engineers. At home you might be interested in the rate at which you can boil an egg or bake a cake. Out-of-doors you might be interested in the rate at which your bike is rusting, the rate at which your lettuces are growing or possibly the rate at which the stonework of buildings is being weathered by acidic gases in the atmosphere.

In the workplace, engineers and other workers are concerned with the rates of chemical reactions in industrial, engineering and farming processes. These might include the rate at which ammonia can be obtained from nitrogen and hydrogen, the rate at which concrete sets or the rate of growth in fruit and vegetable crops.

Industrialists and chemical engineers are not satisfied with merely turning one substance into another. In most cases, they want to obtain products rapidly, easily and as cheaply as possible. Time and money are important in industry. It is often necessary to accelerate reactions using catalysts so that they are economically worthwhile (Section 8.3).

Reaction rates are also important in archaeology. Archaeologists can estimate the age of rocks, fossils and other prehistoric remains by a process known as radioactive dating. This entails measuring the concentration of a decaying radioactive isotope such as $^{14}_{6}C$ in the sample being tested.

Quick Questions

1 a Why does a pressure cooker enable vegetables to be cooked more rapidly?
   b What conditions and processes are used to *slow down* the rate at which perishable foods deteriorate?
   c How do gardeners accelerate the growth of their crops?

8.2 The concept of reaction rate

During a chemical reaction, reactants are being converted to products. The reaction rate (rate of a reaction) tells us how fast the reaction is taking place by indicating how much of a reactant is consumed or how much of a product forms in a given time. Hence,

$$\textbf{Reaction rate} = \frac{\text{change in amount (or concentration) of a substance}}{\text{time taken}}$$

Worked example

When acidified hydrogen peroxide is added to a solution of potassium iodide, iodine is formed.

$$H_2O_2(aq) + 2I^-(aq) + 2H^+(aq) \longrightarrow 2H_2O(l) + I_2(aq)$$

The concentration of iodine rises from 0 to 10^{-5} mol dm^{-3} in 10 seconds.

$$\therefore \quad \text{Reaction rate} = \frac{\text{change in concentration of iodine}}{\text{time taken}}$$

▲ Figure 8.1
This huge limestone lion outside Leeds Town Hall, England, has been slowly weathered by acidic gases in the atmosphere.

▲ Figure 8.2
Finding the optimum temperature, pressure and catalyst for an industrial process is crucial. It could mean the difference between success and failure on the commercial market. This worker is checking the temperature in a cheese-making process.

Figure 8.3
In a pressure cooker, water boils at about 120 °C, instead of the normal 100 °C. This reduces the time needed to cook food.

Figure 8.4
This bag is 100% biodegradable and compostable. The rate at which it degrades is important

Using the symbol Δ to represent the change in a particular quantity, we can write

$$\text{Reaction rate} = \frac{\Delta[I_2]}{\Delta t} = \frac{10^{-5}\,\text{mol dm}^{-3}}{10\,\text{s}}$$

$$= 10^{-6}\,\text{mol dm}^{-3}\,\text{s}^{-1}$$

Strictly speaking, this result gives the *average* reaction rate during the 10 seconds that it took for the concentration of iodine to become $10^{-5}\,\text{mol dm}^{-3}$. By measuring the change in concentration (or amount) over shorter and shorter time intervals we obtain an increasingly accurate estimate of the reaction rate at any moment. This is known as the 'clock' technique.

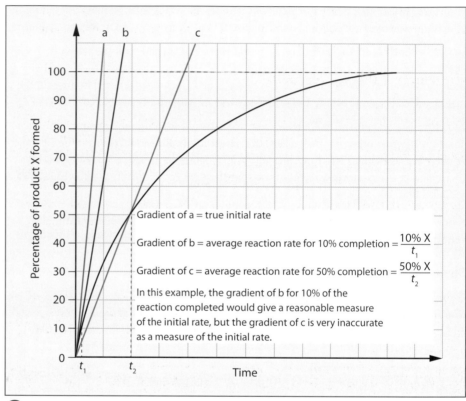

Figure 8.5
Errors in the 'clock technique' for measuring reaction rates

The disadvantages of 'clock' techniques are illustrated more effectively in Figure 8.5. Using 'clock' techniques, the rate is obtained as the inverse of the time for a certain proportion of the reaction to occur. Provided the reaction has gone only a little way towards completion, very little error is introduced as shown by gradient b in Figure 8.5. However, serious errors result if the 'end point' is, say, half-way to completion like that of gradient c in Figure 8.5.

Ideally we should make the time interval almost zero, so Δt tends towards 0 and then we obtain what is effectively the reaction rate at a particular instant, i.e.

$$\frac{\Delta[I_2]}{\Delta t} = \frac{d[I_2]}{dt}$$

In practice, it is usual to plot a graph of the amount or concentration of a particular substance against time. The reaction rate can then be obtained at particular times by drawing tangents to the resulting curve. This technique is illustrated in Section 22.2.

Normally it is convenient to express reaction rates in mol s^{-1} or in $\text{mol dm}^{-3}\,\text{s}^{-1}$, but occasionally it is more convenient to use minutes or even hours as the unit of time.

8.3 Factors affecting the rate of a reaction

Our studies have already indicated several factors which can influence the rate of a reaction.

The availability of reactants and their surface area

Anyone who has built a fire knows that it is easier to start a fire using sticks rather than logs. Similarly, magnesium powder will react much more rapidly than magnesium ribbon with dilute sulfuric acid. In general, the smaller the size of reacting particles, the greater is the total surface area exposed for reaction and, consequently, the faster the reaction. In the case of heterogeneous systems, in which the reactants are in different states, the area of contact between the reacting substances will influence the reaction rate considerably. In homogeneous systems, such as gases and solutions which mix completely, the idea of surface area becomes meaningless.

The concentration of reactants

Increasing the concentration of a reactant normally causes an increase in the rate of a reaction, but this is not always the case. Furthermore, the different reactants can affect the rate of a particular reaction in different ways. For example, when nitrogen oxide reacts with oxygen

$$2NO(g) + O_2(g) \longrightarrow 2NO_2(g)$$

the reaction rate doubles when the oxygen concentration doubles. But doubling the concentration of NO *quadruples* the rate of reaction. The effect of concentration on reaction rates is considered further in Sections 22.1 and 22.3.

The temperature of the reactants

Perishable foods like milk 'go bad' much more rapidly in summer than in winter. In summer, the chemical reactions in the deterioration processes occur more rapidly at the higher temperatures. In general, increasing the temperature increases the rate of chemical reactions.

▲ Figure 8.6
A researcher lifts a specimen of frozen animal tissue from its storage in liquid nitrogen.

Catalysts

Catalysts are substances which alter the rate of chemical reactions without undergoing any overall chemical change themselves. Although catalysts are not used up in reactions, they participate by forming intermediate compounds in the conversion of the reactants to products.

Normally, catalysts are used to accelerate reactions. Certain catalysts can, however, slow reactions down. For example, propane-1,2,3-triol (glycerine) is sometimes added to hydrogen peroxide in order to slow down its rate of decomposition. Catalysts like this are usually called *negative* catalysts.

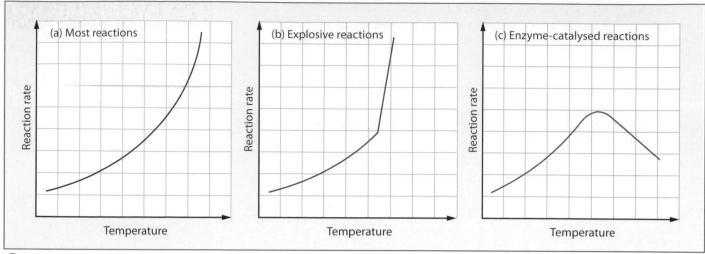

Figure 8.7
Graphs of reaction rate against temperature

Definitions

Catalysis involves substances (catalysts) which alter the rates of chemical reactions without undergoing any overall chemical change themselves.

Enzymes are the catalysts for biological processes. They are composed of proteins.

Catalysts play an important part in industrial processes by enabling reactions to take place which would never occur in their absence. Many industrial processes, including the manufacture of ammonia, sulfuric acid, nitric acid, ethene and polythene rely heavily on the use of catalysts. Biological reactions also rely on catalysts because almost every chemical reaction in animals, plants and micro-organisms requires its own catalyst. Catalysts are involved in simple reactions like the hydrolysis of starch to sugars. They are also involved in highly complex reactions like the replication of DNA which forms the genes in the nucleus of every cell. These biological catalysts, called **enzymes**, are composed of proteins. Some enzymes are so specific that they can only catalyse the reaction of one particular substance. Other enzymes are less specific.

Light

Photosynthesis and photography both involve light-sensitive reactions. The leaves of plants contain a green pigment called chlorophyll. This can absorb sunlight and use this energy to synthesise chemicals and provide food for the plant.

During photosynthesis, plants transform carbon dioxide and water into oxygen and sugars such as glucose. The rate at which this occurs depends on the intensity of the light.

$$6CO_2(g) + 6H_2O(l) \xrightarrow{\text{light}} C_6H_{12}O_6(aq) + 6O_2(g)$$

White silver chloride turns purple and finally dark grey due to a thin layer of silver metal when it is exposed to sunlight. The rate at which this change occurs depends on the intensity of the light.

$$AgCl(s) \xrightarrow{\text{light}} Ag(s) + \tfrac{1}{2}Cl_2(g)$$

The use of silver salts in photography depends on photosensitivity of this kind.

The reactions of halogens with hydrogen (Section 11.5) and with alkanes (Section 14.7) are further examples of photochemical reactions.

8.4 Investigating the effect of concentration changes on the rates of reactions

In a mixture of liquids or gases, the particles (molecules and/or ions) of different substances will be moving about, bumping and colliding with each other. Reactions will only take place if the correct particles collide and this is known as the **collision theory**.

During a reaction, bonds in the reactants must first break so that the atoms can re-arrange themselves and form new bonds in the products. But the right particles must collide and they must collide with enough energy to break bonds in the reactants. This means that:

- some collisions do not result in a reaction and
- reactions can only occur when collisions take place between the appropriate particles with sufficient energy.

In the following experiment we can investigate how the rate of reaction between hydrogen peroxide, hydrogen ions and iodide ions is affected by changes in concentration. We can then explain the results in terms of collisions.

When aqueous hydrogen peroxide is added to a mixture of potassium iodide solution and dilute sulfuric acid, there is a reaction between hydrogen peroxide, hydrogen ions and iodide ions. The products are water and iodine:

$$H_2O_2(aq) + 2H^+(aq) + 2I^-(aq) \longrightarrow 2H_2O(l) + I_2(aq)$$

The rate of reaction can be obtained by measuring the rate of formation of iodine at the start of the reaction. Table 8.1 shows the rate of formation of iodine at the start of seven separate experiments.

Experiment number	Reactant concentrations			Rate of formation of I_2 /mol dm^{-3} s^{-1}
	$[H_2O_2]$/mol dm^{-3}	$[I^-]$/mol dm^{-3}	$[H^+]$/mol dm^{-3}	
1	0.01	0.01	0.10	2×10^{-6}
2	0.02	0.01	0.10	4×10^{-6}
3	0.03	0.01	0.10	6×10^{-6}
4	0.01	0.02	0.10	4×10^{-6}
5	0.01	0.03	0.10	6×10^{-6}
6	0.01	0.01	0.20	2×10^{-6}
7	0.01	0.01	0.30	2×10^{-6}

Table 8.1
The rate of formation of iodine in the reaction between H_2O_2, I^- and H^+

Key Point

The **collision theory** says that reactions can only occur when collisions take place between the appropriate particles with sufficient energy.

Look carefully at experiments 1, 2 and 3 in which the only changes are in the concentration of hydrogen peroxide. As the concentration of hydrogen peroxide, $[H_2O_2]$, doubles and then trebles, the rate of formation of I_2 also doubles and trebles.

When the concentration of hydrogen peroxide is doubled, we can assume that twice as many collisions involving H_2O_2 molecules take place and this results in twice the rate of reaction. Similarly, if the concentration of hydrogen peroxide is trebled, there are three times as many collisions per second involving H_2O_2 molecules and this causes the reaction rate to increase by three times.

Now look at experiments 1, 4 and 5. In these three experiments, the only changes are in the concentration of iodide ions, $[I^-]$. And, as the concentration doubles and then trebles, the rate of formation of I_2 also doubles and trebles.

As before, doubling the concentration of iodide doubles the number of collisions per second involving I^- and this has doubled the reaction rate.

Usually, increasing the concentration of a reactant will increase the reaction rate and this can be explained in terms of an increasing collision rate. However, increasing the concentration of a reactant does not always increase the rate of reaction as you will appreciate by looking at the results of experiments 1, 6 and 7. In this case, doubling and then trebling the concentration of hydrogen ions, $[H^+]$, has no effect on the reaction rate. This is very unusual and we will return to see how the collision theory can explain this result when we look at reaction mechanisms in Section 22.9.

Quick Questions

4 Suppose gas A reacts directly with gas B to form gas C:
$$A(g) + B(g) \longrightarrow C(g)$$
   a How will the collision rate change if $[A(g)]$ doubles?
   b How will the collision rate change if $[B(g)]$ doubles?
   c How would you expect the reaction rate to change if $[B(g)]$ doubles, but $[A(g)]$ stays the same?
   d How would you expect the reaction rate to change if the overall pressure of a mixture of A(g) and B(g) is doubled?

5 Marble chips (calcium carbonate) react with dilute hydrochloric acid to produce carbon dioxide. Suppose the large marble chip in Figure 8.8 is a cube, 2 cm × 2 cm × 2 cm.

  a What is its surface area?

  b Suppose this cube is cut along the dashed lines in Figure 8.8 to make cubes 1 cm × 1 cm × 1 cm. How many smaller cubes will be produced?

  c What is the total surface area of these small cubes?

  d How would you expect the reaction rate with one large cube (2 cm × 2 cm × 2 cm) to compare with that using all of the smaller cubes?

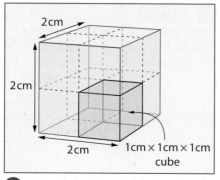

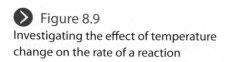
Figure 8.8

8.5 Explaining the effect of changes in concentration, pressure and surface area on the rates of reactions

The collision theory enables us to explain the effect of changes in concentration, pressure and surface area on the rates of reactions.

In any reaction mixture involving gases or aqueous solutions particles are continuously moving and colliding. When they collide, there is always a chance that a reaction will occur. Increasing the concentration of aqueous substances or raising the pressure of gases means that there are more particles in a given volume. This results in more collisions per second and therefore an increase in the reaction rate.

When a reaction takes place between a solid and either a gas or a liquid, the rate of reaction increases when the solid is broken into smaller pieces.

Breaking up the solid increases its surface area in contact with the reactant gas or liquid. This results in more collisions per second between reactant particles and therefore a faster reaction.

8.6 Investigating the effect of temperature change on the rates of reactions

Chemical reactions can be speeded up by raising the temperature and slowed down by lowering the temperature. Most of us have appliances in our homes for speeding up and slowing down chemical reactions. We have a cooker to speed up reactions when we cook food and a fridge to slow down reactions and prevent food from deteriorating.

We can investigate the effect of temperature change on reactions rates using the reaction between sodium thiosulfate solution ($Na_2S_2O_3(aq)$) and dilute hydrochloric acid:

$$Na_2S_2O_3(aq) + 2HCl(aq) \longrightarrow 2NaCl(aq) + H_2O(l) + SO_2(g) + S(s)$$

When the reactants are mixed, a cloudy precipitate of sulfur starts to form, eventually becoming thick and yellow (Figure 8.9).

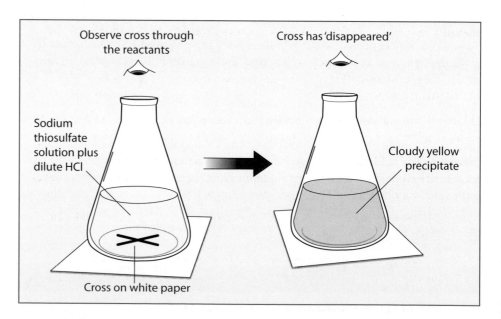

Figure 8.9
Investigating the effect of temperature change on the rate of a reaction

As the precipitate forms and becomes thicker, the ink cross becomes less distinct. We can study the rate of the reaction by looking down through the solution and measuring the time it takes for the cross to disappear.

Table 8.2 shows the results obtained when 10 cm³ of 1.0 mol dm⁻³ dilute HCl were mixed with 50 cm³ of 0.05 mol dm⁻³ Na₂S₂O₃(aq) at different temperatures.

Notice from the results in Table 8.2 that:

▶ the cross disappears more quickly as the temperature rises. This means that the reaction goes faster at higher temperature,
▶ if the temperature rises by 10 °C, the time taken for the reaction is roughly halved. For example, at 298 K, the cross disappears in 120 seconds, whilst at 308 K it disappears in 61 seconds (i.e. about half the time at 298 K).

Using the equation for reaction rate in Section 8.2, we can say:

$$\text{Reaction rate} = \frac{\text{change in amount of sulfur}}{\text{time taken}}$$
$$= \frac{\text{amount of sulfur precipitated}}{\text{time taken}}$$

However, the cross disappears at the same thickness of precipitate each time. So, the amount of sulfur precipitated is the same at each temperature.

$$\therefore \text{ Reaction rate} \propto \frac{1}{\text{time for cross to disappear}}$$

Figure 8.10 shows a graph of this reciprocal against temperature. The graph shows clearly that the reaction rate increases rapidly as temperature increases.

Quick Questions

6 It takes about 10 minutes to fry chips, but about 20 minutes to boil potatoes. Larger potatoes take even longer to boil.
   a Why do larger potatoes take longer to cook than small ones?
   b Give two reasons why chips can be cooked faster than boiled potatoes.
   c Why can boiled potatoes be cooked faster in a pressure cooker?

8.7 Explaining the effect of temperature change on the rates of reactions

For many reactions, a 10 K increase in temperature roughly *doubles* the rate of reaction, so we are led to ask 'How can such a small increase in temperature (10 K in about 300 K) lead to such a large increase in reaction rate?'

Collision theory

At higher temperatures, particles move around faster. So, they collide more often, there are more collisions per second and this causes the rate of reaction to increase.

This increase in collision frequency accounts for an increase in reaction rate as temperature rises, but does it also explain how rapidly the rate increases? The rates of many reactions double for a temperature rise of only 10 K.

From the kinetic theory, we can predict the relative increase in the rate of collisions when the temperature rises by 10 K.

The kinetic energy of a particle is proportional to its absolute temperature:

$$\tfrac{1}{2}mv^2 \propto T$$

but the mass of a given particle remains constant, so $v^2 \propto T$

$$\therefore \quad \frac{v_1^2}{v_2^2} = \frac{T_1}{T_2} \qquad \text{(equation 1)}$$

where v_1 is the velocity at temperature T_1, and v_2 is the velocity at temperature T_2.

 Table 8.2
Investigating the effect of temperature change on the rate of reaction between HCl(aq) and Na₂S₂O₃(aq)

Temperature / K	Time for cross to disappear / s
298	120
302	89
308	61
312	44
318	32

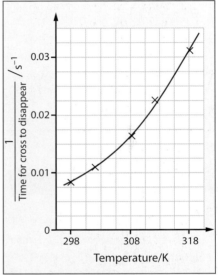

 Figure 8.10
A graph of the reciprocal of the time for the cross to disappear ($\propto$ reaction rate) against temperature

Key Point

The rate of a reaction can be affected by:

▶ the surface area of solids
▶ concentration (partial pressure for gases)
▶ temperature
▶ catalysts
▶ light.

Now, suppose the average speed of a particle is v at 300 K. What will its average speed be at 310 K, i.e. 10 K higher?

Substituting in equation 1,

$$\frac{v_1^2}{v^2} = \frac{310}{300}$$

$$\therefore \quad v_1 = \sqrt{\frac{310}{300}} v = \sqrt{1.033}\, v = 1.016 v$$

$\therefore$ The average speed at 310 K is only 1.016 times greater than that at 300 K, i.e. it has increased by only 1.6%.

Assuming that the frequency of collisions depends on the average speed of the particles, we might expect the rate of collisions and hence the reaction rate to be 1.6% greater at 310 K than at 300 K. In practice, the reaction rate roughly doubles between 300 and 310 K, i.e. it increases by approximately 100%.

Clearly, the *simple* collision theory cannot account fully for the increase in reaction rate as temperature rises. How then do we explain the relatively large increase in reaction rate with temperature?

Activation energy

During a chemical reaction, bonds are first broken and others are then formed. Consequently, energy is required to break bonds and start this process, whether the overall reaction is exothermic or endothermic. Therefore, it is reasonable to assume that *particles do not always react when they collide*. They may not have sufficient energy for the necessary bonds to be broken. In 'soft' collisions the particles simply bounce off each other.

So a reaction will only occur if the colliding particles possess a certain minimum amount of energy. This minimum energy for a reaction to occur between colliding particles is known as the **activation energy**, E_A. The activation energy enables chemical bonds to stretch and break, so that rearrangements of atoms, ions and electrons can then occur as the reaction proceeds. Activation energies explain the fact that reactions go much more slowly than we would expect if every collision resulted in a reaction. In fact, activation energies are closely related to the rates of reactions. If the activation energy is large, only a small proportion of molecules have sufficient energy to react on collision, so the reaction is slow. If, however, the activation energy is very small, most molecules have sufficient energy to react and the reaction is fast.

Figure 8.11 shows the relationship between the activation energy, E_A, and the enthalpy change of reaction, ΔH, as the reaction proceeds for both an exothermic and an endothermic reaction. Diagrams like those in Figure 8.11 are known as **reaction pathways** or **reaction profiles**.

Definition

The **activation energy** is the minimum energy needed for a reaction to occur between the molar amounts shown in the equation for the reaction.

 Figure 8.11

A reaction pathway (reaction profile) for:
(a) an exothermic reaction and
(b) an endothermic reaction

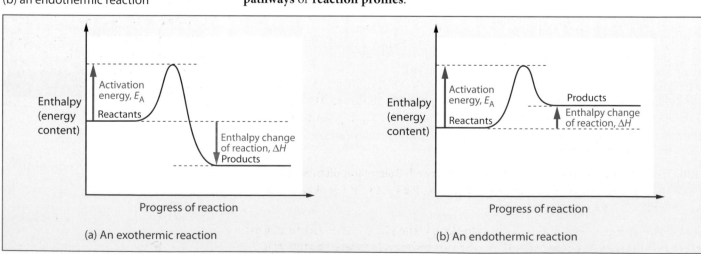

(a) An exothermic reaction

(b) An endothermic reaction

E_A is related closely to reaction rate. It gives an indication of *how fast* the reaction will occur. Unlike E_A, ΔH is related to the equilibrium position of a reaction rather than to the reaction rate. ΔH gives an indication of *how far* a reaction goes towards completion.

The fact that a certain minimum energy is needed to initiate most reactions is well-illustrated by fuels and explosives. These usually require a small input of energy to get started even though they are highly exothermic. Once the reaction has started, it supplies the energy needed to keep itself going.

The idea of activation energy now leads to the next important question. What fraction of the particles possess the activation energy – the minimum energy required for reaction?

Note

A **reaction pathway** (**reaction profile**) is a graph showing how the total enthalpy (energy content) of the substances change during the progress of a reaction from initial reactants to final products.

The Maxwell–Boltzmann distribution

In the mid-nineteenth century, an Englishman, James Clerk Maxwell (1831–1879) and an Austrian, Ludwig Boltzmann (1844–1906) independently worked out the way in which energy was distributed among the molecules of a gas. Using the kinetic theory of gases (Section 4.2) and probability theory, Maxwell and Boltzmann derived equations for the distribution of kinetic energies amongst the molecules of a gas at a particular temperature. From their work, it was possible to draw graphs showing the number of molecules having a particular kinetic energy against the size of the energy as in Figure 8.12.

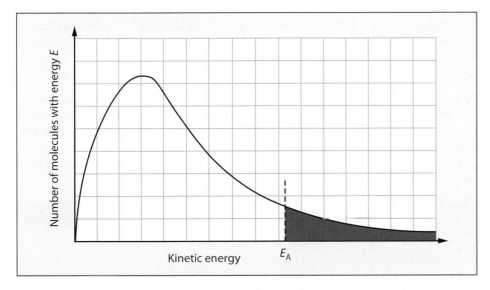

◀ Figure 8.12
Distribution of the kinetic energies for the molecules in a gas at a particular temperature

The spread of kinetic energies for the molecules in the graph is usually described as the **Maxwell–Boltzmann distribution** in honour of the two scientists.

Essentially, the graph is a histogram. It shows the number of molecules in each small range of kinetic energy. So, the area beneath the curve is proportional to the total number of molecules involved.

If E_A is the activation energy, the number of particles with sufficient energy to react is proportional to the red area beneath the curve at energies of E_A and above. Hence, the *fraction* of molecules which can react on collision (with energy > E_A) is given by the ratio:

$$\frac{\text{Red area beneath curve}}{\text{Total area beneath curve}}$$

Using the concept of activation energy and the Maxwell–Boltzmann distribution, we can now answer our original question concerning the relatively large increase in reaction rate with temperature.

Look closely at Figure 8.13. This shows how the kinetic energies of the molecules in a gas might be distributed at T K and $(T + 10)$ K. The area under each curve is proportional

Definition

The **Maxwell–Boltzmann distribution** is the term used to describe the spread of kinetic energies for the molecules in a gas at a particular temperature.

to the total number of molecules. So, as the number of molecules does not change, the areas under the curves must be equal. Consequently, as the temperature rises and kinetic energy increases, the curves must spread to the right and the peak height must fall.

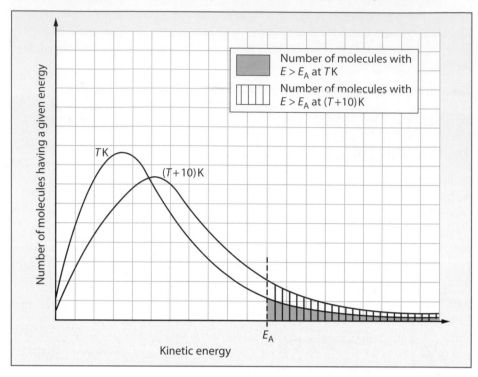

Figure 8.13
Distribution of the kinetic energies of the molecules of a gas at T K and $(T + 10)$ K

Now, let us suppose that colliding molecules must have a kinetic energy of E_A (the activation energy) before a reaction takes place. Notice in Figure 8.13 that only a small number of molecules (indicated by the blue area) have sufficient energy to react at T K. However, when the temperature rises by 10 K to $(T + 10)$ K, the number of molecules with sufficient energy to react (indicated by the red lined area) roughly doubles. This means that the reaction rate doubles as well.

Quick Questions

7 Figure 8.14 shows the volume of hydrogen produced during different reactions between magnesium and hydrochloric acid. Curve X is obtained when 1 g of magnesium ribbon reacts with 100 cm^3 (excess) hydrochloric acid at 30 °C. Which curve would you expect to obtain if:
   a 1 g of Mg ribbon reacts with 100 cm^3 of the same acid at 50 °C,
   b 1 g of Mg ribbon reacts with 100 cm^3 of the same acid at 15 °C,
   c 0.5 g of Mg ribbon reacts with 100 cm^3 of the same acid at 30 °C?

8 Sketch the reaction profile which fits the following data. When compound A is converted to compound C, the reaction proceeds via compound B which can be isolated:

$$A \longrightarrow B \quad \Delta H \text{ is positive.}$$
$$B \longrightarrow C \quad \Delta H \text{ is negative.}$$

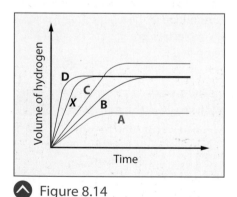

Figure 8.14

8.8 Explaining the effect of catalysts on the rates of reactions

In chemical reactions, existing bonds must break and new bonds must form as reactants are converted to products. In order to break existing bonds, energy is needed and this is often provided in the form of heat. The 'energy barrier' (activation energy) for the uncatalysed reaction between nitrogen and hydrogen to form ammonia, $N_2(g) + 3H_2(g) \longrightarrow 2NH_3(g)$, is shown on the reaction profile in Figure 8.15.

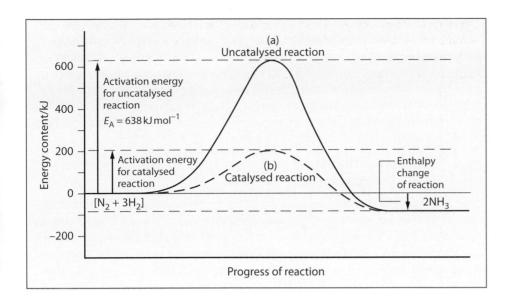

 Figure 8.15
Reaction profiles for the conversion of nitrogen and hydrogen to ammonia (a) uncatalysed and (b) catalysed by tungsten.

Let's follow the progress of the uncatalysed reaction along the black curve from left to right of the reaction profile in Figure 8.15. Molecules of N_2 and H_2 are moving very fast and colliding all the time. As the molecules approach one another, there is little change in their total energy until they get very close to each other. But, when the molecules are within a few nanometres of each other, repulsions between their nuclei and between their negative electron clouds begin to operate. So, the molecules must have sufficient kinetic energy (638 kJ for 1 mol of N_2 and 3 mol of H_2) if they are to overcome these repulsive forces as the molecules get closer.

The peak of the energy profile represents an 'energy barrier'. This has to be surmounted before bonds are stretched and sufficiently broken for products to form.

The height of the 'energy barrier' above the original height of the reactant molecules is the activation energy for the reaction. Only those molecules with enough energy to surmount the energy barrier will be able to react and form products. At the summit of the reaction profile, reactant molecules have a high energy content. They are described as a 'transition state' or an 'activated complex'. This activated complex can either break up and form the product molecules down the right-hand side of the reaction profile or separate into the original reactant molecules and return to the left of the reaction profile.

When nitrogen is mixed with hydrogen, no detectable reaction occurs even at high temperatures and high pressures. As the molecules approach each other, they have insufficient kinetic energy to overcome their mutual repulsions. So they never reach the activated state. They rise part of the way up the left side of the reaction profile, repel one another and separate.

We can, however, speed up the reaction using a catalyst (Section 8.3). Catalysts are neither used up, nor are they chemically changed as a result of their catalysis. This means that a small amount of catalyst is capable of catalysing an infinite amount of reaction.

In industry, the reaction between nitrogen and hydrogen to form ammonia is normally catalysed by iron, although tungsten is more efficient. As Figure 8.15 shows in the dashed red curve, when tungsten is used as the catalyst, the activation energy is reduced to only 200 kJ, less than one-third of that for the uncatalysed reaction. This enables many more molecules to react and the rate of reaction increases substantially. The catalyst works by adsorbing N_2 molecules onto its surface, weakening the $N\equiv N$ bond so it breaks more easily when an H_2 molecule collides with it.

Notice in Figure 8.15 that the energy levels of reactants and products are the same in both the catalysed and uncatalysed reactions.

Figure 8.16
Strong pole-vaulters need lots of kinetic energy in their run-up to convert this into sufficient potential energy and height to clear the cross bar (activation energy barrier).

Figure 8.17
Catalytic converters like this contain platinum or platinum–rhodium alloys as catalysts. In car exhaust systems, catalytic converters can oxidise carbon monoxide to carbon dioxide and reduce oxides of nitrogen (NO and NO_2) to nitrogen.

Key Point

Catalysts usually speed up reactions. They do this by introducing a different reaction path (mechanism) with a lower activation energy than the uncatalysed reaction.

The catalyst has not supplied any extra energy for the reactants, yet the reaction has been speeded up. The catalyst has, in fact, provided a new reaction path for the breaking and rearranging of bonds. This new reaction path has a lower activation energy, so that many more molecules can pass over the energy barrier. The situation in a catalysed reaction can be compared to a pole vaulting event in which the bar has been lowered so that more athletes can get over. Different catalysts work in different ways, but they all provide a new reaction path with lower activation energy.

8.9 Interpreting the action of catalysts in terms of the Maxwell–Boltzmann distribution

The effect of catalysts on reaction rates can be interpreted very neatly in terms of the Maxwell–Boltzmann distribution. A catalyst works by allowing an alternative path for the reaction with a lower activation energy. This does not change the distribution of kinetic energies among the reacting particles in any way. But reducing the activation energy significantly increases the number of particles with sufficient energy to react (Figure 8.18).

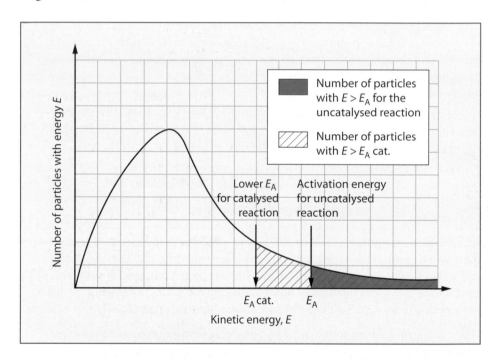

Figure 8.18
Distribution of the kinetic energies of reacting particles and the activation energies for catalysed and uncatalysed reactions. Notice the greater number of particles which have energies greater than the activation energy for the catalysed reaction.

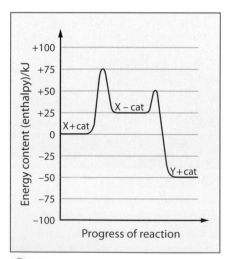

Figure 8.19

Quick Questions

9 In the catalysed conversion of compound X to compound Y, it was found that the reaction proceeded in two stages. In Stage 1, an intermediate, cat–X, was formed involving X and the catalyst (Figure 8.19). In Stage 2, the intermediate broke down forming Y and releasing the catalyst to react with other molecules of X.

$$X + cat \xrightarrow{\text{stage 1}} X\text{–}cat \xrightarrow{\text{stage 2}} Y + cat$$

Use the reaction profile to estimate:
  a the activation energy of Stage 1,
  b the enthalpy change of reaction for Stage 1,
  c the activation energy of Stage 2,
  d the enthalpy change of reaction for Stage 2,
  e the enthalpy change for the overall reaction, $X \longrightarrow Y$,
  f the activation energy for Stage 1 of the reverse reaction.

8.10 The importance of transition metals and their compounds as catalysts

Transition metals and their compounds are important catalysts in industry and in biological systems. Many transition metal ions are required by humans and other living things in minute but definite quantities. Elements required in such small amounts are called trace elements. These trace elements include copper, manganese, iron, chromium and zinc. Some trace elements are essential for the effective catalytic activity of enzymes and these are described as enzyme cofactors. One of the most important enzyme cofactors is the copper(II) ion, Cu^{2+}, for the enzyme cytochrome oxidase. This enzyme catalyses one of the reactions in which energy is obtained from the oxidation of food (Section 5.16). In the absence of copper(II) ions, cytochrome oxidase is completely inhibited and the animal or plant is unable to metabolise food and respire effectively.

Another important enzyme cofactor is the zinc ion, Zn^{2+}, for the enzyme carbonic anhydrase. This enzyme is present in our red blood cells. It catalyses the removal of CO_2 from the blood by converting it to HCO_3^-. Zn^{2+} ions are attached to the enzyme by co-ordinate bonds near the active site. The Zn^{2+} ion with a high charge density assists the breakdown of water molecules into H^+ and OH^- ions. The OH^- ions can then react with CO_2 in the blood forming hydrogencarbonate ions, HCO_3^-.

$$CO_2(aq) + OH^-(aq) \longrightarrow HCO_3^-(aq)$$

A large number of industrial catalysts are either transition metals or their compounds. The most important of these are probably iron in the Haber process to manufacture ammonia, and vanadium(V) oxide in the Contact process to produce sulfur trioxide during the manufacture of sulfuric acid. Table 8.3 lists some of the more important examples of catalysis by transition metals and their compounds.

 Table 8.3
Important examples of catalysis by transition metals and their compounds

Transition element / compound used as catalyst	Reaction / process catalysed
Iron or iron(III) oxide	Haber process to manufacture ammonia: $N_2(g) + 3H_2(g) \longrightarrow 2NH_3(g)$
Vanadium(V) oxide, V_2O_5 or vanadate (VO_3^-)	Contact process to produce sulfur trioxide in the manufacture of sulfuric acid: $2SO_2(g) + O_2(g) \longrightarrow 2SO_3(g)$
Platinum or platinum /rhodium alloys	Conversion of CO to CO_2 and NO / NO_2 to N_2 in catalytic converters: $2CO(g) + 2NO(g) \longrightarrow 2CO_2(g) + N_2(g)$
Nickel, platinum and palladium	Manufacture of margarine and low fat spreads: $RCH = CH_2(g) + H_2(g) \longrightarrow RCH_2CH_3(g)$
Copper(II) ions with cytochrome oxidase	Oxidation of food during respiration

Key Point

Transition metals and their compounds are important catalysts in industry and in biological systems.

Review questions

1 a Give one example of a reaction which occurs instantaneously and one example of a reaction which occurs rapidly at high temperatures, but not at all at room temperature.

   b Explain why these reactions behave as they do.

2 The reaction between marble chips (calcium carbonate) and dilute hydrochloric acid was used by students to investigate the effect of surface area on reaction rate. During the reaction, carbon dioxide escaped from the reacting mixture and this was collected in a graduated syringe. Their results are shown in Figure 8.20.

   In experiment 1, they used five small marble chips (total mass = 2 g) and 50 cm^3 of 0.1 $mol\,dm^{-3}$ hydrochloric acid.

   In experiment 2, they used one large marble chip (mass = 2 g) and 50 cm^3 of the same acid.

   a Write an equation for the reaction involved.

   b Draw and label a diagram of the apparatus used.

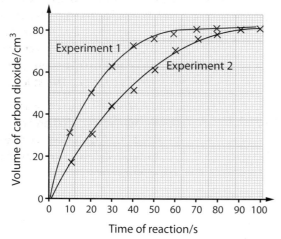

Figure 8.20

   c There is a large excess of marble in both experiments. Why is this?

   d Why is the final volume of carbon dioxide the same in both experiments?

   e Why do the graphs become flat?

   f Use the graphs to estimate a value for the initial reaction rate in each experiment.

   g Explain why the initial reaction rate is different in the two experiments.

3 a Which of the following will affect the rate at which a candle burns?

   i   the temperature of the air,   iii the air pressure,

   ii the shape of the candle,   iv   the length of the wick.

   b Explain your answers to part a.

   c State two other factors that will affect the rate at which a candle burns.

4 Design an experiment to investigate the effect of acid concentration on the weathering of limestone.

5 Solid, black manganese(IV) oxide acts as a catalyst for the decomposition of aqueous hydrogen peroxide, $H_2O_2(aq)$, forming water and hydrogen.

   a What do you understand by the term 'catalyst'?

   b Write an equation for the reaction catalysed by manganese(IV) oxide.

   c Describe an experiment that you could carry out to show that manganese(IV) oxide acts as a catalyst for the decomposition of hydrogen peroxide.

6 a Draw a graph with labelled axes showing the Maxwell–Boltzmann distribution for the energies of the molecules in a gas at 300 K.

   b Repeat part a for molecules in the same gas at 310 K.

   c Explain the shape of the graph at 310 K relative to that at 300 K.

   d Use your graphs to explain why the rates of reactions can double when the temperature rises by only 10 K.

7 The accepted pathway for the exothermic hydrolysis of 2-chloro-2-methylpropane is shown in Figure 8.21.

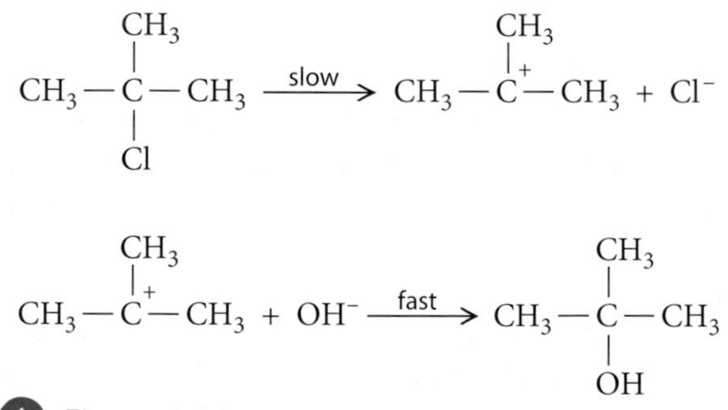

$$CH_3-\underset{\underset{Cl}{|}}{\overset{\overset{CH_3}{|}}{C}}-CH_3 \xrightarrow{\text{slow}} CH_3-\overset{\overset{CH_3}{|}}{\underset{}{C}}{}^+-CH_3 + Cl^-$$

$$CH_3-\overset{\overset{CH_3}{|}}{\underset{}{C}}{}^+-CH_3 + OH^- \xrightarrow{\text{fast}} CH_3-\underset{\underset{OH}{|}}{\overset{\overset{CH_3}{|}}{C}}-CH_3$$

Figure 8.21

   a Sketch a reaction profile for the conversion of $CH_3C(CH_3)ClCH_3$ to $CH_3C(CH_3)OHCH_3$.

   Show the activation energy, the enthalpy change of reaction and the relative positions of reactants, intermediates and products on your reaction profile.

   b When this reaction is investigated, results show that the rate of reaction is unaffected by the concentration of OH^- ions. Suggest a reason for this.

   c Under certain conditions the reaction can be reversed. What conditions would you suggest for this?

9 The periodic table and periodicity

9.1 Families of similar elements

Chemists have always searched for patterns and similarities in the properties and reactions of different substances. From your early studies of chemistry you will appreciate the classification of elements as metals and non-metals and you will have looked at the patterns in reactivity of different metals with oxygen, water and acids. However, there are some elements that cannot be classified easily and unambiguously as either a metal or a non-metal. For example, non-metals are usually volatile and non-conductors of electricity, but graphite (carbon) and silicon (Section 4.13), usually classed as non-metals, have very high melting points and very high boiling points, and both conduct electricity.

Limitations in the overall classification of elements as metals or non-metals led chemists to search for trends and similarities in the properties of much smaller classes of elements. Pairs of similar elements, such as sodium and potassium, calcium and magnesium, and chlorine and bromine, will already be familiar to you.

Quick Questions

1 To what extent do sodium and potassium resemble each other?
   a Do they have similar properties (e.g. melting point, boiling point, hardness, density, lustre)?
   b Do they react in the same way with oxygen in the air?
   c What colour are their oxides, chlorides, sulfates, etc?
   d How do sodium and potassium react with water? Write equations for their reactions with water. Are their reactions and products similar?
   e Are the formulae of their compounds similar? What oxidation number do sodium and potassium have in their compounds?

Döbereiner's triads

In 1829, the German chemist Döbereiner (Figure 9.1) pointed out that many of the known elements could be arranged in groups of three similar elements. He called these families of three elements '**triads**'. Two of Döbereiner's triads are shown in Figure 9.2. These triads were lithium, sodium and potassium (alkali metals) and chlorine, bromine and iodine (halogens). Döbereiner showed that when the three elements in each triad were written in order of relative atomic mass, the middle element had properties in between those of the other two. What is more, the relative atomic mass of the middle element was very close to the average relative atomic masses of the other two elements. For example, in the triad of halogens, the relative atomic mass of bromine (79.9) is close to the average of the relative atomic masses of chlorine and iodine:

$$= \frac{35.5 + 126.9}{2} = \frac{162.4}{2} = 81.2$$

Newlands' octaves

The discovery of triads by Döbereiner and the link between the properties of elements and their relative atomic masses encouraged other chemists to search for patterns.

In 1864, John Newlands, an English chemist, arranged all the known elements in order of their relative atomic masses. He found that one element often had properties like those of the element seven places in front of it in his list. So, Newlands called this the

Alkali metals	Halogens
Li 6.9	Cl 35.5
Na 23.0	Br 79.9
K 39.1	I 126.9

Figure 9.2
Döbereiner's triads of alkali metals and halogens

2 How close is the relative atomic mass of sodium to the average of the relative atomic masses of lithium and potassium?

3 Look at Newlands' octaves in Figure 9.3.
  a Why did Newlands use the word 'octaves'?
  b Name one of Döbereiner's triads which shows up clearly in Newlands' octaves.
  c Name two elements which are grouped together in Newlands' octaves, but which are very different.

'Law of Octaves'. He said that 'the eighth element is a kind of repetition of the first, like the eighth note of an octave in music'.

Figure 9.3 shows the first three of Newlands' octaves.

H	Li	Be	B	C	N	O
F	Na	Mg	Al	Si	P	S
Cl	K	Ca	Cr	Ti	Mn	Fe

Figure 9.3
The first three of Newlands' octaves

The suggestion by Newlands and others that there might be a regular or periodic repetition of elements with similar properties related to their relative atomic masses led to the name **periodic table**.

Unfortunately, Newlands' classification grouped together some elements which were very different and his table of octaves left no spaces for elements that were undiscovered at that time. Because of these problems, Newlands' ideas were criticised and rejected.

9.2 Mendeléev's periodic table

In spite of the criticism of Newlands' table of octaves, chemists continued to search for a pattern linking the properties and relative atomic masses of the elements. In 1869, the Russian chemist, Dmitri Mendeléev, produced new evidence to support the ideas of periodicity which Newlands had suggested five years earlier.

Mendeléev arranged all the elements known to him in order of increasing relative atomic mass and showed that elements with similar properties recurred at regular intervals. Figure 9.4 shows part of Mendeléev's periodic table. Notice that elements with similar properties, such as lithium, sodium and potassium, fall in the same vertical column. These vertical columns of similar elements are called **groups** and the horizontal rows of elements are called **periods**.

Figure 9.4
Part of Mendeléev's periodic table

	Group I	Group II	Group III	Group IV	Group V	Group VI	Group VII	Group VIII
Period 1	H							
Period 2	Li	Be	B	C	N	O	F	
Period 3	Na	Mg	Al	Si	P	S	Cl	
Period 4	K	Ca	*	Ti	V	Cr	Mn	Fe Co Ni
	Cu	Zn	*	*	As	Se	Br	
Period 5	Rb	Sr	Y	Zr	Nb	Mo	*	Ru Rh Pd
	Ag	Cd	In	Sn	Sb	Te	I	

Mendeléev's periodic table was far more successful than Newlands' because of three ingenious steps he took with his table.

▶ In the first place, Mendeléev left gaps in his table in order that similar elements fell in the same vertical group.
▶ Secondly, he suggested that, in due course, elements would be discovered to fill these gaps.
▶ Thirdly, he predicted properties for some of the missing elements from the properties of elements above and below them in his table.

Quick Questions

4 a There are four asterisks in Figure 9.4 to indicate missing elements. Which elements have since been discovered to replace the asterisks?
  b Why did Mendeléev not leave gaps for the noble gases?

Initially, Mendeléev's periodic table was nothing more than a curiosity. But it encouraged chemists to search for further patterns and look for more elements. Because of this, it became a useful tool for chemists. Within 15 years of Mendeléev's predictions three of the missing elements in his table had been discovered and their properties were very similar to his predicitons.

Mendeléev also proposed that Periods 4, 5, 6 and 7 should contain more than just seven elements as in Periods 2 and 3. In order to fit these longer periods into his pattern, he divided them into halves and placed the first half of the elements in the top left-hand corner of their box (e.g. K, Ca, etc., in Period 4), and the second half in the bottom right-hand corners (e.g. Cu, Zn, etc.).

The success of Mendeléev's predictions convinced scientists that his ideas were correct and his periodic table was accepted as a valuable overall summary of the properties of elements. Indeed, Mendeléev's concept of a periodic table has proved very adaptable. It has absorbed more and more new knowledge and the importance and usefulness of his original idea has been demonstrated many times.

9.3 Modern forms of the periodic table

Mendeléev's periodic table arranged the elements in order of relative atomic mass. His Periodic Law stated that *the properties of the elements are a periodic function of their relative atomic masses.* This law fulfilled two important functions.

1 It summarised the properties of elements, putting them into groups with similar properties.
2 It enabled predictions to be made about the properties of known and unknown elements and led to considerable research activity.

Although Mendeléev used the order of relative atomic masses as a basis for his periodic table, he wrote the elements tellurium (Te = 127.6) and iodine (I = 126.9) in the reverse order. Mendeléev found that this was necessary if Te and I were to fall in their correct vertical groups. He argued that iodine must occupy the same group as chlorine and bromine.

Today, we can understand the periodic table in terms of patterns in electronic structure. But Mendeléev's periodic table was proposed long before that. Although the reverse order of Te and I worried scientists at the time of Mendeléev's proposals, it is now explained by modern forms of the Periodic Law. This states that *the properties of elements are a periodic function of their proton (atomic) numbers, not their atomic masses.* (The proton number of Te is 52 and that of I is 53.)

If the elements are numbered along each period from left to right starting at Period 1, then Period 2, etc., the number given to each element is its **proton (atomic) number**.

In modern periodic tables, all the elements are in strict order of their proton numbers. Although there are various forms of the periodic table suitable for one purpose or another, the 'wide form' shown in Figure 9.6 is probably the most useful. In this figure the proton numbers are shown below the symbols for each element.

Those groups with elements in Periods 2 and 3 are numbered from I to VII followed by Group 0. Some of these groups have names as well as numbers. The most common names used for particular groups are:

Group number	Group name
I	alkali metals
II	alkaline-earth metals
VII	halogens
0	noble (inert) gases

Figure 9.5
Dmitri Ivanovich Mendeléev (1839–1907). In 1869, Mendeléev published his periodic table. This was the forerunner of modern periodic tables.

Definition

Transition elements occupy the d- and f-blocks of the periodic table. (A more precise definition of transition elements will be given in Chapter 24.)

 Figure 9.6
The modern periodic table (wide form)

The most obvious difference between the modern periodic table (Figure 9.6) and that proposed by Mendeléev is the removal of the **transition elements** from the simple groups. In Period 4, for example, 10 transition elements (Sc, Ti, V, Cr, Mn, Fe, Co, Ni, Cu, Zn) have been taken out of the simple groups suggested by Mendeléev and placed between Ca in Group II and Ga in Group III.

Sections and blocks in the periodic table

Besides dividing the periodic table into vertical groups of similar elements, it is also useful to split it into five sections of elements with similar properties. These five sections are coloured differently in Figure 9.6.

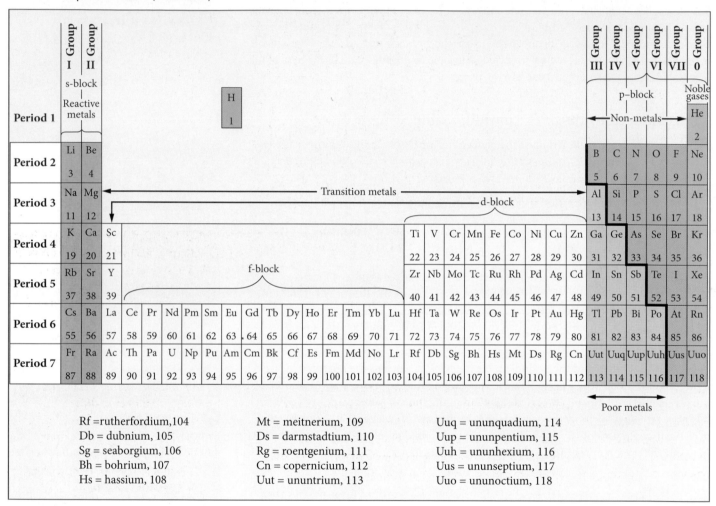

Rf = rutherfordium, 104
Db = dubnium, 105
Sg = seaborgium, 106
Bh = bohrium, 107
Hs = hassium, 108

Mt = meitnerium, 109
Ds = darmstadtium, 110
Rg = roentgenium, 111
Cn = copernicium, 112
Uut = ununtrium, 113

Uuq = ununquadium, 114
Uup = ununpentium, 115
Uuh = ununhexium, 116
Uus = ununseptium, 117
Uuo = ununoctium, 118

Key Point

In the modern periodic tables, all the elements are in strict order of proton (atomic) numbers.

The reactive metals

The elements in Groups I and II form a section of reactive metals. They are sometimes referred to as the 's-block' elements, since the outermost electrons in these metals are in s sub-shells (Section 2.10). These metals (including potassium, sodium, calcium and magnesium) are all high in the activity (electrochemical) series. They have lower densities, lower melting points and lower boiling points than most other metals and they form stable, involatile ionic compounds.

The transition metals

These elements occupy two rectangles between Group II and Group III. Some of the transition metals are called 'd-block' elements, because electrons are being added to d sub-shells across this block of elements (Section 2.10). These metals (including

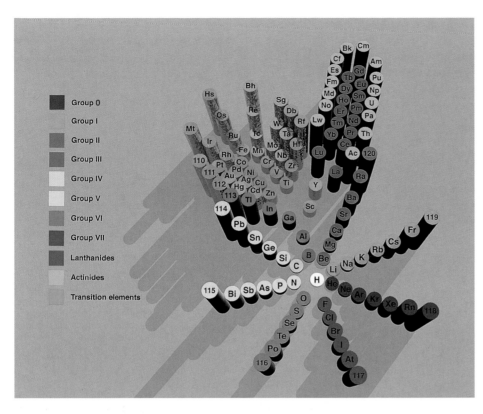

Figure 9.7
An unusual form of the periodic table

Key Point

The periodic table can be split into five sections of elements with similar properties:

▶ reactive metals,
▶ transition metals,
▶ poor metals,
▶ non-metals,
▶ noble gases.

chromium, iron, copper, zinc and silver) are much less reactive than the metals in Groups I and II. In this block of elements there is also a marked horizontal similarity in properties as well as the usual vertical likeness.

The lanthanides and actinides form a sub-section of elements within the transition metals. Indeed, they are sometimes called the inner transition elements. Another name for these elements is the 'f-block' elements, since electrons are being added to f sub-shells across this block of elements. The lanthanides consist of the 14 elements from cerium (Ce) to lutetium (Lu) which come immediately after lanthanum (La) in the periodic table. These elements resemble each other very closely. In fact, the horizontal similarities across this block are so great that chemists experienced considerable difficulty in separating the lanthanides from one another. The actinides are the 14 elements from thorium (Th) to lawrencium (Lr) which follow actinium (Ac) in the periodic table. Only the first three elements in the actinide series (thorium (Th), protactinium (Pa) and uranium (U)) are naturally occurring. All the elements beyond uranium have been synthesised by chemists since 1940. All the actimides are radioactive.

The poor metals

The term 'poor metals' is a useful description for several metals including tin, lead and bismuth. These metals fall in a triangular section of the periodic table to the right of the transition metals. They are usually low in the activity (electrochemical) series and they have some resemblances to non-metals.

The non-metals

These elements also form a triangular section in the periodic table. The elements in this section and the previous one are sometimes called the 'p-block' elements because the outermost electrons in these elements are going into p sub-shells.

The noble gases

The atoms in these elements have completely filled s and p sub-shells of electrons.

Quick Questions

5 How do the transition metals iron and copper compare with the reactive metals sodium and calcium in:
   a their melting points and boiling points,
   b their densities,
   c their reactions with water,
   d the number of different oxidation states which they show in their compounds,
   e the colours of their compounds as solids and in aqueous solution?

6 Draw an outline of the periodic table similar to Figure 9.6. On your outline indicate the position of:
   a metals,
   b metalloids,
   c elements with proton numbers 11 to 18,
   d the alkaline-earth metals,
   e the most reactive non-metal,
   f an element that might be used as a disinfectant,
   g a magnetic element,
   h 'precious' metals.

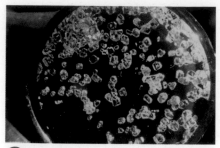

Figure 9.8
Crystals of xenon fluoride, first isolated at the Argonne National Laboratory, USA. This was one of the earliest compounds of a noble gas to be synthesised.

Definition

A **metalloid** is an element with some properties like metals and other properties like non-metals.

They are very unreactive and the first noble gas compound was not made until 1962. Because of their chemical unreactivity these elements were originally called 'inert gases'. Nowadays, several compounds of these elements (mainly oxides and fluorides of xenon and krypton) are known and the adjective 'inert' has been replaced by 'noble'.

9.4 Metals, non-metals and metalloids

Although the periodic table does not classify elements as metals and non-metals, there is a fairly obvious division between the two ('*fairly obvious*', but, not '*clear cut*'). Separating the elements into either metals or non-metals is rather like trying to separate all the shades of grey into either black or white. The fairly obvious division between metals and non-metals is shown by a thick stepped line in Figure 9.6. The 20 or so non-metals are packed into the top right-hand corner above the thick stepped line. Some of the elements next to the thick steps, such as germanium, arsenic and antimony, have similarities to both metals and non-metals and it is difficult to place these in one class or the other. Chemists sometimes use the name **metalloid** for these elements which are difficult to classify one way or the other.

Figure 9.9 shows a classification of elements as metals, metalloids and non-metals on the basis of their electrical conductivity. In this classification:

1 *Metals are good conductors of electricity* with atomic electrical conductivity* greater than 10^{-3} ohm^{-1} cm^{-4}.
2 *Metalloids are poor conductors of electricity* with atomic electrical conductivity usually less than 10^{-3} but greater than 10^{-5} ohm^{-1} cm^{-4}.
3 *Non-metals are virtually non-conductors* (insulators). Their atomic electrical conductivity is usually less than 10^{-10} ohm^{-1} cm^{-4}.

Notice in Figure 9.9 that the cell for carbon is shaded less heavily than those for the other metalloids. This is because carbon exists as three different allotropes – graphite, a poor conductor classed as a metalloid, plus diamond and fullerenes, insulators classed as non-metals.

In spite of problems such as this, the classification of elements into metals, metalloids and non-metals is useful and convenient.

Figure 9.9
Classifying the elements as metals, metalloids and non-metals on the basis of their electrical conductivity. In this figure, metalloids are coloured grey.

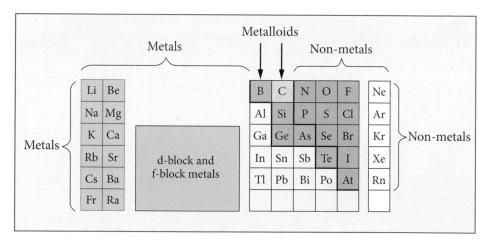

Note

*The atomic electrical conductivity (atomic conductance) is the conductivity of a block of the substance 1 cm^2 in cross-section but long enough to contain one mole of atoms of the element. It is a measure of the conductivity of one mole of atoms of the element.

9.5 Periodic properties

The arrangement of elements in the modern periodic table makes the repeating, periodic pattern of elements with similar properties very clear. The most obvious repeating pattern across the periodic table is from metals on the left, through metalloids with intermediate properties to non-metals on the right. In Period 3, sodium, the left-hand element, is a very reactive metal, whereas chlorine, next to the extreme right is a very reactive non-metal. In between, the elements show a gradual transition from metals to non-metals.

This periodicity in physical and chemical properties of the elements across the table is reflected in a periodic change in their structures and bonding. The structure of the elements varies from metallic, through giant molecular in the metalloids to simple molecular structures in the non-metals.

Key Point

In spite of its limitations, the classification of elements as metals, metalloids and non-metals is very useful. The commonest classification is based on electrical conductivity. Metals are good conductors. Metalloids are poor conductors. Non-metals are non-conductors.

9.6 The periodicity of physical properties

Table 9.1 shows the values of various physical properties and the structure of the elements in the second and third periods of the periodic table.

Table 9.1
Various physical properties and the structure of elements in Periods 2 and 3 of the periodic table

Period 2 / **Property**	**Li**	**Be**	**B**	**C (graphite)**	**N**	**O**	**F**	**Ne**
Melting point /°C	180	1280	2030	3700 sublimes	−210	−219	−220	−250
Boiling point /°C	1330	2480	3930	–	−200	−180	−190	−245
Atomic electrical conductivity × 10^3 /ohm^{-1} cm^{-4}	8	51	–	0.14	–	–	–	–
Type of element	Metals		Metalloids		Non-metals			
Structure	Giant metallic		Giant molecular		Simple molecular N_2	O_2	F_2	Ne
Period 3 / **Property**	**Na**	**Mg**	**Al**	**Si**	**P**	**S**	**Cl**	**Ar**
Melting point /°C	98	650	660	1410	44	119	−101	−189
Boiling point /°C	890	1120	2450	2680	280	445	−34	−186
Atomic electrical conductivity × 10^3 /ohm^{-1} cm^{-4}	10	16	38	0.4	10^{-16}	10^{-22}	–	–
Type of element	Metals			Metalloid	Non-metals			
Structure	Giant metallic			Giant molecular	Simple molecular P_4	S_8	Cl_2	Ar

Notice the following trends across Periods 2 and 3:

1. The melting points rise to the element in Group IV and then fall to low values.
2. The boiling points rise to the element in Group IV and then fall to low values.
3. The electrical conductivity is relatively high for the metals on the left of each period, lower for metalloids in the centre of each period and almost negligible for non-metals on the right.

Melting points and boiling points

The periodicity of the melting points of the elements in Periods 2 and 3 is revealed even more clearly by the graph in Figure 9.10.

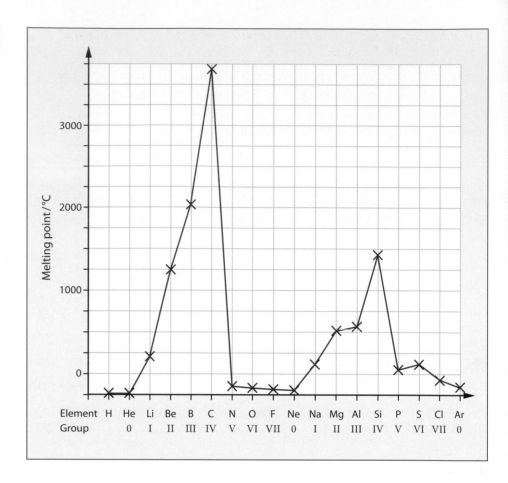

 Figure 9.10

Periodicity of the melting points of the elements in Periods 2 and 3

The melting points of the elements depend closely on their structure and the bonding between atoms. In metals on the left of each period, the bonding between atoms is strong involving the electrostatic attraction of delocalised electrons for positive ions (Section 3.8). So their melting points and boiling points are usually high. The more electrons a metal atom contributes to the delocalised electrons, the stronger the bonding. This results in the rise in melting and boiling points from Group I to Group II and then to Group III in Period 3.

Boron, carbon and silicon, metalloids in Groups III and IV, have giant molecular structures with strong directional covalent bonds linking one atom to another throughout a giant network (Section 4.13). This means that covalent bonds must break before the solid even melts. So this results in very high melting points and boiling points.

Finally, the non-metal elements in Groups V, VI, VII and 0 form simple molecules – monatomic in Ne and Ar, diatomic in N_2, O_2, F_2 and Cl_2, tetratomic in P_4 and octatomic in S_8.

The intermolecular forces between simple molecules in these elements involve only weak induced dipole attractions (Section 4.15). So, their melting points and boiling points are low.

Electrical conductivity

The periodicity of electrical conductivity for the elements in Periods 2 and 3, like the periodicity in melting points and boiling points, depends critically on their structure and bonding.

Metals, such as sodium, magnesium and aluminium, on the left of each period with mobile, delocalised electrons in their structure are good conductors of electricity. When a battery is attached to them, electrons flow out of the metals into the positive terminal of the battery. At the same time, electrons flow into the metal from the negative terminal of the battery.

On the other hand, metalloids and non-metals with molecular structures have no mobile electrons. This means that their electrical conductivity is almost nil.

9.7 The periodicity of atomic properties

Table 9.2 shows the electronic (shell and sub-shell) structures for the elements in Periods 2 and 3.

Elements in the same group of the periodic table have similar electron configurations. For example, Group I elements (Li to Fr) have one electron in their outer shells (ns^1).

Quick Questions

7 a Which elements occur at or near the peaks on the graph in Figure 9.10? What type of structure do these elements have?

 b Suggest a reason why the electrical conductivity rises from Na $\longrightarrow$ Mg $\longrightarrow$ Al.

Period 2	Li	Be	B	C	N	O	F	Ne
Electron shell structure	2, 1	2, 2	2, 3	2, 4	2, 5	2, 6	2, 7	2, 8
Electron sub-shell structure	$1s^22s^1$	$1s^22s^2$	$1s^22s^22p^1$	$1s^22s^22p^2$	$1s^22s^22p^3$	$1s^22s^22p^4$	$1s^22s^22p^5$	$1s^22s^22p^6$

Period 3	Na	Mg	Al	Si	P	S	Cl	Ar
Electron shell structure	2, 8, 1	2, 8, 2	2, 8, 3	2, 8, 4	2, 8, 5	2, 8, 6	2, 8, 7	2, 8, 8
Electron sub-shell structure	$1s^22s^22p^6$ $3s^1$	$1s^22s^22p^6$ $3s^2$	$1s^22s^22p^6$ $3s^23p^1$	$1s^22s^22p^6$ $3s^23p^2$	$1s^22s^22p^6$ $3s^23p^3$	$1s^22s^22p^6$ $3s^23p^4$	$1s^22s^22p^6$ $3s^23p^5$	$1s^22s^22p^6$ $3s^23p^6$

 Table 9.2
The electronic structures of the elements in Periods 2 and 3

Group II elements (Be to Ra) have two electrons in their outer shells (ns^2) and Group VII elements (F to At) have seven outer shell electrons (ns^2np^5). As these electron configurations recur in a periodic pattern, it would not be surprising to find that atomic properties, such as atomic radii, ionic radii and ionisation energies, show similar periodicity.

Key Point

Elements in the same group of the periodic table have similar electron configurations. This results in similar atomic properties.

Atomic radii

The atomic radii of atoms can be obtained from X-ray analysis and electron density maps. Using these techniques, it is possible to measure the distance between the nuclei of atoms and then estimate the radius of individual atoms.

The atomic radii of metals are obtained by measuring the distance between the nuclei of neighbouring atoms in metal crystals (Figure 9.11 (a)). The atomic radius is simply half of the inter-nuclear distance.

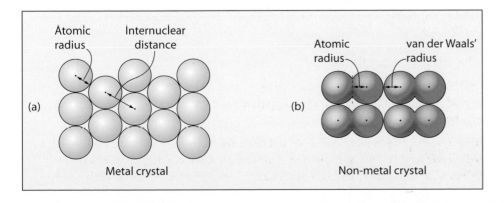

Figure 9.11
Obtaining atomic radii from the distances between the nuclei of atoms

The atomic radii of non-metals are obtained from the distance between the nuclei of similar atoms joined by a covalent bond (Figure 9.11(b)). So, for non-metals, the atomic radius is half of the covalent bond length. Because of this link to covalent bonds, the atomic radii of non-metals are sometimes called covalent radii.

In the simple molecular structures of non-metals, one molecule touches the next and it is sometimes useful to compare the distances between neighbouring atoms which are not chemically bonded. This distance in non-metal crystals is called the **van der Waals radius** (Figure 9.11(b)).

Figure 9.12 shows the atomic radii of the elements from Li to F in Period 2 and from Na to Cl in Period 3. Notice that along each period there is a gradual decrease in atomic radius as electrons are added to the outer shell.

Figure 9.12
Atomic radii of the elements from Li to F and from Na to Cl

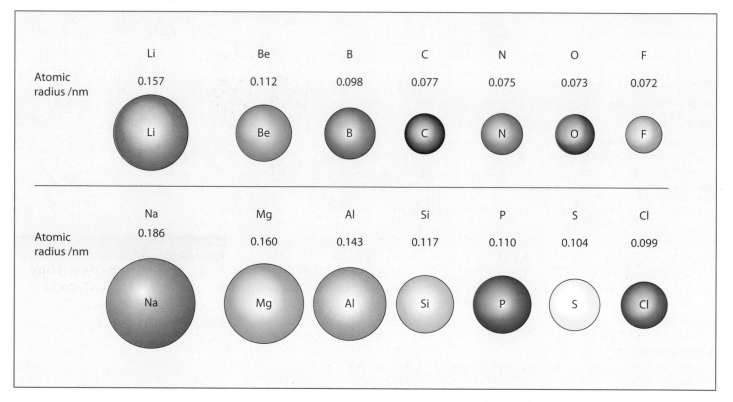

Moving from one element to the next across a period, electrons are added to the same shell. At the same time, protons are added to the nucleus. The electrons are therefore attracted and pulled towards the nucleus by an increasing positive charge, so the radius of the atom decreases. However, the rate of decrease in the radius becomes smaller as more protons are added. The addition of one more proton to the 11 already present in sodium causes a greater proportional increase in the attractive power of the nucleus than the addition of one more proton to the 16 already present in sulfur. Thus, the atomic radius falls by 0.026 nm from Na to Mg, but by only 0.005 nm from S to Cl.

Ionic radii

Ionic radii give us useful information about the structure of some compounds. They provide an accurate measure of the space occupied by particular ions in a crystal lattice. Figure 9.13 shows the radii of the most stable ions of lithium to fluorine in Period 2 and sodium to chlorine in Period 3. It also shows the relative size of the atoms and ions of these elements. Look for the following general patterns in Figure 9.13.

Quick Questions

8 Why do atomic radii increase down each group in the periodic table?

Key Point

Moving from left to right across a period, there is a decrease in atomic radius because the increasingly positive nuclear charge pulls electrons in the same outer shell closer.

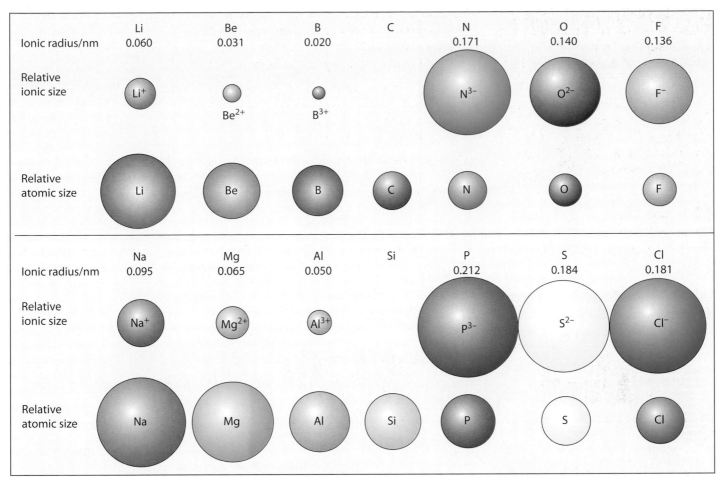

	Li	Be	B	C	N	O	F
Ionic radius/nm	0.060	0.031	0.020		0.171	0.140	0.136

Relative ionic size: Li^+, Be^{2+}, B^{3+}, N^{3-}, O^{2-}, F^-

Relative atomic size: Li, Be, B, C, N, O, F

	Na	Mg	Al	Si	P	S	Cl
Ionic radius/nm	0.095	0.065	0.050		0.212	0.184	0.181

Relative ionic size: Na^+, Mg^{2+}, Al^{3+}, P^{3-}, S^{2-}, Cl^-

Relative atomic size: Na, Mg, Al, Si, P, S, Cl

Figure 9.13
Radii of the most stable ions of Li to F in Period 2 and Na to Cl in Period 3

1 *The ionic radii of positive ions are smaller than their corresponding atomic radii.* For positive ions in the same period (e.g. Na^+, Mg^{2+} and Al^{3+}), the ionic radius decreases across the period from left to right. Now look at Quick question 9.

2 *The ionic radii of negative ions are greater than their corresponding atomic radii.* For negative ions in the same period (e.g. P^{3-}, S^{2-} and Cl^-), the ionic radius decreases with increase in proton number.
When one or more electrons are added to the outer shell of an atom forming a negative ion, there is an increase in the repulsion between negative charge clouds. This results in an overall increase in size of the ion relative to its corresponding atom. What is more, negative ions in the same period (e.g. P^{3-}, S^{2-} and Cl^-) have the same electron structure. So, as the nuclear charge increases with proton number, the electrons are attracted more strongly and the ionic radii become smaller.

3 In *a series of ions with the same electron structure (an isoelectronic series), the ionic radius decreases as proton number increases*. In Figure 9.14 the solid lines, such as that from N^{3-} to Al^{3+}, link those ions which are isoelectronic.

Why does ionic size decrease along these isoelectronic series? The nuclear charge increases along an isoelectronic series (e.g. from N^{3-} to Al^{3+}). This causes the electron cloud to contract because it is pulled in more effectively by an increasing positive charge.

Ionic radii provide a measure of ionic sizes in *solid* crystals, but they give no accurate indication of the size of *aquated* ions in solution. Here the situation can be very different. The increasing ionic size from Li^+ through Na^+ to K^+ might lead us to expect the mobility of the aqueous ions to decrease in the order Li^+, Na^+, K^+. In practice, $K^+(aq)$ is the most mobile and $Li^+(aq)$ is the least mobile of the three aqueous ions. This is due to solvation by water molecules. The small Li^+ ion has the largest charge density per unit of surface area. It therefore attracts polar water molecules around itself more strongly than the Na^+ ion.

Quick Questions

9 a Which electrons are removed from the metals in Groups I, II and III in forming their positive ions?
  b How does the electronic structure of a positive ion (e.g. Na^+) compare with that of the corresponding metal atom (e.g. Na)?
  c Why is the radius of a positive ion less than that of its corresponding atom?
  d Why does the ionic radius decrease from Na^+ to Al^{3+}?

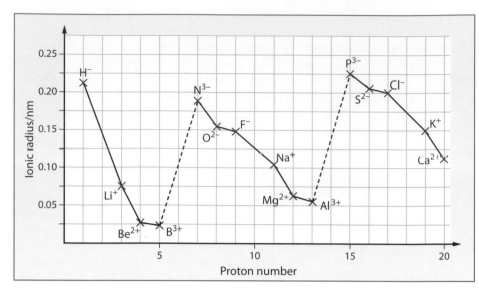

Figure 9.14
Variation in ionic radius with proton number

Quick Questions

10 Which of the elements, Na, Al, P and Cl, has:
  a the largest atomic radius,
  b the smallest atomic radius,
  c the largest ionic radius,
  d the smallest ionic radius?

11 The 2nd, 3rd, 4th and 5th ionisation energies of an element , X, in $kJ\,mol^{-1}$ are 1450, 7740, 10 500, 13 600.
  a To which group of the periodic table does X belong?
  b Explain how you arrived at your answer to part a.

Key Point

Moving from left to right across a period, there is a general increase in the first ionisation energy. This is because the increasing nuclear charge and decreasing atomic radius cause the outermost electron to be held more tightly.

Similarly, Na^+ exerts more attraction on polar water molecules than K^+. Thus, the effective size of the aqueous ions is $Li^+(aq) > Na^+(aq) > K^+(aq)$ and this makes $Li^+(aq)$ least mobile because when it moves it has to drag a larger collection of water molecules with it.

First ionisation energy

The striking periodic variation in the first ionisation energies of the elements, shown in Figure 9.15, is described and then explained fully in Section 2.10.

The first ionisation energy of an element involves removing one electron from each gaseous atom to form one mole of gaseous ions with one positive charge:

$$X(g) \longrightarrow X^+(g) + e^-$$

Moving from left to right across a period, the positive nuclear charge increases and its attraction for the outermost electrons therefore increases. This means that it becomes more difficult to remove an electron from the outermost shell and, in general, the first ionisation energy increases across a period.

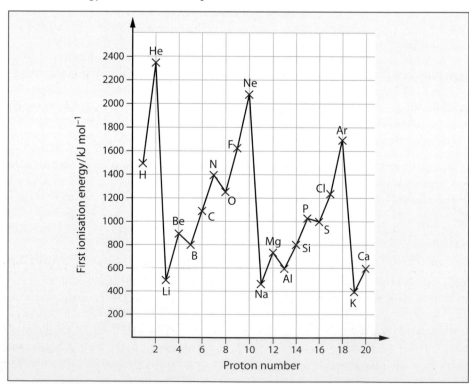

Figure 9.15
The variation in first ionisation energy with proton number

Electronegativity

The electronegativity of an atom provides a numerical measure of the power of that atom in a molecule to attract electrons. The term 'electronegativity' was first introduced in Section 3.9, where we noted that electronegativity values increased from left to right across each period. This trend is emphasised in Figure 9.16.

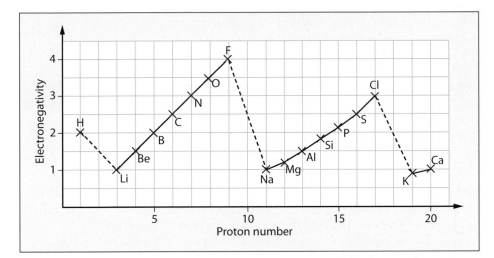

◀ Figure 9.16

Variation in electronegativity with proton number

As expected, the least electronegative elements are Group I metals on the left-hand edge of the periodic table. These metals readily lose electrons and form positive ions. In contrast, the most electronegative elements are the reactive halogens in Group VII near the right-hand edge of the periodic table. These elements readily gain electrons and form negative ions.

Moving from one element to the next across a period, the nuclear charge increases by 1. As the positive charge in the nucleus rises, the atom has an increasing electron-attracting power and therefore an increasing electronegativity.

Key Point

Moving from left to right across each period in the periodic table, there is an increase in the electronegativity of elements.

9.8 The periodicity of chemical properties

Table 9.3 describes the reactions of the elements in Period 3 with oxygen, chlorine and water. Notice how the reactivities of the elements change from sodium to argon, and that the changes in reactivity with both oxygen and chlorine are similar.

 Table 9.3

The reactions of the elements in Period 3 with oxygen, chlorine and water

Element	Heat the element in dry oxygen	Heat the element in dry chlorine	Reaction with water at room temperature
Na	Very vigorous reaction forming $(Na^+)_2O^{2-}(s)$ and $(Na^+)_2O_2^{2-}(s)$	Very vigorous reaction forming $Na^+Cl^-(s)$	Rapid reaction. Na fizzes and skates over the water surface to form $NaOH(aq) + H_2(g)$
Mg	Very vigorous reaction forming $Mg^{2+}O^{2-}(s)$	Vigorous reaction forming $Mg^{2+}(Cl^-)_2(s)$	Slow reaction forming $Mg(OH)_2(s)$ and a few bubbles of $H_2(g)$
Al	Vigorous reaction at first forming $(Al^{3+})_2(O^{2-})_3(s)$ which prevents further attack	Vigorous reaction forming $Al_2Cl_6(s)$	Slow reaction forming a non-porous layer of $(Al^{3+})_2(O^{2-})_3(s)$ which prevents further attack
Si	Slow reaction forming $SiO_2(s)$	Slow reaction forming $SiCl_4(l)$	No reaction
P	Slow reaction forming $P_4O_6(s)$ and $P_4O_{10}(s)$	Slow reaction forming $PCl_3(l)$ and $PCl_5(s)$	No reaction
S	Slow reaction forming $SO_2(g)$ and $SO_3(g)$	Slow reaction forming $SCl_2(g)$ and $S_2Cl_2(l)$	No reaction
Cl	No reaction	No reaction	Very slow reaction forming $HCl(aq)$ and $HClO(aq)$. $HClO(aq)$ then decomposes to $HCl(aq) + O_2(g)$
Ar	No reaction	No reaction	No reaction

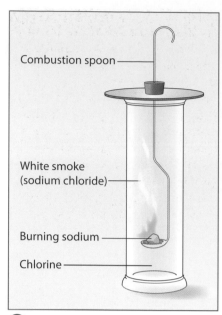

Combustion spoon

White smoke
(sodium chloride)

Burning sodium

Chlorine

△ Figure 9.17
Hot sodium burning in chlorine

Quick Questions

12 a Write an ionic equation for the reaction of sodium with oxygen to form sodium oxide.
b Explain why this is a redox reaction:
  i in terms of electron transfers,
  ii in terms of oxidation numbers.
c When sodium is heated with excess oxygen, it produces sodium peroxide $(Na^+)_2(O_2)^{2-}$ as well as sodium oxide. Draw a 'dot-and-cross' diagram showing only the outermost shells of electrons for Na_2O_2.

13 White phosphorus consists of simple molecules of P_4, in which the four phosphorus atoms occupy the corners of a tetrahedron:
a Draw the structure of a P_4 molecule indicating the covalent bonds between phosphorus atoms and showing any lone pairs of electrons.
b Write an equation for the reaction of white phosphorus with oxygen to form solid phosphorus(III) oxide, P_4O_6.

The reactivity of the elements with both oxygen and chlorine gradually falls from sodium to argon. With water, however, the reactivity falls at first from sodium through magnesium to aluminium. There is no reaction with silicon, phosphorus or sulfur, but then chlorine reacts to form HCl and HClO.

In spite of the variety and differences in the reactions summarised in Table 9.3, they can all be related to the action of the elements as oxidising or reducing agents. The metals (sodium and magnesium) are strong reducing agents. They readily give up electrons to form their corresponding ions:

$$Mg \longrightarrow Mg^{2+} + 2e^-$$

Of course, these two metals never exist freely in nature, but only in compounds as Na^+ and Mg^{2+} ions. As expected, these metals react vigorously on heating with oxygen and chlorine which are strong oxidising agents (Figure 9.17).

$$O_2 + 4e^- \longrightarrow 2O^{2-}$$
$$Cl_2 + 2e^- \longrightarrow 2Cl^-$$

They also react with water (though at different speeds) reducing it to hydrogen:

$$2H_2O + 2e^- \longrightarrow 2OH^- + H_2$$

Aluminium is a less reactive metal and not so strong a reducing agent as sodium and magnesium. It reacts vigorously when heated with either oxygen or chlorine to form an ionic oxide $(Al^{3+})_2(O^{2-})_3$, but a simple molecular chloride, Al_2Cl_6.

$$4Al(s) + 3O_2(g) \longrightarrow 2Al_2O_3(s)$$
$$2Al(s) + 3Cl_2(g) \longrightarrow Al_2Cl_6(s)$$

Aluminium reacts slowly with water forming a non-porous layer of $(Al^{3+})_2(O^{2-})_3$ which protects the metal below from further attack. This is why aluminium, though a reactive metal, can be used to make saucepans, boats and aeroplanes. A tough, thin layer of oxide protects it from attack by air or water.

Silicon in Group IV is a very weak reducing agent. It will react slowly on heating with oxygen and chlorine which are strong oxidising agents, but not with water, a much weaker oxidising agent:

$$Si(s) + O_2(g) \longrightarrow SiO_2(s)$$
$$Si(s) + 2Cl_2(g) \longrightarrow SiCl_4(l)$$

The next two elements, phosphorus and sulfur, can act as either weak reducing agents or weak oxidising agents. They react slowly when heated with oxygen and chlorine (strong oxidising agents), but they have no reaction with water:

$$S(s) + Cl_2(g) \longrightarrow SCl_2(g)$$
$$2S(s) + Cl_2(g) \longrightarrow S_2Cl_2(l)$$
$$S(s) + O_2(g) \longrightarrow SO_2(g)$$
$$2SO_2(g) + O_2(g) \longrightarrow 2SO_3(g)$$

Chlorine, at the other extreme to metals in chemical reactivity, is a strong oxidising agent. So, it has no reaction with oxygen, but it does react with water, oxidising it to HCl and HClO, which decomposes to oxygen and hydrochloric acid:

$$Cl_2(g) + H_2O(l) \longrightarrow HCl(aq) + HClO(aq)$$

then
$$2HClO(aq) \longrightarrow 2HCl(aq) + O_2$$

In this case, the water has acted as a reducing agent:

$$H_2O(l) \longrightarrow \tfrac{1}{2}O_2(g) + 2H^+(aq) + 2e^-$$

On the extreme right of Period 3, argon (a noble gas) shows no reactivity with oxygen, chlorine or water. Other periods show a similar pattern in reactivity to Period 3. Excluding the noble gases, elements change from strong reducing agents on the extreme left to elements which are weak reducing agents and/or weak oxidising agents in the centre and finally to strong oxidising agents.

9.9 Variation in the oxidation numbers of the elements in Periods 2 and 3

Table 9.4 shows the formulae of the chlorides, oxides and hydrides of the elements from lithium to argon in the periodic table.

▼ **Table 9.4**
Formulae of the chlorides, oxides and hydrides of the elements in Periods 2 and 3

Period 2	Li	Be	B	C	N	O	F	Ne
Formula of chloride	LiCl	$BeCl_2$	BCl_3	CCl_4	NCl_3	Cl_2O	ClF	—
Formula of oxide	Li_2O	BeO	B_2O_3	CO CO_2	N_2O NO NO_2 N_2O_3 N_2O_4 N_2O_5	O_2	OF_2	—
Formula of simplest hydride	LiH	BeH_2	BH_3	CH_4	NH_3	H_2O	HF	—

Period 3	Na	Mg	Al	Si	P	S	Cl	Ar
Formula of chloride	NaCl	$MgCl_2$	Al_2Cl_6	$SiCl_4$	PCl_3 PCl_5	SCl_2 S_2Cl_2	Cl_2	—
Formula of oxide	Na_2O Na_2O_2	MgO	Al_2O_3	SiO_2	P_4O_6 P_4O_{10}	SO_2 SO_3	Cl_2O Cl_2O_7	—
Formula of hydride	NaH	MgH_2	AlH_3	SiH_4	PH_3	H_2S	HCl	—

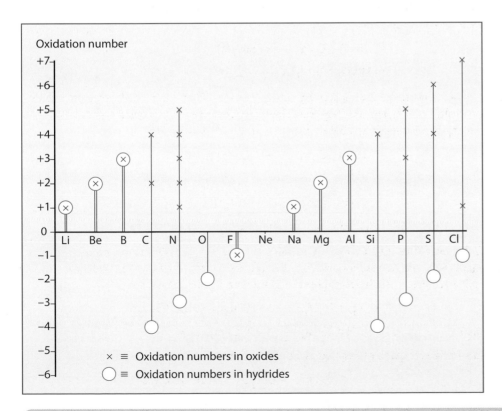

Oxidation number

× ≡ Oxidation numbers in oxides
○ ≡ Oxidation numbers in hydrides

◀ Figure 9.18
Oxidation numbers of the elements from Li to Cl in their oxides and hydrides

Key Point

The change from metals, through metalloids to non-metals across the periodic table is reflected in changing redox behaviour. Excluding the noble gases, elements change from strong reducing agents (e.g. Na, Mg) through weak reducing agents (e.g. Si), then elements which are both weak reducing agents and weak oxidising agents (e.g. P, S) to strong oxidising agents (e.g. Cl_2).

Quick Questions

14 Look at the formulae of the chlorides Li to Ar in Table 9.4.
   a What is the oxidation number of each element in Period 2 in its chloride?
   b How do these oxidation numbers vary across Period 2?
   c Why do the oxidation numbers vary in this way?

The oxidation numbers of the elements from Li to Cl in their oxides and hydrides are represented graphically in Figure 9.18. Oxidation numbers of the elements in compounds with oxygen are indicated by a '×' and in compounds with hydrogen by a 'O'. The oxidation numbers of elements in their oxides are always positive, because oxygen is the most electronegative element apart from fluorine. Fluorine oxide, OF_2, is the only exception, in which the oxidation number of F is –1. The pattern in these oxidation numbers provides further evidence of periodic properties.

The maximum oxidation number of each element (apart from fluorine) in its oxide is the same as its group number. For example, lithium in Group I has an oxidation number of +1, boron in Group III has an oxidation number of +3 and nitrogen in Group V has a maximum oxidation number of +5. These maximum oxidation numbers of each element correspond to the number of electrons in the outermost shell of its atoms.

The oxidation numbers of the elements in their hydrides have been obtained by assuming that the oxidation number of hydrogen in non-metal compounds is +1, whilst its oxidation number in metal hydrides is –1. Another interesting periodic pattern emerges this time. The oxidation numbers of the elements with hydrogen rise from +1 to +3 for the metals in Groups I, II, and III, plunge to –4 for the elements in Group IV and then rise through –3, –2 and –1 for the elements in Groups V, VI and VII.

Many of the oxidation numbers correspond to the loss or gain of enough electrons for the atoms to obtain a stable outer shell with the structure $2s^2 2p^6$ or $3s^2 3p^6$. This tendency is certainly the case for elements in Groups I, II and III which contain 1, 2 and 3 electrons, respectively, in their outer shell. In forming compounds, the elements in these three groups lose 1, 2 or 3 electrons, respectively, forming ions such as Na^+, Mg^{2+} and Al^{3+} with oxidation numbers of +1, +2 and +3. In contrast, the elements in Groups V, VI and VII gain 3, 2 and 1 electrons, respectively, to achieve stable electron structures in ions such as P^{3-}, S^{2-} and Cl^-. Thus, the elements in these groups have oxidation numbers of –3, –2 and –1, respectively.

Key Point

In forming compounds, the elements in Groups I, II and III form positive ions with charges of 1+, 2+ and 3+. In contrast, the elements in Groups V, VI and VII gain electrons forming negative ions with charges of 3–, 2– and 1– respectively.

9.10 Patterns in the properties of chlorides

Table 9.5 shows various properties of the chlorides of the elements in Period 3.

Table 9.5
Properties of the chlorides of elements in Period 3

Formula of chloride	NaCl	MgCl$_2$	Al$_2$Cl$_6$	SiCl$_4$	PCl$_3$ (PCl$_5$)	SCl$_2$ (S$_2$Cl$_2$)	Cl$_2$
State of chloride (at 20 °C)	s	s	s	l	l (s)	l (l)	g
b.pt. of chloride/°C	1465	1418	423	57	74 (164)	59 (136)	–35
Conduction of electricity by chloride in liquid state	good	good	v. poor	nil	nil	nil	nil
Structure of chloride	Giant ionic structures		Simple molecular structures				
Enthalpy change of formation of chloride at 298 K/kJ mol^{-1}	–411	–642	–1408	–640	–320	–20	0
Enthalpy change of formation of chloride at 298 K per mole of Cl atoms/kJ	–411	–321	–235	–160	–107	–10	0
Effect of adding chloride to water	solid dissolves readily		chloride reacts with water, HCl fumes are produced				some Cl$_2$ reacts with water
pH of aqueous solution of chloride	7	6.5	3	2	2	2	2

Structure and bonding

Quick question 15 points to variations which can be explained in terms of the structure and bonding in the chlorides.

NaCl and $MgCl_2$ are giant structures composed of oppositely charged ions attracted to each other by strong electrostatic forces in ionic bonds. This means that the melting points and boiling points of these compounds are high. But the molten substances will conduct electricity because the ions which they contain can move towards the electrode of opposite charge.

All the other chlorides are simple molecular compounds composed of small discrete molecules attracted to each other by relatively weak intermolecular forces. So, the melting points and boiling points of these compounds are low and as liquids they will not conduct electricity.

Enthalpy changes of formation

Elements on the left of the periodic table are highly electropositive metals which readily give up electrons. Elements on the right (excluding the noble gases) are highly electronegative non-metals. Between these extremes, the electronegativity slowly increases from left to right. Consequently, there is a gradual decrease in the enthalpy change when the elements react with 1 mole of chlorine (Cl). Sodium and chlorine react violently with an enthalpy change of 411 kJ per mole of Cl.

$$Na(s) + \tfrac{1}{2}Cl_2(g) \longrightarrow Na^+Cl^-(s) \qquad \Delta H = -411 \, kJ \, mol^{-1}$$

The strong ionic bonding in Na^+Cl^- results in a highly stable product and a reaction which is very exothermic. On the other hand, the reaction between sulfur and chlorine (two electronegative elements) is relatively feeble and much less exothermic:

$$\tfrac{1}{2}S(s) + \tfrac{1}{2}Cl_2(g) \longrightarrow \tfrac{1}{2}SCl_2(l) \qquad \Delta H = -10 \, kJ \, mol^{-1}$$

Reaction with water

When ionic chlorides are added to water, there is an immediate attraction of polar water molecules for ions in the chloride. The solid dissolves forming single aquated ions such as $Na^+(aq)$ and $Cl^-(aq)$ (Figure 9.19). These are separate metal and non-metal ions surrounded by polar water molecules. The solution of sodium chloride is neutral (pH = 7).

$$Na^+Cl^-(s) + aq \longrightarrow Na^+(aq) + Cl^-(aq)$$

Magnesium chloride dissolves readily in water in a similar way to sodium chloride forming the aquated ions, $Mg^{2+}(aq)$ and $Cl^-(aq)$.

In contrast to NaCl and $MgCl_2$, aluminium chloride reacts exothermically with water releasing fumes of hydrogen chloride and producing a very acidic solution (pH = 3). The acidic nature of the solution can be explained in terms of the relatively small, but highly charged Al^{3+} ion surrounded by water molecules. The Al^{3+} ions draw electrons away from their surrounding water molecules causing some of them to give up H^+ ions.

$$[Al(H_2O)_6]^{3+}(aq) \rightleftharpoons [Al(H_2O)_5(OH)]^{2+}(aq) + H^+(aq)$$

The non-metal chlorides in Period 3 all react vigorously and exothermically with water forming acidic solutions and producing fumes of HCl gas.

$$SiCl_4(l) + 2H_2O(l) \longrightarrow SiO_2(s) + \underbrace{4H^+(aq) + 4Cl^-(aq)}_{hydrochloric\ acid}$$

$$PCl_3(l) + 3H_2O(l) \longrightarrow \underset{\substack{phosphonic \\ (phosphorous) \\ acid}}{H_3PO_3(aq)} + \underbrace{3H^+(aq) + 3Cl^-(aq)}_{hydrochloric\ acid}$$

Quick Questions

15 **a** How do the states of the chlorides at 20 °C vary across the third period?
    b How do the boiling points of the chlorides vary across the period?
    c How do the conductivities of the chlorides in the liquid state vary across the period?

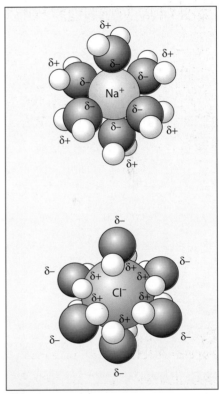

Figure 9.19
The structure of aqueous Na^+ and Cl^- ions

Key Point

Moving from left to right across a period, the **chlorides of elements** change from ionic, involatile metal chlorides to simple molecular, volatile non-metal chlorides.

$$2SCl_2(l) + 2H_2O(l) \longrightarrow S(s) + SO_2(aq) + \underbrace{4H^+(aq) + 4Cl^-(aq)}_{\text{hydrochloric acid}}$$

A similar pattern is found in the chlorides of Period 2 (Review question 7).

9.11 Patterns in the properties of oxides

Tables 9.6 and 9.7 show various properties of the oxides of the elements in Period 2 and Period 3, respectively. There are marked periodic patterns in the structure, bonding and properties of these oxides.

Structure and bonding

In each period, the oxides of metals and metalloids have giant structures, whereas the oxides of non-metals are composed of simple molecules. Thus, lithium oxide, Li_2O; beryllium oxide, BeO; sodium oxide, Na_2O; magnesium oxide, MgO; and aluminium oxide, Al_2O_3, have giant ionic structures. Consequently, they are solids at room temperature, with high melting points and boiling points. These ionic oxides conduct electricity in the molten state.

The metalloids, boron and silicon, form oxides with giant molecular structures. The structure of silicon(IV) oxide, SiO_2, is discussed in Section 4.14. Strong covalent bonds link one atom to the next in these giant structures. They are therefore solids with high melting points and boiling points. The melting point of B_2O_3 is 577 °C and that of SiO_2 is 1700 °C. Unlike ionic solids, giant molecular solids do not conduct electricity in the molten state.

The oxides of non-metals, such as carbon dioxide (CO_2), nitrogen dioxide (NO_2), oxygen difluoride (OF_2), phosphorus(V) oxide (P_4O_{10}), sulfur trioxide (SO_3) and dichlorine heptoxide (Cl_2O_7) consist of discrete small molecules. These simple molecular oxides are much more volatile than the ionic metal oxides and the giant molecular metalloid oxides. They have low melting points and low boiling points and they do not conduct electricity in the liquid state.

Quick Questions

16 The chloride of element X conducts electricity when added to water, but does not conduct as the pure liquid. Which two of the following could be X?
Al   Si   P   K

▼ Table 9.6

Properties of the oxides of elements in Period 2

Formula of oxide	Li_2O	BeO	B_2O_3	CO_2 (CO)	N_2O (NO, NO_2, N_2O_4, N_2O_5)	O_2	OF_2
State of oxide (at 20 °C)	solid	solid	solid	gases	gases (except N_2O_5, a solid)	gas	gas
Conduction of electricity by oxide in liquid state	good	moderate	v. poor	nil	nil	nil	nil
Structure of oxide	Giant structures			Simple molecular structures			
Enthalpy change of formation of oxide at 298 K/kJ mol^{-1}	−596	−611	−1273	−394 (CO$_2$)	+33 (NO$_2$)	–	+22
Enthalpy change of formation of oxide at 298 K per mole of O/kJ	−596	−611	−424	−197	+17	–	+22
Effect of adding oxide to water	Reacts to form LiOH(aq) alkaline solution	BeO does not react with water but it is amphoteric	Reacts to form H_3BO_3, a very weak acid	CO_2 reacts to form H_2CO_3, a weak acid	NO_2 reacts to form an acid solution of HNO_3 and HNO_2	–	OF_2 reacts slowly forming O_2 and an acidic solution of HF
Nature of oxide	**Basic (alkaline)**	**Amphoteric**	**Acidic**	**Acidic**	**Acidic**	–	**Acidic**

Table 9.7
Properties of the oxides of elements in Period 3

Formula of oxide	Na_2O	MgO	Al_2O_3	SiO_2	P_4O_{10} (P_4O_6)	SO_3 (SO_2)	Cl_2O_7 (Cl_2O)
State of oxide (at 20 °C)	solid	solid	solid	solid	solid (solid)	liquid (gas)	liquid (gas)
Conduction of electricity by oxide in liquid state	good	good	good	v. poor	nil	nil	nil
Structure of oxide	Giant structures				Simple molecular structures		
Enthalpy change of formation of oxide at 298 K/kJ mol^{-1}	–416	–602	–1676	–910	–2984 (P_4O_{10})	–395 (SO_3)	+80 (Cl_2O)
Enthalpy change of formation of oxide at 298 K per mole of O/kJ	–416	–602	–559	–455	–298	–132	+80
Effect of adding oxide to water	reacts to form NaOH(aq), alkaline solution	reacts to form Mg(OH)$_2$, weakly alkaline solution	does not react with water but it is amphoteric	does not react with water but it is acidic	P_4O_{10} reacts to form H$_3$PO$_4$, acid solution	SO$_3$ reacts to form H$_2$SO$_4$, acid solution	Cl$_2$O$_7$ reacts to form HClO$_4$, acid solution
Nature of oxide	Basic (alkaline)	Basic (weakly alkaline)	Amphoteric	Acidic	Acidic	Acidic	Acidic

Notice the gradations in structure and bond type across each period from ionic oxides and chlorides to simple molecular oxides and chlorides. The gradations can be correlated with changes in electronegativity across the period from low values on the left to high values on the right. Thus, atoms of low electronegativity, such as Na, Mg and Li, form compounds in which they have given up electrons to either chlorine atoms or oxygen atoms. In contrast to this, compounds formed by the more electronegative atoms (such as Si, P, S and F) with either chlorine or oxygen exist as discrete molecules (e.g. $SiCl_4$, ClF) or as giant covalent structures (e.g. SiO_2).

Thus, bonding in the oxides and chlorides becomes less ionic and more covalent as we move across a period from atoms of low to atoms of high electronegativity.

Enthalpy changes of formation

Tables 9.6 and 9.7 show the standard enthalpy changes of formation of some of the oxides in Periods 2 and 3. Notice that most of the oxides have a negative enthalpy change of formation. This means that oxides are usually very stable. Indeed, only fluorides are generally more stable than oxides.

The standard enthalpy changes of formation of the oxides vary from very large negative values to relatively small positive values. At first sight, there appears to be no distinct pattern in the standard enthalpy changes of formation of the oxides. This is because the number of O atoms in one mole of oxide varies as well as the strength of the bonds. However, when we compare the enthalpy changes of formation *per mole of oxygen atoms*, an obvious pattern appears (Tables 9.6 and 9.7 and Figure 9.20). This value gives the strength of bonds to one mole of oxygen atoms in the oxide. The enthalpy change of formation per mole of oxygen atoms becomes less negative as the proton number increases across a period. This means that, in general, oxygen forms its most stable compounds with elements such as Li, Na, Mg and Al which are furthest removed from it in the periodic table. This is generally the case with other pairs of elements.

Key Point

Moving across Periods 2 and 3, the **oxides of elements** change from ionic, involatile metal oxides, through giant molecular, involatile metalloid oxides to simple molecular, volatile non-metal oxides.

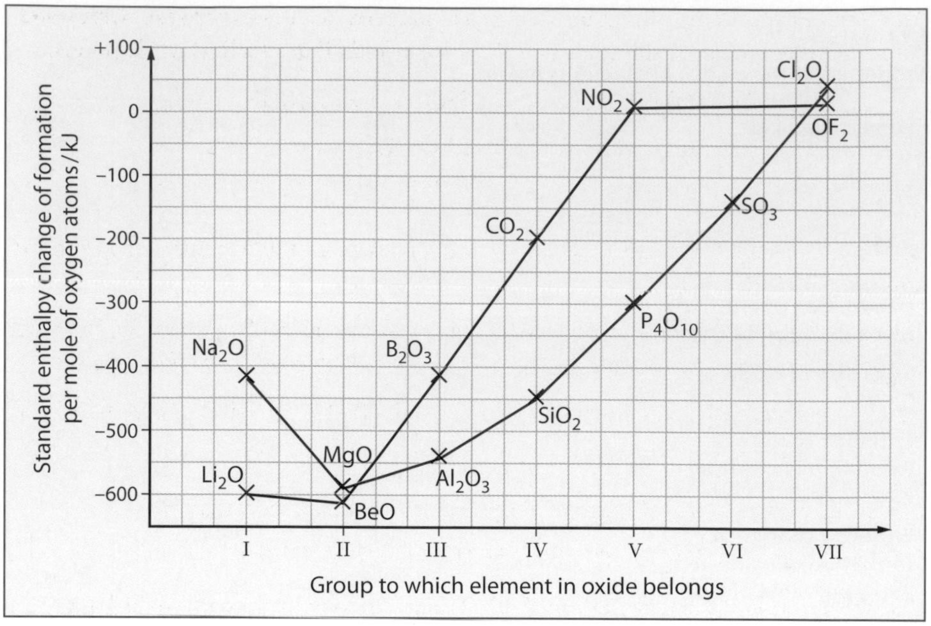

Figure 9.20

Variation in the enthalpy changes of formation of oxides in Periods 2 and 3 per mole of oxygen atoms

Quick Questions

17 Look at Table 9.7 to see how the following oxides react with water: Na_2O, MgO, P_4O_{10}, SO_3 and Cl_2O_7

a Write equations for the reaction of each of these oxides with water.

b How does the acid/base character of the oxides change across Periods 2 and 3?

Definition

Alkalis are soluble bases. So, basic oxides which are soluble in water are also alkalis.

Reaction with water: acid–base character of oxides

Before you read any further try to answer Quick question 17.

As we pass across a period from left to right there is a steady change in the structure of oxides from ionic, through giant molecular to simple molecular. This change in structure leads to a profound difference in the way in which the oxides react with water, acids and alkalis.

The ionic oxides contain O^{2-} ions in the crystal lattice. These O^{2-} ions in Li_2O and Na_2O, the oxides of Group I metals, react vigorously with water to form alkaline solutions.

$$O^{2-}(s) + H_2O(l) \longrightarrow 2OH^-(aq)$$

$$Li_2O(s) + H_2O(l) \longrightarrow 2Li^+(aq) + 2OH^-(aq)$$

$$Na_2O(s) + H_2O(l) \longrightarrow 2Na^+(aq) + 2OH^-(aq)$$

These oxides would react even more vigorously with acids, forming a solution containing cations of the metal.

$$Na_2O(s) + 2H^+(aq) \longrightarrow 2Na^+(aq) + H_2O(l)$$

The oxides of Group II metals do not react so readily with water since the large charge density on the Group II cation holds the O^{2-} ions more firmly.

Thus, MgO is only slightly soluble in water, although it reacts readily with acids to form a solution of magnesium ions.

$$MgO(s) + H_2O(l) \rightleftharpoons Mg^{2+}(aq) + 2OH^-(aq)$$

$$MgO(s) + 2H^+(aq) \longrightarrow Mg^{2+}(aq) + H_2O(l)$$

BeO is insoluble in water, but it shows basic properties by dissolving in acids to form Be^{2+} salts.

$$BeO(s) + 2H^+(aq) \longrightarrow Be^{2+}(aq) + H_2O(l)$$

However, BeO also resembles acidic oxides by reacting with alkalis to form salts called beryllates.

$$BeO(s) + 2OH^-(aq) + H_2O(l) \longrightarrow Be(OH)_4{}^{2-}(aq)$$

tetrahydroxoberyllate(II)
(beryllate)

Aluminium oxide has similar properties to beryllium oxide. It does not react with water, but it will react with both dilute acids ($H^+(aq)$ ions) and dilute alkalis ($OH^-(aq)$ ions).

$$Al_2O_3(s) + 6H^+(aq) \longrightarrow 2Al^{3+}(aq) + 3H_2O(l)$$

$$Al_2O_3(s) + 2OH^-(aq) + 3H_2O(l) \longrightarrow 2[Al(OH)_4]^-(aq)$$
tetrahydroxoaluminate(III)
(aluminate)

Oxides such as BeO and Al_2O_3, which show *both basic and acidic properties*, are called **amphoteric oxides**.

The remaining oxides of the elements in Periods 2 and 3 are all acidic except CO, N_2O and NO which are described as **neutral**. *Neutral oxides show neither acidic nor basic character.*

Boron(III) oxide, B_2O_3, reacts with water to form boric(III) acid, H_3BO_3, a very weak acid.

$$B_2O_3(s) + 3H_2O(l) \longrightarrow 2H_3BO_3(aq)$$

CO_2 dissolves in water and reacts slightly to form weak carbonic acid, H_2CO_3.

$$CO_2(g) + H_2O(l) \rightleftharpoons H_2CO_3(aq)$$

CO_2 will react with the OH^- ions in alkalis to form first hydrogencarbonate, HCO_3^- and then carbonate, CO_3^{2-}.

$$CO_2(g) + OH^-(aq) \longrightarrow HCO_3^-(aq)$$

$$CO_2(g) + 2OH^-(aq) \longrightarrow CO_3^{2-}(aq) + H_2O(l)$$

Silicon(IV) oxide, SiO_2, does not react with water, but it reacts with concentrated alkalis forming silicate, SiO_3^{2-}.

$$SiO_2(s) + 2OH^-(aq) \longrightarrow SiO_3^{2-}(aq) + H_2O(l)$$

NO_2 reacts with water to form a mixture of two acids, HNO_2 and HNO_3.

$$2NO_2(g) + H_2O(l) \longrightarrow HNO_2(aq) + HNO_3(aq)$$

The oxides of P, S and Cl react readily with water to form strong acids.

$$P_4O_{10}(s) + 6H_2O(l) \longrightarrow 4H_3PO_4(aq)$$
phosphorus(V)   phosphoric(V) acid
oxide
(phosphorus pentoxide)

$$SO_3(g) + H_2O(l) \longrightarrow H_2SO_4(aq)$$
sulfur trioxide   sulfuric acid

$$Cl_2O_7(l) + H_2O(l) \longrightarrow 2HClO_4(aq)$$
dichlorine   chloric(VII) acid
heptoxide   (perchloric acid)

$$Cl_2O(g) + H_2O(l) \longrightarrow 2HClO(aq)$$
dichlorine   chloric(I) acid
oxide   (hypochlorous acid)

Notice that *as we cross the periodic table, we move from the ionic oxides of metals, which are basic, to the oxides of metalloids with giant molecular structure, which are weakly basic, weakly acidic or amphoteric, and finally to the simple molecular oxides of non-metals, which are acidic.*

Definitions

Basic oxides are oxides which react with acids to form salts.

Acidic oxides are oxides which react with bases to form salts.

Amphoteric oxides are oxides with both basic and acidic properties. They react with both H^+ ions and OH^- ions to form salts.

Neutral oxides show neither acidic nor basic character.

Key Point

Moving across Periods 2 and 3, the oxides of elements change from giant ionic and basic metal oxides, through giant molecular oxides of metalloids, which are weakly acidic, weakly basic or amphoteric, to simple molecular and acidic non-metal oxides.

Quick Questions

18 The oxide and chloride of an element, X, were mixed separately with water. The two resulting mixtures had different effects on litmus. Which two of the following elements could be X?
Al Si P S

Review questions

1 Consider the elements Na to Cl in the third period.

  a Which of these elements

    i form cations,

    ii form a chloride of empirical formula, XCl_3,

    iii react together to form a compound of formula, XY,

    iv exist as diatomic molecules at room temperature?

  b How do each of the following properties vary for this sequence of elements?

    i the oxidation numbers of the elements in their chlorides,

    ii the boiling points of the chlorides of the elements.

  c Give a brief explanation of the trends in oxidation number in terms of the electronic structures of the atoms of these elements.

  d How do you account for the trend in boiling points of the chlorides of the elements?

2 Use a data book to plot (on the same graph) the melting points and boiling points of the elements from H to Ca against proton number.

  a Which elements occur at or near the peaks on

    i the melting point curves,

    ii the boiling point curves?

  b Which elements occur in the troughs on

    i the melting point curves,

    ii the boiling point curves?

  c What type of structure is found in those elements which occur

    i at or near the peaks,

    ii in the troughs?

  d Which elements are liquid over

    i unusually large temperature ranges,

    ii unusually small temperature ranges?

  e What explanation can you give for the fact that some elements are liquid over:

    i unusually large temperature ranges,

    ii unusually small temperature ranges?

3 Look closely at the boiling points of the elements Na to Ar in Table 9.1.

  a Why do the boiling points rise from Na $\longrightarrow$ Si? Explain the trend in terms of the structure of the elements.

  b Why are the boiling points of the second four elements (P to Ar) much lower than those of the first four?

  c Why is there no clear trend in the boiling points of P, S, Cl and Ar? Draw a sketch graph of the boiling points against relative molecular mass for the simple molecules (P_4, S_8, Cl_2, Ar) of these elements to help you answer this question.

4 An element A reacts with another element X to form a compound of formula AX_2. The element X exists as molecules of formula X_2. Some properties of A, X_2 and AX_2 are tabulated below.

	A	X_2	AX_2
Melting point	High (in the range 700–1200 °C)	Very low (below −50 °C)	Moderately high (in the range 400–700 °C)
Electrical conductivity of the solid	High	Very low	Very low
Electrical conductivity of molten material	High	Very low	High
Electrical conductivity of aqueous solution of the material			High

  a Which particles will move when a potential difference is applied across a sample of: **i** solid A, **ii** molten AX_2?

  b Explain why the electrical conductivity of molten AX_2 is high, whereas that of the solid is very low.

  c Electrolysis of an aqueous solution of AX_2 with platinum electrodes gave A at the cathode and X_2 at the anode. Suggest possible names for the elements A and X consistent with all the above information.

5 The following table shows the melting point and conductivity of five substances.

	Melting point/K	Electrical conductivity in solid state	Electrical conductivity in molten state
Magnesium oxide, MgO	3173	poor	good
Sodium chloride, NaCl	1081	poor	good
Magnesium, Mg	923	good	good
Carbon dioxide, CO_2	217	poor	poor
Silicon(IV) oxide, SiO_2	1883	poor	poor

  a Why is the electrical conductivity of MgO(l) good, but that of MgO(s) poor?

  b Why is the melting point of MgO considerably higher than that of NaCl?

c Why is the electrical conductivity of magnesium good in *both* solid and liquid states?

d Why is the melting point of SiO_2 so much higher than that of CO_2 in spite of the fact that C and Si are both in Group IV of the periodic table?

6 Use the information in the following table to explain the statements below.

	Na	Mg	Al	Si	P	S	Cl
Atomic radius / nm	0.186	0.160	0.143	0.117	0.110	0.104	0.099
Ionic radius / nm	0.095	0.065	0.050			0.184	0.181
First ionisation energy / kJ mol^{-1}	+492	+743	+579	+791	+1060	+1003	+1254

a The atomic radii decreases across the period from Na to Cl.

b The ionic radii of Na^+, Mg^{2+} and Al^{3+} are less than their respective atomic radii, whereas the ionic radii of Cl^- and S^{2-} are greater than their respective atomic radii.

c The first ionisation energies show a general increase from Na to Cl.

d The first ionisation energy of Al is less than that for Mg.

7 The table below shows various properties of the chlorides of the elements in Period 2.

Formula of chloride	LiCl	BeCl$_2$	BCl$_3$	CCl$_4$	NCl$_3$	Cl$_2$O	ClF
State of chloride at 20 °C	s	s	g	l	l	g	g
B.pt. of chloride/°C	1350	487	12	77	71	2	−101
Conduction of electricity by liquid chloride	good	very poor	nil	nil	nil	nil	nil
Structure	giant structures		simple molecular structures				

a Explain the variation in state and boiling point of the chlorides across Period 2.

b LiCl(s) has a giant structure and LiCl(l) is a good conductor of electricity. What type of structure does LiCl(s) have? Why does LiCl(l) conduct electricity?

c How would you expect **i** LiCl, **ii** BCl$_3$ to behave when added to water?

d How is the type of bonding in the chlorides of the elements in Period 2 related to the electronegativity of the element?

8 Consider the elements Li, Be, B, N, F, Ne.

a Which exist as diatomic molecules in the gaseous state at room temperature?

b Which has the highest boiling point?

c Which form a chloride of formula XCl_3?

d Which has the largest first ionisation energy?

e Which has the smallest second ionisation energy?

f Which forms ions with the largest ionic radius?

9 a What is the nature of the bonds in the oxides formed when each of Na, Mg, Al and S react with excess oxygen?

b How do these oxides react with
i water, **ii** dilute acid, **iii** alkali?

c Magnesium chloride is a high melting point solid, whereas silicon tetrachloride is a volatile liquid. Explain the nature of the chemical bonding in these chlorides and show how this accounts for the differences in volatility.

10 Some of the properties of elements Q and R are given below.

Element Q:
i is soft and malleable,
ii floats on water,
iii has a melting point less than 100 °C,
iv forms an ionic hydride of formula QH.

Element R:
i has a density greater than $7 \, g \, cm^{-3}$,
ii has a melting point greater than 2000 K,
iii forms oxides of formulae RO, R_2O_3 and RO_3,
iv forms compounds which are green, orange and violet.

a State the deductions you can make about Q and R from each piece of evidence.

b To which blocks of the periodic table do Q and R belong?

c Identify Q and R as precisely as possible.

10.1 Introduction

Chemists now know of more than one hundred different elements with a bewildering range and variety of properties. Fortunately, their study is made easier because of the periodic table. The patterns and periodicity of similar properties become very evident when we study the elements in each group of the periodic table.

The elements in Group II of the periodic table belong to a family known as the alkaline-earth metals (Table 10.1). The oxides and hydroxides of most Group II metals form *alkaline* solutions and their compounds occur as rocks in the *Earth's* crust. Hence the name alkaline-earth metals (Figure 10.1).

As we study the elements in Group II, look out for two important features:

► the similarities in the elements and in their compounds,
► the gradual trends in the properties of the elements and their compounds down the group.

10.2 Electron structures and atomic properties

All the Group II metals have just two electrons in their outermost shell. These two electrons occupy an s-orbital and the elements belong to the s-block in the periodic table.

> ### Quick Questions
>
> 1 a Write the electron shell structures of Mg and Ca. (The electron shell structure of Be is 2, 2.)
>   b Write the electron sub-shell structures of Mg and Ca. (The electron sub-shell structure of Be is $1s^2$, $2s^2$.)

The s-electrons in Group II metals are situated much further from the nucleus than all their other electrons, so these outer electrons are held only weakly by the positive nucleus. Consequently, the atoms of Group II metals, like those of Group I, readily lose their outermost s-electrons to form stable ions. These ions have a noble gas electron structure with two positive charges, e.g.

$$Mg \longrightarrow Mg^{2+} + 2e^-$$

As we shall see later, this formation of ions with two positive charges dictates most of the properties and reactions of the Group II metals.

Atomic and ionic radii

Table 10.2 summarises the electron structures and important atomic properties of the elements in Group II.

Notice in Table 10.2 and in Figure 10.2 that both the atomic and ionic radii of the elements increase down the group from beryllium to radium.

This increase in the sizes of atoms and ions is not surprising because successive elements have their outermost electrons in a shell further from the nucleus than the previous element.

⊻ Table 10.1
The elements in Group II of the periodic table. (Radium, the last member of Group II is an unstable radioactive element.)

Group II The alkaline-earth metals	
Beryllium	Be
Magnesium	Mg
Calcium	Ca
Strontium	Sr
Barium	Ba
Radium	Ra

⌃ Figure 10.1
The horizontal bands of rock in the Grand Canyon, Arizona, USA, are limestones and sandstones. These rocks contain compounds of calcium and magnesium, two of the alkaline-earth metals in Group II. Limestone is mainly calcium carbonate and sandstone contains calcium and magnesium silicates.

Table 10.2
The electron structures and important atomic properties of the elements in Group II

Element	Be	Mg	Ca	Sr	Ba	Ra
Electron structure	$[He]2s^2$	$[Ne]3s^2$	$[Ar]4s^2$	$[Kr]5s^2$	$[Xe]6s^2$	$[Rn]7s^2$
Atomic radius/nm (metallic radius)	0.112	0.160	0.197	0.215	0.217	0.220
Ionic radius/nm	0.031	0.065	0.099	0.113	0.135	0.140
First ionisation energy /kJ mol^{-1}	900	736	590	548	502	
Second ionisation energy /kJ mol^{-1}	1760	1450	1150	1060	966	
Third ionisation energy /kJ mol^{-1}	14800	7740	4940	4120	3390	

Note

Chemists often use a shortened form for the electron structures of atoms by writing the symbol of the previous noble gas in square brackets to stand for inner shells.

So, using this shortened form, the electron structure of lithium is $[He]2s^1$ and that of sodium is $[Ne]3s^1$.

For each element, the 2+ ion is significantly smaller than its corresponding atom because of the loss of electrons from the outermost shell.

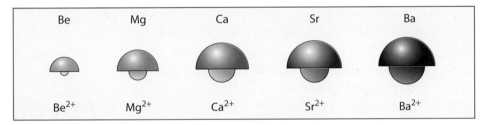

Figure 10.2
Relative sizes of the atoms and ions of the Group II metals

Ionisation energies and oxidation number

Look closely at the first, second and third ionisation energies of the Group II elements in Table 10.2. The second ionisation energy is roughly double that of the first ionisation energy, but the third ionisation energy is between four and eight times higher than the second.

In the case of magnesium, it is approximately twice as difficult to remove the second electron compared to the first, but about five times more difficult to remove the third electron compared to the second.

The first two electrons in an s-orbital can be removed from Group II metals fairly easily. But, the third electron must be removed from an electron configuration similar to that of a noble gas. This third electron is also in a shell closer to the nucleus than the two s-electrons. Its removal is therefore more difficult and requires much more energy. Thus magnesium atoms readily lose two electrons forming Mg^{2+} ions, but never form Mg^{3+} ions. This means that magnesium and the other elements in Group II have only the one oxidation number of +2 in their compounds.

Notice also in Table 10.2 that the first ionisation energies decrease down the group as proton number increases. The same happens with the second and third ionisation energies.

Moving down the group from Be to Ra, the nuclear charge increases, but the shielding effect from inner shells of electrons also increases. We can work out the overall attractive force on the outer electrons from the charge on the nucleus and the charge on electrons in inner shells. This is sometimes called the 'effective nuclear charge'.

Key Point

In the alkaline-earth metals, two electrons occupy the outermost s-orbital. These outermost s-electrons are readily lost forming stable ions with a charge of 2+. As a result, the alkaline-earth metals show only one oxidation number of +2 in all their compounds.

For Be, the effective nuclear charge is:

$+4$	-2	$=$	$+2$
positive charge on nucleus	negative charge on inner electrons		effective nuclear charge

For Mg, the effective nuclear charge is:

$+12$	-10	$=$	$+2$
positive charge on nucleus	negative charge on inner electrons		effective nuclear charge

Can you see that the effective nuclear charge is +2 for *all* the Group II elements because the number of inner shell electrons is always two less than the number of protons? So, down the group from Be to Ra, an effective nuclear charge of +2 is attracting outer s-electrons which are further and further away. This means that the outer electrons are held less strongly by the nucleus and therefore the ionisation energy decreases.

10.3 Physical properties

The relatively low first and second ionisation energies of Group II metals show that their outermost s-electrons are held only weakly by the nucleus. These electrons, like those in Group I metals, can therefore move further from the nucleus than in most other atoms. So, the elements in Groups I and II have larger atomic radii than those elements which follow them in the same period. The large atomic size of Group I and II elements results in weaker forces between neighbouring atoms because there is a reduced attraction of the nuclear charge for shared mobile electrons.

Consequently, the elements in Group II have lower melting points and boiling points than we associate with transition metals like iron and copper. Apart from beryllium, their melting points are less than 840 °C (Table 10.3). In contrast, most of the transition metals melt at temperatures between 1000 and 2500 °C.

⌄ Table 10.3

Physical properties of the elements in Group II

Element	Be	Mg	Ca	Sr	Ba
Melting point /°C	1280	650	838	768	714
Boiling point /°C	2770	1110	1440	1380	1640
Density /g cm^{-3}	1.85	1.74	1.55	2.60	3.51

The reduced interatomic forces in Group II and Group I metals make them relatively soft. All the Group I metals can be cut with a penknife. The Group II metals are not so soft, but they are noticeably softer than transition metals.

The relatively large atomic radii of Group I and Group II metals compared to other elements results in lower densities than other metals. Of the elements listed in Table 10.3, only barium is more dense than aluminium, a metal normally associated with low density. In contrast, most transition metals have a density greater than 7 g cm^{-3}.

10.4 Chemical properties of the Group II elements

All the metals in Group II are high in the activity (electrochemical) series. Hence, these metals are very strong reducing agents losing electrons to form ions with a charge of 2+ when they react. For example: $Mg \longrightarrow Mg^{2+} + 2e^-$

Quick Questions

2 Use the idea of 'effective nuclear charge' to explain why, in any period, the atomic radius of the Group II metal is smaller than that of the Group I metal.

Quick Questions

3 a Why do you think the melting point and boiling point of beryllium are so much higher than other Group II metals?

   b Apart from lithium, all the Group I metals melt below 100 °C. Why are their melting points even lower than those of Group II metals?

Key Point

The s-block metals in Groups I and II have lower melting points, lower boiling points, lower densities and they are softer than transition metals.

Quick Questions

4 Try to think of a reason or reasons why beryllium is a weaker reducing agent than the other Group II metals.

Reactions with oxygen

As strong reducing agents, all the Group II metals, including beryllium, tarnish in air forming a layer of white oxide on top of the silver grey metal. Barium tarnishes so rapidly that it is usually stored under oil like sodium and the other alkali metals:

$$2Be(s) + O_2(g) \longrightarrow 2BeO(s)$$

Apart from beryllium, all the Group II metals burn brightly in oxygen on heating. Beryllium, on heating in air or oxygen, forms a tough layer of beryllium oxide which protects the metal below from further reaction.

Magnesium burns in air with an intense, very bright, white flame forming magnesium oxide:

$$2Mg(s) + O_2(g) \longrightarrow 2MgO(s)$$

Calcium and strontium burn brightly in air with red flames to form their respective oxides.

Barium burns in air with a pale green flame to form barium oxide. In excess air or oxygen, barium produces white ionic barium peroxide, BaO_2, in addition to barium oxide:

$$2Ba(s) + O_2(g) \longrightarrow 2BaO(s)$$

$$Ba(s) + O_2(g) \longrightarrow BaO_2(s)$$

Quick Questions

5 In general, the reactivity of Group II metals with oxygen increases down the group from Be to Ba. Why then do we often get a more intense and rapid reaction when we burn magnesium ribbon and a much weaker flame on burning pieces of calcium?

6 Barium burns in excess air or oxygen to form a mixture of barium oxide and barium peroxide. Draw 'dot-and-cross' diagrams to show the electronic structure of:
   a the oxide ion, O^{2-}
   b the peroxide ion, $^-O{-}O^-$ (O_2^{2-})

7 The bonding in beryllium oxide, $Be^{2+}O^{2-}$, is so strong that it can protect the metal below from further attack when beryllium is heated in oxygen. Suggest a reason for this.

Reactions with water

The reactivity of Group II metals with water gradually increases down the group from Be to Ba.

Beryllium has no apparent reaction with water due to the formation of a thin but very tough and protective oxide layer.

Clean magnesium ribbon reacts very slowly with cold water producing a few tiny bubbles of hydrogen on its surface and magnesium oxide:

$$Mg(s) + H_2O(l) \longrightarrow MgO(s) + H_2(g)$$

Small pieces of calcium react steadily with cold water producing bubbles of hydrogen and a white precipitate of calcium hydroxide:

$$Ca(s) + 2H_2O(l) \longrightarrow Ca(OH)_2(s) + H_2(g)$$

Small pieces of barium react vigorously with cold water producing lots of hydrogen and a solution of barium hydroxide:

$$Ba(s) + 2H_2O(l) \longrightarrow Ba(OH)_2(aq) + H_2(g)$$

▲ Figure 10.3
Beryllium is a light, strong metal with an unusually high melting point. It is therefore used for the construction of high speed aircraft, missiles and space rockets in spite of its high cost.

▲ Figure 10.4
Magnesium powder burns very rapidly with an intense white flame. This has led to its use in fireworks and S.O.S. flares.

Key Point

Group II metals are high in the activity series. They tarnish rapidly in air forming a layer of oxide and reduce water to hydrogen.

Figure 10.5
Calcium hydroxide has been used as a component of mortar in bricklaying for thousands of years. It was used in building the Pyramids in Egypt. Mixed with a little water, $Ca(OH)_2$ slowly hardens by reacting with CO_2 from the air to form water-resistant $CaCO_3$.

Table 10.4
The solubilities of Group II hydroxides at 25 °C

Compound	Solubility /mol per 100 g water at 25 °C
$Mg(OH)_2$	0.20×10^{-4}
$Ca(OH)_2$	15.3×10^{-4}
$Sr(OH)_2$	33.7×10^{-4}
$Ba(OH)_2$	150×10^{-4}

Table 10.5
The solubilities of Group II sulfates at 25 °C

Compound	Solubility /mol per 100 g water at 25 °C
$MgSO_4$	3600×10^{-4}
$CaSO_4$	11×10^{-4}
$SrSO_4$	0.62×10^{-4}
$BaSO_4$	0.009×10^{-4}

10.5 Reactions of the compounds of Group II metals

The common compounds of the alkaline-earth metals are normally white ionic solids, except those compounds in which the anion is coloured, such as manganates and chromates. Another exception is beryllium chloride which is composed of covalently bonded simple molecules.

Reactions of the Group II oxides with water

Beryllium oxide has no reaction with cold water, but magnesium oxide reacts slightly forming magnesium hydroxide.

The oxides of calcium, strontium and barium react vigorously when drops of cold water are added to the dry solids. The mixtures fizz and heat is given out producing the respective hydroxides:

$$CaO(s) + H_2O(l) \longrightarrow Ca(OH)_2(s)$$

Owing to its fast and furious reaction with water, calcium oxide is commonly called 'quicklime' and the calcium hydroxide produced is known as 'slaked lime'.

Reactions of the Group II hydroxides with water

The hydroxides produced when Group II oxides react with water are increasingly soluble from $Mg(OH)_2$ to $Ba(OH)_2$ (Table 10.4). Magnesium hydroxide is only slightly soluble in water producing a very weak alkaline solution.

Calcium hydroxide is sparingly soluble in water. Solutions of calcium hydroxide are usually called limewater. This is used to test for carbon dioxide. If a gas containing carbon dioxide is bubbled into limewater, a milky precipitate of calcium carbonate is produced. This precipitate indicates the presence of carbon dioxide:

$$Ca(OH)_2(aq) + CO_2(g) \longrightarrow CaCO_3(s) + H_2O(l)$$

Strontium hydroxide and barium hydroxide are readily soluble in water forming strongly alkaline solutions.

Reactions of the Group II oxides and hydroxides with acids

All the oxides and hydroxides of Group II metals react with dilute hydrochloric and nitric acids to form salts. For example,

$$MgO(s) + 2HCl(aq) \longrightarrow MgCl_2(aq) + H_2O(l)$$
$$Ca(OH)_2(s) + 2HNO_3(aq) \longrightarrow Ca(NO_3)_2(aq) + 2H_2O(l)$$

Magnesium oxide and hydroxide will also react with dilute sulfuric acid to form a solution of magnesium sulfate. But the oxides of calcium, strontium and barium form an insoluble protective layer of their sulfate which slows down and stops the reaction (Table 10.5).

The oxides and hydroxides of metals, like those in Group II, which react with acids to form salts are described as **basic** (Section 9.11). During reactions, their oxide and hydroxide ions act as bases by accepting H^+ ions (protons) from acids:

$$2H^+(aq) + O^{2-}(s) \longrightarrow H_2O(l)$$
$$H^+(aq) + OH^-(s) \longrightarrow H_2O(l)$$

Those basic hydroxides which dissolve in water to form alkaline solutions are also called **alkalis** (Section 9.11).

Amongst the oxides and hydroxides of Group II, beryllium oxide is unusual. Besides acting as a basic oxide, it can also act as an **acidic oxide**, reacting with bases to form salts called beryllates:

$$BeO(s) + 2OH^-(aq) + H_2O(l) \longrightarrow Be(OH)_4{}^{2-}$$
$$\text{tetrahydroxoberyllate(II)}$$
$$\text{(beryllate)}$$

Oxides, like BeO, which show both basic and acidic properties are described as **amphoteric** (Section 9.11).

Quick Questions

8 a Predict what happens when small pieces of strontium are added to cold water.
  b Write a balanced equation with state symbols for the reaction.
9 Describe and explain what happens when cold water is added dropwise onto barium oxide until water is present in excess.
10 Write balanced equations with state symbols for the following reactions:
  a magnesium oxide with dilute sulfuric acid,
  b barium hydroxide with dilute hydrochloric acid,
  c strontium hydroxide solution with carbon dioxide.

Thermal decomposition of the compounds of Group II metals

Apart from strontium carbonate and barium carbonate, the hydroxides, carbonates and nitrates of Group II metals decompose on heating in a bunsen flame (approx. 900 °C). The products include a solid, ionic metal oxide and a gaseous, simple molecular non-metal oxide.

Hydroxides decompose forming the metal oxide and water vapour:

$$Ca(OH)_2(s) \longrightarrow CaO(s) + H_2O(g)$$

Carbonates decompose forming the metal oxide and carbon dioxide. $SrCO_3$ and $BaCO_3$ need stronger heating to make them decompose:

$$MgCO_3(s) \longrightarrow MgO(s) + CO_2(g)$$

Nitrates decompose forming the metal oxide, brown fumes of nitrogen dioxide and oxygen:

$$Ba(NO_3)_2(s) \longrightarrow BaO(s) + 2NO_2(g) + \tfrac{1}{2}O_2(g)$$

Sulfates decompose on stronger heating:

$$SrSO_4(s) \longrightarrow SrO(s) + SO_3(g)$$

Quick Questions

11 Many rocks throughout the world are composed of limestone (calcium carbonate). What does this suggest about the temperature of these rocks as the Earth was being formed?
12 What mass of solid residue is obtained from the thermal decomposition of 13.0 g of anhydrous barium nitrate?

Key Point

The hydroxides, carbonates, nitrates and sulfates of Group II metals decompose to their oxides on heating.

⌃ Figure 10.6
At the Perrier spring in France, water comes out of the ground naturally sparkling. The limestone rocks beneath the spring are in contact with hot volcanic rocks. These hot rocks cause $CaCO_3$ in the limestone to decompose, producing CO_2 which makes the water naturally sparkling.

Figure 10.7
The only beryllium mineral found in any quantity is the silicate beryl $(3BeSiO_3 \cdot Al_2(SiO_3)_3)$. Emeralds are crystals of beryl, coloured green by traces of chromium(III) ions, Cr^{3+}.

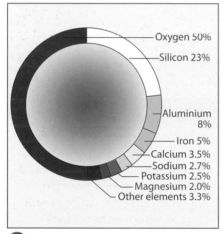

Oxygen 50%
Silicon 23%
Aluminium 8%
Iron 5%
Calcium 3.5%
Sodium 2.7%
Potassium 2.5%
Magnesium 2.0%
Other elements 3.3%

Figure 10.8
The relative abundance of elements on the Earth. (The percentages given relate to the mass of the particular element in the atmosphere and in the Earth's crust to a depth of 40 km).

10.6 Occurrence of the alkaline-earth metals

The alkaline-earth metals occur naturally only as ions with an oxidation number of +2. These 2+ ions occur with anions carrying a charge of 2– as insoluble compounds. So, it is not surprising that Group II elements exist in the Earth principally as solid carbonates, sulfates and silicates.

Magnesium is the eighth most abundant element in the Earth's crust. It is found extensively as the carbonate in magnetite $(MgCO_3)$ and in dolomite $(MgCO_3 \cdot CaCO_3)$. It also occurs as a silicate mineral in asbestos and as aqueous Mg^{2+} ions in sea water.

Calcium is the fifth most abundant element in the Earth's crust and the most abundant s-block element. Vast quantities of calcium occur as the carbonate in chalk, limestone and marble. In fact, calcium carbonate is the second most abundant material in the Earth's crust after silicates like clay, sand and sandstone. Chalk is the softest form of calcium carbonate which has formed from the shells and bones of marine animals that lived millions of years ago. In many places, the chalk was covered by other rocks and put under great pressure. This changed the soft chalk into harder limestone. In other places, the chalk was under both pressure and heat, changing it to marble – the hardest form of calcium carbonate.

Smaller quantities of calcium occur as the sulfate in anhydrite $(CaSO_4)$ and gypsum $(CaSO_4 \cdot 2H_2O)$. These sulfate ores are thought to have resulted from the action of sulfuric acid (produced from the oxidation of sulfide minerals) on limestone.

Strontium and barium are much rarer elements. The commonest ore of strontium is strontianite $(SrCO_3)$ and that of barium is barytes $(BaSO_4)$.

The average abundance of radium (the last element of Group II) in the Earth's crust has been estimated at less than 1 part per million million. Radium exists naturally in compounds as radioactive Ra^{2+} ions. These are formed during breakdown of the nuclei of heavier elements such as uranium.

Figure 10.9
Ever since prehistoric times, people have made designs on the landscape, by removing topsoil to expose deposits of white chalk. This white horse, on a hillside near Uffington, Oxfordshire, England, can be seen for miles.

10.7 Uses of the alkaline-earth metals and their compounds

In spite of being so reactive, magnesium metal has a number of important uses. It is used widely in the preparation of lightweight alloys with a high tensile strength, particularly in the construction of aircraft parts and household goods. Although pure magnesium has poor structural strength, this can be increased by alloying it with aluminium, zinc and manganese. Aluminium increases the tensile strength of the alloy, zinc improves its machining properties and manganese reduces corrosion.

The most commonly used compounds of magnesium are its oxide, MgO and its hydroxide, $Mg(OH)_2$. Magnesium oxide is used as a refractory lining material for high-temperature furnaces because of its exceptionally high melting point and low reactivity.

Magnesium hydroxide is a very weak alkali. Because of this, it is an important constituent of toothpastes. It neutralises acids that might otherwise cause tooth decay. A suspension of $Mg(OH)_2$ in water (commonly called 'milk of magnesia') is used to treat acid indigestion.

Beryllium is too rare and too costly for any large-scale uses, but small amounts are used to harden alloys and other metals such as copper. Beryllium–copper alloys are particularly resistant to sea water corrosion and are therefore used for small parts in boats.

Calcium, strontium and barium metals have very few uses owing to their high reactivity. But the affinity of these elements for oxygen results in their use as deoxidisers in steel production.

Limestone

By far the most commercially important compound of the alkaline-earth metals is limestone. Limestone is one of the most important raw materials for the chemical, agricultural and building industries. The main substance in limestone is calcium carbonate ($CaCO_3$). Quicklime (calcium oxide, CaO) and slaked lime (calcium hydroxide, $Ca(OH)_2$) are readily manufactured from limestone which increases its importance.

Figure 10.10
Compounds of calcium, strontium and barium are often used in fireworks because Ca^{2+}, Sr^{2+} and Ba^{2+} ions emit bright colours on heating to high temperatures.

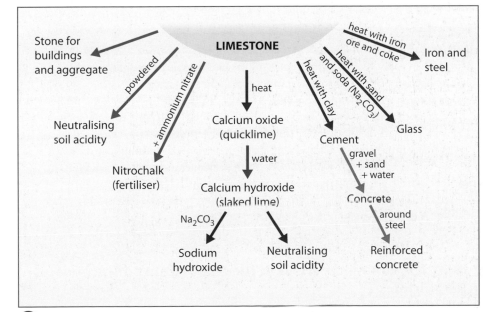

Figure 10.11
Important processes, uses and products of limestone

Quick Questions

13 Fireworks contain a fuel and an oxidising agent. In a particular firework, the fuel is magnesium and the oxidising agent is barium nitrate. Which two of the following substances are produced when the firework is lit?
   a BaO
   b $Ba(NO_2)_2$
   c MgO
   d $Mg(NO_3)_2$

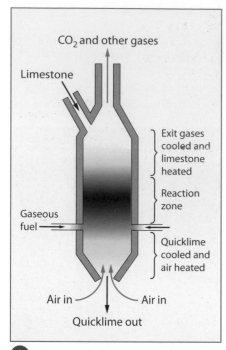

Figure 10.12
A modern gas-fuelled limestone kiln for manufacturing quicklime

The processes and uses of limestone are shown in Figure 10.11. Notice that:

▸ the main uses of limestone itself are in building and construction, in neutralising soil acidity, in fertilisers and in making iron and steel,
▸ the main substances made from limestone are calcium oxide (quicklime), calcium hydroxide (slaked lime), cement and glass.

Calcium oxide (quicklime) is made by the thermal decomposition of limestone at 1000 to 1500 °C (Figure 10.12).

$$CaCO_3(s) \longrightarrow CaO(s) + CO_2(g)$$

Most of the calcium oxide produced from limestone is converted to calcium hydroxide (slaked lime). However, calcium oxide does have an important application in iron-making. When limestone is heated with iron ore and coke, the calcium oxide which forms, reacts with acidic impurities in the ore, such as sand (impure silicon dioxide, SiO_2) forming a molten slag:

$$CaO(s) + SiO_2(s) \xrightarrow{1000\,°C} CaSiO_3(l)$$

Calcium hydroxide (slaked lime) is produced by adding water to calcium oxide:

$$CaO(s) + H_2O(l) \longrightarrow Ca(OH)_2(s)$$

Calcium hydroxide is the cheapest industrial alkali. It is used in water treatment, in neutralising acidic soils and in making sodium hydroxide and bleaching powder.

Cement is made by heating limestone with clay. It contains a mixture of calcium silicate and aluminium silicate. The mixture reacts with water to form hard, interlocking crystals as the cement sets. When cement is used, it is normally mixed with two or three times the volume of sand as well as water.

Concrete is a mixture of cement, sand and water plus aggregate (gravel, broken stones or bricks). As the mixture of cement, sand and water sets around the aggregate, it produces a hard, stone-like building material.

Reinforced concrete is made by allowing concrete to set around a steel framework. This **composite material** has both the hardness of concrete and the flexibility and strength of steel. Because of this, it is used in building large structures like high-rise flats and office blocks.

Definition

A **composite material** contains two different materials in which the particles do not mix, but work together and combine the properties of both materials.

Quick Questions

14 State whether the following processes are exothermic or endothermic:
   a decomposing limestone to produce quicklime,
   b adding water to quicklime to produce slaked lime,
   c neutralising acidity in the soil with slaked lime.
15 Write balanced equations for the following reactions:
   a calcium hydroxide with sodium carbonate to make sodium hydroxide,
   b calcium hydroxide with chlorine to make bleaching powder (calcium chloride + calcium hypochlorite ($Ca(ClO)_2$),
   c limestone with soda (sodium carbonate) and sand (SiO_2) to produce glass (sodium silicate (Na_2SiO_3) and calcium silicate).

10.8 Explaining the trend in thermal stability of carbonates and nitrates

Almost all the compounds of Group II are ionic with a cation which always carries a charge of 2+. So, we would expect Group II compounds to become *more* stable as the 2+ ions become smaller from Ba^{2+} to Be^{2+}. But, this is not the case in practice. Both the carbonates and nitrates of Group II become *less* stable, decomposing more easily from Ba^{2+} compounds to Be^{2+} compounds (Table 10.6).

In order to explain this trend in stability, we need to consider not only the size of ions in the initial carbonates and nitrates, but also the polarisability (Section 5.14) of large anions such as CO_3^{2-} and NO_3^- by cations of varying size from very small Be^{2+} ions to very large Ba^{2+} ions.

The charge density on Mg^{2+} ions is very much larger than that on Ba^{2+} ions. This causes much greater polarisation of the outermost electrons in large CO_3^{2-} and NO_3^- anions. This distorts the CO_3^{2-} and NO_3^- ions, making them less stable. As a result of this, the carbonate and nitrate ions decompose on heating, losing CO_2 and NO_2 respectively and leaving a small oxide ion, O^{2-}, to bond more strongly with the Group II cation.

As polarisation of the large anions by Mg^{2+} is greater than that by Ba^{2+}, MgO forms more easily than BaO and decomposition occurs at a lower temperature. So, the thermal stability of the carbonates and nitrates of Group II elements increases from Mg^{2+} to Ba^{2+} (Figure 10.13).

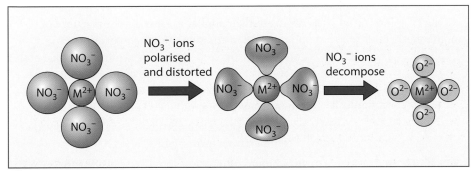

Figure 10.13
The carbonates and nitrates of Group II metals decompose forming their oxides. The smaller the metal ion, the less stable the compound.

10.9 Explaining the variation in solubility of Group II sulfates

In Section 5.15, we used Born–Haber cycles to explain the solubility of ionic compounds in water and established an important link between the enthalpy change of solution, lattice energy and hydration energies.

Table 10.5 (in Section 10.5) and Table 10.7 below show a similar pattern in the solubilities of Group II sulfates and chromates. The solubilities of both the sulfates and chromates decrease as you go down the group from Mg to Ba.

These trends in solubility can be explained in terms of the lattice energies and enthalpy changes of hydration (hydration energies) of the compounds involved.

As we move from $MgSO_4$ through $CaSO_4$ and $SrSO_4$ to $BaSO_4$, the enthalpy change of hydration of the cation becomes much less exothermic. This is because water molecules cannot get so close to the metal ion as its ionic radius increases. This tends to make the salts of larger metal ions like Ba^{2+} less soluble than those of smaller ions, such as Mg^{2+}.

In contrast to their hydration energies, the lattice energies of Group II sulfates do not change greatly from $MgSO_4$ to $BaSO_4$. This is because their lattice energies are determined by the sum of the radii of their constituent ions, $(r_+ + r_-)$. However, the value of $(r_+ + r_-)$ changes relatively slowly from $MgSO_4$ to $BaSO_4$ because r_-, the radius of the SO_4^{2-} ion is much greater than that of any of the positive ions. So, the change in $(r_+ + r_-)$ and therefore in lattice energies is relatively small from $MgSO_4$ to $BaSO_4$.

 Table 10.6
The temperatures at which Group II carbonates and nitrates start to decompose

Compound	Temperature at which decomposition begins / °C
Carbonates	
$BaCO_3$	1360
$SrCO_3$	1280
$CaCO_3$	900
$MgCO_3$	400
$BeCO_3$	cannot exist at room temp.
Nitrates	
$Ba(NO_3)_2$	675
$Sr(NO_3)_2$	635
$Ca(NO_3)_2$	575
$Mg(NO_3)_2$	450
$Be(NO_3)_2$	125

Key Point

The thermal stability of Group II carbonates and nitrates increases from Mg^{2+} to Ba^{2+}.

Key Point

In general, the solubility of Group II compounds containing large anions with a charge of 2– decrease from Mg^{2+} to Ba^{2+}.

Table 10.7
The solubilities of Group II chromates at 25 °C

Compound	Solubility /mol per 100 g water at 25 °C
$MgCrO_4$	8500×10^{-4}
$CaCrO_4$	870×10^{-4}
$SrCrO_4$	5.9×10^{-4}
$BaCrO_4$	0.011×10^{-4}

This means that the variation in solubility of these sulfates is dictated by the trend in enthalpies of hydration of their cations. Also, as these values become less exothermic from Mg^{2+} to Ba^{2+}, their sulfates become less soluble.

Quick Questions

16 a Why are the solubilities in Table 10.7 in mol per 100 g water rather than grams per 100 g water?
   b Why do you think the solubilities of Group II chromates show an even greater decline in solubility than the sulfates?
   c The Group II carbonates do not have the same smooth decline in solubility as the sulfates and chromates. Why is this? (Hint: Think about the relative sizes of the anions involved.)

Review questions

1 a How do each of the following properties of the elements in Group II change with increasing proton number?
   i   Atomic radius
   ii First ionisation energy
   iii Strength as reducing agents
   iv Vigour of reaction with chlorine
   v   Electropositivity.
   b In each case, explain why the five properties change in the way you have suggested.

2 The elements in Group II of the periodic table are barium (Ba), beryllium (Be), calcium (Ca), magnesium (Mg), radium (Ra) and strontium (Sr).
   a Arrange these elements in order of increasing relative atomic mass.
   b Write the electronic structures of any two of these elements except beryllium (e.g. Be would be $1s^2 2s^2$).
   c Draw a sketch graph of ionisation energy against number of electrons removed for the successive ionisation energies of magnesium.
   d Why do all the metals in Group II have an oxidation number of +2 in their compounds?
   e Why do the elements in Group II not have an oxidation number of +1 or +3 in their compounds?

3 a The first member of a group in the periodic table often shows anomalous properties.
   i   State two properties or reactions of beryllium or a compound of beryllium in which the behaviour is anomalous.
   ii State two properties of beryllium or a compound of beryllium which is typical of the elements below it in Group II.

   b Owing to their similar ionic radii, the reactions of lithium and magnesium and their corresponding compounds are very similar. Use your knowledge of the reactions of magnesium and its compounds to decide which one of the following statements concerning lithium and its compounds is correct.
   i   Lithium carbonate decomposes at a relatively low temperature, forming Li_2O and CO_2.
   ii Lithium nitrate decomposes on heating to form lithium nitrite and oxygen.
   iii Lithium burns very slowly in oxygen.
   iv Lithium reacts violently with cold water liberating $H_2(g)$.

4 This question is about the elements in Group II of the periodic table.
   a Copy and complete the table below to show the electronic configuration of calcium atoms and strontium ions, Sr^{2+}.

	1s	2s	2p	3s	3p	3d	4s	4p	4d
Ca	2	2	6						
Sr^{2+}	2	2	6						

   b Explain the following observations:
   i   The atomic radii of Group II elements increase down the group.
   ii The strontium ion is smaller than the strontium atom.
   iii The first ionisation energies of the elements in Group II decrease with increasing proton number.
   c Samples of magnesium and calcium are placed separately in cold water and left for some time. In **each case**, describe what you would see and write a balanced equation for each reaction.

d Strontium nitrate, $Sr(NO_3)_2$ undergoes thermal decomposition:

   i State one observation you would make during this reaction.

   ii Write a balanced equation for this reaction. [4]

Cambridge Paper 2 Q3 June 2007

5 Group II contains the elements Be, Mg, Ca, Sr, Ba.

   a Are these elements metals or non-metals? Give one reason for your answer.

   b Briefly explain how ionisation energies and electrode (reduction) potentials are related to the reactivity of these elements.

   c It is easier to form Mg^+ ions than Mg^{2+} ions from magnesium. In spite of this, Mg^+ is not found in magnesium compounds. Why is this?

   d Group II elements frequently form hydrated salts, while the corresponding compounds of Group I elements are usually anhydrous. Suggest reasons for this difference.

6 Sodium sulfite (Na_2SO_3) is sometimes added to sausage meat to act as a preservative. The amount of Na_2SO_3 present can be determined by boiling a sample of the meat with acid. The quantity of sulfur dioxide which forms is then determined by titration against iodine.

100 g of sausage meat was boiled with 500 cm^3 of 1 $mol\,dm^{-3}$ HCl. The sulfur dioxide evolved was dissolved in water. It was found to require 12.00 cm^3 of 0.025 $mol\,dm^{-3}$ I_2 solution in order to oxidise the SO_2 as in the equation below.

$$SO_2(aq) + 2H_2O(l) + I_2(aq) \longrightarrow$$
$$4H^+(aq) + SO_4{}^{2-}(aq) + 2I^-(aq)$$

In order to check the results of the titration, excess barium chloride is added to the final solution after the titration. The resulting precipitate is collected and weighed. ($Na = 23$, $S = 32$, $O = 16$, $Ba = 137$)

   a How many moles of SO_2 are evolved from 100 g of the sausage meat?

   b How many grams of Na_2SO_3 are present in 100 g of the sausage meat?

   c Government scientists often express the amount of Na_2SO_3 in meat as parts per million (p.p.m.).

   (1 p.p.m. = 1 g of Na_2SO_3 in 10^6 g of meat)

   Express the amount of Na_2SO_3 in the sausage meat in p.p.m.

   d Write an equation for the reaction which occurs when $BaCl_2(aq)$ is added to the solution at the end of the titration.

   e Calculate the mass of precipitate formed when excess $BaCl_2(aq)$ is added to the solution at the end of the titration.

7 Calcium sulfate is found naturally in two forms: anhydrous calcium sulfate ($CaSO_4$, anhydrite) and hydrated calcium sulfate ($CaSO_4{\cdot}2H_2O$ – gypsum). When anhydrite is heated with coke, sulfur dioxide is obtained. This can then be used to manufacture sulfuric acid.

If gypsum is heated at about 125 °C it dehydrates partially forming plaster of Paris, $(CaSO_4)_2{\cdot}H_2O$. When this is mixed with water it changes back to $CaSO_4{\cdot}2H_2O$. The paste expands slightly as it hardens and sets quickly to a firm solid.

When gypsum is heated to 200 °C, it loses all its water of crystallisation. The anhydrous salt which forms sets only very slowly when mixed with water.

   a Explain what is meant by the terms:

   i anhydrous,

   ii hydrated,

   iii water of crystallisation.

   b Write equations for the reactions which occur when:

   i anhydrite is heated with coke,

   ii plaster of Paris is mixed with water,

   iii gypsum is heated at 200 °C.

   c Explain why plaster of Paris is so suitable for:

   i immobilising broken limbs.

   ii making models from moulds.

   d Why is anhydrite not a suitable alternative for the uses mentioned in part c?

11 Group VII – the halogens

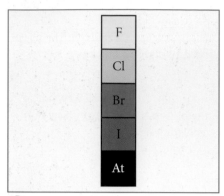

Figure 11.1
Group VII (the halogens)

11.1 Introduction

The halogens are the elements in Group VII of the periodic table – fluorine (F), chlorine (Cl), bromine (Br), iodine (I) and astatine (At) (Figure 11.1). They are known as the *halogens* – a name derived from Greek meaning 'salt formers' – because they combine readily with metals to form salts.

Generally speaking, the halogens comprise the most reactive group of non-metals. In this chapter we shall contrast their properties with those of the reactive metals in Groups I and II of the periodic table. We shall find that *the halogens are strong oxidising agents* whilst the alkali and alkaline-earth metals in Groups I and II are strong reducing agents. In addition, we shall see that the halogens exhibit various oxidation numbers in their compounds. In contrast, the s-block elements have only one oxidation number in their compounds.

Although the halogens show distinct trends in behaviour down Group VII, they also show remarkable similarities. These similarities in their properties and reactions result very largely from their similar electron structures. Each of the halogens has an outer shell containing seven electrons (i.e. ns^2np^5).

Quick Questions

1 **a** Write the electron sub-shell structure for fluorine and chlorine. (For example; that for beryllium would be $1s^2, 2s^2$.)
   b Use the electron structures to explain why fluorine and chlorine are strong oxidising agents.
   c What is the most likely oxidation number of these elements in their compounds?

11.2 Sources of the halogens

The halogens are so reactive that they cannot exist free in nature. Indeed, fluorine is reactive enough to combine directly with almost all the known elements including some of the noble gases. Chlorine reacts directly with all elements except carbon, nitrogen, oxygen and the noble gases.

In the natural environment, halogens always occur as compounds with metals. They are usually present as simple negative ions: fluoride (F^-), chloride (Cl^-), bromide (Br^-) and iodide (I^-). The last element in the Group, astatine (At), occurs in trace amounts in uranium and thorium ores. All the isotopes of astatine are radioactive and the most stable isotope, $^{210}_{85}At$, has a half-life of only 8.3 hours. The name astatine is derived from the Greek word *astatos* meaning 'unstable'.

Fluorine and chlorine are by far the most abundant halogens. The most widespread compounds of fluorine are fluorspar (fluorite), CaF_2 and cryolite, Na_3AlF_6. Unfortunately these extensive deposits of fluoride are dispersed very thinly over the Earth's surface. Only a few sources can be worked economically.

The commonest chlorine compound is, of course, sodium chloride (NaCl) which occurs in sea water and in rock salt. Each kilogram of sea water contains about 30 g of sodium chloride (Figure 11.2).

Figure 11.2
Harvesting salt, sodium chloride, from the sea in Madagascar

Key Points

The halogens are a group of reactive non-metals. They contrast strongly with the alkali and alkaline–earth elements which are reactive metals.

The halogens are so reactive that they occur naturally only in compounds.

The halogens can be obtained by oxidation of halide, Hal⁻ ions.

Bromides and iodides occur in much smaller amounts than either fluorides or chlorides. Sea water contains only small concentrations of bromide, about 70 parts per million by weight, but the extraction of bromine from brine is still economically feasible. Most of the world's production of bromine comes from the Dead Sea in Israel, where the concentration of halide ions is much higher than that usually found in sea water.

Iodides are even scarcer than bromides. Sea water contains traces of iodide (0.05 parts per million by weight). However, the laminarian seaweeds concentrate iodide from the sea to such an extent that fresh wet weed can contain up to 800 parts per million of iodine (Figure 11.3). At one time, the main source of iodine was sodium iodate(v) ($NaIO_3$). This is found only in Chile, mixed with larger proportions of sodium nitrate. Why iodine occurs as iodate(v) in such high concentration in only one part of the world remains a mystery.

Figure 11.3
Iodine can be obtained from laminarian seaweeds. Although sea water contains only traces of iodides, they are concentrated by these seaweeds which can contain up to 800 parts per million.

Figure 11.4
Chilean workmen drilling holes in layers of caliche (impure sodium iodate(v)) prior to blasting

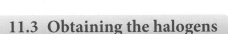

11.3 Obtaining the halogens

The halogens are usually obtained by oxidation of halide ions:

$$2Hal^- \longrightarrow Hal_2 + 2e^-$$

Fluorine is such a powerful oxidising agent that fluoride ions cannot be oxidised by any of the common oxidising agents such as concentrated H_2SO_4, MnO_2 and $KMnO_4$. This means that fluorine must be obtained commercially by electrolysis. But, electrolysis cannot be carried out in aqueous solution, as water or OH^- would be discharged in preference to F^-. And, even if fluorine were produced it would react with water. The electrolyte is hydrogen fluoride dissolved in molten potassium fluoride and the electrodes consist of a graphite anode and a steel cathode (Quick question 2).

The other halogens are much less reactive than fluorine. They can be obtained in the laboratory by oxidising the appropriate halide ions using manganese(IV) oxide, MnO_2 or manganate ions, MnO_4^-.

Quick Questions

2 a Write the symbols, with charges, of all ions present in the electrolyte during the production of fluorine.

b Write an equation to summarise the formation of fluorine at the anode.

c What problems will the formation of fluorine create?

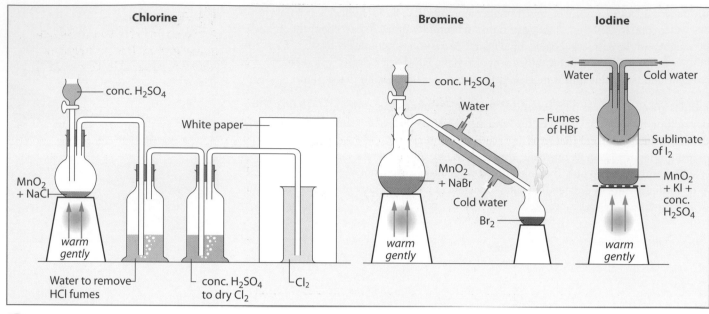

Figure 11.5
Preparing the halogens

Figure 11.5 shows how chlorine, bromine and iodine can be prepared on a small scale in the laboratory. These preparations should be carried out in a fume cupboard. All halogens are poisonous and should be handled with great care. Notice that each preparation is essentially the oxidation of halide ions by MnO_2 in the presence of concentrated H_2SO_4. The apparatus varies from one preparation to the next because chlorine is a gas, bromine is a liquid and iodine is a solid at room temperature.

$$2Hal^- \longrightarrow Hal_2 + 2e^-$$

$$MnO_2 + 4H^+ + 2e^- \longrightarrow Mn^{2+} + 2H_2O$$

Some hydrogen halide is always evolved from the interaction of the halide and concentrated H_2SO_4. The chlorine, for example, will be contaminated with hydrogen chloride fumes. Pure, dry chlorine can be obtained by passing the gas through water to remove the HCl fumes and then through concentrated H_2SO_4 to dry it. Finally, the gas is collected by downward delivery.

Manufacturing chlorine

From an industrial and economic viewpoint, chlorine is by far the most important halogen. At one time, most chlorine was manufactured by the electrolysis of saturated *aqueous* sodium chloride (brine) using cells with a flowing mercury cathode. Serious environmental problems, caused by the escape of relatively small amounts of mercury, have led to the replacement of mercury cells by diaphragm cells (Figure 11.6). The name diaphragm arises because the anode and cathode in each cell are separated by a membrane permeable to $Na^+(aq)$, but not to $Cl^-(aq)$. The process results in the manufacture of hydrogen and sodium hydroxide as well as chlorine.

In the electrolytic process, chlorine is liberated at the titanium anodes and hydrogen is liberated at the nickel cathodes.

Anode (+) (titanium)	$2Cl^-(aq) \longrightarrow Cl_2(g) + 2e^-$	
Cathode (−) (nickel)	$2H_2O(l) + 2e^- \longrightarrow H_2(g) + 2OH^-(aq)$	

As Cl^- ions are removed at the anode, the only ions passing through the permeable membrane into the cathode compartment are $Na^+(aq)$. As water molecules are reduced to hydrogen gas and hydroxide ions at the cathode, this leaves Na^+ and OH^- ions (i.e. effectively sodium hydroxide solution) to flow out of the cell.

Quick Questions

3 **a** Why is conc. H_2SO_4 needed, as well as MnO_2, in preparing the halogens in Figure 11.5?

b Why is conc. H_2SO_4 used rather than dilute H_2SO_4?

c Why is the white paper placed behind the gas jar in which Cl_2 collects?

d The bromine which is collected will be contaminated with fumes of HBr. How could the liquid bromine be purified? (Hint: How are liquids normally purified?)

e Why is cold water passed through the flask on which solid iodine is deposited?

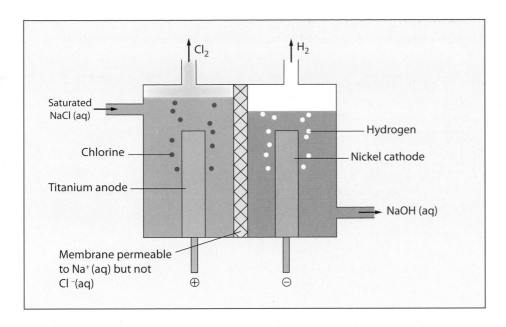

Figure 11.6
A diaphragm cell in which chlorine is manufactured by electrolysis of saturated aqueous sodium chloride

The permeable membrane plays another important part in the process because it keeps the chlorine, produced at the anode, away from the sodium hydroxide, produced in the cathode compartment. If the electrolysis were carried out in a cell without a diaphragm, the Cl_2 and NaOH would react immediately (Section 11.7).

All three products (chlorine, hydrogen and sodium hydroxide) are valuable materials for the chloralkali industry. Chlorine and sodium hydroxide are both in the top 10 manufactured chemicals in terms of the mass produced with worldwide annual productions of more than 10 million tonnes.

The raw materials used for the process – sodium chloride and water, are readily obtainable and cheap, but the electrolytic process is very energy consuming, making high demands on electrical energy.

11.4 Structure and physical properties of the halogens

All the halogens exist as diatomic molecules. The two atoms are linked by a covalent bond (Figure 11.7). These molecules persist in the gaseous, liquid and solid states. Table 11.1 shows some of the atomic and physical properties of the halogens.

All the halogens are coloured – the intensity of colour increasing with increase in proton number. Fluorine is pale yellow, chlorine is pale green, bromine is red–brown and iodine is shiny dark grey, forming a deep violet vapour when heated.

Notice also the change in volatility down the group. Fluorine and chlorine are gases, bromine is a liquid and iodine is a solid. Moving down the group, the volatility of halogens decreases and this is reflected not simply in terms of changes of state, but also in increasing melting points, boiling points, and enthalpy changes of vaporisation. This decreasing volatility is, of course, related to the increasing strength of van der Waals' forces with increasing relative molecular mass. As the halogen molecules get larger, their outer electrons become more polarisable resulting in larger induced dipoles and therefore stronger intermolecular forces (Section 3.11).

All the halogens except fluorine dissolve slightly in water. Fluorine is such a powerful oxidising agent that it converts water to oxygen. The halogens, being non-polar simple molecules, are more soluble in organic solvents, the colour of the solution depending upon the particular halogen and the solvent.

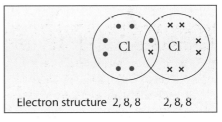

Electron structure 2, 8, 8   2, 8, 8

Figure 11.7
Chlorine, like other halogens, exists as a diatomic molecule. The two chlorine atoms are linked by a covalent bond.

Table 11.1

Atomic and physical properties of the halogens

Element	Fluorine	Chlorine	Bromine	Iodine
Proton number	9	17	35	53
Electron shell structure	2, 7	2, 8, 7	2, 8, 18, 7	2, 8, 18, 18, 7
Outer shell electron configuration	$2s^22p^5$	$3s^23p^5$	$4s^24p^5$	$5s^25p^5$
Relative atomic mass	19.0	35.5	79.9	126.9
State at 20 °C	gas	gas	liquid	solid
Colour	pale yellow	pale green	red–brown	dark grey
Melting point/°C	−220	−101	−7	113
Boiling point/°C	−188	−35	59	183
Enthalpy change of vaporisation /kJ mol^{-1}	+3.3	+10.2	+15	+30
Solubility/g per 100 g of water at 20 °C	reacts readily with water	0.59 (reacts slightly)	3.6	0.018

Quick Questions

4 Astatine is the halogen element immediately below iodine in the periodic table. Its electron structure is 2, 8, 18, 32, 18, 7. Using Table 11.1, predict the following properties of astatine:

  a colour,
  b state at room temperature,
  c relative atomic mass,
  d melting point,
  e enthalpy change of vaporisation,
  f electron sub-shell structure.

Key Point

The halogens show similarities in their properties and reactions with obvious trends in their behaviour. As proton number increases, the halogens become less reactive.

Table 11.2

Standard enthalpy changes of formation of the sodium halides

Compound	Standard enthalpy change of formation /kJ mol^{-1}
NaF	−573
NaCl	−411
NaBr	−361
NaI	−288

In non-polar solvents such as tetrachloromethane and cyclohexane, chlorine is colourless, bromine is red and iodine is violet. In these solvents the elements exist as relatively free molecules as in the gas phase. In polar (electron-donating) solvents such as water, ethanol and propanone (acetone), bromine and particularly iodine give brownish solutions.

11.5 Chemical reactions of the halogens

All the halogens have one electron less than the noble gas which follows them in the periodic table. As expected, their chemistry is dominated by a tendency to gain one electron and obtain an electron structure like a noble gas.

Consequently, *the halogens react with metals to form ionic compounds containing the Hal⁻ ion. With non-metals, they tend to form simple molecular compounds in which the halogen is linked by a single covalent bond(—Hal) to the other element.*

Reactions with metals

The halogens are very reactive with metals. The vigour of reaction depends on

1 the position of the metal in the activity series, and
2 the particular halogen which is reacting.

The reactivity of the halogen decreases with increase in proton number. Fluorine is, in fact, the most reactive of all non-metals. It combines readily and directly with all metals, whereas iodine reacts very slowly even at high temperatures with metals low in the activity series such as silver and gold.

The standard enthalpy changes of formation of the sodium halides (Table 11.2) provide quantitative evidence for the relative reactivity of the halogens. The heat evolved is greatest for the reaction between sodium and fluorine and least for the reaction between sodium and iodine.

The most reactive halogen is fluorine. Chlorine is the next most reactive, then bromine, then iodine. Thus, reactivity decreases as the relative atomic mass of the halogen increases. This is the reverse of the trend in Group II and other groups of metals where reactivity *increases* as the relative atomic mass increases.

Reactions with non-metals

A similar pattern of reactivity emerges here also. Fluorine reacts directly with all non-metals except nitrogen, helium, neon and argon. It will even react with diamond and xenon on heating.

$$C_{(diamond)} + 2F_2(g) \longrightarrow CF_4(l)$$

$$Xe(g) + 2F_2(g) \longrightarrow XeF_4(s)$$

Chlorine and bromine are much less reactive than fluorine. They will not react directly with any of the noble gases, carbon, nitrogen or oxygen. Iodine does not combine with these elements nor with sulfur, but it reacts readily with phosphorus forming the triiodide.

$$P_4(s) + 6I_2(g) \longrightarrow 4PI_3(s)$$

Reaction with hydrogen

The relative reactivity of the halogens with non-metals is well illustrated by their reaction with hydrogen. Fluorine explodes with hydrogen even in the dark at $-200\,°C$. Chlorine and hydrogen explode in bright sunlight but react slowly in the dark. Bromine reacts with hydrogen only on heating and in the presence of a platinum catalyst. Iodine combines only partially and slowly with hydrogen even on heating:

$$H_2(g) + Cl_2(g) \longrightarrow 2HCl(g)$$

$$H_2(g) + I_2(g) \rightleftharpoons 2HI(g)$$

The great reactivity of fluorine with hydrogen and other non-metals is explained partly in terms of its unexpectedly low bond energy (Table 11.3). This means that the initial stage (in which the F—F bond breaks) requires little energy. Hence the activation energy for reactions involving fluorine is low. Fluorine's reactivity is further explained by its ability to form very strong bonds with other elements. Look at Table 5.2, Section 5.10 and compare the C—F and H—F bond energies with those of the C—Cl and H—Cl bonds respectively.

The hydrogen halides (HF, HCl, HBr and HI) are all colourless, simple molecular but polar, compounds which are gases at room temperature.

Most of their properties are determined by the relative strengths of the H—Hal bonds (Figure 11.8). The H—F bond in hydrogen fluoride is so strong that, in spite of its polarity, HF is a relatively weak acid only partly dissociated in aqueous solution:

$$HF(aq) + H_2O(l) \rightleftharpoons H_3O^+(aq) + F^-(aq)$$

Hydrogen chloride, hydrogen bromide and hydrogen iodide are similar in that they:

▸ fume in moist air,
▸ react with water and ionise completely forming strong acids:

$$HCl(aq) + H_2O(l) \longrightarrow H_3O^+(aq) + Cl^-(aq)$$

The decrease in strength of the H—Hal bond energy, D(H—Hal), also explains the relative thermal stabilities of the hydrogen halides.

Neither hydrogen fluoride nor hydrogen chloride are decomposed by a hot glass rod, but hydrogen bromide produces traces of red-brown bromine and iodine gives copious fumes of violet iodine:

$$2HI(g) \longrightarrow H_2(g) + I_2(g)$$

Quick Questions

5 a List three factors which influence the reactivity of the halogen elements.
b How will each of these three factors influence the reactivity of fluorine compared with chlorine?

Table 11.3
Bond energies in the halogen molecules

Element	Bond energy, D (Hal–Hal)/kJ mol^{-1}
F_2	158
Cl_2	244
Br_2	193
I_2	151

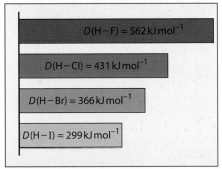

Figure 11.8
H—Hal bond energies, D(H—Hal), determine the relative thermal stabilities of the hydrogen halides and their strengths as acids.

Quick Questions

6 The standard enthalpy changes of formation of HCl and HI are −92 and + 26 kJ mol^{-1}. Which of the following statements is most important in explaining this difference?

a The activation energy, E_A for the H_2/I_2 reaction is smaller than that for the H_2/Cl_2 reaction.

b $D(H—I)$ is smaller than $D(H—Cl)$.

c $D(I—I)$ is smaller than $D(Cl—Cl)$.

 Table 11.4

Standard electrode potentials for the halogens

Electrode reaction at a platinum electrode	Standard electrode potential, $E^{\ominus}$/volts
$F_2(aq) + 2e^- \rightleftharpoons 2F^-(aq)$	+2.87
$Cl_2(aq) + 2e^- \rightleftharpoons 2Cl^-(aq)$	+1.36
$Br_2(aq) + 2e^- \rightleftharpoons 2Br^-(aq)$	+1.07
$I_2(aq) + 2e^- \rightleftharpoons 2I^-(aq)$	+0.54

Figure 11.9

Chlorine acts as an oxidising agent when it bleaches. The piece of printed material on the right was bleached by immersion in chlorine water for a few hours.

Key Point

The chemistry of the halogens is dictated by a tendency to act as oxidising agents in forming halide ions (Hal$^-$). The oxidising power of the halogens is:

$$F_2 > Cl_2 > Br_2 > I_2$$

11.6 The halogens as oxidising agents

When halogens combine with metals or non-metals they normally act as oxidising agents. The elements with which they react have positive oxidation numbers in the resultant compounds.

When halogens combine with metals to form ionic compounds, they gain electrons from the metals to form negative halide ions:

$$2Na(s) + Cl_2(g) \longrightarrow 2Na^+Cl^-(s)$$

The halogens accept electrons during these reactions and act as oxidising agents:

$$Hal_2 + 2e^- \longrightarrow 2Hal^-$$

Fluorine is the most reactive halogen and the most powerful oxidising agent. *The order of decreasing power as oxidising agents is $F_2 > Cl_2 > Br_2 > I_2$.* The electrode potentials (Hal$_2$/Hal$^-$) for the halogens shown in Table 11.4 become less positive from fluorine to iodine. This reflects the decreasing oxidising power.

We have already noticed the different oxidising powers of the halogens with metals, non-metals and hydrogen. The following reactions further emphasise their relative oxidising ability.

Fluorine, chlorine and bromine will all oxidise $Fe^{2+}(aq)$ to $Fe^{3+}(aq)$:

$$2Fe^{2+}(aq) \longrightarrow 2Fe^{3+}(aq) + 2e^-$$

$$Hal_2(aq) + 2e^- \longrightarrow 2Hal^-(aq)$$

Iodine, however, is such a weak oxidising agent that it cannot remove electrons from iron(ii) ions to form iron(iii) ions.

Fluorine and chlorine are such powerful oxidising agents that they can oxidise various coloured dyes to colourless substances. Thus, indicators such as litmus and universal indicator are decolorised when exposed to fluorine and chlorine. Chlorine acts as an oxidising agent when it is used for bleaching (Figure 11.9).

When chlorine water ($Cl_2(aq)$) is added to KI(aq), the solution becomes brown. This is due to the formation of iodine. When bromine water ($Br_2(aq)$) is used in place of chlorine water, iodine is again liberated from KI(aq).

In these two reactions, iodide ions are oxidised to iodine:

$$2I^-(aq) \longrightarrow I_2(aq) + 2e^-$$

The chlorine and bromine act as oxidising agents. They accept electrons from iodide to form chloride and bromide, respectively.

$$Cl_2(aq) + 2e^- \longrightarrow 2Cl^-(aq)$$

$$Br_2(aq) + 2e^- \longrightarrow 2Br^-(aq)$$

Chlorine and bromine can oxidise iodide to iodine, but iodine cannot oxidise chloride or bromide ions. This is because Cl^- and Br^- are so stable they will not release electrons to iodine.

Since chlorine is a stronger oxidising agent than bromine, chlorine can also oxidise bromide ions to bromine. This reaction is used in the manufacture of bromine from bromide ions in sea water:

$$Cl_2(aq) + 2Br^-(aq) \longrightarrow 2Cl^-(aq) + Br_2(aq)$$

So, when chlorine water is added to KBr(aq), yellow–orange bromine is produced.

The reactions of halogens with water

Fluorine and chlorine are strong enough as oxidising agents to oxidise water to oxygen, but bromine and iodine cannot do so.

$$2F_2(g) + 2H_2O(g) \longrightarrow 4HF(g) + O_2(g)$$

The reaction of water with chlorine is very slow. In fact, chloric(I) acid (hypochlorous acid) is first formed, which then decomposes to oxygen and hydrochloric acid.

$$Cl_2(g) + H_2O(l) \longrightarrow H^+(aq) + Cl^-(aq) + HOCl(aq)$$

$$2HOCl(aq) \longrightarrow 2H^+(aq) + 2Cl^-(aq) + O_2(g)$$

11.7 The reactions of halogens with alkalis

Chlorine, bromine and iodine undergo very similar reactions with alkalis. The products depend upon the temperature at which the reaction occurs.

With cold, dilute alkali at 15 °C, the halogen reacts to form a mixture of halide (Hal^-) and halate(I) (hypohalite, $OHal^-$):

$$Hal_2 + 2OH^-(aq) \longrightarrow Hal^-(aq) + OHal^-(aq) + H_2O(l)$$

The $OHal^-$ which is produced in the first reaction then decomposes to form halide and halate(v) ($HalO_3^-$):

$$3OHal^-(aq) \longrightarrow 2Hal^-(aq) + HalO_3^-(aq)$$

For chlorine, this second reaction is very slow at 15 °C, but rapid at 70 °C. Therefore, sodium chlorate(I) (sodium hypochlorite) can be obtained by passing chlorine into sodium hydroxide at 15 °C. Sodium chlorate(v) ($NaClO_3$) is obtained by carrying out the same reaction at 70 °C.

With bromine, both reactions are rapid at 15 °C, but decomposition of OBr^- is slow at 0 °C.

With iodine, decomposition of OI^- occurs rapidly even at 0 °C, so it is difficult to prepare $NaOI$ free from $NaIO_3$.

These two reactions of halogens with alkali involve **disproportionation** – a change in which one particular molecule, atom or ion is simultaneously oxidised and reduced.

When Cl_2 reacts with alkali to form Cl^- and OCl^-, chlorine atoms are both oxidised and reduced. The oxidation number of one Cl atom in Cl_2 changes from 0 to –1 in Cl^- (reduction). The oxidation number of the other Cl atom changes from 0 to +1 in OCl^- (oxidation). Oxidation numbers are shown above each atom in the following equation:

$$\overset{0}{Cl_2}(g) + 2\overset{-2 +1}{OH^-}(aq) \longrightarrow \overset{-1}{Cl^-}(aq) + \overset{-2 +1}{O}\overset{+1 -2}{Cl^-}(aq) + H_2O(l)$$

The halogens, other than fluorine, form compounds in which they have positive oxidation numbers up to +7. Figure 11.10 shows an oxidation number chart for chlorine. The behaviour of halogens in showing several oxidation numbers is, of course, very different from the metals in Groups I and II. These metals show only one oxidation

Quick Questions

7   a   Which halide ions can F_2 oxidise?
    b   Which of the following is the strongest reducing agent?

$$I^- \quad I_2 \quad F_2 \quad F^-$$

Note

According to IUPAC, the acceptable common name for HOCl is hypochlorous acid, although chloric(I) acid is often used.

Oxidation number	Examples
+7	Cl_2O_7, $NaClO_4$
+6	
+5	$NaClO_3$
+4	
+3	$KClO_2$
+2	
+1	Cl_2O, $NaOCl$
0	Cl_2
−1	$NaCl$

Figure 11.10
The range of oxidation numbers for chlorine

Definition

Disproportionation is a reaction in which one molecule, atom or ion is simultaneously oxidised and reduced.

number in their compounds. As we might expect, the most stable oxidation state for halogens is –1. The relative stability of the –1 state decreases as the group is descended and the electronegativity of the halogen decreases.

Fluorine, the most electronegative element, never exhibits a positive oxidation number. This is because it can never form a compound in which it is the less electronegative element. This helps to explain why fluorine reacts with alkalis differently to the other halogens. Fluorine forms a mixture of fluoride and oxygen difluoride in both of which its oxidation number is –1.

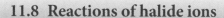

$$2F_2(g) + 2OH^-(aq) \longrightarrow OF_2(g) + 2F^-(aq) + H_2O(l)$$

Quick Questions

8 **a** What is the oxidation number of chlorine in each of OCl^-, Cl^- and ClO_3^-?
  b Write an equation for the disproportionation of OCl^- to Cl^- and ClO_3^-.
  c What is the commonest oxidation number for chlorine in its compounds?
  d Write the formula of a compound containing chlorine in a different oxidation state to that in Cl^-, OCl^- or ClO_3^-.

Key Point

The reaction with $AgNO_3(aq)$ followed by $NH_3(aq)$ or sunlight can be used as a test for halide ions.

11.8 Reactions of halide ions

Some reactions of fluorides, chlorides, bromides and iodides in aqueous solution are summarised in Table 11.5. All common halides are soluble except all lead halides, AgCl, AgBr and AgI. Notice in Table 11.5 that precipitates of these halides are produced when aqueous solutions of halides are treated with either $Pb^{2+}(aq)$ or $Ag^+(aq)$. For example,

$$Ag^+(aq) + Cl^-(aq) \longrightarrow AgCl(s)$$
$$Pb^{2+}(aq) + 2Cl^-(aq) \longrightarrow PbCl_2(s)$$

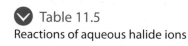 Table 11.5
Reactions of aqueous halide ions

Solution added	F^-(aq)	Cl^-(aq)	Br^-(aq)	I^-(aq)
$Pb(NO_3)_2$(aq)	white ppte of PbF_2	white ppte of $PbCl_2$	yellow ppte of $PbBr_2$	yellow ppte of PbI_2
$AgNO_3$(aq)	no reaction	white ppte of AgCl	cream ppte of AgBr	yellow ppte of AgI
Solubility of AgHal in (a) dil. NH_3(aq) (b) conc. NH_3(aq)	soluble soluble	soluble soluble	insoluble soluble	insoluble insoluble
Effect of sunlight on AgHal	no effect	white AgCl turns purple-grey	cream AgBr turns green-yellow	no effect

Reaction with Ag^+(aq) followed by NH_3(aq)

Silver nitrate solution followed by sunlight or ammonia solution can be used as a test for halide ions.

F^-(aq) forms no precipitate with silver nitrate solution.
Cl^-(aq) gives a white precipitate of AgCl which becomes purple-grey in sunlight and dissolves in dilute NH_3.
Br^-(aq) gives a cream precipitate of AgBr which becomes green-yellow in sunlight and dissolves in concentrated NH_3.
I^-(aq) gives a yellow precipitate of AgI which is unaffected by sunlight and which is insoluble in concentrated NH_3.

The precipitates of AgCl and AgBr dissolve in ammonia solution because they react with it to form soluble complex salts containing the $Ag(NH_3)_2^+(aq)$ ion. For example,

$$AgCl(s) + 2NH_3(aq) \longrightarrow Ag(NH_3)_2^+(aq) + Cl^-(aq)$$

The colour changes which occur when AgCl and AgBr are exposed to sunlight result from conversion of the surface layer of these silver halides to silver and halogen.

$$Ag^+Br^-(s) \longrightarrow Ag(s) + \tfrac{1}{2}Br_2(g)$$

This photochemical change involving AgBr plays an essential part in photography using black and white film. Photographic films contain silver bromide. This decomposes to silver on exposure to light. When the photograph is developed, the film is treated with 'hypo' (sodium thiosulfate solution). This removes excess AgBr as a soluble complex ion $[Ag(S_2O_3)_2]^{3-}$, and the silver remains on the film as a dark shadow.

Figure 11.11
This is a black and white photograph of the first flight over the English Channel by Louis Bleriot in 1909. Black and white photographs depend on a photochemical change involving silver bromide.

Reaction with conc. H_2SO_4

The reactions of solid halides with concentrated sulfuric acid and with concentrated phosphoric(v) acid are summarised in Table 11.6.

Table 11.6
Reactions of solid halides with conc. H_2SO_4 and conc. H_3PO_4

Reagent added	Fluoride	Chloride	Bromide	Iodide
Conc. H_2SO_4	HF(g) produced	HCl(g) produced	HBr(g) + a little red-brown Br_2(g) produced	a little HI(g) + purple I_2(g) produced
Conc. H_3PO_4	HF(g) produced	HCl(g) produced	HBr(g) produced	HI(g) produced

When concentrated H_2SO_4 is added to solid halides the first product is the hydrogen halide. Being relatively volatile, these hydrogen halides are evolved as gases.

$$Hal^-(s) + H_2SO_4(l) \longrightarrow HHal(g) + HSO_4^-(s)$$

However, concentrated sulfuric acid is also an oxidising agent. It is powerful enough to oxidise HBr to Br_2 and III to I_2, but it cannot oxidise HF and HCl.

$$2HBr(g) + H_2SO_4(l) \longrightarrow Br_2(g) + 2H_2O(l) + SO_2(g)$$

$$2HI(g) + H_2SO_4(l) \longrightarrow I_2(g) + 2H_2O(l) + SO_2(g)$$

11.9 Uses of the halogens and their compounds

Apart from chlorine, the halogen elements have few uses, but their compounds are used extensively in industry, in agriculture, in medicine and in the home. Chlorine is used in water purification and as a cheap industrial oxidant in the manufacture of bromine (Section 11.6).

Chlorine in water purification

Since the nineteenth century, the treatment of water supplies with chlorine has helped to radically reduce diseases such as typhoid and cholera. Chlorine disinfects water by reacting with it to form chloric(I) acid, HClO (hypochlorous acid).

$$Cl_2(g) + H_2O(l) \rightleftharpoons HOCl(aq) + HCl(aq)$$

Chloric(I) acid is a strong oxidising agent and a weak acid. It works as a disinfectant because the HOCl molecule can pass through the cell walls of bacteria, unlike OCl⁻ ions. When HOCl molecules get inside a bacterial cell, they oxidise molecules involved in the cell's metabolism and structure, killing the organism.

Quick Questions

9 Look at the reactions of solid halides with conc. H_3PO_4 in Table 11.6:
   a Write an equation for the reaction of conc. H_3PO_4 with NaBr(s).
   b Why does conc. H_3PO_4 produce HBr(g) but no Br_2 like conc. H_2SO_4?

10 Which of the following properties are smaller for chlorine than iodine?
   a solubility of the silver halide in NH_3(aq),
   b strength of intermolecular forces in the element,
   c thermal stability of the hydrogen halide.

Figure 11.12
Swimming pools used to be treated with chlorine gas from cylinders containing liquid chlorine under pressure. Today, swimming pools are usually treated with sodium chlorate(I), NaOCl which produces chloric(I) acid when dissolved in water:

$$OCl^-(aq) + H_2O(l) \rightleftharpoons OH^-(aq) + HOCl(aq)$$

Quick Questions

11 Look carefully at the caption to Figure 11.12.
   a Why do OCl^- ions react with water to form HOCl molecules?
   b Why is it important that HOCl molecules are formed in the water?
   c In sunlight, OCl^- ions are broken down by ultraviolet light.

   $$OCl^-(aq) \overset{uv}{\rightleftharpoons} Cl^-(aq) + \tfrac{1}{2}O_2(g)$$

   Which of the following actions would maintain the highest concentration of HOCl(aq)?
   i   acidify the pool,
   ii add Cl^- ions to the water,
   iii bubble oxygen into the water.

The uses of chlorine compounds

Chlorine is used in the manufacture of many familiar materials. Hydrogen chloride is produced by reacting chlorine with hydrogen, itself a by-product in the electrolytic manufacture of chlorine. Hydrogen chloride made in this way is dissolved in water to produce hydrochloric acid.

$$H_2(g) + Cl_2(g) \longrightarrow 2HCl(g) \xrightarrow{aq} 2H^+(aq) + 2Cl^-(aq)$$
$$\text{hydrochloric acid}$$

Hydrochloric acid is the cheapest industrial acid. It is used in removing (de-scaling) rust from steel sheets before galvanising.

The other by-product of the chloralkali process is sodium hydroxide. Most of this is treated with gaseous chlorine to produce sodium chlorate(I) (sodium hypochlorite, NaOCl):

$$Cl_2(g) + 2NaOH(aq) \longrightarrow NaCl(aq) + NaOCl(aq) + H_2O(l)$$

The solution produced in this reaction, containing both sodium chlorate(I) and sodium chloride, is sold as liquid bleach (Figure 11.13). This is used as a domestic bleach and as a sterilising fluid in dilute solutions. Concentrated solutions are used to bleach paper and wood pulp. The active bleaching agent in the solution is chlorate(I), OCl^-, which acts as a powerful oxidising agent in the bleaching reactions.

VECTA THICK STRONG BLEACH

CONTAINS
Sodium Hypochlorite and
Sodium Hydroxide

WARNING! Do not
use with other products.
May release dangerous
gases (Chlorine).

IRRITANT

Figure 11.13
Sodium chlorate(I), commonly called sodium hypochlorite, is used as a bleach and as a disinfectant.

On warming, sodium chlorate(I) decomposes to sodium chlorate(V), $NaClO_3$ which is used as a weed killer:

$$3NaOCl(aq) \xrightarrow{\text{heat}} 2NaCl(aq) + NaClO_3(aq)$$

Various other organic compounds of chlorine are produced and used commercially. These include solvents such as dichloromethane, plastics such as PVC (Section 27.3) and disinfectants and antiseptics such as Dettol and TCP (Section 25.10).

During the 1940s and 1950s, a range of highly chlorinated organic compounds were developed and used as pesticides. The best known of these compounds are DDT (dichlorodiphenyltrichloroethane) and BHC (benzene hexachloride).

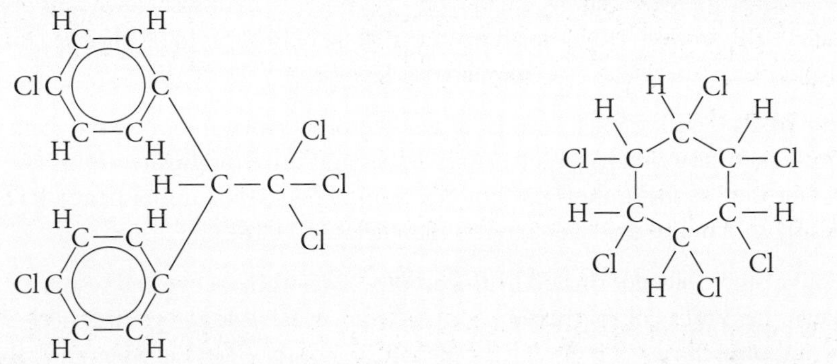

dichlorodiphenyltrichloroethane
(DDT)

benzene hexachloride
(BHC)

⬆ Figure 11.14
The structures of DDT and BHC

DDT and BHC are extremely potent insecticides. They act by changing the structure of nerve cell walls, allowing sodium and potassium ions to escape. Eventually, nerve action ceases and the insect is paralysed.

The spraying of large areas of arable land with these chlorine-containing pesticides led to enormous improvements in the yield of crops. They helped to eliminate many insect-borne diseases such as malaria, yellow fever and sleeping sickness. The World Health Organisation (WHO) estimated that 5 million lives were saved in the first eight years of using DDT.

Unfortunately, however, these chlorinated compounds are so stable that they remain unchanged on the crops and accumulate in the soil. The compounds are also fat-soluble but not water-soluble. This means that they concentrate in the fatty tissues of animals that ingest them. During the 1950s, unusually large numbers of seed-eating birds began to die in the cereal-growing areas of Western Europe. Then, in the 1960s, birds of prey and amphibians became much less common in Europe and North America. Their eggs and corpses were analysed and found to contain abnormally high levels of DDT.

Tests showed that insects can develop a tolerance to small, non-lethal doses of DDT. But, these insects are eaten by small carnivores which concentrate the DDT in their own fatty tissue. Larger predators which eat these small carnivores concentrate the DDT even further. Ultimately, the DDT reaches toxic levels, particularly in birds of prey and amphibians several stages down the food chain.

Experiments also showed that DDT has less harmful effects in mammals which have enzymes that convert DDT to DDE (dichlorodiphenyldichloroethene). The different

⬆ Figure 11.15
During the 1950s and 1960s, peregrine falcons became extinct in some areas of Europe and North America owing to the use of DDT and other toxic insecticides. The birds are now re-establishing themselves in these areas following restrictions on the use of DDT imposed in the 1970s.

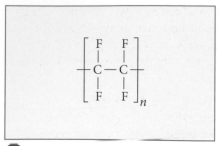

Figure 11.16
The structure of poly(tetrafluoroethene)

Figure 11.17
The running surfaces of skis are coated with PTFE

shape of DDE reduces its interference with the cell wall of nerves. However, the identification of high levels of DDT in human milk led to increasing concerns and the world consumption of DDT fell from 400 000 tonnes in 1963 to only about half that quantity in 1971.

Since 1972, the governments of many countries throughout the world have imposed severe restrictions on the use of DDT which is now only used for the most difficult control schemes.

Uses of the other halogens and their compounds

Small quantities of fluorine are used in rocket propulsion. However, much larger quantities are used to make uranium hexafluoride, UF_6, for the separation of ^{238}U and ^{235}U isotopes for use in nuclear power stations. By allowing gaseous UF_6 to diffuse slowly, it is possible to separate $^{238}UF_6$ from $^{235}UF_6$. This allows the mixture of isotopes to be 'enriched', increasing the proportion of the useful ^{235}U compared to less useful ^{238}U.

Fluorine is also used to make a wide range of fluoroalkanes and other halogenated compounds for use as refrigerants, anaesthetics and fire extinguisher fluids (Section 16.3). One of the most important fluorocarbons is poly(tetrafluoroethene), PTFE, frequently sold under the trade name Teflon (Figure 11.16).

PTFE, like other fully fluorinated hydrocarbons, is very unreactive and resists almost all corrosive chemicals. For this reason, it is used for valves, seals and gaskets in chemical plants and laboratories. It is also an excellent insulator for electrical wires and cables.

PTFE has an extremely low coefficient of friction and strong anti-stick properties. Because of this, thin layers of it are coated on the running surface of skis and on the cooking surface of non-stick saucepans.

Bromine and iodine are used in much smaller quantities than either chlorine or fluorine. Bromine is used in making a range of halogenated organic products including flame retardants, medicines and dyes. Iodine, like bromine, is also used in the manufacture of medicines and dyes. A small amount of iodine is essential in our diet in order to produce thyroxin, a hormone involved in controlling metabolism. Shortage of thyroxin causes a condition called goitre. In many regions of the world, sodium iodide is added to table salt to supplement iodine in the diet.

Review questions

1 **a** Chlorine and sodium hydroxide are manufactured by the electrolysis of brine. Write equations to summarise what happens during the electrolysis when:

　　i precautions are taken to prevent the products mixing with each other,

　　ii the main products (chlorine and sodium hydroxide) are deliberately mixed with one another.

　b Bromine is produced commercially by concentrating sea water and then reacting it with chlorine:

　　i What method could be used to concentrate the sea water?

　　ii Write an equation for the production of bromine from concentrated sea water by reaction with chlorine.

　c Fluorine can be obtained by the electrolysis of HF dissolved in molten KF.

　　i Write equations for the reactions at the anode and cathode during this electrolysis.

　　ii Fluorine cannot be obtained by the electrolysis of *aqueous* KF. Why not?

2 From your knowledge of the halogens, predict what happens in the following situations involving astatine, At_2 and astatides. Write equations for any reactions which take place. (Ignore the radioactive nature of astatine.)

　a Astatine vapour is mixed with hydrogen at 100 °C.

　b Astatine is added to aqueous sodium hydroxide.

　c Concentrated sulfuric acid is added to solid sodium astatide.

　d Aqueous silver nitrate is added to aqueous sodium astatide.

3 The halogens (F, Cl, Br and I) form a well-defined group of elements in the periodic table.

  a Explain how the following support this statement:

    i electron structure,

    ii redox behaviour,

    iii physical properties of the elements,

    iv usual oxidation state

  b Describe three specific properties that show a clear gradation as the group is descended from F to I.

  c Fluorine and its compounds show some properties which are not typical of the rest of the group. Mention two of these properties and suggest a reason or reasons for the difference.

4 Explain the following observations:

  a A mass spectrograph of chlorine shows five peaks with relative masses of 35, 37, 70, 72 and 74.

  b As Group VII of the periodic table is descended, the halogens become weaker oxidising agents.

  c In its compounds, fluorine shows only one oxidation state, whereas chlorine shows several.

  d Hydrogen fluoride has a higher boiling point than hydrogen chloride and hydrogen iodide.

5 This question is about the elements of Group VII, the halogens.

  a Copy and complete the following table:

Halogen	Colour	Physical state at room temperature
chlorine		
bromine		
iodine		

  b Concentrated sulfuric acid is added to separate solid samples of magnesium chloride, magnesium bromide and magnesium iodide.

    i Describe, in *each* case, *one* observation you would be able to make.

    ii Give an equation for the reaction of concentrated sulfuric acid with magnesium chloride.

c When dilute nitric acid and aqueous silver nitrate are added to a solution of a magnesium halide, MgX_2, a pale cream precipitate is formed. This precipitate is soluble in conc. aqueous ammonia but not in dilute aqueous ammonia.

    i What is the identity of the precipitate?

    ii Give an equation, with state symbols, for the reaction of the precipitate with concentrated aqueous ammonia.

d A hot glass rod is plunged into separate gas jars, one containing hydrogen chloride and one containing hydrogen iodide.

    i For *each* gas, state what you would observe, if anything, and write an equation for any reaction that takes place.

    ii Explain your answer to part i in terms of enthalpy changes.

    iii What is the role of the hot glass rod in any reaction that occurs?

Cambridge Paper 2 Q3 June 2006

6 The percentage of copper in a sample of brass was determined as follows. 2.0 g of the brass was converted to 200 cm^3 of a solution of copper(II) nitrate free from nitric or nitrous acid and acidified with ethanoic (acetic) acid. 20.0 cm^3 of this solution liberated sufficient iodine from potassium iodide solution to react with 25.0 cm^3 of 0.1 mol dm^{-3} sodium thiosulfate solution. (Cu = 64; the reaction between $Cu^{2+}(aq)$ and $I^-(aq)$ can be represented as $2Cu^{2+}(aq) + 4I^-(aq) \longrightarrow 2CuI(s) + I_2(aq)$.)

  a Write an equation or half-equations for the reaction between iodine and sodium thiosulfate.

  b How many moles of I_2 were liberated by 20 cm^3 of the copper(II) nitrate solution?

  c How many moles of copper are there in 200 cm^3 of the copper(II) nitrate solution?

  d What is the percentage by mass of copper in the brass?

12.1 The properties and reactions of nitrogen and sulfur

Figure 12.1
Piles of yellow sulfur at a chemical plant near Vancouver, Canada. Sulfur used to be mined from underground deposits, but today, most sulfur for industry is recycled from the sulfur removed from fossil fuels, especially crude oil.

Physical properties

Nitrogen and sulfur are both members of the p-block of elements in the periodic table but their physical properties are very different. Nitrogen exists as a colourless gas composed of diatomic N_2 molecules under normal conditions. Its melting point and boiling point are very low (Table 12.1). On the other hand, sulfur exists as a yellow solid composed of octatomic S_8 molecules. It melts at about 115 °C and boils at 445 °C.

Table 12.1
The physical properties of nitrogen and sulfur

Property	Nitrogen	Sulfur
State at room temperature	Gas	Solid
Colour	Colourless	Yellow
Melting point /°C	−210	115
Boiling point /°C	−196	445

The very different properties of nitrogen and sulfur are readily explained in terms of their structures. In each nitrogen molecule, the two atoms are held together by a triple covalent bond. In sulfur molecules, the eight atoms form a puckered crown-like ring in which the sulfur atoms are linked by simple covalent bonds (Figure 12.2).

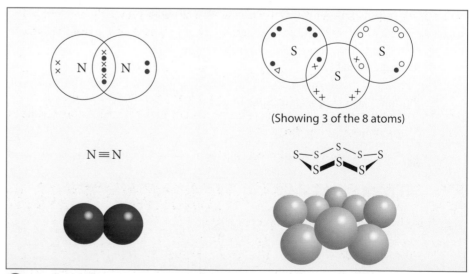

(Showing 3 of the 8 atoms)

Figure 12.2
The bonding in molecules of nitrogen and sulfur

The differences in melting and boiling points of nitrogen and sulfur are readily explained by their different relative molecular masses. Molecules of nitrogen, N_2, have a relative molecular mass of only 28.0, whereas those of sulfur, S_8, have a relative molecular mass of 256.8.

Both N_2 and S_8 are symmetrical non-polar molecules. They are held together by weak induced dipole attractions, but molecules of S_8 are nearly ten times heavier. With almost ten times the number of electrons, S_8 molecules have significantly stronger induced dipole attractions and therefore a higher melting point and boiling point.

Key Point

Nitrogen exists as gaseous N_2 molecules. Sulfur exists as solid S_8 molecules. Sulfur is much less volatile than nitrogen.

Chemical reactions

Owing to the triple covalent bond in N_2 molecules, nitrogen is remarkably inert and unreactive. All three bonds must be broken before nitrogen can take part in any reactions and this requires a very large input of energy, 994 kJ per mole of N_2 molecules. Not surprisingly, nitrogen is slow to react. Compare this with 496 kJ mol^{-1} for the double bond in O_2 molecules and only 264 kJ mol^{-1} for the single bond between S atoms in S_8 molecules.

The only element to react with nitrogen at room temperature is lithium, forming lithium nitride:

$$6Li(s) + N_2(g) \longrightarrow 2Li_3N(s)$$

However, when magnesium burns in the air, it reacts with nitrogen as well as oxygen forming magnesium nitride in addition to magnesium oxide:

$$3Mg(s) + N_2(g) \longrightarrow Mg_3N_2(s)$$

At high temperatures and pressures and in the presence of an iron catalyst, nitrogen will react with hydrogen to form ammonia, NH_3:

$$N_2(g) + 3H_2(g) \longrightarrow 2NH_3(g)$$

This reaction is used in the manufacture of ammonia by the Haber process, which we will study in more detail in Sections 12.6 and 12.7.

Nitrogen will also react with oxygen at high temperatures to form nitrogen monoxide, NO. In this case, nitrogen is oxidised by the more electronegative oxygen:

$$N_2(g) + O_2(g) \longrightarrow 2NO(g)$$

Key Point

The triple bond in N_2 molecules results in a high activation energy for reactions of nitrogen and makes it very unreactive.

Compared with nitrogen, sulfur is a very reactive element. With elements of lower electronegativity, such as metals, sulfur acts as an oxidising agent forming sulfides. For example, when iron filings and powdered sulfur are heated together, the mixture glows red hot even when the external heating is removed. A very exothermic reaction occurs forming iron(II) sulfide:

$$8Fe(s) + S_8(s) \longrightarrow 8FeS(s)$$

With elements of higher electronegativity, such as chlorine and oxygen, sulfur acts as the reducing agent and is oxidised. Sulfur burns steadily in oxygen with a blue flame to form chokingly pungent sulfur dioxide:

$$S_8(s) + 8O_2(g) \longrightarrow 8SO_2(g)$$

Quick Questions

1 Show, by calculating oxidation numbers, that:
   a Nitrogen is reduced when it reacts with metals.
   b Nitrogen is oxidised when it reacts with oxygen.
2 By calculating oxidation numbers, show that sulfur reacts as an oxidising agent with iron, but a reducing agent with oxygen.

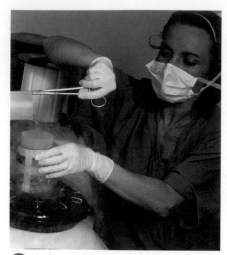

Figure 12.3
Using liquid nitrogen at very low temperatures, it is possible to store not only frozen eggs and sperm but also human embryos. One of these embryos can be implanted in a woman's uterus years later and it will develop and eventually be born as a healthy baby.

Key Point

The most important use of atmospheric nitrogen is in manufacturing ammonia. Most of the other uses of nitrogen depend on its lack of reactivity. The most important use of elemental sulfur is in manufacturing sulfuric acid.

Figure 12.4
During a lightning flash, nitrogen reacts with oxygen in the air to form nitrogen monoxide, NO. The nitrogen monoxide then reacts with more oxygen to form nitrogen dioxide, NO_2. This combines with water in the atmosphere to produce nitrous and nitric acids causing acid rain.

12.2 The uses of nitrogen and sulfur

Most of the uses of nitrogen depend upon its inertness and lack of reactivity. The worldwide production of nitrogen amounts to about 25 million tonnes each year. This nitrogen is obtained by the fractional distillation of liquid air.

Most of this nitrogen is used to create inert atmospheres for chemical processes where the presence of oxygen would be hazardous. The largest application of this kind is in oil recovery, where nitrogen under pressure is used to force oil to the surface from underground deposits. Another use of pure nitrogen is to provide a blanketing layer of unreactive gas in the manufacture of electronic components. A small-scale but extremely important use of nitrogen, this time as a liquid, is in the storage of materials at very low temperatures (Figure 12.3). The applications of *pure* nitrogen are minor, however, compared with the use of nitrogen in the air to manufacture more than 160 million tonnes of ammonia worldwide each year (Section 12.7).

Like those of nitrogen, the uses of sulfur are also concentrated on the manufacture of one major product – this time sulfuric acid (Section 12.11). About 200 million tonnes of sulfuric acid are produced each year throughout the world. It can be produced relatively cheaply and is used as a dehydrating agent, an oxidising agent and an acid. For these reasons, it is sometimes claimed that almost every manufactured product has required the use of sulfuric acid at some stage.

Another important use of sulfur is in the vulcanisation (strengthening) of rubber. In this process, sulfur is used to form cross-links between the rubber molecules (Section 15.7). Sulfur is also used in the production of sulfur dioxide and sodium sulfite. These two compounds are used as food preservatives in fruit bottling, brewing and wine making because of their anti-microbial properties.

12.3 Pollution from the oxides of nitrogen and sulfur

When gasoline (petrol) and diesel fuel are burnt at high temperatures in vehicle engines, nitrogen and oxygen from the air are taken into the cylinders and react to form nitrogen monoxide:

$$N_2(g) + O_2(g) \longrightarrow 2NO(g)$$

The same reaction also occurs in the furnaces of power stations and during an electric storm, when N_2 and O_2 in the air are exposed to very high temperatures due to sparks or electrical discharges. As soon as the nitrogen monoxide has formed, it reacts with more oxygen in the air to form an equilibrium mixture with red-brown nitrogen dioxide, NO_2:

$$2NO(g) + O_2(g) \rightleftharpoons 2NO_2(g)$$

NO and NO_2 are interconverted rapidly in the atmosphere, so they are collectively referred to as NO_x. As these oxides of nitrogen pass into the atmosphere, they contribute to atmospheric pollution and the problems of acid deposition ('acid rain').

Sulfur dioxide, like nitrogen monoxide and nitrogen dioxide is another cause of acid rain. The sulfur dioxide is released when sulfur compounds, which contaminate fossil fuels, are burnt in power stations, homes and vehicles. The problems of acid rain, created by sulfur dioxide, have been reduced in many countries by removing sulfur from fuels during their production.

Sulfur dioxide, SO_2, is not oxidised spontaneously to sulfur trioxide, SO_3, by oxygen in the air, but it is oxidised readily in the presence of nitrogen monoxide, NO. When NO is present in the air, it reacts with oxygen rapidly forming nitrogen dioxide, NO_2:

$$2NO(g) + O_2(g) \longrightarrow 2NO_2(g)$$

As soon as NO$_2$ forms, it reacts with SO$_2$ producing SO$_3$:

$$SO_2(g) + NO_2(g) \longrightarrow SO_3(g) + NO(g)$$

At the same time, NO is regenerated ready to form NO$_2$ again and then oxidise more SO$_2$. Therefore, NO acts as a catalyst in converting oxygen and sulfur dioxide to sulfur trioxide.

SO$_2$ and SO$_3$, collectively known as SO$_x$, react readily with water in the atmosphere to form sulfurous and sulfuric acids respectively. These acids eventually reach the ground as acid deposition in rain, mist or snow.

Both NO$_x$ and SO$_x$ are major causes of atmospheric pollution. Health problems such as asthma and bronchitis are made worse, plants can be affected adversely and even the stonework of buildings is damaged. The problems created by NO$_x$ and SO$_x$ in the atmosphere and the ways and means by which scientists have helped to reduce their impact is discussed further in Section 14.8.

Quick Questions

3 Write equations for the reactions which occur when water reacts with:
   a NO$_2$ to produce nitric acid, HNO$_3$, and nitrous acid, HNO$_2$,
   b SO$_2$ to produce sulfurous acid, H$_2$SO$_3$,
   c SO$_3$ to produce sulfuric acid, H$_2$SO$_4$.
4 Write an equation for the reaction of nitric acid in rainwater with calcium carbonate in the limestone of a building.
5 Once NO$_2$ is present in the atmosphere, it is photochemically decomposed to NO in a reaction that also produces oxygen atoms. In a second reaction, the oxygen atoms react with O$_2$ molecules to form ozone (O$_3$). Write equations for these two reactions.

◀ Figure 12.5
In cities, such as Los Angeles with high populations, millions of vehicles and a warm, sunny climate, exhaust gases can lead to the formation of photochemical smog. This contains a mixture of NO$_x$, SO$_x$ and ozone, all of which irritate our respiratory systems. This can lead to coughing, sore throats, runny noses and can cause severe illness in the elderly and in people with heart or lung diseases. The red-brown colour of the smog is due to NO$_2$.

12.4 The hydrides of nitrogen and sulfur

The most important hydrides of nitrogen and sulfur are ammonia, NH$_3$, and hydrogen sulfide, H$_2$S, respectively. Both of these are colourless gases at room temperature, but with very different smells and chemical characters.

Ammonia has a strong, sharp smell. It is basic and extremely soluble in water forming an alkaline solution containing OH$^-$ ions:

$$NH_3(aq) + H_2O(l) \rightleftharpoons NH_4^+(aq) + OH^-(aq)$$

In contrast to ammonia, hydrogen sulfide is very poisonous and has a disgusting smell like bad eggs. It is acidic, dissolving in water to form a weakly acidic solution containing H_3O^+ ions:

$$H_2S(aq) + H_2O(l) \rightleftharpoons HS^-(aq) + H_3O^+(aq)$$

Ammonia is, in fact, the only common alkaline gas. So, it is often convenient to test for ammonia by showing that it will turn damp red litmus paper blue.

Key Points

Ammonia is basic and very soluble in water.
Hydrogen sulfide is weakly acidic and soluble in water.

Test for ammonia:
Ammonia is the only common gas to turn damp red litmus paper blue.

Quick Questions

6 When ammonia is bubbled into aqueous sulfur dioxide, two salts are formed. Which of the following formulae are correct for the two salts?
   a NH_4SO_3 b $(NH_4)_2SO_3$
   c NH_4HSO_3 d $(NH_4)_2SO_4$

12.5 Ammonia as a base

Ammonia is by far the most important nitrogen compound. Its industrial synthesis from inert nitrogen gas makes an abundant supply of nitrogen available to industry and agriculture. Most ammonia is used to manufacture ammonium salts for fertilisers and nitric acid for the production of more fertilisers, synthetic fibres and explosives. To understand these uses we must look at the properties of ammonia itself.

Ammonia acts as a base in many of its reactions, bonding to H^+ ions to form ammonium ions, NH_4^+ (Figure 12.7). This reaction is reversible.

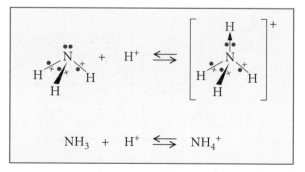

Figure 12.7
The formation and structure of an ammonium ion

Reaction with water

Ammonia is extremely soluble in water because NH_3 molecules form hydrogen bonds with H_2O molecules, and also react with H_2O molecules to form a solution containing ammonium ions and hydroxide ions:

$$NH_3(aq) + H_2O(l) \rightleftharpoons NH_4^+(aq) + OH^-(aq)$$

This reaction is incomplete, and NH_3 is only partly dissociated into NH_4^+ and OH^- ions. Therefore, ammonia solution is only a weak electrolyte.

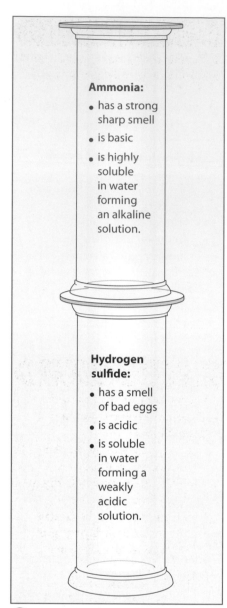

Figure 12.6
Contrasting the key properties of ammonia and hydrogen sulfide

Ammonia:
- has a strong sharp smell
- is basic
- is highly soluble in water forming an alkaline solution.

Hydrogen sulfide:
- has a smell of bad eggs
- is acidic
- is soluble in water forming a weakly acidic solution.

Reaction with acids

Ammonia reacts with acids to form ammonium salts. This is how solid fertilisers, such as ammonium nitrate, NH_4NO_3 (Nitram) and ammonium sulfate, $(NH_4)_2SO_4$, are manufactured from ammonia:

$$NH_3(g) + HNO_3(aq) \longrightarrow NH_4NO_3(aq)$$

$$2NH_3(g) + H_2SO_4(aq) \longrightarrow (NH_4)_2SO_4(aq)$$

An acid–base reaction also occurs when ammonia gas reacts with hydrogen chloride producing a white smoke. The white smoke is tiny particles of solid ammonium chloride suspended in the air. This reaction is sometimes used as a test for ammonia:

$$NH_3(g) + HCl(g) \longrightarrow NH_4Cl(s)$$

Making ammonia on a small scale

The easiest way to prepare a small amount of ammonia is to remove H^+ ions from NH_4^+ ions in an ammonium salt, like ammonium chloride, NH_4Cl. Alkalis containing OH^- ions will do this. Figure 12.8 shows how dry ammonia can be made by heating a mixture of ammonium chloride and calcium hydroxide. In effect, the reaction involves the displacement of ammonia from one of its salts.

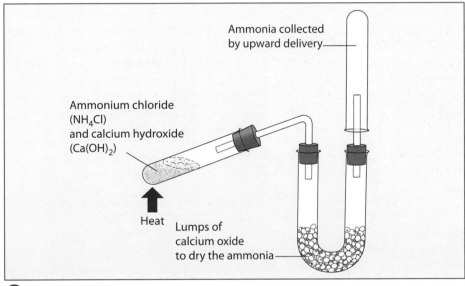

Ammonia collected by upward delivery

Ammonium chloride (NH₄Cl) and calcium hydroxide (Ca(OH)₂)

Heat

Lumps of calcium oxide to dry the ammonia

Figure 12.8
Making ammonia on a small scale

12.6 The manufacture of ammonia

Towards the end of the 19th century, the agricultural industry began to require increasing supplies of nitrogenous fertilisers to increase the yield from crops and feed rapidly increasing populations, particularly in Europe and North America.

At the same time, the chemical industry also required increasing quantities of nitrogen compounds to produce nitric acid for dyes and explosives such as TNT and dynamite.

Agriculture and industry were competing with each other for dwindling supplies of nitrogenous raw material from places such as Chile, where there were naturally occurring deposits of nitrates. An alternative supply of nitrogen, such as ammonia or nitrate, had to be found. Otherwise, people would starve or the chemical industry would stagnate. Ironically, the problem was solved by war preparations in Germany

Figure 12.9

Fritz Haber (1868–1934). Haber was born in Breslau, Germany. After studying chemistry at university, he obtained a post as lecturer at the technical college in Karlsruhe. Whilst working there, Haber discovered a method of synthesising ammonia. In 1918, he was awarded the Nobel Prize in Chemistry for this discovery.

between 1909 and 1914. German military leaders realised that once war was declared, their country would be subjected to a strict blockade. The importation of raw materials might then cease, and Germany would have to meet its own demands for nitrogenous fertilisers and nitric acid to make the explosives needed for warfare.

In 1909, the leading German chemical company BASF (Badische Anilin und Soda Fabrik) turned its research expertise and financial resources towards the problem. BASF started to investigate the possibility of manufacturing ammonia from nitrogen in the air. In the previous year, a young German chemist, Fritz Haber, had discovered that nitrogen and hydrogen could be made to react and form an equilibrium mixture containing ammonia. But the reaction needed a suitable catalyst at 600 °C and a pressure of 200 atmospheres:

$$N_2(g) + 3H_2(g) \rightleftharpoons 2NH_3(g) \qquad \Delta H^\ominus = -92\,kJ\,mol^{-1}$$

BASF bought from Haber the rights to his ammonia process. Then, they spent more than two million US dollars in transforming Haber's simple apparatus (Figure 12.10) into a giant industrial plant capable of producing 10 000 tonnes of ammonia per year. By 1913, German production of nitrogen compounds (mainly fertilisers and explosives) was about 120 000 tonnes per year. Without these materials, Germany would have run out of food and explosives and the First World War would have ended before 1918.

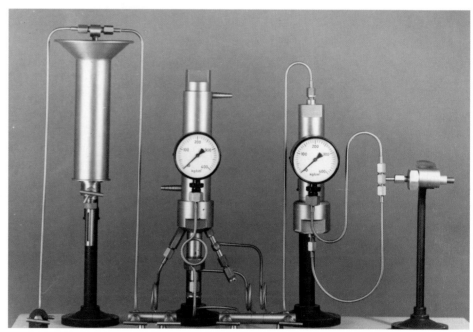

Figure 12.10

Haber's apparatus for the synthesis of ammonia

Choosing conditions for the manufacture of ammonia

The commercial success of any process depends on the speed, efficiency and economy with which the products can be obtained from the starting materials. Many people (including managers, economists and engineers) will be involved in decisions about the methods and materials for any large industrial process. However, the major problem confronting chemists is to convert reactants to products:

▶ as quickly as possible and
▶ as completely as possible.

The first of these requirements is one of kinetics – reaction rates. It involves finding the conditions to get maximum product as quickly as possible.

Quick Questions

9 a What conditions did Haber employ to increase the reaction rate in synthesising ammonia?

b Haber's experiment produced a reaction mixture containing only 8% by volume of ammonia. What conditions would increase the yield of ammonia at equilibrium? (Hint: Remember Le Chatelier's principle, Section 7.10.)

Definition

Le Chatelier's principle says that, if a system in equilibrium is subjected to a change, processes (reactions) occur which tend to remove the change imposed.

The second requirement is one of equilibrium. It involves finding conditions which will maximise the proportion of product in the equilibrium mixture. In this respect, Le Chatelier's principle can be used to predict the conditions for maximum yield of product at equilibrium.

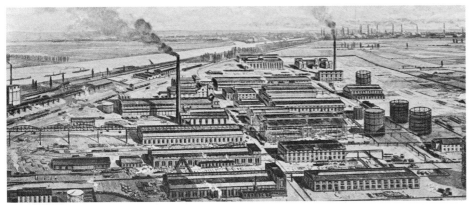

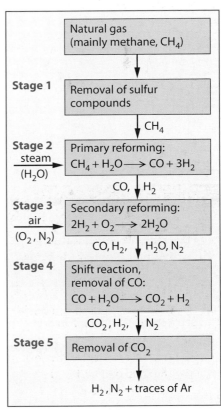

Figure 12.11
The world's first ammonia manufacturing plant. This was built by BASF at Oppau in Germany and opened in 1912.

Quick Questions

10 a What conditions of temperature and pressure (high, moderate or low) would you choose to synthesise ammonia:
   i as fast as possible?
   ii to get the highest yield of ammonia?
   b What compromise in conditions will be necessary in the commercial manufacture of ammonia?

12.7 The modern Haber process

The present manufacture of ammonia involves two main processes:

▶ The production of a purified mixture of nitrogen and hydrogen.
▶ The synthesis of ammonia – the Haber process.

The production of a purified mixture of nitrogen and hydrogen

The purified mixture of nitrogen and hydrogen is produced in a complex series of reactions from natural gas (mainly methane), water and air (Figure 12.12).

The first stage involves removal of hydrogen sulfide and organic sulfur compounds from natural gas. Otherwise these sulfur compounds will poison the catalysts used later in the production of the nitrogen/hydrogen mixture.

In the second stage (primary reforming), methane in the natural gas reacts with steam in a catalysed reaction to produce carbon monoxide and hydrogen:

$$CH_4(g) + H_2O(g) \longrightarrow CO(g) + 3H_2(g) \qquad \Delta H^\ominus = +210\,kJ\,mol^{-1}$$

In the third stage (secondary reforming), the products from Stage 2 are mixed with a controlled volume of air (78% nitrogen, 21% oxygen and 1% argon) so that some of the hydrogen can react with oxygen to form steam:

$$2H_2(g) + O_2(g) \longrightarrow 2H_2O(g) \qquad \Delta H^\ominus = -482\,kJ\,mol^{-1}$$

At this point, the gas contains carbon monoxide, hydrogen, steam, nitrogen and traces of argon.

In the fourth stage (the shift reaction), carbon monoxide reacts with the steam produced in Stage 3 using a catalyst at high temperature to produce carbon dioxide and more hydrogen:

$$CO(g) + H_2O(g) \longrightarrow CO_2(g) + H_2(g) \qquad \Delta H^\ominus = -42\,kJ\,mol^{-1}$$

The fifth stage involves removing unwanted carbon dioxide by passing the gases through an alkaline solution, such as concentrated potassium carbonate or an organic base.

Finally the gases are dried to produce a mixture containing 74% hydrogen and 25% nitrogen with traces of argon.

Figure 12.12
A flow diagram for the production of the nitrogen/hydrogen mixture from methane, water and air

11 a What conditions of temperature and pressure favour the formation of the products in the primary reforming process?

b How does the ratio of N_2 to H_2 in the final reactants mixture compare with their ratio in the equation for the synthesis of ammonia?

c Why is it unnecessary to remove traces of argon from the final nitrogen/hydrogen mixture?

d Write an equation for the removal of CO_2 in Stage 5 using an alkaline solution containing OH^- ions.

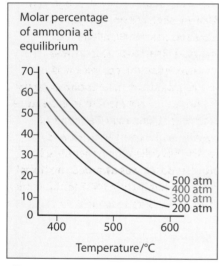

Molar percentage of ammonia at equilibrium

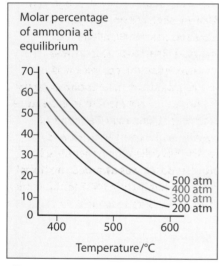

Figure 12.14

The percentage of ammonia in the equilibrium mixture obtained from a 1:1 mixture of N_2 and H_2 at different temperatures and pressures

Key Point

The Haber Process, $N_2(g) + 3H_2(g) \longrightarrow 2NH_3(g)$, uses an iron catalyst at 400 °C and 100 atm pressure.

The synthesis of ammonia – the Haber process

Once a mixture has been produced with the correct proportions of N_2 and H_2, it is used to manufacture NH_3 in the Haber process. Figure 12.13 shows a flow diagram for the modern Haber process.

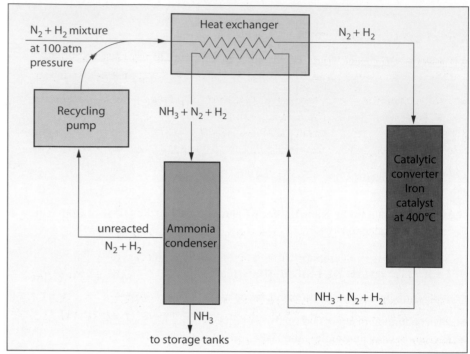

Figure 12.13

A flow diagram for the Haber process

Notice that hot product gases from the catalytic converter are used to warm up the incoming mixture of compressed gases before the reaction between nitrogen and hydrogen in the converter:

$$N_2(g) + 3H_2(g) \rightleftharpoons 2NH_3(g) \qquad \Delta H^{\ominus} = -92\,\text{kJ}\,\text{mol}^{-1}$$

Le Chatelier's principle predicts that high pressure and low temperature will increase the proportion of ammonia at equilibrium and these predictions are borne out by the results in Figure 12.14.

High pressure obviously gives a higher yield of ammonia but, the higher the pressure, the greater the cost and maintenance of equipment. Although pressures up to 600 atm have been used, the favoured pressure nowadays is a compromise of 100 atm.

In contrast to pressure, the temperature must be low to give a high yield of ammonia. However, at low temperatures the rate of reaction is so slow that it makes the process uneconomic. In practice, the operating temperature is usually a compromise of about 400 °C.

In addition to temperature and pressure, the catalyst is a vitally important component in this industrial process. A more efficient catalyst permits lower operating temperatures.

Experience has shown that the catalytic activity of the iron is improved by mixing it with small amounts of potassium hydroxide as a promotor.

The hot gases leaving the converter pass through the heat exchange system and are then cooled to −50 °C in the condenser. The ammonia liquifies (b.pt. −33 °C) and collects in the storage tanks. Unreacted nitrogen and hydrogen are recycled.

Quick Questions

12 Which of the following statements are correct?

K_p for the Haber process increases if

a pressure increases, **b** temperature decreases,

c a catalyst is used, **d** catalyst promoters are used.

The uses and importance of ammonia

Ammonia ranks second to sulfuric acid as the chemical with the largest annual worldwide production. About 160 million tonnes of it are manufactured each year.

Ammonia forms the basis of the nitrogen industry. It will react with acids to give ammonium salts. It can also be oxidised to nitric acid which in turn produces nitrates. Both ammonium salts and nitrates are used as fertilisers, which provide the outlet for about 85% of ammonia. Of the remaining 15%, one-third is used in the production of polyamides including nylon.

12.8 From ammonia to nitric acid

The manufacture of nitric acid from ammonia involves three stages.

1 Catalytic oxidation of ammonia to nitrogen monoxide (NO).

$$4NH_3(g) + 5O_2(g) \rightleftharpoons 4NO(g) + 6H_2O(g) \qquad \Delta H^{\ominus} = -950\,kJ\,mol^{-1}$$

2 Oxidation of nitrogen monoxide (NO) to nitrogen dioxide (NO_2).

$$2NO(g) + O_2(g) \longrightarrow 2NO_2(g) \qquad \Delta H^{\ominus} = -114\,kJ\,mol^{-1}$$

3 Reaction of nitrogen dioxide with water to form nitric acid.

$$3NO_2(g) + H_2O(l) \longrightarrow 2HNO_3(aq) + NO(g) \qquad \Delta H^{\ominus} = -117\,kJ\,mol^{-1}$$

Quick Questions

13 Look closely at Stage 1 above in the manufacture of nitric acid.

a What conditions of temperature and pressure would give the maximum yield of nitrogen monoxide (NO) at equilibrium?

b In practice, the first stage is carried out by passing dry ammonia and air at 7 atm over a 90% platinum/10% rhodium catalyst at 900 °C. Explain why the conditions for the process differ from those predicted in your answer to part a.

▲ Figure 12.15

Carl Bosch (1874–1940). Bosch was the son of a plumber. After studying chemistry at university, Bosch joined BASF in 1899 and quickly gained a reputation as a brilliant chemical engineer. Bosch was responsible for developing the complex and sophisticated chemical engineering for the first industrial plant to manufacture ammonia at Oppau in Germany. In honour of his work the process is sometimes called the Haber–Bosch process. Bosch was awarded the Nobel Prize for Chemistry in 1931 for his work on high-pressure reactions.

◀ Figure 12.16

This technician is holding a sheet of the platinum/rhodium catalyst used in Stage 1 of the manufacture of nitric acid from ammonia. The sheet of catalyst is a gauze woven from wire of the platinum alloy to give a high surface area.

Key Point

Nitrogen shows a range of oxidation states in its oxides and oxoacids.

Quick Questions

14 a Why does a deficiency of nitrogen cause plants to become stunted?

b Why does a deficiency of nitrogen cause yellowing of the leaves?

15 a What is the percentage by mass of nitrogen in ammonium nitrate, NH_4NO_3? (N = 14, H = 1, O = 16)

b How do you think ammonium nitrate is produced industrially?

$$CH_2-O-NO_2$$
$$|$$
$$CH-O-NO_2$$
$$|$$
$$CH_2-O-NO_2$$

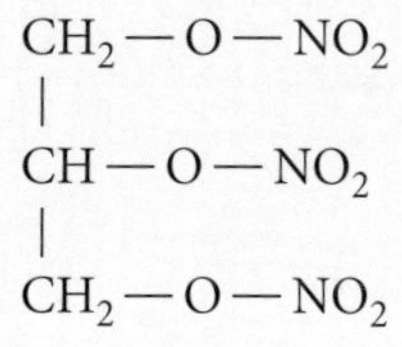

propane-1,2,3-triyl trinitrate
(nitroglycerine)

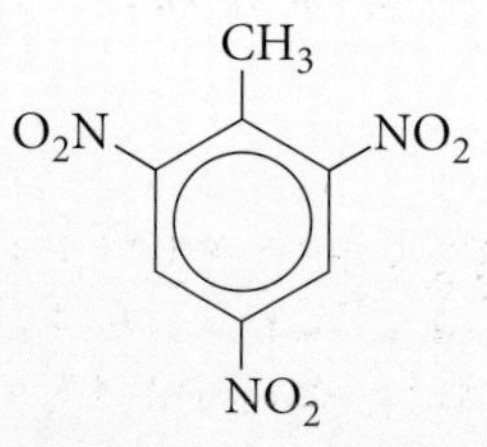

methyl-2,4,6-trinitrobenzene
(trinitrotoluene, TNT)

Figure 12.17
Explosives manufactured from nitric acid

After Stage 1, the product gases are cooled to 25 °C. They are then mixed with more air so that colourless nitrogen monoxide is immediately oxidised to red–brown nitrogen dioxide:

$$2NO(g) + O_2(g) \longrightarrow 2NO_2(g)$$

In the third stage, nitrogen dioxide reacts with water in large absorption towers. These are designed to ensure thorough mixing of the ascending gases and the descending solution. The final product contains about 60% nitric acid. More concentrated acid can be obtained by distilling the 60% solution with concentrated sulfuric acid.

12.9 Fertilisers and explosives from nitric acid

Nitric acid plays an important part in the production of fertilisers and explosives which consume 75% and 15% of its production, respectively.

Fertilisers

Nitrogen is an essential element for plant growth. It is required for the formation of proteins, chlorophyll and nucleic acids. Plants suffering from nitrogen deficiency become stunted with yellow leaves.

The intensive cropping of agricultural land means that nitrogen is removed from the soil and must be replaced by the application of fertilisers. Otherwise the soil will become barren and infertile. In some countries, ammonia is injected directly into the soil, but nitrogenous fertilisers are usually nitrates or ammonium salts. Indeed, ammonium nitrate (sold commercially as 'NITRAM') is the most widely used fertiliser in many countries, because of its relatively high percentage of nitrogen.

Nitrochalk consisting of ammonium nitrate crystals coated with chalk (calcium carbonate) contains about 26% nitrogen. It is non-deliquescent, unlike ammonium nitrate crystals and it provides a convenient way of liming the soil at the same time.

Fertilisers can be used as single compounds such as ammonium nitrate or as mixtures of compounds containing nitrogen, phosphorus and potassium ('NPK' fertilisers). The proportions of nitrogen, phosphorus and potassium in NPK fertilisers are usually shown as % nitrogen (N), % phosphorus(v) oxide (P_2O_5) and % potassium oxide (K_2O). These percentages vary in different NPK fertilisers depending on their precise purpose.

Explosives

Although 90% of ammonium nitrate is used in fertilisers, a large proportion of the remaining 10% goes towards the production of explosives. Most explosives are manufactured by processes which use concentrated nitric acid.

The formulae of two of these explosives are shown in Figure 12.17. Both of these substances decompose explosively. The explosions produce large amounts of gas and heat which causes a sudden increase in pressure. Carbon and hydrogen in the explosives are oxidised by oxygen from the nitro-groups forming carbon dioxide and water (steam). For example, the explosion of 'nitroglycerine' (made by the action of concentrated nitric acid on glycerol (propane-1,2,3-triol)) can be summarised by the equation:

$$4C_3H_5(NO_3)_3(l) \longrightarrow 12CO_2(g) + 10H_2O(g) + 6N_2(g) + O_2(g)$$

Notice that in this case, there is more than enough oxygen within the 'nitroglycerine' molecule to ensure complete oxidation to CO_2 and $H_2O(g)$. This is not the case with aromatic nitro-compounds such as TNT. These explosives are frequently blended with compounds containing a high percentage of oxygen, such as chlorates and nitrates, to ensure complete oxidation during explosion.

Nitroglycerine is very unstable and may explode unexpectedly. To make it safer, the Swedish industrialist, Alfred Nobel (1833–96) invented a method of stabilising nitroglycerine by absorbing it into an inert material such as 'Fuller's earth' (dried clay) or 'Kieselguhr' (powdered silicon(IV) oxide). The resulting explosive is called dynamite. The subsequent manufacture of dynamite earned Nobel a fortune, which he used to establish the world famous prizes named after him.

The manufacture of explosives requires elaborate and very special safety precautions. You should *never* attempt to make any explosive yourself.

◀ Figure 12.18
Controlled demolition, using explosives, of a block of flats in London, UK

12.10 Problems with the over-use of fertilisers

Millions of tonnes of *inorganic* fertilisers, made from non-living materials such as rocks, minerals and nitrogen in the air, are used to improve crop yields every year. They include compounds such as ammonium nitrate, ammonium sulfate, potassium nitrate and nitrochalk. They are widely used because they are easy to spread and their action is rapid. However, they do not improve the quality or the structure of the soil which is important for healthy plants.

In contrast, *organic* fertilisers are obtained from rotting plants and animal faeces. They include seaweed, manure and compost. Leguminous crops such as clover and peas are sometimes ploughed into the soil as organic fertilisers, because they are rich in nitrogen. As these organic fertilisers contain a lot of plant fibre, they improve the structure of the soil. Farming methods which use only organic fertilisers and no artificial pesticides are described as organic and the organic foods produced are often more expensive.

◀ Table 12.2
Comparing organic and inorganic fertilisers

	Organic fertilisers	**Inorganic fertilisers**
Examples	Manure, compost	Ammonium sulfate, potassium sulfate, ammonium phosphate
Cost	Relatively cheap	Expensive
Concentration of nutrients	Low	High
Ease of use	Awkward, bulky and sticky	Easy to use as powder or granules
Speed of action	Slow	Rapid
Effect on ecosystem	Provide food for decomposers (earthworms, ants, bacteria, fungi) and improve soil structure and aeration	May change soil pH May harm decomposers and other insects Allow elements not required by plants to accumulate May cause eutrophication in water courses

Figure 12.19
Eutrophication resulting in dead and rotting materials in streams and rivers is becoming increasingly common due to the over-use of inorganic fertilisers.

Inorganic fertilisers disrupt an ecosystem because they can harm insects and decomposers, change the soil pH and interfere with healthy plant growth.

The over-use of inorganic fertilisers has also caused problems in rivers and lakes. Inorganic fertilisers, particularly ammonium salts and nitrates all of which are soluble, dissolve in rainwater and get washed into streams and rivers. This abnormal enrichment of waterways with mineral nutrients is called **eutrophication**. Once in the waterways, the fertilisers encourage excessive growth of bacteria, algae and water plants. These organisms deplete oxygen dissolved in the water and eventually all the oxygen is used up. Without oxygen, the bacteria, algae, plants and fish die and decay and the waterway becomes a stinking sewer.

Despite these problems, we cannot manage at present without inorganic fertilisers – we could not grow enough crops to feed the ever increasing world population.

12.11 The manufacture of sulfuric acid

Sulfuric acid is arguably the world's most important industrial chemical. A greater mass of sulfuric acid is manufactured every year than any other chemical. It is used to make hundreds of different products needed throughout the whole range of global industries.

There are three stages in the manufacture of sulfuric acid and these are shown diagrammatically in Figure 12.20.

1 Production of sulfur dioxide.
2 Oxidation of sulfur dioxide to sulfur trioxide – the Contact process.
3 Absorption of sulfur trioxide.

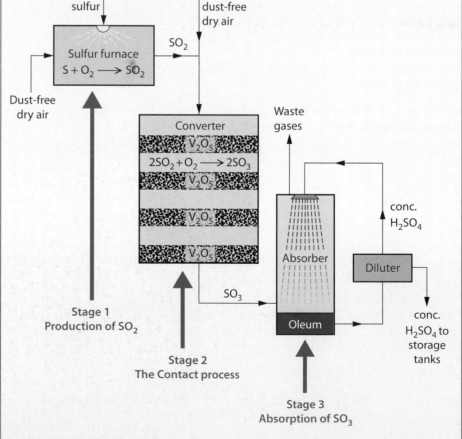

Figure 12.20
A flow diagram for the manufacture of sulfuric acid

Stage 1 Production of sulfur dioxide

If elemental sulfur is the starting point, it is melted and then sprayed into a furnace. Here it reacts with oxygen from a blast of dust-free dry air to produce sulfur dioxide. The reaction is very exothermic:

$$S_8(l) + 8O_2(g) \longrightarrow 8SO_2(g) \qquad \Delta H = -2376\,\text{kJ}\,\text{mol}^{-1}$$

The most important source of sulfur is its recovery from natural gas and crude oil. These contain organic sulfur compounds and hydrogen sulfide, which must be removed before they are used as fuels or chemical feedstock.

An alternative source of sulfur dioxide in some countries comes from the roasting in air of sulfide ores, such as zinc sulfide and lead sulfide, prior to the extraction of constituent metals:

$$2ZnS(s) + 3O_2(g) \longrightarrow 2ZnO(s) + 2SO_2(g)$$

About one-third of the sulfur dioxide required is obtained from roasting sulfide ores, and this is increasing. Plants which once expelled SO_2 into the atmosphere are now recovering it as sulfuric acid. In particular, China now produces most of its sulfuric acid from iron pyrites, FeS_2.

Stage 2 Oxidation of sulfur dioxide to sulfur trioxide – the Contact process

In Stage 2, additional supplies of dust-free air are added to the sulfur dioxide. The mixture then passes through beds of vanadium(v) oxide catalyst at a temperature of about 450 °C.

Here the sulfur dioxide and oxygen react exothermically to form sulfur trioxide:

$$2SO_2(g) + O_2 \rightleftharpoons 2SO_3(g) \qquad \Delta H = -197\,\text{kJ}\,\text{mol}^{-1}$$

Stage 3 Absorption of sulfur trioxide

Direct absorption of SO_3 in water is unsatisfactory, because the heat evolved vaporises the H_2SO_4 produced forming a mist of tiny droplets which is slow to settle out as a liquid. Instead, the SO_3 is first absorbed in concentrated H_2SO_4, then diluted.

So, in Stage 3, the sulfur trioxide passes into an absorber tower where it reacts with concentrated sulfuric acid, producing a liquid called oleum:

$$SO_3(g) + H_2SO_4(l) \longrightarrow \underset{\text{oleum}}{H_2S_2O_7(l)}$$

Finally, the oleum is diluted with water to produce concentrated sulfuric acid containing 98% H_2SO_4:

$$H_2S_2O_7(l) + H_2O(l) \longrightarrow 2H_2SO_4(l)$$

Figure 12.21
The sulfuric acid plant in Maharashtra State, Western India

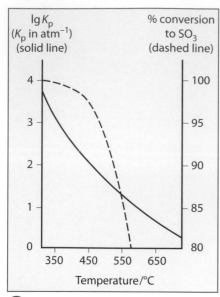

Figure 12.22
The effect of temperature on the equilibrium constant and on the percentage conversion to SO_3 at equilibrium for a typical mixture of SO_2 and O_2 at 1 atm pressure

Key Point

The Contact process,
$$2SO_2(g) + O_2(g) \longrightarrow 2SO_3(g),$$
uses a vanadium(v) oxide catalyst at 450 °C and atmospheric pressure.

Choosing conditions for the Contact process

The bottleneck in production of sulfuric acid is the conversion of SO_2 to SO_3 in the Contact process:

$$2SO_2(g) + O_2(g) \rightleftharpoons 2SO_3(g) \qquad \Delta H = -197\,kJ\,mol^{-1}$$

However, in addition to using a catalyst, the rate of SO_3 production can be improved by:

▸ increasing the temperature and
▸ increasing the concentration (pressure) of O_2 and/or SO_2.

As the reaction is exothermic and three moles of reactants form two moles of product, Le Chatelier's principle predicts that the maximum yield of SO_3 at equilibrium will be obtained at:

▸ low temperature and
▸ high pressure.

Notice how reaction rate and equilibrium considerations conflict in the choice of temperature for the Contact process. Maximum yield of SO_3 would be obtained at low temperatures, but under these conditions the reaction rate would be very slow. In practice, a compromise temperature of about 450 °C is chosen. There are two other reasons for keeping the temperature as low as possible – fuel costs and corrosion of the reaction chambers, which both increase rapidly as the temperature of the process rises. Figure 12.22 shows the effect of temperature on the equilibrium constant and on the percentage conversion to SO_3 at equilibrium for a typical gas mixture.

Notice how the percentage conversion to SO_3 falls very rapidly above 450 °C. At 450 °C, conversion to SO_3 is 97%, but at 550 °C conversion to SO_3 falls to 86%. Because of the high conversion to SO_3 at 450 °C and atmospheric pressure, it is unnecessary to carry out the process at increased pressure.

There is one other aspect of the Contact process worth consideration. As the reaction in the converter proceeds, heat is evolved in the exothermic reaction and this raises the catalyst to a higher temperature. At this higher temperature, the percentage conversion to SO_3 is much reduced (Figure 12.22). Thus, it is necessary to cool the product gases between successive beds of the catalyst.

By clever use of heat exchangers to cool the catalyst and heat the incoming gases, the operating temperature can be maintained at about 450 °C without external heating.

Quick Questions

17 Which one of the following effects does V_2O_5 have on the equilibrium of the following reaction?
$$2SO_2(g) + O_2(g) \rightleftharpoons 2SO_3(g)$$
 a It speeds up the forward reaction.
 b It slows down the reverse reaction.
 c It increases the value of K_p.
 d It increases $[SO_3(g)]$ at equilibrium.

18 Which of the following statements help to explain the use of heat exchangers in the Contact process?
 a To warm the incoming gas.
 b To avoid overheating the catalyst.
 c To give more SO_3 at equilibrium.
 d To speed up the reaction.

12.12 The importance and uses of sulfuric acid

The annual world production of sulfuric acid is about 200 million tonnes. Figure 12.23 shows the main uses of sulfuric acid, together with approximate percentages for the different products and purposes.

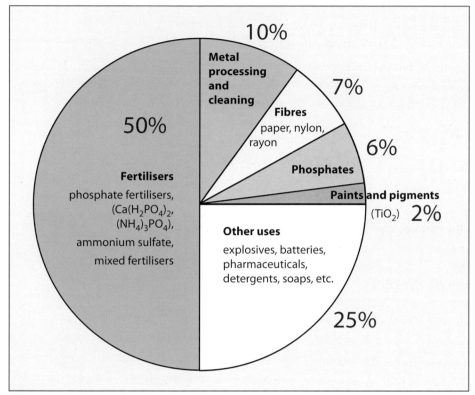

10%

Metal processing and cleaning

7%

Fibres
paper, nylon, rayon

6%

50%

Phosphates

Fertilisers
phosphate fertilisers,
$(Ca(H_2PO_4)_2)$,
$(NH_4)_3PO_4$),
ammonium sulfate,
mixed fertilisers

Paints and pigments
(TiO_2) **2%**

Other uses
explosives, batteries,
pharmaceuticals,
detergents, soaps, etc.

25%

▲ Figure 12.23
The uses of sulfuric acid

Notice that sulfuric acid is required for the manufacture of many basic materials including fertilisers, metals, fibres, paints, pigments, pharmaceuticals and detergents. Because of this, its level of production can be used as a guide to a country's industrial activity. Also, problems with the bulk storage of sulfuric acid mean that its production must respond quickly to any changes in consumption.

By far the largest amount of sulfuric acid is used in the manufacture of various fertilisers, particularly phosphates and ammonium sulfate which is important for sulfur-deficient soils. It is also widely used in metal processing, including the manufacture of copper and zinc and in cleaning the surface of sheet steel prior to galvanising (coating with zinc) and covering with a thin layer of tin for use in food cans.

> **Key Point**
>
> Sulfuric acid is used in many manufacturing processes and is an extremely important industrial chemical.

Review questions

1 a Explain why traces of sulfur dioxide are emitted from oil-burning furnaces.

 b Write an equation to show how *one* of the following chemicals could be used to reduce the sulfur dioxide in part a. $CaCl_2$ $Ca(OH)_2$ $CaSO_4$

 c Draw a 'dot-and-cross' diagram to explain why molecules of sulfur dioxide are described as non-linear or V-shaped.

2 a Ammonia and chlorine react in the gas phase to form nitrogen and ammonium chloride.

 i Write a balanced equation for this reaction.

 ii Explain why ammonia acts as both a base and a reducing agent in the reaction.

 b Draw 'dot-and-cross' diagrams to show the bonding and structure of:

 i ammonia,   ii ammonium chloride.

 c In some parts of the world, liquid ammonia is injected directly into the soil as a fertiliser. Suggest one advantage and one disadvantage of using ammonia in this way.

3 a Write balanced equations with state symbols for the following reactions:

 i the synthesis of ammonia from nitrogen and hydrogen,

 ii the reaction of ammonia with oxygen to produce nitrogen monoxide and steam,

 iii the reaction of nitrogen dioxide with water to produce nitric and nitrous acids.

 b For each of the reactions in part a:

 i State the changes in oxidation number of nitrogen.

 ii Classify the changes to nitrogen as oxidation, reduction or disproportionation.

4 Ammonium nitrate is an important fertiliser:

 a Explain why ammonium nitrate can be used as a fertiliser.

 b Ammonium nitrate can be mixed with chalk (calcium carbonate) to produce 'nitrochalk'.

 i What benefit does 'nitrochalk' provide for agriculture which pure ammonium nitrate does not?

 ii Suppose you are supplied with two fertilisers, both of which are known to contain ammonium nitrate, but only one of which contains chalk. Describe simple chemical tests that you could carry out to show that both fertilisers contain an ammonium salt but only one contains a carbonate.

 c The uncontrolled use of nitrate fertilisers is causing problems in some inland waterways:

 i How do nitrates get into watercourses?

 ii How do nitrates affect living things in watercourses and cause problems?

5 Explain the following:

 a The eight-membered ring in S_8 molecules is not planar.

 b Nitrogen is a gas at room temperature, whereas sulfur is a solid.

 c Sulfur is much more reactive than nitrogen.

 d Sulfur atoms with six electrons in their outermost shell can form sulfur hexafluoride, SF_6, but nitrogen with five electrons in its outermost shell cannot form nitrogen pentafluoride, NF_5.

6 Hydrogen can be obtained from natural gas by partial oxidation with steam. This involves the following endothermic reaction.

$$CH_4(g) + H_2O(g) \rightleftharpoons CO(g) + 3H_2(g)$$

 a Write an expression for K_p for this reaction.

 b How will the value of K_p be affected by:

 i increasing the pressure,

 ii increasing the temperature,

 iii using a catalyst?

 c How will the composition of the equilibrium mixture be affected by:

 i increasing the pressure,

 ii increasing the temperature,

 iii using a catalyst?

7 The first step in the manufacture of nitric acid from ammonia involves the exothermic oxidation of ammonia to nitrogen monoxide (NO) and steam.

 a Write an equation for the reaction of ammonia with oxygen to form nitrogen monoxide and steam.

 b Predict, qualitatively, the conditions of temperature and pressure for maximum yield of nitrogen monoxide in the equilibrium mixture.

 c The industrial manufacture of nitrogen monoxide from ammonia employs high temperature and a pressure of 7 atm. How and why are these industrial conditions different from those you predicted in **b** for maximum yield of nitrogen monoxide at equilibrium?

 d Describe, with equations, how nitrogen monoxide produced by this process is converted to nitric acid.

13 Introduction to organic chemistry

13.1 Carbon – a unique element

There are millions of compounds containing carbon and hydrogen whose formulae are known to chemists – far more than the number of compounds of all the other elements put together. Why does carbon form such an enormous number of compounds?

There are three important properties of carbon that enable it to form so many stable compounds.

a Carbon has a fully shared octet of electrons in its compounds

In methane the outer-shell electrons of carbon are shared by four hydrogen atoms as shown in Figure 13.1. This means that the carbon atoms have no lone pairs or empty orbitals in their outer shells, so they are unable to form any more bonds. The inability of carbon to bond any further once it has an octet of electrons is responsible for the kinetic stability of its compounds (see below).

b Carbon can form strong single, double and triple bonds to itself

The stability of the single C—C bond can be seen by comparing the bond energies in Table 13.1.

It is worth comparing carbon and silicon here since we might expect silicon to show similarities to carbon, being the next member of Group IV. Note the strength of the C—C bond compared with that of the Si—Si bond. Note too the high strength of the C—H bond. Almost all carbon compounds also contain hydrogen.

If carbon compounds are to be stable they must be stable under normal conditions, and that means in the presence of air. In fact, compounds containing carbon and hydrogen are *not* stable relative to their oxidation products, carbon dioxide and water. We would therefore expect them to react with oxygen exothermically, which, of course, they do. For example, methane:

$$CH_4(g) + 2O_2(g) \longrightarrow CO_2(g) + 2H_2O(g) \qquad \Delta H^\ominus = -890\,kJ\,mol^{-1}$$

Now, although methane is energetically unstable relative to its combustion products, it does not react with air until heated to quite high temperatures. In other words, it needs lighting before it burns. This is because the reaction between methane and oxygen has a high activation energy which must be supplied before the reaction will proceed. So methane, like most compounds containing carbon and hydrogen, is energetically unstable in the presence of air, but kinetically stable.

Compounds containing silicon and hydrogen are also energetically unstable relative to their combustion products:

$$SiH_4(g) + 2O_2(g) \longrightarrow SiO_2(s) + 2H_2O(g) \qquad \Delta H^\ominus = -1428\,kJ\,mol^{-1}$$

This reaction is much more exothermic than the corresponding one for methane, largely because of the very high energy of the Si—O bond. But, unlike methane, silane is not kinetically stable in the presence of oxygen. The activation energy of the above reaction is quite low, and silane bursts into flame spontaneously in air. Other compounds containing silicon and hydrogen behave similarly, so silicon does not form a huge range of compounds like carbon. However, the high strength of the Si—O bond relative to the Si—Si and Si—H bonds means that silicon exists naturally as highly stable silicon(IV) oxide (sand) and silicate minerals like clay and granite.

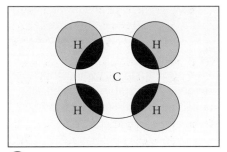

▲ Figure 13.1
Electron sharing in methane

▼ Table 13.1
Some average bond energies

Bond	Bond energy /kJ mol^{-1}
C—C	346
C=C	610
C≡C	835
Si—Si	226
Si=Si	318 (estimated)
S—S	272
C—O	360
Si—O	464
C—H	413
Si—H	318

Key Point

Organic compounds, though kinetically stable in air, are energetically unstable relative to their combustion products.

Figure 13.2

In the presence of air, hydrocarbons are energetically unstable, but kinetically stable. They can, therefore, be safely stored at room temperature …

Figure 13.3

… but at high temperatures they burn rapidly.

The ability of carbon to form strong bonds to itself means that it can form chains and rings of varying size. This is called catenation. These chains and rings are the basis of carbon's many stable compounds.

The kinetic stability of hydrocarbons in air is very important to society. It means they can be stored and, barring accidents, the energy of their oxidation can be released when it is required. This makes hydrocarbons, of which oil is our major source, the most important modern fuels.

c Carbon can form four covalent bonds

The bond energies in Table 13.1 suggest that sulfur, like carbon, should be able to form reasonably stable bonds to itself. However, sulfur forms only two bonds, so a chain of sulfur atoms cannot have side-groups attached to it. In contrast, carbon can form *four* bonds. This means a chain of carbon atoms can have many different groups attached, leading to a wide diversity of compounds.

Figure 13.4

Like carbon atoms, these skydivers can form four bonds each. This means they can form chains and rings.

13.2 Organic chemistry

The diversity of carbon chemistry is responsible for the diversity of life itself. You yourself are a unique individual because you contain unique DNA and unique proteins. Only carbon can form a range of compounds diverse enough to provide a different set for every individual.

The major source of compounds containing carbon and hydrogen is living or once-living material: animals, plants, coal, oil and gas. For this reason, it was originally thought that only living organisms could produce these compounds. This has since been shown to be untrue, but the name 'organic' is still applied to that branch of chemistry concerned with the study of compounds containing C—H bonds. This includes the vast majority of carbon compounds, although CO, CO_2 and carbonates are usually considered to belong to the field of inorganic chemistry.

The study of organic chemistry is of central importance in understanding the chemistry, and therefore the biology, of living systems. (The chemical study of living systems is called **biochemistry**.) A knowledge of organic chemistry enables chemists to develop and manufacture medicines, agricultural chemicals, anaesthetics and other chemicals whose effects on life processes are important to humans.

Many other organic chemicals are important to modern society. These include, for example, the many polymers (polythene, nylon) whose properties of flexibility and elasticity are a direct result of carbon's unique ability to form chains.

It is clearly important to understand organic chemistry, but with so many compounds to consider, we need some means of simplifying and systematising our study.

13.3 Functional groups

The ability of carbon to form strong bonds to itself and to hydrogen leads to the formation of stable compounds. **Hydrocarbons** contain *only* hydrogen and carbon. The simplest hydrocarbons, containing only single bonds, are the **alkanes.** Butane is an example:

$$H-\overset{\displaystyle H}{\underset{\displaystyle H}{C}}-\overset{\displaystyle H}{\underset{\displaystyle H}{C}}-\overset{\displaystyle H}{\underset{\displaystyle H}{C}}-\overset{\displaystyle H}{\underset{\displaystyle H}{C}}-H$$

But consider another compound, butan-1-ol:

$$H-\overset{\displaystyle H}{\underset{\displaystyle H}{C}}-\overset{\displaystyle H}{\underset{\displaystyle H}{C}}-\overset{\displaystyle H}{\underset{\displaystyle H}{C}}-\overset{\displaystyle H}{\underset{\displaystyle H}{C}}-OH$$

Butane and butan-1-ol have very different properties. Butane is a gas, butan-1-ol is a liquid. Butane has no effect on sodium, but butan-1-ol reacts to give hydrogen. Clearly, the —OH group in butan-1-ol has a big effect on the properties of the unreactive butane skeleton to which it is attached.

The —OH group in butan-1-ol is an example of a **functional group.** A given functional group, such as —OH, has much the same effect whatever the size and shape of the hydrocarbon skeleton it is attached to. This greatly simplifies the study of organic compounds because all molecules containing the same functional group can be considered as members of the same family, with similar properties. As the hydrocarbon chain gets bigger it increasingly dominates the properties of the compound. Because of this, members of the family show a steady gradation (gradual change) of physical and chemical properties as the size of the hydrocarbon part of the molecule increases.

A family of compounds containing the same functional group is called a **homologous series.** Butan-1-ol is a member of the homologous series of **alcohols,** all of which contain the —OH group. The first two alcohols are methanol and ethanol:

$$H-\overset{\displaystyle H}{\underset{\displaystyle H}{C}}-OH \qquad\qquad H-\overset{\displaystyle H}{\underset{\displaystyle H}{C}}-\overset{\displaystyle H}{\underset{\displaystyle H}{C}}-OH$$

methanol   ethanol

For any homologous series we can write a general formula in terms of the number of carbon atoms present. For example, the general formula of the alcohols is $C_nH_{2n+1}OH$.

The main functional groups and homologous series considered in this book are shown in Table 13.2. The methods of naming the different compounds will be explained as we go along.

The idea of functional groups can also be applied to compounds containing more than one group. Thus, the properties of the molecule as a whole can be predicted by considering the effect of each functional group.

Figure 13.5
These identical twins are made from identical sets of carbon compounds.

Definitions

Hydrocarbons are compounds containing only carbon and hydrogen.

A **functional group** is the atom or group of atoms which gives an organic compound its characteristic properties.

Key Point

Organic compounds have a basic hydrocarbon skeleton with functional groups attached. Compounds with the same functional group form a family with similar properties called a homologous series.

Quick Questions

1 To which homologous series does each of the following belong? Look at Table 13.2:
   a propanal, CH_3CH_2CHO,
   b propene, $CH_3CH=CH_2$,
   c propanoic acid, CH_3CH_2COOH.

Table 13.2
Common functional groups

Functional group	Name of homologous series	Example
—OH	alcohols	CH_3OH, methanol
—NH$_2$	amines	CH_3NH_2, methylamine
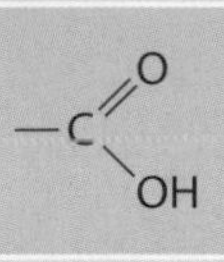 (carboxylic acid structure)	carboxylic acids	CH_3COOH, ethanoic acid
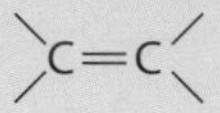 (C=C structure)	alkenes	$H_2C=CH_2$, ethene
—Halogen	halogeno compound	CH_3Cl, chloromethane
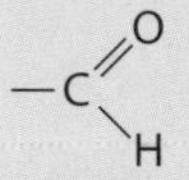 (aldehyde structure)	aldehydes	CH_3CHO, ethanal
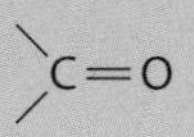 (ketone structure)	ketones	CH_3COCH_3, propanone
—O—	ethers	CH_3OCH_3, methoxymethane

13.4 Finding the formulae of organic compounds

To predict the properties of a compound, we need to know its **structural formula,** showing the position and nature of its functional groups.

Finding empirical formulae

Chapter 1 explains how the empirical formula of a compound may be found from its percentage composition. The composition of organic compounds can be found by combustion analysis. A known mass of a compound is burned and the carbon dioxide and water formed are collected and measured. Other elements that may be present, such as nitrogen and halogens, can also be estimated. From the masses of the combustion products the empirical formula can be calculated. In modern laboratories this combustion analysis is performed automatically by machines.

Quick Questions

2 In a combustion analysis machine, the compound is burned in pure oxygen rather than air. Suggest a reason why.
3 What products would you expect to be formed when the following compounds are burned completely in a combustion analysis machine?
   a ethane, CH_3CH_3,
   b ethanol, CH_3CH_2OH,
   c ethylamine, $CH_3CH_2NH_2$.

Worked example

A compound **X** containing only carbon, hydrogen and oxygen was subjected to combustion analysis. 0.1 g of the compound on complete combustion gave 0.228 g of carbon dioxide and 0.0931 g of water. Calculate the empirical formula of the compound.

First, calculate the mass of C and H in 0.1 g of the compound.

44 g of CO_2 contains 12 g of C $\Rightarrow$ mass of C in 0.1 g of **X** $= \frac{12}{44} \times 0.228 = 0.0621$ g

18 g of H_2O contains 2 g of H $\Rightarrow$ mass of H in 0.1 g of **X** $= \frac{2}{18} \times 0.0931 = 0.0103$ g

Mass of C + H in 0.1 g of **X** $= 0.0621 + 0.0103 = 0.0724$ g

The remainder of the 0.1 g of **X** must be oxygen.

$\Rightarrow$ Mass of O in 0.1 g $= 0.1 - 0.0714 = 0.0276$ g

$\therefore$ Ratio by mass C : H : O is $\quad 0.0621 \quad : \quad 0.0103 \quad : \quad 0.0276$

$\Rightarrow$ Ratio by moles C : H : O is $\quad \frac{0.0621}{12} \quad : \quad \frac{0.0103}{1} \quad : \quad \frac{0.0276}{16}$

$\qquad\qquad\qquad = \quad 0.005\,18 \quad : \quad 0.0103 \quad : \quad 0.001\,73$

$\qquad\qquad\qquad = \qquad 3 \qquad : \qquad 6 \qquad : \qquad 1$

$\therefore$ Empirical formula of **X** is C_3H_6O.

Finding molecular formulae

Once we have found the empirical formula, we can find the molecular formula of the compound, provided we know its relative molecular mass. There are several methods for doing this, but it is usually done by mass spectrometry (Section 1.2).

When the molecules of an organic compound pass through a mass spectrometer, they get broken up into fragments. This means that the mass spectrum of the compound contains several peaks, corresponding to the different fragments. But a few molecules pass through intact. These intact molecules give a peak showing the relative molecular mass of the compound.

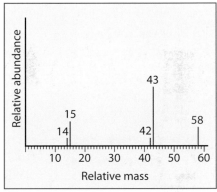

⌃ Figure 13.6

Mass spectrum of compound **X**, molecular formula C_3H_6O. Most of the peaks are caused by molecular fragments. The highest mass peak at 58 is the molecular ion, so the relative molecular mass of X is 58.

Worked example

The mass spectrum obtained for the compound **X** in the worked example above is shown in Figure 13.6.

The relative mass of the heaviest particle recorded in the spectrum is 58. We can assume that this corresponds to the intact molecule with a single positive charge, i.e. $\mathbf{X}^+$, the **molecular ion, M.** This means that the relative molecular mass of **X** must be 58.

With an empirical formula of C_3H_6O, **X** could have molecular formula C_3H_6O, $C_6H_{12}O_2$, $C_9H_{18}O_3$ and so on. But since its relative molecular mass is known to be 58, the only possible molecular formula is C_3H_6O.

Section 13.7 explains how the fragments can tell you the molecule's structure.

Definitions

The **empirical formula** shows the simplest whole number ratio for the atoms of each element in a compound.

The **molecular formula** shows the actual number of atoms of each element in one molecule of the compound.

The **structural formula** shows how the different atoms are joined together, and the positions of functional groups.

Quick Questions

4 The structural formula of ethanoic acid is given in Table 13.2.
   a What is its empirical formula?
   b What is its molecular formula?
   c Write one other structure with the same molecular formula.
5 A hydrocarbon has an empirical formal CH_2, and its relative molecular mass is 56. What is the molecular formula of the hydrocarbon?

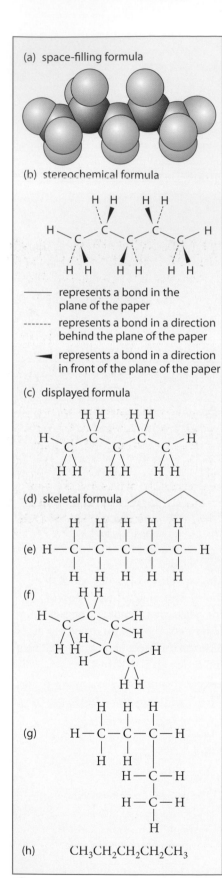

(a) space-filling formula

(b) stereochemical formula

—— represents a bond in the plane of the paper

------- represents a bond in a direction behind the plane of the paper

◄ represents a bond in a direction in front of the plane of the paper

(c) displayed formula

(d) skeletal formula

(e)

(f)

(g)

(h) $CH_3CH_2CH_2CH_2CH_3$

Figure 13.7
Representations of the structural formula of pentane, C_5H_{12}

Finding structural formulae

The molecular formula of a compound gives the number of atoms of the different elements in one molecule of the compound, but it gives no information about the way the atoms are arranged. A compound with molecular formula C_3H_6O, for example, could have the structural formulae:

$$CH_3CH_2CHO \qquad \text{or} \qquad CH_3COCH_3$$
$$\text{propanal} \qquad\qquad\qquad\qquad \text{propanone}$$

To find the exact structural formula of a compound, we need more information to decide which functional groups it contains. This information can be found by various methods.

a Instrumental methods

These are described in Sections 13.7 to 13.9.

b Physical properties of the compound, such as boiling point

Physical properties are dependent on structure and provide information that can help us work out the structural formula. For example propanal and propanone are both liquids, but they have different boiling points and smell different.

c Chemical properties of the compound

Each functional group has certain chemical characteristics. For example, aldehydes such as propanal are reducing agents, but ketones such as propanone are not. Thus a knowledge of the chemical properties of a compound can provide clues about the functional groups it contains.

13.5 Writing structural formulae

The structural formula of a compound shows how the atoms are joined together. This information is crucial in deciding the properties of the compound. The three-dimensional shape of the molecule also affects its properties. It is helpful if the structural formula can give some indication of the shape, i.e. stereochemistry of the molecule.

The methane molecule is tetrahedral in shape (Section 3.7). This tetrahedral arrangement of bonds is common to all saturated carbon atoms (that is, carbon atoms bonded to four other groups). The tetrahedral shape creates a problem in representing structural formulae, because it is difficult to show a three-dimensional shape on two-dimensional paper.

Figure 13.7 shows eight ways of representing the structural formula of pentane, C_5H_{12}. The most accurate representation is the space-filling model (Figure 13.7(a)). Try building this model yourself. Space-filling models show the extent of the electron cloud of each atom.

Figure 13.7(b) attempts to represent the tetrahedral carbon atom by showing the directions of the bonds. This is called a stereochemical formula. This system can be simplified to Figure 13.7(c), which is called a displayed formula.

Sometimes a skeletal formula (Figure 13.7(d) is used, in which only carbon–carbon bonds and functional groups are shown.

A common and easy way of writing formulae is shown Figure 13.7(e) but this has drawbacks because it represents the bond angles at each carbon atom as 90° instead of 109°. This type of diagram shows its limitations when an attempt is made to represent rotation about a single bond: compare Figure 13.7(g) with Figure 13.7(f). It is easy to see that (f) can be obtained from (c) by rotating about the bond between the second and third carbon atoms. On the other hand, (g) appears to be a different compound to (e), though both in fact represent pentane.

Finally, Figure 13.7 (h) shows a shortened way of writing a structural formula.

These different ways of representing structural formulae will be used as appropriate in different parts of this book.

Quick Questions

6 Write a displayed formula for: $CH_3CH_2CHOHCH_2CH_3$

7 Write a skeletal formula for:

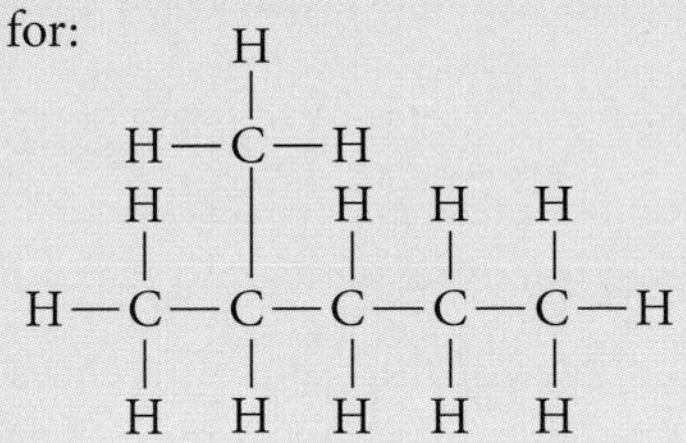

Hint: The skeletal formula for

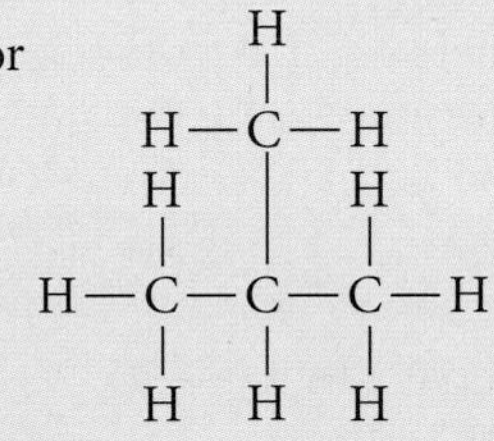

is

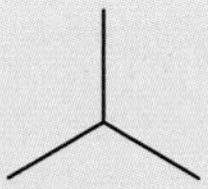

8 Write a displayed formula for:

a   b

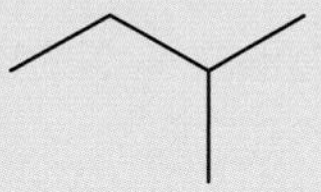

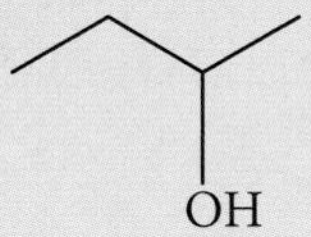

Hint: The skeletal formula for 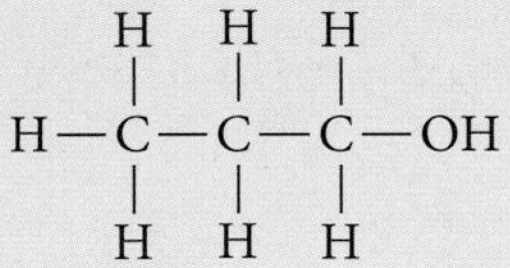   is

Definitions

A **displayed formula** shows all the atoms and all the bonds between them.

A **skeletal formula** omits all C and H atoms, showing only carbon–carbon bonds and functional groups.

Isomers are compounds with the same molecular formula but with their atoms arranged in different ways.

Structural isomers have the same molecular formula, but they differ in which atom is attached to which.

13.6 Isomerism

Compounds possessing the same molecular formula but with their atoms arranged in different ways, are called **isomers**. Isomerism is very common in organic chemistry. All carbon compounds with four or more carbon atoms, and many with less, show isomerism. There are 4.11×10^9 isomers with the molecular formula $C_{30}H_{62}$. Isomerism also occurs, though less commonly, in inorganic chemistry. In this book we are concerned with three main types of isomerism: structural, optical and *cis–trans*.

Structural isomerism

Structural isomers differ in which atom is attached to which. Propanal and propanone (Table 18.1, Section 18.2) are structural isomers. In propanal, the C=O group is at the end of the molecule; in propanone it is in the middle. Structural isomers can be members of different homologous series, like the pair just mentioned, or they may be members of the same series like butane and methylpropane below, which are both alkanes:

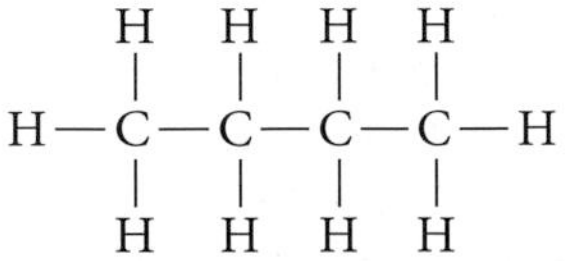

butane

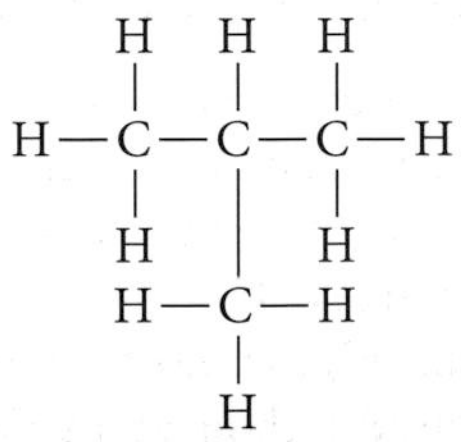

methylpropane

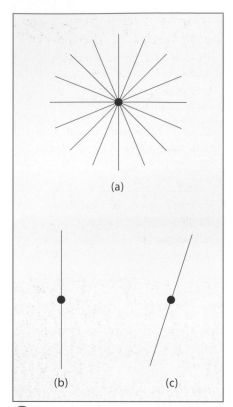

Figure 13.8

(a) Normal light ray travelling towards the observer: each line represents a wave seen 'end-on'

(b) Plane-polarised light: only waves in a single plane are present.

(c) Plane-polarised light from (b) after passing through an optically active solution

Figure 13.9
The different flavours of oranges and lemons are caused by the two different optical isomers of the same molecule, called limonene.

Note that

$$
\begin{array}{c}
\quad\ \text{H}\quad\ \text{H}\quad\ \text{H} \\
\quad\ |\quad\ \ |\quad\ \ | \\
\text{H}-\text{C}-\text{C}-\text{C}-\text{H} \\
\quad\ |\quad\ \ |\quad\ \ | \\
\quad\ \text{H}\quad\ \text{H}\quad\ \text{C}-\text{H} \\
\quad\qquad\qquad\ | \\
\quad\qquad\qquad\ \text{H}
\end{array}
$$

is *not* a third isomer of C_4H_{10}, because it can be formed from butane simply by rotating about a single bond. It is possible to rotate a structure freely about any single carbon–carbon bond. Structures that can be interconverted by rotating about a bond in this way are not isomeric. The structure above appears at first sight to be a separate isomer, but this is due to representing the carbon bond angles as 90° instead of 109°. Try experimenting with a molecular model to make this point clear.

Structural isomers usually show considerable differences in their physical and chemical properties, even if they are members of the same homologous series. Thus, the boiling point of butane is 273K while that of methylpropane is 261K. It is not really surprising that structural isomers differ. Any functional group is influenced by its environment and has its properties modified by the atoms to which it is attached.

> ### Quick Questions
>
> **9** How many isomers are there of molecular formula C_5H_{12}? Write their structural formulae.
>
> **10** Write the structural formulae of all the alcohols (i.e. the compounds with the —OH functional group) of molecular formula $C_4H_{10}O$.

Optical isomerism

Optical isomers are identical to one another, except in one property – their effect on polarised light.

Polarised light

Light is a form of electromagnetic radiation consisting of waves. A ray of normal light has waves which vibrate in many directions at right angles to the direction of travel of the ray (Figure 13.8(a)).

Certain materials have the ability to remove from normal light all waves except those vibrating in a single plane (see Figure 13.8(b)). The light is then said to be **plane polarised.** It is rather like the light being combed as it passes through the polariser. A well known polariser is polaroid, which is used in the lenses of some sunglasses.

Certain chemical substances have the ability to rotate the plane of polarised light. That is, if polarised light is shone through a solution of the substance in a suitable solvent, the plane in which the polarised light vibrates will be rotated either clockwise or anticlockwise (Figure 13.8(c)). These substances are said to be **optically active.** An example is 2-hydroxypropanoic acid (lactic acid, $CH_3CHOHCOOH$), the substance responsible for the sour taste in sour milk.

Chemists have isolated two forms of lactic acid: one of these rotates plane polarised light clockwise and is called the (+) isomer, the other rotates it anticlockwise and is called the (−) isomer (Figure 13.10). For a given concentration of solution the two forms rotate light to exactly the same extent, but in opposite directions. In addition, crystals of the two forms are found to be mirror images of one another. Apart from these differences, the forms are physically and chemically identical. Since the two forms of lactic acid are chemically identical, they must contain exactly the same groups attached to each other in the same way. Their difference can only lie in the way the groups are arranged relative to each other in space. And they must be exact opposites in this respect since their effect on polarised light is equal and opposite.

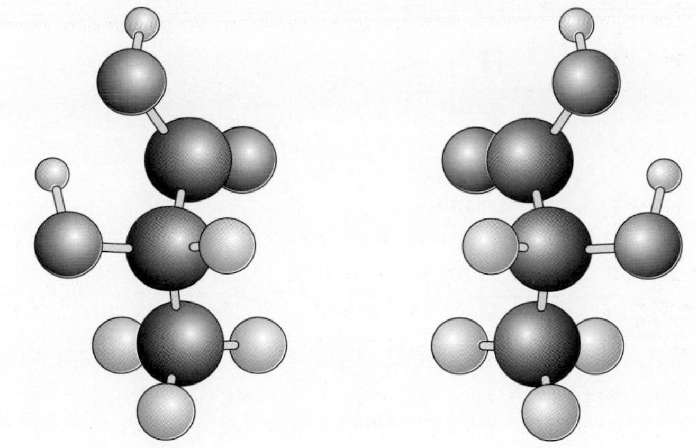

Figure 13.10
The optical isomers of lactic acid. The two isomers are images of one another reflected in an imaginary mirror placed between them.

Definitions

Optical isomers are molecules with identical properties, except they rotate plane polarised light in opposite directions.

Optical isomers have **chiral** molecules, which means they are asymmetric.

Chiral molecules have two forms which are mirror images of one another. The two forms are called **enantiomers**.

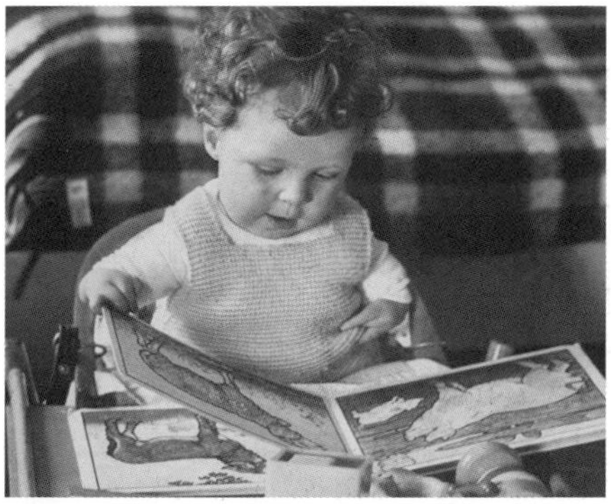

Figure 13.11
The thalidomide tragedy in the 1960s was caused by a drug that was prescribed to pregnant mothers to treat morning sickness. One of the enantiomers of thalidomide is a safe and effective medicine; unfortunately the other enantiomer is a powerful foetus-deformer. If chemists had known this at the time, they might have been able to prevent the tragedy.

Chiral molecules

When we look at the molecular structure of different optically active substances we see they have one thing in common. They all have *asymmetric* molecules; this means that their molecules have no centre, line or plane of symmetry. As a result they are different from their own mirror image. Optical activity is shown by all substances with asymmetric molecules. Such substances have two isomeric forms which are mirror images of one another and which rotate polarised light in opposite directions.

Molecules that are asymmetric are said to be **chiral.** The simplest type of chiral molecule is one in which four different groups are attached to the same carbon atom; lactic acid is such a molecule. The asymmetric carbon atom is called a **chiral centre.** Figure 13.10 shows the two forms, called (+)-lactic acid and (−)-lactic acid.

Try making models of each of the forms to satisfy yourself that they are indeed different molecules and that they are mirror images. You can show they are different by trying to superimpose the two molecules on one another: it is impossible to arrange them so that all the groups correspond in position. Note, though, that in both (+)- and (−)-lactic acid all the groups are attached together in the same way, and that the spacings of the various groups are the same in each isomer. This is why the two forms of lactic acid have identical chemical properties.

The relationship between optical isomers such as (+)- and (−)-lactic acid is like the relationship between your right and your left hand. The lengths of the different fingers and the distances between them are the same for both hands, but they are mirror images of one another and cannot be superimposed on each other – try putting your right glove on the left hand. Structures that are mirror images of one another are called **enantiomers.**

Key Point

If a molecule contains a carbon atom to which four different groups are attached, it will have optical isomers.

Many optically active compounds have complex structures and it is difficult to tell quickly whether or not their molecules are chiral. For simpler molecules it is safe to say that if the molecule contains a carbon atom to which four different groups are attached, it will show optical isomerism.

Figure 13.12 shows an example of a simple molecule with a chiral centre.

11 Would you expect CH_2ClF to show optical isomerism? Explain.

12 Which of the isomeric alcohols whose structures you wrote in Quick question 10 would show optical isomerism? Identify the chiral centre and write the structures of the enantiomers.

13 Write the structure of the first alkane to show optical isomerism.

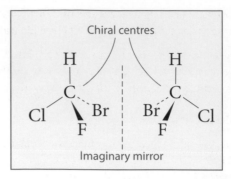

Figure 13.12
CHBrClF is a simple molecule with a chiral centre. The two enantiomers are shown here.

Many organic compounds and some inorganic ones have chiral molecules and therefore show optical isomerism. Optically active compounds are very common in nature. Almost all amino acids are optically active, as are all the sugars. The interesting thing is that most naturally occurring optically active compounds occur as one isomer only. Thus all naturally occurring glucose is (+)-glucose. The enzymes that produce and break down these substances are able to recognise them because of their asymmetric molecules. The active site of the enzyme is also asymmetric, and exactly fits the asymmetric substrate. This is why enzymes are so specific in their action.

Lactic acid can be extracted from natural sources such as sour milk or muscle tissue and the acid from these sources always shows optical activity. It is also possible to prepare lactic acid in the laboratory from fairly simple starting materials, but the acid prepared in this way shows no optical activity at all.

This is because most laboratory methods for preparing lactic acid produce the (+)- and (–)-isomers in equal amounts. The two isomers cancel each other out in their effect on polarised light.

A mixture of optical isomers containing equal molar quantities of (+) and (–) isomers is called a racemic mixture or a racemate. Because of the identical chemical properties of the isomers, it is very difficult to separate them from a racemic mixture, though methods do exist for doing so. This is important in the preparation of pure pharmaceutical products.

Cis–trans isomerism

This type of isomerism occurs in some molecules with double bonds. It is described in Section 15.3.

13.7 Instrumental methods of analysis

Modern chemistry laboratories use sophisticated instruments to find the structure of organic chemicals. These instruments are very sensitive and they usually need only very small amounts of the chemical to work on. Laboratories in industry and universities are equipped with a range of instruments, but they tend to be very expensive, so you don't find many of them in schools. The most important methods of instrumental analysis are mass spectrometry, spectroscopy and chromatography. In this section we will look in some detail at mass spectrometry; then, in Section 13.8, spectroscopy, and finally, in Section 13.9, chromatography.

Mass spectrometry

The mass spectrometer is an instrument which turns atoms and molecules into ions and measures their mass. The working of a simple mass spectrometer is described in Section 1.2.

When an organic compound passes through a mass spectrometer, its molecules get broken into positively charged fragments. These fragments provide useful information.

Each fragment gives a corresponding line in the mass spectrum. From the position of the line, we can find the relative mass of the fragment, and use this to work out its formula. By piecing together the fragments we can deduce the structure of the parent molecule.

Using fragments to deduce structures

Look at Figure 13.13. The mass spectrum for compound **X**, molecular formula C_3H_6O, has prominent peaks at 15 and 43, and these give strong clues to the compound's structure. They correspond to CH_3^+ and CH_3CO^+, respectively, and this suggests the structural formula CH_3COCH_3 (propanone). The other possible structure for **X**, CH_3CH_2CHO (propanal) is ruled out because this would give a strong peak at mass 29, corresponding to the two fragments $CH_3CH_2^+$ and CHO^+, both of which we would expect CH_3CH_2CHO to form.

We can also find out a lot by looking at the *differences* between the masses of peaks in the spectrum. For example, in the mass spectrum of **X** (Figure 13.13), the peaks at mass 58 and 43 differ in mass by $(58 - 43) = 15$. This suggests that the peak at 43 has been formed by the loss of a CH_3 group from the molecular ion (the peak at 58).

Key Point

Certain fragments are very common in mass spectra. They include:

Mass	Fragment
15	CH_3^+
28	CO^+ or $C_2H_4^+$
29	$CH_3CH_2^+$
43	$CH_3CH_2CH_2^+$
77	$C_6H_5^+$ (benzene ring)

Definition

A **molecular ion**, M^+ is formed by removing one electron from the intact molecule to give an ion with a single + charge.

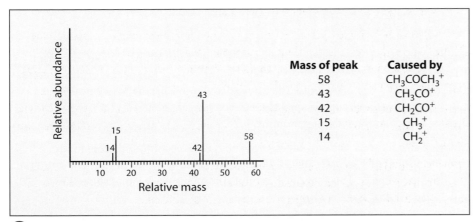

Mass of peak	Caused by
58	$CH_3COCH_3^+$
43	CH_3CO^+
42	CH_2CO^+
15	CH_3^+
14	CH_2^+

Figure 13.13

Mass spectrum of a compound **X**, molecular formula C_3H_6O. The table identifies the fragment causing the major peaks. (There are several smaller peaks which have been omitted.)

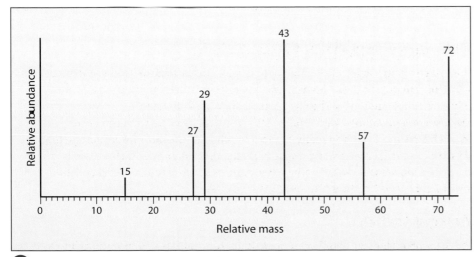

Figure 13.14

The mass spectrum of butanone

Quick Questions

Look at Figure 13.14, which shows the mass spectrum of butanone, $CH_3CH_2COCH_3$

14 a Identify the molecular ion peak M^+. What is the relative molecular mass of butanone?

 b Identify the fragments responsible for the peaks at 15 and 29.

 c Work out the difference in mass between the peaks at 72 and 57. What fragment must have been lost from the molecular ion to create the peak at 57?

 d Work out the difference in mass between the peaks at 72 and 43. What fragment must have been lost from the molecular ion to create the peak at 43?

Patterns from isotopes

Some isotopes give tell-tale messages in mass spectra. Two of these tell-tale messages result in what are called M + 1 and M + 2 peaks.

M + 1 peaks

Carbon has two stable isotopes, carbon-12 and carbon-13. C-12 makes up 98.9 % of all the carbon atoms and C-13 makes up the other 1.1 %.

If you look to the right of the molecular ion, $\mathbf{M^+}$, you can usually see a much smaller peak, called an M+1 peak. This is caused by the molecular ion with a C-13 atom replacing *one* of the C-12 atoms.

If the compound has just one C atom, this extra peak will be just 1.1 % of the height of the main molecular ion peak, because only 1.1 % of the molecules will have a C-13 atom. But if the compound has *two* C atoms, the extra peak will be 2.2 % of the height of the main molecular ion peak, because there are two chances that the molecule has a C-13 atom in it. If there are *three* C atoms, the extra peak will be 3.3 % of the height, and so on.

You can use these M + 1 peaks to work out the number of C atoms in the parent molecule (see Figure 13.15).

M + 2 peaks

Bromine has two stable isotopes, ^{79}Br and ^{81}Br, which are almost equally abundant. This means that a molecule containing one Br atom shows *two* peaks for the molecular ion, separated by two mass units. The M peak contains ^{79}Br; the M + 2 peak contains ^{81}Br. The peaks are of roughly equal heights, so equal-sized M + 2 peaks are a tell-tale sign for compounds containing bromine.

Chlorine also has two stable isotopes, ^{35}Cl and ^{37}Cl. So compounds containing chlorine also give M and M + 2 peaks, but in this case the M + 2 peak is one-third the height of the M peak. This is because natural chlorine is made up of about 75 % ^{35}Cl and 25 % ^{37}Cl, that is a 3 : 1 ratio.

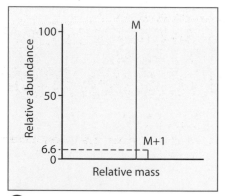

Figure 13.15
The M+1 peak is 6.6% of the height of the M peak. This indicates that the compound contains 6.6/1.1 = 6 C atoms.

Quick Questions

Look at Figure 13.16.
15 a Identify the M and M + 2 peaks.
    b Find the two peaks corresponding to Br^+.
    c What is the peak at mass = 43 due to?

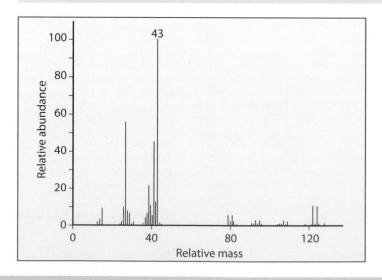

Figure 13.16
The mass spectrum for
1-bromopropane, $CH_3CH_2CH_2Br$

13.8 Spectroscopy

When electromagnetic radiation, such as light or infrared, shines on a chemical, the chemical may interact with the radiation in some way. The commonest example is visible light: colour is produced when chemicals emit or absorb visible light of a particular frequency.

The way a particular chemical interacts with radiation gives chemists information about its molecules. A **spectroscope** is an instrument which allows radiation to interact with a sample of a chemical and then analyses the changes.

Different kinds of radiation interact with chemicals in different ways (Figure 13.18). The effects are summarised in Table 13.3. In this section we will look at two types of spectroscopy that are particularly useful to chemists: infrared and nuclear magnetic resonance.

⬥ Figure 13.17
The yellow colour of sodium vapour street lamps is due to the visible emission spectrum of sodium.

Type of radiation	Frequency range/s^{-1}	Effect on molecule	Type of spectroscopy
ultraviolet	10^{15}–10^{17}	excites the electrons	ultraviolet/visible spectroscopy
visible light	10^{14}–10^{15}	excites the electrons	ultraviolet/visible spectroscopy
infrared	10^{11}–10^{12}	makes bonds vibrate	infrared spectroscopy
microwaves	10^{9}–10^{11}	makes molecules rotate	microwave spectroscopy
radio waves	10^{6}–10^{8}	changes the magnetic alignment of the nuclei of some atoms	nuclear magnetic resonance

⬥ Table 13.3
How different types of radiation interact with chemicals

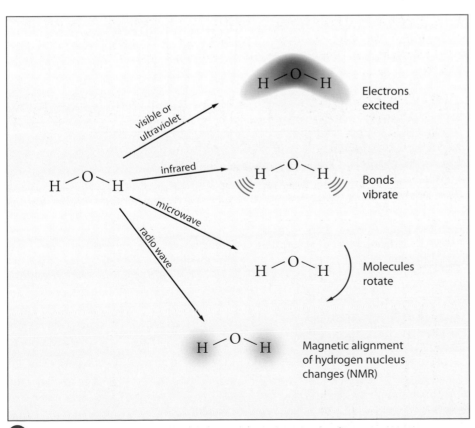

⬥ Figure 13.18
The effect of different types of radiation on a water molecule

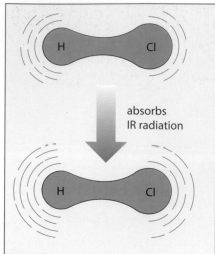

Figure 13.19
When an HCl molecule absorbs infrared radiation, it vibrates more energetically. The frequency of radiation absorbed is $7.21 \times 10^{13}\,\text{s}^{-1}$, and this frequency is characteristic of the H—Cl bond.

Figure 13.20
The infrared spectrum of propanone. Notice that the horizontal axis shows 'wavenumber'. This is the conventional way of representing the radiation absorbed. It is the reciprocal of the wavelength measured in cm. Notice too that the vertical axis shows 'transmittance', the percentage of the radiation that passes through the sample without being absorbed.

Table 13.4
Characteristic IR absorptions of some common bonds

Infrared spectroscopy

Infrared spectroscopy makes use of the fact that molecules absorb infrared (IR) radiation. IR has a wavelength longer than visible light, between about 2500 nm and 25 000 nm. The energy of the radiation is absorbed in making the bonds vibrate (Figure 13.19). When the molecule absorbs the radiation, the bonds vibrate more energetically.

Different bonds absorb radiation of different frequencies. The frequency is characteristic of the particular bond concerned. So, we can use IR absorption to identify the bonds, and therefore the functional groups, in an organic molecule.

The spectrometer produces an **infrared spectrum** on a chart recorder, like that in Figure 13.20 – this is the IR spectrum of propanone, CH_3COCH_3.

You will see that the spectrum is quite complicated, even though propanone is a simple molecule with only three types of bond. The complexity arises because the bonds can each vibrate in a number of different ways, and the vibrations can interact with each other. Nevertheless, it is possible to see some characteristic peaks of absorption which we can use to identify functional groups in the molecule.

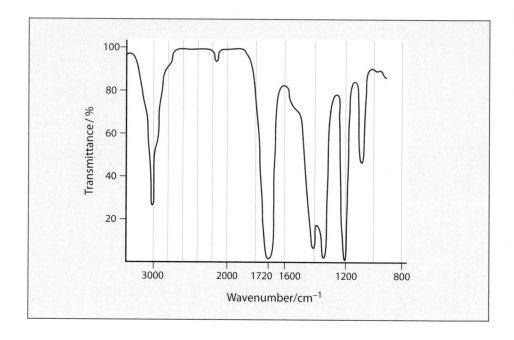

Bond	Found in	Wavenumber of absorption/ cm^{-1}	Intensity (M = medium, S = strong)
C—H	alkanes, alkenes, arenes	2840 to 3095	M/S
C=C	alkenes	1610 to 1680	M
C=O	aldehydes, ketones, acids, esters	1680 to 1750	S
C—O	alcohols, esters	1000 to 1300	S
C≡N	nitriles	2200 to 2280	M
C—Cl	chloro compound	700 to 800	S
O—H	'free' hydrogen-bonded in alcohols, phenols hydrogen-bonded in acids	3580 to 3670 3230 to 3550 2500 to 3300	S S (broad) M (broad)
N—H	primary amines	3100 to 3500	S

Table 13.4 gives the characteristic wavenumbers of absorptions of some common bonds. In the spectrum of propanone in Figure 13.20, the strong peak at about $1720\,\mathrm{cm}^{-1}$ corresponds to the $C={=}O$ bond. The weaker absorption at $3000\,\mathrm{cm}^{-1}$ corresponds to the $C-H$ bond. This peak is weaker even though there are several H atoms in the molecule: in IR spectroscopy the strength of the peak is a characteristic of the bond itself, not of the number of bonds present.

Most of the interesting parts of an IR spectrum are found in the region beyond $1500\,\mathrm{cm}^{-1}$. The peaks outside this region are less useful, but they are helpful in showing the **fingerprint** of the compound: the characteristic pattern of its IR spectrum. The fingerprint can be used to compare the compound's spectrum with IR spectra of known compounds given in reference books.

Infrared spectra are very useful, particularly in identifying the functional groups in an unknown compound.

Quick Questions

16 Figure 13.21 shows the IR spectrum of ethanol, CH_3CH_2OH. Look closely at the spectrum.
  a What bond gives rise to the peak at just below $3000\,\mathrm{cm}^{-1}$?
  b What bond gives rise to the peak at about $3400\,\mathrm{cm}^{-1}$?

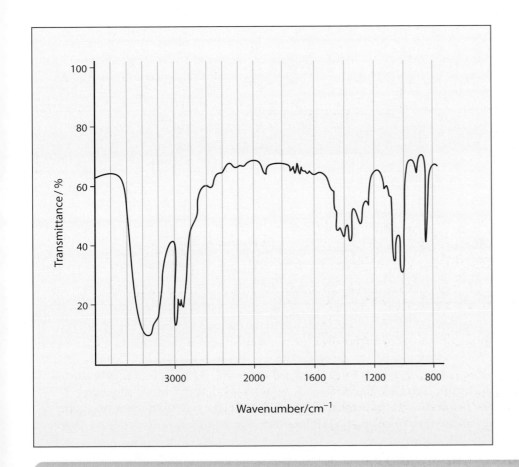

◀ Figure 13.21
The infrared spectrum of ethanol, CH_3CH_2OH

Figure 13.22

A chemist using a nuclear magnetic resonance spectrometer. NMR is probably the most valuable technique available to organic chemists.

Nuclear magnetic resonance (NMR) spectroscopy

When certain atoms are placed in a strong magnetic field, their nuclei behave like tiny bar magnets and align themselves with the field. Electrons behave like this too, and for this reason both electrons and nuclei are said to possess 'spin', since any spinning electric charge has an associated magnetic field.

Just as electrons with opposite spin pair up together (Section 2.8), a similar thing happens with the protons and neutrons in the nucleus. If a nucleus has an even number of protons and neutrons (e.g. $^{12}_{6}C$), their magnetic fields cancel out and it has no overall magnetic field. But if the number of protons and neutrons is odd (e.g. $^{13}_{6}C$, $^{1}_{1}H$), the nucleus has a magnetic field.

If the substance is now placed in an external magnetic field, the nuclear 'magnets' line up with the field, in the same way as a compass needle lines up with a magnetic field (Figure 13.23).

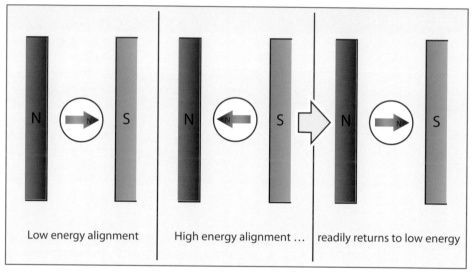

Low energy alignment | High energy alignment … | readily returns to low energy

Figure 13.23

Alignment of a compass needle in a magnetic field

The nuclear 'magnet' can have two alignments, of low and high energy (Figure 13.24). To make the nucleus change to the high energy alignment, a quantity of energy, ΔE must be supplied.

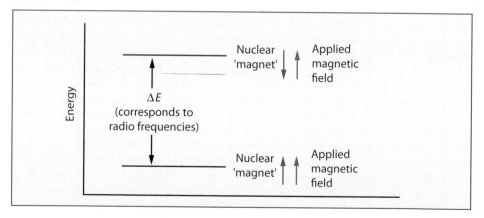

Figure 13.24

Two alignments of the nuclear 'magnet' in an external magnetic field. The energy difference, ΔE, between the two orientations is the basis of the technique of NMR.

It happens that the energy, ΔE, absorbed in the transition from low to high energy corresponds to radio frequencies. The precise frequency depends on the environment of the nucleus; in other words, on the other nuclei and electrons in its neighbourhood. So, by placing the sample being examined in a strong magnetic field and measuring the frequencies of radiation it absorbs, information can be obtained about the environments of nuclei in the molecule. The technique is called **nuclear magnetic resonance spectroscopy**, or **NMR** for short. Figure 13.25 shows a simplified diagram of an NMR spectrometer.

Key Point

Nuclear magnetic resonance, spectroscopy, NMR, involves placing a compound in a strong magnetic field so certain nuclei can be in a high energy or low energy state. Transition between the two states involves absorption of energy of characteristic frequency.

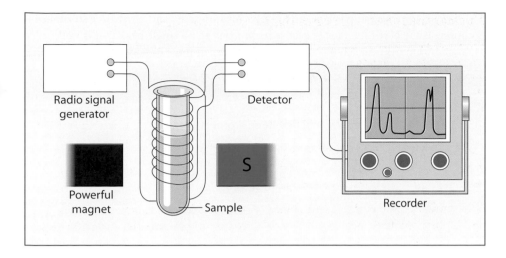

Figure 13.25
A simplified NMR spectrometer. The sample is dissolved in a solvent, such as ^2H_2O or CCl_4 which do not have nuclear magnetic properties.

The technique of NMR is particularly useful in identifying the number and type of hydrogen atoms (^1H) in a molecule. This is called **proton NMR**, because ^1H nuclei are protons. It is also used to find the positions of carbon atoms. The common isotope of carbon, ^{12}C, does not have a nuclear 'magnet', but natural carbon contains 1% of the ^{13}C isotope, which does show magnetic behaviour and can be identified using NMR.

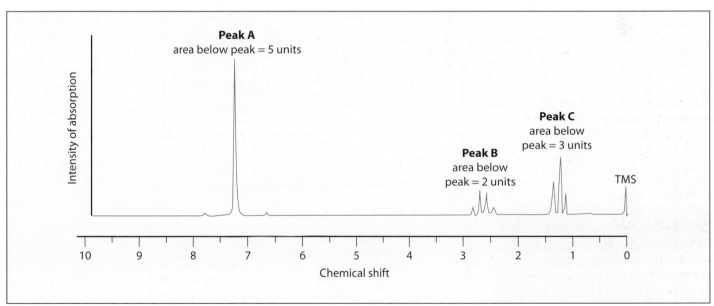

Figure 13.26
The proton NMR spectrum of ethylbenzene, $C_6H_5CH_2CH_3$

Figure 13.26 shows the proton NMR spectrum of ethylbenzene, $C_6H_5CH_2CH_3$. Notice that there are three major peaks, of differing heights. Each peak corresponds to an H atom in a different molecular environment. The area under each peak is proportional to the number of that type of H atom in the molecule. The largest peak (A) corresponds to the 5 H atoms in C_6H_5, the benzene ring. The second largest (C) corresponds to the 3 H atoms in the —CH_3 group, and the third peak (B) corresponds to the 2 H atoms in the CH_2 group.

H atoms in the same environment have similar positions in the NMR spectrum. Normally, this position is measured as a **chemical shift**, δ, from a fixed reference point. The reference point normally used is the absorption of a substance known as TMS. The chemical shift of TMS is set at zero.

Definitions

The **chemical shift** for an NMR peak is the distance of the peak from a fixed reference point, usually TMS.

Proton NMR is used to find the molecular environment of ^1H nuclei (protons).

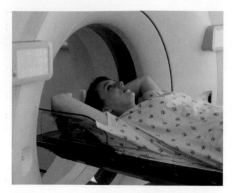

Figure 13.27
Nuclear magnetic resonance is the principle behind Magnetic Resonance Imaging (MRI) body scanners, which detect protons in water molecules in the body. Unlike X-ray investigation, the technique is believed to be completely harmless to the patient. The patient lies inside the tube, surrounded by a powerful magnetic field.

TMS stands for tetramethylsilane, $Si(CH_3)_4$. This non-toxic and unreactive substance is chosen as the NMR reference because its protons give a single peak that is well separated from the peaks found in the NMR spectra of most organic compounds.

Table 13.5 gives the chemical shifts for some common proton environments.

⌄ Table 13.5
Chemical shifts for protons , ^1H, in different environments

Type of proton	Chemical shift, δ, in region of
R—CH$_3$	0.9
R—CH$_2$—R	1.3
R—CH—R (with R above)	2.0
—C—CH$_2$— (C double bond O)	2.3
—O—CH$_3$	3.8
—O—CH$_2$—R	4.0
—O—H	5.0
⬡—H	7.5
—C=O with H	9.5
—C=O with O—H	11.0

Figure 13.29 shows a simplified proton NMR spectrum for ethanol, CH_3CH_2OH. It has been simplified by removing some of the detail, so the peaks appear more clearly. Notice that it includes the peak for TMS. Notice too that an integrated trace is shown. This gives the relative areas under each of the peaks.

Quick Questions

17 a Estimate the chemical shift for each of the peaks in Figure 13.28.
  b Use Table 13.5 to identify each of the peaks.
  c Explain the relative areas under the peaks.

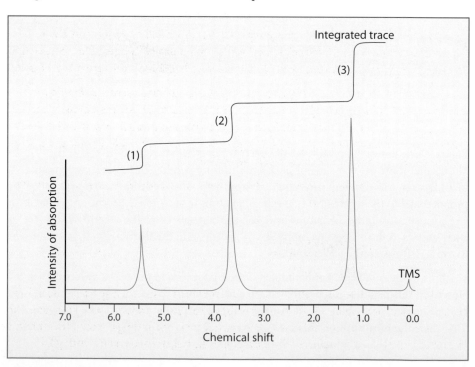

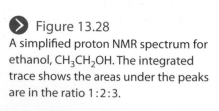

Figure 13.28
A simplified proton NMR spectrum for ethanol, CH_3CH_2OH. The integrated trace shows the areas under the peaks are in the ratio 1:2:3.

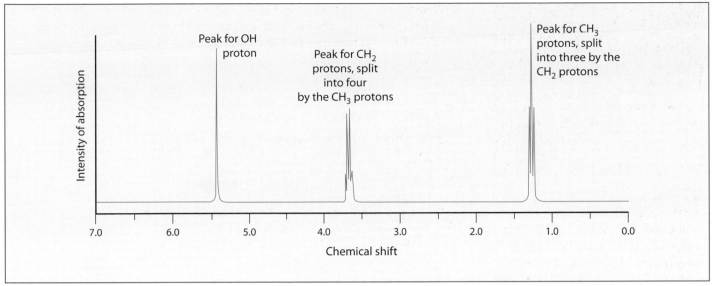

Figure 13.29
A detailed high-resolution NMR spectrum for ethanol

Spin-spin coupling

The simplified NMR spectrum of ethanol shown in Figure 13.28 shows three single peaks. The smallest peak corresponds to the single OH proton; the middle peak corresponds to the two CH_2 protons and the largest peak corresponds to the three CH_3 protons. A detailed, **high-resolution** spectrum of ethanol shows that the CH_2 and CH_3 peaks are in fact split into a number of subsidiary peaks (Figure 13.29). This splitting is caused by **spin-spin coupling** between protons on neighbouring carbon atoms.

Here is what happens. One of the carbon atoms in ethanol has two protons on it (CH_2), the other has three (CH_3). Consider the CH_3 protons, which act like tiny magnets. When the external magnetic field is applied, these three tiny magnets can arrange themselves in four different ways.

1 All three aligned *with* the magnetic field
2 All three aligned *against* the magnetic field
3 Two aligned with the field and one aligned against it
4 One aligned with the field and two aligned against it.

Each of these four arrangements gives a slightly different overall magnetic field. Each different field interacts with the neighbouring CH_2 protons slightly differently, so these CH_2 protons give four different peaks, very close to one another. These peaks are called a **quartet**.

Similarly, the two protons on the CH_2 group can arrange themselves differently in the external magnetic field. This time there are three different arrangements – see if you can work out what they are. Each of these different fields interacts with the neighbouring CH_3 protons slightly differently, so we see three different peaks close to one another. These three peaks are called a **triplet**.

As a general rule: *a group carrying* n *protons will cause the protons on a neighbouring group to split into* n + 1 *peaks*.

13.9 Chromatography

What is chromatography?

In simple **paper chromatography**, a mixture of coloured substances is spotted onto a sheet of absorbent paper (Figure 13.30). The paper dips into a solvent and the solvent separates the components of the mixture. This is a simple and quick way of finding out the components of a mixture of coloured substances (chromatography literally means 'colour writing').

Note

Interpreting high-resolution NMR Spectra

When you are interpreting a high-resolution spectrum like the one in Figure 13.29, you can

► Use the position of each overall peak to identify the type of proton causing the peak.
► Use the integrated trace to find the numbers of protons causing each peak.
► Use the 'n + 1' rule to get information about the numbers of protons on neighbouring groups.

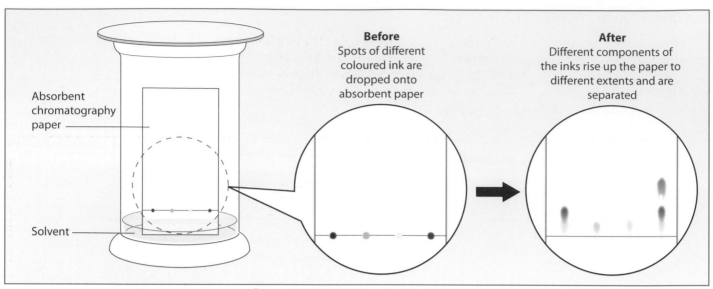

 Figure 13.30

Simple paper chromatography. You can see that the purple ink at the right-hand side is actually made up of yellow, red and blue dyes.

Definition

The **stationary phase** in chromatography is a solid or a liquid held on a solid support. The **mobile phase** is a liquid or gas that moves through the stationary phase.

Note

There is a difference between adsorption and absorption. **Adsorption** occurs when a substance sticks to the surface of a solid. **Absorption** occurs when the substance is distributed throughout the solid, like water in a sponge.

There are several types of chromatography – they do not all involve paper, or even coloured substances.

All chromatography involves separating mixtures by using a **stationary phase** and a **mobile phase**. In paper chromatography, the stationary phase is the paper and the mobile phase is the solvent. The different components of the mixture are attracted to different extents to the two phases and get **partitioned** between the two phases. In the example in Figure 13.30, the blue dye is partitioned more into the mobile solvent phase than the stationary paper phase, so it travels further up the paper.

When a component is present in both phases, it is said to be **partitioned** between the phases.

Table 13.6 summarises the different types of chromatography that we look at in this section.

Table 13.6

The different types of chromatography

Type	Stationary phase	Mobile phase	Used for
Paper chromatography	Paper	Solvent rising up the paper	Separating mixtures to identify their components
Thin layer chromatography (TLC)	An adsorbent solid spread thinly on a metal or plastic plate	Solvent rising up the plate	Separating mixtures to identify their components
Column chromatography	An adsorbent solid packed in a column	Solvent poured down the column	Separating mixtures to identify their components. Using the separated components for further chemical reactions
High performance liquid chromatography (HPLC)	An adsorbent solid packed in a column	Solvent forced through the column under pressure	Separating mixtures to identify their components. Using the separated components for further chemical reactions
Gas–liquid chromatography (GLC)	Organic liquid spread on surface of an unreactive column, heated in an oven	Unreactive gas forced through the column under pressure	Separating mixtures of gases or volatile liquids to identify their components

Thin layer chromatography (TLC)

Figure 13.31 shows a typical TLC set-up. The adsorbent solid on the plate is a very fine powder, giving a high surface area for adsorption. The adsorbent solid on the plate is often a thin layer of silicon(IV) oxide (silica, SiO_2). As the solvent rises up the plate, the components get partitioned between the solid silica and the liquid solvent. Different components travel different distances.

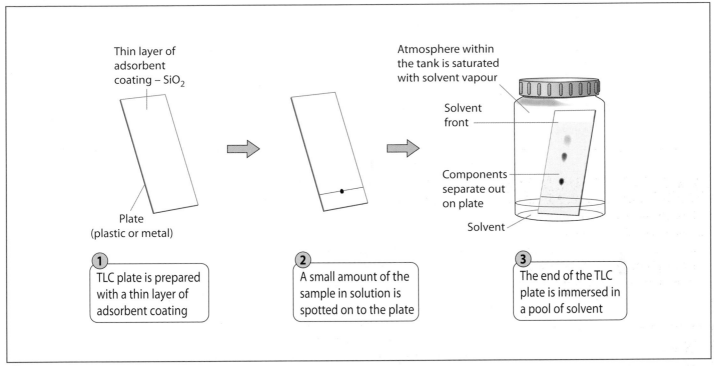

1 TLC plate is prepared with a thin layer of adsorbent coating

2 A small amount of the sample in solution is spotted on to the plate

3 The end of the TLC plate is immersed in a pool of solvent

⌃ Figure 13.31
A typical arrangement for thin layer chromatography

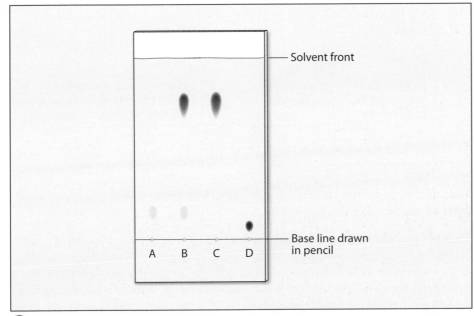

⌃ Figure 13.32
The result of a TLC experiment to investigate food dyes. A: yellow dye; B: green dye; C: blue dye; D: purple dye

Quick Questions

18 Figure 13.32 shows the result of an investigation of four food dyes, A, B, C and D. Look closely at Figure 13.32.
   a What do the results tell you about the green food dye B?
   b What can you say about the way the purple food dye D is partitioned between the stationary and mobile phases?
   c Why is it important that the base line is drawn in pencil, not ink?
   d Calculate the R_f values for dyes A and C (see box for explanation of R_f values).

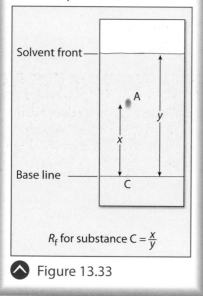

Note

R_f values

R_f values are a way of measuring how far a component has travelled, relative to the solvent front (Figure 13.33).

R_f values are constant for a given component and a given set of chromatography conditions: in particular the solvent, the stationary phase and the temperature. If two spots on a chromatogram have the same R_f value, they are probably chemically identical.

Solvent front

Base line

R_f for substance $C = \dfrac{x}{y}$

▲ Figure 13.33

▶ Figure 13.34
Column chromatography

TLC is an improvement on paper chromatography, because it is fast and gives clearer, sharper separation of the spots. It is used widely in research laboratories, for example to check the products of a reaction, or to see whether a product contains impurities.

The components separated by TLC do not have to be coloured. Colourless spots can be shown up by using a *locating agent* to make them coloured. Another method of showing up the spots involves shining ultraviolet light on the plate.

Column chromatography and HPLC

In column chromatography, the stationary phase is an adsorbent solid (often silica) packed into a column (Figure 13.34). The mobile phase is a solvent poured through the column and the separated components flow out at the bottom and can be collected. So column chromatography can be used to separate and identify components (as in TLC), but it can also be used to collect the separated components and do further experiments on them. This is very useful when you are synthesising organic chemicals.

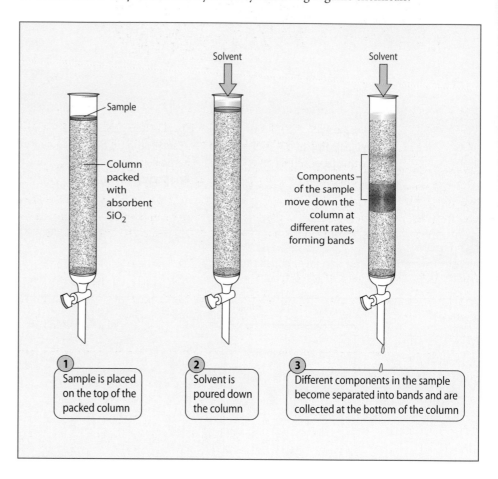

You can also use column chromatography *quantitatively* – to measure the *quantities* of different products formed in a reaction.

High performance liquid chromatography (HPLC) is an advanced version of column chromatography. To get good separation, you need very fine solid particles in the column, giving a big surface area. The solid (often silica) is powdered into very small particles, about a tenth of the width of a human hair. A pump is needed to force the solvent through the densely packed column – at pressures up to 500 atmospheres. Figure 13.35 shows a modern HPLC machine. The biggest part of the machine is the electronics that control it and detect the output from the column. The column itself is quite small – about the size of a pencil.

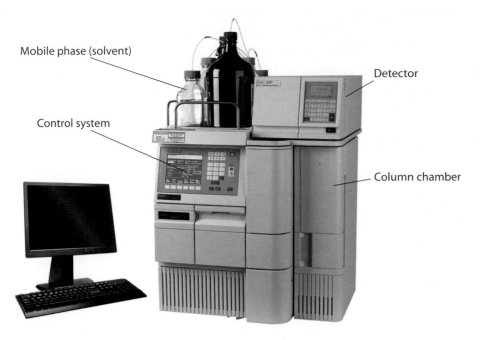

Mobile phase (solvent)

Control system

Detector

Column chamber

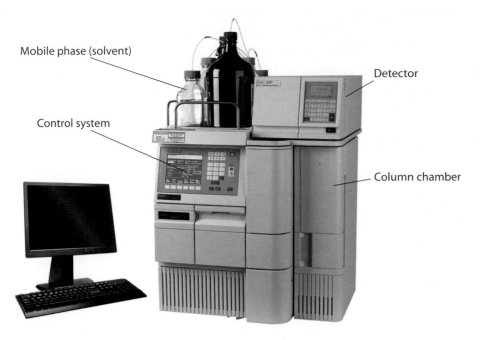

Figure 13.35
An HPLC system

The separated components are detected as they come out of the column at different times. The time each component spends in the column (the **retention time**) can be used to compare unknown components with a known reference substance.

HPLC is widely used to separate small quantities of pure compounds from mixtures. This is useful in the pharmaceutical industry, for example, where small but valuable quantities of a new drug may be separated from a reaction mixture. When scientists are studying biochemical reactions, they often need to separate compounds which are very similar to one another. The efficient separation of HPLC makes it very useful for this.

Definition

Retention time is the time that a component spends in the column in HPLC or GLC before it reaches the detector.

Gas–liquid chromatography (GLC)

In GLC, the mobile phase is a *gas*, not a liquid. Figure 13.36 shows the arrangement in outline.

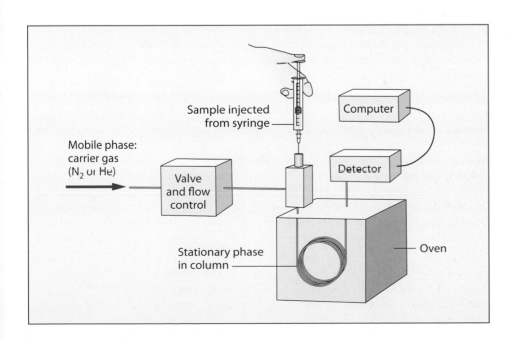

Sample injected from syringe

Computer

Mobile phase: carrier gas (N₂ or He)

Valve and flow control

Detector

Stationary phase in column

Oven

Figure 13.36
An outline of gas chromatography

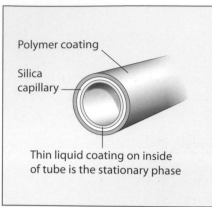

The inert carrier gas, usually nitrogen or helium, flows through a column inside a heated oven. The inside of the column is coated with a thin layer of a stable, non-volatile liquid, such as silicone oil (Figure 13.37).

The sample being analysed could be a gas or a volatile liquid, which vaporises in the oven. When it is injected into the gas stream, the sample gets carried along through the column. The components are partitioned between the gas phase and the liquid in the column. Components that are absorbed most strongly by the liquid stay longer in the column and come out last. Components that are adsorbed least strongly by the liquid come out first. The time that a component spends in the column is its retention time.

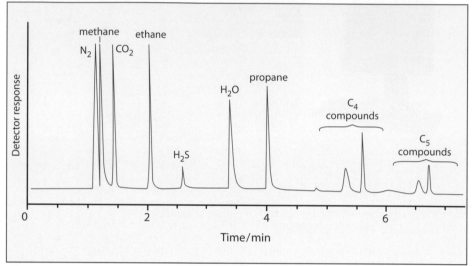

Figure 13.38
A gas chromatogram of a sample of natural gas

Figure 13.38 shows a gas chromatogram for a sample of natural gas from a refinery. Notice the following points.

▸ There are several peaks, because natural gas is a mixture of gases. Each peak represents a different compound.
▸ The peaks differ in height. The height of each peak is approximately proportional to the relative quantity of the component in the mixture.
▸ Gases like methane and N_2, with small molecules, have low retention times and come out of the column early. Gases with larger molecules have longer retention times.

GLC is a valuable method for separating mixtures of gases and volatile liquids and identifying their components. It is very sensitive and can detect tiny quantities. It also gives a measure of the *quantity* of components as well as detecting their presence.

GLC is widely used in research, industry and environmental monitoring. It is very useful in the petroleum industry for monitoring the composition of crude oil, gasoline and other fractions. Each sample of crude oil has a characteristic GLC 'fingerprint' so you can use GLC to find the source of oil pollution. Police use GLC to measure the concentration of alcohol in a car driver's breath.

Quick Questions

19 Look at Figure 13.38 and answer these questions.
  a Estimate the retention times of: **i** CO_2, **ii** H_2S under these conditions.
  b Suggest a reason why compounds like methane and N_2 have low retention times – in other words, they come out of the column more quickly than compounds like propane.
  c Use the peak heights to estimate the ratio of methane to propane in this sample of natural gas.
  d Suggest a reason why it is important to be able to detect the presence of H_2S in natural gas.
  e The chromatogram shows peaks for *two* C_4 alkanes. Draw structural formulae for each.

Review questions

1 **A**, **B** and **C** are isomeric compounds of molecular formula C_3H_8O. Two of the compounds are members of the same homologous series. The table gives some data about **A**, **B** and **C**.

	A	**B**	**C**
Boiling point/K	370	284	356
Density/g cm^{-3}	0.80	0.72	0.79

   a Which two compounds are members of the same homologous series?

   b Write the structural formulae of all three possible isomers of C_3H_8O.

   c Use Table 13.2 to decide to which homologous series each of the isomers you have drawn belongs.

   d Is it possible from the data given to say which of the isomers you have drawn is **A**, which is **B** and which is **C**? Explain your answer.

2 Use Table 13.2 to decide to which homologous series each of the following compounds belongs.

   a $CH_3CH_2CH_2OH$

   b $CH_3CH_2COCH_3$

   c CH_3CH_2Cl

   d CH_3CH_2COOH

   e $CH_3CH{=}CHCH_3$

   f H_2NCH_2COOH

   g $CH_3CH_2OCH_2CH_2CH_3$

3 Write the full structural formulae of all the isomers of the following, stating which type of isomerism is involved:

   a C_3H_7Cl

   b C_6H_{14}

   c $C_2H_3Cl_2Br$

4 Fluorocarbons are compounds analogous to alkanes but containing fluorine instead of hydrogen, e.g. CF_4, CF_3CF_3, etc. They are extremely unreactive, being quite stable in air even at high temperatures. Long-chain fluorocarbons such as PTFE ($-CF_2-CF_2-CF_2-CF_2-CF_2-$) are used as unreactive corrosion-resistant materials for gaskets and protective coatings.

Use the bond energies given below to explain why fluorocarbons are stable in air while hydrocarbons are energetically unstable.

Bond	Bond energy/kJ mol^{-1}
C—H	413
C—F	485
H—O	463
F—O	234

5 The Earth's crust contains only 0.036% by mass of carbon, compared with 49% oxygen and 26% silicon. Despite its relatively low abundance on Earth, carbon is the element that forms the basis of all living things. What are the properties of carbon that make it so suitable for this role?

6 An organic compound was subjected to combustion analysis. 1.0 g of the compound formed 1.37 g carbon dioxide, 1.12 g water and no other products.

   a Calculate the percentage by mass of carbon and of hydrogen in the compound.

   b What other element must be present?

   c Calculate the empirical formula of the compound.

   d The mass spectrum of the compound is shown in Figure 13.39.

   i Use this to find the relative molecular mass of the compound, and thus its molecular formula.

   ii Using the fragments shown in the mass spectrum, deduce the structural formula of the compound.

   iii Give the formula of the fragment to which each peak can be attributed.

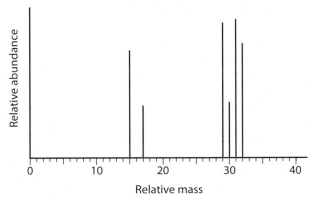

▲ Figure 13.39

7 Figure 13.40 shows the simplified NMR spectrum of 1-phenylbutan-2-one.

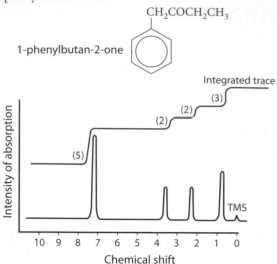

Figure 13.40
The simplified NMR spectrum of 1-phenylbutan-2-one

a Table 13.5 shows the chemical shifts for some types of protons (H atoms). Use the table to identify the protons responsible for the peaks at

   i   7.2

   ii 2.3

   iii 0.9

b Using your answer to part a, and a process of elimination, identify the protons responsible for the peak at 3.6.

8 Ethoxyethane, $CH_3CH_2OCH_2CH_3$, and butan-1-ol, $CH_3CH_2CH_2CH_2OH$ are isomers. Figure 13.41 shows the infrared spectra of these two compounds. The spectra are labelled A and B.

a Use Table 13.4 to decide which spectrum belongs to which compound.

b Identify the bonds responsible for the peaks marked 1, 2, 3 and 4 on Figure 13.41.

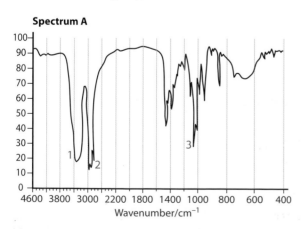

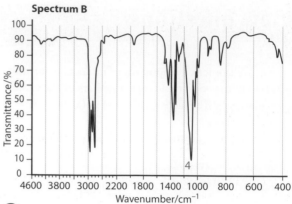

Figure 13.41

9 At one time, bromomethane, CH_3Br, was widely used to control insect pests in agricultural crops and timber. It is now known to break down in the stratosphere and contribute to the destruction of the ozone layer.

   Samples can be screened for traces of bromomethane by subjecting them to mass spectrometry.

a Which peak(s) would show the presence of bromine in the compound?

b How would you tell by studying the M and M + 2 peaks that the compound contained bromine rather than chlorine?

Cambridge Paper 4 Q9 June 2008

10 Amino acids may be separated by using two-dimensional paper chromatography. This involves putting a spot of the mixture on the corner of a piece of chromatography paper and allowing a solvent to soak up the paper. This paper is then dried, turned through 90° and placed in a second solvent. This method gives better separation than a one-solvent method.

a Paper chromatography relies on partition between the solvent applied and another phase. What is this second phase?

b The table below shows the R_f values for some amino acids in two different solvents.

Amino acid	R_f solvent 1	R_f solvent 2
A	0.1	0.2
B	0.0	0.4
C	0.3	0.0
D	0.8	0.9
E	0.6	0.5

Plot the positions of the amino acids after two-dimensional paper chromatography on some graph paper.

   i   Which amino acid travelled fastest in *both* solvents?

   ii Which amino acid did not move at all in solvent 2?

Cambridge Paper 4 Q8c June 2007

14 Alkanes

14.1 Crude oil

As well as supplying a large part of our energy needs, crude oil is the source of most organic chemicals. Together with natural gas, it is the most important of all modern raw materials.

Crude oil is a complex mixture of hydrocarbons which has no uses in its raw form. To provide useful products, its components must be partly separated and if necessary modified. The fundamental process is primary distillation, the details of which are summarised diagrammatically in Figure 14.2.

⌃ Figure 14.1
An oil refinery

Fractions from primary distillation

Refinery gas (1–2% of crude oil) is similar in composition to natural gas. It contains those hydrocarbons that are gases at normal temperatures. These include the alkanes with one to four carbon atoms in their molecules, with methane as the major component. The main use of refinery gas is as a gaseous fuel, but like natural gas it can also be used as a starting point for making petrochemicals, since most organic chemicals are built up from small molecules containing one, two or three carbon atoms.

Gasoline (15–30%) is a complex liquid mixture of hydrocarbons containing mainly C_5–C_{10} compounds whose boiling points range from 40 °C to 180 °C. The major use of gasoline is of course as a fuel in internal combustion engines (Section 14.7).

A considerable proportion of this fraction is also used in the manufacture of chemicals, after the cracking processes described in Section 14.7. The part of the gasoline fraction used to produce chemicals is called **naphtha.**

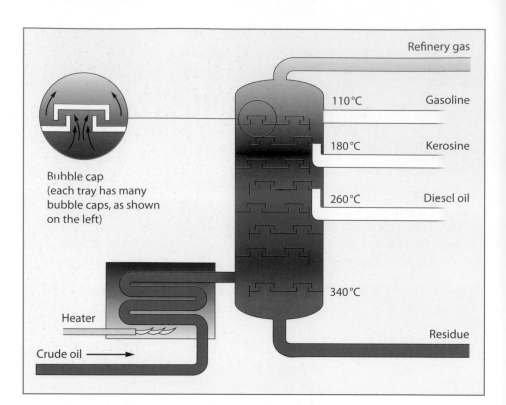

Figure 14.2
Primary distillation of crude oil

Refinery gas

110 °C Gasoline

180 °C Kerosine

260 °C Diesel oil

340 °C

Residue

Bubble cap
(each tray has many
bubble caps, as shown
on the left)

Heater

Crude oil →

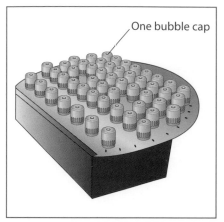

One bubble cap

▲ Bubble tray in a primary distillation column

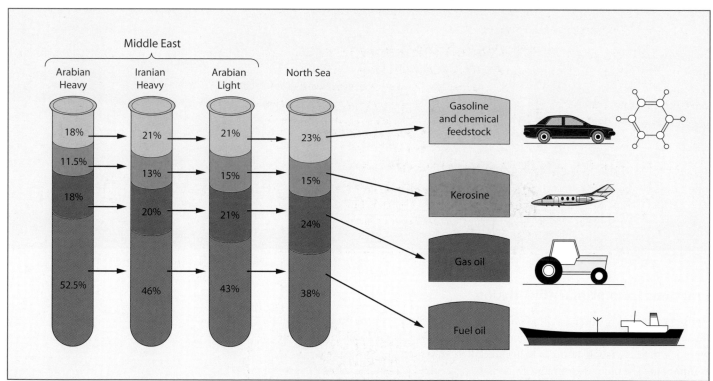

Figure 14.3
Variation of composition of crude oil from different sources. Different crudes contain varying proportions of the different fractions. The output of the refinery can be arranged to suit market demands by blending different crudes and by converting heavy fractions into lighter ones by cracking.

Kerosine (10–15%) consists mainly of C_{11} and C_{12} hydrocarbons, with boiling points from 160 °C to 250 °C. It is used as a fuel in jet engines and for domestic heating. It can also be cracked to produce extra gasoline (Section 14.7).

Diesel oil or **gas oil**, (15–20%) containing C_{13}– C_{25} compounds, boils in the range 220 °C–350 °C. It is used in diesel engines, where the fuel is ignited by compression rather than a spark and also in furnaces for industrial heating. Like kerosine, it can be cracked to produce extra gasoline.

Figure 14.4
Refuelling planes in mid-flight. Jet fuel
is made from kerosine.

Residue (40–50%) The residual oil from primary distillation boils above 350 °C and is a highly complex mixture of involatile hydrocarbons. Most of it is used as fuel oil in large furnaces such as those in power stations or big ships. A proportion of it, however, is used to make **lubricating oils** and **waxes**.

To obtain lubricating oil and paraffin wax from the residue, the appropriate hydrocarbons must be distilled off. The distillation is done in a vacuum to avoid the high temperatures needed at atmospheric pressure, because high temperatures would crack the hydrocarbons. The solid left after vacuum distillation is an involatile tarry material called **bitumen** or **asphalt**. It is used to surface roads and to waterproof materials.

14.2 The composition of crude oil

The distillation of crude oil can be carried out on a small scale in the laboratory: you may be familiar with the experiment represented in Figure 14.5. Table 14.1 gives the properties of the fractions collected over different temperature ranges.

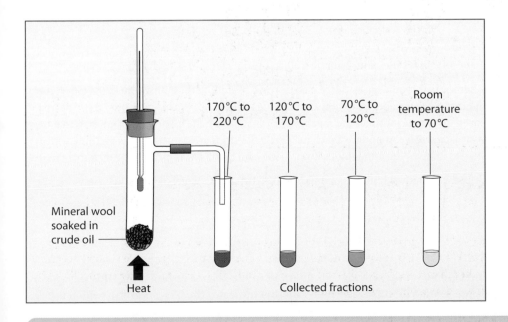

Figure 14.5
The laboratory distillation of crude oil.
Fractions are collected over several
different boiling ranges.

Table 14.1

Properties of fractions obtained in the laboratory distillation of crude oil

Table 14.1

	Colour	Viscosity	Behaviour when ignited
Room temp. to 70 °C	pale yellow	runny	burns readily: clean yellow flame
70 °C–120 °C	yellow	fairly runny	easily ignited: yellow flame, some smoke
120 °C–170 °C	dark yellow	fairly viscous	harder to ignite: quite smoky flame
170 °C–220 °C	brown	viscous	hard to ignite: smoky flame

Quick Questions

1   Look again at Table 14.1.
    a   Why are the first fractions easiest to ignite?
    b   Why do the higher fractions burn with the smokiest flames?
    c   Suggest a reason why the fractions become increasingly viscous.

Notice that all the properties show a steady gradation: there are no sudden changes in the properties of the different fractions. This suggests that crude oil is a mixture of many components and that the components have similar properties, many of them belonging to the same homologous series.

Hydrocarbons are the main and the most important components of crude oil. Three different homologous series of hydrocarbons are present:

▶ **Alkanes**, which we will look at in more detail later in this chapter.
▶ **Cycloalkanes**, considered in Section 14.3.
▶ **Aromatics** (compounds containing benzene-type rings), considered in Chapter 25. Aromatic hydrocarbons are sometimes called **arenes**.

The proportions of the different types of compounds present depend on the source of the oil. Alkanes and cycloalkanes form the large majority, with aromatics making only about 10% of the hydrocarbon total.

Definition

A **saturated organic compound** is one which contains only single bonds.

All the volatile fractions of crude oil are made up of hydrocarbons. The increasing boiling ranges of the fractions correspond to the increasing size of the alkane molecules they contain.

14.3 Naming alkanes

The alkanes are **saturated hydrocarbons.** 'Saturated' means that they contain the maximum content of hydrogen possible, with no double or triple bonds between carbon atoms. The general formula of the alkanes is C_nH_{2n+2}. It is possible to have alkanes with straight or with branched chains, for example:

$$CH_3-CH_2-CH_2-CH_2-CH_3 \qquad \text{straight chain}$$

$$CH_3-CH_2-\underset{\underset{CH_3}{|}}{CH}-CH_3 \qquad \text{branched chain}$$

Table 14.2 gives the names and formulae of some simple straight-chain alkanes.

After the first four alkanes, the name is formed by adding the suffix **-ane** to the Greek root indicating the number of carbon atoms in the molecule (e.g. *pent-* five, *hex-* six). The first part of the name indicates the number of carbon atoms; the ending -ane indicates that it is an alkane. All compounds based on alkane skeletons are named by this method, with a stem to indicate the number of carbon atoms and a suffix to indicate the functional group.

Branched-chain alkanes are named as straight-chain alkanes with side groups attached. For example:

$$CH_3-CH_2-\underset{\underset{CH_3}{|}}{CH}-CH_3$$

is regarded as butane with CH_3 attached to the *second* carbon atom. It is therefore called 2-methylbutane. The prefix **methyl** indicates the CH_3- group, which is

Table 14.2
Straight-chain alkanes

Formula	Name
CH_4	methane
CH_3CH_3	ethane
$CH_3CH_2CH_3$	propane
$CH_3CH_2CH_2CH_3$	butane
$CH_3(CH_2)_3CH_3$	pentane
$CH_3(CH_2)_4CH_3$	hexane
$CH_3(CH_2)_8CH_3$	decane
$CH_3(CH_2)_{18}CH_3$	eicosane

just methane with a hydrogen atom removed so it can be attached to another atom. Similarly CH_3CH_2- is called **ethyl,** $CH_3CH_2CH_2-$ **propyl,** and so on. Side groups of this kind are called **alkyl groups;** their general formula is C_nH_{2n+1}. The symbol **R** is often used to represent a general alkyl group. A side group containing an aromatic hydrocarbon is called an **aryl** group.

Another example of a branched-chain alkane is:

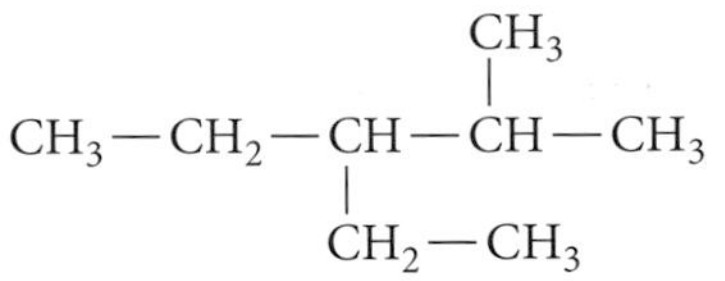

$$CH_3-CH_2-\overset{\displaystyle \overset{CH_3}{|}}{CH}-\overset{\displaystyle \underset{CH_2-CH_3}{|}}{CH}-CH_3$$

This molecule can be regarded as a five-carbon chain with a methyl and an ethyl group attached to the second and third carbon atoms, respectively. It is therefore called 3-ethyl-2-methylpentane. The side groups are listed in alphabetical order (ethyl before methyl). If the carbon atoms were numbered from the left, the name would be 3-ethyl-4-methylpentane. This name is not used because the convention is to use the name that includes the lowest numbers. Note also that for each side group the name includes a number showing the carbon atom to which the group is attached. Thus

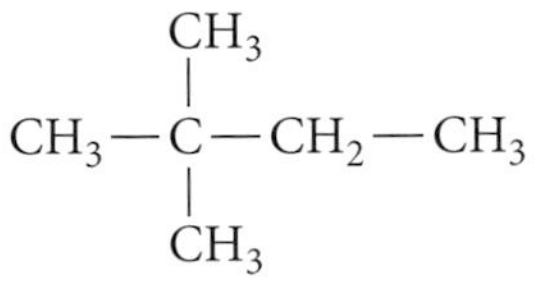

$$CH_3-\overset{\displaystyle \overset{CH_3}{|}}{\underset{\displaystyle \underset{CH_3}{|}}{C}}-CH_2-CH_3 \quad \text{and} \quad CH_3-\overset{\displaystyle \overset{CH_3}{|}}{\underset{\displaystyle \underset{CH_3}{|}}{C}}-\overset{\displaystyle \underset{CH_2-CH_3}{|}}{CH}-CH_2-CH_2-CH_2-CH_3$$

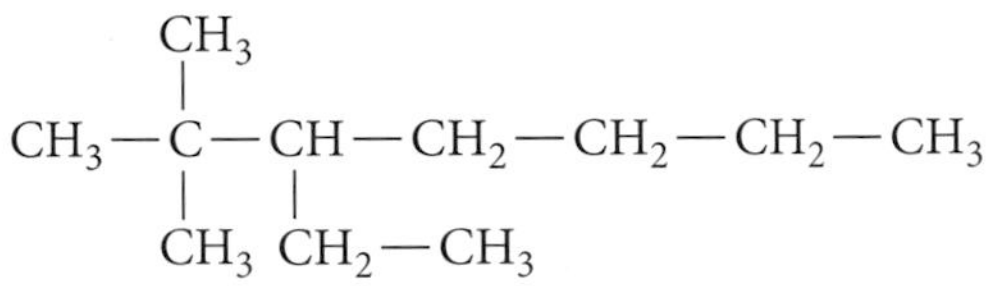

is called 2,2-dimethylbutane   is called 2,2-dimethyl-3-ethylheptane

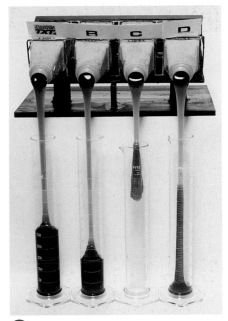

To name an alkane from its structural formula:

1 Look for the longest unbranched (main) chain in the molecule.
2 Look for the side groups attached to the main chain and the numbers of the carbon atoms to which they are attached.
3 Write down the name of the longest unbranched chain, and add the names of the side groups and the numbers of the carbon atoms to which they are attached.

The system of nomenclature described above is called the IUPAC system (IUPAC stands for International Union of Pure and Applied Chemistry). This system can be extended to all organic compounds. Using this system, if you know the name of a compound, you can write its structural formula.

The IUPAC system is gradually replacing older, less systematic ways of naming organic compounds. IUPAC names will be used wherever possible in this book. The only disadvantage of the system is that for complicated molecules the name may become very cumbersome. In such cases systematic nomenclature is often abandoned and a more easily spoken name is used. Thus the name 2,3,4,5,6-pentahydroxyhexanal is dropped in favour of 'glucose'.

Key Point

Alkanes are saturated hydrocarbons with general formula:

$$C_nH_{2n+2}$$

Quick Questions

2 Name the following alkanes:
a

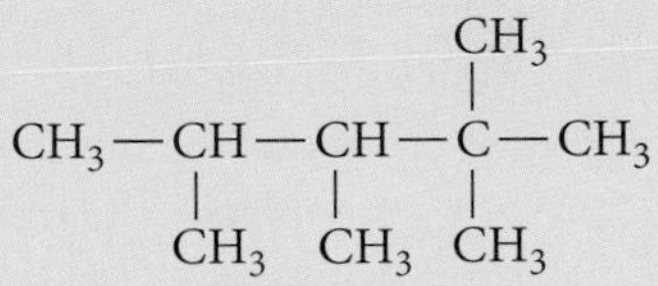

$$CH_3-CH-CH-\overset{\displaystyle \overset{CH_3}{|}}{C}-CH_3$$
$$\underset{\displaystyle \underset{CH_3}{|} \quad \underset{CH_3}{|} \quad \underset{CH_3}{|}}{}$$

b

$$CH_3-\overset{\displaystyle \overset{CH_3}{|}}{\underset{\displaystyle \underset{CH_3}{|}}{C}}-CH_3$$

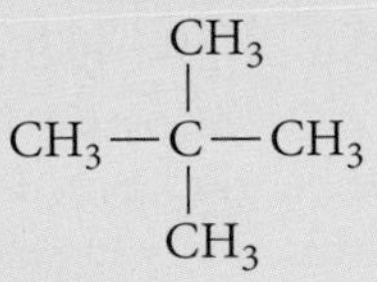

c $CH_3(CH_2)_6CH_3$

3 Write the formulae of the following alkanes:
a 2,2,4-trimethylhexane,   c 5-ethyldecane.
b methylpropane,

Note

Alkyl groups are named using the following prefixes:

methyl CH_3-

ethyl CH_3CH_2-

propyl $CH_3CH_2CH_2-$

R $C_nH_{2n+1}-$

Cycloalkanes

The alkanes considered so far have all had open-chain molecules, that is molecules which come to an end at some point. It is also possible for alkane molecules to form rings and these compounds are named by using the prefix **cyclo-**. Some cycloalkanes are shown below.

	Structural formula	Skeletal formula

(a) cyclopropane, C_3H_6

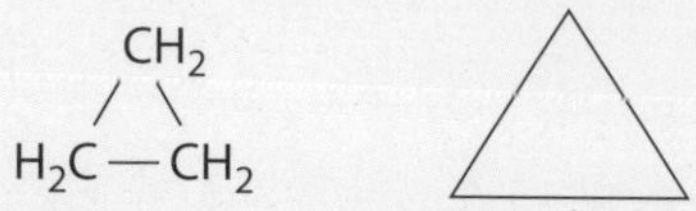

(b) cyclobutane, C_4H_8

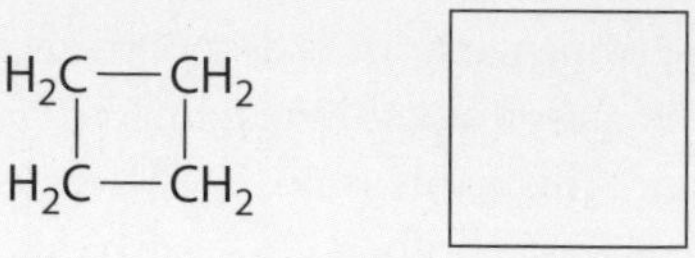

(c) cyclopentane, C_5H_{10}

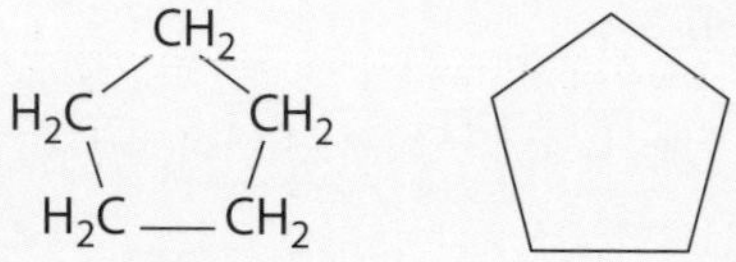

(d) cyclohexane, C_6H_{12}

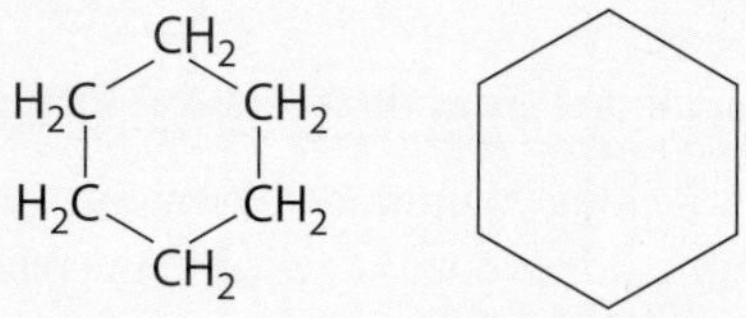

(This skeletal formula shows cyclohexane as planar, which is inaccurate, but easier to draw.)

Cyclopropane, cyclobutane and cyclopentane all have planar molecules. The natural bond angles around a saturated carbon atom are 109° (the angle at the centre of a tetrahedron). In cyclopropane the bond angle is only 60°, since the carbon atoms within the molecule form an equilateral triangle. Hence there is considerable **ring strain** in this molecule, so cyclopropane is unstable and reactive, tending to break open its ring. Cyclohexane and the higher cycloalkanes can relieve ring strain by puckering, so that their rings are no longer planar. These cycloalkanes are therefore stable and very similar in properties to open-chain alkanes.

Quick Questions

4 Suggest explanations for these facts:
   - **a** Cyclobutane is less reactive than cyclopropane, but more reactive than open-chain alkanes.
   - **b** Cyclopentane is no more reactive than open-chain alkanes.

14.4 Physical properties of alkanes

The alkanes form a homologous series. As with all homologous series the members show a gradual change in physical properties as the number of carbon atoms in their molecules increases. Some properties of individual straight-chain alkanes are shown in Table 14.3. Notice that these properties show a gradual change as the number of carbon atoms in the molecules increases. Figure 14.7 shows graphically the smooth increase of boiling point with increasing number of carbon atoms.

Steady variation in physical properties is characteristic of all homologous series. This means we can predict the properties of a compound from the properties of other members in the same homologous series.

As Table 14.3 shows, the first four straight-chain alkanes (C_1–C_4) are gases. The next twelve (C_5–C_{16}) are liquids, and the remainder are waxy solids.

Number of carbon atoms	Formula	Name	State (at 298 K)	Boiling point/K	Melting point/K	Density /g cm^{-3}
1	CH_4	methane	g	112	90	0.424
2	C_2H_6	ethane	g	184	101	0.546
3	C_3H_8	propane	g	231	85	0.501
4	C_4H_{10}	butane	g	273	138	0.579
5	C_5H_{12}	pentane	l	309	143	0.626
6	C_6H_{14}	hexane	l	342	178	0.657
7	C_7H_{16}	heptane	l	371	182	0.684
8	C_8H_{18}	octane	l	399	216	0.703
9	C_9H_{20}	nonane	l	424	219	0.718
10	$C_{10}H_{22}$	decane	l	447	243	0.730
11	$C_{11}H_{24}$	undecane	l	469	247	0.740
12	$C_{12}H_{26}$	dodecane	l	489	263	0.749
15	$C_{15}H_{32}$	pentadecane	l	544	283	0.769
20	$C_{20}H_{42}$	eicosane	s	617	310	0.785

◀ Table 14.3
Physical properties of straight-chain alkanes

Quick Questions

Use the data in Table 14.3 to predict:
5 a the density of tridecane ($C_{13}H_{28}$),
  b the boiling point of tetradecane ($C_{14}H_{30}$),
  c the formula of the first alkane to be solid at 20 °C.

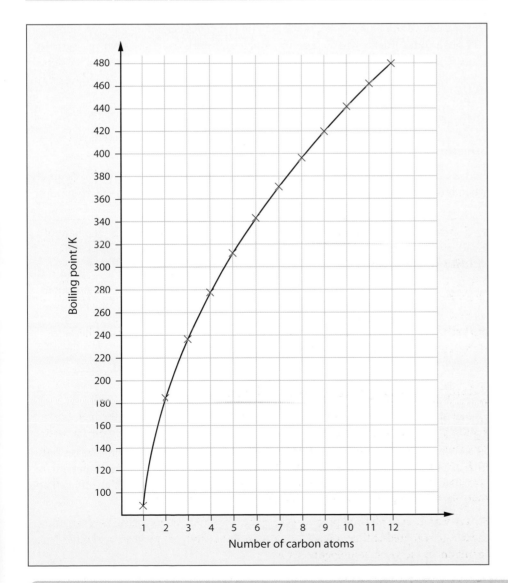

◀ Figure 14.7
Variation of boiling point of straight-chain alkanes with number of carbon atoms

▲ Figure 14.8
An explosion on the Deepwater Horizon oil rig in 2010 led to a race to stop escaping crude oil reaching beaches in the Gulf of Mexico.

Alkanes are colourless when pure. The viscosity of liquid alkanes increases with increasing molecular mass. Alkanes are all less dense than water and therefore float on it. This can cause major environmental hazards when oil leaks occur at sea (Figure 14.8).

The steady change in physical properties of alkanes is caused by steadily increasing molecular size. The increase in boiling point is due to the increasing forces of attraction (induced dipoles) between molecules of increasing size (Section 3.11). The higher viscosity of the higher alkanes is due to the tendency of the long molecules to become 'tangled up' with one another.

The trends in physical properties considered so far have been those among the straight-chain alkanes. Branched-chain alkanes do not show the same steady gradation of properties as straight-chain alkanes. On the whole, the effect of branching is to increase volatility and reduce density. Thus the boiling point of pentane is 309 K while that of its highly branched isomer 2,2-dimethylpropane is 283 K.

14.5 Reaction types and reaction mechanisms in organic chemistry

One of the things that makes organic chemistry enjoyable is the fact that although there are many, many different organic compounds, the reactions of these compounds are easy to classify.

In the next section we will look at the characteristic reactions of alkanes, but in this section we will look at some of the general principles behind organic reactions. Although some organic reactions go in a single step, most reactions involve a series of steps. The sequence of steps in a reaction is called the **reaction mechanism.**

Types of organic reaction

There are three important types of organic reaction.

Substitution reactions

In this type of reaction one atom or group of atoms is substituted by another. For example, when bromoethane reacts with ammonia, bromine atoms are replaced by $-NH_2$ groups:

$$CH_3CH_2Br + NH_3 \longrightarrow CH_3CH_2NH_2 + HBr$$
bromoethane   ethylamine

Addition reactions

In these reactions, two molecules react together to form a single molecule. A double bond is often involved. For example, when bromine reacts with ethene, the bromine adds across the double bond to form a single product, 1, 2-dibromoethane:

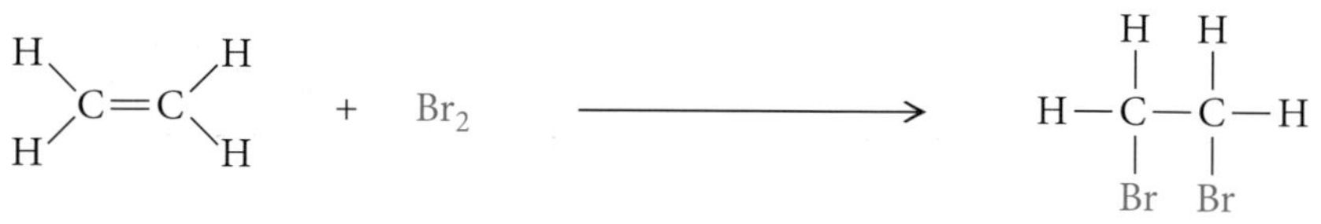

Elimination reactions

In an elimination reaction, a small molecule is removed (eliminated) from a larger one. This usually results in the formation of a double bond. For example, when bromoethane is heated with a solution of hydroxide ions in ethanol, HBr is eliminated, forming ethene:

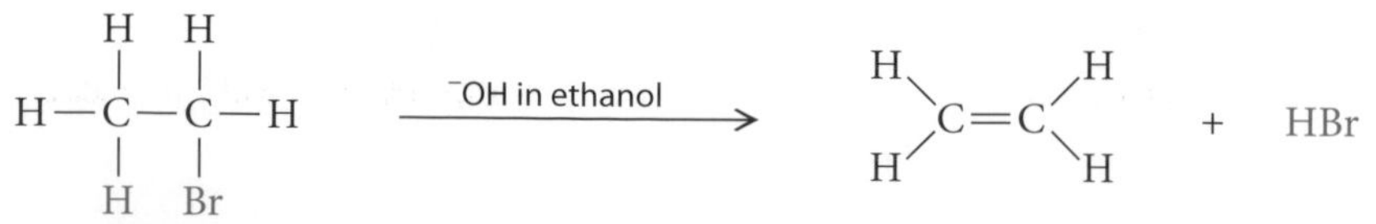

> **Definitions**
>
> In **substitution reactions**, one atom or group is replaced by another.
>
> In **addition reactions**, two molecules react together to form a single molecule.
>
> In **elimination reactions** a small molecule is removed from a larger one.

6 Look at the reactions represented by the equations below. Classify each reaction as
   A Substitution   **B** Addition   **C** Elimination
   a $CH_2{=}CH_2 + H_2 \longrightarrow CH_3CH_3$
   b $CH_3CH_2Br + {}^-OH \longrightarrow CH_3CH_2OH + Br^-$
   c $CH_3CH_2OH \longrightarrow CH_2{=}CH_2 + H_2O$
   d $CH_4 + Cl_2 \longrightarrow CH_3Cl + HCl$

Ways of breaking bonds

All reactions involve the breaking, and remaking, of bonds. Breaking bonds is sometimes called **bond fission.** The way bonds break has an important influence on reactions.

In a covalent bond, a pair of electrons is shared between two atoms. For example, in the HCl molecule:

$$H \div Cl$$

When the bond breaks, these electrons get redistributed between the two atoms. There are two ways this redistribution can happen.

Homolytic fission

In this type of bond fission, one of the two shared electrons goes to each atom. In the case of HCl:

$$H \div Cl \longrightarrow H\cdot + Cl\cdot$$

The dot · beside each atom represents the unpaired electron that the atom has gained from the shared pair in the bond. The atoms have no overall electric charge, because each one has an equal number of protons and electrons. But the atoms are highly reactive, because the unpaired electron has a strong tendency to pair up with another electron from another substance (Figure 14.9). These highly reactive atoms or groups of atoms with unpaired electrons are called **free radicals.**

A hydrogen radical

$$H \, \bullet \qquad\Rightarrow\qquad H \, \bullet \quad Cl \stackrel{\bullet}{\div} Cl \qquad\Rightarrow\qquad H \stackrel{\bullet}{\div} Cl \quad Cl\bullet$$

This has an unpaired electron which would be much more stable if paired with another electron

The hydrogen radical meets a chlorine molecule

The unpaired electron on the hydrogen pairs up with an electron on one of the chlorine atoms. This forms a bond. It leaves the other chlorine as a reactive free radical

◄ Figure 14.9
Why free radicals are reactive.

Another example of free radical formation occurs when a C—H bond in methane is broken:

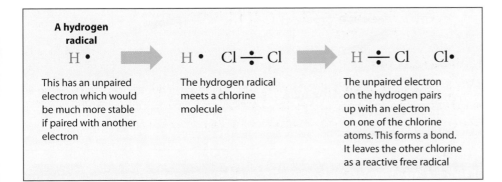

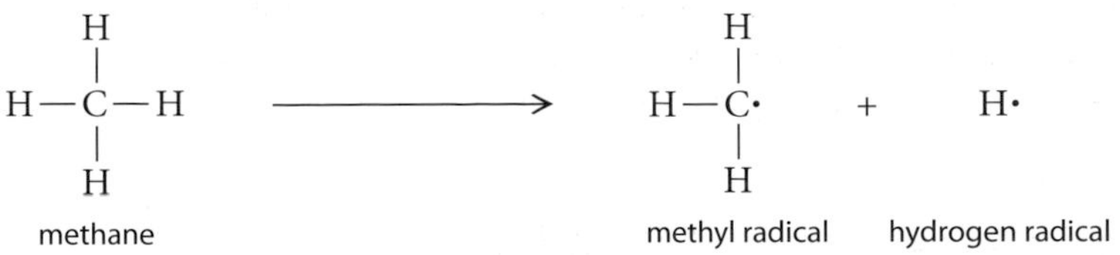

methane   methyl radical   hydrogen radical

⬢ Figure 14.10
One of the causes of ageing is believed to be the damaging effect of free radicals on tissues such as skin.

7 Look at the groups a to f listed below. Classify each as

    A carbocation
    B carbanion
    C free radical
    D none of these

    a $CH_3CH_2\cdot$
    b $CH_3CH_2CH_2^+$
    c CH_3^+
    d $CH_3CH_2^-$
    e ^-OH
    f $\cdot OH$

8 Which type of bond fission, homolytic or heterolytic, would you expect for
    a a bond between identical atoms,
    b a bond between atoms whose electronegativities differ widely,
    c a bond between atoms whose electronegativities are similar?

Definitions

A **carbocation** has a positive charge on a carbon atom.

A **carbanion** has a negative charge on a carbon atom.

Electrophiles are attracted to groups that can donate electron pairs.

Nucleophiles are attracted to groups that can accept electron pairs.

Figure 14.11

Free radicals are most commonly formed when the bond being broken has electrons that are fairly equally shared, in a non-polar bond. When the electrons are unequally shared, in a polar bond, heterolytic fission is more likely.

Heterolytic fission

In heterolytic bond fission, *both* of the shared electrons go to just *one* of the atoms when the bond breaks. This atom becomes negatively charged, because it has one more electron than it has protons. The other atom becomes positively charged. In the case of HCl:

$$H \div Cl \longrightarrow H^+ + :Cl^-$$

Heterolytic fission is more common where a bond is already polar. For example, the bromoalkane shown below contains a polar C — Br bond. Under certain conditions this can break heterolytically:

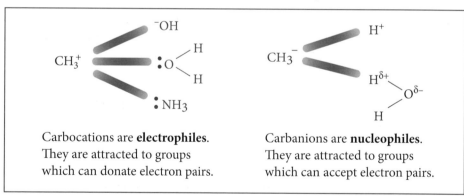

Notice that the first ion contains a *positively* charged carbon atom. It is an example of a **carbocation**. An ion containing a *negatively* charged carbon atom is called a **carbanion**. Both these types of ions are unstable and highly reactive, so they only exist as short-lived reaction intermediates.

Electrophiles and nucleophiles

Like all ions, carbocations and carbanions are attracted to groups which carry an opposite electric charge. Carbocations are **electrophiles** ('electron-lovers'): they are short of electrons, so they are attracted to groups which can donate electron pairs. Carbanions are **nucleophiles** ('nucleus-lovers'): they are electron-rich, so they are attracted to groups which can accept electron pairs. This idea is illustrated in Figure 14.11.

Carbocations are **electrophiles**. They are attracted to groups which can donate electron pairs.

Carbanions are **nucleophiles**. They are attracted to groups which can accept electron pairs.

Quick Questions

9 Look at the ions and groups labelled **a** to **h** below. Classify each as:
    A electrophile,   **B** nucleophile,   **C** could act as electrophile or nucleophile.

    a Cl^-
    b NO_2^+
    c CH_3^+
    d HI
    e ^-CN
    f H^+
    g CH_3NH_2
    h CH_3OH

The terms electrophile and nucleophile don't apply just to ions. Many organic compounds are polar: they carry partial charges, though not full positive or negative charges. These partial charges can also make a group electrophilic or nucleophilic, like this:

$$\overset{\delta+}{CH_3} \rule[0.5ex]{2em}{0.4pt} \overset{\delta-}{Br}$$

electrophilic   nucleophilic

Describing reaction mechanisms

We can use the ideas and terms outlined above to describe the mechanisms of the important classes of organic reactions that we will meet in this book.

Curly arrows

When reaction mechanisms are being described, a 'curly arrow' is sometimes used to show the **movement of a pair of electrons.** The beginning of the arrow shows where the electron pair starts from and the arrow head shows where the pair ends up. Here is an an example.

The arrow shows a pair of electrons moving from the Br^- ion to the region between the bromine and the carbon, where the pair forms a covalent bond between the two atoms.

The same reaction is shown again below, with all the bonding electrons indicated.

A *curly half-arrow* is used to show the *movement of a single electron* in reactions involving free radicals. The beginning of the arrow shows where the single electron starts from, and the half-arrow head shows where it ends up.

For example, the reaction

$$H^\bullet + Cl\rule[0.5ex]{1em}{0.4pt}Cl \longrightarrow H\rule[0.5ex]{1em}{0.4pt}Cl + Cl^\bullet$$

would be shown as

This is shown again below, with all the bonding electrons indicated.

Figure 14.12 summarises the way curly arrows and half-arrows are used.

Note

Important types of reaction mechanism, and where you will find them in this book:

Free radical substitution:
   Section 14.6

Electrophilic addition:
   Section 15.4

Nucleophilic substitution:
   Section 16.4

Nucleophilic addition:
   Section 18.1

Electrophilic substitution:
   Section 25.5

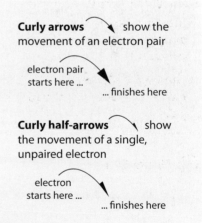

Curly arrows show the movement of an electron pair

electron pair starts here ...

... finishes here

Curly half-arrows show the movement of a single, unpaired electron

electron starts here ...

... finishes here

▲ Figure 14.12

Quick Questions

10 Draw curly arrows or half-arrows to show the movement of electrons in these reactions:

   a $CH_3\rule[0.5ex]{1em}{0.4pt}Br + {}^-OH \longrightarrow CH_3\rule[0.5ex]{1em}{0.4pt}OH + Br^-$
   b $CH_4 + Br^\bullet \longrightarrow CH_3\rule[0.5ex]{1em}{0.4pt}Br + H^\bullet$

Table 14.4
The effect of common reagents on hexane

Reagent	Effect
air (oxygen)	no effect cold. Burns when heated
sodium hydroxide solution	no effect hot or cold
concentrated sulfuric acid	no effect hot or cold
potassium manganate(VII) (potassium permanganate) solution	no effect hot or cold
bromine	no effect in dark. Bromine is slowly decolorised in sunlight

Definitions

The stages of a free-radical chain reaction are:

Initiation, when free-radicals are formed and get the reaction started,

Propagation, where a free-radical reacts with a neutral molecule, forming another free-radical to keep the reaction going,

Termination, where two free radicals combine, ending the reaction chain.

14.6 Reactivity of alkanes

The effects of various reagents on hexane, a typical alkane, are shown in Table 14.4.

If you look at the table, you can see that hexane is clearly unreactive. It is unaffected by acids, alkalis, dehydrating agents and aqueous oxidising agents. A close look at the substances in the table with which hexane *does* react, namely bromine and oxygen, shows two things. First, neither of these reagents has any centre of electrical charge in their molecule: they are non-polar. This contrasts with the other substances in the table, which are all polar reagents. Second, before reaction can occur, energy must be supplied: heat in the case of oxygen, light in the case of bromine.

Ions and polar molecules have no effect on alkane molecules, because the C—H bond is non-polar. Carbon and hydrogen are very close in electronegativity, so the electron pair in the covalent bond between carbon and hydrogen is fairly evenly shared. This means there is little polarity in the C—H bond. The electron pair in the C—C bond is also evenly shared, so this bond is non-polar too.

Thus there are no polar bonds in alkane molecules, and so no centres of electrical charge to act as electrophiles or nucleophiles and attract normally reactive species such as ^{-}OH, H^+ and MnO_4^{-}.

So, alkanes are unreactive towards polar and ionic reagents, but will react with non-polar substances such as oxygen and bromine. The mechanism of these reactions involves free radicals.

Free-radical reactions of alkanes

Hexane and bromine react in sunlight. A similar reaction occurs between methane and chlorine and we will study this simple example in more detail.

Methane and chlorine do not react at all in the dark, but in sunlight an explosive reaction occurs, forming chloromethane and hydrogen chloride:

$$CH_4(g) + Cl_2(g) \xrightarrow{light} CH_3Cl(g) + HCl(g) \quad \Delta H = -98\,kJ\,mol^{-1}$$

This is a substitution reaction. Like all reactions between covalent molecules, it involves breaking some bonds, for which energy must be supplied, and making new bonds, from which energy is released.

The reaction between methane and chlorine does not proceed in the dark at room temperature because there is not enough energy available to break bonds and start the reaction.

$$Cl_2 \longrightarrow Cl\cdot + Cl\cdot \quad \Delta H = +242\,kJ\,mol^{-1}$$
$$CH_4 \longrightarrow CH_3\cdot + H\cdot \quad \Delta H = +435\,kJ\,mol^{-1}$$

These figures show that the Cl—Cl bond is easier to break than the C—H bond. When light is shone on the reaction mixture, chlorine molecules are supplied with enough energy to split them into atoms. This stage is called **initiation.** Using curly half-arrows, we can show the movement of electrons in this reaction:

$$Cl\overset{\frown}{}Cl \longrightarrow Cl\cdot + Cl\cdot$$

The chlorine atoms, being free radicals, are highly reactive. When they collide with a methane molecule they combine with one of its hydrogen atoms, forming a new free radical:

$$CH_4 + Cl\cdot \longrightarrow CH_3\cdot + HCl \quad \Delta H = +4\,kJ\,mol^{-1}$$

Using curly half-arrows, we can show the movement of the single electrons in this reaction:

$$H-\underset{\underset{H}{|}}{\overset{\overset{H}{|}}{C}}-H \quad Cl\cdot \longrightarrow \quad H-\underset{\underset{H}{|}}{\overset{\overset{H}{|}}{C}}\cdot \quad H-Cl$$

The $CH_3\cdot$ free radical then reacts with another chlorine molecule to form a $Cl\cdot$ radical:

$$CH_3\cdot + Cl_2 \longrightarrow CH_3Cl + Cl\cdot \qquad \Delta H = -97\,kJ\,mol^{-1}$$

and so the process continues. These two reactions enable a **chain reaction** to occur: they are **propagation** steps. Note that each propagation step involves both the breakage and the formation of a bond: the net energy change is therefore relatively small.

The reaction chain ends when two free radicals collide and combine; this is called **termination** and is highly exothermic:

$$Cl\cdot + Cl\cdot \longrightarrow Cl_2 \qquad \Delta H = -242\,kJ\,mol^{-1}$$
$$CH_3\cdot + Cl\cdot \longrightarrow CH_3Cl \qquad \Delta H = -339\,kJ\,mol^{-1}$$
$$CH_3\cdot + CH_3\cdot \longrightarrow C_2H_6 \qquad \Delta H = -346\,kJ\,mol^{-1}$$

Each chain may go through 100 to 10 000 cycles before termination occurs. The processes are extremely rapid, so the reaction is explosive.

The net result of the reaction above is the formation of large amounts of CH_3Cl and HCl plus small amounts of C_2H_6. (In the presence of excess chlorine, further substitution may occur, forming CH_2Cl_2, $CHCl_3$ and CCl_4, as explained in the next section.) The overall energy change of $-98\,kJ\,mol^{-1}$ represents the difference between the energy released in forming C—Cl and H—Cl bonds and the energy absorbed in breaking C—H and Cl—Cl bonds. The activation energy is provided by light which initiates the reaction.

Practically all the reactions of alkanes involve free-radical mechanisms, with high activation energies and a tendency to proceed rapidly in the gas phase.

14.7 Important reactions of alkanes

Alkanes show little reactivity towards all the common polar and ionic reagents. Indeed the alkanes were once known as the **paraffins**, from the Latin words *parum* (little) and *affinitas* (affinity). There are only a few reactions of alkanes, but these are of great importance.

Burning

Alkanes are *kinetically* stable in the presence of oxygen, but they are *energetically* unstable with respect to their oxidation products. When they are ignited, the necessary activation energy is supplied and combustion occurs. The reaction involves a free-radical mechanism, which occurs rapidly in the gas phase. Because it is a gas-phase reaction, liquid and solid alkanes must first be vaporised. This is why less volatile alkanes burn less readily. If the oxygen supply is plentiful, the combustion products are carbon dioxide and water. For example:

$$C_7H_{16}(g) + 11O_2(g) \longrightarrow 7CO_2(g) + 8H_2O(l) \qquad \Delta H = -4854\,kJ\,mol^{-1}$$

If the oxygen supply is limited, the products may include carbon monoxide and carbon (soot).

The combustion of hydrocarbons occurs in power stations, furnaces, domestic heaters, candles, gas heaters, internal combustion engines and many other devices. An understanding of the combustion of hydrocarbons is vital to the design of all these devices.

∧ Figure 14.13
The combustion of alkanes is a gas-phase reaction. Before it can burn, candle wax must vaporise. The wick provides a surface from which the molten wax vaporises.

Key Point

Free-radical reactions are chain reactions, in which each reaction initiates the next one. Chain reactions are fast and often explosive.

Quick Questions

11 a Write an equation for the complete combustion of octane, C_8H_{18}.
   b Estimate the enthalpy change of combustion of octane given that the enthalpy changes of combustion of hexane and heptane are $-4195\,kJ\,mol^{-1}$ and $-4854\,kJ\,mol^{-1}$, respectively.

12 Write an equation for the combustion of octane to form carbon monoxide and water.

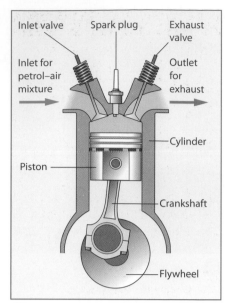

Figure 14.14

How a four-stroke gasoline engine works. The compression stroke is shown here. As the piston compresses the gasoline–air mixture, a spark makes the mixture explode, pushing the piston down and turning the crankshaft.

An important example is the gasoline engine (Figure 14.14). In the cylinder of a motor-car engine a mixture of gasoline vapour (which contains mostly C_5 to C_{10} alkanes) and air is ignited by an electric spark. The rate at which this reaction occurs and the ease with which it is initiated are very important for the efficiency of the engine. If the explosion starts too early the pistons are subjected to harmful jarring. 'Knocking' or 'pinking' of the engine then occurs. This is more likely to occur in the engines of high-performance cars.

Reducing the knocking problem

Gasoline mixtures that are rich in straight-chain alkanes such as heptane ignite very readily and explode rapidly, causing 'knocking' and inefficient combustion. The combustion of branched-chain alkanes such as 2,2,4-trimethylpentane (iso-octane) is much smoother and more controlled, so gasoline mixtures rich in branched-chain alkanes are more efficient fuels and less likely to cause knocking. The **octane rating** of 2,2,4-trimethylpentane is set at 100, and that of heptane is set at 0. Gasoline mixtures rich in branched-chain alkanes have high octane numbers and burn smoothly and efficiently in high-performance engines.

There are two ways of meeting the demand of modern high-performance engines for fuels with high octane numbers. One is to produce artificial gasoline mixtures that are rich in branched-chain alkanes, using cracking and isomerisation (see below). The other is to add an anti-knock compound to gasoline. For many years, this involved a lead compound $Pb(CH_2CH_3)_4$, which resulted in toxic lead being released into the environment.

Section 14.8 has more about the environmental effects of burning motor fuels.

Reaction with halogens

The reaction of methane with chlorine (Section 14.6) produces several products:

$$CH_4 + Cl_2 \longrightarrow CH_3Cl + HCl$$
chloromethane

$$CH_3Cl + Cl_2 \longrightarrow CH_2Cl_2 + HCl$$
dichloromethane

$$CH_2Cl_2 + Cl_2 \longrightarrow CHCl_3 + HCl$$
trichloromethane

$$CHCl_3 + Cl_2 \longrightarrow CCl_4 + HCl$$
tetrachloromethane

The proportions of these different products depend on the proportions of chlorine and methane used. All four products are useful in industry, but they have to be separated by distillation.

Similar reactions occur between alkanes and fluorine, chlorine and bromine. The reaction with fluorine occurs in the absence of sunlight and is explosive.

Cracking

When alkanes are heated to high temperatures their molecules vibrate strongly enough to break and form smaller molecules. One of these molecules is usually an alkane. These reactions are known as **cracking**. For example:

$$C_{11}H_{24} \longrightarrow C_9H_{20} + H_2C{=}CH_2$$
undecane   nonane   ethene

By using a catalyst, cracking can be made to occur at fairly low temperatures. This is known as **catalytic cracking** (Figure 14.16).

Figure 14.15

A catalytic cracker ('cat cracker') unit in an oil refinery

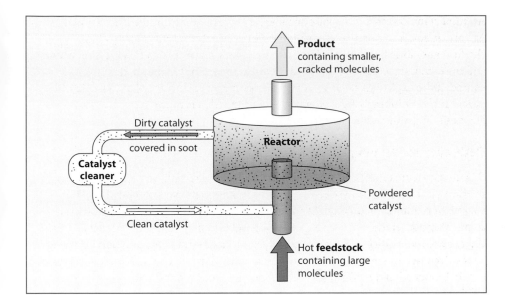

Quick Questions

13 Look at the equation for the cracking of $C_{11}H_{24}$.
a In which petroleum fraction would $C_{11}H_{24}$ be found?
b In which fraction would C_9H_{20} be found?
c Why must one of the products of cracking always be unsaturated?
14 Write an equation, similar to the undecane equation, for the cracking of decane, $C_{10}H_{22}$.

Figure 14.16
How a catalytic cracker works

Cracking involves *free radical* reactions. Heating alkanes makes them break up into free radicals. These radicals then react with one another to give a mixture of products, including branched-chain alkanes and alkenes.

Cracking is very important in the petroleum industry. It is used to convert heavy fractions into higher value products. In particular, it is used to provide extra gasoline and as a source of alkenes.

a To provide extra gasoline

The example above shows how undecane, a member of the kerosine fraction, can be cracked to produce nonane, a component of gasoline. So, heavier fractions can be cracked to produce extra gasoline. Cracking tends to produce branched-chain rather than straight-chain alkanes, so the gasoline produced this way also has a high octane rating. Processes similar to cracking can be used to convert low-grade gasoline to high-grade fuel. This process of breaking up straight-chain alkanes and reassembling them as branched-chain isomers is called isomerisation. It is very important in the production of unleaded gasoline.

b As a source of alkenes

Because they are so unreactive, alkanes are not a good starting point from which to make organic chemicals from crude oil. Alkenes, with their reactive double bonds, are a better starting point. The petrochemical industry uses vast quantities of ethene and propene as units for building larger organic molecules.

Cracking reactions produce alkenes, and under the right conditions large yields of ethene and propene can be obtained. Alkenes are therefore produced as by-products from cracking heavy fractions (such as $C_{11}H_{24}$) and by cracking some of the gasoline fraction (which is called naphtha when used as a source of alkenes). Alkenes are also made by cracking gaseous alkanes (ethane, propane and butane).

The reactions of alkenes and the products obtained from them are discussed in Chapter 15.

14.8 The environmental impact of motor vehicle fuels

Motor vehicles produce millions of tonnes of exhaust gases every year. Some of these are harmful to humans and the environment. Table 14.5 gives the composition of typical gasoline engine exhaust for cars with and without catalytic converters.

Quick Questions

15   **a**   What single gas omitted from Table 14.5 makes up most of the remainder of car exhaust?
    b   Why does exhaust contain carbon monoxide?
    c   Why does exhaust contain oxides of nitrogen?
    d   Why does exhaust contain hydrocarbons?

Table 14.5

The percentage of gases in typical gasoline engine exhaust, percentage by volume

Gas	Non-catalyst cars	Catalyst cars
carbon dioxide	14	15
oxygen	0.7	0
hydrogen	0.25	0
carbon monoxide	1.0	0.2
hydrocarbons	0.06	0.01
aldehydes	0.004	0.002
nitrogen oxides	0.2	0.02
sulfur dioxide	0.005	0.005
lead (as solids if leaded petrol is used)	4 mg m^{-3}	not applicable

Effects on humans

The most dangerous pollutants in car exhaust are carbon monoxide, hydrocarbons and oxides of nitrogen. Carbon monoxide and hydrocarbons are present in the exhaust because of incomplete combustion of the fuel. Carbon monoxide is very toxic because it forms a stable compound with haemoglobin in the blood, making the haemoglobin unable to transport oxygen. In a confined space the carbon monoxide in exhaust gases can be fatal. In the open air, the danger is less, though the carbon monoxide concentration in a busy street is undesirably high.

Acid deposition and photochemical smog

Acid deposition describes all the different ways in which acid materials get deposited on the Earth, including rain, snow, mist and solid particles. Acid deposition has serious environmental effects, including the destruction of life in acidified lakes, the destruction of trees and the erosion of stonework on buildings. Acid deposits contain mainly sulfuric acid, nitric acid, sulfates and nitrates, and automobile exhausts are thought to play a major part in their formation. Acidic oxides of sulfur and nitrogen are heavily involved (Section 12.3).

Exhaust gases are also involved in the formation of **photochemical smog**. In many cities, strong sunlight interacts with pollutant gases from millions of motor vehicles. This causes a complex series of photochemical reactions which produce a choking mixture of ozone, NO_x and other gases (Section 12.3).

Removing pollutant gases from car exhaust

The most significant pollutant gases in car exhaust are CO, NO_x and unburnt hydrocarbons, C_xH_y. As Figure 14.18 shows, governments have passed laws requiring that these gases are steadily reduced in vehicle exhausts.

There are several ways that the exhaust pollution problem can be tackled.

Using oxygenated fuels

'Oxygenates' are fuels containing oxygen in their molecules. The most important are the alcohols methanol and ethanol (Chapter 17). Adding these oxygenates to petrol makes combustion more complete and reduces the quantities of CO and C_xH_y that are produced.

Controlling the quantity of air mixed with the fuel

Before entering the combustion chamber of a vehicle, gasoline is mixed with air in the carburettor or fuel-injection system. The 'stoichiometric' quantity of air is the quantity

Figure 14.17
Car exhausts are analysed to make sure they meet government standards for emissions

Key Point

The major pollutants in car exhaust are:

▸ carbon monoxide, CO,
▸ nitrogen oxides, NO_x (mainly NO and NO_2),
▸ unburned hydrocarbons, C_xH_y.

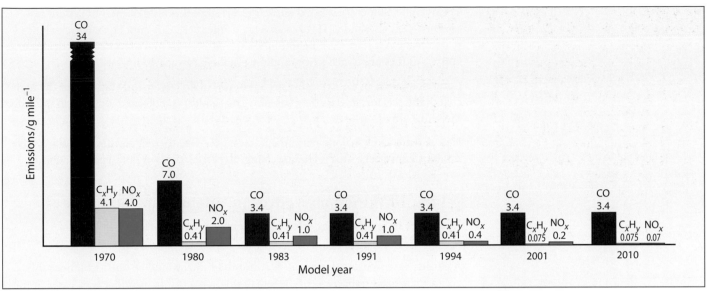

Figure 14.18

Exhaust emission limits set by the US Federal government for new vehicles. (Since 2001, the hydrocarbon limit, C_xH_y, has been for hydrocarbons other than methane.)

that is exactly enough to combine with the fuel, with none left over. A 'rich' mixture has an excess of fuel: this is what you get when the engine is 'choked' for a cold start. Rich mixtures produce large quantities of CO and C_xH_y so cold winter mornings are a bad time for exhaust pollution. A 'lean' mixture has an excess of air, and this produces less CO and C_xH_y. 'Lean burn' engines are one way that car manufacturers can meet the demands for reduced emissions of pollutant gases.

Using catalytic converters

A catalytic converter removes pollutant gases from the exhaust by oxidising or reducing them. The exhaust gases pass through a converter containing a precious metal catalyst, usually an alloy of platinum and rhodium. Several reactions may take place. NO_x and CO may take part in a redox reaction which removes both of them at the same time. NO_x oxidises CO to CO_2, and is itself reduced to harmless nitrogen gas.

$$2NO(g) + 2CO(g) \longrightarrow N_2(g) + 2CO_2(g)$$

CO and C_xH_y are oxidised by air:

$$2CO(g) + O_2(g) \longrightarrow 2CO_2(g)$$

$$C_7H_{16}(g) + 11O_2(g) \longrightarrow 7CO_2(g) + 8H_2O(g)$$

(using C_7H_{16} to represent a typical hydrocarbon).

For all three of these reactions to happen, you need a 'three-way converter' (Figure 14.19). An oxygen monitor is fitted to the engine. This checks the quantity of oxygen going into the engine to make sure there is enough to carry out the oxidation reactions.

Quick Questions

16 Look at Table 14.5.
   a Which exhaust gases are *significantly* lower in catalyst cars than in non-catalyst cars?
   b Which exhaust gas is *higher* in catalyst cars than in non-catalyst cars?
   c Suggest a reason for your answer in part b.

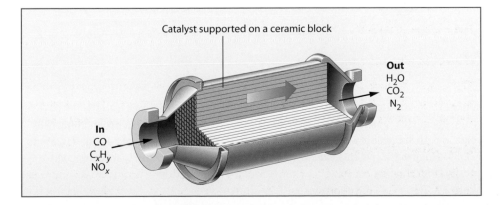

Catalyst supported on a ceramic block

In
CO
C_xH_y
NO_x

Out
H_2O
CO_2
N_2

Figure 14.19
A three-way catalytic converter

The overall result of passing exhaust gases through this kind of catalyst system is to convert NO_x, CO, and C_xH_y to relatively harmless N_2, CO_2 and H_2O. The catalytic reactions do not start working until the catalyst has reached a temperature of about 200 °C, so they are not effective until the engine has warmed up.

Catalyst systems of this type cost several hundred dollars, mainly because of the high cost of the precious metals they contain. The catalyst is 'poisoned' by lead, so unleaded fuel must always be used.

Unfortunately, the major environmental impact of motor vehicles cannot be solved by catalytic converters. This is climate change, which is covered in the next section.

14.9 The greenhouse effect and global climate change

Climate change is by far the most serious environmental problem linked to hydrocarbon fuels.

A garden greenhouse keeps plants warmer than they would be outside. It does this because the glass traps some of the Sun's radiant energy. In a similar way, gases in the atmosphere help to keep the Earth warm, by absorbing some of the Sun's energy that would otherwise be reflected back into space. Without this **greenhouse effect**, the average surface temperature of the Earth would be about 33 °C lower, that's –18 °C instead of 15 °C.

The Earth absorbs visible and ultraviolet (UV) radiation from the Sun and then re-emits part of this energy as infrared (IR). IR radiation is absorbed by molecules in the atmosphere, making them vibrate faster and increasing their energy. The energy is passed on to other molecules, making the atmosphere hotter.

Some gases are better than others at absorbing the Sun's energy and keeping the Earth warm. Oxygen and nitrogen are not very effective, but carbon dioxide is. This is because CO_2 readily absorbs infrared radiation of the frequency that the Earth emits. Figure 14.20 shows how carbon dioxide warms the atmosphere by the greenhouse effect.

> Figure 14.20
Why carbon dioxide is a greenhouse gas

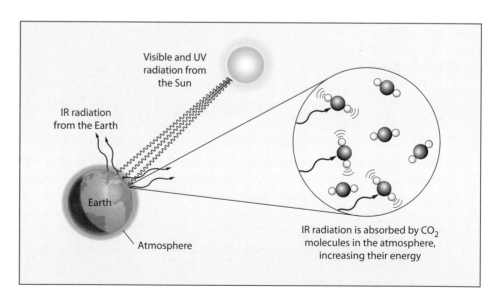

Visible and UV radiation from the Sun

IR radiation from the Earth

Earth

Atmosphere

IR radiation is absorbed by CO_2 molecules in the atmosphere, increasing their energy

Definitions

The **greenhouse effect** is the natural warming of the atmosphere caused by the absorption of radiation by certain gases such as CO_2.

The **enhanced greenhouse effect** is the additional warming due to gases that have been released into the atmosphere as a result of human activity.

Whenever fossil fuels like gasoline are burned, carbon dioxide goes into the atmosphere (see Quick question 17). Since the industrial revolution began about 200 years ago, the concentration of CO_2 in the atmosphere has risen from about 0.028% to 0.038% (Figure 14.21). This causes an **enhanced greenhouse effect,** absorbing extra energy from the Sun.

Nearly all climate scientists are convinced that the enhanced greenhouse effect is causing the Earth's climate to get warmer. Figure 14.22 shows how the temperature of the Earth's surface has changed between 1850 and 2007. The temperature differences are small – there is only about 1 °C difference between the coldest year and the warmest. But small temperature changes can make a big difference to the environment. Many environmental scientists believe that changes can already be seen in melting glaciers and increased frequency of hurricanes.

A few scientists are sceptical, saying the changes to average temperatures are part of a natural cycle and not a long term trend caused by humans burning fossil fuels. However, the large majority of climate scientists say that the trend is caused by humans, and that the situation is set to get a lot worse unless the release of carbon dioxide by burning fossil fuels is drastically reduced.

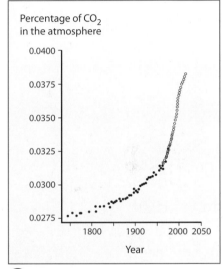

 Figure 14.21
The changing percentage of CO_2 in the atmosphere since 1750. The data before 1958 (closed circles) were obtained by analysis of bubbles of gas trapped below the surface of the Antarctic ice-cap. The data after 1958 (open circles) were obtained by analysing the atmosphere at the Mauna Loa observatory in Hawaii.

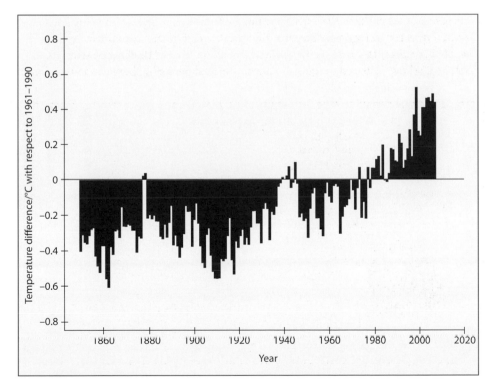

◄ Figure 14.22
Changes to the Earth's surface temperature over the period 1850–2007. The red bars show the temperature differences compared with the average for 1961–1990.

Controlling CO_2

The only way to reduce the emission of CO_2 is to burn less fossil fuels. This is much easier said than done. The world economy depends on fossil fuels, and the consumption of oil, gas and coal is increasing very rapidly in many developing countries. Here are some of the main approaches to reducing CO_2 emissions. There isn't a single way to fix it – the problem will only be solved by a combination of many different measures.

- Increase efficiency, with more fuel-efficient cars, better insulated homes and less wasted energy. (But notice that putting catalytic converters on cars does nothing to reduce the amount of CO_2 they produce.)
- Use nuclear energy and renewable energy sources, such as hydropower, wind and solar power, to generate electricity.
- Use fossil fuels that produce less CO_2. Table 14.6 shows the 'CO_2 load' of different fuels. The table shows that, for each kilojoule of energy produced, natural gas produces about three-quarters as much CO_2 as gasoline, and less than half as much CO_2 as coal.

Table 14.6

The 'CO_2 load' of different fuels. This is the mass of CO_2 produced per kJ of energy released by the fuel.

Fuel	'CO_2 load'/$g\,kJ^{-1}$
Coal (calculated as pure C)	0.11
Natural gas (calculated as pure CH_4)	0.049
Gasoline (calculated as pure C_8H_{18})	0.064
Hydrogen (H_2)	0

Quick Questions

18 a Why is the CO_2 load of hydrogen zero?
   b Why is the CO_2 load of natural gas so much less than that of coal?
19 Calculate the CO_2 load of undecane, $C_{11}H_{24}$, a constituent of diesel oil, by following steps a to c. (The enthalpy change of combustion of undecane, $\Delta H_c = -7431\,kJ\,mol^{-1}$.)
   a Write a balanced equation for the complete combustion of undecane.
   b Calculate the mass of CO_2 formed when one mole of undecane burns.
   c Use ΔH_c to calculate the mass of CO_2 formed per kJ released when undecane burns. This is the CO_2 load.
   d Compare your answer for the CO_2 load of undecane with the other values in Table 14.6. Which is it closest to? Explain why.

Review questions

1 Name the following alkanes:

a $CH_3—CH_2—CH_2—CH_2—CH_2—CH_3$

b $CH_3—CH—CH_2—CH—CH_2—CH_3$
     | |
     CH_3 CH_2CH_3

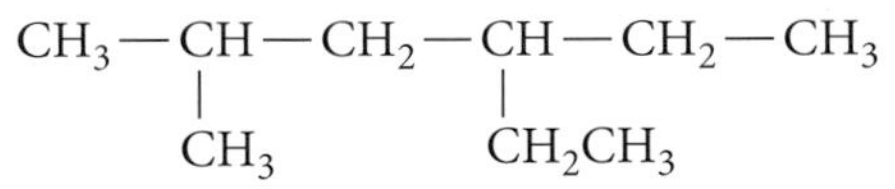

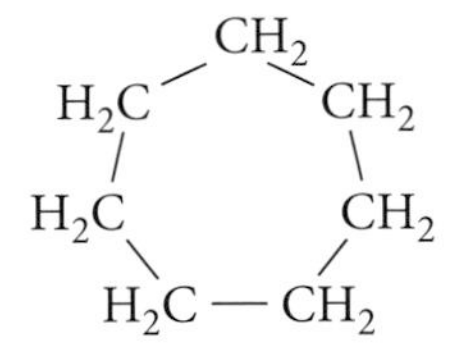

c

d $CH_3—CH_2—CH_2—CH_2$
                      |
                      CH_2
                      |
                      CH_3

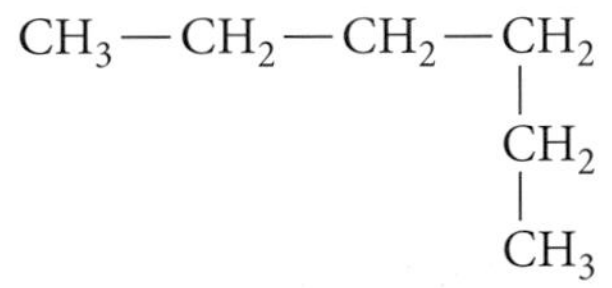

e

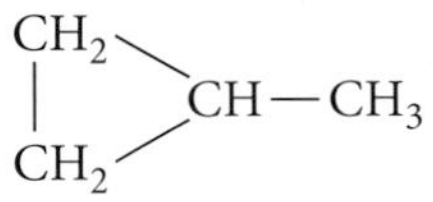

f

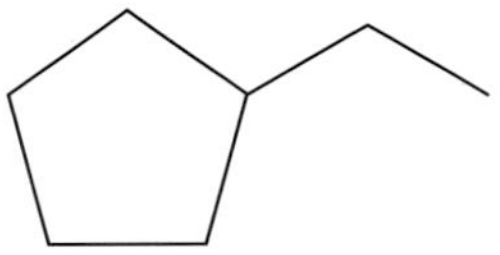

2 Consider the following alkanes:

A $CH_3(CH_2)_{18}CH_3$

B

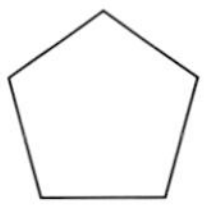

C $CH_3(CH_2)_6CH_3$

D $CH_3CH(CH_2)_4CH_3$
        |
        CH_3

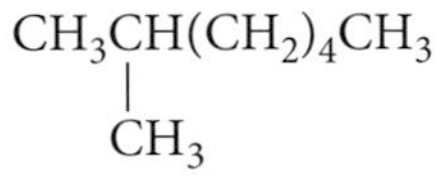

E $CH_3CH_2CH_3$

a In which crude oil fraction would each be found?

b Which two alkanes are isomers?

c Which alkane might have been formed by cracking another of the alkanes shown above? (Identify *both* alkanes.)

d Arrange the alkanes in order of increasing boiling point.

e Which alkane is a solid at 298 K?

f Which alkane's formula does *not* fit the general formula C_nH_{2n+2}?

3 Write **a** structural formulae, **b** molecular formulae for the following alkanes:

   i ethylcyclohexane

   ii 1,2-dimethylcyclopentane

   iii 2,2,3-trimethylbutane

   iv 3,4-diethyl-2,2-dimethylheptane.

4 Study the table below, which gives the enthalpy changes of combustion and the relative molecular masses of some alkanes.

Alkane	No. of carbon atoms, n	Relative molecular mass	Enthalpy change of combustion ΔH_c/kJ mol^{-1}
methane	1	16	−890
ethane	2	30	−1560
propane	3	44	−2220
butane	4	58	−2877
pentane	5	72	−3509
methylbutane	5	72	−3503
2,2-dimethylpropane	5	72	−3517
hexane	6	86	−4195

a Plot a graph of ΔH_c against n for the first six straight-chain alkanes.

b Does ΔH_c increase by approximately the same amount for each extra carbon atom in an alkane chain?

c What is the average increase in ΔH_c per carbon atom?

d Your answer to part c represents the enthalpy change of combustion of which structural group?

e Compare the ΔH_c values for the three isomeric alkanes with five carbon atoms. Why are they so similar?

f Work out the enthalpy change of combustion per gram of butane, pentane and hexane, respectively. Comment on your result.

5 This question is about octane (C_8H_{18}) and cyclooctane (C_8H_{16}).

a Draw the structural formula of octane.

b Draw the skeletal formula of cyclooctane.

c Suppose octane and cyclooctane were each reacted with chlorine so that only one hydrogen atom in each molecule was substituted by chlorine. Draw structural formulae for all the products you would expect in each case.

6 The table below shows the enthalpy changes of combustion of some cycloalkanes.

Name	Formula	Enthalpy change of combustion ΔH_c/kJ mol^{-1}
cyclopropane	$(CH_2)_3$	−2090
cyclobutane	$(CH_2)_4$	−2740
cyclopentane	$(CH_2)_5$	−3320
cyclohexane	$(CH_2)_6$	−3948
cycloheptane	$(CH_2)_7$	−4635
cyclooctane	$(CH_2)_8$	−5310
cyclononane	$(CH_2)_9$	−5980
cyclodecane	$(CH_2)_{10}$	−6630

a For each compound, calculate the enthalpy change of combustion per CH_2 group.

b Would you expect the enthalpy change of combustion per CH_2 group to be constant among: **i** open-chain alkanes, **ii** unstrained cycloalkanes, **iii** cycloalkanes with ring strain?

c Explain your answers and comment on the figures you obtained in part a.

7 If a few drops of bromine are added to hexane a deep red solution is formed. No reaction occurs if the mixture is kept in the dark, but in sunlight the red colour slowly disappears and a misty gas is given off.

a Identify the misty gas formed in this reaction.

b Write an overall equation for the reaction of one mole of bromine with one mole of hexane under these conditions.

c Why does the reaction only occur in sunlight?

d This is a free-radical chain reaction. Write equations for:

  i the initiation stage,

  ii a propagation stage,

  iii a termination stage.

8 Write the structural formulae of all the products you would expect to be formed when ethane reacts with excess chlorine in sunlight.

9 For each of the following, state whether or not you would expect any significant reaction to occur. Write equations for any reactions which you think will occur.

a Chlorine is bubbled through hexane in sunlight.

b Sodium metal is added to warm hexane.

c Hexane is boiled with acidified potassium dichromate(VI) solution.

d Chlorine and hexane vapour are heated in the dark.

e Hexane and hydrogen are heated in sunlight.

f Hexane vapour is heated strongly on its own, in the absence of air.

10 Methane can be used as an alternative to gasoline as a fuel for motor cars. What modifications would be necessary to an ordinary car to enable it to run on methane? What would be the advantages of using methane instead of gasoline?

11 Natural gas from the Texas Panhandle has the percentage composition by volume shown in the table below.

Component	Percentage by volume
methane	80.9
ethane	6.8
propane	2.7
butane and higher alkanes	1.6
nitrogen	7.9
carbon dioxide	0.1

The natural gas is processed in the following way. Carbon dioxide is first removed, then ethane, propane, butane and higher alkanes are separated off and mainly used to make alkenes, especially ethene.

a How might the carbon dioxide be removed from the gas?

b How might ethane and the higher alkanes be separated off?

c What would the remaining natural gas be used for?

d Why is it not necessary to remove the nitrogen from the gas?

e Why is ethene an important industrial chemical?

f Ethene is produced from ethane by a cracking reaction that eliminates hydrogen. Write an equation for this reaction. What conditions would be needed to make the reaction occur?

g In Europe ethene is usually manufactured from naphtha (part of the gasoline fraction from crude oil) rather than from ethane. Suggest a reason for this difference.

12 The engines of modern motor cars have exhaust systems which are fitted with catalytic converters in order to reduce atmospheric pollution from substances such as NO.

a State *three more* pollutants, other than CO_2 and H_2O, that are present in the exhaust gases of a car engine.

b What is the active material present in the catalytic converter?

c Write *one* balanced equation to show how NO is removed from the exhaust gases of a car engine by a catalytic converter.

Cambridge Paper 4 Q3b June 2008

15 Alkenes

Why alkenes are important

What do PVC raincoats and antifreeze have in common? Or polythene bottles and adhesives? Like many other things in everyday use, they are made from ethene, the simplest alkene and the most versatile organic compound in use today. Ethene, $CH_2=CH_2$, with its reactive double bond, can be used as a building block to prepare complex organic molecules. Propene, $CH_3CH=CH_2$, is used in a similar way, though on a smaller scale. Large quantities of these alkenes are manufactured by cracking processes described in Section 14.7.

15.1 Naming alkenes

Ethene and propene are the first two members of the homologous series of **alkenes.** All members of this series contain a double carbon–carbon bond, $C=C$. They have two atoms of hydrogen less than the corresponding alkane and their general formula is C_nH_{2n}. Because they have less than the maximum content of hydrogen they are said to be **unsaturated**.

Physically, alkenes are similar to alkanes, with boiling points generally a little lower (e.g. ethane 185 K, ethene 169 K; propane 231 K, propene 225 K).

Alkenes are named using the same general rules described for alkanes in Section 14.3. The suffix **-ene** is used instead of -ane, together with a number indicating the position of the double bond in the chain. Thus the molecule $CH_3CH=CHCH_3$ is named but-2-ene. Although the double bond joins carbon atoms 2 and 3, the number 2 is used because this is the lower. Table 15.1 gives the formulae and names of some more alkenes.

 Table 15.1
Naming alkenes

Formula	Name
$CH_3CH_2CH_2CH=CH_2$	pent-1-ene
$CH_3CH_2CH=CHCH_3$	pent-2-ene

$CH_3C=CHCH_3$
    |
   CH_3
	2-methylbut-2-ene

$CH_2=CHCH_2CH=CH_2$	penta-1,4-diene
(hexagon ring)	cyclohexene

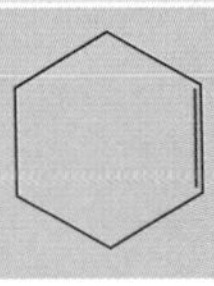

15.2 The nature of the double bond

When ethene is bubbled through bromine, the bromine is decolorised. No sunlight is needed. A colourless liquid, immiscible with water, is formed, but no hydrogen bromide is produced. This reaction is clearly different in nature to the reaction of bromine with alkanes such as ethane, in which sunlight is needed to make the reaction occur

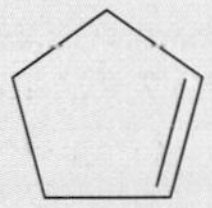

⌃ Figure 15.1
Ethene is used to speed up the ripening of fruit such as bananas. This 'Ethylene Gas Generator' produces ethene by the dehydration of ethanol over a heated catalyst. What might the catalyst be?

Table 15.2

Bond energies and bond lengths of
C—C and C=C

	C—C (in ethane)	C=C (in ethene)
Bond energy/ kJ mol^{-1}	346	598
Bond length/nm	0.154	0.134

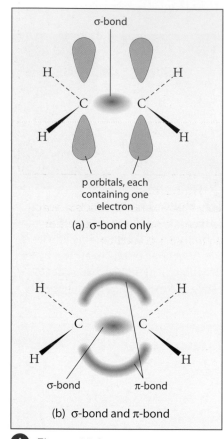

(a) σ-bond only

(b) σ-bond and π-bond

Figure 15.2
σ– and π–bonds in ethene. Note that the plane of the molecule is at right angles to the page.

Definitions

A **sigma bond (σ)** is formed by the overlap of orbitals between two atoms. It lies on the axis between the two atoms.

A **pi (π)** bond is formed by the sideways overlap of two p orbitals. It has two regions, one above and one below the plane of the molecule.

and hydrogen bromide is produced. The reaction between ethene and bromine is an **addition** reaction. One of the two carbon–carbon bonds in ethane breaks, enabling bonds to be formed to bromine:

1,2-dibromoethane
(a colourless liquid)

Addition reactions are characteristic of all alkenes, which are much more reactive than alkanes. This reactivity might be a little surprising at first: we might expect a double bond to be stronger than a single one, and therefore more stable.

Look at the bond energies in Table 15.2. You can see that the bond energy of a double bond is greater than that of a single bond, though not twice as great. This suggests that the two bonds in C=C may not be identical. In fact, two kinds of covalent bond are involved.

▶ A bond situated symmetrically between the two carbon atoms, formed by the overlap of two orbitals. This is called a **sigma (σ) bond** (Figure 15.2(a)).
▶ A bond formed by the 'sideways' overlap of two 2p orbitals. Because each p orbital has two lobes, this bond has two regions, one above and one below the plane of the molecule. It is called a **pi (π) bond** (Figure 15.2(b)).

The two electrons of the π-bond are not situated on the axis between the carbon atoms. This means they are not 'on average' as close to the nuclei of the carbon atoms as the electron pair in the σ-bond. Therefore they do not attract the nuclei so strongly, so the π-bond is not as strong as the σ-bond. Nevertheless, the π-bond and σ-bond together are stronger than the single σ-bond that links the carbon atoms in ethane. Consequently, the C=C bond in ethene is stronger and shorter than the C—C bond in ethane.

If double bonds are stronger than single bonds, why are alkenes more reactive than alkanes? When bromine adds across the double bond in ethene the π-bond is broken: this requires energy. The energy used in breaking the one π-bond, however, is more than repaid by the energy released when two *new* bonds are made to bromine atoms. There is more about the mechanism of addition reactions in Section 15.4.

15.3 Cis–trans isomerism

X-ray diffraction evidence shows that the ethene molecule is planar. This agrees with our ideas on the shapes of molecules (Section 3.7), since ethene has no lone pairs. The three negative centres around each carbon atom are arranged trigonally, at approximately 120° to one another, so the ethene molecule is drawn like this:

Unlike the CH_3 groups in ethane, the CH_2 groups in the ethene molecule cannot be rotated about the bond between the carbon atoms. It is possible to rotate about a σ-bond, because this does not affect the orbital overlap. With a double bond though, rotation would involve breaking the π-bond and this requires more energy than is available at normal temperatures.

Now consider the compound 1,2-dibromoethene, BrCH=CHBr. This can be made by addition of bromine to ethyne (HC≡CH). The product is a colourless non-polar liquid, boiling point 381 K, melting point 266 K and immiscible with water.

However, another compound is known with the same formula BrCH=CHBr. This second substance is also a colourless liquid, immiscible with water, but its boiling point is 383 K, its melting point is 220 K and its molecule is quite polar. These two compounds exhibit a kind of isomerism called **cis–trans** or **geometric** isomerism. It arises from the lack of free rotation about a double bond. The first compound (b.pt. = 381 K, m.pt. = 266 K, non-polar) has the structure:

$$\underset{Br}{\overset{H}{\diagdown}} C = C \underset{H}{\overset{Br}{\diagup}}$$

This is called *trans*-1,2-dibromoethene because the bromine atoms are on opposite sides of the double bond (*trans*, Latin: opposite). The second compound (b.pt. = 383 K, m.pt. = 220 K, polar) has the structure:

$$\underset{Br}{\overset{H}{\diagdown}} C = C \underset{Br}{\overset{H}{\diagup}}$$

This is called *cis*-1,2-dibromoethene (*cis*, Latin: on the same side), both bromine atoms being on the same side of the double bond. Since free rotation about the double bond is prevented, these isomers cannot be readily converted to one another, unless the temperature is high enough to make available sufficient energy to break the π-bond. Figure 15.3 shows the three-dimensional shapes of the two isomers.

Cis–trans isomerism is common in compounds containing double bonds. The isomers normally have similar chemical properties but often their physical properties are markedly different. *Cis–trans* isomerism is not limited to compounds with C=C double bonds. It can arise wherever rotation about a bond is restricted, for example in ring compounds.

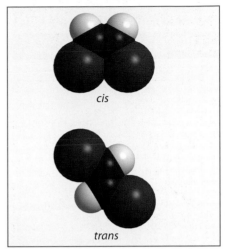

Figure 15.3
Computer-generated models showing the three-dimensional shapes of *cis*- and *trans*-dibromoethene

Quick Questions

3 **a** Why is *cis*-1,2-dibromoethene polar while its *trans* isomer is not?
  b Suggest a reason for the large difference in melting point between the two isomers.
  c Would you expect but-2-ene, $CH_3CH=CHCH_3$, to have *cis* and *trans* isomers? If so, write their structures.

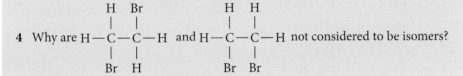

4 Why are $H-\overset{\displaystyle H}{\underset{\displaystyle Br}{C}}-\overset{\displaystyle Br}{\underset{\displaystyle H}{C}}-H$ and $H-\overset{\displaystyle H}{\underset{\displaystyle Br}{C}}-\overset{\displaystyle H}{\underset{\displaystyle Br}{C}}-H$ not considered to be isomers?

15.4 Mechanism of addition to a double bond

Ethene and bromine undergo an addition reaction to form 1, 2-dibromoethane. The reaction occurs in the dark at room temperature, which suggests that a free-radical mechanism is *not* involved. The mechanism involves heterolytic rather than homolytic fission.

The ethene molecule has a region of high electron density caused by the π-electrons in its double bond. As the bromine molecule approaches the ethene, it becomes polarised by this negative charge:

$$\underset{H}{\overset{H}{\diagdown}} \underset{\underset{\displaystyle Br}{|}}{\overset{C-C}{\cdots}} \underset{H}{\overset{H}{\diagup}} \longrightarrow \underset{H}{\overset{H}{\diagdown}} \underset{\underset{\displaystyle Br^{\delta-}}{\overset{\displaystyle Br^{\delta+}}{|}}}{\overset{C-C}{\cdots}} \underset{H}{\overset{H}{\diagup}}$$

Definition

In **electrophilic addition**, an electrophilic (electron-loving) atom or group of atoms accepts electrons from a double bond. The double bond breaks and two new bonds form.

A loose association forms between the ethene and bromine molecules. Negative charge moves from the double bond towards the positively charged bromine atom. At the same time, electrons in the Br—Br bond are repelled towards the negatively charged Br atom. The result is the formation of a C—Br bond and the production of two ions, a carbocation and a Br^- ion:

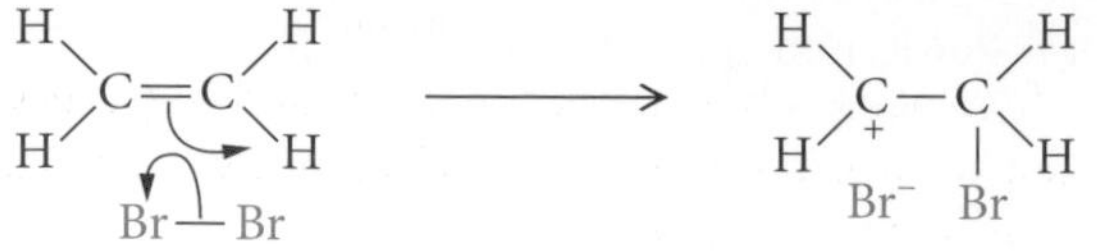

Notice the curly arrows in this equation. They are used to show the movement of electron pairs (Section 14.5).

Quick Questions

5 **a** What product would you expect from an addition reaction between ethene and hydrogen chloride? Write its formula.

b In the reaction between ethene and HCl, which end of the HCl molecule would you expect to attack the double bond initially?

c Draw the structure of the intermediate carbocation in this reaction.

The carbocation is very unstable and quickly combines with the Br^- to form 1,2-dibromoethane:

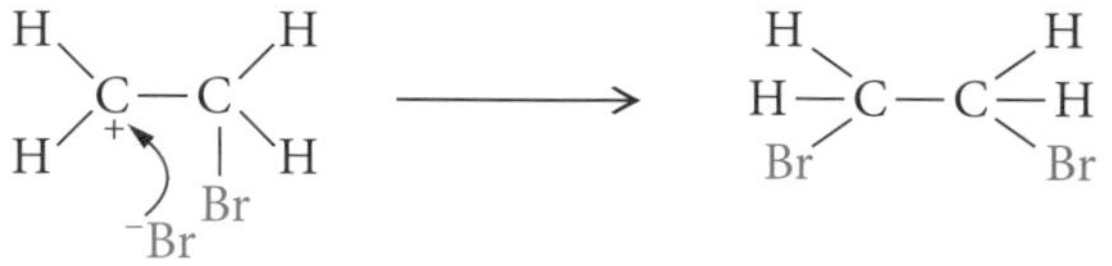

This mechanism is described as **electrophilic addition** because the Br_2 molecule acts as an electrophile when it is attracted to the electron-rich double bond.

Evidence for this mechanism comes from observing what happens when the reaction is carried out in the presence of Cl^- ions. When ethene reacts with bromine in the presence of Cl^- ions, it is found that 2-bromo-1-chloroethane, $ClCH_2—CH_2Br$, is formed as well as 1,2-dibromoethane. This suggests that the intermediate carbocation has indeed been formed and has reacted with Cl^- as well as Br^- ions.

15.5 Important reactions of alkenes

Nearly all the important reactions of alkenes are addition reactions. Many have mechanisms similar to the reaction with bromine. You can often predict the product of an addition reaction. Look carefully at the molecule which is being added to ethene. Try to decide how it will split into a positive and negative part. These two parts will then join on either side of the double bond.

Key Point

The characteristic reaction of alkenes is electrophilic addition in general:

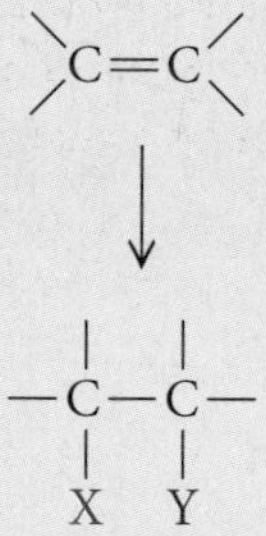

The reactions given below all involve ethene, which is industrially the most important alkene. The reactions of ethene are typical of alkenes in general, because the alkenes are a well-graded homologous series.

Many of these reactions can be carried out in the laboratory, though some require special conditions normally only available on an industrial scale. Industrially, ethene is produced by cracking light hydrocarbon fractions. In the laboratory it can be prepared by the dehydration of ethanol (Section 17.5).

Reaction with halogens

We have already seen that bromine reacts with ethene under normal conditions to form 1,2-dibromoethane. As you might expect, chlorine reacts in a similar way:

$$CH_2{=}CH_2 + Cl_2 \longrightarrow \underset{\text{1,2-dichloroethane}}{CH_2Cl—CH_2Cl}$$

Quick Questions

6 Predict the structure of the products of addition reactions between ethene and:

a H_2O

b HOCl

c H_2

The reaction occurs under ordinary conditions. The product, 1,2-dichloroethane, is used to manufacture chloroethene ('vinyl chloride'), from which PVC is made.

Fluorine reacts explosively with ethene, but the reaction with iodine is rather slow.

Decolorisation of yellow-orange bromine water is a useful test-tube reaction to detect a C=C double bond.

Reaction with hydrogen

Hydrogen and ethene do not react under normal conditions, but in the presence of a finely divided metal catalyst, usually nickel at about 140 °C, ethane is produced:

$$H_2C=CH_2 + H_2 \xrightarrow{Ni} CH_3CH_3$$
$$\text{ethane}$$

This reaction is useful when analysing organic compounds. By measuring the number of moles of hydrogen absorbed by one mole of a hydrocarbon, the number of double (or triple) bonds in its molecule can be found.

Hydrogenation of double bonds is used to convert edible oils into margarine and other spreads. Vegetable oils such as palm oil consist of esters of long-chain carboxylic acids which contain double bonds. Treatment of these oils with hydrogen in the presence of a nickel catalyst saturates the carbon chains by adding hydrogen across the double bonds. This raises the melting point of the oil so that it becomes solid, i.e. a fat, at room temperature. By controlling the degree of hydrogenation, you can make the margarine as soft or hard as required.

There is now considerable medical evidence that saturated fats, i.e. those containing few double bonds, are dangerous to health. Doctors believe they contribute to heart and circulatory disease. Animal fats such as cream, butter and pork fat tend to have a high proportion of saturated fats plus cholesterol, which is also dangerous in large amounts. Vegetable fats, including soft margarine made from vegetable oil, contain more unsaturated fats and less cholesterol. They are therefore thought to be healthier.

Reaction with hydrogen halides

Ethene reacts with gaseous hydrogen halides or concentrated aqueous solutions of hydrogen halides, such as concentrated HCl, in the cold:

$$CH_2=CH_2 + HX \longrightarrow CH_3CH_2X$$

This is another example of electrophilic addition. Hydrogen ions act as electrophiles, attacking the double bond and forming an intermediate carbocation:

This ion then reacts with halide ions X⁻ to form the product:

For example, the reaction of hydrogen chloride with ethene produces chloroethane.

When propene reacts with a hydrogen halide such as HCl, there are two possible products:

1-chloropropane **A**

2-chloropropane **B**

Key Point

Test for a C=C double bond
Compounds containing a C=C double bond decolorise yellow-orange bromine water and purple acidified $KMnO_4(aq)$.

Figure 15.4
Margarine contains less saturated fat than is present in butter.

Quick Questions

7 Predict the formula of the product of the reaction between:
   a propene and chlorine,
   b but-2-ene and hydrogen.
8 A hydrocarbon P contains four carbon atoms. One mole of P reacts with two moles of hydrogen. Write a structural formula for P.

The inductive effect

Carbon is not very electronegative. Many functional groups attract electrons more than carbon, so they pull electrons away from the carbon atom they are bonded to. For example

$$C \longrightarrow Cl$$

This is called the **negative inductive (–I) effect**.

A few groups tend to push electrons towards carbon. For example

$$C \longleftarrow CH_3$$

This is called the **positive inductive (+I) effect**. All alkyl groups have a small positive inductive effect.

9 Predict the major products of reactions between:
 a $CH_3CH_2CH\!=\!CH_2$ and HBr
 b $CH_3C\!=\!CHCH_3$ and HI
      |
      CH_3

When this reaction is carried out, much more **B** is formed than **A**. We can find a reason for this by looking at the mechanism of the reaction.

When hydrogen ions attack the double bond in propene, two different carbocations, **C** and **D**, can be formed:

D is the more stable of these two ions. This is because of a very important property of an alkyl group:

an alkyl group tends to push electrons slightly towards any carbon atom to which it is attached.

This is a positive inductive effect.

In ion **D** there are two methyl groups pushing electrons onto the positively charged carbon atom; in **C** there is only one ethyl group doing so. As a result the positive charge is stabilised slightly more in **D** than in **C**, because the donated electrons tend to cancel out the charge. **D** is therefore the more stable of the two possible intermediate carbocations, though still very unstable. It therefore tends to persist longer than **C**, making it more likely to combine with Cl⁻ to form the product **B**. This means that **B** is the major product, though a certain amount of **A** is also formed.

The major products of asymmetric addition reactions like this can be predicted by a useful general rule, known as **Markovnikoff's rule**. It says that:

when a molecule HA adds to an asymmetric alkene, the major product is the one in which the hydrogen atom attaches itself to the carbon atom already carrying the larger number of hydrogen atoms.

For example, in the reaction of 2-methylpropane with HCl:

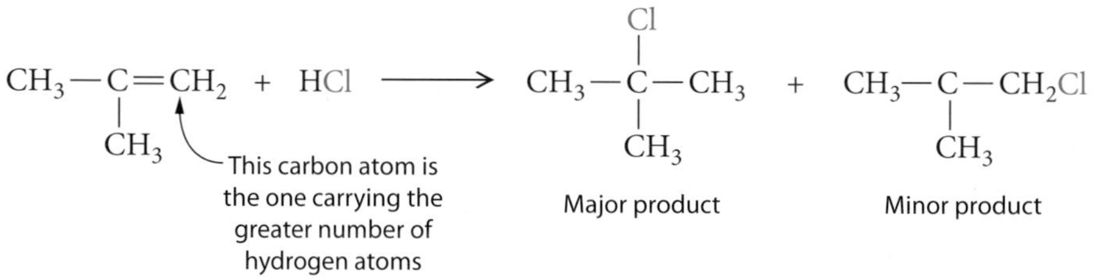

Reaction with H_2O

Ethene undergoes addition with H_2O.

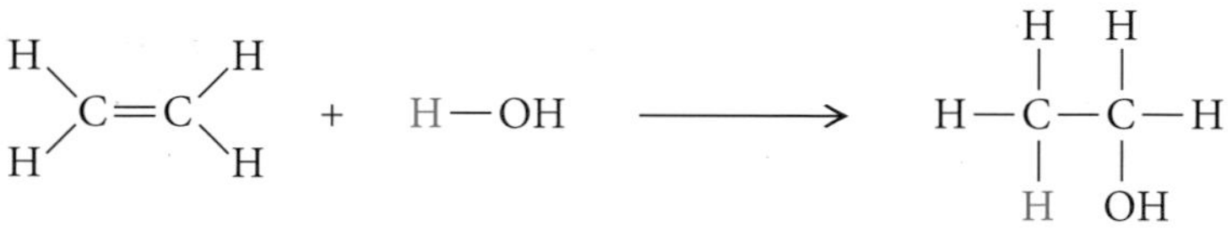

The reaction only occurs in the vapour phase, using a catalyst at high temperature and pressure. This is one of the major methods for manufacturing ethanol. Phosphoric acid, H_3PO_4, is used as the catalyst:

$$CH_2\!=\!CH_2(g) + H_2O(g) \xrightarrow[330\,°C,\ 60\ atm]{H_3PO_4} CH_3CH_2OH(g)$$
ethanol

Reaction with potassium manganate(VII)

Cold, dilute manganate(VII)

Ethene will decolorise cold acidified dilute potassium manganate(VII) (KMnO$_4$). This reaction is a useful test for a C=C double bond.

The reaction with manganate(VII) ions (MnO$_4^-$) is complicated and involves both addition and oxidation. The product is ethane-1,2-diol, HOCH$_2$CH$_2$OH, which is itself further oxidised if excess manganate(VII) is present.

$$CH_2=CH_2 \xrightarrow{\text{cold, dilute } MnO_4^-/H^+} HOCH_2-CH_2OH$$

The balanced equation for this reaction, like all redox reactions in organic chemistry, can be written using the half-equation method described in Sections 6.2 and 6.3:

$$CH_2=CH_2 + 2H_2O \longrightarrow HOCH_2CH_2OH + 2H^+ + 2e^- \qquad (1)$$
$$\text{ethane-1,2-diol}$$

$$MnO_4^- + 8H^+ + 5e^- \longrightarrow Mn^{2+} + 4H_2O \qquad (2)$$

Multiplying (1) by 5 and (2) by 2 and adding, we get

$$5CH_2=CH_2 + 2H_2O + 2MnO_4^- + 6H^+ \longrightarrow 5HOCH_2CH_2OH + 2Mn^{2+}$$

Hot, concentrated manganate(VII)

When an alkene is heated with acidified concentrated potassium manganate(VII), the double bond is completely broken. This is called oxidative cleavage.

The products are carbonyl compounds: aldehydes or ketones (Section 18.1). If aldehydes are formed, they get further oxidised to carboxylic acids.

For example, with hex-2-ene:

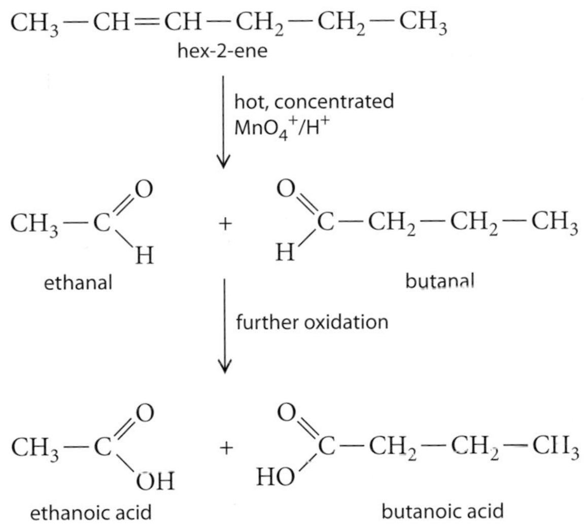

$$CH_3-CH=CH-CH_2-CH_2-CH_3$$
hex-2-ene

hot, concentrated
MnO$_4^+$/H$^+$

CH$_3$—C$\langle$O / H + O$\rangle$C—CH$_2$—CH$_2$—CH$_3$ / H

ethanal   butanal

further oxidation

CH$_3$—C$\langle$O / OH + O$\rangle$C—CH$_2$—CH$_2$—CH$_3$ / HO

ethanoic acid   butanoic acid

This reaction is a useful way to find the position of a double bond in larger alkene molecules. By identifying the products of oxidative cleavage, you can work out the original position of the double bond.

Reaction with itself

Perhaps the most important industrial reaction of ethene is with itself, to form poly(ethene). This is described in Section 15.6.

All the other reactions described in this section are applicable to alkenes in general, not just to ethene.

Some important addition reactions of a double bond are summarised in Figure 15.5.

Quick Questions

10 Deduce the structural formulae of alkenes which give the following products on oxidative cleavage with hot, concentrated MnO$_4^-$.

   a Alkene X, which gives two molecules of CH$_3$COOH.

   b Alkene Y, which gives one molecule of

   CH$_3$—C—CH$_3$
   ‖
   O

   and one molecule of CH$_3$CH$_2$COOH.

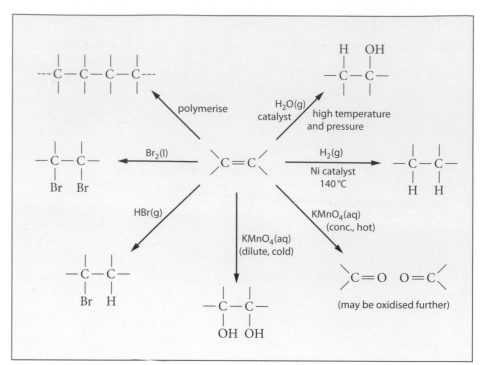

Figure 15.5
Some reactions of a C=C double bond

Quick Questions

11 Predict the outcome of the following reactions:
   a $CH_3CH{=}CH_2 + H_2$ over Ni catalyst
   b $CH_3CH{=}CHCH_3 + H_2O(g)$
   c Excess $CH_3CH{=}CH_2 +$ cold, dilute $KMnO_4$ in acid
   d $CH_3CH{=}CHCH_3 + Br_2$

15.6 Addition polymerisation

Chapter 27 covers polymers in detail. Here we will look at one important type of polymerisation, which occurs when small molecules containing a double bond, such as ethene, join together. The double bond in each molecule can be thought of as opening up and enabling the free bonds to link with one another forming a chain.

Thus with ethene:

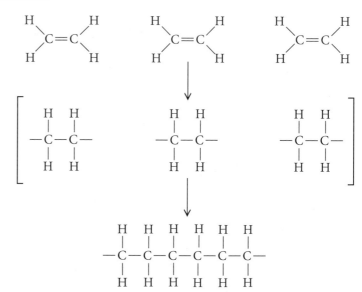

Figure 15.6
These two bottles, one made from high-density polythene and the other from low-density polythene, were heated in an oven. Which is which?

The product, **poly(ethene)** or **polythene**, is the most common synthetic polymer. Notice from the last equation that the empirical formula of ethene is the same as that of its monomer, CH_2. This is always the case with addition polymerisation, but is not so with condensation polymerisation (Section 27.3).

Polythene

Polythene was discovered in 1933. The original method involved heating ethene at about 200 °C and a pressure of 1200 atm in the presence of traces of oxygen. This is a free-radical reaction, in which oxygen acts as an initiator. Organic peroxides, R—O—O—R, are also used to initiate the reaction.

The polythene produced in this way has branched chains which cannot pack closely together (Figure 15.7(a)). It is therefore fairly readily melted and easily deformed. The polymer melts at about 105 °C and softens in boiling water. The majority of polythene manufactured is of this form: it is called **low-density polythene** (LDPE). It is used for making film and sheeting for bags and wrappers and for making moulded articles such as washing-up bowls and 'squeezy' bottles.

Another method of making polythene was developed by Ziegler in the 1950s. This process uses catalysts at low temperatures and pressure (about 60 °C and 1 atm). The molecules produced have few branches and can pack closely together (Figure 15.7(b)). This form of polythene, called **high-density polythene** (HDPE), is more rigid and melts at a higher temperature (about 135 °C) than the low-density form. High-density polythene is used for moulding rigid articles such as bleach bottles and milk-bottle crates.

Poly(propene) (polypropylene)

Propene, which like ethene is readily available from crude oil, can be polymerised by the Ziegler process:

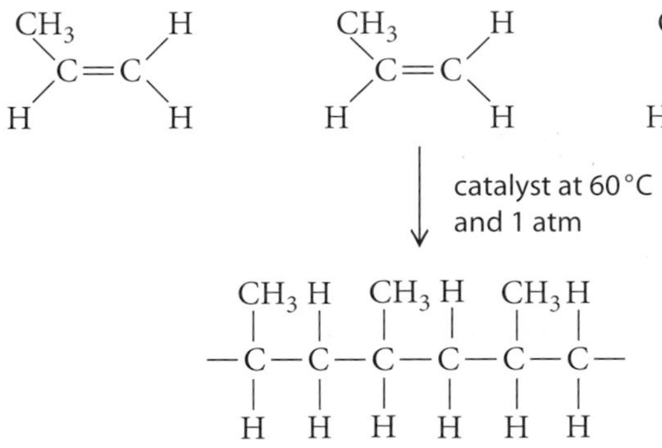

Poly(propene) produced in this way has the —CH$_3$ side groups arranged in a regular way. Its chains pack together closely, producing a material similar to high-density polythene. It is used in mouldings and film and can be made into a fibre.

Poly(chloroethene) (PVC)

Chloroethene (vinyl chloride) polymerises to form poly(chloroethene), also called polyvinyl chloride, PVC:

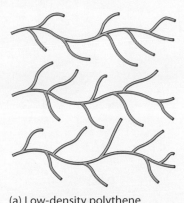

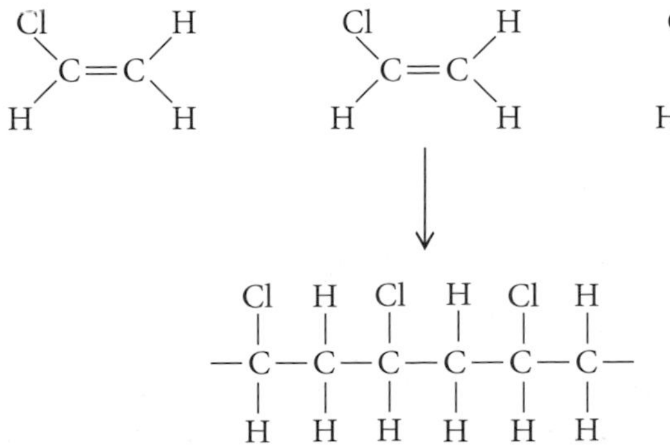

(a) Low-density polythene
Many branched chains, irregularly packed. Largely amorphous with few crystalline regions. Low tensile strength, low melting point

(b) High-density polythene
Few branched chains, regularly packed, with extensive crystalline regions. Higher tensile strength, high melting point

⌃ Figure 15.7
Low-density and high-density polythene

⌃ Figure 15.8
Polypropene is used to make rope. Unlike rope made from natural fibres, it does not rot and it can be any colour.

(a)

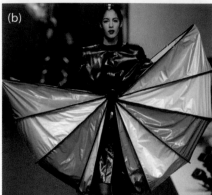

(b)

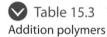
Figure 15.9

Pure PVC is rigid, because of strong intermolecular forces. It can be made more flexible by adding a plasticiser.

(a) Guttering made from unplasticised PVC

(b) Clothing made from plasticised PVC

Key Point

Many compounds containing double bonds undergo addition polymerisation. The most important of these is ethene, which polymerises to polythene.

Table 15.3

Addition polymers

The polar C—Cl bond in PVC results in considerable intermolecular attraction between the polymer chains, making PVC a fairly strong material. Its best feature is its versatility. On it own, PVC is a fairly rigid plastic, but additives called plasticisers can be used to make it flexible. It is used among other things for coating fabrics and for covering wires and cables.

PVC is a versatile polymer, but its disposal can present problems. Burning waste PVC produces hydrogen chloride gas, which is toxic. It must be removed from the combustion gases by reaction with a base before they are released into the atomsphere.

Some other addition polymers, their monomers and their uses are shown in Table 15.3.

Figure 15.10
Furniture made from Perspex

Polymer: systematic name	poly(phenylethene)	poly(tetrafluoroethene)	poly(methyl-2-methyl-propenoate)	poly(propenenitrile)
Polymer: common name	polystyrene	PTFE, 'Teflon', etc.	'Perspex', acrylic	'Acrilan', etc.
Monomer: systematic name	phenylethene	tetrafluoroethene	methyl 2-methyl-propenoate	propenenitrile
Monomer: formula	$C_6H_5CH{=}CH_2$	$CF_2{=}CF_2$	CH_3 C—$COOCH_3$ CH_2	$CH_2{=}CHCN$
Monomer: common name	styrene	tetrafluoroethylene	methyl methacrylate	acrylonitrile
Properties	brittle (but cheap)	very stable, low friction, anti-stick properties	transparent	strong, fibre properties
Uses	expanded polystyrene for insulation, plastic toys, etc.	non-stick coatings on pans. Insulators	as a substitute for glass	making textiles (wool substitute)

15.7 Rubber – a natural addition polymer

Rubbers are **elastomers** – polymers with elastic properties (Figure 15.11).

Many forms of synthetic rubber are available, most of them cheaper and more useful than natural rubber, but the natural polymer is still much used. Natural rubber is as an *addition* polymer. The monomer is 2-methylbuta-1,3-diene (isoprene), whose formula is shown in Figure 15.11.

When dienes of this sort undergo addition polymerisation, one of the double bonds remains in the resulting polymer. The structure of poly(2-methylbuta-1,3-diene) is also shown in Figure 15.11.

Natural rubber comes from the Para rubber tree. When the outer bark is cut, a milky liquid called **latex** oozes out (Figure 15.12). This contains small globules of rubber suspended in water. Some adhesives are made from latex with ammonia added as a preservative. Raw rubber is readily precipitated from latex, but is sticky and soft. In 1832 Charles Mackintosh used rubber to waterproof cotton coats, but these became sticky in hot weather.

It was not until Charles Goodyear discovered vulcanisation in 1839 that rubber became really useful for footwear, tyres, hoses and foam rubber, etc. When raw rubber is heated with sulfur it becomes harder, tougher and less temperature-sensitive. It is said to be **vulcanised**. Sulfur atoms form cross-links between rubber chains, as shown in Figure 15.13.

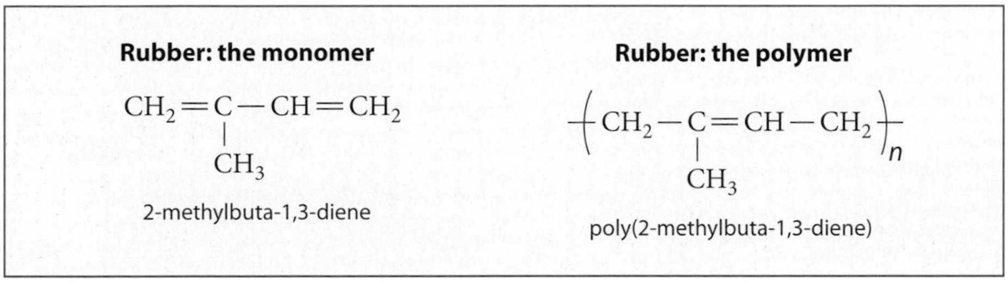

Quick Questions

Use Table 15.3 to answer these questions.

12 Draw a section of the chain structures of poly(phenylethene) and poly(propenenitrile).

13 Suggest a reason why poly(phenylethene) is brittle.

14 Suggest a reason why poly(propenenitrile) is a strong material, suitable for making fibres.

▲ Figure 15.12
Tapping latex from a rubber tree

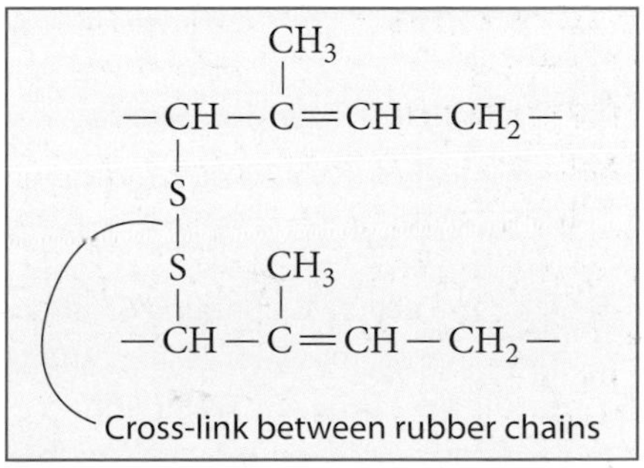

Rubber: the monomer

$$CH_2 = C - CH = CH_2$$
$$\quad\quad |$$
$$\quad\quad CH_3$$

2-methylbuta-1,3-diene

Rubber: the polymer

$$\left(CH_2 - C = CH - CH_2 \right)_n$$
$$\quad\quad\quad |$$
$$\quad\quad\quad CH_3$$

poly(2-methylbuta-1,3-diene)

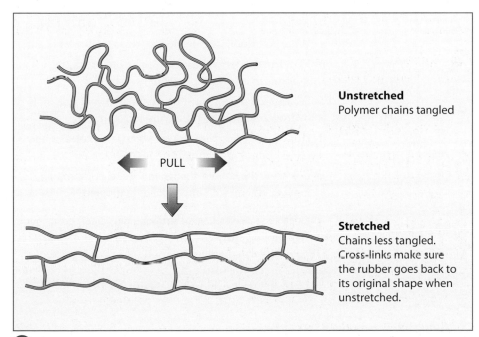

Unstretched
Polymer chains tangled

PULL

Stretched
Chains less tangled. Cross-links make sure the rubber goes back to its original shape when unstretched.

▲ Figure 15.11
Why rubber is elastic

$$CH_3$$
$$\quad |$$
$$- CH - C = CH - CH_2 -$$
$$|$$
$$S$$
$$|$$
$$S \quad\quad CH_3$$
$$| \quad\quad\quad |$$
$$- CH - C = CH - CH_2 -$$

Cross-link between rubber chains

▲ Figure 15.13
Vulcanising involves heating rubber with sulfur to introduce cross-links.

Review questions

1 a Name the following compounds:

   i $CH_3CH_2CH_2CH=CH_2$

   ii

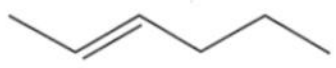

   iii

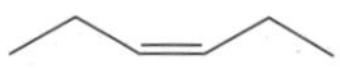

   iv $CH_3C\equiv CCH_3$

   v

b Write structural formulae for these compounds:

   i   propene

   ii *cis*-pent-2-ene

   iii cycloocta–1,3,5,7-tetraene

   iv 2-methylpent-2-ene.

2 1,1-dichloroethene, $Cl_2C=CCH_2$ readily undergoes addition polymerisation.

a What is the systematic name of the polymer?

b Write the structure of a section of the polymer chain.

c Would you expect the polymer to have a higher or a lower melting point than poly(chloroethene) (PVC)? Explain your answer.

3 a For each of the following compounds, say whether you would expect them to show *cis–trans* isomerism. For the compounds that show *cis–trans* isomerism, draw the structures of all the possible isomers.

   i

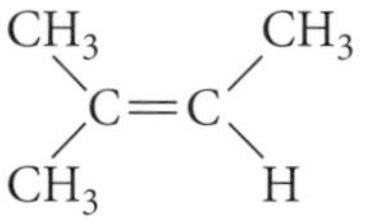

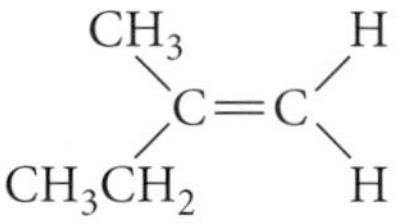

   ii

   iii

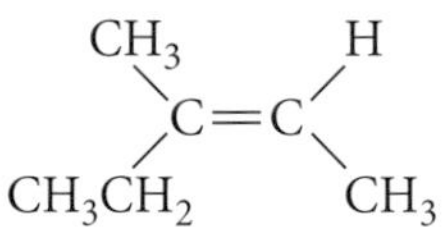

   iv

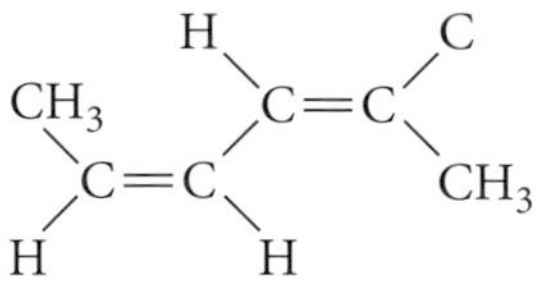

b 1,2-dichlorocyclopropane, CH_2

                 / \\
           $ClCH-CHCl$

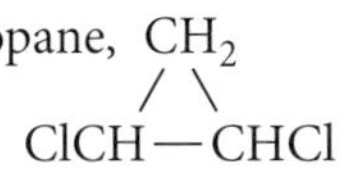

has *cis* and *trans* isomers.

Explain why this is so, and draw the structure of the two isomers.

4 A hydrocarbon **A** contains 87.8% carbon and 12.2% hydrogen by mass. Its relative molecular mass is 82. **A** decolorises bromine water and in the presence of a nickel catalyst it reacts with hydrogen to form **B**. 0.1 g of **A** was found to absorb 27.3 cm^3 of hydrogen (measured at s.t.p.). **B** does not decolorise bromine water.

The molar volume of a gas at s.t.p. is 22.4 dm^3.

a What is the empirical formula of **A**?

b What is the molecular formula of **A**?

c How many moles of hydrogen react with 1 mole of **A**?

d How many double bonds does **A** have in its molecule?

e What is the molecular formula of **B**?

f Suggest structural formulae for **A** and **B**.

5 a How would you distinguish between hex-1-ene and cyclohexane, using simple test-tube reactions?

b Predict the structures of the products of the following reactions.

   i   **ii**

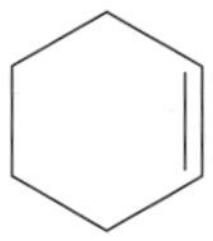

       + $KMnO_4$(aq)   + HI(g)
        (cold, dilute)

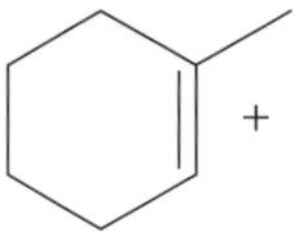

   iii $CH_3CH=CHCH_3$ reacts with itself, i.e. polymerises.

6 Some bond energies are given below.
$C=C$ (in ethene) 598 kJ mol^{-1}
$C-C$ (in ethane) 346 kJ mol^{-1}
$C-H$ (general) 413 kJ mol^{-1}
$H-H$ 436 kJ mol^{-1}

Consider the reaction of ethene with hydrogen:

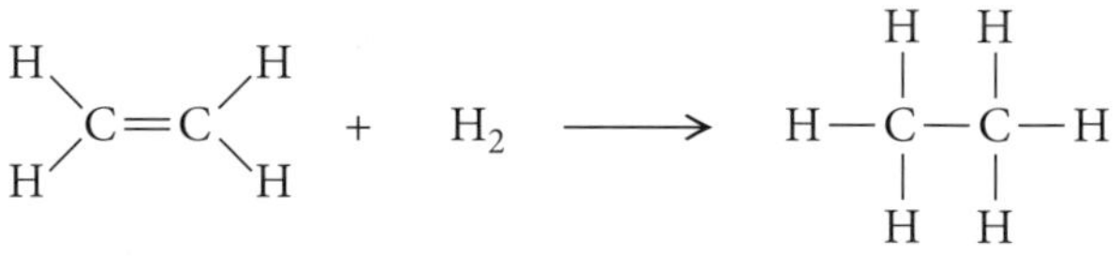

a What conditions are needed for this reaction to occur?

b How much energy must be supplied to break the π-bonds in one mole of ethene?

c How much energy must be supplied to split one mole of hydrogen molecules into hydrogen atoms?

d How much energy is released when two moles of new $C-H$ bonds are formed?

e Calculate the energy change for the reaction of one mole of ethene with one mole of hydrogen.

7 Alkenes such as ethene and propene have been described as the building blocks of the organic chemical industry. Discuss this statement, giving examples. What particular features of the chemistry of alkenes make them suitable for this role and why are alkanes less suitable?

8 Consider the following compounds.

A hex-2-ene $CH_3CH_2CH_2CH{=}CHCH_3$

B hex-1-ene, $CH_3CH_2CH_2CH_2CH{=}CH_2$

C hexane $CH_3CH_2CH_2CH_2CH_2CH_3$

D cyclohexane

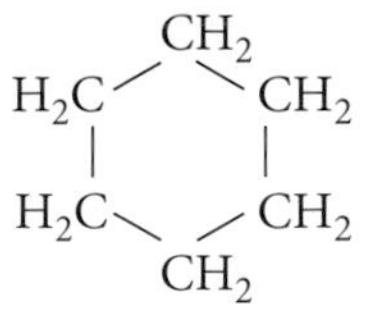

The answers to the following questions may be one or more than one of the compounds **A–D**.

a Which would decolorise bromine in the absence of sunlight?

b Which would react with chlorine, but only when heated or exposed to light?

c Which would absorb 1 mole of hydrogen per mole of the compound in the presence of a nickel catalyst?

d Which has *cis* and *trans* isomers?

e Which are unsaturated?

9 When ethene is reacted with HCl, C_2H_5Cl is the only product.

a Using structural formulae, give an equation for the reaction between ethene and HCl.

b What type of reaction occurs between HCl and ethene?

c Explain why there is no further reaction between C_2H_5Cl and HCl.

Cambridge Paper 2 Q1d June 2007

10 Polychloroethene (polyvinyl chloride, PVC), which is commonly used in water pipes and food wrap, is formed from chloroethene (vinyl chloride ($H_2C{=}CHCl$) in an addition polymerisation reaction.

a Explain the term 'polymerisation'.

b Draw a reaction scheme for the formation of PVC from its monomer, showing three repeat units.

c What characteristics does the monomer need to have to undergo an addition polymerisation reaction?

d From what monomer would the following polymer be formed?

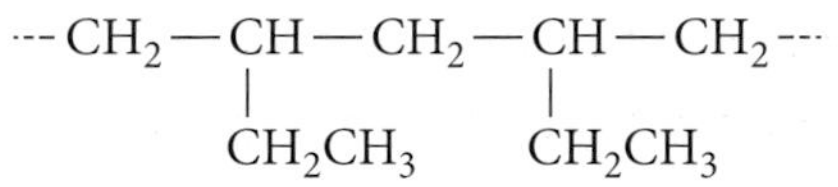

$$\cdots CH_2{-}CH{-}CH_2{-}CH{-}CH_2\cdots$$
$$\quad\quad\ |\quad\quad\quad\quad |$$
$$\quad CH_2CH_3\quad\ CH_2CH_3$$

16 Organic halogen compounds

16.1 Anaesthetics

Before anaesthetics, surgery was a savage and primitive affair. Anaesthetics enabled surgery to develop from crude carpentry to its present sophisticated form.

Three of the most important early anaesthetics were nitrous oxide (dinitrogen oxide, N_2O), ether (ethoxyethane, $CH_3CH_2OCH_2CH_3$) and chloroform (trichloromethane, $CHCl_3$). Nitrous oxide is non-toxic and non-flammable, but it only produces light anaesthesia. Ether is an effective anaesthetic but it is highly flammable and therefore dangerous. Chloroform produces deep anaesthesia and is non-flammable, but it is toxic and carries the risk of liver damage.

The ideal inhalant anaesthetic must be a gas or volatile liquid, so that it can be inhaled and absorbed via the lungs. It must be non-flammable. It must produce deep anaesthesia and it must be non-toxic. In 1951, chemists began the search for a new anaesthetic. They decided to look at **halogenoalkanes**.

Chemists knew that the substitution of chlorine atoms into an alkane increased its anaesthetic properties, but also made it toxic. Thus dichloromethane, CH_2Cl_2, is a fairly weak anaesthetic with little toxicity, trichloromethane (chloroform), $CHCl_3$, is stronger and more toxic, while tetrachloromethane is a very strong anaesthetic and also very toxic.

The introduction of halogen atoms into an alkane skeleton also tends to make it non-flammable. Fluorine is useful in this respect as the C—F bond is very stable. Fluoroalkanes are inert, non-flammable and non-toxic. Chemists therefore looked for a short-chain halogenoalkane containing fluorine for inertness and chlorine for anaesthetic properties, with a suitable boiling point in the range 40 °C to 60 °C. They concentrated on two-carbon molecules, and after many trials produced the compound shown in Figure 16.1.

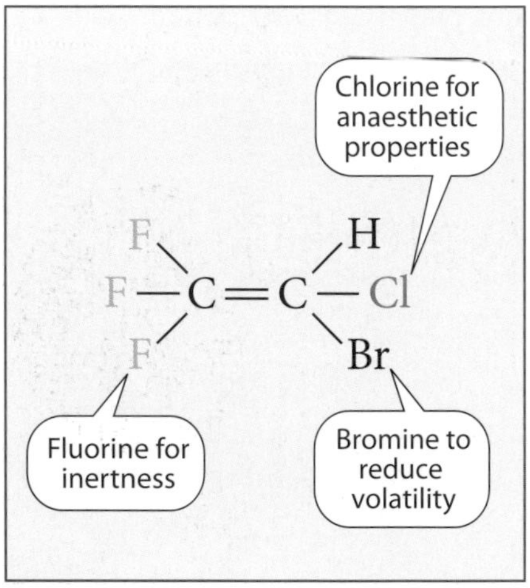

This substance was given the name Halothane. Following its discovery in 1956 it became widely used in hospitals. It is still widely used across the world, but is being replaced by safer anaesthetics based on fluorocarbons.

Halothane illustrates several of the important properties of halogenoalkanes. It shows the increasing reactivity of the C—Hal bond as we go from F to I; the decreasing volatility of R—Hal as we move in the same direction; and the effect of halogen atoms in reducing the flammability of a hydrocarbon.

Although organic halogen compounds are uncommon in nature, a study of their properties is important to chemists as they have many uses in industry and in the laboratory.

![Chlorine for anaesthetic properties; Fluorine for inertness; Bromine to reduce volatility — structural diagram of Halothane]

⌃ Figure 16.1
Halothane
(2-bromo-2-chloro-1,1,1-trifluoroethane)

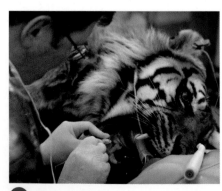

⌃ Figure 16.2
A tiger having a false tooth fitted under Halothane anaesthetic

16.2 Naming halogen compounds

Organic halogen compounds contain the **halogeno** functional group: —F, —Cl, —Br, —I. They are named using the prefixes **fluoro-**, **chloro-**, **bromo-** and **iodo-**. Numbers are used if necessary to indicate the position of the halogen atom in the molecule. Thus,

$$CH_3CH_2Cl \qquad \text{chloroethane,}$$

$$CH_3CHBrCH_3 \qquad \text{2-bromopropane}$$

If the molecule contains more than one halogen atom of the same kind, the prefixes **di-**, **tri-**, etc., are used. Thus,

$$CH_2ClCH_2Cl \qquad \text{1,2-dichloroethane}$$

$$CHCl_2CHClCH_3 \qquad \text{1,1,2-trichloropropane}$$

Quick Questions

1 Name these compounds:
 a $CH_3CH_2CHICH_3$
 b CH_3CHCl_2
2 Write formulae for these compounds:
 a 1,3,5-tribromocyclohexane
 b 1,2-dibromo-3-chloropropane

Acyl halides contain the $-\overset{\displaystyle\parallel}{\underset{\displaystyle O}{C}}-Hal$ group. They are covered in Section 18.6.

16.3 The nature of the carbon–halogen bond

Unreactive halogenoalkanes: the C—F bond

The C—F bond is very strong: compare its bond energy of 485 kJ mol^{-1} with 435 kJ mol^{-1} for C—H and 327 kJ mol^{-1} for C—Cl. The C—F bond is thus very unreactive. So fluorocarbons, compounds containing fluorine and carbon only, are extremely inert. The C—Cl bond is more reactive than C—F, but highly chlorinated compounds such as CCl_4 and $CHCl_3$ are, nevertheless, fairly inert. In particular they are non-flammable.

Compounds containing carbon, hydrogen and fluorine are called **fluoroalkanes** or hydrofluorocarbons (HFCs). Because of their low reactivity, they are very useful as fire extinguishers, aerosol propellants and refrigerants. An example is 1, 1, 1, 2–tetrafluoroethane:

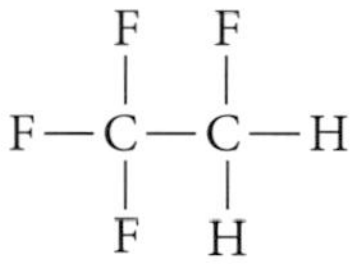

$$\overset{\displaystyle F}{\underset{\displaystyle F}{F-C}}-\overset{\displaystyle F}{\underset{\displaystyle H}{C}}-H$$

With a boiling point of –26 °C, it is an excellent refrigerant.

Chlorofluorocarbons and the ozone layer

Chlorofluorocarbons (CFCs) contain chlorine as well as fluorine. An example is CCl_2F_2, dichlorodifluoromethane.

At one time, CFCs were used widely as refrigerants, solvents and aerosol propellants. It is now known that CFCs are responsible for serious damage to the stratospheric ozone layer which absorbs most of the Sun's harmful ultraviolet radiation before it can reach us on Earth.

The trouble with CFCs is that they are *too* unreactive. Once they have entered the atmosphere, from an aerosol spray, an old refrigerator or whatever, they stay there and do not break down for many years. Slowly the molecules of CFCs diffuse upwards

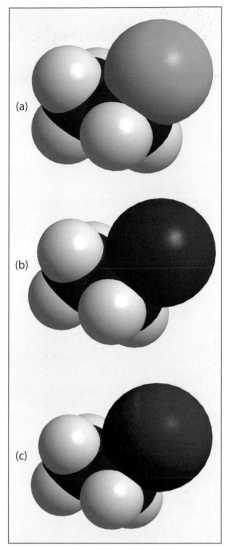

(a)

(b)

(c)

⌃ Figure 16.3
Molecules of (a) chloroethane, (b) bromoethane and (c) iodoethane.

Key Point

The C—F bond is very strong, so fluoroalkanes are very stable and unreactive.

Figure 16.4
Babies' skin is very sensitive to ultraviolet radiation. They can be protected by anti-UV hats and clothing.

Figure 16.5
This large plastic balloon filled with helium will carry instruments for measuring the ozone concentration over the Arctic.

Key Point

CFCs break down in the stratosphere to produce reactive Cl· radicals, which destroy ozone.

Table 16.1
Boiling points of some halogen compounds

Compound	State at 298 K	Boiling point/K
CH_3F	g	195
CH_3Cl	g	249
CH_3Br	g	277
CH_3I	l	316
CH_2Cl_2	l	313
$CHCl_3$	l	335
CCl_4	l	350
C_6H_5Cl	l	405

until they reach the upper atmosphere (the stratosphere). This part of the atmosphere is constantly bombarded with powerful ultraviolet radiation, some of which is just the right frequency to break the relatively weak C—Cl bond in CFCs. This results in homolytic fission of the C—Cl bond, producing Cl· radicals. It is these free radicals that are responsible for damage to the ozone layer.

This is what happens:

Ozone, O_3, is continuously formed in the stratosphere by reaction between O· radicals and O_2 molecules. First, O· radicals are formed when ultraviolet radiation splits up O_2 molecules. Then these radicals react with more O_2 molecules to make O_3. The two reactions are

$$O_2(g) \longrightarrow 2O\cdot(g) \tag{1}$$

Then

$$O\cdot(g) + O_2(g) \longrightarrow O_3(g) \tag{2}$$

This is a natural process that goes on all the time. It produces O_3 that absorbs harmful ultraviolet radiation before it reaches the Earth.

However, when Cl· radicals are present in the stratosphere, they react with ozone, forming O_2 and ClO· radicals.

$$O_3(g) + Cl\cdot(g) \longrightarrow O_2(g) + ClO\cdot(g) \tag{3}$$

The ClO· radicals then react with O· radicals

$$ClO\cdot(g) + O\cdot(g) \longrightarrow Cl\cdot(g) + O_2(g) \tag{4}$$

If you look closely at reactions (3) and (4), you will see that together they have the effect of removing both O_3 and O·. What is more, the Cl· radicals get regenerated by reaction (4). In effect they never get used up, so they are acting as a catalyst in converting O_3 and O· to O_2.

In fact, one Cl· radical can catalyse the breakdown of about 1 million O_3 molecules, so you can see why CFCs have such a devastating effect on the ozone layer.

During the 1980s, measurements of the concentration of ozone in the stratosphere showed that the ozone layer was thinning and developing a 'hole', especially over the Antarctic. Chemists became increasingly convinced that this 'hole' was due to the effects of CFCs, and in 1990 more than 60 countries signed an agreement to phase out the use of CFCs by the year 2000. Even so, because CFCs persist in the atmosphere for so long, scientists estimate that it will be many more years before the ozone layer returns to the condition it was in before the trouble started. CFCs have now been replaced by other compounds, especially fluoroalkanes.

Compounds containing C—F bonds are not typical of the halogenoalkanes as a whole, which tend to be much more reactive than fluoroalkanes. The reactions of halogenoalkanes are considered in the next section.

Physical properties of halogenoalkanes

The C—Hal bond is polar, but not polar enough to have an appreciable effect on the physical properties of organic halogen compounds. They are all immiscible with water. As Table 16.1 shows, their volatility is determined by the size and number of halogen atoms they contain. Thus iodomethane is a liquid at room temperature while chloromethane is a gas, even though chloromethane has a more polar molecule.

16.4 Nucleophilic substitution

Reactive halogenoalkanes: the C—Br bond

To get some idea of the nature of the C—Hal bond, we will consider some bromine-containing compounds.

Table 16.2 compares the effect of adding silver nitrate solution to sodium bromide and 1-bromobutane. In the case of 1–bromobutane, vigorous shaking is necessary since it does not mix appreciably with the aqueous phase.

Table 16.2

The effect of aqueous silver nitrate on bromine-containing compounds at room temperature

Sodium bromide	1-bromobutane
pale yellow precipitate appears immediately	no reaction at first: faint precipitate appears after several minutes

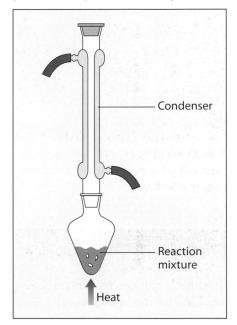

Figure 16.6

Reflux apparatus used for boiling an organic reaction mixture. In this case, the reactants are bromoethane and aqueous sodium hydroxide. The condenser prevents escape of volatile reagents.

— Condenser

— Reaction mixture

Heat

Quick Questions

3 Look at Table 16.2 and answer these questions.
   a What is the pale yellow precipitate produced in the reaction between silver nitrate and sodium bromide?
   b Write an ionic equation for this reaction.
   c Why does silver nitrate produce no immediate precipitate with 1-bromobutane, even though it contains bromine?
   d Suggest a reason why a precipitate appears after several minutes.

The C—Br bond in 1-bromobutane is covalent. 1–bromobutane therefore contains no Br^- ions, so it does not produce a precipitate of silver bromide with silver nitrate. The slow appearance of a precipitate of silver bromide suggests that Br^- ions are slowly being produced. Why? To explain this, we need to look at the nature of the C—Br bond.

Bromine is more electronegative than carbon, so the C—Br bond is polar:

$$\overset{\delta+}{\underset{}{C}}-\overset{\delta-}{Br}$$

The partial positive charge on the carbon atom tends to attract **nucleophiles**, such as NH_3, ^-OH and H_2O, with their lone pairs of electrons. Water molecules from the aqueous silver nitrate can act as nucleophiles since their oxygen atoms carry two lone pairs and a partial negative charge:

$$\overset{\delta+}{\underset{H}{\overset{H}{\diagdown}}}O\text{:}\;\;\delta-$$

The water attacks the partially positive carbon atom in 1-bromobutane and a substitution reaction takes place, releasing bromide ions:

$$CH_3CH_2CH_2\overset{}{\underset{H}{\overset{H}{\diagdown}}}\overset{\delta+}{C}\overset{\delta-}{-Br} \longrightarrow CH_3CH_2CH_2\overset{}{\underset{H}{\overset{H}{\diagdown}}}\overset{+}{C}-O\overset{H}{\diagup}_H \longrightarrow CH_3CH_2CH_2\overset{}{\underset{H}{\overset{H}{\diagdown}}}C-OH + H^+ + Br^-$$

butan-1-ol

This is an example of a **nucleophilic substitution** reaction. Such reactions are typical of halogenoalkanes and their mechanism is discussed further below.

The reaction of halogenoalkanes with $^-OH(aq)$

The reaction of halogenoalkanes with H_2O is slow, but if ^-OH is used as a nucleophile instead of H_2O it is quicker. Thus bromoethane forms ethanol quite rapidly when heated under reflux with aqueous sodium hydroxide (Figure 16.6).

$$CH_3CH_2Br + {}^-OH \longrightarrow CH_3CH_2OH + Br^-$$

bromoethane   ethanol

Key Point

The typical reactions of halogenoalkanes involve **nucleophilic substitution**.

Table 16.3
Average bond energies. The weaker the bond between the carbon and halogen atoms, the easier it is to hydrolyse.

Bond	Average bond energy/kJ mol^{-1}
C—F	484
C—Cl	338
C—Br	276
C—I	238

Quick Questions

4 Predict the organic product of the reactions when aqueous sodium hydroxide is boiled with:
   a 2-chloropropane
   b 1,2-dibromoethane

The substitution reaction outlined above between CH_3CH_2Br and ^-OH is general for all halogenoalkanes. It can be used as a method for preparing alcohols:

$$RHal + {}^-OH \longrightarrow ROH + Hal^-$$

For a given alkyl group — R, the iodo compound reacts most readily, the bromo compound less so and the chloro compound reacts least readily. This is because the C—Hal bond becomes progressively stronger from I to Cl (Table 16.3).

Testing for halogenoalkanes

We can use alkaline hydrolysis to find out which halogen is present in a halogenoalkane. The halogenoalkane is boiled with aqueous sodium hydroxide, then cooled and neutralised with dilute nitric acid. Then silver nitrate solution is added. A white precipitate of silver chloride indicates a chloroalkane; a pale yellow precipitate indicates a bromoalkane and a yellow precipitate indicates an iodoalkane. For example:

$$RCl + {}^-OH \longrightarrow ROH + Cl^-$$

$$Ag^+(aq) + Cl^-(aq) \longrightarrow AgCl(s)$$

The mechanism of nucleophilic substitution

Chemists have studied the mechanism of reactions between halogenoalkanes and hydroxide ions extensively. It involves nucleophilic attack by the ^-OH ion, e.g.

The actual mechanism is more complex than the simplified version shown here. Chemists believe that there are two different mechanisms possible:

▸ a step-by-step, two-stage mechanism known as S_N1
▸ a single step mechanism, known as S_N2.

The evidence for the proposed mechanisms, and the reasons for the terms 'S_N1' and 'S_N2', are explained below (see 'How do we know?').

We will look at the two mechanisms in turn, using a generalised substitution reaction:

(R^1, R^2 and R^3 are alkyl groups or hydrogen atoms.)

The step-by-step mechanism (S_N1)

In this case, a two-stage mechanism is involved:

The first step, which is relatively slow, involves breaking the C—Br bond to form the intermediate carbocation. The carbocation is very unstable and reactive, so the second step is fast. The overall rate of the reaction is determined by the slow first step – the **rate determining step** (Section 22.9).

The single-step mechanism (S_N2)

Here the reaction occurs in a single step. The ^-OH ion is attracted to the central carbon atom, and as it moves in, it repels the Br atom. At some 'middle' stage in the reaction, Br and OH are both partially bonded to the carbon, the OH on its way in and the Br on its way out. This is the **transition state** of the reaction. It is *not* a reaction intermediate which exists independently, but simply the middle stage in a continuous process during which the ^-OH moves in and the Br^- moves out.

$$
\begin{array}{c}
\underset{Br}{\overset{R^1}{\underset{R^2}{\big|}}}\!\!C\!\!\underset{R^3}{}\ +\ ^-OH
\end{array}
\longrightarrow
\left[Br\text{----}\underset{R^2\quad R^3}{\overset{R^1}{C}}\text{----}OH \right]
\longrightarrow
Br^-\ +\ \underset{R^2}{\overset{R^1}{}}\!\!C\!\!\underset{R^3\ OH}{}
$$

transition state

Notice that the transition state involves *five* groups around the central carbon atom.

Figure 16.7 illustrates the difference between S_N1 and S_N2 in another way.

Key Point

S_N1 mechanisms involve a two-step nucleophilic substitution process.

S_N2 mechanisms involve a one-step process.

Quick Questions

5 In the S_N1 mechanism, suggest a reason why the second stage is faster than the first stage.

6 In the S_N2 mechanism, what difference might it make if R^1, R^2 and R^3 are:
   a big, bulky alkyl groups,
   b H atoms?

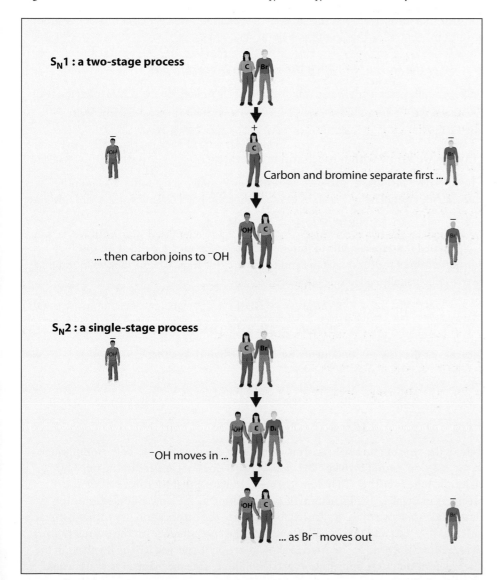

S_N1 : a two-stage process

Carbon and bromine separate first ...

... then carbon joins to ^-OH

S_N2 : a single-stage process

^-OH moves in ...

... as Br^- moves out

Figure 16.7
Breaking up is hard to do ...

Which mechanism operates in practice?

In many nucleophilic substitution reactions, both S_N1 and S_N2 mechanisms operate at the same time, but in most cases one proceeds much faster than the other. Which of the mechanisms is dominant depends on a number of factors. Two of the most important factors are the structure of the halogenoalkane and the nature of the solvent.

Structure of the halogenoalkane

In the two-stage S_N1 mechanism, the rate of the reaction is determined by the ease with which the intermediate carbocation forms. If the substituent groups R^1, R^2, and R^3 are all alkyl groups rather than H atoms, they will tend to donate electrons by the inductive effect (Section 15.5). This stabilises the carbocation and favours its formation.

$$R^2 - \underset{\underset{R^3}{|}}{\overset{\overset{R^1}{|}}{C}} - Br \longrightarrow R^2 - \underset{\underset{R^3}{|}}{\overset{\overset{R^1}{|}}{C^+}} + Br^-$$

If some or all of R^1, R^2 and R^3 are hydrogen atoms, the formation of the carbocation will be less favoured, so the rate of the reaction will be slower. Thus, the two-step S_N1 mechanism is favoured by the presence of substituent alkyl groups.

On the other hand, the single step S_N2 mechanism is favoured by the presence of substituent hydrogen atoms. The small hydrogen atoms make it easier to fit five groups around the central carbon in the transition state.

Nature of the solvent in which the reaction is carried out

Polar solvents, particularly water, stabilise ions by hydrating them. This favours the two-step S_N1 mechanism. Conversely, the single-step mechanism, S_N2, is favoured in non-polar solvents.

How do we know which mechanism operates?

Chemists have investigated the mechanism of nucleophilic substitution reactions by studying the rates of many reactions between different bromoalkanes and hydroxide ions.

$$RBr + {}^-OH \longrightarrow ROH + Br^-$$

They carried out experiments to find how the reaction rate depended on the concentration of the bromoalkane and the hydroxide ions. Chemists have found that two results (usually called rate laws) are possible.

(A) Rate $= k[RBr]$ (i.e. rate law does not involve ^-OH)

(B) Rate $= k[RBr]\,[^-OH]$ (i.e. rate law includes ^-OH)

It turns out that the reaction involving 2,2-dimethylbromoethane

$$CH_3 - \underset{\underset{CH_3}{|}}{\overset{\overset{CH_3}{|}}{C}} - Br + {}^-OH \longrightarrow CH_3 - \underset{\underset{CH_3}{|}}{\overset{\overset{CH_3}{|}}{C}} - OH + Br^-$$

follows the type of rate law in **A**. This means that the slowest, rate-determining step of the reaction cannot involve ^-OH at all, since it does not appear in the rate law equation, Rate $= k[RBr]$. Therefore, chemists conclude that for this reaction the mechanism involves the formation of a carbocation as the slow, rate-determining step – this is the two-stage S_N1 mechanism. Once the carbocation has formed, it reacts quickly with ^-OH and this quick second stage has no influence on the overall rate of the reaction. The term S_N1 is used to show that only **1** species, RBr, is involved in the rate-determining step: **Substitution Nucleophilic 1**. This mechanism is as we would expect for 2,2-dimethylbromoethane, because the three methyl groups stabilise the intermediate carbocation.

However, chemists found that the corresponding reaction involving bromomethane

$$CH_3Br + {}^-OH \longrightarrow CH_3OH + Br^-$$

is second order, with the type of rate law in **B**. This means that the rate-determining step must involve both ^-OH and RBr, since both appear in the rate law equation. Therefore, we conclude that for this reaction the mechanism involves a single step in which the two reactants form a transition state: an S_N2 mechanism. The term S_N2 is used to show that **2** species, RBr and ^-OH, are involved in the rate-determining step – **S**ubstitution **N**ucleophilic **2**.

Chemists have used such rate law studies to investigate the reaction mechanisms for reactions involving many different types of compounds and there is more about this in Chapter 22.

Quick Questions

7 Consider the reaction between bromobutane and I^- ions:
$$CH_3CH_2CH_2CH_2Br + I^- \longrightarrow CH_3CH_2CH_2CH_2I + Br^-$$
The reaction is carried out in a propanone solvent. The rate law for this reaction is found to be
Rate = $k[CH_3CH_2CH_2CH_2Br]\,[I^-]$
a Which mechanism, S_N1 or S_N2, operates in this reaction? How do you know?
b Describe the mechanism of the reaction.

16.5 Important substitution reactions of halogenoalkanes

Organic halogen compounds have important uses. Several of these have already been mentioned, and their use as insecticides is considered in Section 11.9.

Organic halogen compounds are uncommon in nature so they have to be synthesised. They are usually manufactured from alkenes or alkanes. Chloroethane, for example, is made by the addition reaction between ethene and hydrogen chloride. In the laboratory, halogenoalkanes are usually prepared from alcohols (Section 17.5).

As well as being useful in their own right, halogenoalkanes are important intermediates in synthesis. They can be used to introduce a reactive site into a hydrocarbon molecule. The reactive halogen can then be substituted by another group which could not be introduced directly. Examples of groups that can be introduced in this way are discussed later in this section.

This kind of synthesis is important in small-scale preparations such as those carried out in the laboratory or in the manufacture of pharmaceuticals. Many drugs have complicated organic molecules and their synthesis involves building up a complex molecule from a simple starting compound. These syntheses often involve many steps.

The synthetic methods described in this chapter are used for small-scale operations, but they are relatively expensive and therefore not used in large-scale, high-tonnage industrial operations. On this scale high-pressure, high-temperature catalytic processes, very different from those used in the laboratory, tend to be chosen. These processes involve expensive plant and equipment, but cheap reagents. In the long run they are cheaper than small-scale laboratory processes.

Halogenoalkanes undergo substitution reactions with a wide range of nucleophiles. Species that can act as nucleophiles include not only those carrying a full negative charge (^-OH, ^-CN, CH_3COO^-, $CH_3CH_2O^-$) but also neutral molecules carrying an unshared pair of electrons (H_2O, NH_3).

Figure 16.8
Small-scale synthesis of a pharmaceutical. Laboratory methods are used and the reagents are often expensive.

Quick Questions

8 Predict the structure of the compound formed by the reaction of bromoethane with each of the following nucleophiles. Hints to the answers are given below, but try not to look at these until you have predicted the structures yourself.
a ^-CN
b $CH_3CH_2O^-$
c NH_3

Figure 16.9
Chloroalkanes are excellent solvents. Dichloromethane will even dissolve dried paint, so it is used as paint stripper. Unlike trichloromethane and tetrachloromethane, its toxicity is low, so it is safer to use.

Quick Questions

9 Describe the reagents and conditions you would use to prepare:
  a propane-1,2-diol from 1,2-dibromopropane,
  b 2-methylpropanenitrile from 2-iodopropane,
  c methylamine from bromomethane.

Reaction with hydroxide ion

This has already been dealt with (Section 16.4).

Reaction with water

This has already been dealt with (Section 16.4). In the case of bromoethane the product is ethanol, though the reaction is slow even when heated:

$$CH_3CH_2Br + H_2O \longrightarrow CH_3CH_2OH + HBr$$
ethanol

Reaction with cyanide ion

When bromoethane is heated under reflux with a solution of potassium cyanide in ethanol, propanenitrile is formed:

$$CH_3CH_2Br + {}^-CN \longrightarrow CH_3CH_2CN + Br^-$$
propanenitrile

This reaction is useful in synthesis as a means of increasing the length of a carbon chain.

Reaction with ammonia

Ammonia is a nucleophile because of the lone pair of electrons on its nitrogen atom. When bromoethane is heated with a concentrated aqueous solution of ammonia in a sealed tube, ethylamine is formed:

$$CH_3CH_2Br \, (l) + NH_3 \, (aq) \longrightarrow CH_3CH_2NH_2 \, (aq) + HBr \, (aq)$$
ethylamine

The nitrogen atom in ethylamine still possesses a lone pair, so it can still act as a nucleophile. With excess bromoethane, further substitution therefore occurs:

$$CH_3CH_2Br + CH_3CH_2NH_2 \longrightarrow CH_3CH_2NHCH_2CH_3 + HBr$$
ethylamine   diethylamine

And so on.

The reactions mentioned above have all been illustrated by reference to bromoethane, but they are applicable to halogenoalkanes generally. Many different compounds can be synthesised from halogenoalkanes.

16.6 Elimination reactions

In the reactions we have considered so far, $^-$OH has acted as a nucleophile in its reactions with halogenoalkanes. For example, when 2-bromopropane is heated with aqueous sodium hydroxide, the hydroxide ion acts as a nucleophile and substitution occurs:

$$CH_3-\underset{\underset{\text{2-bromopropane}}{}}{\overset{\overset{Br}{|}}{CH}}-CH_3 \, (aq) + {}^-OH \, (aq) \longrightarrow CH_3-\underset{\underset{\text{propan-2-ol}}{}}{\overset{\overset{OH}{|}}{CH}}-CH_3 \, (aq) + Br^- \, (aq)$$

Under certain conditions, however, $^-$OH can act as a *base* instead of a nucleophile. When $^-$OH behaves in this way, it removes H^+ from a halogenoalkane. The C—Br bond breaks at the same time as $^-$OH removes H^+ from the neighbouring carbon atom, and an alkene is formed. For example, when bromoethane is heated with a solution of sodium hydroxide in *ethanol* instead of in water, ethene is formed:

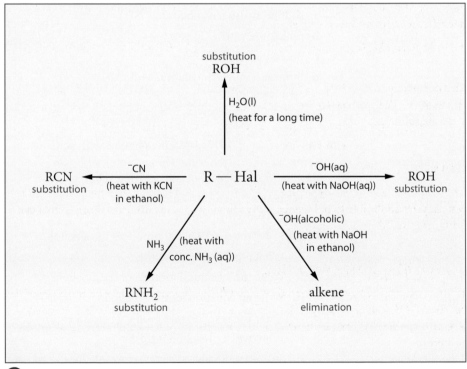

i.e. $CH_3CH_2Br + {}^-OH \longrightarrow CH_2{=}CH_2 + H_2O + Br^-$

The overall effect of this reaction is to eliminate HBr from the molecule of bromoethane, so this is an **elimination reaction**.

Notice that the reagent, sodium hydroxide, is the same for both substitution and elimination reactions. In the substitution reaction, ^-OH acts as a nucleophile; in the elimination reaction it acts as a base. By altering the conditions, we can alter the manner in which it acts: in aqueous solution it behaves as a nucleophile, causing substitution; in ethanolic solution it behaves as a base, causing elimination. Under each set of conditions, both substitution and elimination can in fact occur, but by controlling the conditions we can ensure that one particular reaction occurs to a greater extent than the other.

Figure 16.10 summarises the typical reactions of halogenoalkanes.

Key Point

^-OH ion can react with halogenoalkanes in two ways:

▶ as a nucleophile, substituting the halogen atom with ^-OH, in a substitution reaction,
▶ as a base, removing H^+ and causing an elimination reaction.

Quick Questions

10 Try to predict the effect of an elimination reaction in a halogenoalkane that contains *two* halogen atoms. For example, what would be the product if 1,2-dibromoethane were heated under reflux with a solution of sodium hydroxide in ethanol?

▲ Figure 16.10
Typical reactions of halogenoalkenes (Hal=Cl, Br, or I).

Review questions

1 Consider the following compounds.

   A CCl_3F

   B CF_3-CF_3

   C CH_3CH_2Cl

   D $CH_3CHBrCH_3$

   a Name each compound.

   b Which compound would react *most* readily with aqueous sodium hydroxide?

   c Which compound would react *least* readily with aqueous sodium hydroxide?

   d Which compound(s) would undergo an elimination reaction when heated under reflux with NaOH in ethanol?

   e Which would be suitable for use as a refrigerant?

2 This question concerns the hydrolysis of three different halogenoalkanes:

1-chlorobutane

1-bromobutane

1-iodobutane.

Four drops of each halogenoalkane are added separately to three separate tubes standing in a water-bath at 60 °C. Each tube contains 1 cm^3 of 0.1 mol dm^{-3} silver nitrate solution. The results are as follows:

1-Chlorobutane – slight cloudiness after three minutes. Still only slightly cloudy after 15 minutes.

1-Bromobutane – slightly cloudy after one minute, opaque after three minutes, coagulation and precipitation after six minutes.

1-Iodobutane – immediately opaque, yellow precipitate within first minute.

   a What is the precipitate formed in each case?

   b Why does the precipitate form?

   c In each case a substitution reaction is occurring between the halogenoalkane and a nucleophile. What is the nucleophile involved?

   d Which halogenoalkane undergoes substitution most readily and which least readily?

   e Explain your answer to part d, given the following bond energies in kJ mol^{-1}:

     C—Cl (in chloroethane) 339

     C—Br (in bromoethane) 284

     C—I (in iodoethane) 218

3 Predict the products of reactions between the following pairs of substances.

   a $CH_3CH_2CH_2Br$ and NH_3

   b $CH_2ICH_2CH_2I$ and $^-OH(aq)$

   c CH_3Br and KCN (ethanolic)

   d 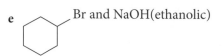 Br and NaOH(aq)

   e Br and NaOH(ethanolic)

4 Use your knowledge of nucleophilic substitution reactions to predict the structural formulae of the products of reactions between bromoethane, CH_3CH_2Br, and the following.

   a sodium hydrogensulfide ($Na^{+-}SH$)

   b potassium nitrite ($K^+NO_2^-$)

   c potassium chloride

   d lithium tetrahydridoaluminate, $LiAlH_4$ (nucleophile = H^-)

   e sodamide ($Na^{+-}NH_2$).

5 The reaction of 3-chloro-3-ethylpentane with aqueous sodium carbonate at room temperature yields two products, one an alcohol and one an alkene.

   a Write the structural formula of:
3-chloro-3-ethylpentane

   b Consider the reaction that produces the alcohol.

     i What type of reaction is this?

     ii Write the structural formula of the alcohol.

     iii Write an equation for the reaction.

   c Consider the reaction that produces the alkene.

     i What type of reaction is this?

     ii Suggest a possible structural formula for the alkene.

     iii Write an equation for the reaction.

   d What changes in reaction conditions might favour the formation of the alkene?

6 Identify the compounds **A** to **D** and the reagents **P** and **Q** in the reaction scheme below.

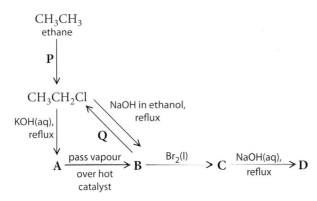

7 'Teflon' is used as a non-stick coating for cookware. To withstand the high temperatures, Teflon must be non-reactive and its structure includes very strong carbon–halogen bonds. Use the bond energies in Table 16.3 to answer the following questions:

a Which halogen do you think is most likely to be found in the structure of Teflon?

b Which halogen is least likely to be found in Teflon?

Teflon was produced accidently whilst trying to synthesise a new refrigerant. Hydrofluoroalkanes (HFAs) are commonly used as refrigerants and propellants.

c What properties of HFAs made them advantageous as refrigerants?

d Why are HFAs more beneficial to use than chlorofluorocarbons (CFCs)?

8 Nucleophilic substitution reactions occur either via an S_N1 or S_N2 mechanism.

a What is the difference between an S_N1 and an S_N2 mechanism?

b Which step of an S_N1 reaction occurs very quickly and why?

c Reactions taking place in water usually proceed via an S_N1 mechanism. Why is this?

d In an S_N2 reaction, which of the following species would be the most reactive? Explain your answer.

   A CH_3Cl **B** CH_3Br

   C CH_3I **D** CH_3F

9 Halogenoalkanes can undergo either nucleophilic substitution reactions or elimination reactions when they react with hydroxide ions, such as in NaOH(aq).

a ⁻OH acts as a nucleophile in nucleophilic substitution reactions. How does it behave in an elimination reaction?

b If you wanted to perform an elimination reaction, what would be an appropriate solvent?

c Draw a diagram to illustrate the mechanism of the elimination reaction when KOH in ethanol reacts with $CH_3CH_2CH_2Cl$. Indicate with curly arrows the movement of electrons.

d Discuss is the difference in reactivity between primary and tertiary halogenoalkanes.

Figure 17.1
Yeast on a vat of fermenting beer. The yeast has multiplied ten-fold and the best is retained for the next brewing. The remainder is sold to manufacturers of yeast extract.

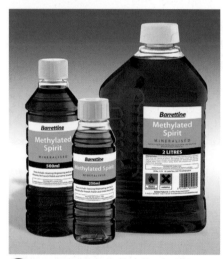

Figure 17.2
Methylated spirit is ethanol with methanol added to make it undrinkable. A dye is often added too.

Figure 17.3
A computer-generated model of a molecule of ethanol

17.1 Fermentation

When Noah left the ark he promptly planted a vineyard and was soon drinking its produce. People have been fermenting grape juice for at least ten thousand years, and probably fermenting honey for even longer. The reason for doing this is that fermentation of sugar by yeast produces ethanol, the well-known intoxicant. Ethanol is a member of the homologous series of alcohols and is commonly just called 'alcohol'.

$$C_6H_{12}O_6(aq) \longrightarrow 2CH_3CH_2OH(aq) + 2CO_2(g)$$

a sugar   ethanol

This is an exothermic reaction that provides the yeast with energy.

Alcoholic fermentation occurs naturally wherever sugar-containing materials such as fruit are allowed to decay.

In the manufacture of alcoholic drinks, fermentation is carried out under more controlled conditions. The source of sugar varies: it may be grapes (for wine), honey (for mead), malted grain (for beer), apples (for cider) or indeed any sugar-containing fruit or plant.

From the standpoint of yeast, the ethanol produced in fermentation is a toxic waste product, which kills the yeast at concentrations greater than about 15% by volume. It is therefore impossible to produce alcoholic drinks containing more than 15% alcohol by fermentation alone. To get a higher concentration, fermented liquids are sometimes distilled to make *spirits*. For example, wine is distilled to produce brandy. A typical spirit contains about 40% ethanol.

Alcohols are toxic

All the members of the homologous series of alcohols are toxic to a greater or lesser extent. The first member of the series, methanol, is much more toxic than ethanol. It is added to industrial alcohol to make it undrinkable. It is then called methylated spirit. Unfortunately a few people drink it nevertheless, which can cause blindness or even death.

The higher alcohols are moderately toxic, unpleasant-tasting compounds. A mixture of these compounds is produced in small amounts during fermentation: the mixture is called *fusel oil* (*fusel* is German for 'bad liquor').

Ethanol itself is intoxicating in small amounts but toxic in large amounts. In small amounts it makes people feel relaxed. In large amounts it has a serious effect on mental and physical performance and its use can be very dangerous, particularly to drivers of motor vehicles. It is addictive if regularly taken in large quantities. Ethanol is firmly established in many societies as an accepted social drug, but there is little doubt that if it were newly introduced today and its properties were known, it would be banned as a dangerous drug. In many Islamic societies drinking alcohol is unacceptable.

Manufacturing ethanol

Ethanol is an important industrial chemical. It is used as a solvent for perfumes, lacquers and inks, and as a fuel (Section 17.5).

Ethanol can be manufactured by two methods.

▶ *By fermentation.* Industrial ethanol can be manufactured by fermentation of cheap carbohydrates such as grain or molasses, followed by distillation. This method has the advantage that it uses renewable raw materials, but these materials could also be used to feed people. This method is used mainly in tropical countries where the raw materials can be easily and cheaply grown.

▶ *From ethene.* In countries where the raw materials for fermentation are relatively expensive, ethanol is normally manufactured by the catalytic hydration of ethene (Section 15.5).

17.2 Naming alcohols

In this chapter we shall be considering organic **hydroxy compounds,** containing the —OH group. A study of the properties of the —OH group is important to chemists because of the industrial importance of compounds containing this functional group and because of its wide occurrence in biological molecules.

Aliphatic and aromatic hydroxy compounds differ considerably in their properties. They are, therefore, regarded as two distinct groups of compounds: aliphatic hydroxy compounds are called **alcohols** and aromatic ones are called **phenols**.

Alcohols are named using the suffix **-ol**, preceded if necessary by a number to indicate its position in the carbon skeleton. Thus, CH_3OH is methanol, $CH_3CH_2CH_2OH$ is propan-1-ol, $CH_3CHOHCH_3$ is propan-2-ol and

$$CH_3-\underset{\underset{OH}{|}}{\overset{\overset{CH_3}{|}}{C}}-CH_3 \text{ is 2-methylpropan-2-ol.}$$

Alcohols with structures of the form RCH_2OH, where R is an alkyl or aryl group or hydrogen, are called **primary** alcohols. Methanol and propan-1-ol are primary alcohols.

Those with the structure $\overset{R^1}{\underset{R^2}{\diagdown}}CHOH$ are called **secondary** alcohols, for example

propan-2-ol. Alcohols with the structure $R^2-\overset{R^1}{\underset{R^3}{\diagup}}COH$ are called **tertiary** alcohols, for

example 2-methylpropan-2-ol. Primary, secondary and tertiary alcohols have some important differences in chemical reactivity (Section 17.5).

Some alcohols, particularly biologically occurring ones, contain more than one —OH group in their molecule. They are known as **polyhydric** alcohols. They are named using the suffixes **-diol**, **-triol**, etc., depending on how many —OH groups they contain. Thus,

$HOCH_2CH_2OH$ ethane-1,2-diol,

$HOCH_2CHOHCH_2OH$ propane-1,2,3-triol.

Compounds in which the —OH group is attached to an aromatic ring are called phenols. The simplest and most important is phenol itself.

Phenols have the —OH group attached *directly* to the ring, and we will study them in more detail in Chapter 25.

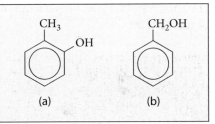

Key Point

Alcohols are aliphatic compounds containing the —OH group.

They are named using the suffix -ol.

Definitions

Primary alcohols have the structure RCH_2OH (R can be H).

Secondary alcohols have the structure: $R_1 R_2 CHOH$

Tertiary alcohols have the structure: $R_1 R_2R_3 COH$

Figure 17.4
A phenol and an alcohol

Quick Questions

1 One of the two compounds shown in Figure 17.4 is an alcohol and one is a phenol. Which is which?
2 Name the following:
   a $CH_3CHOHCH_2CH_3$
   b $H-\underset{\underset{OH}{|}}{\overset{\overset{H}{|}}{C}}-\underset{\underset{OH}{|}}{\overset{\overset{CH_3}{|}}{C}}-CH_3$
3 Write formulae for the following:
   a butane-1,2,4-triol
   b 2-methylpentan-2-ol
4 From questions 2 and 3, classify those alcohols containing a single —OH group as primary, secondary or tertiary.

Hydrogen bonding...

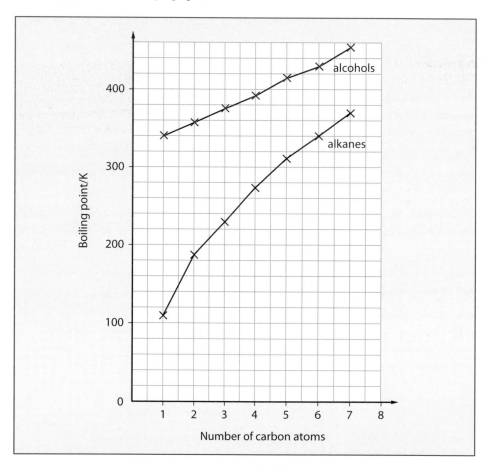

(a) between water molecules

(b) between ethanol molecules

(c) between ethanol and water

⌃ Figure 17.5
Hydrogen bonding in ethanol and water

⌄ Figure 17.6
Boiling points of the first seven straight-chain alkanes and straight-chain primary alcohols

Quick Questions

5 Look at the graphs in Figure 17.6 and answer the following questions.

  a Why are the boiling points of alcohols higher than those of the corresponding alkanes?

  b Why do the differences in boiling points between corresponding alcohols and alkanes get less as the number of carbon atoms increases?

  c Where would the two graphs intersect, and what is the physical significance of the point of intersection?

17.3 Alcohols as a homologous series

The alcohols illustrate very well the steady change in physical properties when a homologous series is ascended. The —OH group has a big effect on the physical properties of any molecule of which it is a part.

Hydrogen bonding in ethanol

Hydrogen bonding (Section 3.12) has a big influence on the properties of ethanol. Hydrogen bonds form between —OH groups of adjacent ethanol molecules (Figure 17.5(b)). This gives ethanol relatively high intermolecular forces and therefore relatively low volatility. Hydrogen bonding between ethanol molecules and water molecules explains why ethanol is soluble in water in all proportions (Figure 17.5(c)).

Pure ethanol has a strong affinity for water and tends to absorb it from the atmosphere. Furthermore, it is impossible to separate pure water from an ethanol–water mixture by distillation alone. The two liquids form a constant-boiling mixture containing 95.6% ethanol which distils over unchanged.

To produce 100% ethanol (called **absolute alcohol**) the remaining 4.4% water must be removed by a chemical drying agent such as calcium oxide.

Figure 17.6 shows the boiling points of the first seven straight-chain primary alcohols and the first seven straight-chain alkanes. As the homologous series of alcohols is ascended, the influence of the —OH group becomes less and less important compared with that of the increasingly large hydrocarbon chain. The properties of the higher alcohols therefore tend more and more towards those of the corresponding alkane. This trend can be seen in solubility as well as volatility, as Table 17.1 shows.

Table 17.1
Solubility of alcohols in water

Name	Formula	Solubility/g per 100 g of water
methanol	CH_3OH	infinite (miscible in all proportions)
ethanol	CH_3CH_2OH	infinite (miscible in all proportions)
propan-1-ol	$CH_3CH_2CH_2OH$	infinite (miscible in all proportions)
butan-1-ol	$CH_3CH_2CH_2CH_2OH$	8.0
pentan-1-ol	$CH_3CH_2CH_2CH_2CH_2OH$	2.7
hexan-1-ol	$CH_3CH_2CH_2CH_2CH_2CH_2OH$	0.6

Alcohols as solvents

The lower alcohols such as ethanol tend to be good solvents for polar as well as non-polar solutes. This is because they contain both a highly polar —OH group and a non-polar hydrocarbon chain. For example, both sodium hydroxide (ionic) and hexane (molecular) dissolve well in ethanol. This property makes ethanol, and other alcohols like methanol and propanol, valuable solvents in the laboratory and in industry. One example is the use of ethanol as a base for perfumes: the ethanol dissolves both the water-insoluble oils which provide the aroma and the water which makes up the bulk of the preparation.

Hydrogen bonding between —OH groups also has an effect on the viscosity of alcohols, particularly those with more than one —OH group in their molecule. Thus ethanol has a viscosity of $1.06 \times 10^{-3}\,N\,s\,m^{-2}$ at 298 K, about the same as water. But propane-1,2,3-triol (commonly called glycerine) is very thick and sticky, with a viscosity of $942 \times 10^{-3}\,N\,s\,m^{-2}$ at the same temperature. This is because of extensive hydrogen bonding between its molecules, which carry three —OH groups each.

17.4 The amphoteric nature of alcohols

We can think of alcohols as being derived from water, by replacing one hydrogen atom by an alkyl group.

$$H—O—H \qquad\qquad R—O—H$$
water an alcohol

We might therefore expect alcohols to show some similarity to water, and this is true of their physical properties (Section 17.3).

Water is an amphoteric compound. It can act as an acid, donating a proton, or as a base, accepting a proton:

as an acid: $\qquad H_2O \longrightarrow {}^-OH + H^+$

as a base: $\qquad H_2O + H^+ \longrightarrow H_3O^+$

overall: $\qquad H_2O + H_2O \longrightarrow {}^-OH + H_3O^+$
acid   base

Alcohols also show amphoteric behaviour:

as an acid: $\qquad ROH \longrightarrow RO^- + H^+$

as a base: $\qquad ROH + H^+ \longrightarrow ROH_2^+$

overall: $\qquad ROH + ROH \longrightarrow RO^- + ROH_2^+$
acid   base

When an alcohol acts as an acid, it cleaves at the O—H bond. When it acts as a base, it attracts H^+ ions (protons) to lone pairs of electrons on its O atom and can

Figure 17.7
Windscreen de-icing fluid contains propan-2-ol. This alcohol mixes completely with water and has a low freezing point.

subsequently cleave at the R—O bond. Both these forms of bond cleavage are characteristic of alcohols, and we will consider them separately. In doing so we will take ethanol as a typical alcohol.

Reactions involving cleavage of the O—H bond

Reaction with sodium

When a small piece of sodium is added to ethanol, the sodium sinks and a steady stream of hydrogen is given off. Compare this with the much more vigorous reaction of sodium with water.

In this reaction, ethanol is behaving as a very weak acid, releasing H^+ ions which are reduced to hydrogen by Na. The rate of the reaction shows that ethanol is an even weaker acid than water.

Ethanol is converted to ethoxide ion, so the two products are hydrogen and sodium ethoxide:

$$2Na + 2CH_3CH_2OH \longrightarrow 2CH_3CH_2O^-Na^+ + H_2$$
sodium ethoxide

Compare this reaction to that of sodium with water, forming hydrogen and sodium hydroxide.

Esterification

In the presence of an acid catalyst, alcohols react with carboxylic acids to form esters. Water is eliminated. The alcohol cleaves at the O—H bond:

$$CH_3-\underset{\underset{O}{\|}}{C}-OH \ + \ CH_3CH_2-O-H \ \xrightarrow{H^+} \ CH_3-\underset{\underset{O}{\|}}{C}-O-CH_2CH_3 \ + \ H_2O$$

ethanoic acid   ethanol   ethyl ethanoate (an ester)

This reaction is considered more fully in Section 19.3.

Another way to convert an alcohol to an ester is by acylation, using an acyl chloride (Section 19.5)

Reactions involving cleavage of the C—O bond

Reaction with halide ions

In the presence of a strong acid, the O atom of alcohols becomes protonated. A lone pair of electrons on the O bonds to H^+:

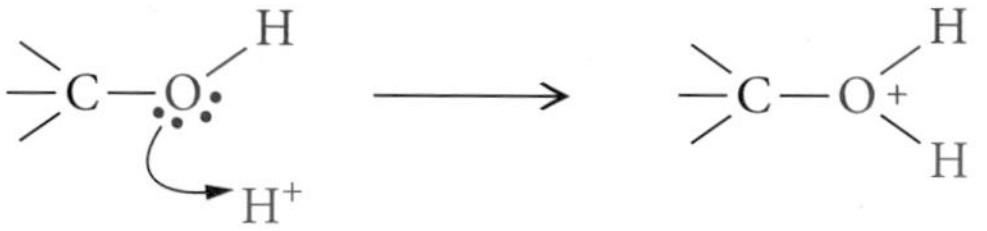

The oxygen atom then carries a positive charge, and tends to attract electrons very strongly from the carbon atom next to it (Figure 17.8). This causes a large positive charge on the carbon, making it attractive to nucleophiles such as halide ions. So, *in the presence of concentrated sulfuric acid*, ethanol reacts with Br^- to form bromoethane:

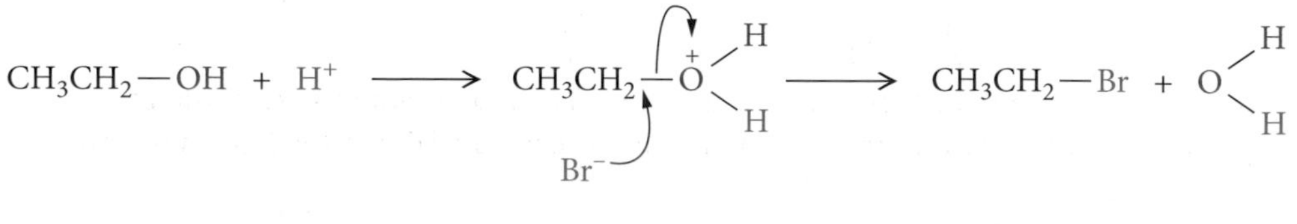

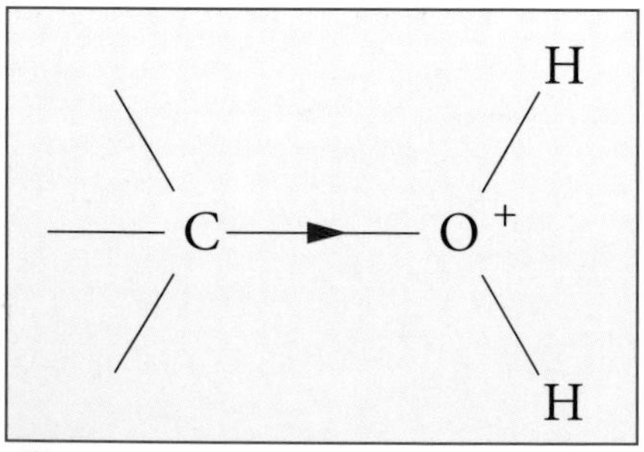

$$CH_3CH_2-OH \ + \ H^+ \ \longrightarrow \ CH_3CH_2-\overset{+}{O}\diagdown^H_H \ \longrightarrow \ CH_3CH_2-Br \ + \ O\diagup^H_H$$

Key Point

Ethanol cleaves at the O—H bond when it reacts with:

▶ sodium, to release hydrogen,
▶ carboxylic acids to form esters.

In these reactions, ethanol acts as an acid (proton donor), although weaker than water.

Quick Questions

6 Predict the formulae of the products of reactions between:
  a propan-2-ol and sodium,
  b propan-1-ol and propanoic acid, in the presence of an acid catalyst.

▲ Figure 17.8
The effect of protonation on polarisation of the C—O bond in alcohols. The arrow indicates the direction of displacement of electrons in the bond.

Overall, this amounts to the reaction of the alcohol with HBr:

$$CH_3CH_2OH + HBr \longrightarrow CH_3CH_2Br + H_2O$$
bromoethane

In some cases an intermediate carbocation may actually be formed before the Br^- attacks. Cleavage of the C—O bond is greatly helped by protonation because it is much easier for the molecule to lose a neutral H_2O molecule than a charged ^-OH ion. We say that H_2O is a better *leaving group* than ^-OH.

This is another example of a nucleophilic substitution reaction. Compare it with the substitution reactions of halogenoalkanes (Section 16.4). The reaction with ethanol is normally carried out by heating it under reflux with potassium bromide and concentrated sulfuric acid. Similar reactions occur between alcohols and Cl^-, Br^- and I^- in acid conditions. Cl^- is the least reactive and I^- the most reactive.

Other halogenation reactions

There are several other reagents that are useful for replacing an —OH group with a halogen atom. Some of the more important are:

phosphorus pentachloride: $ROH + PCl_5 \longrightarrow RCl + HCl + POCl_3$

phosphorus tribromide or triiodide (in practice a mixture of red phosphorus and bromine or iodine is used): $3ROH + PBr_3 \longrightarrow 3RBr + H_3PO_3$

17.5 Reactions involving the carbon skeleton

So far we have considered reactions of alcohols in which part or all of the —OH group is replaced. Alcohols also undergo reactions which involve *both* the carbon skeleton *and* the —OH group.

Dehydration to form alkenes

Consider again the protonated form of ethanol that we met in the last section. This ion can readily lose water, forming a carbocation:

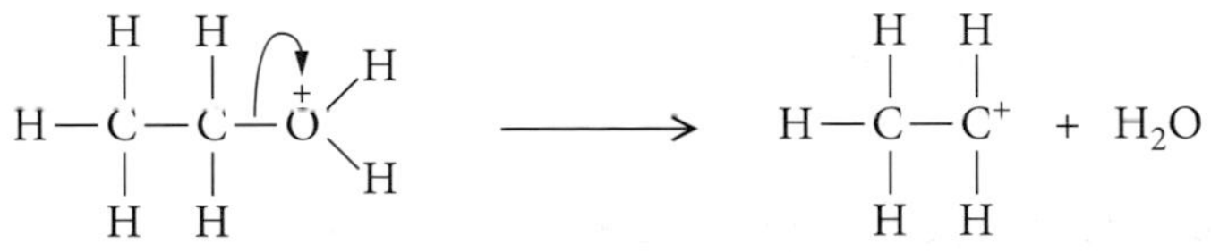

The carbocation is an unstable intermediate. In the reaction with HBr, it reacts rapidly with Br^- to form bromoethane. In the absence of any nucleophile like Br^-, however, a different reaction may occur. The ion may lose H^+ and form ethene:

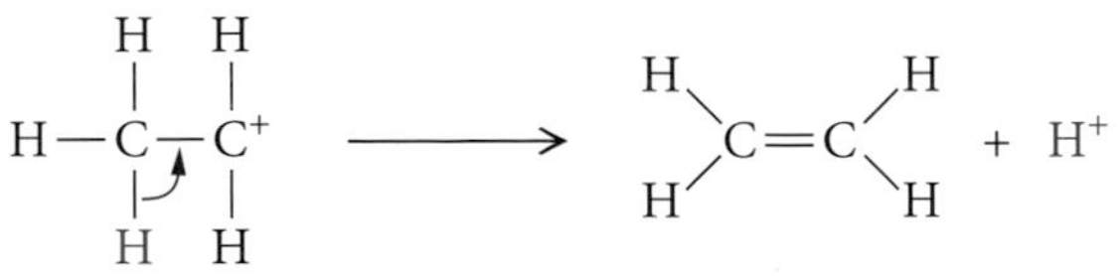

Thus in the presence of strong acids, ethanol forms ethene. This is an elimination reaction:

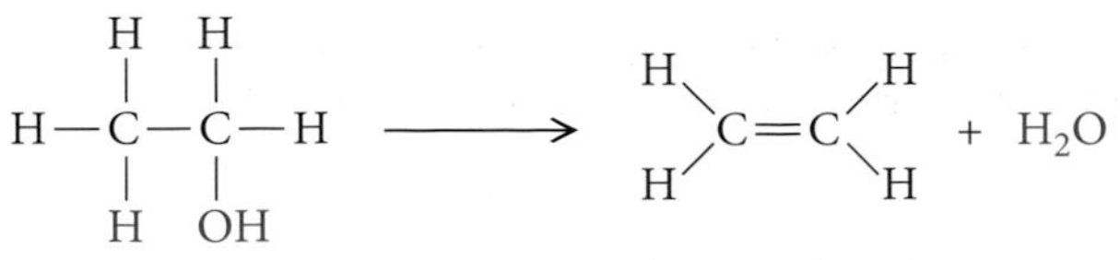

Since water is eliminated, this can also be thought of as a dehydration reaction. In practice the intermediate carbocation may never form: the H^+ and the H_2O may leave simultaneously.

Key Point

Alcohols can be converted to halogenoalkanes by heating with concentrated sulfuric acid and halide ions. The reaction involves cleavage of the C—O bond.

Key Point

Alcohols can be dehydrated to form alkenes, by heating the alcohol with concentrated sulfuric acid, or by passing the alcohol vapour over a heated catalyst.

Quick Questions

7 Predict the formulae of the main products of the following reactions.
   a butan-2-ol is heated with excess concentrated sulfuric acid,
   b propan-2-ol vapour is passed over heated aluminium oxide.

The reaction can be carried out in the laboratory by heating ethanol at 170 °C with excess concentrated sulfuric acid. Ethene is evolved and can be collected over water. The concentrated sulfuric acid can be thought of as a dehydrating agent, removing water from ethanol.

Another way of preparing ethene from ethanol is by catalytic dehydration of ethanol vapour. A heated catalyst of aluminium oxide or pumice stone is used, in the apparatus shown in Figure 17.9.

Dehydration reactions like these apply to alcohols in general, not just ethanol.

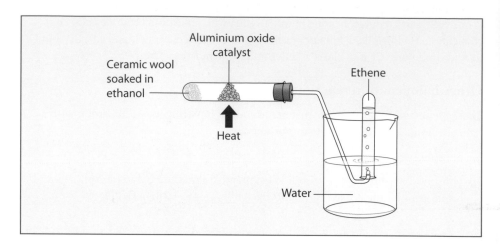

▶ Figure 17.9
Catalytic dehydration of ethanol

Oxidation to aldehydes and ketones

Table 17.2 shows the effect of warming three different alcohols with acidified potassium dichromate(VI).

▼ Table 17.2
Effect of warming alcohols with acidified potassium dichromate(VI)

Name	Formula	Observation
propan-1-ol	$CH_3CH_2CH_2OH$	orange dichromate(VI) slowly turns green
propan-2-ol	$CH_3CHOHCH_3$	orange dichromate(VI) slowly turns green
2-methylpropan-2-ol	CH_3 $\vert$ CH_3-C-CH_3 $\vert$ OH	no change

Quick Questions

8 Look at Table 17.2 and answer these questions.
   a Classify each of the three alcohols as primary, secondary or tertiary.
   b When potassium dichromate(VI) is reduced in acid solution, green $Cr^{3+}(aq)$ ions are formed. Which of the three alcohols is/are oxidised by acidified potassium dichromate(VI)?

Primary and secondary alcohols are readily oxidised by a variety of oxidants. Acidified dichromate(VI), acidic or alkaline manganate(VII) or air in the presence of a catalyst are all suitable oxidants. The initial product is a **carbonyl** compound (Chapter 18) which in the case of a primary alcohol is an **aldehyde**:

propan-1-ol → propanal, an aldehyde + $2H^+$ + $2e^-$ (electrons accepted by oxidiser)

Key Point

Primary alcohols can be oxidised to aldehydes, then to acids. Secondary alcohols can be oxidised to ketones. Tertiary alcohols cannot be oxidised without breaking up the molecule.

Aldehydes are themselves readily oxidised to acids, thus:

$$H-\underset{\underset{H}{|}}{\overset{\overset{H}{|}}{C}}-\underset{\underset{H}{|}}{\overset{\overset{H}{|}}{C}}-C\overset{O}{\underset{H}{\diagdown}} + H_2O \longrightarrow H-\underset{\underset{H}{|}}{\overset{\overset{H}{|}}{C}}-\underset{\underset{H}{|}}{\overset{\overset{H}{|}}{C}}-C\overset{O}{\underset{OH}{\diagdown}} + 2H^+ + 2e^-$$

propanal   propanoic acid

The product of oxidising a primary alcohol is therefore usually an acid, unless the aldehyde is distilled from the reaction mixture as it forms.

In the case of a secondary alcohol the product of oxidation is a **ketone**:

$$H-\underset{\underset{H}{|}}{\overset{\overset{H}{|}}{C}}-\underset{\underset{OH}{|}}{\overset{\overset{H}{|}}{C}}-\underset{\underset{H}{|}}{\overset{\overset{H}{|}}{C}}-H \longrightarrow H-\underset{\underset{H}{|}}{\overset{\overset{H}{|}}{C}}-\underset{\underset{O}{||}}{C}-\underset{\underset{H}{|}}{\overset{\overset{H}{|}}{C}}-H + 2H^+ + 2e^-$$

propan-2-ol   propanone,
                                     a ketone

Ketones are not readily oxidised, so the reaction stops at this point.

Tertiary alcohols cannot be readily oxidised because they have no hydrogen atom which can be removed from the carbon atom carrying the —OH group. With strong oxidants, their molecules break up, giving a mixture of oxidation products.

Other ways of oxidising ethanol

Some bacteria can oxidise ethanol to ethanoic acid (acetic acid), and this has always been a problem for wine producers. The bacterium, *Acetobacter*, uses air to oxidise ethanol in wine, producing a weak solution of ethanoic acid called vinegar. The bacterium uses this oxidative process as a source of energy. Once a bottle of wine has been opened it will turn to vinegar fairly quickly because of the considerable number of these bacteria in the air. One way of preventing this is to add extra alcohol to the wine so that the bacteria cannot tolerate the higher concentration of ethanol. Wine treated in this way is said to be *fortified*: sherry and port are examples.

The human body gets rid of ethanol by oxidation. It is oxidised in the liver, first to ethanal and eventually to CO_2 and water. The liver can oxidise about 8 g of ethanol an hour, but it cannot cope with very high concentrations. Excessive quantities of ethanol can damage the liver and cause a condition called cirrhosis.

Alcohols as fuels

Ethanol, like all alcohols, burns to give CO_2 and H_2O. Ethanol has a clean, smokeless flame and is sometimes used as a fuel, for example in stoves burning methylated spirits. Its use as a fuel is limited by its cost in temperate countries, but in some tropical countries, where sugar is easily grown, ethanol is produced cheaply by fermentation. In Brazil, for example, ethanol is an important motor fuel because it can be produced at a cost comparable to that of gasoline.

Methanol is also an important fuel. It is quite cheap and can be blended with gasoline to increase its octane number and reduce pollution.

The tri-iodomethane (iodoform) reaction

When ethanol is warmed with iodine and an alkali, a yellow solid with a characteristic smell is formed. This solid is tri-iodomethane (iodoform), CHI_3. The reaction is explained in more detail in Section 18.6.

Figure 17.10
The oxidation of ethanol is used to detect alcohol in motorists' breath. This roadside 'Alcolmeter' contains an electrochemical cell in which alcohol is oxidised at one of the electrodes. The greater the concentration of alcohol, the higher the voltage of the cell and the larger the reading on the meter.

Quick Questions

9 Predict the products, if any, of oxidising the following alcohols with acidified dichromate(VI):
   a ethanol (product distilled off immediately),
   b ethanol (reagents heated together under reflux for some time),
   c 2-methylbutan-2-ol,
   d butan-2-ol.

Figure 17.11
Wine fermenting in a barrel. The curved tube at the top is a trap to keep out bacteria. It allows carbon dioxide to escape from the fermenting wine, but it does not let in bacteria which might turn the wine to vinegar.

The characteristic reactions of ethanol are summarised in Figure 17.13.

△ Figure 17.12
Ethanol is added to gasoline to produce a cleaner fuel with higher performance. This is more common in tropical countries where food crops, which can be fermented to produce ethanol, grow quickly.

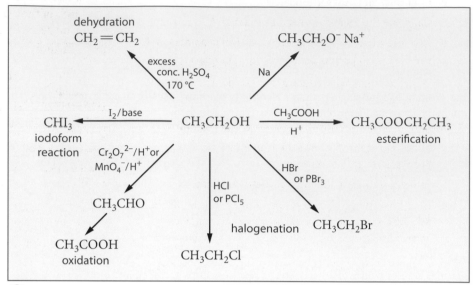

△ Figure 17.13
A summary of the reactions of ethanol

Review questions

1 Consider the following compounds.

A

$$CH_3CH_2 - \underset{\underset{OH}{|}}{\overset{\overset{CH_3}{|}}{C}} - CH_3$$

B $CH_3CH_2CHOHCH_3$

C CH_3OH

D $CH_3CH_2CH_2CH_2OH$

a Name each compound.

b Which is/are primary alcohol(s)?

c Which is/are tertiary alcohol(s)?

d Which is/are secondary alcohol(s)?

2 Refer again to the compounds in question 1.

a Which react with sodium metal?

b Which could be oxidised to an aldehyde?

c Which could be oxidised to a ketone?

d Which would form an alkene when heated with excess concentrated sulfuric acid?

e Which has the lowest boiling point?

3 Table 17.3 gives some physical properties of water, ethanol and hexane.

 Table 17.3
Some physical properties of water, ethanol and hexane

Name	water	ethanol	hexane
Formula	H_2O	CH_3CH_2OH	$CH_3(CH_2)_4CH_3$
M_r	18	46	86
Boiling point/°C	100	78	69
Density at 273 K/g cm^{-3}	1.00	0.79	0.66
Surface tension at 293 K/N m^{-1}	7.28	2.23	1.84

Suggest explanations in terms of intermolecular forces, why

a the boiling point of ethanol is greater than that of hexane,

b the density of water is greater than that of ethanol,

c the surface tension of water is greater than that of hexane.

4 An organic liquid **A** contains carbon, hydrogen and oxygen only. On combustion 0.463 g of **A** gave 1.1 g of carbon dioxide and 0.563 g of water. When vaporised, 0.1 g of **A** occupy 54.5 cm^3 at 208 °C and 98.3 kPa (740 mm Hg). (Standard pressure = 101 kPa (760 mm Hg); 1 mole of a gas occupies 22.4 dm^3 at s.t.p.)

a What is the percentage composition of **A**?

b Find the empirical formula of **A**.

c Calculate the relative molecular mass of **A**.

d Give the structures of possible non-cyclic isomers of **A**.

e Which isomers will react with sodium to give hydrogen?

f Which isomers will reduce acidified dichromate(vi) ion to green Cr^{3+}?

5 Suggest explanations for the following observations.

a Butan-1-ol is much more soluble in 5 mol dm^{-3} hydrochloric acid than in water.

b Ethanol is more acidic than 2-methylpropan-2-ol.

c Heating butan-2-ol with excess concentrated sulfuric acid produces a mixture of three isomeric alkenes. Draw their structures.

6 Give the reagents and conditions you would use to carry out the following conversions. One or more steps may be involved in the conversions.

a ethanol to ethyl ethanoate (using ethanol as the only organic starting material),

b ethanol to 1,2-dibromoethane,

c ethene to ethanoic acid,

d ethanol to propanenitrile.

7 Predict the structure of the organic products of the following reactions.

a Butan-2-ol is warmed with acidified potassium manganate(vii).

b Propan-2-ol is warmed with excess concentrated sulfuric acid.

c Methanol is treated with phosphorus pentachloride.

8 For each of the common uses of alcohols given below, explain which property or properties of the alcohols involved make them suitable for that use.

a Ethanol is used to clean surfaces before applying adhesive.

b Ethane-1,2-diol is used as an antifreeze.

c Ethanol is used as the base for many perfumes.

d Vinegar is manufactured from a weak solution of ethanol.

e Fruits, such as peaches, are sometimes preserved in brandy.

f 'Breathalysers' containing potassium dichromate(vi) can be used to test for ethanol in motorists' breath.

9 Alcohols may be classified into primary, secondary and tertiary. Some reactions are common to all three types of alcohol. In other cases, the same reagent gives different products depending on the nature of the alcohol.

Re-draw the grid below. In the empty squares give the structural formula of the organic compound formed in each of the reactions indicated. If no reaction occurs, write 'no reaction' in the space.

Alcohol Reagent(s) and conditions	$CH_3CH_2CH_2CH_2OH$	$CH_3CH_2CH(OH)CH_3$	$(CH_3)_3COH$
Red phosphorus and iodine Heat under reflux			
Concentrated H_2SO_4 Heat			
$Cr_2O_7{}^{2-}/H^+$ Heat under reflux			

Cambridge Paper 2 Q2d June 2007

10 Compounds containing the allyl group, $CH_2{=}CHCH_2{-}$, have pungent smells and are found in onions and garlic.

Allyl alcohol, $CH_2{=}CHCH_2OH$, is a colourless liquid which is soluble in water.

a Allyl alcohol behaves as an alkene and as a primary alcohol.

Give the structural formula of the organic compound formed when allyl alcohol is

i reacted with Br_2,

ii heated under reflux with an acidified solution of $Cr_2O_7{}^{2-}$ ions.

b When allyl alcohol is reacted with MnO_2 at room temperature, propenal, $CH_2{=}CHCHO$ is formed. What type of reaction is this?

c Allyl alcohol may be converted into propanal, CH_3CH_2CHO, by using a ruthenium(iv) catalyst in water:

$$CH_2{=}CHCH_2OH \xrightarrow{\text{ruthenium(iv) catalyst}} CH_3CH_2CHO$$

The reactant and the product are isomers. What form of isomerism do they display?

d Allyl alcohol can be converted into propanal in two steps *without* the use of a ruthenium(iv) catalyst:

$$CH_2{=}CHCH_2OH \xrightarrow{\text{Step I}} CH_3CH_2CH_2OH \xrightarrow{\text{Step II}} CH_3CH_2CHO$$

What reagents and conditions would be used for *each* step?

e By considering your answers to parts b and d, suggest what is unusual about the single-step reaction in part c.

f Suggest the structural formula of the organic compound formed when allyl alcohol is:

i reacted with cold, dilute $MnO_4{}^-$ ions,

ii heated under reflux with acidified $MnO_4{}^-$ ions.

Cambridge Paper 2 Q4 June 2006

18 Carbonyl compounds

Figure 18.1
Aldehydes and ketones contribute to the flavour of some fruit. The smell of ethanal is reminiscent of apples.

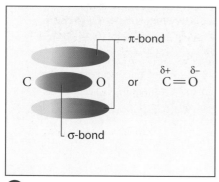

Figure 18.2
Electron distribution in the C=O bond

18.1 The carbonyl group

In Chapter 15 we looked at compounds containing C=C double bonds and saw that their typical reactions involves electrophilic addition. In Chapter 17 we looked at compounds containing the C—O single bond, which is polar and tends to bring about substitution reactions. This chapter is about the **carbonyl** group, C=O, which is the functional group in aldehydes and ketones. We might expect the reactions of this group to show similarity to the reactions of both

$$\ce{>C=C<} \quad \text{and} \quad \ce{>C-O-}$$

The double bond between C and O in the carbonyl group, like the double bond in alkenes, can be considered to consist of a σ-bond and a π-bond. Unlike the C=C group, however, the carbonyl group does not have an even distribution of electrons between the two atoms. There is a greater electron density over the more electronegative oxygen atom, as shown in Figure 18.2.

This electron distribution makes the carbon atom attractive towards nucleophiles. Nucleophiles tend to attack and bond to this carbon, breaking the π-bond and resulting in addition. With a general nucleophile $\overset{\bullet\bullet}{X}$—Y, the overall reaction is

$$\ce{>\overset{\delta+}{C}=\overset{\delta-}{O}} + \overset{\bullet\bullet}{X}-Y \longrightarrow \ce{>C<^{OY}_{X}}$$

The exact mechanism depends on the nature of X and Y. In most cases, Y is hydrogen. This sort of reaction, which is typical of the carbonyl group, is called **nucleophilic addition**. Compare it with the *electrophilic* addition which is typical of alkenes. The mechanism of the nucleophilic addition reaction involving HCN is described in Section 18.3.

The reactions of the carbonyl group are important because the group is common in biological molecules, particularly carbohydrates (Section 18.7). Carbonyl compounds also have considerable industrial significance, for example as solvents and in the manufacture of plastics.

Quick Questions

1 **a** What feature of the carbonyl group makes it attractive to nucleophiles?
   b Why is the double bond in alkenes, C=C, attractive to electrophiles rather than nucleophiles?

18.2 Aldehydes and ketones – nature and naming

Aldehydes and ketones both contain the carbonyl group, but they differ in its position in the hydrocarbon skeleton. **Aldehydes** have the carbonyl group at the end of a chain. Their general formula is therefore $\ce{R-\underset{O}{\overset{\|}{C}}-H}$, usually written RCHO, where R is an alkyl or aryl group or hydrogen. **Ketones** have the carbonyl group in a non-terminal position in the chain. Their general formula is $\ce{R^1-\underset{O}{\overset{\|}{C}}-R^2}$, usually written R^1COR^2, where R^1 and R^2 are alkyl or aryl groups.

Thus aldehydes and ketones are structurally quite similar, but their properties differ considerably and they form different homologous series.

Aldehydes are named using the suffix **-al** after a prefix indicating the number of carbon atoms (including the one in the carbonyl group). Thus CH_3CHO is called ethanal. Ketones are named using the suffix **-one** after a prefix indicating the number of carbon atoms, together with a number, if necessary, to indicate the position of the carbonyl group in the chain. Thus $CH_3COCH_2CH_2CH_3$ is called pentan-2-one. Table 18.1 gives the names and formulae of some important aldehydes and ketones.

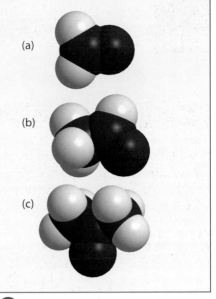

(a)

(b)

(c)

Figure 18.3
Computer-generated models of molecules of (a) methanal, (b) ethanal and (c) propanone

Quick Questions

2 Name the compounds:
   a $CH_3CH_2CH_2CHO$,
   b $CH_3CH_2COCH_2CH_2CH_3$
3 Give the formula of:
   a hexan-2-one,
   b 3–methylpentanal
4 Why is there no such compound as ethanone?

Table 18.1
Some important aldehydes and ketones

Formula	Systematic name	Other name	State at room temp.	Boiling point/K	Solubility in water
HCHO	methanal	formaldehyde	g	254	soluble
CH_3CHO	ethanal	acetaldehyde	l	294	infinite*
CH_3CH_2CHO	propanal	propionaldehyde	l	321	soluble
CH_3COCH_3	propanone	acetone	l	329	infinite*
$CH_3COCH_2CH_3$	butanone	methylethyl ketone	l	353	very soluble
$CH_3CH_2COCH_2CH_3$	pentan-3-one		l	375	very soluble
⬡—CHO	benzaldehyde		l	451	slightly soluble
⬡—$COCH_3$	phenylethanone	acetophenone	l	475	insoluble

* miscible in all proportions

Physical properties

The polarity of the $C=O$ group has a big influence on the physical properties of aldehydes and ketones. The earlier members of both series are considerably less volatile than alkanes of corresponding relative molecular mass. Thus ethanal (CH_3CHO), with a boiling point of 21 °C, is a liquid (though a very volatile one) at room temperature. Propane, $CH_3CH_2CH_3$, has the same relative molecular mass but is a gas at room temperature (boiling point, –42 °C).

The polar $—C=O$ group has less effect on intermolecular forces than the $—OH$ group in alcohols, which is able to participate in hydrogen bonding. Compare the boiling points of ethanal and ethanol, which are 20 °C and 78 °C, respectively.

The early members of the aldehydes and ketones are soluble in water. They act as solvents for both polar and non-polar solutes. Propanone (acetone), for example, is a widely used industrial solvent.

Figure 18.4
Figure 18.4
Propanone is widely used as a solvent. Nail varnish remover contains propanone.

Key Point

Aldehydes are made by oxidising primary alcohols. Ketones are made by oxidising secondary alcohols.

Quick Questions

5 a Consider the cases when —R^1 and —R^2 are
  i both —H,
  ii —CH_3 and —H,
  iii both —CH_3.
  Place **i**, **ii** and **iii** in order according to the size of positive charge you would expect on the carbonyl carbon in each of the three compounds.
  (Remember that alkyl groups such as —CH_3 tend to donate electrons to groups to which they are attached.)
  b Place **i**, **ii** and **iii** in order of readiness to undergo nucleophilic addition.
  c Name compounds **i**, **ii** and **iii**.

As expected, the polar —C=O group has less and less influence on the physical properties of carbonyl compounds as the homologous series are ascended. Table 18.1 gives some physical properties of carbonyl compounds.

Methanal, the simplest aldehyde, is toxic and carcinogenic.

Making aldehydes and ketones

In the laboratory these compounds can be made by the oxidation of alcohols, as described in Section 17.5. In fact, the name *aldehyde* comes from *alcohol dehydro*genate. Aldehydes are made by oxidising primary alcohols, ketones by oxidising secondary alcohols. In the laboratory, the oxidising agent is usually acidified dichromate(VI), $Cr_2O_7^{2-}/H^+$ (Figure 18.5). Some ethanal is manufactured by the oxidation of ethanol, but most is made by direct oxidation of ethene (see review question 3 at the end of this chapter).

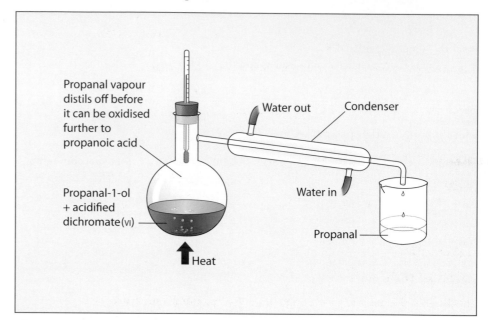

Figure 18.5
Propanal can be prepared in the laboratory by oxidising propan-1-ol using acidified dichromate(VI).

18.3 Addition reactions of carbonyl compounds

The characteristic reaction of compounds containing the carbonyl group is nucleophilic addition:

$$\begin{array}{c} R^1 \\ \diagdown \overset{\delta+}{} \overset{\delta-}{} \\ C=O \\ \diagup \\ R^2 \end{array} + \overset{\cdot\cdot}{X}-Y \longrightarrow \begin{array}{c} R^1 \qquad OY \\ \diagdown \diagup \\ C \\ \diagup \diagdown \\ R^2 \qquad X \end{array}$$

The readiness with which this reaction occurs is partly determined by the size of the partial positive charge on the carbon atom of the carbonyl group. The larger the δ^+ charge, on this atom, the more attractive it is to nucleophiles.

Aldehydes tend to be more reactive than ketones in nucleophilic addition reactions. Methanal is the most reactive aldehyde.

The important stage of addition to C=O is attack by a nucleophile. Compare this with C=C, where the important stage of addition is attack by an *electrophile*. Carbonyl compounds do react with some of the compounds that react with alkenes, but many of their reactions are different. Some examples are now given.

Reaction with HCN

In most addition reactions of carbonyl compounds, the molecule adding across the $C{=}O$ double bond is of the form HX. A good example is HCN.

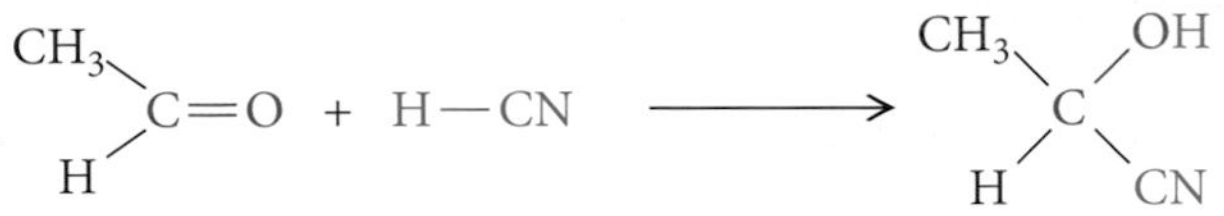

$$CH_3\!\!\diagdown\!\!C{=}O \;+\; H{-}CN \;\longrightarrow\; CH_3\!\!\diagup\!\!C\!\!\diagdown\!\!\begin{smallmatrix}OH\\CN\end{smallmatrix}$$

The mechanism of this reaction is as follows. Hydrogen cyanide is a weak acid which dissociates to form ⁻CN ions:

$$HCN(aq) + H_2O(l) \rightleftharpoons H_3O^+(aq) + {}^-CN(aq)$$

The ⁻CN is a nucleophile and attacks the carbon atom of the carbonyl group:

This intermediate ion then reacts with water to form the product:

The reaction occurs with ketones as well as aldehydes:

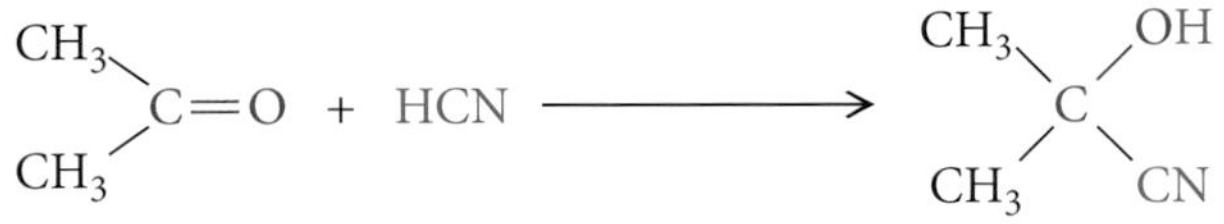

$$CH_3\!\!\diagdown\!\!C{=}O \;+\; HCN \;\longrightarrow\; CH_3\!\!\diagup\!\!C\!\!\diagdown\!\!\begin{smallmatrix}OH\\CN\end{smallmatrix}$$

Reduction

Reduction with hydrogen

Like the $C{=}C$ bond, the $C{=}O$ bond undergoes addition with hydrogen in the presence of a metal catalyst such as platinum or nickel. Aldehydes give primary alcohols, ketones give secondary alcohols.

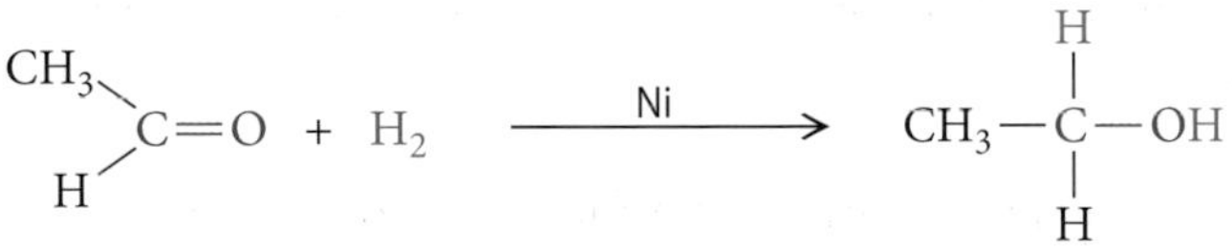

$$CH_3\!\!\diagdown\!\!C{=}O \;+\; H_2 \xrightarrow{\;Ni\;} CH_3{-}\overset{\displaystyle H}{\underset{\displaystyle H}{C}}{-}OH$$

ethanal   ethanol

$$CH_3\!\!\diagdown\!\!C{=}O \;+\; H_2 \xrightarrow{\;Ni\;} CH_3{-}\overset{\displaystyle H}{\underset{\displaystyle OH}{C}}{-}CH_3$$

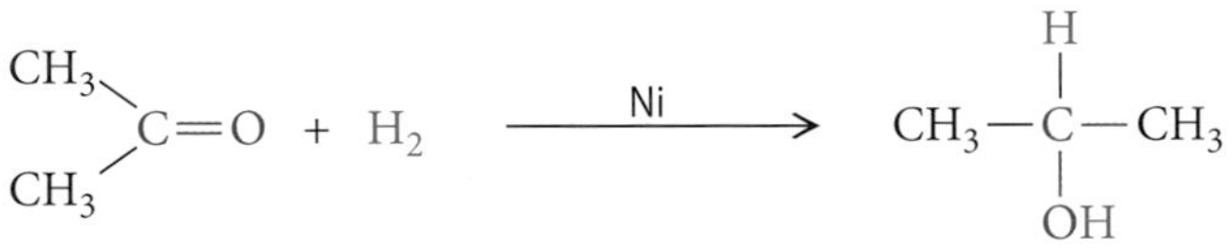

propanone   propan-2-ol

Reduction with metal hydrides

Metal hydrides contain the hydride ion, ⁻H. They are useful reducing agents in organic chemistry. The ⁻H ion acts as a nucleophile, attacking the carbon atom of the carbonyl group. With ethanal:

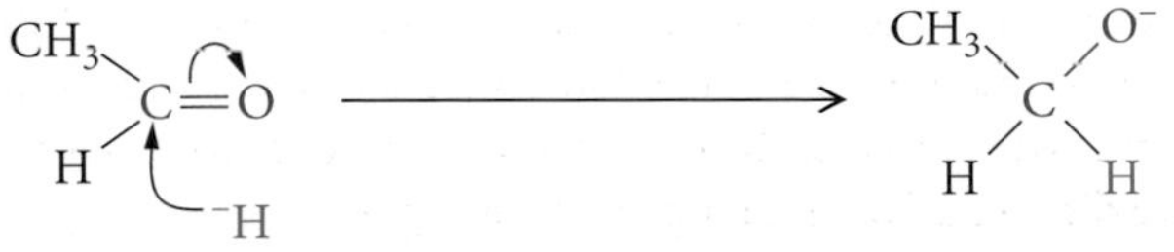

> ### Quick Questions
>
> 6 Predict the product of a nucleophilic addition reaction between ethanal and:
> a H_2O (think of it as H—OH),
> b CH_3CH_2OH.

> ### Key Point
>
> Carbonyl compounds can be reduced to alcohols, using
>
> a hydrogen and a catalyst, or
> b hydrides, especially $LiAlH_4$ and $NaBH_4$.

Quick Questions

7 Write the formulae of the products formed when the following compounds are reduced using NaBH$_4$ in water:
   a butanal,
   b butanone,
   c cyclopentanone,

Definition

In a **condensation reaction** a small molecule (usually H$_2$O) is eliminated between two large molecules, as the larger molecules join together. Condensation reactions are also called **addition–elimination**.

⌃ Figure 18.6
Methanal is very toxic. Small quantities of it are released into the air from the plastics used in some kinds of wall insulation. However, certain plants absorb methanal from the air – this photo shows the spider plant, *Chlorophytum comosum*, which is particularly effective at removing methanal vapour.

Key Point

The products of condensation reactions between 2,4-dinitrophenylhydrazine and carbonyl compounds are useful in identifying individual carbonyl compounds.

The intermediate ion then reacts with water to give the alcohol:

$$\text{CH}_3\text{CH(O}^-\text{)H} \xrightarrow{\text{H}_2\text{O}} \text{CH}_3\text{CH(OH)H} + \text{}^-\text{OH}$$

Two hydrides can be used to reduce carbonyl compounds to alcohols. Lithium tetrahydridoaluminate, LiAlH$_4$, also known as lithium aluminium hydride, is a powerful reducing agent. It will reduce carboxylic acids, esters and amides as well as carbonyl compounds. LiAlH$_4$ is easily hydrolysed, so it must be used in a dry ether solvent.

Sodium tetrahydridoborate, NaBH$_4$, also known as sodium borohydride, is a less powerful reducing agent. It reduces carbonyl compounds to alcohols, but does not affect the other functional groups that are reduced by LiAlH$_4$. It is therefore useful for the *selective* reduction of carbonyl groups, leaving other functional groups unaffected. NaBH$_4$ is more convenient to use than LiAlH$_4$, because it is not easily hydrolysed, so it can be used in solution in water or alcohol.

18.4 Condensation reactions of carbonyl compounds

Sometimes addition reactions of carbonyl compounds are followed by elimination of a molecule of water. Many of these addition–elimination reactions involve derivatives of ammonia with the general form X—NH$_2$.

$$\text{R}^1\text{R}^2\text{C}=\text{O} + \text{H}_2\text{N}-\text{X} \longrightarrow \left[\text{R}^1\text{R}^2\text{C(OH)(N(X)H)} \right] \longrightarrow \text{R}^1\text{R}^2\text{C}=\text{N}-\text{X} + \text{H}-\text{O}-\text{H}$$

This is an **addition–elimination** reaction, also called a **condensation** reaction.

An important condensation reaction is the reaction of carbonyl compounds with 2,4–dinitrophenylhydrazine (2,4–DNPH). For example, with ethanal:

ethanal   2,4-dinitrophenylhydrazine

$$\text{CH}_3\text{CH}=\text{O} + \text{H}_2\text{N}-\text{NH}-\text{C}_6\text{H}_3(\text{NO}_2)_2 \longrightarrow \text{CH}_3\text{CH}=\text{N}-\text{NH}-\text{C}_6\text{H}_3(\text{NO}_2)_2$$

The products of condensation reactions between 2,4-dinitrophenylhydrazine and carbonyl compounds are all orange crystalline solids with well-defined melting points. They are useful in identifying individual carbonyl compounds. The condensation product is prepared, its melting point is measured accurately and the compound is identified from tables of melting points.

Quick Questions

8 Ethanal undergoes a condensation reaction with hydroxylamine, NH$_2$OH. Predict the structure of the product of this reaction.

18.5 Oxidation of carbonyl compounds

Table 18.2 shows the effects of some oxidising agents on different carbonyl compounds.

 Table 18.2
The effect of warming different carbonyl compounds with oxidising agents

Carbonyl compound	Oxidising agent	
	Complexed Ag^+ in alkaline solution (Tollen's reagent)	Complexed Cu^{2+} in alkaline solution (Fehling's or Benedict's solution)
ethanal	silver mirror formed on walls of tube	red precipitate of copper(I) oxide formed
propanone	no reaction	no reaction

Aldehydes have a hydrogen atom attached to the C atom in their carbonyl group. This hydrogen is activated by the carbonyl group and is readily oxidised to —OH. Aldehydes are therefore readily oxidised to carboxylic acids. For example, propanal is oxidised by Ag^+:

$$CH_3CH_2-\underset{\underset{O}{\|}}{C}-H + H_2O \longrightarrow CH_3CH_2-\underset{\underset{O}{\|}}{C}-OH + 2H^+ + 2e^-$$

propanal   propanoic acid

$$Ag^+ + e^- \longrightarrow Ag$$

Overall:

$$CH_3CH_2-\underset{\underset{O}{\|}}{C}-H + H_2O + 2Ag^+ \longrightarrow CH_3CH_2-\underset{\underset{O}{\|}}{C}-OH + 2H^+ + 2Ag$$

In general, aldehydes are oxidised to caboxylic acids even by mild oxidising agents such as Ag^+ and Cu^{2+}. Other oxidising agents, including acidified manganate(VII) (MnO_4^-/H^+) and acidified dichromate(VI)($Cr_2O_7^{2-}/H^+$), are also effective.

Ketones are not readily oxidised at all. They have no effect on these mild oxidising agents such as Ag^+ and Cu^{2+}. This is because they have no oxidisable hydrogen atom joined to the carbonyl group. Strong oxidising agents such as hot concentrated nitric acid can oxidise ketones, but the effect is to break up the molecule, forming at least two smaller molecules of carboxylic acid.

Tests for the aldehyde group

The oxidising agents in Table 18.2 are commonly used to test for the aldehyde group.

Fehling's solution is made by mixing copper(II) sulfate solution with an alkaline solution containing 2,3-dihydroxybutanedioate ions (tartrate ions). These complex the Cu^{2+} ions and prevent precipitation of copper(II) hydroxide from the alkaline solution. *Benedict's solution* is similar. Aldehydes reduce blue Cu^{2+} ions in Fehling's solution and Benedict's solution to a red precipitate of copper(I) oxide.

Tollens' reagent is made by adding excess ammonia solution to a solution of silver(I) ions. It contains the $[Ag(NH_3)_2]^+$ ion in alkaline solution. Complexing of Ag^+ by NH_3 prevents precipitation of silver hydroxide. Aldehydes reduce Ag^+ ions to solid Ag on warming, giving a silver mirror.

Many other oxidising agents can be used to convert aldehydes to carboxylic acids. In the laboratory, acidified dichromate(VI) is commonly used to prepare carboxylic acids from aldehydes.

Quick Questions

9 Look at Table 18.2 and answer the questions.
   a Write a half-equation for the reduction of Ag^+ to Ag.
   b Which of the carbonyl compounds in the table are oxidised by Ag^+?
   c Suggest a compound to which ethanal might be oxidised.
   d To what oxidation state is Cu^{2+} reduced by ethanal?
   e Which is the more powerful reducing agent, ethanal or propanone?

Key Points

▶ Aldehydes are readily oxidised to carboxylic acids. Ketones are not readily oxidised.
▶ The aldehyde group can be detected by its reducing action on mild oxidising agents such as Ag^+ and Cu^{2+}.

(a)

(b)

▲ Figure 18.7
Tests for aldehydes as reducing agents:

(a) Benedict's solution: blue Cu^{2+} ions are reduced to red Cu_2O by aldehydes. Fehling's solution is similar.

(b) Tollens' reagent: colourless Ag^+ ions are reduced to a silver mirror by aldehydes.

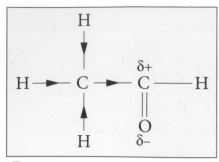

Figure 18.8
The polarising effect of the carbonyl group in ethanal. The polar C=O group draws electrons away from the C atom joined to it, and this makes the C—H bonds more polar than normal.

Quick Questions

10 Which of the following compounds would give a yellow precipitate of triiodomethane (iodoform) when heated with a solution of iodine in alkaline aqueous sodium carbonate?
  a CH_3CH_2CHO
  b $CH_3CHOHCH_3$
  c $CH_3CH_2CH_2OH$
  d $CH_3CH_2COCH_3$
  e $HCHO$

Key Point

Table 18.3 summarises some of the important reactions of ethanal and propanone.

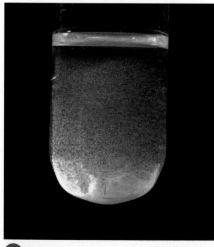

Figure 18.9
Triiodomethane (iodoform) crystals formed by the reaction of ethanal with I_2 and an alkali

18.6 Effect of the carbonyl group on neighbouring atoms

The carbonyl group is polar with a considerable partial positive charge on the carbon atom. This charge has the effect of withdrawing electrons from neighbouring carbon atoms, as shown in Figure 18.8.

The result is to make the C—H bonds of the neighbouring carbon atom more polar than normal. As a result, the hydrogen atoms are more readily replaced than those in alkanes. For example, ethanal reacts readily with iodine even in the dark, forming triiodoethanal:

$$CH_3CHO + 3I_2 \longrightarrow CI_3CHO + 3HI$$
ethanal   triiodoethanal

If an alkali is also present, the triiodoethanal then reacts with the base to form triiodomethane, commonly known as iodoform:

$$CI_3CHO + {}^-OH \longrightarrow CHI_3 + HCOO^-$$
triiodoethanal   triiodomethane

The iodoform reaction

This kind of reaction is not limited to ethanal. Any compound of the general formula CH_3COR, where R is an alkyl group or hydrogen, will form CHI_3 when warmed with iodine and an alkali.

Triiodomethane is a yellow solid with a characteristic smell. Being insoluble in water, it appears as a yellow crystalline precipitate (Figure 18.9). This reaction, known as the **iodoform reaction,** is a useful test for compounds of the form CH_3COR. The reaction is also given by compounds with the formula CH_3CHOHR, since these are themselves oxidised by the iodine to CH_3COR. For example, ethanol, CH_3CH_2OH, gives the iodoform reaction.

Key Point

Compounds containing the group $CH_3CO—$ or $CH_3CHOH—$ give a yellow crystalline precipitate of triiodomethane (iodoform) when heated with iodine and an alkali.

 Table 18.3
Important reactions of ethanal and propanone

Reagent	Reaction type	Product from ethanal	Product from propanone
HCN	addition	CH_3CH with OH and CN	CH_3CCH_3 with OH and CN
$LiAlH_4$, $NaBH_4$ or H_2/Ni	addition/reduction	CH_3CH_2OH	$CH_3CHOHCH_3$
$NH_2NH—DNP^*$	condensation	CH_3CH ‖ $NNH—DNP$	CH_3CCH_3 ‖ $NNH—DNP$
Ag^+(complexed) (Tollens' reagent)	oxidation	CH_3COOH Ag^+ reduced to Ag	no reaction
Cu^{2+} (complexed) (Fehling's or Benedict's solution)	oxidation	CH_3COOH Cu^{2+} reduced to Cu_2O	no reaction
I_2/base	iodoform reaction	yellow crystalline ppte of CHI_3	yellow crystalline ppte of CHI_3

* DNP stands for the 2,4-dinitrophenyl group (Section 18.4)

18.7 Sugars – naturally occurring carbonyl compounds

Sugars are sweet-tasting soluble carbohydrates. Carbohydrates derive their name from the fact that they are composed of carbon, hydrogen and oxygen with H and O in the ratio of 2 : 1, as in water. **Monosaccharides** such as glucose (Figure 18.10) are usually **pentoses** or **hexoses**, i.e. they contain 5 or 6 carbon atoms in their molecules. **Disaccharides** such as sucrose consist of two monosaccharide molecules joined by the elimination of a molecule of water. **Polysaccharides** such as starch are made up of many monosaccharide units joined together. Table 18.4 gives some examples of common carbohydrates. Notice that the monosaccharides all have asymmetric molecules, so they show optical isomerism.

The most obvious feature of the structures of the monosaccharides and disaccharides is the presence of large numbers of —OH groups. These give them a large capacity for hydrogen bonding, so they are involatile solids, soluble in water.

As well as showing the properties of polyhydroxy compounds, sugars show many properties in solution that are typical of carbonyl compounds. For example, glucose gives a crystalline condensation compound with 2,4-dinitrophenylhydrazine (Section 18.4). This is surprising since the structure of glucose shown in Figure 18.10 contains no carbonyl group.

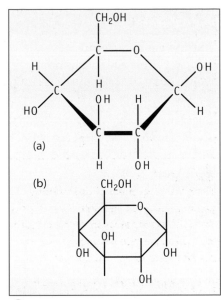

▲ Figure 18.10
The structure of glucose: (a) full structural formula (b) skeletal formula. The hexagonal ring should be thought of as *perpendicular* to the paper, with the —OH groups projecting above or below it.

▼ Table 18.4
Some common carbohydrates

Name	Type	Structure	Occurrence
glucose	monosaccharide, aldose, hexose		occurs abundantly in plants and animals
fructose	monosaccharide, ketose, hexose		in fruit and honey
ribose	monosaccharide, aldose, pentose		component of the molecules of ribonucleic acid (RNA) and vitamin B_{12}
sucrose	disaccharide	glucose — fructose	sugar cane, sugar beet (commonly simply called 'sugar')
maltose	disaccharide	glucose — glucose	malt
lactose	disaccharide	glucose — galactose	milk
starch	polysaccharide	chains of glucose units	plant energy storage organs, e.g. potato, wheat grain, rice
cellulose	polysaccharide	chains of glucose units (linked differently to those in starch)	structural material of plants

Key Points

Sugars (monosaccharides) such as glucose and fructose are polyhydroxy compounds usually containing five or six C atoms.

Sugars exist in ring- and open-chain form. The open chain form contains either an aldehyde or a ketone group.

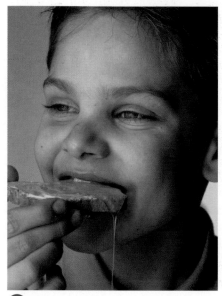

▲ Figure 18.11
Honey is a concentrated solution of sugars including glucose and fructose. Hydrogen bonding between the sugars and water makes the honey viscous.

The carbonyl properties of glucose arise from the fact that it can exist as an 'open-chain' form as well as its normal ring form.

ring form   open-chain form

The two forms are readily interconverted and in aqueous solution about 1% of glucose molecules exist in the open-chain form. This form carries an aldehyde group, so glucose has several properties typical of an aldehyde. Because of this, it is sometimes called an **aldose**. For example, glucose shows the reducing properties typical of an aldehyde. The reduction of Fehling's solution (or Benedict's solution) is a standard test for glucose and other reducing sugars.

The open-chain form of fructose is

As this contains a typical ketone group, fructose is described as **ketose.**

Why does the open-chain form of glucose and other sugars change to the ring form? The answer lies in the tendency of the carbonyl group to undergo nucleophilic addition. The nucleophile involved is the oxygen atom of one of the —OH groups in the same molecule. An internal nucleophilic addition reaction occurs, forming a ring.

glucose: open-chain form   glucose: ring form

This reaction occurs spontaneously. Under normal conditions in aqueous solution the two forms exist in equilibrium, with the ring form predominating.

An understanding of the chemistry of carbohydrates is vital to an understanding of biology. These molecules occur in all living organisms, as structural materials (e.g. cellulose), energy storage compounds (e.g. starch) and primary energy sources (e.g. glucose).

Figure 18.12
Sucrose dissolves because its —OH groups form hydrogen bonds to water.

Quick Questions

11 Consider the sugars below labelled **A**, **B**, and **C**. In each case the normal ring form is shown, with the open-chain form below it.

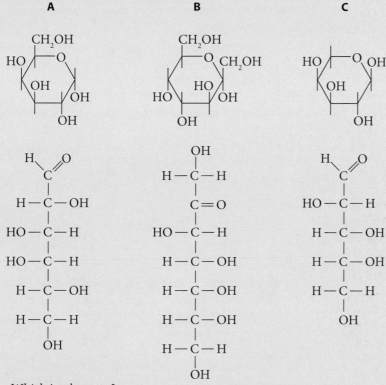

  a Which is a heptose?
  b Which would reduce Tollens' reagent to silver?
  c Which is a ketose?
  d Which would undergo a condensation reaction with 2,4-dinitrophenylhydrazine?

Review questions

1 Consider the following compounds:

A $CH_3COCHCH_2CH_3$
    |
    CH_3

B $CH_3CH_2CH_2CH_2CHO$

C $CH_2OH(CHOH)_4CHO$

D (benzene ring with CHO)

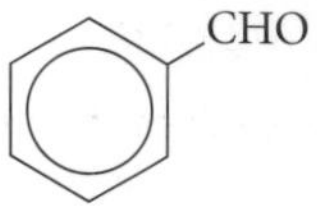

E (cyclohexanone)

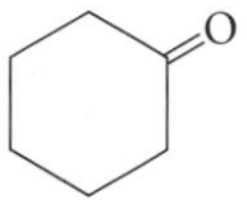

a Which are aldehydes?

b Which are ketones?

c Which is a hexose?

d Name compounds **A**, **B** and **E**.

e Which would produce a red precipitate of copper(I) oxide when boiled with Fehling's solution?

f Which would give a yellow crystalline precipitate when reacted with iodine in alkaline solution?

g Which would be reduced to a secondary alcohol by $NaBH_4$?

2 'The $-C=C-$ and $-C=O$ groups might be expected to show chemical similarity, but in fact they show very little.' Is this true? Illustrate and discuss the statement, referring to a range of examples and to the underlying chemical principles.

3 This question is about the manufacture of ethanal. Read the passage below and then answer the questions on it.

One method for manufacturing ethanal uses the oxidation of ethene. Ethene and oxygen are bubbled together through an aqueous solution containing a $PdCl_2$ catalyst:

$$CH_2=CH_2(g) + \tfrac{1}{2}O_2(g) \xrightarrow{PdCl_2(aq)} CH_3CHO(aq)$$

The mechanism of the reaction involves initial attack on the ethene molecule by water. Since H_2O is a nucleophile, it is necessary to make the ethene, which is normally subject only to electrophilic attack, attractive to nucleophiles. This is done by the Pd^{2+} catalyst, which forms a complex with the ethene, decreasing the electron density in the double bond and making the molecule prone to nucleophilic attack.

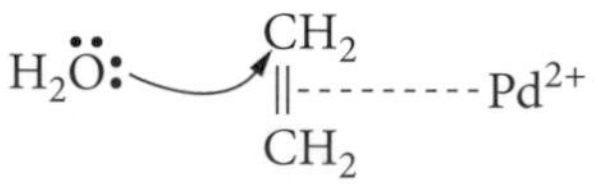

The complex formed in this way breaks down by a number of stages to form ethanal and palladium metal. The palladium metal is oxidised back to Pd^{2+} by oxygen.

a The suggested mechanism involves the formation of a complex between ethene and Pd^{2+}. How do you think ethene is bonded to Pd^{2+} in this complex?

b Suggest a reason why the formation of this complex renders ethene attractive to nucleophiles rather than electrophiles.

c Could this method be used to manufacture *propanone*? If so, what starting material would be used instead of ethene?

4 Predict the formulae of the products of the following reactions.

a $CH_3COCH_2CH_3 + H_2 \xrightarrow{Ni}$

b $C_6H_5COCH_3 + NH_2NH-DNP$

c $CH_3CH_2COCH_2CH_3 + HCN \longrightarrow$

d $C_6H_5CHO \xrightarrow{MnO_4^-/H^+}$

e $CH_3CH_2CHO + Cl_2 \longrightarrow$

5 Write structural formulae for all compounds of molecular formula C_4H_8O containing a carbonyl group. How would you distinguish between the different compounds, using simple chemical tests?

6 a A compound **A** has molecular formula $C_4H_6O_2$. **A** reacts with HCN to form compound **B**, $C_6H_8O_2N_2$. **A** is readily oxidised by acidified potassium dichromate(VI) to an acidic compound **C**, $C_4H_6O_4$. When 1.0 g of **C** is dissolved in water and titrated with 1.0 mol dm^{-3} sodium hydroxide, 16.9 cm^3 of sodium hydroxide are required for neutralisation.

Suggest structural formulae for **A**, **B** and **C** and explain the above reactions.

b A compound **X** contains 64.3% C, 7.1% H, and 28.6% O by mass. Its relative molecular mass is 56. **X** reduces Fehling's solution to copper(I) oxide. **X** reacts with hydrogen in the presence of a nickel catalyst: 0.1 g of **X** was found to react with 80 cm^3 of hydrogen (measured at s.t.p.). (1 mol of gas occupies 22 400 cm^3 at s.t.p.)

Suggest a structural formula for **X** and explain the above reactions.

7 Suggest explanations for the following.

  a Five different oxidation products of ethane-1,2-diol are known (excluding carbon dioxide and water).

  b When ethanal is added to heavy water (D_2O) containing a small amount of base, trideuteroethanal (CD_3CHO) is formed. (Deuterium (D) is an isotope of hydrogen, sometimes written as ^2H. Its chemical properties are identical to those of normal hydrogen, but its atoms have twice the mass.)

  c Trichloroethanal (CCl_3CHO) undergoes addition reactions far more readily than ethanal.

8 This question is about Grignard reagents, compounds that are of great use in organic synthesis for forming carbon—carbon bonds.

Grignard reagents are compounds of general formula RMgX where X = Br or I. They are very reactive, and they contain the highly nucleophilic ion R^-. When a solution of a Grignard reagent in dry ether is added to a carbonyl compound, the R^- attacks the carbonyl group:

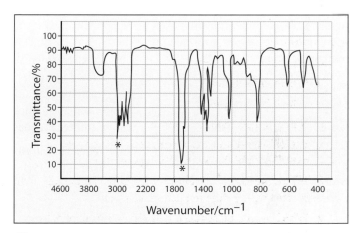

If water is now added to this product, an alcohol is formed:

Overall:

$$\text{C=O} + RMgBr + H_2O \longrightarrow \text{C} \overset{OH}{\underset{R}{}} + MgBrOH$$

Predict the formulae of the compounds formed when the following are treated with the Grignard reagent methyl magnesium bromide, CH_3MgBr, followed by water.

  a methanal,

  b ethanal,

  c propanone,

  d carbon dioxide.

9 **X** is a carbonyl compound with three carbon atoms in its molecule.

Figure 18.13 shows the infrared spectrum of **X**.
Figure 18.14 shows the NMR spectrum of **X**.
Use Table 13.4, giving the characteristic IR absorptions of some common bonds, and Table 13.5, giving the chemical shifts for some types of protons, when you answer this question.

  a Use the spectra to identify **X**. Give its name and formula.

  b Identify the bonds which are responsible for the peaks marked * on the IR spectrum (Figure 18.13).

  c Identify the protons responsible for each of the peaks in the NMR spectrum (Figure 18.14).

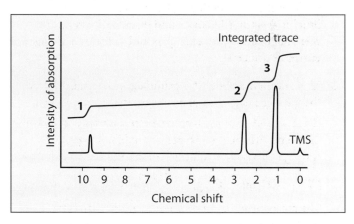

Figure 18.13
The infrared spectrum of **X**

Figure 18.14
The NMR spectrum of **X**

19 Carboxylic acids and their derivatives

19.1 Carboxylic acids

Why is an ant like a pickled onion?

They both owe their powerful effect to carboxylic acids. Ants sting by spraying methanoic acid onto the skin when they bite; pickled onions taste sharp because of ethanoic acid in vinegar.

Carboxylic acids and their derivatives occur widely in nature. They are also present in many manufactured products, such as soaps and polyesters.

Carboxylic acids contain the **carboxyl** group, $-\overset{\displaystyle \|}{\underset{\displaystyle O}{C}}-OH$. Their general formula is usually written as RCOOH, where R is an alkyl or aryl group or hydrogen. They are named using the suffix **-oic acid** after a prefix indicating the number of carbon atoms (including the one in the carboxyl group). The first two members of the series, whose systematic names are methanoic acid and ethanoic acid, are often called by their traditional names, formic acid and acetic acid.

When two carboxyl groups are present, the acid is dibasic and the suffix **-dioic acid** is used. Table 19.1 gives some information about five important carboxylic acids.

Figure 19.1

An ant bites a termite. The ant makes a wound with its jaws, then sprays on methanoic acid.

Table 19.1
Some important carboxylic acids

Formula	Systematic name	Traditional name	Occurrence and uses
HCOOH	methanoic acid	formic acid (from Latin *formica*, an ant)	used in textile processing and as a grain preservative; ants use it as a poison, spraying it when they bite their victims
CH_3COOH	ethanoic acid	acetic acid (from Latin *acetum*, vinegar)	in vinegar; used in making artificial textiles
CH_3CH_2COOH	propanoic acid	propionic acid	calcium propanoate is used as an additive in bread manufacture
COOH (benzene ring)	benzoic acid		food preservative
COOH COOH	ethanedioic acid	oxalic acid	in rhubarb leaves

In the carboxyl group, $-C=O$ and $-O-H$ are so close together that they affect one another's properties a great deal. Carboxylic acids therefore have many reactions that are different from those of both alcohols which contain the $O-H$ bond and carbonyl compounds containing the $C=O$ bond.

The properties of the $-COOH$ group are modified only slightly when it is attached to a benzene ring. As a result, aromatic carboxylic acids have many properties in common with aliphatic ones.

Physical properties

Ethanoic acid melts at 17 °C and boils at 118 °C. This means that although it is normally a liquid in the laboratory, it freezes in cold weather. Because of the readiness with which it freezes, and the similarity of solid ethanoic acid to ice, pure ethanoic acid is sometimes described as 'glacial'. Dilute solutions of the acid freeze at about the same temperature as water.

The boiling point of ethanoic acid is higher than that of either ethanol (78 °C), which has the same number of carbon atoms, or propan-1-ol (97 °C), which has the same relative molecular mass. The relatively high boiling points of carboxylic acids are due to hydrogen bonding. Carboxylic acids form stronger hydrogen bonds than alcohols. This is because their —OH group is more polarised due to the presence of the electron-withdrawing —C=O group:

Carboxylic acids can also form doubly hydrogen-bonded dimers (Figure 19.3). Carboxylic acids in the liquid and solid states exist mostly in this dimer form.

Because of their capacity for hydrogen bonding, the early members of the carboxylic acids are miscible with water in all proportions. As with all homologous series, solubility decreases with increasing molecular size. Table 19.2 shows the physical properties of some carboxylic acids.

Methanoic, ethanoic and propanoic acids have strong, sharp vinegary odours. The C_4 to C_8 acids (butanoic to octanoic) have very strong, unpleasant odours. Butanoic acid is responsible for the smell of rancid butter and is present in human sweat. Its smell can be detected at concentrations of 10^{-11} mol dm^{-3} by humans and at concentrations of 10^{-17} mol dm^{-3} by dogs (Figure 19.4).

▼ Table 19.2
Some properties of carboxylic acids

Acid	Formula	State at room temperature	Boiling point/°C	Solubility in water
methanoic	HCOOH	l	101	infinite*
ethanoic	CH_3COOH	l	118	infinite*
propanoic	CH_3CH_2COOH	l	141	infinite*
butanoic	$CH_3CH_2CH_2COOH$	l	164	infinite*
octanoic	$CH_3(CH_2)_6COOH$	l	237	slightly soluble
benzoic	C_6H_5COOH	s	249	slightly soluble
ethanedioic	$(COOH)_2$	s		soluble

*miscible in all proportions

Making carboxylic acids

Carboxylic acids can be prepared in the laboratory by several methods. Two important methods are described below.

By oxidation of primary alcohols

Primary alcohols are converted to carboxylic acids by *prolonged* oxidation on refluxing with acidified dichromate(VI). The alcohol is converted first to an aldehyde, which is further oxidised to the acid. There are further details of these reactions in Section 17.5.

▲ Figure 19.2
Both these bottles contain pure ethanoic acid (acetic acid). Pure ethanoic acid freezes at 17 °C. The bottle on the left has been kept in the fridge.

▲ Figure 19.3
Carboxylic acids can form hydrogen-bonded dimers

$$\text{RCH}_2\text{OH} \xrightarrow{\text{Cr}_2\text{O}_7{}^{2-}\text{(aq)/H}^+\text{(aq)}} \text{RCHO} \xrightarrow{\text{Cr}_2\text{O}_7{}^{2-}\text{(aq)/H}^+\text{(aq)}} \text{RCOOH}$$

primary alcohol   aldehyde   carboxylic acid

By hydrolysis of nitriles

Nitriles contain the —C≡N group (Section 26.2). Nitriles are hydrolysed to carboxylic acids by heating with strong acid or to carboxylic acid salts by heating with strong alkali, e.g.

$$\text{RCN} + 2\text{H}_2\text{O} + \text{H}^+ \longrightarrow \text{RCOOH} + \text{NH}_4{}^+$$

$$\text{RCN} + 2\text{H}_2\text{O} + \text{OH}^- \longrightarrow \text{RCOO}^- + \text{NH}_3 + \text{H}_2\text{O}$$

Nitriles are themselves prepared from halogenoalkanes (Section 16.5).

19.2 Some important reactions of carboxylic acids

Formation of salts

Like all acids, carboxylic acids react with inorganic bases to form salts and water. For example:

$$\text{CH}_3\text{COOH(aq)} + \text{NaOH(aq)} \longrightarrow \text{CH}_3\text{COO}^-\text{Na}^+\text{(aq)} + \text{H}_2\text{O(l)}$$

ethanoic acid   sodium hydroxide   sodium ethanoate

Salts of carboxylic acids (also called carboxylates) have very different properties from the free acids, because they are ionic rather than molecular substances. All carboxylate salts are solids, and most are soluble in water. Because carboxylic acids are only weakly acidic, their salts are basic. For example, a $1.0\,\text{mol}\,\text{dm}^{-3}$ solution of sodium ethanoate has a pH of about 9.5.

As the hydrocarbon chain becomes longer, the properties of carboxylate salts tend more and more towards those of molecular hydrocarbons rather than ionic salts.

Soap

Carboxylate salts with medium-length chains have the properties of both hydrocarbons and salts. This gives them detergent properties, which means they help oils and grease to mix with water. They are called **soaps**.

Soap was the first detergent and it has been used for thousands of years. Soaps are a mixture of the salts of medium- and long-chain carboxylic acids. They are made by boiling fats with a strong alkali, like sodium or potassium hydroxide.

A typical salt present in household soap is sodium octadecanoate (sodium stearate):

$$\text{CH}_3(\text{CH}_2)_{16}\text{COO}^-\,\text{Na}^+$$

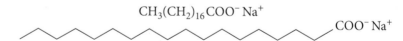

COO⁻ Na⁺

Quick Questions

2   Look carefully at the structure of sodium octadecanoate.
    a   Which end of the structure would you expect to mix well with water?
    b   Which end would mix well with grease?
    c   Explain how this substance would help to remove grease from a surface, so it can be washed away by water.
    d   Suggest reasons why hot water is better for washing in than cold water.

Key Points

Carboxylic acids react with inorganic bases to form carboxylate salts. Carboxylate salts with medium to long hydrocarbon chains are **soaps**.

Detergents are substances which are attracted to both polar and non-polar materials, helping them to mix.

Figure 19.4
The ability of a dog to track a person is due to its ability to detect carboxylic acids in the sweat from the person's feet. Each person's sweat glands produce a characteristic blend of carboxylic acids which can be detected and recognised by the sensitive nose of the dog.

Key Point

Carboxylic acids can be prepared by a) the oxidation of primary alcohols or b) the hydrolysis of nitriles.

Figure 19.5
Rhubarb leaves are poisonous because they contain ethanedioic acid (oxalic acid). This acid is poisonous, because of the toxic nature of its anion, the ethanedioate ion. The correct way to treat this kind of poisoning is to prescribe an alkali which can form an insoluble salt with ethanedioic acid. For example, magnesium hydroxide can be used, because magnesium ethanedioate is insoluble.

Figure 19.6 shows what happens when soap chains interact with grease.

Figure 19.7 shows the effect of detergent on the ability of water to wet a fabric.

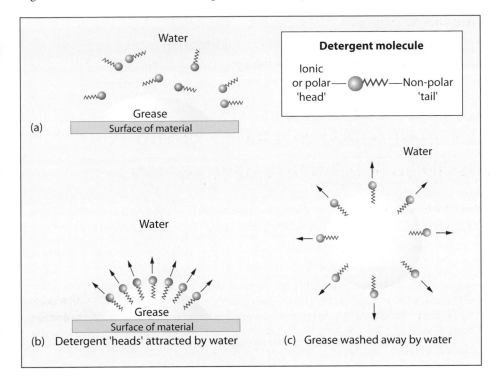

Figure 19.6
Effect of detergent on grease

Figure 19.7
The effect of soap on the wetting ability of water. The droplets at the top are pure water; those at the bottom are a soap and water solution.

Reaction with bases

This has already been considered earlier in this section.

Formation of acyl halides

See Section 19.5

Esterification

This important reaction is considered in the next section.

19.3 Esters

Figure 19.8 illustrates an experiment in which ethanol and ethanoic acid are heated in the presence of concentrated sulfuric acid. Study it, then answer Quick question 3.

Figure 19.8
Reaction between ethanol and ethanoic acid in the presence of concentrated sulfuric acid

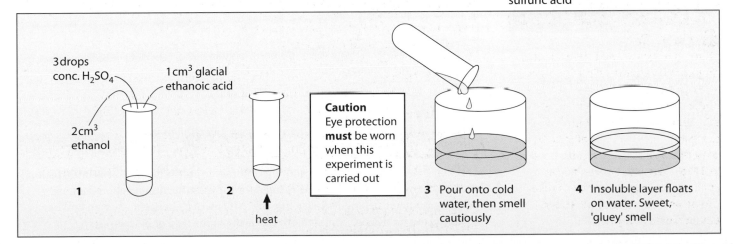

Quick Questions

3 Look at Figure 19.8.

a Does **i** ethanol, **ii** ethanoic acid, **iii** sulfuric acid, mix with water?

b Do any of the substances in part a have a sweet, 'gluey' smell?

c What can you say about the product of this reaction between ethanol and ethanoic acid?

d Why do you think the reaction mixture is poured onto cold water before smelling?

Key Point

Esters are the product of the reaction of an alcohol and carboxylic acid. The reaction is often called **esterification**. Water is also formed, and an acid catalyst is needed.

Figure 19.10
Beeswax contains esters made from long-chain alcohols and long-chain acids. A typical ester in beeswax is $C_{15}H_{31}COOC_{30}H_{61}$

Ethanol and ethanoic acid react together to form an ester, ethyl ethanoate:

$$CH_3\underset{\underset{O}{\|}}{C}-OH + CH_3CH_2OH \longrightarrow CH_3\underset{\underset{O}{\|}}{C}-OCH_2CH_3 + H_2O$$

ethanoic acid   ethanol   ethyl ethanoate

This reaction proceeds extremely slowly under ordinary conditions. However, it goes at an appreciable rate in the presence of a strong acid catalyst such as sulfuric or hydrochloric acid. This is a general method of preparing esters, and can be applied to any combination of acid and aliphatic alcohol:

$$R^1-\underset{\underset{O}{\|}}{C}-OH + R^2OH \longrightarrow R^1-\underset{\underset{O}{\|}}{C}-OR^2 + H_2O$$

Notice that the bridging oxygen atom between the R^1CO and R^2 groups in the ester could have come either from the alcohol or from the acid. As the equation stands, there is no way of telling from which molecule it originated. To put the problem another way, does the oxygen atom in the alcohol molecule end up in the ester or in the water?

The answer to this question was found by two American chemists, Roberts and Urey, using a technique known as **isotopic labelling**. The method is outlined in Figure 19.9.

---

1 Prepare methanol 'labelled' with the oxygen isotope ^{18}O

$$CH_3{}^{18}OH$$

2 React this with benzoic acid in the presence of an acid catalyst

$$\bigcirc\!\!-COOH + CH_3{}^{18}OH$$

3 Two sets of products are possible according to which bond breaks in $CH_3{}^{18}OH$

$\bigcirc\!\!-CO^{18}OCH_3 + H_2O$ or $\bigcirc\!\!-COOCH_3 + H_2{}^{18}O$

    $(M_r = 138)$ $(M_r = 136)$

4 Separate the ester and measure its relative molecular mass in a mass spectrometer

5 M_r of the ester is found to be 138, corresponding to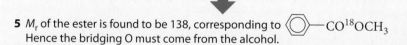
Hence the bridging O must come from the alcohol.

Figure 19.9
Investigating the fate of oxygen atoms in an esterification reaction

This experiment shows that the bridging O comes from the alcohol. This information enabled chemists to propose a mechanism for the esterification reaction. Isotopic labelling is often used to investigate the mechanism of a chemical reaction and has advanced our knowledge considerably, especially in the field of biochemistry.

Naming esters

Esters are the products of condensation reactions (Section 18.4) between alcohols and carboxylic acids. They can also be prepared by reacting alcohols with acyl chlorides (Section 19.5). Esters contain the $-\overset{\|}{\underset{O}{C}}-O$ functional group.

Esters are named by regarding them as alkyl (or aryl) derivatives of carboxylic acids. Thus the name is obtained from a stem indicating the alcohol from which the ester is derived, with the suffix **-yl**, followed by a stem indicating the acid, with the suffix-**oate.** So the ester derived from methanol and propanoic acid is called methyl propanoate. You can work out the name of an ester from its formula in four stages:

1 Divide the formula into two portions by mentally drawing a line after the bridging O of the $-COO$ group, e.g.

$$CH_3CH_3C-O\overset{|}{\underset{O}{\|}}CH_2CH_2CH_2CH_3$$

2 Name the portion that does *not* carry the $-COO$ group. This portion is an alkyl group – *butyl* in this example.
3 Name the portion that carries the $-COO$ group. This portion is a carboxylate group – *propanoate* in this example.
4 Combine **2** and **3** to give the name of the ester – *butyl propanoate*.

Properties and uses of esters

Unlike the acids and alcohols from which they are derived, esters have no free $-OH$ groups so they cannot form hydrogen bonds. They are therefore volatile compared with acids and alcohols of similar molecular mass and they are not very soluble in water. Ethyl ethanoate, for example, is a liquid, boiling point 77 °C. At 25 °C its solubility is 8.5 g per 100 g of water.

Volatile esters have characteristically pleasant, fruity smells. The flavour and fragrance of many fruits and flowers are due to mixtures of compounds, many of them esters (Table 19.3). Artificial fruit flavourings are made by mixing synthetic esters to give the approximate flavour (raspberry, pear, cherry, etc.) required. Organic acids are usually added to give the sharp taste characteristic of fruit. Artificial fruit flavours can only approximate to the real thing, because it would be too costly to include all the components of the complex mixture present in the real fruit.

Some adhesives smell of ethyl ethanoate. Polystyrene cement, for example, consists of polystyrene dissolved in ethyl ethanoate. When the cement is applied, the ethyl ethanoate evaporates, leaving behind the solid plastic which binds together the surfaces being joined.

Name of ester	Skeletal formula	Flavour
ethyl butanoate		pineapple
octyl ethanoate		orange
methyl butanoate		apple
pentyl ethanoate		banana
ethyl methanoate		rum

◀ Table 19.3
Some esters and their flavours

Quick Questions

Try these questions to test your understanding of how esters are named and prepared.

4 Give the name of
   a $CH_3CH_2COOCH_3$,
   b $HCOOCH_2CH_3$.
5 Write the formula of
   a heptyl decanoate,
   b phenyl benzoate.
6 How would you prepare propyl propanoate?
7 Write the formula of the product formed when ethanoic acid and butan-2-ol are warmed together in the presence of an acid catalyst.

Key Points

Esters are formed by a condensation reaction between an alcohol and a carboxylic acid, usually in the presence of an acid catalyst.

Esters have the general formula R^1COOR^2.

Ester hydrolysis is the reverse of esterification and involves boiling an ester with an acid or alkali.

8 a Write an expression for the equilibrium constant for the esterification of ethanol and ethanoic acid as in the equation alongside.

b If one mole of ethanoic acid and one mole of ethanol are allowed to reach equilibrium at 25 °C, how many moles of ethyl ethanoate are present at equilibrium? (Assume $K_c = 4$ at 25 °C.)

9 What would you expect to happen if ethyl ethanoate were refluxed with an excess of water for a long time?

△ Figure 19.11
Pyrethrum flowers contain a natural insecticide, called pyrethrin. It is an ester, which is hydrolysed quickly when eaten by mammals, making it harmless to them, but is toxic to insects. Pyrethroid insecticides can be manufactured which work in the same way but are safe in the environment.

Polyesters

Polyesters are polymers joined together by ester linkages. Polyesters are covered in Section 27.4.

Esterification as an equilibrium reaction: hydrolysis

The reaction of a carboxylic acid with an alcohol to form an ester is fairly slow, even in the presence of an acid catalyst. As well as being kinetically quite slow, esterification is also an equilibrium reaction that does not normally reach completion.

If known quantities of ethanol, ethanoic acid and hydrochloric acid catalyst are sealed together and left for two or three weeks, an equilibrium is reached.

Ethanol, ethanoic acid, ethyl ethanoate and water are all present, as well as unchanged acid catalyst.

$$CH_3CH_2OH + CH_3COOH \underset{}{\overset{H^+}{\rightleftharpoons}} CH_3COOCH_2CH_3 + H_2O$$
$$\text{ethanol} \qquad \text{ethanoic acid} \qquad\qquad \text{ethyl ethanoate}$$

If the reaction mixture is now titrated with standard sodium hydroxide, the amount of ethanoic acid present at equilibrium can be found.

From this and the starting amounts of reactants and catalyst, the amounts of all the other components of the equilibrium mixture can be worked out. Using these results the equilibrium constant K_c for the esterification reaction can easily be found: its value is about 4 at 25 °C.

A mixture of one mole of ethanol and one mole of ethanoic acid brought to equilibrium at 25 °C contains two-thirds of a mole of both ethyl ethanoate and water and one-third of a mole of both ethanol and ethanoic acid. The esterification reaction is thus far from complete, and it is readily reversed.

Hydrolysis of esters

The reverse of esterification is **ester hydrolysis**, in which an ester reacts with water to form an alcohol and a carboxylic acid. Like esterification, ester hydrolysis is a slow reaction, but it is speeded up by an acid catalyst. It is also catalysed by alkali, but in this case the carboxylic acid formed by hydrolysis reacts with excess alkali to form the carboxylate salt. This removes the carboxylic acid from the equilibrium mixture as it is formed, which means that ester hydrolysis can proceed to completion in the presence of alkali. This cannot happen when an acid catalyst is used.

For example:

$$CH_3COOCH_2CH_3 + {}^-OH \longrightarrow CH_3COO^- + CH_3CH_2OH$$
$$\text{ethyl ethanoate} \qquad\qquad\qquad \text{ethanoate ion} \qquad \text{ethanol}$$

or

$$CH_3COOCH_2CH_3 + NaOH \longrightarrow CH_3COO^-Na^+ + CH_3CH_2OH$$

Esters are therefore hydrolysed more effectively in alkaline than in acidic solution. Even with an acid or alkali catalyst, the reaction is quite slow and the mixture must be boiled under reflux. Strictly speaking, the alkali is not acting catalytically in this reaction because, as the equation shows, the hydroxide ions, ^-OH, are used up.

Ester hydrolysis is important in biological systems. The photograph in Figure 19.11 shows an important example.

19.4 The carboxyl group and acidity

Like water and alcohols, carboxylic acids contain the —OH group, but they are much stronger acids than water or alcohols. Most of the acids we taste in our food, such as ethanoic acid in vinegar and citric acid in lemons, are carboxylic acids. They seem to have a very strong acid taste, although they are only present in fairly low concentrations – vinegar is only about 7% ethanoic acid, and the rest is mainly water. Carboxylic acids are weak compared to many inorganic acids such as hydrochloric acid. (Remember the distinction between concentration and strength of an acid in Section 7.14.)

When a carboxylic acid dissociates in water, it forms a carboxylate anion $RCOO^-$:

$$H_2O + R-C\overset{O}{\underset{OH}{\diagup}} \longrightarrow H_3O^+ + R-C\overset{O}{\underset{O^-}{\diagup}}$$

carboxylic acid   carboxylate anion

X-ray diffraction studies of the anion show that the two carbon–oxygen bonds are of equal length, which implies that the two bonds are identical. This suggests that the negative charge and double bond character are distributed evenly over the whole carboxylate group, delocalising the negative charge:

$$R-C\overset{O}{\underset{O}{\lessgtr}}^-$$

Each oxygen atom effectively carries half a negative charge. This delocalisation of charge makes the carboxylate anion less likely to join up with H^+ again. As a result, the equilibrium

$$RCOOH(aq) + H_2O(l) \rightleftharpoons RCOO^-(aq) + H_3O^+(aq)$$

is much further to the right than the equivalent equilibrium for alcohols. So, carboxylic acids are stronger acids than alcohols. Nevertheless, the majority of the acid is still in the un-ionised form, and a solution of ethanoic acid of concentration $0.1\,mol\,dm^{-3}$ is only 0.3% ionised.

The effect of neighbouring groups on acid strength

In general, the higher the density of negative charge on the $RCOO^-$ ion, the more it will attract H^+ and reform the un-ionised RCOOH. So we can say that the more negative the charge density on $RCOO^-$, the weaker the acid will be, and vice-versa.

The nature of the neighbouring group to which —COOH is attached can have a considerable effect on the strength of a carboxylic acid. In general, an electron-withdrawing group reduces the density of negative charge on —COO^-, and so increases the strength of the acid. An electron-donating group does the opposite.

Table 19.4 gives the K_a values for different carboxylic acids. (Section 21.10 deals with K_a values: the point you need to know here is that the larger the K_a value, the stronger the acid.)

◀ Table 19.4
The strength of different carboxylic acids (the higher the K_a value, the stronger the acid)

Carboxylic acid	Formula	K_a at 25°C/mol dm^{-3}
methanoic acid	HCOOH	1.6×10^{-4}
ethanoic acid	CH_3COOH	1.7×10^{-5}
propanoic acid	CH_3CH_2COOH	1.3×10^{-5}
butanoic acid	$CH_3CH_2\,CH_2COOH$	1.5×10^{-5}
octanoic acid	$CH_3(CH_2)_6COOH$	1.4×10^{-5}
chloroethanoic acid	$ClCH_2COOH$	1.3×10^{-3}
dichloroethanoic acid	$Cl_2CHCOOH$	5.0×10^{-2}
trichloroethanoic acid	Cl_3CCOOH	2.3×10^{-1}
benzoic acid	C_6H_5COOH	6.4×10^{-5}
ethanedioic (oxalic) acid	$(COOH)_2$	3.5×10^{-2} (first dissociation) 4.0×10^{-5} (second dissociation)

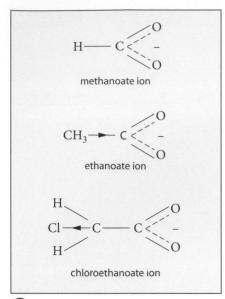

Figure 19.12
Anions formed from methanoic, ethanoic and chloroethanoic acids

Definition

Acyl halides have the general formula RCOHal, where R can be an alkyl or aryl group. They are named using the suffix -oyl halide.

Quick Questions

11 What reagents would you use to prepare:
   a propanoyl chloride,
   b butanoyl bromide?

The electron-donating —CH_3 group in ethanoic acid tends to increase the negative charge density on the carboxylate group in the ethanoate ion. Ethanoic acid is therefore weaker than methanoic acid, which carries no electron-donating alkyl group. On the other hand, in chloroethanoic acid, the electron-withdrawing Cl atom reduces the charge density on the chloroethanoate ion, so chloroethanoic caid is stronger than ethanoic acid (Figure 19.12).

The electron-donating effect of all alkyl groups is roughly equal, so carboxylic acids with more than one carbon atom (ethanoic, propanoic, butanoic, etc.) are roughly equal in strength.

Quick Questions

10 Explain why:
   a dichloroethanoic acid is a stronger acid than chloroethanoic acid,
   b benzoic acid is a stronger acid than ethanoic acid,
   c ethanedoic acid is a stronger acid for its first dissociation than for its second dissociation.

19.5 Acyl chlorides

Acyl halides are derivatives of carboxylic acids in which the –OH group is replaced by a halogen atom. The commonest acyl halides are the **acyl chlorides**, which contain the group:

$$-\overset{\displaystyle}{\underset{\displaystyle O}{C}}-Cl$$

An important example is ethanoyl chloride, $CH_3-\overset{}{\underset{O}{C}}-Cl$, which is usually written CH_3COCl.

Acyl halides have the general formula **RCOHal**, where R can be an alkyl or an aryl group. They are named using the suffix **-oyl halide** after a prefix indicating the number of carbon atoms in the molecule, including that in the —COCl group. So, for example, $CH_3CH_2CH_2COCl$ is named butanoyl chloride, and the acyl chloride formed from benzoic acid, C_6H_5COOH, is named benzoyl chloride, C_6H_5COCl.

Preparing acyl chlorides

Acyl chlorides are prepared by reacting carboxylic acids with the same chlorinating agents that are used to convert alcohols to chloroalkanes (Section 17.4). Phosphorus pentachloride is often used, as in the example below, but $SOCl_2$ can also be used.

$$\underset{\text{ethanoic acid}}{CH_3\underset{O}{\overset{\|}{C}}OH} + PCl_5 \longrightarrow \underset{\text{ethanoyl chloride}}{CH_3\underset{O}{\overset{\|}{C}}Cl} + HCl + POCl_3$$

To prepare acyl bromides, PBr_3 is often used as the brominating agent.

The reactivity of acyl chlorides

Like halogenoalkanes, acyl halides react with nucleophiles, but they are much more reactive than halogenoalkanes.

Reaction with water: hydrolysis

When water is added to ethanoyl chloride (which is a liquid), a violent reaction occurs. The liquid boils and fumes of hydrogen chloride are evolved:

$$CH_3COCl(l) + H_2O(l) \longrightarrow CH_3COOH(aq) + HCl(g)$$

This reaction is far more vigorous than the hydrolysis of chloroalkanes. Chloroethane, for example, is hydrolysed only very slowly by water, even when heated. Aryl chlorides such as chlorobenzene are even slower to hydrolyse.

The reason for the rapid hydrolysis of acyl halides is that they react with nucleophiles by a different mechanism to that of halogenoalkanes. The first stage of the reaction involves *addition* of the nucleophile across the $C=O$ double bond (Figure 19.13). This is similar to the nucleophilic addition reactions of carbonyl compounds, described in Section 18.1. This addition reaction is immediately followed by *elimination* of HCl from the intermediate ion.

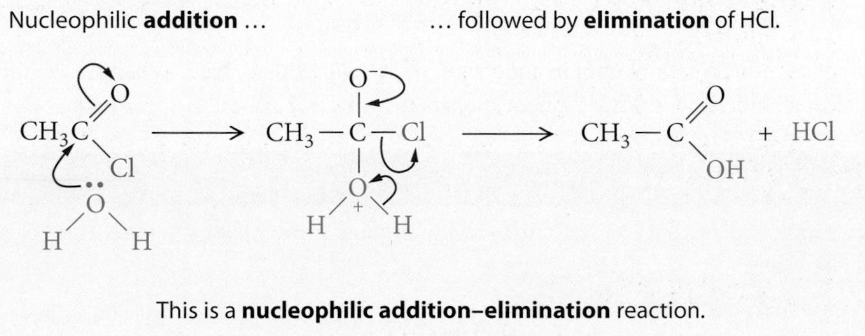

Nucleophilic **addition** ...   ... followed by **elimination** of HCl.

This is a **nucleophilic addition–elimination** reaction.

◀ Figure 19.13
The mechanism of acylation. This example shows the reaction of ethanoyl chloride with H_2O to form ethanoic acid.

This different reaction mechanism enables acyl halides to react with nucleophiles much more readily than halogenoalkanes. As you might expect, acyl halides react with all the nucleophiles that will substitute halogenoalkanes.

Hydrolysis of acyl halides is not a useful reaction, because the product is a carboxylic acid, which is what acyl halides are prepared from in the first place. However, acyl halides react with other nucleophiles to give useful products.

Like halogenoalkanes, acyl halides are important in synthesis. Just as RHal is used to attach an alkyl group R to a nucleophile, RCOHal is used to attach an acyl group, RCO. Because of their greater reactivity, acyl halides react readily with those nucleophiles, such as ammonia and alcohols, that react only slowly with halogenoalkanes.

Some useful reactions of acyl halides

Although ethanoyl chloride is used in the following examples, the reactions are general for acyl halides.

With ammonia and primary amines

Ethanoyl chloride reacts violently with an aqueous solution of ammonia at room temperature. The product is ethanamide, an amide. Amides have the general formula $RCONH_2$, and should not be confused with amines, RNH_2.

$$CH_3COCl(l) + NH_3(aq) \longrightarrow CH_3CONH_2(aq) + HCl(aq)$$
ethanamide

Primary amines can be thought of as ammonia in which one H atom is substituted by an alkyl or aryl group. They react vigorously at room temperature with acyl halides in a similar way to ammonia. For example:

$$CH_3COCl(l) + CH_3CH_2NH_2(aq) \longrightarrow CH_3CONHCH_2CH_3(aq) + HCl(aq)$$
N-ethylethanamide

The product is a secondary amide. The *N* in the name indicates that the amide is substituted on the nitrogen atom, rather than a carbon atom.

With alcohols and phenols

Ethanoyl chloride reacts vigorously with alcohols at room temperature to produce esters. For example, with ethanol and ethanoyl chloride, the product is ethyl ethanoate:

$$CH_3COCl(l) + CH_3CH_2OH(l) \longrightarrow CH_3COOCH_2CH_3(l) + HCl(g)$$
$$\text{ethyl ethanoate}$$

This is an important method for making esters. It is particularly useful for making esters of phenol, C_6H_5OH, because phenol does not react directly with carboxylic acids. Phenol does, however, react readily with acyl halides. The ester phenyl benzoate can be made by the reaction of phenol with benzoyl chloride:

$$C_6H_5COCl + C_6H_5OH \longrightarrow C_6H_5COOC_6H_5 + HCl$$
$$\text{benzoyl chloride} \quad \text{phenol} \qquad \text{phenyl benzoate}$$

In practice, a solution of phenol in aqueous sodium hydroxide is used. When this solution is shaken with benzoyl chloride, phenyl benzoate forms as a crystalline precipitate.

Key Points

The characteristic reaction mechanism of acyl halides is nucleophilic addition–elimination.

Acyl halides react with all the nucleophiles that halogenoalkanes react with, but they react much more readily.

Quick Questions

12 Name and predict the structures of the organic products of the reactions between:
   a benzoyl chloride and ethanol,
   b propanoyl chloride and methylamine, CH_3NH_2,
   c ethanoyl chloride and phenol.

Review questions

1 Consider the following compounds:

A CH₃CHCOOH
     |
     CH₃

B $CH_3(CH_2)_{16}COO^-Na^+$

C $CH_3CH_2COOCH_3$

D $HOOCCH_2CH_2COOH$

a Which is an ester?

b Which is a dibasic acid?

c Which is a carboxylate salt?

d Name each compound.

e Which would be almost insoluble in water, but would slowly dissolve when boiled with sodium hydroxide solution?

f Which would form a pleasant-smelling liquid when warmed with ethanol and concentrated sulfuric acid?

g Which would have detergent properties?

2 Predict the formulae of the products of the following reactions:

a $CH_3CH_2CH_2OH(l) \xrightarrow{Cr_2O_7{}^{2-}(aq)/H^+(aq)}$

b $CH_3CN(l) + HCl(aq) \xrightarrow{boil}$

c $CH_3COO(CH_2)_4CH_3(l) + NaOH(aq) \xrightarrow{boil}$

d $CH_3COOH(aq) + Ca(OH)_2(aq) \longrightarrow$

e $CH_3COCl(l) + CH_3CH_2NH_2(g) \longrightarrow$

3 How would you carry out the following conversions in the laboratory? One or more steps may be involved in each case.

a CH_3CHO to $CH_3COOCH_2CH_3$,

b $CH_2 = CH_2$ to CH_3COOH,

c CH_3CH_2CN to $CH_3CH_2COO^-Na^+$,

d CH_3CH_2OH to $CH_3COO^-Na^+$.

4 What simple *chemical* tests would you use to distinguish one compound from the other in the following pairs:

a CH_3CH_2CHO and CH_3CH_2COOH,

b $CH_3COCH_2CH_3$ and $CH_3COOCH_2CH_3$,

c CH_3COOCH_3 and $CH_3COOCH_2CH_3$,

d $CH_2FCOO^-Na^+$ and $CH_3COO^-Na^+$.

5 Write structural formulae for all acids and esters of molecular formula $C_4H_8O_2$. What simple chemical tests would you use to distinguish between the different esters?

6 This question is about an investigation into the mechanism of hydrolysis of ethyl ethanoate:

$$CH_3COOCH_2CH_3 + H_2O \xrightarrow[\text{catalyst}]{\text{HCl}} CH_3COOH + CH_3CH_2OH$$

ethyl ethanoate   ethanoic acid   ethanol

Two experiments, **A** and **B**, were carried out. Read the accounts of the experiments, then answer the questions which follow.

Experiment A Ethyl ethanoate was refluxed with deuterated water, D_2O, containing deuterium chloride, DCl (D = $_1^2$H). The two hydrolysis products were separated and purified and their relative molecular masses were measured using a mass spectrometer.

The alcohol formed had $M_r = 47$, and the acid had $M_r = 61$.

a Compare M_r for the alcohol formed in this experiment with the value expected from the equation above. Account for any difference.

b Compare M_r for the acid formed in this experiment with the value expected from the equation above. Account for any difference.

c What information, if any, does experiment **A** give about the mechanism of the hydrolysis reaction?

Experiment B Ethyl ethanoate was refluxed with $H_2^{18}O$ containing HCl. The two hydrolysis products were separated and purified and their relative molecular masses were measured using a mass spectrometer. The alcohol formed had $M_r = 46$, and the acid had $M_r = 62$.

d and **e** Repeat steps **a** and **b** above, this time using the results from experiment **B**.

f What information, if any, does experiment **B** give about the mechanism of the hydrolysis reaction?

g Compare your answer with the information given in Section 19.3.

7 Acyl halides (RCOHal) are very reactive and are important in organic synthesis, as they can be used to attach an RCO group onto a nucleophile. Generally, they are thought of as intermediates which can be used to form a wide range of chemical compounds.

a What type of reaction does an acyl halide undergo?

b Give the structure of the product formed when ethanoyl chloride (CH_3COCl) reacts with:

  i water,

  ii ammonia,

  iii butan-1-ol,

  iv ethylamine.

c Why are acyl halides so reactive?

d Are acyl halides more or less reactive than halogenoalkanes? Give a reason for your answer.

8 Suggest explanations for the following observations.

a Butane, propan-1-ol, propanal and ethanoic acid all have approximately the same relative molecular mass, but their boiling points are 273 K, 371 K, 322 K and 391 K, respectively.

b When chloroethanoyl chloride, $ClCH_2COCl$, is added to water, one chlorine atom is quickly substituted to give $ClCH_2COOH$, but the other Cl atom is substituted only slowly, even on boiling.

Definitions

Oxidation is the loss of electrons or an increase in oxidation number.

Reduction is the gain of electrons or a decrease in oxidation number.

⌃ Figure 20.1
Rows of copper cells at a copper refinery. This photo shows one row of anodes withdrawn from its cell.

Key Points

Electrochemical cells generate electric currents from chemicals. They transform chemical energy into electrical energy.

Electrolytic cells use electric currents to produce chemicals. They transform electrical energy into chemical energy. The process is called electrolysis.

Definitions

Redox reactions involve transfer of electrons.

Oxidising agents accept electrons.

Reducing agents donate electrons.

20.1 What is electrochemistry?

Electrochemistry involves a study of the interconversion of electrical energy and chemical energy. So, it covers the use of electricity to produce new materials by electrolysis in electrolytic cells as well as the use of chemicals to generate electric currents in electrochemical cells and batteries.

Some important aspects of electrochemistry, including electrolysis and the redox reactions involved in electrochemical processes, have already been discussed in Chapter 6. During electrolysis, the reactions at the electrodes involve redox. For example, in the purification of impure copper (Figure 20.1), metallic copper atoms undergo **oxidation** by loss of electrons at the anode, while copper(II) ions, Cu^{2+}, undergo **reduction** by gaining electrons and forming copper metal at the cathode:

Anode (+): impure copper $\qquad Cu(s) \longrightarrow Cu^{2+}(aq) + 2e^-$

Cathode (−): pure copper $\qquad Cu^{2+}(aq) + 2e^- \longrightarrow Cu(s)$

20.2 Electrochemical cells

Redox reactions are involved at the terminals of electrochemical cells in which electric currents are generated, as well as being involved at the electrodes of electrolytic cells when electric currents provide the energy to bring about chemical reactions.

When zinc is added to copper(II) sulfate solution, a redox reaction occurs in which zinc goes into solution as zinc ions, $Zn^{2+}(aq)$ and copper(II) ions, $Cu^{2+}(aq)$ are converted to copper metal. The blue solution due to copper(II) ions gradually fades and a deposit of red-brown copper is produced.

The equation for the reaction is:

$$Zn(s) + \underbrace{Cu^{2+}(aq) + SO_4^{2-}(aq)}_{\text{copper sulfate solution}} \longrightarrow \underbrace{Zn^{2+}(aq) + SO_4^{2-}(aq)}_{\text{zinc sulfate solution}} + Cu(s)$$

Ignoring the sulfate ions which take no part in the reaction we get:

$$Zn(s) + Cu^{2+}(aq) \longrightarrow Zn^{2+}(aq) + Cu(s)$$

During the reaction, zinc metal, $Zn(s)$, loses electrons forming aqueous zinc ions, $Zn^{2+}(aq)$, while aqueous copper ions, $Cu^{2+}(aq)$, accept these electrons forming copper metal, $Cu(s)$.

Using the apparatus in Figure 20.2 the overall process can be separated into two distinct half-reactions:

$$Zn(s) \longrightarrow Zn^{2+}(aq) + 2e^- \qquad \text{at the zinc rod}$$

electron transfer

$$Cu^{2+}(aq) + 2e^- \longrightarrow Cu(s) \qquad \text{at the copper rod}$$

At the zinc rod, zinc atoms give up electrons and form zinc ions which go into solution as $Zn^{2+}(aq)$. The electrons flow from the zinc rod through the external circuit to the copper rod where they combine with Cu^{2+} ions to produce copper atoms. As electrons flow through the connecting wire they form an electric current and cause the bulb to light up.

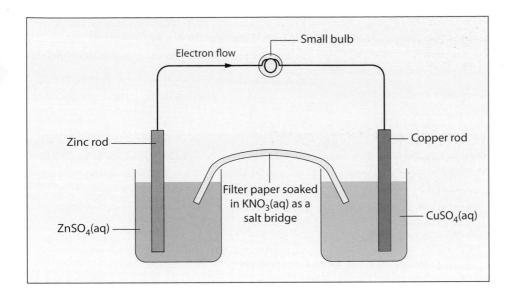

Small bulb

Electron flow

Zinc rod

Copper rod

Filter paper soaked
in KNO_3(aq) as a
salt bridge

$ZnSO_4$(aq)

$CuSO_4$(aq)

Figure 20.2
Electron transfer between zinc and
copper(ɪɪ) sulfate solution

Quick Questions

1 a Will electrons flow from
      positive to negative or from
      negative to positive?
   b Will the zinc rod be positive
      or negative?
   c What happens when the salt
      bridge is removed?

The overall movement of charge around the circuit in Figure 20.2 is shown in
Figure 20.3.

When the circuit is complete, the zinc rod 'dissolves away' and the concentration of
Zn^{2+}(aq) in the left-hand beaker rises. If this net increase in positive charge is not
'neutralised', the reaction will soon stop because the excess positive charge in the
solution will prevent any more Zn^{2+} ions from entering it.

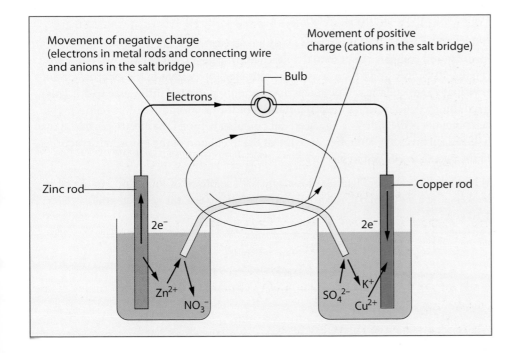

Movement of negative charge
(electrons in metal rods and connecting wire
and anions in the salt bridge)

Movement of positive
charge (cations in the salt bridge)

Bulb

Electrons

Zinc rod

Copper rod

$2e^-$

$2e^-$

Zn^{2+}

NO_3^-

SO_4^{2-}

K^+

Cu^{2+}

Figure 20.3
Movement of charge around a circuit

The excess positive charge from Zn^{2+}(aq) in the solution is counteracted by nitrate
ions, NO_3^-(aq), moving out of the salt bridge into the solution or by zinc ions moving
into the salt bridge (Figure 20.3).

In the right-hand beaker, copper ions are converted to copper atoms leaving an excess
of negative charge (sulfate ions, SO_4^{2-}(aq)) in solution. This excess negative charge is
balanced by SO_4^{2-}(aq) moving into the salt bridge or potassium ions, K^+(aq), diffusing
out of the salt bridge into the solution.

Notice how the salt bridge serves two main functions.

It completes the circuit by allowing ions carrying charge to move one way or the other *without the solutions in the beakers mixing*. When the salt bridge is removed, the current ceases and the bulb goes out because charge (whether it is ions or electrons) can no longer flow around the circuit.

It provides anions and cations to replace those consumed or balance those produced at the electrodes.

Quick Questions

2 Write an overall balanced equation and then two ionic half-equations for each of the following redox reactions. In each case, state which species is oxidised and which is reduced.
   a Magnesium with silver nitrate solution.
   b Aqueous chlorine with sodium iodide solution.

20.3 Cell potentials

In an electrochemical cell such as that in Figure 20.2, a redox reaction occurs as two half-reactions in separate **half-cells**. Electrons flow from one half-cell to the other through a wire connecting the electrodes. A salt bridge connects the solutions of the two half-cells and completes the circuit.

The simplest salt bridge is just a strip of filter paper soaked in potassium nitrate solution and dipping into the solution in each beaker. More permanent salt bridges consist of porous solids such as sintered glass or inverted 'U' tubes filled with agar jelly containing potassium nitrate solution.

If the bulb in Figure 20.2 is replaced by a high resistance voltmeter, it is possible to measure and compare the potential difference (voltage) created by the two half-cells (Figure 20.4). By using a high resistance voltmeter, the current in the external circuit is virtually zero and the cell produces its maximum **potential difference** (p.d.). This maximum p.d. is known as the **cell potential** or the electromotive force (e.m.f.). The cell potential is the difference in electrical potential between the electrodes in the two half-cells. It gives a quantitative measure of the likelihood of the redox half-reactions in the two half-cells taking place.

When the apparatus in Figure 20.4 is set up with solutions of $1\,mol\,dm^{-3}$ in the two beakers, the maximum p.d. recorded by the voltmeter (i.e. the cell potential) at 25 °C is 1.10 volts.

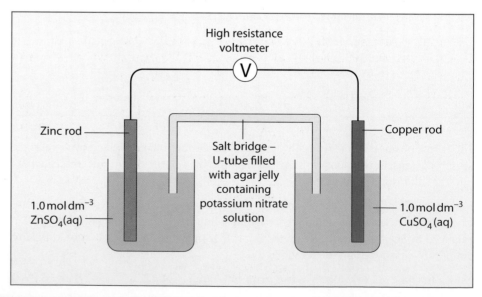

Figure 20.4
Measuring the cell potential for a cell composed of a $Zn(s)|Zn^{2+}(aq)$ half-cell and a $Cu(s)|Cu^{2+}(aq)$ half-cell

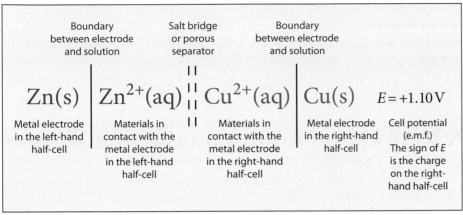

Figure 20.5
A cell diagram for the cell composed of the half-cells, $Zn(s)|Zn^{2+}(aq)$ and $Cu(s)|Cu^{2+}(aq)$

In order to describe different cells, chemists use a convenient shorthand called a **cell diagram**. The combination of half-cells in Figure 20.4 and the resulting e.m.f. is summarised as a cell diagram in Figure 20.5 with an explanation below each item.

The sign and magnitude of the voltage shown in the cell diagram above relates to changes indicated by the order of items in the diagram, i.e.

$$Zn(s) \longrightarrow Zn^{2+}(aq) + 2e^-$$

and

$$Cu^{2+}(aq) + 2e^- \longrightarrow Cu(s)$$

So, for the change,

$$Zn(s) + Cu^{2+}(aq) \longrightarrow Zn^{2+}(aq) + Cu(s) \quad E_{cell} = +1.10\,V$$

and for the reverse process,

$$Zn^{2+}(aq) + Cu(s) \longrightarrow Zn(s) + Cu^{2+}(aq) \quad E_{cell} = -1.10\,V$$

Notice that this convention leads to a positive value of E_{cell}, the cell potential, if the reaction tends to go from left to right as in the cell diagram. A negative value of E indicates that the cell reaction as written will not take place, but will tend to go in the opposite direction.

When the $Zn(s)|Zn^{2+}(aq)$ half-cell in Figure 20.4 is replaced by other metal/metal ion half-cells, the results in Table 20.1 are obtained.

The first three metals in Table 20.1 (zinc, iron and lead) all produce positive cell potentials, indicating that the half-reactions go from left to right as written in the cell diagram (Figure 20.6).

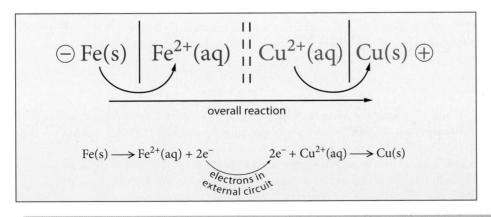

$$Fe(s) \longrightarrow Fe^{2+}(aq) + 2e^- \qquad 2e^- + Cu^{2+}(aq) \longrightarrow Cu(s)$$

electrons in external circuit

> **Definition**
>
> A **cell diagram** is a convenient shorthand which summarises the potential, the electrodes and other materials of an electrochemical cell.

Table 20.1
The cell potentials developed between various metal/metal ion half-cells and the $Cu(s)|Cu^{2+}(aq)$ half-cell

Metal/metal ion half-cell	Cell potential /volts
$Zn(s)|Zn^{2+}(aq)$	+1.10
$Fe(s)|Fe^{2+}(aq)$	+0.78
$Pb(s)|Pb^{2+}(aq)$	+0.47
$Cu(s)|Cu^{2+}(aq)$	0.00
$Ag(s)|Ag^+(aq)$	−0.46

Figure 20.6
The directions of reactions in a cell composed of the half-cells $Fe(s)|Fe^{2+}(aq)$ and $Cu(s)|Cu^{2+}(aq)$

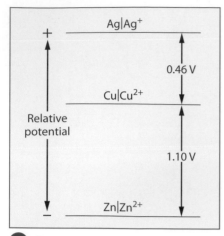

Figure 20.7
Relative cell potentials

When the cells deliver a current, all three metals (zinc, iron and lead) form the negative terminal with respect to copper. These metals are stronger reducing agents than copper and go into solution as their ions. On the other hand, the $Ag(s)\,|\,Ag^+(aq)$ half-cell produces a negative potential relative to the copper half-cell. This means that the silver electrode is positive with respect to copper and in this case copper is acting as the reducing agent. The cell reaction can be summarised as

$$Ag(s)\,|\,Ag^+(aq) \mathbin{\vdots\vdots} Cu^{2+}(aq)\,|\,Cu(s) \qquad E_{cell} = -0.46\ \text{V}$$

So, the reaction

$$2Ag(s) + Cu^{2+}(aq) \longrightarrow 2Ag^+(aq) + Cu(s)$$

has a negative cell potential of –0.46 volts. It is the reverse reaction which actually occurs when a current flows. Figure 20.7 shows diagrammatically the relative potentials for the copper, zinc and silver half-cells.

Notice that the cell potentials in Table 20.1 can be used to compare the relative tendencies of the metals to release electrons and form ions. Thus zinc, at the head of the table, is a strong reducing agent, releasing electrons readily, whilst silver at the bottom of the table is a very poor reducing agent. The cell potentials also give a quantitative measure of the position of a metal in the electrochemical (redox) series.

Quick Questions

3 a What factors, besides the particular metals and ions, will affect the value of cell potentials measured using apparatus like that in Figure 20.4?

b Look at Figure 20.7. Write half-equations for reactions in the half-cells and calculate the cell potential for the following cell:

$$Zn(s)\,|\,Zn^{2+}(aq) \mathbin{\vdots\vdots} Ag^+(aq)\,|\,Ag(s)$$

20.4 Standard electrode potentials

In measuring cell potentials, it is impossible to obtain the potential (e.m.f.) of a single half-cell because voltages can only be measured for a complete circuit with two electrodes. However, it would be extremely useful if we could summarise all cell data by giving each half-cell a characteristic value. This can be done by choosing one particular half-cell and giving it a **standard electrode potential** of zero. The electrode potentials of other half-cells can then be compared with this standard.

The standard chosen for electrode potentials is not a metal but hydrogen. The standard hydrogen half-cell, usually called the **standard hydrogen electrode** is shown in Figure 20.8. This half-cell can be summarised as $Pt\,|\,H_2(g),\ 2H^+(aq) \mathbin{\vdots}$ with a standard electrode potential, $E^{\ominus} = 0.00\ \text{V}$ at 25 °C.

The standard hydrogen electrode consists of hydrogen gas at one atmosphere pressure and 25 °C bubbling around a platinised platinum electrode. The electrode is immersed in a $1.0\ \text{mol dm}^{-3}$ solution of H^+ ions. It is simultaneously bathed in $H^+(aq)$ and $H_2(g)$. Hydrogen is adsorbed on the platinum and an equilibrium is established between the adsorbed layer of H_2 gas and $H^+(aq)$ ions in the solution.

$$H_2(g) \rightleftharpoons 2H^+(aq) + 2e^-$$

The platinised platinum electrode has three functions.

1 It acts as an *inert* metal connection to the $H_2\,|\,H^+$ system. (There is no tendency for Pt to react in any way.)

2 It allows H_2 gas to be adsorbed onto its surface.

3 It is covered by a loosely deposited layer of finely divided platinum (i.e. it is *platinised*). This increases its surface area so that it can establish an equilibrium between $H_2(g)$ and $H^+(aq)$ as rapidly as possible.

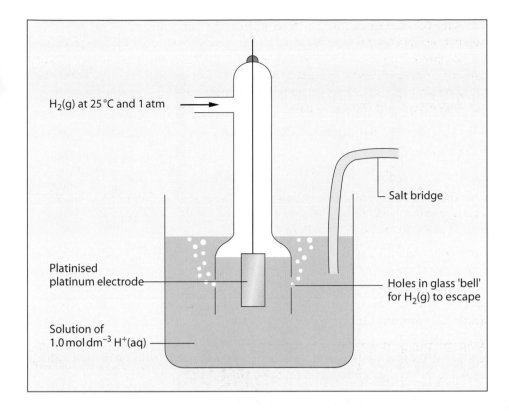

Figure 20.8
The standard hydrogen half-cell

Standard conditions

The potential of a standard hydrogen electrode (half-cell) will depend on the temperature, the concentration of H^+ ions and the pressure of the $H_2(g)$. So, in measuring and comparing electrode potentials we must choose the same **standard conditions** for all measurements. The standard conditions chosen are similar to those for thermochemical measurements (Chapter 5).

▶ All solutions have a concentration of $1\,mol\,dm^{-3}$.
▶ Any gases involved have a pressure of 1 atmosphere.
▶ The temperature is 25 °C (298 K).
▶ Platinum is used as the electrode when the half-cell system does not include a metal.

When cell potentials and electrode potentials are measured under standard conditions the values obtained are called *standard* cell potentials and *standard* electrode potentials with the symbols $E^{\ominus}_{cell}$ and $E^{\ominus}$ respectively.

20.5 Measuring standard electrode potentials

Figure 20.9 shows a standard $Cu^{2+}(aq)\,|\,Cu(s)$ half-cell and a standard $Fe^{3+}(aq)\,|\,Fe^{2+}(aq)$ half-cell. The $Fe^{3+}(aq)\,|\,Fe^{2+}(aq)$ system does not involve a metal so platinum must be used as the electrode. Notice also that the solution in this half-cell is $1.0\,mol\,dm^{-3}$ with respect to both $Fe^{3+}(aq)$ and $Fe^{2+}(aq)$.

The standard electrode potentials of all half-cells are measured relative to a standard hydrogen electrode. The standard electrode potential of the $Cu^{2+}(aq)\,|\,Cu(s)$ half-cell is therefore the potential difference between the electrodes of a cell consisting of the standard hydrogen half-cell and the standard $Cu^{2+}(aq)\,|\,Cu(s)$ half-cell (Figure 20.10). When this cell is set up the potential is +0.34 volts and copper is the positive terminal. The reactions occurring at the electrodes are:

$$H_2(g) \longrightarrow 2H^+(aq) + 2e^- \qquad \text{and}$$
$$Cu^{2+}(aq) + 2e^- \longrightarrow Cu(s)$$

> **Key Point**
>
> The standard conditions for electrochemical measurements are similar to those for thermochemical measurements.

> **Note**
>
> **Platinised** platinum is only used as the electrode in the standard hydrogen electrode where it is important to maintain an equilibrium between H_2 gas and aqueous H^+ ions. Other half-cells requiring an inert metal electrode use ordinary platinum.

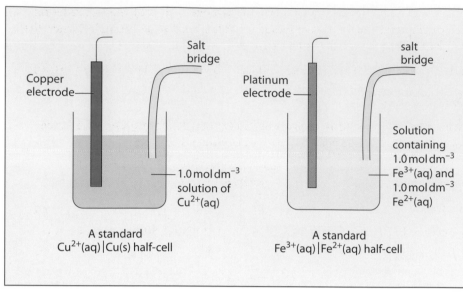

Figure 20.9
Two standard half-cells

As the standard electrode potential for the system $H_2(g) \longrightarrow 2H^+(aq) + 2e^-$ is arbitrarily taken as zero, the standard electrode potential for the system $Cu^{2+}(aq) + 2e^- \longrightarrow Cu(s)$ is $+0.34$ V. The plus sign is used because the copper electrode is positively charged with respect to the standard hydrogen electrode. So we can write:

$$Cu^{2+}(aq) + 2e^- \longrightarrow Cu(s) \qquad\qquad E^\ominus = +0.34\,V$$

and

$$Pt(s)\,|\,H_2(g), 2H^+(aq)\,\vdots\vdots\,Cu^{2+}(aq)\,|\,Cu(s) \qquad E^\ominus_{cell} = +0.34\,V$$

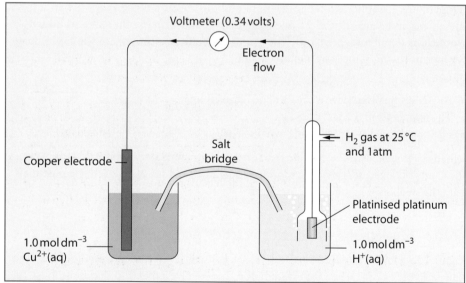

Figure 20.10
Measuring the standard electrode potential of the $Cu^{2+}(aq)\,|\,Cu(s)$ half-cell

Definition

The **standard electrode potential**, $E^\ominus$ of a standard half-cell is the potential of that half-cell relative to a standard hydrogen electrode under standard conditions.

$E^\ominus$ relates to the reduction half-equation and has the same sign as the half-cell relative to the standard hydrogen half-cell.

Key Point

Standard electrode potentials are sometimes called standard reduction potentials or standard redox potentials, because they relate to reduction half-reactions.

By convention, the oxidised form is written first when a particular redox half-equation and its standard electrode potential are being referred to. Thus, $Cu^{2+}(aq)\,|\,Cu(s)$, $E^\ominus = +0.34$ V means that the half-cell reaction

$$Cu^{2+}(aq) + 2e^- \longrightarrow Cu(s)$$

has a standard electrode potential of $+0.34$ V. **Standard electrode potentials** are sometimes called **standard reduction potentials** because they relate to the *reduction of the more oxidised species.*

When inert electrodes are present (as in the hydrogen half-cell), the reduced form of the components of the system should be written next to the electrode. Hence, the standard hydrogen electrode is written as:

$2H^+(aq), H_2(g) \mid Pt(s)$ or as $Pt(s) \mid H_2(g), 2H^+(aq)$, but *not* as $Pt(s) \mid 2H^+(aq), H_2(g)$.

Worked example

When a standard zinc half-cell is connected to a standard hydrogen half-cell, the voltage produced is 0.76 volts and the zinc electrode is negative.

a What is the standard electrode potential of the $Zn(s) \mid Zn^{2+}(aq)$ half-cell?
b Write equations for the half-reactions which occur in each half-cell.

Answer

a The standard electrode potential of the $Zn(s) \mid Zn^{2+}(aq)$ half-cell relates to the reduction of Zn^{2+} ions to Zn:

$$Zn^{2+}(aq) + 2e^- \longrightarrow Zn(s)$$

As the zinc electrode is negative compared to the standard hydrogen electrode, its standard electrode potential will have a negative value. So, we can write:

$$Zn^{2+}(aq) + 2e^- \longrightarrow Zn(s) \qquad E^\ominus = -0.76\,V$$

b As the zinc electrode is negative, electrons flow in the external circuit from the zinc half-cell to the hydrogen half-cell. So, the half reaction in the zinc half-cell is:

$$Zn(s) \longrightarrow Zn^{2+}(aq) + 2e^-$$

and that in the hydrogen half-cell is:

$$2H^+(aq) + 2e^- \longrightarrow H_2(g)$$

Remember that a current will only flow in the external circuit if the overall *cell* potential is positive and, in this example, the appropriate $E^\ominus$ values are:

$$Zn(s) \longrightarrow Zn^{2+}(aq) + 2e^- \qquad E^\ominus = +\,0.76\,V$$

$$2H^+(aq) + 2e^- \longrightarrow H_2(g) \qquad E^\ominus = 0.00\,V$$

overall: $\quad Zn(s) + 2H^+(aq) \longrightarrow Zn^{2+}(aq) + H_2(g) \qquad E^\ominus_{cell} = +0.76\,V$

Notice also in this case that the reaction at the hydrogen electrode and the current in the external circuit are in the opposite direction to those which occurred when a $Cu(s) \mid Cu^{2+}(aq)$ half-cell was connected to a standard hydrogen electrode.

20.6 Relative strengths of oxidising and reducing agents

Standard electrode potentials provide a direct measure of the relative oxidising and reducing power of different species.

The standard electrode potentials of the zinc, hydrogen and copper half-cells are listed in Table 20.2 in order to emphasise the relative strengths of the different oxidising and reducing agents. Cu^{2+} accepts electrons more readily than H^+ or Zn^{2+}. It is a stronger oxidising agent than either H^+ or Zn^{2+}. On the other hand, Zn is a more powerful reducing agent than either H_2 or Cu since it will reduce H^+ to H_2 and Cu^{2+} to Cu.

Notice that *the relative size of $E^\ominus$ gives a measure of the strengths of both oxidants and reductants.* When standard electrode (reduction) potentials are listed from the most positive to the most negative value, Cu^{2+}, the more powerful oxidising agent is always above Zn^{2+}.

4 True or false: The statement $Cu^{2+}(aq) \mid Cu(s)$, $E^\ominus = +0.34$ volts means that:
  a Copper is more positive than Cu^{2+} ions.
  b A solution of $Cu^{2+}(aq)$ is 0.34 volts more positive than a Cu electrode immersed in it.
  c The reaction $Cu^{2+}(aq) + 2e^- \longrightarrow Cu(s)$ has a standard electrode potential of +0.34 volts.
  d The copper electrode of a standard $Cu^{2+}(aq) \mid Cu(s)$ half-cell is 0.34 volts more positive than the platinum electrode in a standard hydrogen half-cell to which it is connected.

5 Write the half-equation and the standard electrode potential for the right-hand electrode in each of the following cells:
$Pt \mid H_2(g), 2H^+(aq) \mid\mid$
$Pb^{2+}(aq) \mid Pb(s)$
$E^\ominus_{cell} = -0.13\,V$
$Pt \mid H_2(g), 2H^+(aq) \mid\mid Fe^{3+}(aq),$
$Fe^{2+}(aq) \mid Pt$
$E^\ominus_{cell} = +0.77\,V$

6 The standard electrode potential for the $Zn^{2+}(aq) \mid Zn(s)$ electrode is $-0.76\,V$.
The standard cell potential, $E^\ominus_{cell}$, for the cell,
$Zn(s) \mid Zn^{2+}(aq) \mid\mid Sn^{2+}(aq) \mid Sn(s)$
is $+0.62\,V$.
  a Write half-equations for the reactions at the electrodes when the cell delivers a current.
  b What is the standard electrode potential of the $Sn^{2+}(aq) \mid Sn(s)$ electrode?

Table 20.2
Standard electrode potentials of the zinc, hydrogen and copper half-cells

	Oxidising agent		Reducing agent		$E^{\ominus}$/V
increasing strength of oxidising agent	$Cu^{2+}(aq) + 2e^-$	$\longrightarrow$	$Cu(s)$	increasing strength of reducing agent	+0.34
	$H^+(aq) + e^-$	$\longrightarrow$	$\frac{1}{2}H_2(g)$		0.00
	$Zn^{2+}(aq) + 2e^-$	$\longrightarrow$	$Zn(s)$		−0.76

However, for the reverse reactions to those in Table 20.2,

$$Zn(s) \longrightarrow Zn^{2+}(aq) + 2e^-, \qquad E^{\ominus} = +0.76\,V$$

and

$$Cu(s) \longrightarrow Cu^{2+}(aq) + 2e^-, \qquad E^{\ominus} = -0.34\,V$$

In this case, Zn, the more powerful reducing agent is always below Cu and this is reflected in a more positive value of $E^{\ominus}$ for $Zn(s) \longrightarrow Zn^{2+}(aq) + 2e^-$.

The standard electrode potentials for a large number of redox half-reactions are shown in Table 20.3. The table is arranged so that the strongest oxidising agent, F_2, with the most positive value for $E^{\ominus}$ is at the top of the list. The weakest oxidising agent, K^+, with the most negative value for $E^{\ominus}$ is at the bottom. So, F_2 is a better competitor for electrons than any other oxidising agent in the list and K^+ is the weakest competitor for electrons.

Key Point

The more positive the value of $E^{\ominus}$, the more likely (energetically favourable) is the reaction.

Table 20.3
Standard electrode potentials

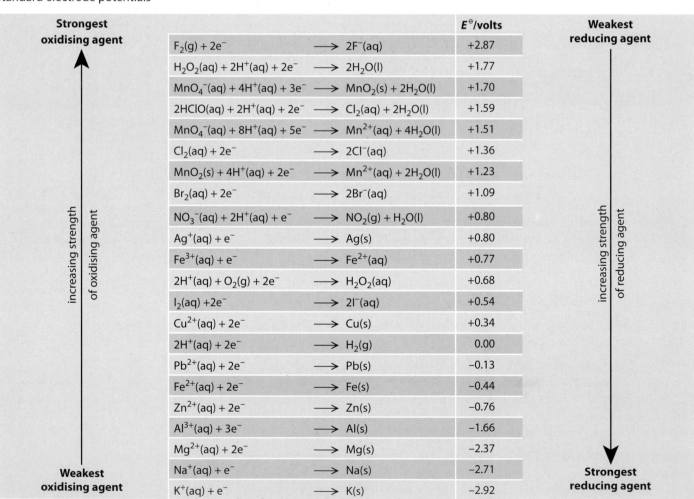

Strongest oxidising agent		$E^{\ominus}$/volts	Weakest reducing agent
$F_2(g) + 2e^-$ $\longrightarrow$ $2F^-(aq)$		+2.87	
$H_2O_2(aq) + 2H^+(aq) + 2e^-$ $\longrightarrow$ $2H_2O(l)$		+1.77	
$MnO_4^-(aq) + 4H^+(aq) + 3e^-$ $\longrightarrow$ $MnO_2(s) + 2H_2O(l)$		+1.70	
$2HClO(aq) + 2H^+(aq) + 2e^-$ $\longrightarrow$ $Cl_2(aq) + 2H_2O(l)$		+1.59	
$MnO_4^-(aq) + 8H^+(aq) + 5e^-$ $\longrightarrow$ $Mn^{2+}(aq) + 4H_2O(l)$		+1.51	
$Cl_2(aq) + 2e^-$ $\longrightarrow$ $2Cl^-(aq)$		+1.36	
$MnO_2(s) + 4H^+(aq) + 2e^-$ $\longrightarrow$ $Mn^{2+}(aq) + 2H_2O(l)$		+1.23	
$Br_2(aq) + 2e^-$ $\longrightarrow$ $2Br^-(aq)$		+1.09	
$NO_3^-(aq) + 2H^+(aq) + e^-$ $\longrightarrow$ $NO_2(g) + H_2O(l)$		+0.80	
$Ag^+(aq) + e^-$ $\longrightarrow$ $Ag(s)$		+0.80	
$Fe^{3+}(aq) + e^-$ $\longrightarrow$ $Fe^{2+}(aq)$		+0.77	
$2H^+(aq) + O_2(g) + 2e^-$ $\longrightarrow$ $H_2O_2(aq)$		+0.68	
$I_2(aq) + 2e^-$ $\longrightarrow$ $2I^-(aq)$		+0.54	
$Cu^{2+}(aq) + 2e^-$ $\longrightarrow$ $Cu(s)$		+0.34	
$2H^+(aq) + 2e^-$ $\longrightarrow$ $H_2(g)$		0.00	
$Pb^{2+}(aq) + 2e^-$ $\longrightarrow$ $Pb(s)$		−0.13	
$Fe^{2+}(aq) + 2e^-$ $\longrightarrow$ $Fe(s)$		−0.44	
$Zn^{2+}(aq) + 2e^-$ $\longrightarrow$ $Zn(s)$		−0.76	
$Al^{3+}(aq) + 3e^-$ $\longrightarrow$ $Al(s)$		−1.66	
$Mg^{2+}(aq) + 2e^-$ $\longrightarrow$ $Mg(s)$		−2.37	
$Na^+(aq) + e^-$ $\longrightarrow$ $Na(s)$		−2.71	
Weakest oxidising agent $K^+(aq) + e^-$ $\longrightarrow$ $K(s)$		−2.92	Strongest reducing agent

increasing strength of oxidising agent

increasing strength of reducing agent

On the other hand, potassium metal (K), the most powerful reducing agent in Table 20.3, is at the bottom of the list whilst the fluoride ion (F^-), the weakest reducing agent, is at the top.

This illustrates the conjugate character of an oxidising agent and its corresponding reducing agent. The stronger the oxidising agent, the weaker is its corresponding (conjugate) reducing agent and vice versa.

Notice also in Table 20.3 that the order of metals, as reducing agents according to their $E^\ominus$ values, is exactly the same as the order of metals in the reactivity series. This is why the reactivity series is also known as the electrochemical series.

Quick Questions

7 Using Table 20.3, arrange the following in order of decreasing strength:
   a oxidising agents: Al^{3+}, Br_2, NO_3^-, Fe^{3+}, MnO_4^-
   b reducing agents: Ag, H_2O_2, I^-, Mg, NO_2

20.7 The use of standard electrode potentials

Using standard electrode potentials to calculate standard cell potentials

Using standard electrode potentials, it is possible to calculate the standard cell potential of any cell by combining the standard electrode potentials of the two half-cells that make up the full cell.

The size and sign of any standard electrode potential tells you how feasible or how likely it is that the half-reaction will occur. The more positive the standard electrode potential, the more likely it is that the half-reaction will occur. So, as we have seen already, the half-reaction $Cu^{2+}(aq) + 2e^- \longrightarrow Cu(s)$ with a standard electrode potential of $+ 0.34$ volts is more likely to occur than the half-reaction $Zn^{2+}(aq) + 2e^- \longrightarrow Zn(s)$ with a standard electrode potential of -0.76 volts.

One of the first practical cells to be used was the Daniell cell invented in 1836 (Figure 20.11).

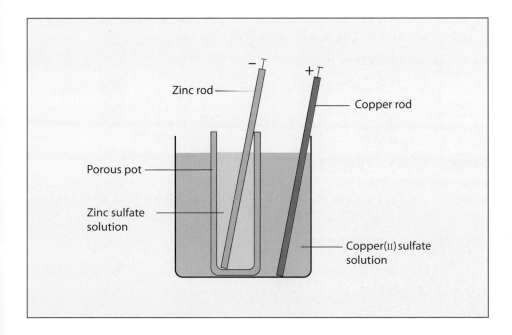

◀ Figure 20.11
A Daniell cell

Notice how closely the Daniell cell resembles the arrangement in Figure 20.4 combining the $Zn^{2+}(aq)\,|\,Zn(s)$ and $Cu^{2+}(aq)\,|\,Cu(s)$ half-cells. The only significant difference is that the Daniell cell uses a rigid porous pot as a salt bridge to allow movement of ions between the two solutions, whereas the arrangement in Figure 20.4 uses agar gel containing potassium nitrate.

When a Daniell cell is operating under standard conditions, the reaction in the copper half-cell is:

$$Cu^{2+}(aq) + 2e^- \longrightarrow Cu(s) \qquad\qquad E^\ominus = +0.34\,V$$

The half-cell reaction goes in this direction because the $Cu^{2+}(aq)|Cu(s)$ half-cell has the more positive electrode potential. The process in the zinc half-cell is therefore the *reverse* of

$$Zn^{2+}(aq) + 2e^- \longrightarrow Zn(s) \qquad\qquad E^\ominus = -0.76\,V$$
$$(i.e.\ Zn(s) \longrightarrow Zn^{2+}(aq) + 2e^- \qquad E^\ominus = +0.76\,V)$$

Combining these two processes, we have:

$$Cu^{2+}(aq) + 2e^- \longrightarrow Cu(s) \qquad\qquad E^\ominus = +0.34\,V$$
$$Zn(s) \longrightarrow Zn^{2+}(aq) + 2e^- \qquad\qquad E^\ominus = +0.76\,V$$

overall: $\quad Cu^{2+}(aq) + Zn(s) \longrightarrow Cu(s) + Zn^{2+}(aq) \qquad E^\ominus_{cell} = +1.10\,V$

Key Point

Standard electrode (reduction) potentials are sometimes listed from the most negative to the most positive rather than from the most positive to the most negative. But whatever the order of listing, the half-cell with the most positive electrode potential has the strongest tendency to gain electrons and therefore acts as the positive electrode when the half-cells are connected.

Therefore, when the cell delivers a current, the cell potential is $+1.10\,V$. The positive sign for $E^\ominus_{cell}$ indicates that the reaction goes spontaneously as written. Zinc goes into solution as zinc ions, electrons flow from the zinc electrode through the external circuit to the copper electrode where copper(II) ions accept these electrons and deposit as copper. The zinc electrode is negative, the copper electrode is positive and the cell diagram is:

$$Zn(s)\,|\,Zn^{2+}(aq)\,\vdots\,\vdots\,Cu^{2+}(aq)\,|\,Cu(s) \qquad\qquad E^\ominus_{cell} = +1.10\,V$$

In calculating cell potentials and predicting reactions at the electrodes, it is important to appreciate that $E^\ominus$ values relate to the feasibility or likelihood of the reaction occurring and *not* to the quantity of materials reacting.

The reaction is just as likely (or unlikely) to occur whether we have a milligram or a kilogram, one mole or two moles, a thimbleful or a bucketful, so the cell potential (e.m.f.) is independent of the number of electrons transferred.

So, for example, if $E^\ominus$ for the reaction

$$Fe^{3+}(aq) + e^- \longrightarrow Fe^{2+}(aq)\ is + 0.77\,V,$$
$E^\ominus$ for the reaction $\quad 2Fe^{3+}(aq) + 2e^- \longrightarrow 2Fe^{2+}(aq)$ is also $+0.77\,V$ and *not* $+1.54\,V$.

Worked example

One of the commonest cells in use today is the alkaline cell (Figure 20.12). This uses a zinc half-cell ($Zn^{2+}(aq) + 2e^- \longrightarrow Zn(s)$, $E = -0.76\,V$) in combination with a half-cell containing a carbon (graphite) electrode in contact with moist manganese(IV) oxide ($2MnO_2(s) + H_2O(l) + 2e^- \longrightarrow Mn_2O_3(s) + 2OH^-(aq)$, $E = +0.74\,V$).

a Write half-equations for the reactions at each electrode.
b Calculate the cell potential.
c Say which electrode is positive.
d Draw the cell diagram.

Answer

a At the carbon electrode, with the more positive electrode potential:

$$2MnO_2(s) + H_2O(l) + 2e^- \longrightarrow Mn_2O_3(s) + 2OH^-$$

At the zinc electrode, $\qquad Zn(s) \longrightarrow Zn^{2+}(aq) + 2e^-$

b The overall cell potential, $E_{cell} = +0.74\,V + 0.76\,V = +1.5\,V$

c In the external circuit, electrons flow from the zinc electrode to the carbon electrode, so the carbon electrode is positive.

d The cell diagram showing the direction in which reactions occur is:

$$Zn(s)|Zn^{2+}(aq) \vdots\vdots 2MnO_2(s) + H_2O(l), Mn_2O_3(s) + 2OH^-(aq)|C_{(graphite)} \quad E_{cell} = +1.5\,V$$

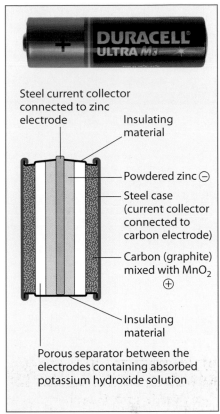

Quick Questions

8 Use Table 20.3 to calculate the standard cell potential, say which direction a current will flow and write the cell diagram for the following pairs of half-reactions:

a $H_2O_2(aq) + 2H^+(aq) + 2e^- \rightleftharpoons 2H_2O(l)$;
$\qquad Cu^{2+}(aq) + 2e^- \rightleftharpoons Cu(s)$

b $Fe^{3+}(aq) + e^- \rightleftharpoons Fe^{2+}(aq)$;
$\qquad Br_2(aq) + 2e^- \rightleftharpoons 2Br^-(aq)$

▲ Figure 20.12

The structure of an alkaline cell with a photo of an actual cell above

Using standard electrode potentials to predict the feasibility of reactions

As we have seen already, $E^\ominus$ values provide an indication of the feasibility of half-reactions and $E^\ominus_{cell}$ values indicate the direction in which reactions occur in electrochemical cells.

A positive value of $E^\ominus_{cell}$ for the reaction:

$$Zn(s) + Cu^{2+}(aq) \longrightarrow Zn^{2+}(aq) + Cu(s); \qquad E^\ominus_{cell} = +1.10\,V$$

indicates that the changes shown in the overall equation are likely to take place. And from experience, we know that the reaction does in fact occur.

On the other hand, a negative value of $E^\ominus_{cell}$ for the reverse reaction

$$Zn^{2+}(aq) + Cu(s) \longrightarrow Zn(s) + Cu^{2+}(aq); \qquad E^\ominus_{cell} = -1.10\,V$$

indicates that the changes shown in the equation will *not* occur. In fact, for reactions in which $E^\ominus_{cell}$ is negative, a reaction is likely to occur in the reverse direction.

So, a positive value for a cell potential tells us that the half-reactions that could take place at the electrodes when a cell is set up are likely to occur. This suggests that the half-reactions are also likely to occur if the reactants are mixed together in a test tube. The more positive the cell potential, the more feasible is the reaction.

Quick Questions

9 a Use Table 20.3 to predict whether a reaction is feasible between:
  i $F_2(g)$ and $Na(s)$
  ii $F^-(aq)$ and $Na^+(aq)$
  iii $F_2(g)$ and $Na^+(aq)$
  iv $Cl_2(aq)$ and $Fe^{2+}(aq)$
  v $I_2(aq)$ and $Br^-(aq)$
  vi $Cl_2(aq)$ and $I^-(aq)$

b Write the overall balanced equations for those reactions in part **a** which you predict are feasible.

20.8 Limitations to the predictions from standard electrode potentials

Although deductions about the feasibility of reactions from standard electrode potentials are normally reliable, the predictions do not always work out in practice. There are two main reasons for this:

▶ a slow rate of reaction and
▶ reaction conditions which are non-standard.

Predictions from $E^\ominus$ data and reaction rates

Although an overall positive value of $E^\ominus_{cell}$ for a pair of redox half-equations suggests that a reaction should take place, in practice the reaction may be too slow to show any detectable change. The important point is that $E^\ominus_{cell}$ relates to the relative stabilities of the reactants and products. It is, in fact, related to $\Delta H^\ominus_{reaction}$. Therefore, it indicates the *energetic* feasibility of the reaction. But, $E^\ominus_{cell}$ gives no indication about the rate of a reaction or its *kinetic* feasibility.

For example, the $E^\ominus$ values below predict that $Cu^{2+}(aq)$ should oxidise $H_2(g)$ to $H^+(aq)$:

$$Cu^{2+}(aq) + 2e^- \longrightarrow Cu(s) \qquad\qquad E^\ominus = +0.34\,V$$

$$H_2(g) \longrightarrow 2H^+(aq) + 2e^- \qquad\qquad E^\ominus = 0.00\,V$$

overall: $Cu^{2+}(aq) + H_2(g) \longrightarrow Cu(s) + 2H^+(aq) \qquad E^\ominus_{cell} = +0.34\,V$

But nothing happens when hydrogen is bubbled into copper(II) sulfate solution because the activation energy of the reaction is so high and the rate of reaction is effectively zero.

Predictions from $E^\ominus$ data under non-standard conditions

Another important point to remember about $E^\ominus$ values is that they relate *only* to standard conditions. Changes in temperature, pressure and concentration of either the reactants or the products in a redox half-equation will affect the values of an electrode potential, and this may affect whether the reaction is feasible or not.

Quick Questions

10 The half equation for a redox reaction represents an equilibrium between two sides of an equation such as:

$$Cu^{2+}(aq) + 2e^- \rightleftharpoons Cu(s) \qquad E^\ominus = +0.34\,V$$

a Bearing in mind that $E^\ominus$ values relate to the feasibility of a reaction, will the conversion of $Cu^{2+}(aq)$ to $Cu(s)$ become more or less feasible if the concentration of $Cu^{2+}(aq)$:
   i increases, ii decreases?

b How do you think the value of $E^\ominus$ will change if the concentration of $Cu^{2+}(aq)$:
   i increases, ii decreases?

In fact, all electrode potentials for metal ion | metal systems, such as $Cu^{2+}(aq)|Cu(s)$, become more positive when the concentration of the metal ion is increased. This is because the reduction of the metal ions is more likely to occur if their concentration is higher. Conversely, electrode potentials become less positive when the concentration of metal ions is reduced.

For example, the standard electrode potential of the $Cu^{2+}(aq)|Cu(s)$ half-cell is +0.34 V, but the electrode potential falls to +0.31 V if the concentration of $Cu^{2+}(aq)$ is reduced from $1.0\,mol\,dm^{-3}$ to $0.1\,mol\,dm^{-3}$.

Similarly, the standard electrode potential of the $Zn^{2+}(aq)|Zn(s)$ half-cell is $-0.76\,V$, but the electrode potential falls to $-0.79\,V$ if the concentration of $Zn^{2+}(aq)$ is only $0.1\,mol\,dm^{-3}$.

The effect of non-standard conditions on electrode potentials is nicely illustrated by the reaction between solid manganese(IV) oxide, $MnO_2(s)$ and hydrochloric acid. Under standard conditions, MnO_2 will not oxidise $1.0\,mol\,dm^{-3}$ HCl(aq) to chlorine.

$$MnO_2(s) + 4H^+(aq) + 2e^- \longrightarrow Mn^{2+}(aq) + 2H_2O(l) \qquad E^\ominus = +1.23\,V$$

$$2Cl^-(aq) \longrightarrow Cl_2(g) + 2e^- \qquad E^\ominus = -1.36\,V$$

$$MnO_2(s) + 4H^+(aq) + 2Cl^-(aq) \longrightarrow Mn^{2+}(aq) + 2H_2O(l) + Cl_2(g) \qquad E^\ominus_{cell} = -0.13\,V$$

But, when manganese(IV) oxide is *warmed* with *concentrated* hydrochloric acid, the concentration of $H^+(aq)$ and $Cl^-(aq)$ are increased. The electrode potential of each half-equation becomes more positive and the overall cell potential becomes positive. Hence, chlorine is produced.

20.9 Commercial cells

Cells and batteries provide an economic way of storing and, when required, using the energy from chemical reactions. Today, we rely on cells and batteries for the electrical energy to power a wide range of gadgets and machines from wrist watches and mobile phones to computers and cars.

However, the convenient and portable cells and batteries which we use in our everyday machines and gadgets are very different from the experimental and impractical cells discussed in the previous sections of this chapter.

Alkaline cells

Alkaline cells, described in the worked example of Section 20.7 and in Figure 20.12 are probably the commonest, cheapest and most convenient cells in use today. They are used in a wide range of electrical gadgets such as clocks, torches and radios.

The cell diagram for an alkaline cell is:

$$Zn(s)|Zn^{2+}(aq) \vdots \vdots 2MnO_2(s) + H_2O(l), \; Mn_2O_3(s) + 2OH^-(aq)|C_{(graphite)} \qquad E = +1.5\,V$$

The negative terminal in the cell is zinc which goes into solution as zinc ions.

$$Zn(s) \longrightarrow Zn^{2+}(aq) + 2e^-$$

However, as the $Zn^{2+}(aq)$ ions form, they react with hydroxide ions which diffuse out of the porous separator to form insoluble zinc oxide and water:

$$Zn^{2+}(aq) + 2OH^-(aq) \longrightarrow ZnO(s) + H_2O(l)$$

So, overall the reaction at the zinc electrode is:

$$Zn(s) + 2OH^-(aq) \longrightarrow ZnO(s) + H_2O(l) + 2e^-$$

▲ Figure 20.13
Various types of alkaline cells used in radios, torches, clocks and toys

At the positive carbon (graphite) electrode, the reaction is:

$$2MnO_2(s) + H_2O(l) + 2e^- \longrightarrow Mn_2O_3(s) + 2OH^-(aq)$$

Adding these two half-equations together gives the overall cell reaction:

$$Zn(s) + 2MnO_2(s) \longrightarrow ZnO(s) + Mn_2O_3(s)$$

A single dry cell can produce a voltage of about +1.5 V, although batteries of these cells giving 100 V and more are sometimes used. Alkaline cells can produce a higher voltage and a steady current over a longer time than many other cells.

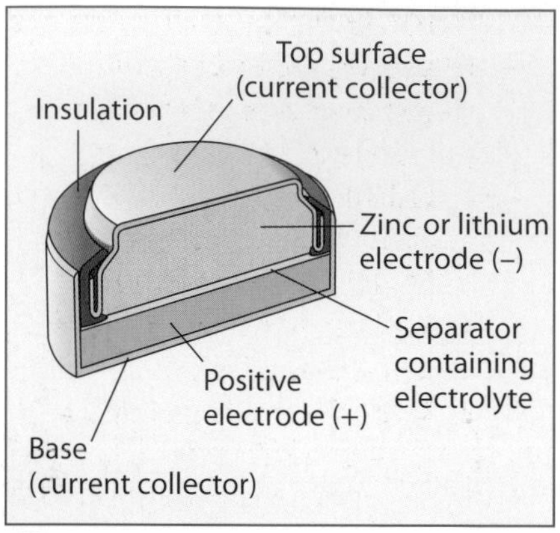

Miniature ('button') cells

Although very small alkaline cells are made and sold commercially, miniature button cells often use lithium instead of zinc as the negative electrode:

$$Li(s) \longrightarrow Li^+(aq) + e^- \qquad E^\ominus = +3.04\,V$$

Button cells are used in devices, such as wrist watches and heart pace-makers, where a steady voltage is required for the life of the cell.

In laboratory cells, the electrolyte in the salt bridge is selected so that it takes no part in the cell reaction. However, in commercial alkaline and button cells, the electrolyte often takes an important part in the reactions. In addition to connecting the two half-cells, it may also be used to remove products of the half-reactions or maintain the concentration of reactants.

Figure 20.14
A typical 'button' cell

- Insulation
- Top surface (current collector)
- Zinc or lithium electrode (–)
- Separator containing electrolyte
- Positive electrode (+)
- Base (current collector)

Definitions

Primary cells are those which cannot be recharged and used again.

Secondary cells are those which can be recharged and used again.

Rechargeable (secondary) cells

Alkaline cells cannot provide continuous supplies of electrical energy indefinitely. Nor is there any means of restoring or *recharging* these cells so that they can be used again. Cells such as these which cannot be recharged and re-used are called **primary cells**. In contrast, cells which can be recharged and used again are called **secondary cells**.

When a secondary cell is recharged, an electric current is passed through it in the opposite direction to that which flows when the cell is generating electricity. Chemical reactions occur at the terminals and the original substances reform. This recharging is an example of electrolysis and is the reverse of discharge.

Lead–acid cells (car batteries)

Lead–acid cells are probably the most widely used secondary cells. Most car batteries are composed of six lead–acid cells connected in series. This gives a total voltage of approximately 12 V.

Car batteries do not normally power the vehicle except in hybrid or electric cars. They simply provide electricity for the lights and windscreen wipers plus power for the starter motor.

The negative terminal in a lead–acid cell is a lead plate. This gives up electrons forming lead(II) ions when the cell is producing a current.

Figure 20.15
This battery-powered Volvo is being recharged

$$Pb(s) \longrightarrow Pb^{2+}(aq) + 2e^-$$

The positive terminal is lead coated with lead(IV) oxide. When the cell is working, the lead(IV) oxide reacts with H^+ ions in the sulfuric acid electrolyte, also forming lead(II) ions plus water.

$$PbO_2(s) + 4H^+(aq) + 2e^- \longrightarrow Pb^{2+}(aq) + 2H_2O(l)$$

The Pb^{2+} ions formed during discharge react with sulfate ions, SO_4^{2-}, in the electrolyte forming insoluble lead(II) sulfate on both terminals:

$$Pb^{2+}(aq) + SO_4^{2-}(aq) \longrightarrow PbSO_4(s)$$

It is important that the discharge of a lead–acid cell does not continue for too long, otherwise the solid lead(II) sulfate which forms on the terminals becomes coarser and thicker. This acts as an insulator on the terminals and prevents the process being reversed when the cell is recharged.

In the past, expensive and heavy lead–acid batteries with relatively low voltage and limited mileage between recharges, have made battery-powered vehicles impractical. In recent years, however, the rising cost of fossil fuels, the political and social pressures for a greener environment plus improved smaller batteries with lower mass and higher voltage have made the battery-powered motor car more viable. At the same time, several European governments are launching programmes to help manufacturers develop electric vehicles and encourage people to buy them. Ahead of all this, some car makers are already producing hybrid cars operated partly by batteries and partly by fossil fuels.

Quick Questions

13 Why are lead–acid batteries unsuitable for powering electric cars?

14 Lead–acid batteries have been used successfully to power some delivery vehicles. Explain why.

15 a Use Table 20.3 to state the value of $E^\ominus$ for a $Pb^{2+}(aq)\,|\,Pb(s)$ half-cell.
   b What is the approximate voltage provided by one lead–acid cell?
   c Calculate the approximate electrode potential of the $Pb(s)\,|\,PbO_2(s) + 4H^+(aq),$ $Pb^{2+}(aq) + 2H_2O(l)$ half-cell in a lead–acid cell.
   d Write the cell diagram for a lead–acid cell.

20.10 Fuel cells

Fuel cells are electrochemical cells which convert the chemical energy of a fuel directly to electrical energy. Fuel cells differ from conventional electrochemical cells, such as alkaline cells and lead–acid cells, in having a continuous supply of reactants from which to produce an electric current. Fuel cells use a variety of fuels including hydrogen, hydrocarbons and alcohols.

The hydrogen / oxygen fuel cell

One of the most important fuel cells is the hydrogen / oxygen fuel cell (Figure 20.16).

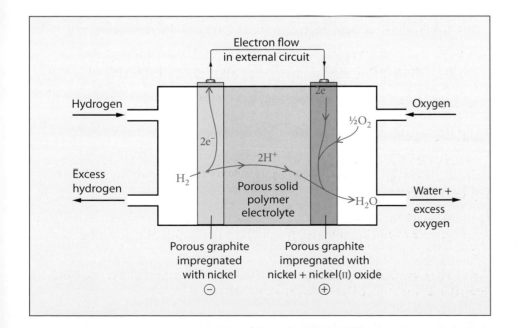

◀ Figure 20.16
A hydrogen / oxygen fuel cell

Figure 20.17

This is the Hyundai Tucson hydrogen fuel cell car

In this cell the negative terminal is porous graphite impregnated with nickel, and the positive terminal is porous graphite impregnated with nickel and nickel(II) oxide. The nickel and nickel(II) oxide act as catalysts splitting hydrogen and oxygen molecules into single atoms.

Hydrogen is passed onto the negative terminal. Here, hydrogen molecules, H_2, split into single H atoms in contact with the nickel catalyst and then lose electrons, forming H^+ ions.

Negative terminal : $\qquad H_2 \longrightarrow 2H \longrightarrow 2H^+ + 2e^-$

The H^+ ions diffuse into and migrate through the porous solid electrolyte while electrons flow into the external circuit as an electric current.

Oxygen is passed onto the positive terminal. Here the oxygen molecules, O_2, split into single O atoms in contact with the nickel / nickel(II) oxide catalyst. They then combine with H^+ ions from the electrolyte and electrons from the external circuit to form water.

Positive terminal: $\quad O + 2H^+ + 2e^- \longrightarrow H_2O$

By adding together the equations for reactions at both terminals, the overall reaction in the hydrogen / oxygen fuel cell is:

$$H_2 + O \longrightarrow H_2O$$

Starting from H_2 and O_2, the overall equation becomes:

$$2H_2(g) + O_2(g) \longrightarrow 2H_2O(l)$$

In principle, this cell is similar to typical electrochemical cells. The key difference is that fresh reactants (H_2 and O_2) are continuously fed into the cell and the product (H_2O) is drawn off. The steady flow, of reactants into the cell and product out of the cell, allows the voltage to stay at 1.2 V and power is maintained all the time.

Fuel cells are also very efficient with 70% or more of the chemical energy being converted into electricity. By comparison, the most modern power plants and vehicle engines using fossil fuels can only achieve a conversion of chemical energy to electrical energy of 45%. In addition, hydrogen / oxygen fuel cells emitting water vapour as the only exhaust gas are pollution-free.

In recent years, compact lightweight H_2/O_2 fuel cells have been developed together with the technology to store large quantities of hydrogen more safely. Scientists are already working on the development of catalysts which could be used in vehicles to produce hydrogen from fossil fuels. The race to develop and manufacture a compact, family-sized fuel cell car is definitely on.

20.11 Measuring the quantity of electricity (electric charge)

We will now return to electrolytic cells and study the process of electrolysis in more detail. The energy that causes the chemical changes during electrolysis is provided by an electric current. An electric current, symbol I, is simply a flow of electrons.

Knowing the charge on one electron, we could measure the quantity of electricity (electric charge) that has flowed along a wire by counting the number of electrons which pass a certain point. The charge on one electron is, however, much too small to be used as a practical unit in measuring the quantity of electric charge.

The symbol for electric charge is Q and the practical unit used in measuring electric charge is the coulomb, C. An electron carries a charge of -1.6×10^{-19} C. So, six million, million, million (6×10^{18}) electrons carry one coulomb of negative charge.

If one coulomb of charge passes along a wire in one second, then the rate of flow of charge (i.e. the electric current) is 1 coulomb per second or 1 **ampere**, A.

If Q coulombs flow along a wire in t seconds, the electric current, I, is given by:

$$I = \frac{Q}{t}$$

This equation can be rearranged to give:

$$\underset{\text{charge in coulombs}}{Q} = \underset{\text{current in amps}}{I} \times \underset{\text{time in seconds}}{t}$$

Key Point

When an electric current flows:

$$\underset{\text{charge}}{Q} = \underset{\text{current}}{I} \times \underset{\text{time}}{t}$$

Quick Questions

17 The current in a small torch bulb is 0.25 A. How much electric charge flows if the torch is used for 20 minutes?

20.12 How much electric charge is needed to deposit one mole of copper during electrolysis?

Using the apparatus in Figure 20.18, we can find the amount of electric charge required to deposit one mole of copper (63.5 g) on the cathode during electrolysis. The rheostat (variable resistor) is used to keep the current constant and quite low. If the current is too large, the copper deposits too fast and flakes off the cathode.

The copper cathode is cleaned, dried and weighed. The circuit is connected up and a current of about 0.15 A is passed for at least 45 minutes.

After electrolysis, the cathode is removed and washed, first with distilled water and then with propanone. When it is completely dry, it is reweighed.

Here are the results of one experiment.

Mass of copper cathode before electrolysis	=	54.07 g
Mass of copper cathode after electrolysis	=	54.25 g
Mass of copper deposited	=	0.18 g

Time of electrolysis = 60 min = 3600 s
Current = 0.15 A
Quantity of electric charge used = $I \times t$
     = 0.15 A × 3600 s
     = 540 C

From these results:

0.18 g of copper is produced by 540 C

∴ 1 mol of copper (63.5 g) is produced by $\dfrac{540\,C \times 63.5\,g}{0.18\,g}$ = 190 500 C

Accurate experiments show that 1 mol of copper is deposited by 193 000 C. This quantity of electricity would operate a 2 kW electric kettle for about 6 hours.

Quick Questions

18 a Why must the cathode be clean and dry when it is weighed before electrolysis?
  b Why is the cathode washed in distilled water and then propanone after electrolysis?
  c A similar experiment to the last one was carried out using silver electrodes in silver nitrate solution.
    0.60 g of silver was deposited in 60 minutes using a current of 0.15 A.
    Calculate the quantity of charge needed to deposit 1 mol (108.0 g) of silver.

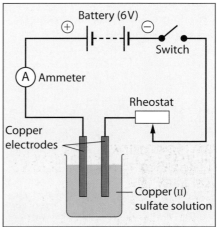

 Figure 20.18
Finding the amount of charge required to deposit 1 mol of copper

 Table 20.4
The charge required to produce 1 mol of seven different elements

Element	Charge required to produce 1 mol/C
copper	193 000
silver	96 500
sodium	96 500
aluminium	289 500
lead	193 000
hydrogen	96 500
oxygen	193 000

Figure 20.19
Michael Faraday was one of the first scientists to study electrolysis. Here he is giving the Christmas Lecture at the Royal Institution, London, in 1855. Prince Albert, Consort to Queen Victoria, is in the audience.

The Faraday constant

Table 20.4 shows the charge required to produce 1 mol of seven different elements.

Notice in Table 20.4 that when molten liquids and aqueous solutions are electrolysed, the quantity of electric charge needed to produce 1 mol of an element is always a multiple of 96 500 C (i.e. 96 500 C or 193 000 C (2 × 96 500 C) or 289 500 C (3 × 96 500 C)).

Because of this, 96 500 coulombs is called the **Faraday constant** (*F*), in honour of Michael Faraday (1791–1867). Faraday was one of the first scientists to measure the masses of elements produced during electrolysis (Figure 20.19).

During electrolysis, 1 mole of sodium ions requires 1 mole of electrons in order to form one mole of sodium atoms.

$$Na^+ + e^- \longrightarrow Na$$

These electrons carry the 96 500 C of charge which are required to deposit one mole of sodium. So, 96 500 C is the charge on 1 mole of electrons and the Faraday constant is defined as 96 500 coulombs per mole (9.65×10^4 C mol^{-1}).

Copper ions (Cu^{2+}) carry twice as much charge as sodium ions (Na^+), so it is not surprising that twice as much charge ($2 \times 96 500$ C = 193 000 C) is required to deposit one mole of copper.

20.13 Determining the Avogadro constant by an electrolytic method

During electrolysis, the positive charge on 1 mol of sodium ions is 'neutralised' by one Faraday of charge (96 500 C) from 1 mol of electrons:

$$Na^+ + e^- \longrightarrow Na$$

Therefore, we can say: $F = L \times e$

In this equation,   *F* is the Faraday constant,

   L is the Avogadro constant and

   e is the charge on one electron.

This equation enables us to determine the value of the Avogadro constant, *L*, by an electrolytic method. If we know the values of *F* and *e*, we can calculate the value of *L*.

During the early part of the nineteenth century, scientists began to measure the quantity of electric charge needed to deposit one mole of different elements. They used electrolytic methods similar to that described in Section 20.12. As a result of these experiments, scientists agreed a value of 96 500 C mol^{-1} for the Faraday constant, *F*.

Then in 1910, R. A. Millikan determined the charge on one electron by measuring the tiny charges due to electrons on individual oil droplets. The accepted value for the negative charge on one electron, *e*, is 1.60×10^{-19} C.

So, rearranging the equation above,

$$L = \frac{F}{E} = \frac{96\,500\,\text{C mol}^{-1}}{1.60 \times 10^{-19}\,\text{C}} = 6.03 \times 10^{23}\,\text{mol}^{-1}$$

Today, the accepted value of *L*, the Avogadro constant is 6.02×10^{23} mol^{-1}.

20.14 Calculating the amounts of substances produced during electrolysis

From your earlier study of Section 6.5, you should be able to identify the substances liberated at the electrodes during the electrolysis of different molten and aqueous substances. But, can we calculate the *amounts* of substances liberated during electrolysis?

The amount of a substance produced during electrolysis can be calculated provided we know:

▸ the number of faradays required to produce 1 mol of the product and
▸ the quantity of electric charge in faradays that has passed to the electrodes.

Using the molar mass of the product or the molar volume ($22.4\,dm^3\,mol^{-1}$ at s.t.p. or $24\,dm^3\,mol^{-1}$ under room conditions), it is then possible to calculate the mass or volume of the product.

Worked example

An electric current of 1.93 A passed through an electrolytic cell composed of copper electrodes in concentrated copper(II) chloride solution for 50 minutes:

a How many faradays will produce 1 mol of: **i** copper, Cu, **ii** chlorine, Cl_2?
b How many faradays of charge passed during the experiment?
c What amount (in moles) of: **i** copper, Cu, **ii** chlorine, Cl_2, was produced?
d What mass of copper, Cu, was produced? (Cu = 63.5)
e What volume of Cl_2 gas was produced under room conditions?

Answer

a During electrolysis, the following reactions occur:

At the cathode (−): $Cu^{2+}(aq) + 2e^- \longrightarrow Cu(s)$

At the anode (+): $2Cl^-(aq) \longrightarrow Cl_2(g) + 2e^-$

∴ 1 mol of copper, Cu, is produced by $2\,F = 2 \times 96\,500\,C = 193\,000\,C$ and

1 mol of chlorine, Cl_2, is produced by $2\,F = 2 \times 96\,500\,C = 193\,000\,C$

b Charge passed, $\quad Q = \underset{\substack{\text{current in} \\ \text{amperes}}}{I} \times \underset{\substack{\text{time in} \\ \text{seconds}}}{t} = 1.93\,A \times (50 \times 60)\,s$

$$= 5790\,C$$

c i $193\,000\,C$ produce 1 mol of Cu

∴ $5790\,C$ produce $\dfrac{1\,mol}{193\,000\,C} \times 5790\,C = 0.03\,mol\,Cu$

ii $193\,000\,C$ produce 1 mol of Cl_2

∴ $5790\,C$ produce $\dfrac{1\,mol}{193\,000\,C} \times 5790\,C = 0.03\,mol\,Cl_2$

d Mass of copper produced = $0.03\,mol \times 63.5\,g\,mol^{-1} = 1.905\,g$

e Volume of Cl_2 produced under room conditions = $0.03\,mol \times 24\,dm^3\,mol^{-1}$
$$= 0.72\,dm^3$$

Quick Questions

19 How much aluminium is produced at the cathode when a current of 100 kA passes through molten aluminium oxide dissolved in cryolite during a 3 hour shift in the industrial production of aluminium? (Al = 27.0)

20 What volumes of gases are produced at the anode and cathode when a current of 0.25 A passes for 40 minutes through aqueous sodium sulfate, $Na_2SO_4(aq)$?

Review questions

1 **a** Explain what you understand by the terms oxidation and reduction.

b Illustrate your answer to part a by writing separate half-equations involving electrons for the oxidation and reduction processes when the following substances react:

  i Mg with Cl_2,

  ii MnO_2 with conc. HCl,

  iii H_2O_2 with I^-,

  iv Fe^{3+} with I^-.

c Suppose you were required to determine the standard electrode potential of a $Zn^{2+}(aq)|Zn(s)$ half-cell.

Draw a fully-labelled diagram showing how the apparatus would be assembled for the determination.

2 Chlorine gas and iron(II) ions react together in aqueous solution as follows:

$$Cl_2(g) + 2Fe^{2+}(aq) \longrightarrow 2Cl^-(aq) + 2Fe^{3+}(aq)$$

a Figure 20.20 shows the apparatus needed to measure the $E^{\ominus}_{cell}$ for the above reaction.

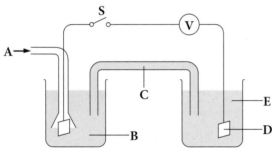

⌃ Figure 20.20

  i Identify what the five letters A–E in Figure 20.20 represent.

  ii Use Table 20.3 to calculate the $E^{\ominus}_{cell}$ for this reaction, and decide which direction (left or right, or right to left) electrons will flow through the voltmeter, V, when switch S is closed.

b Iron(III) chloride readily dissolves in water:

$$FeCl_3(s) \longrightarrow Fe^{3+}(aq) + 3Cl^-(aq)$$

  i Use the following data to calculate the standard enthalpy change for this process:

Species	$\Delta H^{\ominus}_f / kJ\,mol^{-1}$
$FeCl_3(s)$	−399.5
$Fe^{3+}(aq)$	−48.5
$Cl^-(aq)$	−167.2

  ii A solution of iron(III) chloride is used to dissolve unwanted copper from printed circuit boards. When a copper-coated printed circuit board is immersed in $FeCl_3(aq)$ the solution turns pale blue. Suggest an equation for the reaction between copper and iron(III) chloride and use Table 20.3 to calculate the $E^{\ominus}$ for the reaction.

Cambridge Paper 4 Q1 June 2008

3 **a** Draw a fully labelled diagram of a cell incorporating the following two electrode reactions:

$$H_2O_2(aq) + 2H^+(aq) + 2e^- \longrightarrow 2H_2O(l) \quad E^{\ominus} = +1.77\,V$$
$$Sn^{4+}(aq) + 2e^- \longrightarrow Sn^{2+}(aq) \qquad\qquad E^{\ominus} = +0.15\,V$$

b Show the direction in which electrons flow in the external circuit when the cell is used to generate electricity.

c Write equations for the reactions at each electrode when the cell produces an electric current.

d Deduce a value for the cell potential when the cell operates under standard conditions.

e How would the potential of the cell be affected if:

  i the concentration of H^+ ions was increased,

  ii the concentration of Sn^{4+} ions was increased?

Explain your reasoning in each case.

4 The following data should be used in answering this question:

Electrode reaction	$E^{\ominus}/V$
$Zn(s) \longrightarrow Zn^{2+}(aq) + 2e^-$	+0.76
$Fe(OH)_2(s) + OH^-(aq) \longrightarrow Fe(OH)_3(s) + e^-$	+0.56
$Fe(s) \longrightarrow Fe^{2+}(aq) + 2e^-$	+0.44
$Sn(s) \longrightarrow Sn^{2+}(aq) + 2e^-$	+0.14
$4OH^-(aq) \longrightarrow 2H_2O(l) + O_2(aq) + 4e^-$	−0.40

When iron rusts, redox occurs and the iron forms hydrated iron(III) oxide, $Fe_2O_3 \cdot xH_2O$

a Explain, with appropriate equations, what happens when iron rusts.

b Iron may be protected from rusting by coating with zinc or tin.

Explain why zinc protects iron more effectively than tin, once the protective metal coating has been scratched to expose the iron below.

c Why do you think tin is used rather than zinc to coat and protect the inside of 'tin' cans, whereas zinc is used rather than tin to galvanise and protect buckets?

5 The following table gives the standard electrode potentials for a number of half-reactions.

	$E^{\ominus}/V$
$Zn^{2+}(aq) + 2e^- \longrightarrow Zn(s)$	-0.76
$Fe^{2+}(aq) + 2e^- \longrightarrow Fe(s)$	-0.44
$I_2(aq) + 2e^- \longrightarrow 2I^-(aq)$	$+0.54$
$Fe^{3+}(aq) + e^- \longrightarrow Fe^{2+}(aq)$	$+0.77$
$Ce^{4+}(aq) + e^- \longrightarrow Ce^{3+}(aq)$	$+1.61$

a Which half-equation is used as the standard for these electrode potentials?

b Which of the substances in the table above is:

  i the strongest oxidising agent,

  ii the strongest reducing agent?

c Which substance(s) in the table could be used to convert iodide ions to iodine? Write balanced equations for any possible conversions.

d A half-cell is constructed by putting a platinum electrode in a solution which is $1.0\,mol\,dm^{-3}$ with respect to both Fe^{2+} and Fe^{3+} ions. This half-cell is then connected by means of a 'salt bridge' to another half-cell containing an iron electrode in a $1.0\,mol\,dm^{-3}$ solution of Fe^{2+} ions.

  i What is the potential of this cell?

  ii If the two electrodes are connected externally, what reactions take place in each half-cell?

  iii In which direction do electrons flow in the external circuit?

e Write an equation for the reaction you would expect to occur when an iron nail is placed in a solution of iron(III) sulfate.

6 Sodium is manufactured by electrolysis of molten sodium chloride.

a Write equations for the reactions at the anode and cathode when the process occurs.

b How much electric charge is required to produce 1 mol of sodium, Na?

c How long will it take to produce 1 kg of sodium when a current of 100 kA passes through an industrial electrolytic cell containing molten sodium chloride? (Na = 23.0)

21.1 Introduction

Equilibria involving ions in aqueous solutions are important in industrial, analytical and biological processes. The principles and characteristics of these ionic equilibria are very similar to those of other systems in chemical equilibrium.

In this chapter, we shall concentrate on two important types of ionic equilibria:

▶ the equilibrium between an undissolved solid and its dissolved ions in solution, i.e. **solubility equilibria**;

In most cases, these solubility equilibria involve sparingly soluble ionic solids in water, that is, solids which are only slightly soluble.

▶ the equilibrium between a dissolved undissociated molecule and its dissociated ions, i.e. **dissociation equilibria**.

In most cases, these dissociation equilibria involve either acids or bases.

21.2 The solubility of sparingly soluble ionic solids in water

When increasing quantities of a sparingly soluble ionic solid are added to water, a saturated solution is eventually formed. Ions in the saturated solution are in equilibrium with the excess undissolved solute:

$$MX(s) \rightleftharpoons M^+(aq) + X^-(aq)$$

In this section, we will study the relationship between the concentrations of the aqueous ions and the undissolved solute at equilibrium.

Table 21.1 shows the equilibrium concentrations of $Ag^+(aq)$ and $BrO_3^-(aq)$ in contact with undissolved $AgBrO_3$ when different initial volumes of $0.1\,mol\,dm^{-3}$ $AgNO_3$ and $0.1\,mol\,dm^{-3}$ $KBrO_3$ are added to $200\,cm^3$ of distilled water at $16\,°C$.

Figure 21.1

As a coral reef grows, the concentration of Ca^{2+} and CO_3^{2-} ions from reef-building organisms in the surrounding sea-water must be sufficient to precipitate calcium carbonate which forms the reef.

Initial volume of $0.1\,mol\,dm^{-3}$ $AgNO_3$ /cm^3	Initial volume of $0.1\,mol\,dm^{-3}$ $KBrO_3$ /cm^3	Concentration of $Ag^+(aq)$ at equilibrium/mol dm^{-3}	Concentration of $BrO_3^-(aq)$ at equilibrium/mol dm^{-3}	$[Ag^+]_{eq} \times [BrO_3^-]_{eq}$ /mol^2 dm^{-6}
40	10	0.0144	0.0024	3.45×10^{-5}
30	20	0.0081	0.0041	3.32×10^{-5}
25	25	0.0058	0.0058	3.36×10^{-5}
20	30	0.0042	0.0082	3.44×10^{-5}
10	40	0.0033	0.0102	3.37×10^{-5}

Table 21.1

Concentrations of $Ag^+(aq)$ and BrO_3^- (aq) in contact with undissolved $AgBrO_3$ when different initial volumes of $0.1\,mol\,dm^{-3}$ $AgNO_3$ and $0.1\,mol\,dm^{-3}$ $KBrO_3$ are added to $200\,cm^3$ of distilled water

Notice, in the final, right-hand column in Table 21.1, that the product of the concentrations of $Ag^+(aq)$ and $BrO_3^-(aq)$ is more or less constant with a mean value of $3.39 \times 10^{-5}\,mol^2\,dm^{-6}$. So, for this equilibrium we can write

$$[Ag^+(aq)]_{eq}[BrO_3^-(aq)]_{eq} = \text{a constant at a given temperature}$$
$$= 3.39 \times 10^{-5}\,mol^2\,dm^{-6} \text{ at } 16\,°C.$$

In other words, *the product of the concentrations of $Ag^+(aq)$ and BrO_3^- (aq) is independent of the amount of $AgBrO_3$ present*, provided there is some undisssolved $AgBrO_3$ in contact with the solution.

Explaining the constant value of $[Ag^+(aq)][BrO_3^-(aq)]$ at equilibrium

When equilibrium between pure $AgBrO_3$ and its solution is reached, we have:

$$AgBrO_3(s) \rightleftharpoons Ag^+(aq) + BrO_3^-(aq)$$

Hence we can write an equilibrium constant expression as:

$$K_c = \frac{[Ag^+(aq)][BrO_3^-(aq)]}{[AgBrO_3(s)]}$$

But, $[AgBrO_3(s)]$, which represents the concentration of a pure solid, is constant (see note in margin). So, rearranging the last equation, we can write:

$$[Ag^+(aq)][BrO_3^-(aq)] = K_c[AgBrO_3(s)] = \text{a new constant}$$

This new constant is known as the **solubility product** and is given the symbol, $K_{s.p.}$.

Using the general formula A_xB_y for a sparingly soluble salt, we can deduce a general expression for the solubility product as follows.

At equilibrium,

$$A_xB_y(s) \rightleftharpoons xA^{y+}(aq) + yB^{x-}(aq)$$

Hence

$$K_c = \frac{[A^{y+}(aq)]^x[B^{x-}(aq)]^y}{[A_xB_y(s)]}$$

But $[A_xB_y(s)]$ is constant. Therefore

$$[A^{y+}(aq)]^x[B^{x-}(aq)]^y = K_{s.p.}, \text{ the solubility product of } A_xB_y$$

The solubility products of some common compounds are given in Table 21.2.

 Table 21.2
The solubility products of some common compounds at 25 °C

Compound	Solubility product	Compound	Solubility product
Barium sulfate, $BaSO_4$	$1.0 \times 10^{-10}\,mol^2\,dm^{-6}$	Lead(II) sulfate, $PbSO_4$	$1.6 \times 10^{-8}\,mol^2\,dm^{-6}$
Calcium carbonate, $CaCO_3$	$5.0 \times 10^{-9}\,mol^2\,dm^{-6}$	Lead(II) sulfide, PbS	$1.3 \times 10^{-28}\,mol^2\,dm^{-6}$
Calcium fluoride, CaF_2	$4.0 \times 10^{-11}\,mol^3\,dm^{-9}$	Nickel sulfide, NiS	$4.0 \times 10^{-21}\,mol^2\,dm^{-6}$
Calcium sulfate, $CaSO_4$	$2.0 \times 10^{-5}\,mol^2\,dm^{-6}$	Silver bromate, $AgBrO_3$	$6.0 \times 10^{-5}\,mol^2\,dm^{-6}$
Copper(II) sulfide, CuS	$6.3 \times 10^{-36}\,mol^2\,dm^{-6}$	Silver bromide, $AgBr$	$5.0 \times 10^{-13}\,mol^2\,dm^{-6}$
Lead(II) bromide, $PbBr_2$	$3.9 \times 10^{-5}\,mol^3\,dm^{-9}$	Silver chloride, $AgCl$	$2.0 \times 10^{-10}\,mol^2\,dm^{-6}$
Lead(II) chloride, $PbCl_2$	$2.0 \times 10^{-5}\,mol^3\,dm^{-9}$	Silver iodide, AgI	$8.0 \times 10^{-17}\,mol^2\,dm^{-6}$
Lead(II) iodide, PbI_2	$7.1 \times 10^{-9}\,mol^3\,dm^{-9}$	Zinc sulfide, ZnS	$1.6 \times 10^{-24}\,mol^2\,dm^{-6}$

21.3 Calculating solubility products and solubilities

The solubility product of a salt can be obtained from its solubility. The following example shows how this is done.

A saturated solution of silver chloride contains $1.46 \times 10^{-3}\,g\,dm^{-3}$ at 18 °C. What is the solubility product of silver chloride at this temperature?

The solubility of silver chloride at 18 °C $= 1.46 \times 10^{-3}\,g\,dm^{-3}$

$$= \frac{1.46 \times 10^{-3}}{143.5}\,mol\,dm^{-3} \quad (M_r(AgCl) = 143.5)$$

$$= 1 \times 10^{-5}\,mol\,dm^{-3}$$

Now at equilibrium,

$$AgCl(s) \rightleftharpoons Ag^+(aq) + Cl^-(aq)$$

Note

We can calculate the concentration of a pure solid, such as $AgBrO_3$, from its density. For example:

Density of $AgBrO_3$ = 5.21 g/cm³

$$= 5.21 \times 1000\,g\,dm^{-3}$$

$$\therefore [AgBrO_3(s)] = \frac{5210\,g\,dm^{-3}}{235.9\,g\,mol^{-1}}$$

$$= 22.1\,mol\,dm^{-3}$$

Key Point

The solubility product, $K_{s.p.}$, of the sparingly soluble salt, A_xB_y, is given by:

$$K_{s.p.} = [A^{y+}]^x[B^{x-}]^y$$

$K_{s.p.}$ is constant at constant temperature.

Quick Questions

1 Write an expression for the solubility product of:
   a $AgCl$,
   b PbI_2,
   c Bi_2S_3.

2 Why is the experimental value for the solubility product of silver bromate at 16 °C less than that given in Table 21.2?

Figure 21.2
Stag's horn coral growing off the shore of the Seychelle Islands in the Indian Ocean

Quick Questions

3 Why must the temperature be stated with all solubility product values?

4 The units for the solubility product of lead(II) chloride, $PbCl_2$, are different to those of lead(II) sulfide, PbS. Why is this?

5 The concentration of calcium carbonate in river water, saturated with the solid at 20 °C, was found to be $7 \times 10^{-5}\,mol\,dm^{-3}$. What is the solubility product of $CaCO_3$ at 20 °C?

6 If the solubility of Bi_2S_3 at 25 °C is $x\,mol\,dm^{-3}$, which one of the following gives the numerical value of its solubility product at 20 °C?

  a x^2,
  b $4x^3$,
  c $27x^5$,
  d $108x^5$.

$$\therefore \qquad [Ag^+(aq)] = 1 \times 10^{-5}\,mol\,dm^{-3}$$

$$\text{and} \qquad [Cl^-(aq)] = 1 \times 10^{-5}\,mol\,dm^{-3}$$

$$\Rightarrow \qquad K_{s.p.}(AgCl) = [Ag^+(aq)][Cl^-(aq)] = 1 \times 10^{-5}\,mol\,dm^{-3} \times 1 \times 10^{-5}\,mol\,dm^{-3}$$

$$= 1 \times 10^{-10}\,mol^2\,dm^{-6}$$

so, the solubility product of silver chloride at 18 °C is $1 \times 10^{-10}\,mol^2\,dm^{-6}$.

Conversely, the solubility of a salt can be obtained from its solubility product. The following calculation shows how this is done.

The solubility product of silver carbonate at 20 °C is $8 \times 10^{-12}\,mol^3\,dm^{-9}$. What is its solubility at this temperature?

Let us suppose the solubility is $s\,mol\,dm^{-3}$. At equilibrium,

$$Ag_2CO_3(s) \rightleftharpoons 2Ag^+(aq) + CO_3^{2-}(aq)$$

$\therefore$ if the solubility of Ag_2CO_3 is $s\,mol\,dm^{-3}$

$$[Ag^+(aq)] = 2s \qquad \text{and} \qquad [CO_3^{2-}(aq)] = s$$

$$\therefore \qquad K_{s.p.}(Ag_2CO_3) = [Ag^+(aq)]^2[CO_3^{2-}(aq)] = 8 \times 10^{-12}\,mol^3\,dm^{-9}$$

$$= (2s)^2 s = 8 \times 10^{-12}\,mol^3\,dm^{-9}$$

$$\therefore \qquad 4s^3 = 8 \times 10^{-12}\,mol^3\,dm^{-9}$$

$$\Rightarrow \qquad s^3 = 2 \times 10^{-12}\,mol^3\,dm^{-9}$$

$$\therefore \qquad s = 1.26 \times 10^{-4}\,mol\,dm^{-3}$$

i.e. the solubility of silver carbonate at 20 °C is $1.26 \times 10^{-4}\,mol\,dm^{-3}$.

Notice that the concentration of Ag^+ is both doubled and squared in the solubility product expression relative to the solubility of Ag_2CO_3. This is because the balanced solubility equation shows that the concentration of Ag^+ will be double the solubility (i.e. $2s$ not s) and the expression for $K_{s.p.}$ requires this concentration of Ag^+ to be squared (i.e. $(2s)^2$ not $2s$).

21.4 Limitations to the solubility product concept

The solubility product concept is valid only for saturated solutions in which the total concentration of ions is no more than about $0.01\,mol\,dm^{-3}$. For concentrations greater than this, the value of $K_{s.p.}$ is no longer constant. This means that it is quite inappropriate to use the solubility product concept for soluble compounds such as NaCl, $CuSO_4$ and $AgNO_3$. As a consequence of this, the numerical values of solubility products are always very small, rarely exceeding 10^{-4}. For substances of extremely low solubility, $K_{s.p.}$ may be less than 10^{-40}.

The solubility product of a sparingly soluble salt is essentially a modified equilibrium constant. Like other equilibrium constants, its value will change with temperature. Consequently, the temperature at which a solubility product is measured should always be specified.

21.5 Using the solubility product concept

The common ion effect

Although the solubility product of a salt is constant at constant temperature, the concentrations of the individual ions may vary over a very wide range. When a saturated solution is obtained by dissolving the pure salt in water, the concentrations of the ions are in a ratio determined by the formula of the compound. For example, the concentrations of Ca^{2+} and F^- ions in pure saturated calcium fluoride solution must be in the ratio 1:2. However, when a saturated solution is obtained by mixing two solutions containing a **common ion**, there may be a big difference in the concentration of the ions of any sparingly soluble electrolyte. In these cases, solubility products can be used to determine the concentration of ions in the solution. We can illustrate this by considering the solubility of $BaSO_4$, first in water and then in $0.1\,mol\,dm^{-3}$ sodium sulfate solution ($K_{s.p.}(BaSO_4) = 1 \times 10^{-10}\,mol^2\,dm^{-6}$).

Solubility of BaSO₄ in water

Suppose the solubility of $BaSO_4$ in water $= s\,mol\,dm^{-3}$

$$BaSO_4(s) \rightleftharpoons Ba^{2+}(aq) + SO_4^{2-}(aq)$$

$$\therefore \quad K_{s.p.}(BaSO_4) = [Ba^{2+}][SO_4^{2-}]$$

$$\Rightarrow \quad 1 \times 10^{-10} = s \times s = s^2$$

$$\Rightarrow \quad s = 10^{-5}\,mol\,dm^{-3}$$

$\therefore$ solubility of $BaSO_4$ in water $= 10^{-5}\,mol\,dm^{-3}$.

Solubility of BaSO₄ in 0.1 mol dm⁻³ Na₂SO₄ (aq)

Suppose the solubility of $BaSO_4$ in $0.1\,mol\,dm^{-3}\,Na_2SO_4(aq) = s'\,mol\,dm^{-3}$

$$BaSO_4(s) \rightleftharpoons Ba^{2+}(aq) + SO_4^{2-}(aq)$$

$$Na_2SO_4(s) \longrightarrow 2Na^+(aq) + SO_4^{2-}(aq)$$

In this case, $\quad [Ba^{2+}(aq)] = s'\,mol\,dm^{-3}$

but, $\quad [SO_4^{2-}(aq)] = (s' + 0.1)\,mol\,dm^{-3}$

$$\Rightarrow \quad K_{s.p.} = [Ba^{2+}][SO_4^{2-}] = s'(s' + 0.1)\,mol^2\,dm^{-6}$$

$$\Rightarrow \quad s'(s' + 0.1) = 1 \times 10^{-10}\,mol^2\,dm^{-6}$$

but, as s' is much less than $0.1\,mol\,dm^{-3}$,

we can say $\quad (s' + 0.1) \simeq 0.1\,mol\,dm^{-3}$

$$\Rightarrow \quad s' \times 0.1 = 1 \times 10^{-10}\,mol^2\,dm^{-6}$$

$$\Rightarrow \quad s' = 10^{-9}\,mol\,dm^{-3}$$

$\therefore$ solubility of $BaSO_4$ in $0.1\,mol\,dm^{-3}\,Na_2SO_4 = 10^{-9}\,mol\,dm^{-3}$

This calculation illustrates the important generalisation known as the **common ion effect**. This says that *in the presence of either A^+ or B^- from a second source, the solubility of the salt A^+B^- is reduced.*

Predicting precipitation

Another important application of solubility products is that they enable chemists to predict the maximum concentrations of ions in a solution. Hence, it is possible to tell whether or not precipitation will occur.

Suppose we mix a $10^{-3}\,mol\,dm^{-3}$ solution of $Ca^{2+}(aq)$ with an equal volume of a $10^{-3}\,mol\,dm^{-3}$ solution of $SO_4^{2-}(aq)$ at $25\,°C$. Will a precipitate of $CaSO_4$ form?

The solubility product for calcium sulfate is $2 \times 10^{-5}\,mol^2\,dm^{-6}$ at $25\,°C$.

So, $\quad K_{s.p.}(CaSO_4) = [Ca^{2+}][SO_4^{2-}] = 2 \times 10^{-5}\,mol^2\,dm^{-6}$

Immediately after mixing equal volumes of the two solutions and before any precipitation has occurred,

$$[Ca^{2+}] = [SO_4^{2-}] = \frac{10^{-3}}{2}\,mol\,dm^{-3} = 5 \times 10^{-4}\,mol\,dm^{-3}$$

Note

In the context of solubility, a **common ion** is an ion which two solutions share in common.

Key Point

The solubility product concept is invalid for saturated solutions in which the total concentration of ions is greater than $0.01\,mol\,dm^{-3}$.

Quick Questions

7 Calculate the solubility of silver chloride in:
   a water
   b $0.1\,mol\,dm^{-3}\,NaCl(aq)$
   ($K_{s.p.}(AgCl) = 2.0 \times 10^{-10}\,mol^2\,dm^{-6}$)

Key Point

As a result of the **common ion effect**, the solubility of the salt, A^+B^-, is reduced if $A^+(aq)$ or $B^-(aq)$ ions are present from a second source.

Ionic product refers to the product of multiplying together the concentrations of two aqueous ions raised to the appropriate powers, *before* any precipitate has formed.

Solubility product refers to the product of the concentrations raised to the same appropriate powers *after* a precipitate has formed and the concentrations have reached their equilibrium values.

(The concentration of each ion is halved because each solution is diluted by mixing with the other.) Hence, the ionic product for $CaSO_4$ immediately after mixing,

$$= [Ca^{2+}][SO_4^{2-}] = 5 \times 10^{-4} \times 5 \times 10^{-4} \, mol^2 \, dm^{-6}$$

$$= 25 \times 10^{-8} = 2.5 \times 10^{-7} \, mol^2 \, dm^{-6}$$

This ionic product is less than the value of $K_{s.p.}$ for $CaSO_4$, so no precipitate will form.

Let us now suppose that we mix equal volumes of $10^{-2} \, mol \, dm^{-3}$ solutions.

Immediately after mixing,

$$[Ca^{2+}] = [SO_4^{2-}] = 5 \times 10^{-3} \, mol \, dm^{-3}$$

and the ionic product,
$$[Ca^{2+}][SO_4^{2-}] = 5 \times 10^{-3} \times 5 \times 10^{-3} \, mol^2 \, dm^{-6}$$

$$= 25 \times 10^{-6} \, mol^2 \, dm^{-6}$$

$$= 2.5 \times 10^{-5} \, mol^2 \, dm^{-6}$$

In this case, the ionic product is greater than the solubility product. Therefore precipitation of $CaSO_4$ occurs. As a result of precipitation, the concentrations of aqueous Ca^{2+} and SO_4^{2-} ions are lowered by the reaction,

$$Ca^{2+}(aq) + SO_4^{2-}(aq) \longrightarrow CaSO_4(s)$$

and the product $[Ca^{2+}][SO_4^{2-}]$ falls from 2.5×10^{-5} to $2.0 \times 10^{-5} \, mol^2 \, dm^{-6}$.

The precipitation of solids from aqueous solution is of great importance in nature and industry. Stalagmites and stalactites precipitate slowly from water in which the concentrations of $Ca^{2+}(aq)$ and $CO_3^{2-}(aq)$ have an ionic product greater than the solubility product of calcium carbonate (Figure 21.3). Coral reefs grow in a similar fashion. In this case, the concentration of $Ca^{2+}(aq)$ and $CO_3^{2-}(aq)$ around the coral must be large enough to precipitate calcium carbonate from the surrounding sea water.

> **Figure 21.3**

Stalactites growing from the roof of caverns at Marianna, Florida, USA. The stalactites form when calcium carbonate precipitates from a saturated solution dripping from the roof.

Key Point

Precipitation of an insoluble salt occurs when:
  ionic product > solubility product.

In this case, precipitation occurs until:
  ionic product = solubility product

Selective precipitation

The different solubilities of salts can be used to separate different substances from each other by carefully selected precipitation reactions. Suppose we have a solution containing magnesium chloride, calcium chloride and barium chloride. How can we separate the three different metal ions? Both magnesium and calcium chromate(VI) are soluble but barium chromate(VI) is insoluble. So, addition of a solution of $K_2CrO_4(aq)$ will precipitate $BaCrO_4(s)$, which can then be removed by filtration.

The remaining solution now contains $Mg^{2+}(aq)$ and $Ca^{2+}(aq)$, However, $MgSO_4$ is soluble, while $CaSO_4$ is insoluble. So, addition of $Na_2SO_4(aq)$ to the mixture will precipitate $CaSO_4$ and leave $Mg^{2+}(aq)$ in solution. Finally, the $Mg^{2+}(aq)$ can be removed as solid $MgCO_3$ by adding Na_2CO_3 solution.

Notice that the order in which reagents are added is important. If we added Na_2SO_4 solution before adding K_2CrO_4 solution, a mixture of $BaSO_4$ and $CaSO_4$ would be precipitated. On the other hand, if Na_2CO_3 solution was added to the mixture of the three cations, $MgCO_3$, $CaCO_3$ and $BaCO_3$ would all be precipitated. Thus, both the precipitating reagents and their order of addition must be chosen carefully.

▲ Figure 21.4
As an oyster grows, it must adjust conditions so that the concentration of carbonate ions and calcium ions is large enough to precipitate calcium carbonate from sea water to form its shell.

Quick Questions

8 The solubility of silver bromide in water is $7 \times 10^{-7}\,mol\,dm^{-3}$ at $25\,°C$. Calculate its solubility product.

9 $K_{s.p.}((Ag^+)_2C_2O_4{}^{2-}) = 5 \times 10^{-12}\,mol^3\,dm^{-9}$ at $25\,°C$
   a Write an expression for the solubility product of silver ethanedioate, $(Ag^+)_2C_2O_4{}^{2-}$.
   b Calculate the concentration of silver ethanedioate in a saturated aqueous solution at $25\,°C$.
   c How would you expect the value of a solubility product to vary with temperature? Explain your answer.

21.6 The strengths of acids and bases

The strengths of different acids and bases can be compared using conductivity measurements.

Strong electrolytes, such as hydrochloric acid and sodium hydroxide, are virtually completely dissociated into ions in aqueous solutions. Therefore, they are better conductors than weak electrolytes, such as ethanoic (acetic) acid, which are only partially dissociated:

$$HCl(aq) \longrightarrow H^+(aq) + Cl^-(aq) \qquad \text{complete dissociation}$$

$$CH_3COOH(aq) \rightleftharpoons H^+(aq) + CH_3COO^-(aq) \qquad \text{only partial dissociation}$$

The simple descriptive terms 'strong' and 'weak' are much too limited and inaccurate as a method of comparing the strengths of acids and bases. So, chemists looked for a more accurate and quantitative comparison. In the case of acids, relative strengths can be compared by measuring the concentration of H^+ ions or by measuring their **pH,** for a fixed concentration of acid. The 'p' in pH comes from the German word 'potenz' meaning power and the 'H' from $[H^+]$.

The pH of a solution is the negative logarithm to base ten of the hydrogen ion concentration in $mol\,dm^{-3}$, i.e.,

$$pH = -lg\,[H^+(aq)]$$

Hydrogen ion concentrations in aqueous solution range from about $10^{-15}\,mol\,dm^{-3}$ to $10\,mol\,dm^{-3}$. So, it is convenient to have a scale that is both negative and logarithmic in order to show the relative strengths of acids. The negative sign produces positive pH values for almost all solutions and practical situations. The logarithmic scale reduces the extremely wide variation in $[H^+(aq)]$ to a narrow range of pH from about 15 to –1.

▲ Figure 21.5
A scientist measuring the pH of glacial melt water

Definition

The **pH** of a solution provides a quantitative measure of its H^+ ion concentration.

$$pH = -lg[H^+(aq)]$$

Note

The accepted abbreviation for logarithm to base 10 is lg, but log_{10} and lg_{10} are often used.

See the note headed 'Logarithm to base 10' alongside Section 2.9.

Worked examples

a What is the pH of $10^{-1}\,\mathrm{mol\,dm^{-3}}$ HCl?

Answer

Since HCl is fully dissociated and monobasic,

$$[\mathrm{H^+(aq)}] \text{ in } 10^{-1}\,\mathrm{mol\,dm^{-3}}\text{ HCl} = 10^{-1}\,\mathrm{mol\,dm^{-3}}$$

$$\therefore \qquad \mathrm{pH} = -\lg[\mathrm{H^+(aq)}] = -\lg[10^{-1}] = -(-1) = +1$$

b What is the pH of $10^{-3}\,\mathrm{mol\,dm^{-3}}$ H_2SO_4?

Answer

Since H_2SO_4 is fully dissociated and dibasic,

$$[\mathrm{H^+(aq)}] \text{ in } 10^{-3}\,\mathrm{mol\,dm^{-3}}\,H_2SO_4 = 2 \times 10^{-3}\,\mathrm{mol\,dm^{-3}}$$

$$\therefore \qquad \mathrm{pH} = -\lg[\mathrm{H^+(aq)}] = -\lg(2 \times 10^{-3})$$

$$= -(+0.30 - 3.00)$$

$$= -(-2.70) = +2.70$$

c The pH of pure water at 25 °C is 7. What is its hydrogen ion concentration?

Answer

$$\mathrm{pH} = -\lg[\mathrm{H^+(aq)}]$$

$$\Rightarrow \qquad 7 = -\lg[\mathrm{H^+(aq)}]$$

$$\therefore \qquad \lg[\mathrm{H^+(aq)}] = -7$$

$$\therefore \qquad [\mathrm{H^+(aq)}] = 10^{-7}\,\mathrm{mol\,dm^{-3}}$$

Quick Questions

10 Suppose the acid, HX, is a weak electrolyte. What happens to the extent of dissociation of HX in aqueous solution if:
  a water is added,
  b gaseous HCl is bubbled in,
  c solid NaX is added?

11 What are the pH values of the following solutions:
  a $10^{-3}\,\mathrm{mol\,dm^{-3}}$ HCl,
  b $1.0\,\mathrm{mol\,dm^{-3}}$ HCl,
  c $3.0\,\mathrm{mol\,dm^{-3}}$ HX which is only 50% dissociated?

Quick Questions

12 a Why is $[\mathrm{H^+}] = [\mathrm{OH^-}]$ in pure water?
  b At 25 °C, $K_w = 10^{-14}\,\mathrm{mol^2\,dm^{-6}}$. What is the value of K_c for the reaction:
  $H_2O(l) \rightleftharpoons \mathrm{H^+(aq)}$
  $+ \mathrm{OH^-(aq)}$ at 25 °C?
  c Look at the information in Table 21.3. How does the value of K_w change as the temperature increases?
  d Explain the effect of temperature on K_w using Le Chatelier's principle
  ($H_2O(l) \rightleftharpoons \mathrm{H^+(aq)}$
  $+ \mathrm{OH^-(aq)}$;
  $\Delta H = +58\,\mathrm{kJ\,mol^{-1}}$).

21.7 The dissociation of water

When water is purified by repeated distillation, its conductivity falls to a constant, low value. Even the purest water has a tiny electrical conductivity. This is further evidence that water dissociates to form ions, i.e.

$$H_2O(l) \rightleftharpoons \mathrm{H^+(aq)} + \mathrm{OH^-(aq)}$$

Obviously, the concentration of ions is very small, as the pH of pure water shows. Although the equilibrium in this reaction lies far to the left, we can write an equilibrium constant for the dissociation of water as

$$K_c = \frac{[\mathrm{H^+}][\mathrm{OH^-}]}{[H_2O]}$$

Only a minute trace of the water is ionised, so each cubic decimetre of water will contain virtually 1000 g of H_2O. And, as 1 mole of water weighs 18 g, we can say,

$$[H_2O] \text{ in water} = \frac{1000}{18} = 55.56\,\mathrm{mol\,dm^{-3}}, \text{ which is constant.}$$

Thus, we can incorporate this constant for $[H_2O]$ in the value of K_c, just as we did with the concentration of undissolved solute in considering the solubility product of a sparingly soluble salt, i.e.

$$K_c[H_2O] = [\mathrm{H^+}][\mathrm{OH^-}] = \text{a new constant, } K_w$$

constant   constant

This overall constant, K_w, is called the **ionic product for water**.

At 25 °C, $\quad [\mathrm{H^+}] = [\mathrm{OH^-}] = 10^{-7}\,\mathrm{mol\,dm^{-3}}$

hence $\quad K_w = [\mathrm{H^+}][\mathrm{OH^-}] = 10^{-7}\,\mathrm{mol\,dm^{-3}} \times 10^{-7}\,\mathrm{mol\,dm^{-3}} = 10^{-14}\,\mathrm{mol^2\,dm^{-6}}$

21.8 The pH scale

In pure water and in neutral solutions such as aqueous sodium chloride, H^+ and OH^- ions arise only from the ionisation of water.

Hence in water and in neutral solutions,

$$[H^+] = [OH^-]$$

At 25 °C, $[H^+][OH^-] = 10^{-14} \, mol^2 \, dm^{-3}$

$\therefore$ $[H^+] = [OH^-] = 10^{-7} \, mol \, dm^{-3}$ $\therefore pH = 7$

Notice, however, in Table 21.3 that K_w, which equals $[H^+][OH^-]$, rises with temperature.

At 50 °C, for example,

$$K_w = [H^+][OH^-] = 5.47 \times 10^{-14} \, mol^2 \, dm^{-6}$$

So, at 50 °C,

$$[H^+] \text{ in neutral solution} = (5.47 \times 10^{-14} \, mol^2 \, dm^{-6})^{1/2}$$

$$= 2.34 \times 10^{-7} \, mol \, dm^{-3}$$

$\therefore$ pH of water at 50 °C $= -lg[2.34 \times 10^{-7}] = -(+0.4 - 7.0)$

$$= 6.6$$

The pH of pure water and neutral solutions is 7.0 only at 25 °C. At 0 °C, the pH of a neutral solution is 7.5, at 50 °C it is 6.6 and at 100 °C it is 6.1.

So far, we have been considering neutral solutions in which H^+ and OH^- ions can only come from water. In acidic and alkaline solutions H^+ and OH^- ions may arise from sources other than water. Nevertheless, the system

$$H_2O(l) \rightleftharpoons H^+(aq) + OH^-(aq)$$

is still in equilibrium. So, the product $[H^+][OH^-]$ remains constant for all solutions at the same temperature. Therefore, it is possible to determine both the $[H^+]$ and the $[OH^-]$ in any solution. For example, in $10^{-2} \, mol \, dm^{-3}$ HCl,

$$[H^+] = 10^{-2} \, mol \, dm^{-3} \text{ and } \therefore pH = 2$$

but since $[H^+][OH^-] = 10^{-14} \, mol^2 \, dm^{-6}$ for this solution, we can say:

$$10^{-2} \times [OH^-] = 10^{-14} \, mol^2 \, dm^{-6}$$

$\therefore$ $[OH^-]$ in $10^{-2} \, mol \, dm^{-3}$ HCl $= \dfrac{10^{-14}}{10^{-2}} = 10^{-12} \, mol \, dm^{-3}$

Likewise, in $10^{-1} \, mol \, dm^{-3}$ NaOH,

$$[OH^-] = 10^{-1} \, mol \, dm^{-3}$$

but since $[H^+][OH^-] = 10^{-14} \, mol^2 \, dm^{-6}$ for this solution, $[H^+] \times 10^{-1} = 10^{-14} \, mol^2 \, dm^{-6}$

$\therefore$ $[H^+]$ in $10^{-1} \, mol \, dm^{-3}$ NaOH $= 10^{-13} \, mol \, dm^{-3}$ and $\therefore pH = 13$

These results will help you to appreciate the following important generalisations.

For neutral solutions,

$$[H^+] = [OH^-] = 10^{-7} \, mol \, dm^{-3} \text{ and } pH = 7 \text{ at } 25 °C.$$

For acidic solutions,

$$[H^+] > [OH^-] \text{ and } pH < 7 \text{ at } 25 °C.$$

For basic (alkaline) solutions,

$$[H^+] < [OH^-] \text{ and } pH > 7 \text{ at } 25 °C.$$

 Table 21.3

Values of K_w at various temperatures

Temperature/°C	$K_w/mol^2 \, dm^{-6}$
0	0.11×10^{-14}
10	0.30×10^{-14}
20	0.68×10^{-14}
25	1.00×10^{-14}
50	5.47×10^{-14}
100	51.3×10^{-14}

Figure 21.6

The colour of hydrangeas varies with the pH of the soil provided aluminium ions are present. In alkaline soils, hydrangea petals are pink as Al^{3+} ions are 'trapped' in the soil as insoluble aluminium hydroxide. In acid soils, Al^{3+} ions are free to become absorbed through the roots and react with the pink constituent in the petals to form a blue complex ion.

Key Points

At 25 °C, the product, $[H^+][OH^-] = 10^{-14} \, mol^2 \, dm^{-6}$ for all aqueous solutions.

For neutral solutions, $[H^+] = [OH^-] = 10^{-7} \, mol \, dm^{-3}$ and pH = 7.

In acidic solutions, $[H^+] > [OH^-]$ and pH < 7.

In alkaline solutions, $[H^+] < [OH^-]$ and pH > 7.

Figure 21.7 relates the pH scale to the hydrogen ion concentration and to changing acidity and alkalinity.

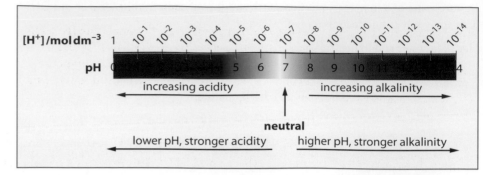

Figure 21.7
The pH scale

Quick Questions

13 Assuming $K_w = 10^{-14} \, mol^2 \, dm^{-6}$, calculate [H$^+$] and the pH of:
a 1.0 mol dm^{-3} HCl,
b 0.10 mol dm^{-3} H$_2$SO$_4$,
c 1.0 mol dm^{-3} NaOH,
d 0.10 mol dm^{-3} NaOH
e 1.0 mol dm^{-3} Ca(OH)$_2$.

Key Point

A **pH meter** uses an Ag/AgCl electrode inside a glass membrane that is permeable only to aqueous H$^+$ ions. This is sometimes called a **glass electrode**.

21.9 Measuring the hydrogen ion concentration and pH

One way of measuring the hydrogen ion concentration of a solution, and hence its pH, would be to use a hydrogen electrode (Section 20.4) connected to a reference electrode which is unaffected by H$^+$ ions. Under standard conditions, the hydrogen electrode is assigned an electrode potential of 0.00 V. Therefore, the overall cell potential will be that of the reference electrode. But, if the concentration of H$^+$ ions in the hydrogen half-cell changes, its electrode potential is no longer 0.00 V and the overall cell potential (voltage) will also change.

Using solutions of known [H$^+$], a calibration curve can be constructed showing the overall cell potential against the [H$^+$] in the hydrogen half-cell. It is then possible to place the combined hydrogen and reference electrodes in any solution and obtain its pH from the resulting voltage.

Unfortunately, the hydrogen half-cell is awkward and inconvenient to use. It requires a bulky hydrogen cylinder, it is difficult to maintain a steady flow of hydrogen, it takes time to reach equilibrium and the platinised platinum electrode is easily 'poisoned' (affected adversely) by impurities in the gas or in the solution.

In practice, it is more convenient to use a **glass electrode** in place of a hydrogen electrode. This consists of silver coated with silver chloride in a solution of 0.1 mol dm^{-3} hydrochloric acid inside a thin glass membrane which is permeable only to aqueous H$^+$ ions. This glass electrode is then combined with a reference electrode and a sensitive voltmeter (calibrated to read pH) in order to form a **pH meter**.

Figure 21.8
pH meters, such as this one, allow the accurate determination of the pH of any solution. The delicate glass membrane is protected by a plastic sheath.

21.10 Dissociation constants of acids and bases

The pH of a solution is sometimes used to indicate the strength of a constituent acid or base. The use of pH is, however, very limited in this context because its value will change as the concentration of the acid or base changes. This led chemists to look for a more useful means of representing the strengths of acids and bases. They found this by considering the dissociation of these substances in aqueous solution.

When the weak acid HA is dissolved in water, we can write:

$$HA(aq) \rightleftharpoons H^+(aq) + A^-(aq)$$

$$\text{Hence, } K_c = \frac{[H^+(aq)][A^-(aq)]}{[HA(aq)]}$$

In dealing with acids, K_c is usually replaced by the symbol, K_a, which is known as the **acid dissociation constant** or simply the dissociation constant of the acid. By eliminating the (aq) state symbols, we can simplify the last equation to

$$K_a = \frac{[H^+][A^-]}{[HA]}$$

Since dissociation constants are effectively equilibrium constants, they are unaffected by concentration changes. Because of this, the numerical value of K_a provides a better measure of the strength of an acid and its extent of dissociation than pH.

The stronger the acid, the greater the extent of dissociation, the greater are $[H^+]$ and $[A^-]$ and the larger is K_a. The values of K_a for some acids are given in Table 21.4. Notice that the values of K_a in Table 21.4 are all measured at 25 °C. This serves to emphasise that the values of acid dissociation constants, like those of all equilibrium constants, are influenced by changes in temperature.

▼ Table 21.4
The values of K_a for some acids

Acid	Equilibrium in aqueous solution	K_a at 25 °C /mol dm^{-3}
Sulfuric acid	$H_2SO_4(aq) \rightleftharpoons H^+(aq) + HSO_4^-(aq)$	very large
Nitric acid	$HNO_3(aq) \rightleftharpoons H^+(aq) + NO_3^-(aq)$	40
Trichloroethanoic acid	$CCl_3COOH(aq) \rightleftharpoons H^+(aq) + CCl_3COO^-(aq)$	2.3×10^{-1}
Dichloroethanoic acid	$CHCl_2COOH(aq) \rightleftharpoons H^+(aq) + CHCl_2COO^-(aq)$	5.0×10^{-2}
Sulfurous acid	$H_2SO_3(aq) \rightleftharpoons H^+(aq) + HSO_3^-(aq)^-$	1.6×10^{-2}
Chloroethanoic acid	$CH_2ClCOOH(aq) \rightleftharpoons H^+(aq) + CH_2ClCOO^-(aq)$	1.3×10^{-3}
Nitrous acid	$HNO_2(aq) \rightleftharpoons H^+(aq) + NO_2^-(aq)$	4.7×10^{-4}
Methanoic acid	$HCOOH(aq) \rightleftharpoons H^+(aq) + HCOO^-(aq)$	1.6×10^{-4}
Benzoic acid	$C_6H_5COOH(aq) \rightleftharpoons H^+(aq) + C_6H_5COO^-(aq)$	6.4×10^{-5}
Ethanoic acid	$CH_3COOH(aq) \rightleftharpoons H^+(aq) + CH_3COO^-(aq)$	1.7×10^{-5}
Hydrated aluminium ion	$[Al(H_2O)_6]^{3+}(aq) \rightleftharpoons H^+(aq) + [Al(H_2O)_5OH]^{2+}(aq)$	1.0×10^{-5}
Carbonic acid	$H_2CO_3(aq) \rightleftharpoons H^+(aq) + HCO_3^-(aq)$	4.5×10^{-7}
Hydrogen sulfide	$H_2S(aq) \rightleftharpoons H^+(aq) + HS^-(aq)$	8.9×10^{-8}
Boric acid	$H_3BO_3(aq) \rightleftharpoons H^+(aq) + H_2BO_3^-(aq)$	5.8×10^{-10}
Hydrogen peroxide	$H_2O_2(aq) \rightleftharpoons H^+(aq) + HO_2^-(aq)$	2.4×10^{-12}
Water	$H_2O(l) \rightleftharpoons H^+(aq) + OH^-(aq)$	5.6×10^{-16}

decreasing acid strength ↓

Determination of dissociation constants

The value of K_a for a weak acid can be determined using the equation above if we know the concentration of the acid and its pH. The following example shows how this is done.

Worked example

The pH of 0.01 mol dm^{-3} ethanoic (acetic) acid (CH_3COOH) is 3.40 at 25 °C. What is the dissociation constant of ethanoic acid at this temperature?

$$CH_3COOH(aq) \rightleftharpoons H^+(aq) + CH_3COO^-(aq)$$

$$\Rightarrow \quad K_a = \frac{[H^+][CH_3COO^-]}{[CH_3COOH]}$$

Before substituting any values in this equation for K_a, we can make two important assumptions:

▶ The concentration of H^+ ions arising from the dissociation of water molecules in the solution of ethanoic acid is very much smaller than that from the acid itself,
∴ $[H^+] \approx [CH_3COO^-]$

▶ Ethanoic acid is a weak acid. This means that the proportion of ethanoic acid which dissociates is negligible compared with the undissociated acid. So, in a 0.01 mol dm^{-3} solution of CH_3COOH,
$[CH_3COOH] \approx 0.01$ mol dm^{-3}

> ### Key Point
>
> For a weak acid, dissolved in water,
> $$K_a = \frac{[H^+][A^-]}{[HA]}$$
> K_a, the **acid dissociation constant** is constant at constant temperature. The value of K_a gives a quantitative measure of the strength of the acid and its extent of dissociation.
>
> K_a, like $K_{s.p.}$, is a modified equilibrium constant.

Figure 21.9
The coloured material in red cabbage can be used as an indicator. The purple coloured extract shown in the middle tube is red in dilute hydrochloric acid and yellow in dilute sodium hydroxide solution

Key Point

$pK_a = -\lg K_a$

Quick Questions

15 Why is the pH of an acid unsatisfactory in deciding whether the acid is strong or weak?

16 Nitric acid is a strong acid and ethanoic acid is a weak acid, yet it takes the same amount of sodium hydroxide to neutralise one mole of both these acids. Why is this?

17 Calculate K_a for methanoic acid (HCOOH), given that the pH of $0.10 \, mol \, dm^{-3}$ HCOOH is 2.4.

18 What is the value of pK_a for:
   a chloroethanoic acid, $K_a = 1.3 \times 10^{-3} \, mol \, dm^{-3}$,
   b benzoic acid, $K_a = 6.4 \times 10^{-5} \, mol \, dm^{-3}$?

Answer

Bearing in mind these two assumptions, we can now calculate $[H^+]$ from the given pH and then substitute values for $[H^+]$, $[CH_3COO^-]$ (assumed to be the same as $[H^+]$) and $[CH_3COOH]$ (assumed to be $0.01 \, mol \, dm^{-3}$) in the equation for K_a.

$$pH = -\lg[H^+] = 3.40$$

Hence

$$\lg[H^+] = -3.40 = (-4.00 + 0.60)$$

∴

$$[H^+] = 4.0 \times 10^{-4} \, mol \, dm^{-3}$$

Now, assuming

$$[CH_3COO^-] = [H^+] = 4.0 \times 10^{-4} \, mol \, dm^{-3}$$

and

$$[CH_3COOH] = 0.01 \, mol \, dm^{-3},$$

$$K_a = \frac{[H^+][CH_3COO^-]}{[CH_3COOH]} = \frac{4 \times 10^{-4} \times 4 \times 10^{-4}}{0.01} \frac{mol \, dm^{-3} \times mol \, dm^{-3}}{mol \, dm^{-3}}$$

$$K_a = 1.60 \times 10^{-5} \, mol \, dm^{-3}$$

By reversing this calculation, it is possible to calculate the pH of a solution of a weak acid provided we know its concentration.

Quick Questions

14 This question will help you to calculate the pH of $1.0 \, mol \, dm^{-3}$ benzoic acid (C_6H_5COOH). $K_a(C_6H_5COOH) = 6.4 \times 10^{-5} \, mol \, dm^{-3}$
   a Write an equation for the dissociation of C_6H_5COOH.
   b Using this equation, write an expression for the dissociation constant of benzoic acid.
   c Assuming $[H^+] = [C_6H_5COO^-]$ and $[C_6H_5COOH] \gg [H^+]$, calculate $[H^+]$ for $1.0 \, mol \, dm^{-3}$ benzoic acid.
   d Finally, calculate the pH of $1.0 \, mol \, dm^{-3}$ benzoic acid.

The K_a values of weak acids vary from about 10^{-2} to 10^{-10}. In order to compare the strengths of different acids using more accessible and more manageable values, the term pK_a is often used instead of K_a. $pK_a = -\lg K_a$ (i.e. the negative log to base 10 of K_a). This gives pK_a values from 2 to 10.

Just as acid dissociation constants can be used to compare the strengths of different acids, we can also use base dissociation constants for bases. If the base BOH is in equilibrium with its dissociated ions,

$$BOH \rightleftharpoons B^+ + OH^-$$

and

$$K_b = \frac{[B^+][OH^-]}{[BOH]}$$

where K_b is known as the dissociation constant of the base.

21.11 Acid–base indicators

Acid–base indicators such as methyl orange, phenolphthalein and bromothymol blue are substances which change colour according to the hydrogen ion concentration of the solution to which they are added (Figure 21.10). Consequently, they are used to test for acidity and alkalinity and to detect the end point in acid–base titrations.

Most indicators can be regarded as weak acids of which either the undissociated molecule or the dissociated anion, or both, are coloured. If we take methyl orange as an example and write the undissociated molecule as HMe,

$$HMe(aq) \rightleftharpoons H^+(aq) + Me^-(aq)$$

red   colourless   yellow

Addition of acid (i.e. H^+ ions) displaces this equilibrium to the left. When this happens $[HMe] \gg [Me^-]$ and the solution becomes red.

On the other hand, when alkali (containing OH^- ions) is added to methyl orange it reacts with H^+ ions forming water. The equilibrium in the above system moves to the right in order to replace some of the H^+ ions. In this case, $[Me^-] \gg [HMe]$, and the methyl orange becomes yellow.

methyl orange (yellow form)

phenolphthalein (colourless form)

Indicators can be regarded as weak acids, so it is possible to determine their acid dissociation constants. Using HIn for the undissociated form of the indicator, we can write

$$HIn(aq) \rightleftharpoons H^+(aq) + In^-(aq)$$

$$\Rightarrow \qquad K_a(HIn) = \frac{[H^+][In^-]}{[HIn]}$$

The numerical values of these dissociation constants can be obtained by measuring the pH of a solution of known concentration for each indicator (Section 21.10). The dissociation constants for some indicators are shown in Table 21.5.

Using indicators in titrations

The aim of any titration is to determine the **equivalence point** when the amount of one reactant added from a burette is just sufficient to react with all of a second reactant in the flask. The **end point** of the titration is the point at which the titration is stopped. Ideally, the end point should coincide with the equivalence point for the two reacting solutions. In order to achieve this in acid–base titrations, the indicator should change colour sharply at the equivalence point on adding a single drop of either acid or alkali. At the end point of the titration, the colour of the indicator will be mid-way between the acid colour of HIn and the alkaline colour of In^-, i.e. $[HIn] = [In^-]$.

Now, $\qquad K_a(HIn) = \dfrac{[H^+][In^-]}{[HIn]}$

But, at the end point, $\qquad [HIn] = [In^-]$

So, at the end point, $\qquad K_a(HIn) = [H^+]$

$\therefore \qquad$ pH at end point $= -\lg[H^+] = -\lg K_a(HIn) = pK_a(HIn)$

Therefore, the pH at the end point of an acid–base titration will be equal to pK_a of the indicator used.

The pHs at the end point for the four indicators listed in Table 21.5 are shown in Table 21.6.

Quick Questions

19 The dissociation of phenolphthalein in aqueous solution can be represented as:
$$HPh(aq) \rightleftharpoons H^+(aq) + Ph^-(aq)$$
Phenolphthalein is pink in strongly alkaline solution, and colourless in strongly acidic solution.
what are the colours of:
a HPh,
b Ph^- for phenolphthalein?

 Figure 21.10
The formulae of two common indicators

Key Point

Acid–base indicators change colour according to the $[H^+]$ of the solution to which they are added.

They are used to test for acidity and alkalinity and to detect the end point in acid–base titrations.

 Table 21.5
The values of K_a for some indicators

Indicator	K_a at 25 °C /mol dm^{-3}
Phenolphthalein	7×10^{-10}
Bromothymol blue	1×10^{-7}
Litmus	3×10^{-7}
Methyl orange	2×10^{-4}

 Table 21.6
The pH at the end point for some indicators

Indicator	pH at end point $= -\lg K_a(HIn)$ $= pK_a(HIn)$
Phenolphthalein	9.1
Bromothymol blue	7.0
Litmus	6.5
Methyl orange	3.7

The range of an indicator

The colour change of an indicator is due to the change from one coloured form to another. Near the end point, both coloured forms will be present in appreciable quantities. Experience shows that our eyes cannot judge precisely when the two forms are in equal concentrations, and therefore the exact end point, as indicators effectively change colour over a range of about 2 pH units.

The values in Tables 21.5 and 21.6 show that the dissociation constants of indicators differ widely. Consequently, they change colour over widely differing ranges of pH. This last point is illustrated in Figure 21.11.

> Figure 21.11
> The pH ranges of some indicators

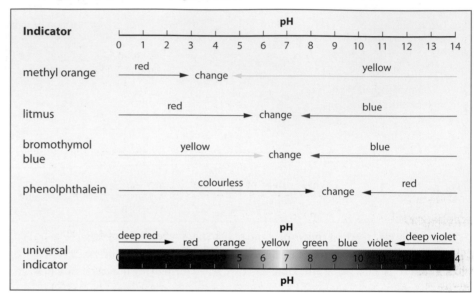

Some indicators, like phenolphthalein, are only slightly soluble in water. They are therefore often used as solutions in alcohol or in a mixture of alcohol and water. Because of this, the colour of an indicator may change when it is added to water, independently of its colour change with pH.

As you would expect, the end point of each indicator is in the centre of its pH range. Figure 21.11 also shows that some indicators change colour over a range of pH well away from 7. This is an important point. Many students misunderstand the use of indicators and expect them all to change colour at pH 7. In fact it is because different indicators change colour at different pH values that acid–base titrations have such a wide application in industry and in the laboratory.

21.12 pH changes during acid–base titrations

During any acid–base titration there is a change in pH as base is added to acid or vice versa. In most cases the pH changes sharply by several units at the equivalence point and this allows us to detect the end point of the titration with an indicator.

The actual change in pH during the course of a titration depends largely upon the strength of the acid and base used.

Titrating strong acid against strong base

The graph in Figure 21.12 shows how the pH changes during the titration of $50\,cm^3$ of $0.1\,mol\,dm^{-3}$ HCl(aq) with $0.1\,mol\,dm^{-3}$ NaOH(aq). As the alkali is added, the pH changes slowly at first. It does, however, change very rapidly from about 3.5 to 9.5 at the equivalence point. Thus, any indicator which changes colour between pH 3.5 and 9.5 will identify the equivalence point. This means that any one of methyl orange, bromothymol blue or phenolphthalein could be used.

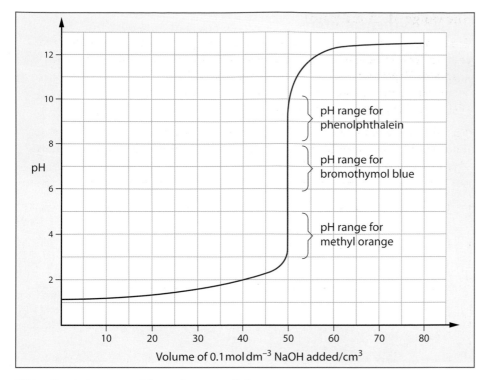

◀ Figure 21.12
pH changes during the titration of
50 cm^3 of 0.1 mol dm^{-3} HCl(aq) with
0.1 mol dm^{-3} NaOH(aq)

Quick Questions

22 Look carefully at Figure 21.12.
   a At what pH is the exact
      equivalence point?
   b What should the precise
      initial pH be before any
      NaOH(aq) is added?
   c i   Over what pH range would
          methyl orange change
          colour at the end point?
      ii How much 0.1 mol dm^{-3}
          NaOH has been added
          when this happens?

Titrating strong acid against weak base

The graph in Figure 21.13 shows how the pH changes when 0.1 mol dm^{-3} NH$_3$(aq) (weak base) is added to 50 cm^3 of 0.1 mol dm^{-3} HCl(aq). As before, there is little variation in pH when the base is first added. But, at the equivalence point the pH changes rapidly from about 3.5 to 7.0. Thus, any indicator which changes colour between pH 3.5 and 7.0 will identify the equivalence point accurately. In this case, methyl orange is a very suitable indicator. Bromothymol blue may also be used, but phenolphthalein is useless because it does not begin to change colour until about pH 8.

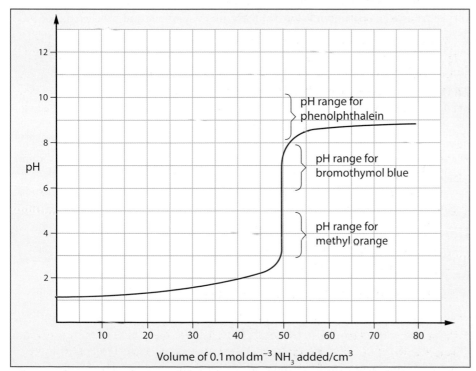

◀ Figure 21.13
pH changes during the titration of
50 cm^3 of 0.1 mol dm^{-3} HCl(aq) with
0.1 mol dm^{-3} NH$_3$(aq)

Quick Questions

23 a Why does the initial pH start
      at 1.0 in both Figure 21.12
      and Figure 21.13?
   b Why is the final pH in Figure
      21.12 at about 12.5, but that
      in Figure 21.13 at about 9?

Changes in pH during the titration of weak acid with strong base and during the titration of weak acid with weak base are discussed in Review question 7 at the end of this chapter.

 Figure 21.14

The cells of animals and plants are protected against small amounts of acid or alkali by buffers. However serious damage results when too much acid or alkali is involved. The conifers in this photo have been damaged by acid rain.

Figure 21.15

Most shampoos contain a buffer solution so that they don't irritate the skin. They are sometimes marketed as 'pH balanced shampoos'.

Definition

Buffers are solutions which resist changes in pH on addition of acid or alkali and on dilution.

When $0.1\ cm^3$ of $1.0\ mol\ dm^{-3}$ HCl(aq) is added to $1\ dm^3$ of water or sodium chloride solution, the pH changes sharply from 7.0 to 4.0 (i.e. by 3 units of pH). This shows that the pH values of water and sodium chloride solution are extremely sensitive to even small additions of acid or alkali. If this happened when small amounts of acid or alkali are added to biological systems, living organisms would be killed instantly. Fortunately, animals and plants are protected against sharp changes in pH by the presence of **buffers**.

Buffers are solutions which resist changes in pH when acid or alkali are added to them.

Buffers are also important in many industrial processes where the pH must not deviate very much from an optimum value. Similarly, many synthetic and processed foods must be prepared in a buffered form so that they can be eaten and digested in our bodies without undue change in pH. The pH of blood is normally 7.4. Under most circumstances, a change of only 0.5 units in the pH of blood would be fatal.

All this leads to the question 'How do buffers act and how do they resist changes in pH when acid or alkali is added?'

Buffer solutions usually consist of:

▶ a solution containing a weak acid and one of its salts (e.g. ethanoic acid and sodium ethanoate, carbonic acid and sodium hydrogencarbonate),

▶ a solution containing a weak alkali and one of its salts (e.g. ammonia solution and ammonium chloride).

How does a buffer solution work?

In order to understand how a buffer solution works, we can consider the hypothetical weak acid, HA, in a solution with its salt MA. In this solution, HA will be slightly dissociated whilst MA is fully dissociated. Leaving out state symbols for clarity and convenience, this can be written as:

$$HA \rightleftharpoons H^+ + A^-$$

$$MA \longrightarrow M^+ + A^-$$

So, the mixture contains a relatively high concentration of un-ionised HA (an acid) and a relatively high concentration of A^- (its conjugate base).

If an acid is suddenly added to this system, H^+ ions from the acid combine with A^- ions to form un-ionised HA. Provided there is a large reservoir of A^- ions in the buffer, nearly all the added H^+ ions are removed. Thus, $[H^+]$ changes very little and the pH is only slightly altered.

If an alkali, such as sodium hydroxide, is suddenly added to the system, OH^- ions from the alkali combine with H^+ ions to form water. This reduces the concentration of H^+ ions in the buffer, but more HA dissociates to restore the equilibrium and the $[H^+]$ rises almost to its original value. Provided there is a large reservoir of HA in the buffer, the $[H^+]$ changes very little and again the pH is only slightly altered. By having these reserves of both HA and A^- in the buffer mixture, changes in the pH resulting from the addition of acid or alkali can be minimised.

Essentially, the stable pH of a buffer is due to

- a high $[A^-]$ which traps added H^+ ions (acid), and
- a high [HA] which can supply H^+ ions to trap added OH^- ions (base).

Solutions of this kind can counteract the effects of adding acid or alkali.

Calculating the pH of buffer solutions

In a buffer solution composed of the weak acid HA and its salt MA,

$$HA \rightleftharpoons H^+ + A^-$$

and

$$MA \longrightarrow M^+ + A^-$$

We can write an expression for the dissociation constant of HA as:

$$K_a = \frac{[H^+][A^-]}{[HA]}$$

$$\therefore \quad [H^+] = K_a \frac{[HA]}{[A^-]}$$

In the buffer mixture, [HA] is effectively the concentration of acid used in the buffer solution ([acid]) because the acid will be only very slightly dissociated. Similarly, $[A^-]$ is effectively the concentration of salt in the buffer ([salt]) because the salt is fully dissociated into ions.

Thus we can write:

$$[H^+] \simeq K_a \times \frac{[acid]}{[salt]}$$

This last equation explains why $[H^+]$, and therefore the pH of a buffer, is also affected very little by dilution. This is because the acid and its salt are diluted to exactly the same extent and therefore the ratio [acid]/[salt] remains constant.

Using the last equation, notice the special case which applies when [acid] = [salt]. In this case, $[HA^+] = K_a$ and $pH = pK_a$.

The pH of the buffer is numerically equal to pK_a for the acid.

The following example will help you to understand the points which have just been discussed.

Worked example

A buffer solution was made by adding 3.28 g of sodium ethanoate to 1 dm^3 of 0.01 mol dm^{-3} ethanoic acid. What is the pH of this buffer? ($K_a(CH_3COOH) = 1.7 \times 10^{-5}\,mol\,dm^{-3}$)

Answer

$$K_a = \frac{[H^+][CH_3COO^-]}{[CH_3COOH]}$$

Quick Questions

24 a How many moles of NaOH does 0.1 cm^3 of 1.0 mol dm^{-3} NaOH contain?

b What is the $[OH^-]$ if the amount of NaOH in part a dissolves in 1 dm^3 of water?

c What is the pH of the solution in part b? ($K_w = 10^{-14}\,mol^2\,dm^{-6}$)

d What is the pH *change* if 0.1 cm^3 of 1.0 mol dm^{-3} NaOH is added to 1 dm^3 of water?

25 Why does the pH change very little if 0.1 cm^3 of 1.0 mol dm^{-3} NaOH is added to a solution containing both ethanoic acid and sodium ethanoate?

Key Point

The $[H^+]$ and pH of a buffer containing the weak acid, HA, and its salt, MA can be calculated using:

$$[H^+] = K_a \frac{[HA]}{[A^-]} \simeq K_a \frac{[acid]}{[salt]}$$

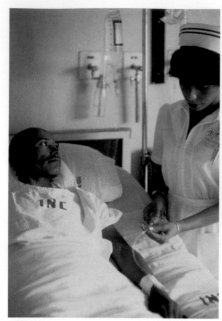

Figure 21.16
Injections and drips into a patient's body must be carefully buffered so that the pH values of body fluids do not change too much.

$$\therefore \qquad [H^+] = K_a \frac{[CH_3COOH]}{[CH_3COO^-]}$$

But, $\quad [CH_3COOH] = [\text{acid}] = 0.01 \, \text{mol dm}^{-3}$

and $\quad [CH_3COO^-] = [\text{salt}] = \dfrac{3.28}{82} = 0.04 \, \text{mol dm}^{-3}$

$$\therefore \qquad [H^+] = 1.7 \times 10^{-5} \times \frac{0.01}{0.04} = 4.25 \times 10^{-6} \, \text{mol dm}^{-3}$$

$$\therefore \qquad \text{pH of the buffer} = -\lg[H^+] = -\lg(4.25 \times 10^{-6})$$
$$= -(+0.63 - 6.00) = -(-5.37)$$

$$\Rightarrow \qquad \text{pH of the buffer} = 5.37$$

In order to appreciate the action of a buffer, let's calculate the change in pH when $1 \, \text{cm}^3$ of $1.0 \, \text{mol dm}^{-3}$ NaOH(aq) is added to $1 \, \text{dm}^3$ of the buffer in the last example.

Calculating the change in pH when alkali is added to a buffer

Worked example

Calculate the change in pH when $1 \, \text{cm}^3$ of $1.0 \, \text{mol dm}^{-3}$ NaOH(aq) is added to $1 \, \text{dm}^3$ of buffer in the last example.

$$\text{pH of buffer initially} = 5.37$$

When NaOH is added to the buffer it reacts with CH_3COOH and forms CH_3COONa (i.e.$[CH_3COOH]$ falls and $[CH_3COO^-]$ rises). $1 \, \text{cm}^3$ of $1.0 \, \text{mol dm}^{-3}$ NaOH contains $0.001 \, \text{mol}$ of OH^-. This removes $0.001 \, \text{mol}$ of CH_3COOH and forms $0.001 \, \text{mol}$ of CH_3COONa.

$\therefore$ After adding $0.001 \, \text{mol}$ of NaOH,

$$[CH_3COOH] = 0.010 - 0.001 = 0.009 \, \text{mol dm}^{-3}$$
$$[CH_3COO^-] = 0.040 + 0.001 = 0.041 \, \text{mol dm}^{-3}$$

(We have made the approximation that the total volume of solution remains $1 \, \text{dm}^3$ after adding $1 \, \text{cm}^3$ of NaOH(aq).)

$$\Rightarrow \qquad [H^+] = K_a \frac{[\text{acid}]}{[\text{salt}]} = 1.7 \times 10^{-5} \times \frac{0.009}{0.041} \, \text{mol dm}^{-3}$$
$$[H^+] = 3.73 \times 10^{-6} \, \text{mol dm}^{-3}$$

$\therefore$ pH after adding NaOH $= 5.43$

Therefore, the pH of the buffer changes by only 0.06 units, which illustrates the pH-stabilising effect of the buffer very well.

The main use of buffers in the laboratory is in preparing solutions of known and constant pH. Buffer solutions cannot be made by simply preparing acid or alkaline solutions of a given concentration. The pH of such solutions will vary slightly as gases from the atmosphere, such as CO_2, dissolve in them, or as traces of alkali dissolve from the glass vessel.

Buffer solutions are also important in medicine and in agriculture because the pH values of living organisms must be maintained at certain critical values. Because of this, intravenous injections must be carefully buffered so as not to change the pH of the blood from its normal value of 7.4. In living systems, the buffering action is usually provided by:

▶ carbonic acid, H_2CO_3, and hydrogencarbonate, HCO_3^-,
▶ dihydrogenphosphate, $H_2PO_4^-$, and hydrogenphosphate, HPO_4^{2-} and
▶ various proteins which can both accept and donate H^+ ions.

In healthy people, the blood pH is stabilised by buffer action involving carbonic acid from dissolved carbon dioxide and hydrogencarbonate ions:

$$H_2O(l) + CO_2(g) \longrightarrow H_2CO_3(aq) \rightleftharpoons H^+(aq) + HCO_3^-(aq)$$

$H_2PO_4^-/HPO_4^{2-}$ and proteins in the blood also contribute to the buffering action.

During respiration, CO_2 is produced which diffuses into the blood where it forms weak carbonic acid and hydrogencarbonate ions. Excess carbon dioxide can be removed from the blood as it passes through the lungs and excess hydrogencarbonate ions are removed by the kidneys.

Quick Questions

27 A blood sample contains $1.8 \times 10^{-3}\,mol\,dm^{-3}$ H_2CO_3 ($K_a = 4.5 \times 10^{-7}\,mol\,dm^{-3}$) and $2.0 \times 10^{-2}\,mol\,dm^{-3}$ HCO_3^-.
   a Assuming that H_2CO_3 and HCO_3^- provide the only buffering effect, calculate the pH of this blood.
   b Calculate the pH if 1 drop ($0.1\,cm^3$) of $1.0\,mol\,dm^{-3}$ HCl contaminates $1\,dm^3$ of this blood.
   c What is the pH change in part b?

Review questions

1 Monuments made of marble or limestone, such as the Taj Mahal in India and the Mayan temples in Mexico, are suffering erosion by acid rain. The carbonate stone is converted by the acid rain into the relatively more soluble sulfate:

$$\underset{\text{acid rain}}{CaCO_3(s) + H_2SO_4(aq)} \longrightarrow CaSO_4(s) + H_2O(l) + CO_2(g)$$

   a i Write an expression for the solubility product, $K_{s.p.}$, of $CaSO_4$, stating its units.
   ii The $K_{s.p.}$ of $CaSO_4$ has a numerical value of 3×10^{-5}. Use your expression in i to calculate $[CaSO_4]$ in a saturated solution.
   iii Hence calculate the maximum loss in mass of a small statue if $100\,dm^3$ of acid rain falls on it. Assume the statue is made of pure calcium carbonate, and that the acid rain becomes saturated with $CaSO_4$.

   b The life of such monuments is now being extended by treating them with a mixture of urea and barium hydroxide solutions. After soaking into the pores of the carbonate rock, the urea gradually decomposes to ammonia and carbon dioxide. The carbon dioxide then reacts with the barium hydroxide to form barium carbonate:

$$\underset{\text{urea}}{(NH_2)_2CO(aq) + H_2O(l)} \longrightarrow 2NH_3(g) + CO_2(g)$$

$$Ba(OH)_2(aq) + CO_2(g) \longrightarrow BaCO_3(s) + H_2O(l)$$

Acid rain then converts the barium carbonate to its sulfate:

$$BaCO_3(s) + H_2SO_4(aq) \longrightarrow BaSO_4(s) + H_2O(l) + CO_2(g)$$

Barium sulfate is much less soluble than calcium sulfate. A saturated solution contains $[Ba^{2+}] = 9.0 \times 10^{-6}\,mol\,dm^{-3}$.
   i Explain why barium sulfate is less soluble than calcium sulfate.
   ii Write an expression for the $K_{s.p.}$ of barium sulfate and use the data to calculate its value.

Cambridge Paper 4 Q2a–b June 2006

2 The solubility product of lead(II) sulfate, $PbSO_4$, in water, is $1.6 \times 10^{-8}\,mol^2\,dm^{-6}$.
   a Calculate the solubility of lead(II) sulfate in
   i pure water,
   ii $0.1\,mol\,dm^{-3}$ $Pb(NO_3)_2$ solution,
   iii $0.01\,mol\,dm^{-3}$ Na_2SO_4 solution.
   b Why is lead(II) sulfate more soluble in water than in any solution containing either $Pb^{2+}(aq)$ or SO_4^{2-} (aq)?
   c Use your understanding of solubility product to explain the 'common ion' effect.

   Illustrate your answer with an example.

3 Chloric(I) (hypochlorous) acid, HOCl, is a weak acid
   $K_a(HOCl) = 3.2 \times 10^{-8}\,mol\,dm^{-3}$
   a Calculate the $[H^+]$ and $[OH^-]$ in $1.25 \times 10^{-2}\,mol\,dm^{-3}$ HOCl.
   b What is the pH of $1.25 \times 10^{-2}\,mol\,dm^{-3}$ HOCl?

4 Explain the following observations:

   a When $1.0 \, mol \, dm^{-3}$ hydrochloric acid is diluted, the pH rises. Eventually, the pH reaches a static value and does not change on further dilution.

   b Meythyl orange can be used to determine the end point in the titration of nitric acid with ammonia solution, but phenolphthalein is no use in this case.

   c The pH of $10^{-8} \, mol \, dm^{-3}$ HCl is not 8.

5 Assuming that the pH of blood is maintained at 7.4 by the acid $H_2PO_4^-$ and its salt, HPO_4^{2-}, calculate the ratio of the concentration of $H_2PO_4^-$ to that of HPO_4^{2-} in blood.

   $(K_a(H_2PO_4^-) = 6.4 \times 10^{-8} \, mol \, dm^{-3})$

6 Calculate the pH of:

   a $10^{-4} \, mol \, dm^{-3} \, HCl(aq)$,

   b $10^{-4} \, mol \, dm^{-3} \, Ba(OH)_2 \, (aq)$,

   c $1.0 \, mol \, dm^{-3} \, H_2X \, (aq)$ which is only 50% dissociated,

   d $0.01 \, mol \, dm^{-3}$ propanoic acid,

     $(K_a = 1.45 \times 10^{-5} \, mol \, dm^{-3})$

   e $1.0 \, mol \, dm^{-3} \, NH_3(aq). \, (K_b = 1.7 \times 10^{-5} \, mol \, dm^{-3})$

7 Figure 21.17 shows how the pH changes when $0.1 \, mol \, dm^{-3}$ CH_3COOH is titrated against $0.1 \, mol \, dm^{-3}$ NaOH (curve a). Curve b shows what happens when $0.1 \, mol \, dm^{-3}$ CH_3COOH is titrated against $0.1 \, mol \, dm^{-3} \, NH_3$.

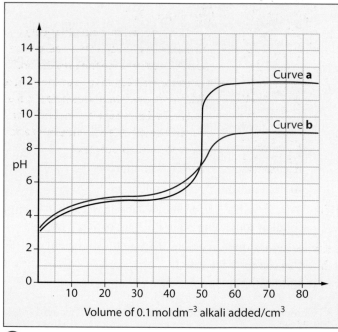

a The pH of $0.1 \, mol \, dm^{-3}$ HCl is 1.0. Why is the pH of $0.1 \, mol \, dm^{-3} \, CH_3COOH$ about 2.8 rather than 1.0?

b Is methyl orange a suitable indicator to use when titrating $0.1 \, mol \, dm^{-3} \, CH_3COOH$ against $0.1 \, mol \, dm^{-3}$ NaOH? Explain.

c Is phenolphthalein a suitable indicator to use when titrating $0.1 \, mol \, dm^{-3} \, CH_3COOH$ against $0.1 \, mol \, dm^{-3}$ NaOH? Explain.

d No indicator will detect the equivalence point accurately when titrating $0.1 \, mol \, dm^{-3} \, CH_3COOH$ against $0.1 \, mol \, dm^{-3} \, NH_3$. Why is this?

e In view of the statement in d, how could you determine the equivalence point in titrating a weak acid against a weak alkali?

Figure 21.17
pH changes during the titration of $50 \, cm^3$ of $0.1 \, mol \, dm^{-3}$ $CH_3COOH(aq)$ with $0.1 \, mol \, dm^{-3}$ NaOH(aq) (curve a) and during the titration of $50 \, cm^3$ of $0.1 \, mol \, dm^{-3} \, CH_3COOH(aq)$ with $0.1 \, mol \, dm^{-3} \, NH_3(aq)$ (curve b)

22.1 Introducing reaction kinetics

In Chapter 8, we introduced the concept of **reaction rate** and studied factors affecting the rates of reactions such as concentration, pressure, surface area, temperature and catalysts. We then went on to explain the effect of changes in concentration, pressure and surface area on the rates of reactions using the collision theory. However, the simple collision theory could not adequately explain the effects of temperature and catalysts. In order to obtain a full explanation, it was necessary to introduce a more sophisticated version of the collision theory involving the concept of activation energy and the Maxwell–Boltzmann distribution of kinetic energies amongst the molecules of a gas.

In this chapter, we shall again study the rates of chemical reactions quantitatively, but in more detail. In particular, we will study the effect of changing the concentration of different reactants on reaction rates. Then, we will use the results to gain an understanding of reaction mechanisms – the steps by which bonds break in the reactants and then new bonds form in the products.

Definitions

Reaction kinetics, sometimes called chemical kinetics, is the detailed quantitative study of the rates of chemical reactions.

Reaction rate (rate of a reaction) is the rate of change in the amount or concentration of a particular reactant or product.

◀ Figure 22.1
By studying reactions quantitatively, chemists can deduce equations, create models to summarise their results and then make predictions about the effects of different conditions. This helps in the design of safer, more efficient and more environmentally friendly industrial and pharmaceutical processes.

22.2 Techniques for studying the rates of reactions

The rate of a chemical reaction can be obtained by following a property which changes as the reaction occurs. By following this property and analysing the reaction mixture at suitable intervals, it is possible to determine the concentration of both reactants and products. Hence, we can obtain a measure of the reaction rate (i.e. the rate at which the concentration of a particular substance changes with time).

The method used to analyse the reaction mixture depends on the reaction under consideration. The following techniques illustrate four possible approaches.

Titration techniques

This method is particularly suitable for reactions in solution such as that between iodine and propanone catalysed by acid:

$$CH_3COCH_3(aq) + I_2(aq) \xrightarrow{H^+} CH_2ICOCH_3(aq) + H^+(aq) + I^-(aq)$$

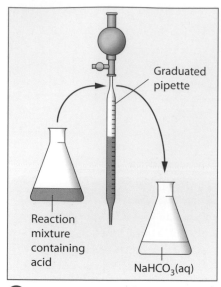

Figure 22.2

Removing a measured sample of the propanone/iodine reaction mixture and quenching it in sodium hydrogencarbonate, $NaHCO_3(aq)$, before titrating the unreacted iodine

Figure 22.3

A simplified diagram of a colorimeter

Quick Questions

1 In a study of the propanone/iodine reaction the concentration of iodine fell from $0.0100 \, mol \, dm^{-3}$ to $0.0095 \, mol \, dm^{-3}$ in 600 s.
   a What was the average rate of reaction in $mol \, dm^{-3} \, s^{-1}$ of iodine during that time?
   b Why is the reaction *not* quenched by neutralising the acid catalyst with $NaOH(aq)$?

2 When the decomposition of hydrogen peroxide solution, $H_2O_2(aq)$, was studied at room temperature, its concentration fell from $2.00 \times 10^{-2} \, mol \, dm^{-3}$ to $0.20 \times 10^{-2} \, mol \, dm^{-3}$ in 10^5 seconds.
   a Write a balanced equation for the decomposition of $H_2O_2(aq)$.
   b What is the rate of decomposition of H_2O_2?

3 Suggest one disadvantage of titration techniques in measuring reaction rates?

Titration techniques can also be used to analyse the alkaline hydrolysis of an ester (such as methyl methanoate):

$$HCOOCH_3(l) + OH^-(aq) \longrightarrow HCOO^-(aq) + CH_3OH(aq)$$

In these cases, the reaction can be followed by removing and analysing small portions of the reaction mixture at intervals. Very often, the removed portion must be added to a reagent which will 'quench' the reaction (i.e. stop it). This prevents further changes in concentration while the analysis is carried out. For example, in studying the reaction between iodine and propanone, portions of the reaction mixture can be pipetted into sodium hydrogencarbonate solution. This 'quenches' the reaction by neutralising the acid catalyst (Figure 22.2). The quenched mixture can then be analysed carefully by titrating the unreacted iodine against a standard solution of sodium thiosulfate.

Colorimetric analysis

This method is especially convenient for systems in which one of the substances is coloured (e.g. the reaction of iodine with propanone or the reaction of bromine with methanoic acid).

The intensity of colour can be followed during the reaction using a photoelectric colorimeter (Figure 22.3). From these measurements the concentration of the coloured substance can be obtained at different times.

In a colorimeter, a narrow beam of light passes through the reaction mixture towards a sensitive photocell. In many colorimeters, it is possible to select the most appropriate colour of light by choosing a particular filter.

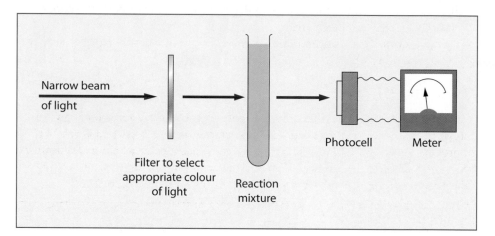

The current generated in the photocell is proportional to the amount of light transmitted by the reaction mixture. This, in turn, depends upon its colour intensity. Thus, the current from the photocell will be greatest when the light transmitted by the reaction mixture is the greatest. This occurs when the coloured substance is most dilute. However, the meter is usually calibrated to show not the fraction of light *transmitted* but the fraction of light *absorbed*. This will be proportional to the concentration of the coloured substance in the reaction mixture.

Pressure measurements

This technique is particularly suitable for gaseous reactions which involve changes in pressure. For example, the gaseous decomposition of 2-methyl-2-iodopropane can be followed conveniently by measuring the pressure at suitable time intervals.

Conductimetric analysis

Many reactions in aqueous solution involve changes in the ions present as the reaction proceeds. So, the electrical conductivity of the solution will change during the reaction. This can be used to determine the changing concentrations of reactants and products with time. Essentially, this method involves immersing two inert electrodes in the reaction mixture and then following the change in electrical conductivity of the solution with time (Figure 22.4).

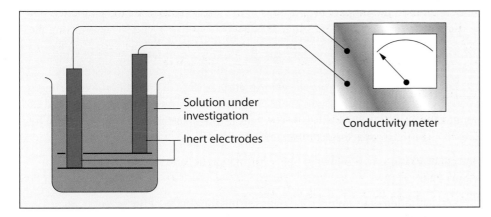

Solution under investigation

Inert electrodes

Conductivity meter

◀ Figure 22.4
Following the change in electrical conductivity of a solution with time

The last three methods have two advantages over titration techniques.

▶ They do not require the removal of samples from the reacting mixture. In these three cases, the extent of the reaction is determined at regular intervals without disturbing the reaction mixture.
▶ The time at which the concentration of a reactant or product is being measured can be taken more accurately.

Calculating the rate of reaction

It is important to appreciate that measurements on a reacting system do not give the rate of reaction directly. They simply give the concentration of a particular reactant or product, X, at a given time, t. By plotting a graph of the concentration of X against time, it is possible to determine the rate of reaction at time t. This involves drawing a tangent to the graph at time t and then calculating its gradient (Figure 22.5).

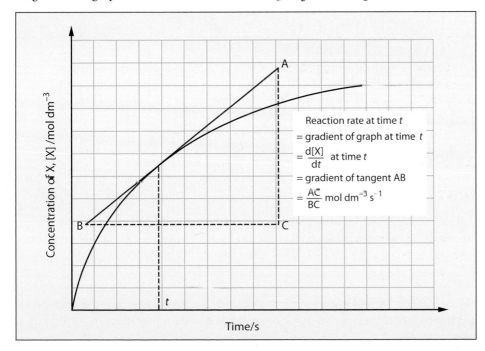

Concentration of X, [X] /mol dm^{-3}

Time/s

Reaction rate at time t
= gradient of graph at time t
= $\dfrac{d[X]}{dt}$ at time t
= gradient of tangent AB
= $\dfrac{AC}{BC}$ mol dm^{-3} s^{-1}

Quick Questions

4 Suggest a suitable technique for measuring the rate of each of these reactions:
   a $BaCO_3(s)$
       $\rightarrow BaO(s) + CO_2(g)$
   b $2NO(g) + 2CO(g)$
       $\rightarrow N_2(g) + 2CO_2(g)$
   c $C_4H_9Br(l) + NaOH(aq)$
       $\rightarrow C_4H_9OH(l) + NaBr(aq)$
   d $Br_2(aq) + HCOOH(aq)$
       $\rightarrow 2Br^-(aq) + 2H^+(aq) + CO_2(g)$

◀ Figure 22.5
Obtaining the rate of reaction at a given time from a graph of concentration against time

Table 22.1

Results for the kinetic study of the reaction between bromine and methanoic acid

Time/s	$[Br_2]$/mol dm^{-3}
0	0.0100
30	0.0090
50	0.0080
80	0.0073
120	0.0067
180	0.0052
240	0.0044
350	0.0029
470	0.0020
600	0.0012

22.3 Investigating the effect of concentration on reaction rate more fully – rate equations

We will now consider the influence of concentration on reaction rates in more detail. A convenient reaction to study is that between bromine and methanoic acid, HCOOH, in aqueous solution.

The reaction is catalysed by acid:

$$Br_2(aq) + HCOOH(aq) \xrightarrow{H^+} 2Br^-(aq) + 2H^+(aq) + CO_2(g)$$

The reaction can be followed colorimetrically by measuring the intensity of the red–brown bromine at suitable time intervals. By plotting a calibration curve of known bromine concentrations against colorimeter readings, it is possible to deduce the concentrations of bromine from the colorimeter readings obtained in the experiment. Some typical results are shown in Table 22.1. The concentration of methanoic acid was about 1.0 mol dm^{-3}. So, it is virtually constant throughout the experiment because it is present in such large excess compared to Br_2.

The concentrations of bromine in Table 22.1 are plotted graphically against time in Figure 22.6.

The concentration of bromine, $[Br_2]$, falls during the course of the reaction. So, the rate of the reaction can be expressed in terms of the rate at which the bromine concentration changes.

$$\text{Reaction rate} = -\frac{\text{change in bromine concentration}}{\text{time taken}}$$

$$= -\frac{d[Br_2]}{dt} \text{ at one point on the graph in Figure 22.6}$$

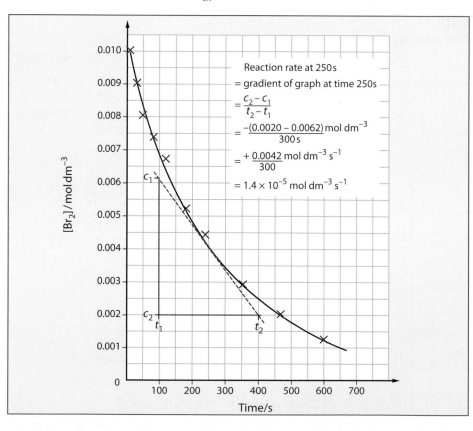

Figure 22.6

The variation of bromine concentration with time in the reaction between methanoic acid and bromine

Notice the negative sign in the last expression. $d[Br_2]$ is negative because the bromine is being used up. So, the negative sign is necessary to give the rate of reaction, $-d[Br_2]/dt$, a positive value.

372

22.3 Investigating the effect of concentration on reaction rate more fully – rate equations

In order to obtain the reaction rate at any given time, we must draw a tangent to the curve at this particular time and measure its gradient. This has been done on Figure 22.6 for the reaction rate at 250 seconds. Values of the reaction rate for different bromine concentrations at various times are shown in Table 22.2. These values of reaction rate are plotted vertically against bromine concentration in Figure 22.7.

Quick Questions

5 Look carefully at Figures 22.6 and 22.7.
  a How does the bromine concentration change with time?
  b How does the reaction rate change with time?
  c Is the rate of reaction affected by the bromine concentration?
  d How does the rate of reaction depend on the bromine concentration?
  e Write a mathematical expression relating reaction rate to bromine concentration.

 Table 22.2

Values of the reaction rate at different bromine concentrations. (Values of $[Br_2]$ are taken from Figure 22.6 at the times shown in the table. Values of the reaction rate were obtained from the gradients of tangents drawn to the graph in Figure 22.6.)

Time/s	$[Br_2]/$ $mol\,dm^{-3}$	Reaction rate $(-d[Br_2]/dt)$ $/mol\,dm^{-3}\,s^{-1}$
50	0.0085	2.9×10^{-5}
100	0.0068	2.3×10^{-5}
200	0.0048	1.7×10^{-5}
250	0.0041	1.4×10^{-5}
300	0.0034	1.1×10^{-5}
400	0.0025	0.8×10^{-5}
500	0.00175	0.5×10^{-5}

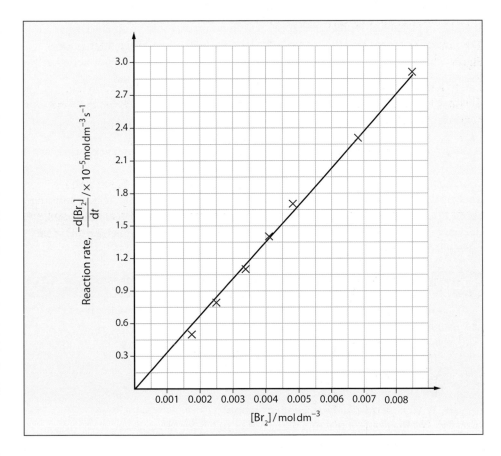

Figure 22.7

Variation of reaction rate with bromine concentration

The graph in Figure 22.7 shows that the reaction rate is directly proportional to the bromine concentration. So, we can write:

$$\text{reaction rate} \propto [Br_2]$$

$$\rightarrow \quad \text{reaction rate} = k[Br_2]$$

where k is a constant, known as the **rate constant** for the reaction.

22.4 Order of reaction and rate equations

Experiments show that the rates of most reactions can be related to the concentrations of individual reactants by an equation of the form

$$\text{Rate} = k[X]^n$$

Note

A **rate equation** is sometimes called a rate law.

Definitions

A **rate equation** shows how the rate of a reaction varies with the concentrations of those substances taking part in the reaction.

The **order of a reaction** with respect to a given reactant is the power of that reactant's concentration in the rate equation.

The **overall order of a reaction** is the sum of the orders of the individual reactants in the rate equation.

Key Points

For the general reaction,

$$xA + yB \longrightarrow \text{products},$$

it is found experimentally that the reaction rate can be expressed as:

$$\text{Rate} = k[A]^m[B]^n$$

This is known as the rate equation for the reaction. k is called the rate constant and its value is constant at a particular temperature. When all the reactant concentrations are $1.0\,\text{mol}\,\text{dm}^{-3}$, k is numerically equal to the reaction rate.

The rate equation can only be obtained by experiment. It cannot be deduced from the balanced equation.

This expression in which [X] is the concentration of the reactant being investigated and n is usually 0, 1 or 2, is known as the **rate equation**. The value of n gives the **order of the reaction with respect to X**.

If we take the more general case of a reaction,

$$xA + yB \longrightarrow \text{products}$$

the rate equation can be written as

$$\text{Rate} = k[A]^m[B]^n$$

The indices m and n are known as the orders of the reaction with respect to A and B respectively. The **overall order** of the reaction is $(m + n)$. It is therefore important to be clear whether the overall order of the reaction or the order with respect to an individual reactant is being discussed.

In the rate equation, k is known as the rate constant, but k is only constant at one particular temperature. In almost all cases, k increases as the temperature increases. It is also important to appreciate that the rate equation can only be obtained by experiment; it cannot be deduced from the balanced equation. Notice in the rate equation just above that the powers of the concentrations m and n are not necessarily the same as the coefficients a and b in the balanced chemical equation.

The reaction between propanone and iodine in aqueous solution,

$$CH_3COCH_3(aq) + I_2(aq) \longrightarrow CH_2ICOCH_3(aq) + H^+(aq) + I^-(aq)$$

is catalysed by aqueous H^+ ions. Experiments show that the rate can be expressed as:

$$\text{Rate} = k[CH_3COCH_3][H^+]$$

This particular reaction shows very clearly that the balanced equation tells us nothing about the rate equation. I_2 is an important reactant in the balanced equation, but it does not even appear in the rate equation. Furthermore, H^+ ions which feature in the rate equation do not appear as reactants in the balanced equation.

Quick Questions

6 Look at the rate equation for the reaction of propanone with iodine in acid solution just above.
   a What is the order of the reaction with respect to:
      i propanone,
      ii iodine,
      iii H^+ ions?
   b What is the overall order of the reaction?
7 What is the order of reaction with respect to Br_2 for the reaction of methanoic acid with Br_2 which we studied in Section 22.3?

First order reactions

Figure 22.7 in the previous section shows a graph of the reaction rate against concentration for a typical first order reaction. In this case, the rate of reaction is proportional to the concentration of the reactant under investigation. So, the concentration term of the reactant is raised to the power of one in the rate equation:

$$\text{Rate} = k[A]^1 = k[A]$$

Doubling the concentration of A, doubles the reaction rate. Tripling the concentration of A, triples the rate and so on, giving a straight line through the origin in the plot of rate against concentration.

8   Look again at Figure 22.6 showing a typical concentration–time graph for a first order reaction. How many seconds does it take for the concentration of bromine, $[Br_2]$, to fall from:
    a $0.010\,mol\,dm^{-3}$ to $0.005\,mol\,dm^{-3}$,
    b $0.005\,mol\,dm^{-3}$ to $0.0025\,mol\,dm^{-3}$,
    c $0.0025\,mol\,dm^{-3}$ to $0.001\,25\,mol\,dm^{-3}$,
    d $0.008\,mol\,dm^{-3}$ to $0.004\,mol\,dm^{-3}$?

At any time during the reaction it takes about 200 seconds for the concentration of bromine to halve (i.e. to fall from a concentration of say $x\,mol\,dm^{-3}$ to $x/2\,mol\,dm^{-3}$).

The time taken for the concentration of a reactant to fall to half its original value is called the **half-life of the reaction**, $t_{1/2}$. And, experiments at constant temperature show that all first order reactions have constant half-lives. In other words, the half-life of a first order reaction is independent of concentration. The half-life is the same wherever it is read off the graph. This is illustrated in Figure 22.8. The constant half-life provides a useful method of showing that a reaction is first order.

The units of k for a first order reaction

Rates of reaction are usually given in $mol\,dm^{-3}\,s^{-1}$ and reactant concentrations are in $mol\,dm^{-3}$. So the units of k will depend upon the order of the reaction.

For a first order reaction:

$$Rate = k[A]^1 = k[A]$$

So, $k = \dfrac{rate}{[A]}$ and the units of k are therefore $\dfrac{mol\,dm^{-3}\,s^{-1}}{mol\,dm^{-3}} = s^{-1}$

9   **a** Assume the half-life for the first order decomposition of compound X is 100 seconds. How long will it take for [X] to fall from:
        i $0.100\,mol\,dm^{-3}$ to $0.050\,mol\,dm^{-3}$,
        ii $0.080\,mol\,dm^{-3}$ to $0.010\,mol\,dm^{-3}$?
    b Using the rate equation for a first order reaction, it is possible to deduce mathematically that
        $t_{1/2} = \dfrac{0.69}{k}$. What is the value of k for the reaction described in part a?

10 **a** Knowing that rate $= k[A]$ for a first order reaction, explain briefly how the rate constant, k, can be found from a graph of rate plotted against concentration.
    b Calculate the rate constant for the rate against concentration graph in Figure 22.7.

Second order reactions

For a reaction that is second order with respect to a particular reactant, the reaction rate is proportional to the square of the concentration of that reactant.

$$Rate = k[A]^2$$

In this case, doubling [A] increases the reaction rate by $2^2 = 4$ times and tripling [A] increases the rate by $3^2 = 9$ times.

The **half-life of a reaction** is the time taken for the concentration of one of the reactants to fall by half.

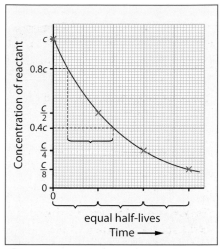

Figure 22.8
A concentration–time graph for a first order reaction. The half-life is the same wherever it is read off the graph.

A reaction is first order with respect to a particular reactant if a concentration–time graph for that reactant shows constant half-lives.

Figure 22.9 shows a typical concentration–time graph for a second order reaction. In this case, the half-life is *not* constant. In fact, the time taken for the concentration to fall from c to $c/2$ is half the time it takes the concentration to fall from $c/2$ to $c/4$.

As before, tangents can be drawn to the curve at intervals in order to calculate the rate for different concentrations of the reactant. This produces a graph similar to that shown in Figure 22.10. This time, the rate–concentration graph is not a straight line but a curve with increasing gradient.

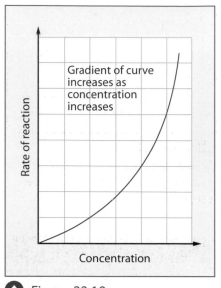

▲ Figure 22.10

A typical rate–concentration graph for a second order reaction

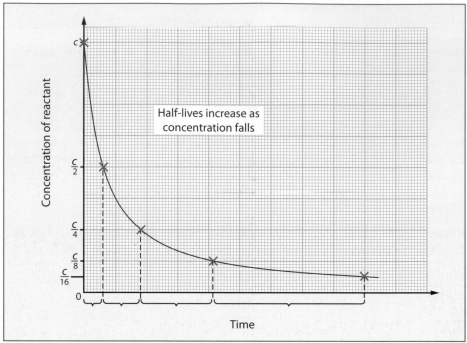

▲ Figure 22.9

Variation in the concentration of reactant against time for a second order reaction

The units of k for a second order reaction

For a second order reaction, Rate $= k[A]^2$

So, $\quad k = \dfrac{\text{rate}}{[A]^2}$

and the units of k are therefore $\dfrac{\text{mol dm}^{-3}\,\text{s}^{-1}}{(\text{mol dm}^{-3})^2} = \dfrac{\text{s}^{-1}}{\text{mol dm}^{-3}} = \text{mol}^{-1}\,\text{dm}^3\,\text{s}^{-1}$

Quick Questions

11 The rate equation for the hydrolysis of an ester in alkali is
   Rate $= k[\text{ester}][\text{OH}^-]$
   a What is the order of the reaction with respect to
     i the ester,
     ii OH^-?
   b What is the overall order of the reaction?
   c The rate of the reaction is $0.001\,\text{mol dm}^{-3}\,\text{s}^{-1}$ at 25 °C when $[\text{ester}] = 0.10\,\text{mol dm}^{-3}$ and $[\text{OH}^-] = 0.05\,\text{mol dm}^{-3}$. Calculate the value of k, the rate constant and show its units.
   d Calculate the reaction rate at 25 °C when:
     i $[\text{ester}] = 0.05\,\text{mol dm}^{-3}$ and $[\text{OH}^-] = 0.10\,\text{mol dm}^{-3}$
     ii $[\text{ester}] = 0.20\,\text{mol dm}^{-3}$ and $[\text{OH}^-] = 0.02\,\text{mol dm}^{-3}$

Zero order reactions

If a reaction is zero order with respect to a particular reactant, the rate of the reaction is unaffected by changes in the concentration of that reactant. This means that the reaction continues at the same rate even when the concentration of the reactant changes. In this case, the gradient of the concentration–time graph stays the same as the concentration of the reactant falls (Figure 22.11).

In the rate equation for a zero order reaction, the concentration of the reactant under investigation is raised to the power zero.

$$\text{Rate} = k[A]^0$$

Any number raised to the power zero equals one. So for a zero order reaction we can write

$$\text{Rate} = k$$

The rate of reaction is numerically equal to the rate constant and, at a given temperature, the reaction rate is constant.

The decomposition of ammonia to N_2 and H_2 using a hot platinum catalyst is a good example of a zero order reaction. Provided there is enough ammonia to be adsorbed over the whole surface of the platinum, the rate of reaction remains constant whatever the concentration of ammonia. In this case, the reaction rate is determined by the surface area of the catalyst which is constant. Now try Quick question 12.

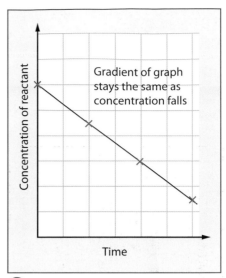

Figure 22.11
Variation in the concentration of reactant against time for a zero order reaction

Quick Questions

12 **a** Write a balanced equation with state symbols and a rate equation for the decomposition of ammonia using a platinum catalyst.
   b Draw sketch graphs for the reaction in part a to show:
   i $[NH_3(g)]$ against time (with time on the horizontal axis),
   ii rate of reaction against $[NH_3(g)]$ (with $[NH_3(g)]$ on the horizontal axis).
   c In a particular experiment at 150 °C, the rate of reaction was 0.002 mol dm^{-3} s^{-1} when the concentration of ammonia was 0.20 mol dm^{-3}. What is the value of the rate constant at this temperature and what are its units?

Key Point

A reaction is zero order with respect to a particular reactant if a concentration–time graph for that reactant produces a straight line (constant gradient).

22.5 The initial rates method for determining rate equations

The initial rates method is a variation of the method described in Section 22.3 to determine the rate equation and the rate constant of a reaction. As before, values of the concentration of a reactant at different times are used to plot a concentration–time graph. But this time, the reaction is only followed for a short time – just long enough to allow one tangent to be drawn at $t = 0$. The gradient of this tangent gives the instantaneous rate of reaction at the start of the reaction – the **initial rate**. The experiment is then repeated several times using different initial concentrations.

For example, the decomposition of dinitrogen pentoxide,

$$2N_2O_5(g) \longrightarrow 4NO_2(g) + O_2(g)$$

was studied by carrying out four experiments, each with a different concentration of N_2O_5. In each experiment, the concentration of N_2O_5 was carefully monitored for 40 seconds in order to plot a graph of $[N_2O_5(g)]$ against time.

For each graph, only the tangent at $t = 0$ is drawn. The gradient of this tangent then gives the initial rate of the reaction for the initial concentration of N_2O_5.

The results from one experiment and the concentration–time graph plus the tangent at $t = 0$ are shown in Figure 22.12.

Definition

The **initial rate of a reaction** is the instantaneous reaction rate at the moment the reaction starts.

Table 22.3

Initial reaction rates and initial N_2O_5 concentrations in experiments to study the decomposition of N_2O_5

Initial reaction rate, $d[N_2O_5(g)]/dt$ /mol dm^{-3} s^{-1}	Initial concentration of $N_2O_5(g)$ /mol dm^{-3}
0.69×10^{-9}	2.55×10^{-8}
0.46×10^{-9}	1.70×10^{-8}
0.23×10^{-9}	0.85×10^{-8}

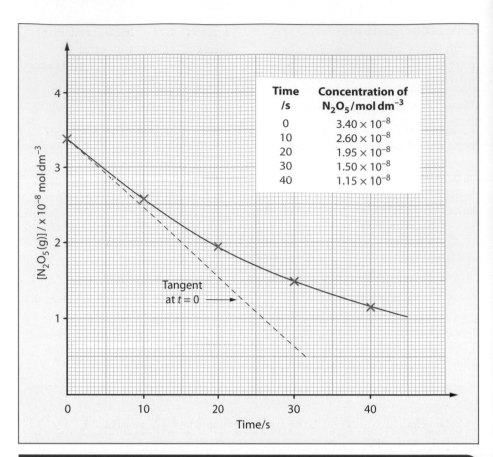

Time /s	Concentration of N_2O_5/mol dm^{-3}
0	3.40×10^{-8}
10	2.60×10^{-8}
20	1.95×10^{-8}
30	1.50×10^{-8}
40	1.15×10^{-8}

Figure 22.12

The results of one experiment and the concentration–time graph plus the tangent at $t = 0$ for the decomposition of N_2O_5

Key Point

The initial rates method involves measuring the initial rate for different concentrations of a given reactant, then deducing the order of the reaction from the way initial rate varies with concentration.

Quick Questions

13 **a** Use the tangent in Figure 22.12 to calculate the initial reaction rate in terms of the rate of change in N_2O_5 concentration, $d[N_2O_5(g)]/dt$, at $t = 0$.
   b The initial reaction rates of the other three experiments and their corresponding initial N_2O_5 concentrations are listed in Table 22.3. Use these results and that from part a to deduce the order of the reaction with respect to N_2O_5.
   c Write a rate equation for the decomposition of N_2O_5.
   d Use your rate equation to calculate the rate constant for the decomposition of N_2O_5.

22.6 Investigating the reaction between hydrogen and nitrogen monoxide using the initial rates method

Table 22.4 shows data concerning the reaction between hydrogen and nitrogen monoxide at 800 °C.

$$2H_2(g) + 2NO(g) \longrightarrow 2H_2O(g) + N_2(g)$$

Table 22.4

Data showing the initial concentrations and initial rates of six experiments involving the reaction between hydrogen and nitrogen monoxide at 800 °C

Experiment number	Initial concentration of nitrogen monoxide /mol dm^{-3}	Initial concentration of hydrogen /mol dm^{-3}	Initial rate of production of nitrogen /mol dm^{-3} s^{-1}
1	6×10^{-3}	1×10^{-3}	3×10^{-3}
2	6×10^{-3}	2×10^{-3}	6×10^{-3}
3	6×10^{-3}	3×10^{-3}	9×10^{-3}
4	1×10^{-3}	6×10^{-3}	0.5×10^{-3}
5	2×10^{-3}	6×10^{-3}	2.0×10^{-3}
6	3×10^{-3}	6×10^{-3}	4.5×10^{-3}

In experiments 1, 2 and 3, the initial concentration of hydrogen doubles and then trebles while the initial concentration of nitrogen monoxide stays the same. In experiments 4, 5 and 6, the initial concentration of nitrogen monoxide doubles and then trebles while the initial concentration of hydrogen is the same.

Now answer Quick question 14.

From Quick question 14 you should be able to write a rate equation for the reaction between hydrogen and nitrogen monoxide as:

$$\text{Rate} = k[H_2(g)]\,[NO(g)]^2$$

By substituting the results from experiment 1 in the rate equation, we can now determine the value of k, the rate constant:

$$3 \times 10^{-3}\,\text{mol dm}^{-3}\,\text{s}^{-1} = k \times 1 \times 10^{-3}\,\text{mol dm}^{-3} \times (6 \times 10^{-3}\,\text{mol dm}^{-3})^2$$

$$\Rightarrow \quad k = \frac{3 \times 10^{-3}\,\text{mol dm}^{-3}\,\text{s}^{-1}}{1 \times 10^{-3}\,\text{mol dm}^{-3}} \times \frac{1}{(6 \times 10^{-3}\,\text{mol dm}^{-3})^2}$$

$$= \frac{10^6\,\text{mol dm}^{-3}\,\text{s}^{-1}}{12(\text{mol dm}^{-3})^3}$$

$$= 8.33 \times 10^4\,\text{mol}^{-2}\,\text{dm}^6\,\text{s}^{-1}$$

As the rate constant is constant at a fixed temperature, we can write the value of k as:

$$k = 8.33 \times 10^4\,\text{mol}^{-2}\,\text{dm}^6\,\text{s}^{-1} \text{ at } 800\,°C.$$

22.7 The effect of temperature change on reaction rates

In Sections 8.6 and 8.7, we studied the effect of temperature change on reaction rates. We found that almost all reactions can be speeded up by increasing the temperature. In many cases, the reaction rate roughly doubles for a temperature rise of only 10 K.

In order to explain this rapid increase in reaction rates with temperature, we argued that the reactant particles would not react unless they collided with at least a certain minimum amount of energy. This minimum energy for a reaction is known as the activation energy, E_A.

Using probability theory and the kinetic theory of gases, James Clerk Maxwell and Ludwig Boltzmann derived equations for the distribution of kinetic energies amongst the molecules of a gas at different temperatures. From their work, it is possible to draw graphs showing the number of molecules with a particular kinetic energy against the size of the kinetic energy (Figure 22.14). The spread of kinetic energies for the molecules in these graphs was named the Maxwell–Boltzmann distribution in honour of the two scientists. From their graphs of the distributions of kinetic energies, Maxwell and Boltzmann calculated that the fraction of molecules with an energy greater than the activation energy, E_A in J mol^{-1} is given by $e^{-E_A/RT}$. In this expression, e is the exponential function and the base of natural logarithms equal to 2.718, R is the gas constant $(8.3\,\text{J K}^{-1}\,\text{mol}^{-1})$ and T is the absolute temperature.

This suggests that at a given temperature, T, the reaction rate should be proportional to $e^{-E_A/RT}$.

Now, since the rate constant, k, is a measure of the reaction rate when the concentration of all reactants is $1\,\text{mol dm}^{-3}$, we can write

$$k \propto e^{-E_A/RT}$$

$$\Rightarrow \quad k = Ae^{-E_A/RT}$$

This last expression is sometimes called the **Arrhenius equation** because it was first predicted by the Swedish chemist Svante Arrhenius in 1889.

Quick Questions

14 a Look at the data for experiments 1, 2 and 3 in Table 22.4.
       What happens to the initial rate when the initial concentration of H_2 is:
       i doubled,
       ii trebled?
    b How does the reaction rate depend on $[H_2]$?
    c What is the order of reaction with respect to hydrogen?
    d Look at the data for experiments 4, 5 and 6.
       What happens to the initial rate when the initial concentration of NO is:
       i doubled,
       ii trebled?
    e How does the reaction rate depend on $[NO]$?
    f What is the order of reaction with respect to nitrogen monoxide?
    g What is the overall order of the reaction?

Figure 22.13
The metabolic rate in an animal's body is very dependent on the temperature of its surroundings. Many animals like these dormice hibernate in the winter when food is scarce. Their body temperature falls to just a few degrees above the surroundings and metabolic processes take place very slowly using stored fats or carbohydrates.

> Figure 22.14

Graphs showing the distribution of kinetic energies amongst the molecules of a gas at $T\,K$ and $(T+10)\,K$.

Quick Questions

15 a The activation energies of many reactions fall between $100\,\text{kJ}\,\text{mol}^{-1}$ and $300\,\text{kJ}\,\text{mol}^{-1}$. What is the significance of this range of values in relation to the bonds which need to be broken in reactions?

b Use the equation

$$\ln k = \ln A - \frac{E_A}{RT}$$

to explain why the rate constant, k, increases as temperature increases.

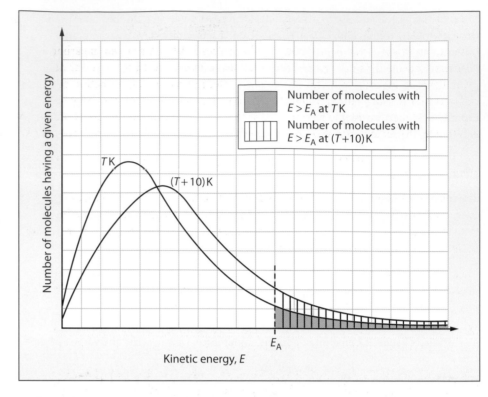

> **Note**
>
> **Logarithms to base e**
>
> Logarithms to base e are called **natural logarithms**. The correct symbol for 'logarithm to base e' is 'ln'. Logarithms to base e follow similar rules to logarithms to base 10.
>
> Remember, the logarithm of a number to a given base is the power to which the base must be raised to equal the number.
>
> So, $\lg 10 = 1$, $\lg 10^2 = 2$
>
> $\therefore$ $\ln e = 1$, $\ln e^2 = 2$
>
> and $\ln e^{-E_A/RT} = -\dfrac{E_A}{RT}$

In the Arrhenius equation, A (the Arrhenius constant), can be regarded as a **collision frequency and orientation factor** in the reaction rate, whilst $e^{-E_A/RT}$ represents an **activation state factor**. Thus, A is determined by the total number of collisions per unit time and the orientation of molecules when they collide, i.e. the way they are lined up with one another on collision. In contrast, $e^{-E_A/RT}$ is determined by the fraction of molecules with sufficient energy to react.

If we take logarithms to base e in $k = Ae^{-E_A/RT}$

$$\ln k = \ln A + \ln e^{-E_A/RT}$$

$$\Rightarrow \quad \ln k = \ln A - \frac{E_A}{R} \times \frac{1}{T}$$

Comparing the last equation with

$$y = c + mx$$

a graph of $\ln k$ against $1/T$ should be a straight line with gradient $-E_A/R$ (Figure 22.15).

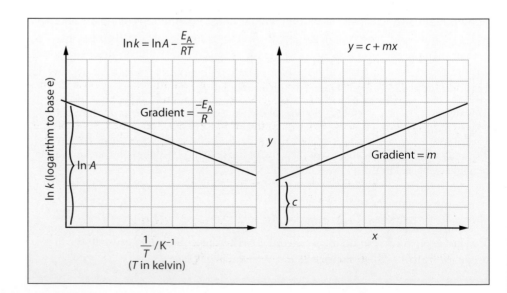

> Figure 22.15

Comparing the equation $\ln k = \ln A - E_A/RT$ with $y = c + mx$

22.8 The effect of catalysts on reaction rates

Catalysts and their effect on reaction rates have already been discussed in earlier chapters. Catalysts, including enzymes, were introduced in Section 8.3 as one of several factors affecting the rates of reactions. Later in Section 8.8, the action of catalysts in providing a different reaction path with a lower activation energy was explained using the example of ammonia synthesis. This effect on the activation energy was then interpreted in terms of the Maxwell–Boltzmann distribution in Section 8.9.

In Chapter 12 we discussed the importance of catalysis in the Haber process and the Contact process.

The reactions in both the Haber process and the Contact process involve gases passing over a solid catalyst. Reactions of this kind in which the reactants are in a different state to the catalyst are said to involve **heterogeneous catalysis**. Usually, heterogeneous catalysts are solids while the reactants are gases or in solution.

In contrast to this, catalysed reactions in which the reactants and catalyst are mixed together in the same state are said to involve **homogeneous catalysis**. Generally in this case, the catalyst and reactants are dissolved in the same solution.

Many important industrial processes involve heterogeneous catalysis whereas the actions of enzymes in biological systems usually involve homogeneous catalysis.

Quick Questions

16 Classify the following catalysts as heterogeneous or homogeneous:
   a H^+ ions in the reaction of I_2 with propanone,
   b the enzyme catalase in the decomposition of hydrogen peroxide in plant cells,
   c nickel in the hydrogenation of unsaturated fats,
   d salivary amylase in the breakdown of starch.
17 Suggest three reasons why catalysts are so important in industry.
18 What is the big advantage of heterogeneous catalysis over homogeneous catalysis for industrial processes?

Very often, homogeneous catalysis operates through the formation of an intermediate involving the catalyst and a reactant. This intermediate can then react more easily than the original reactant.

Consider the hypothetical reaction between substances A and B to form C, catalysed by X. Substance A might first combine with the catalyst to form AX which then reacts more easily with B, forming C and regenerating the catalyst X.

$$A + X \longrightarrow AX$$
$$\text{then} \quad AX + B \longrightarrow C + X$$

Let's apply this hypothetical reaction scheme to the conversion of atmospheric sulfur dioxide, SO_2, to sulfur trioxide, SO_3, catalysed by nitrogen monoxide, NO.

SO_2 does not readily react with oxygen in the air to form SO_3. But when NO is present, this reacts immediately with oxygen to form NO_2 as an intermediate. This NO_2 then reacts readily with SO_2 to form SO_3 and regenerate the catalyst NO.

$$O_2(g) + 2NO(g) \xrightarrow{\text{catalyst}} 2NO_2(g)$$

$$2NO_2(g) + 2SO_2(g) \xrightarrow{\quad\quad} 2SO_3(g) + 2NO(g) \atop \text{catalyst}$$

$$\text{Overall:} \quad O_2(g) + 2SO_2(g) \longrightarrow 2SO_3(g)$$

Key Point

The Arrhenius equation, $k = Ae^{-E_A/RT}$ relates the rate constant to temperature. It can be used to calculate the activation energy, E_A. E_A is the activation energy in $J\,mol^{-1}$ and R is the gas constant $(8.3\,J\,K^{-1}\,mol^{-1})$. So, by plotting a graph of the logarithms to base e of rate constants against the reciprocal of the absolute temperature, we can measure the gradient, E_A/R. From this gradient, we can calculate the activation energy, E_A.

▲ Figure 22.16
The original process for manufacturing the herbicide 'Roundup' required methanol and hydrogen cyanide – two unpleasant and very poisonous chemicals. Fortunately, a new process using a copper catalyst has now been developed by Monsanto which requires much less toxic raw materials.

Key Point

There are two distinct types of catalysis:

▶ **heterogeneous catalysis** in which the reactants are in a different physical state to the catalyst
▶ **homogeneous catalysis** in which the reactants and catalyst are in the same physical state.

Figure 22.17
Ibuprofen is an important analgesic which reduces swelling and pain. Its manufacture neatly illustrates the effectiveness of catalysts. At one time, ibuprofen was manufactured in a five-step process using a homogeneous catalyst that could not be recovered, leaving 60% of the mass of the products as waste. Today, ibuprofen is produced in just two steps, using heterogeneous catalysts which can be recovered easily, leaving no waste.

Many industrial processes involve heterogeneous catalysis in which gas molecules interact with and become attached to the surface of a solid catalyst. This process is called adsorption and the attachment of reactant molecules to surface atoms of the catalyst can involve covalent bonds, permanent-dipole and induced-dipole attractions. Adsorption increases the reaction rate by holding the reactants close together on the catalyst surface, making a reaction more likely.

When ammonia is synthesised from nitrogen and hydrogen with a hot iron catalyst in the Haber process, chemists believe the following sequence of reactions occur:

▶ Reactant gases are first adsorbed as molecules on the catalyst surface:

$$H_2(g) \longrightarrow H_{2\,adsorbed}$$

$$N_2(g) \longrightarrow N_{2\,adsorbed}$$

▶ The adsorbed molecules are atomised on the hot catalyst.

$$H_{2\,adsorbed} \longrightarrow 2H_{adsorbed}$$

$$N_{2\,adsorbed} \longrightarrow 2N_{adsorbed}$$

▶ Nitrogen and hydrogen atoms then react on the catalyst surface forming ammonia:

$$N_{adsorbed} + H_{adsorbed} \longrightarrow NH_{adsorbed}$$

$$NH_{adsorbed} + H_{adsorbed} \longrightarrow NH_{2\,adsorbed}$$

$$NH_{2\,adsorbed} + H_{adsorbed} \longrightarrow NH_{3\,adsorbed}$$

▶ Finally ammonia desorbs, leaving the surface of the catalyst:

$$NH_{3\,adsorbed} \longrightarrow NH_3(g)$$

Further examples of homogeneous and heterogeneous catalysis involving transition metals and their compounds are discussed in Section 24.13.

Quick Questions

19 **a** At high pressure and with limited surface area of catalyst, the reaction rate of nitrogen with hydrogen to form ammonia is independent of the concentrations of $N_2(g)$ and $H_2(g)$.
  i Why is this?
  ii What is the overall order of reaction?
  b At lower pressure and excess surface area of catalyst, the reaction rate is proportional to $[N_2(g)]$ but independent of $[H_2(g)]$.
  i Which step in the sequence of reactions is now dictating the reaction rate?
  ii Why is this the dictating reaction?
  iii What is the overall order of reaction?

22.9 The importance of reaction rate studies

Reaction rate studies give us detailed information about the rates of chemical reactions. This is important for industrial processes in which time and the efficient use of resources are crucial. The study of reaction rates also helps us to understand biological processes and many other reactions, such as rusting and burning, which affect our everyday lives. In general, the most exhaustive studies of reaction rates have been done on industrial processes. In these cases chemists, engineers and economists endeavour to obtain maximum product from the minimum amount of raw material, using minimum fuel in the minimum possible time.

Reaction mechanisms

Rate studies also allow us to interpret reactions on a molecular level. By considering the order of a reaction with respect to the different reactants, we can speculate about the sequence in which bonds break and atoms rearrange. From these ideas, it is possible to suggest a **reaction mechanism**. We have already looked in detail at the reaction mechanisms involved in the nucleophilic substitution of halogenalkanes with hydroxide ions in Section 16.4. So, let's now look at some other examples.

The reaction of hydrogen bromide with oxygen at 700 K

$$4HBr(g) + O_2(g) \longrightarrow 2H_2O(g) + 2Br_2(g)$$

This balanced equation indicates that four HBr molecules react with one O_2 molecule. If the reaction were to take place in a single step, these five molecules would need to collide with each other simultaneously. This is extremely improbable. Since the reaction occurs quite rapidly at 700 K, it is likely that it proceeds by a series of steps rather than by a single step involving the simultaneous collision of five molecules. In fact, it is important to appreciate that most chemical reactions take place in a series of simple steps and not in a single step as suggested by the balanced equation.

Quantitative studies of the reaction between HBr and O_2 show that the reaction rate is proportional to both the concentration of HBr and the concentration of O_2. This means that the reaction is first order with respect to both HBr and O_2. We can summarise this information in a rate equation as

$$\text{Rate} = k[HBr][O_2]$$

Notice that the balanced equation involves HBr and O_2 in molar proportions of $4:1$, yet the rate equation suggests proportions of $1:1$. How can we explain this?

The overall reaction must take place in a series of simple steps. These must satisfy both the rate equation and the balanced equation.

The following mechanism has been proposed for the reaction:

HBr + O_2 ⟶ H̶B̶r̶O̶O̶			Step 1	Slow
H̶B̶r̶O̶O̶ + HBr ⟶		2̶H̶B̶r̶O̶	Step 2	Fast
H̶B̶r̶O̶ + HBr ⟶		$H_2O + Br_2$	Step 3	Fast
H̶B̶r̶O̶ + HBr ⟶		$H_2O + Br_2$	Step 4	Fast
Overall	4HBr + O_2 ⟶	$2H_2O + 2Br_2$		

Notice that each step in the reaction mechanism involves the collision of only two molecules. These are much more likely events than the simultaneous collision of five molecules. Step 3 and step 4, although identical, are written separately because each of the two HBrO molecules, formed in step 2, react separately with HBr molecules. Notice also the suggestion that the first step in the mechanism is slow whilst the others are fast. This explains why the reaction rate is proportional to both [HBr] and [O_2], since one molecule of each of these is involved in the slow first step. This proposed reaction mechanism is consistent with the rate equation and the observed reaction kinetics.

The slow first step producing HBrOO is very much a 'bottleneck' in the oxidation of hydrogen bromide. HBrOO forms, but it is immediately consumed in the fast second step by reaction with HBr. The HBrO molecules which form also react rapidly with more HBr. Although the second, third, and fourth steps are very rapid, they produce water and Br_2 only as fast as the slowest stage in the sequence. Hence, those factors that determine the rate of formation of HBrOO determine the overall rate of reaction. The formation of HBrOO is the one step which dictates the rate because it is the slowest stage in the reaction mechanism with the highest activation energy. The slowest stage in a mechanism is called the **rate-determining step** or the **rate-limiting step**.

The reaction of iodine with propanone

20 a Write a rate equation for
    the reaction of iodine with
    propanone in acid solution.
  b Which substances are
    probably involved in the slow,
    rate-determining step of the
    reaction?

Iodine reacts with propanone in acid solution as follows:

$$I_2(aq) + CH_3COCH_3(aq) \xrightarrow{H^+(aq)} CH_2ICOCH_3(aq) + H^+(aq) + I^-(aq)$$

The reaction is first order with respect to CH_3COCH_3, zero order with respect to I_2 and first order with respect to H^+. Now look at Quick question 20.

The rate-determining step in the reaction must involve propanone and H^+ ions because the concentrations of these substances affect the reaction rate. Iodine is not involved in the rate-determining step because its concentration does not influence the reaction rate. The suggested mechanism for the reaction is shown in Figure 22.18.

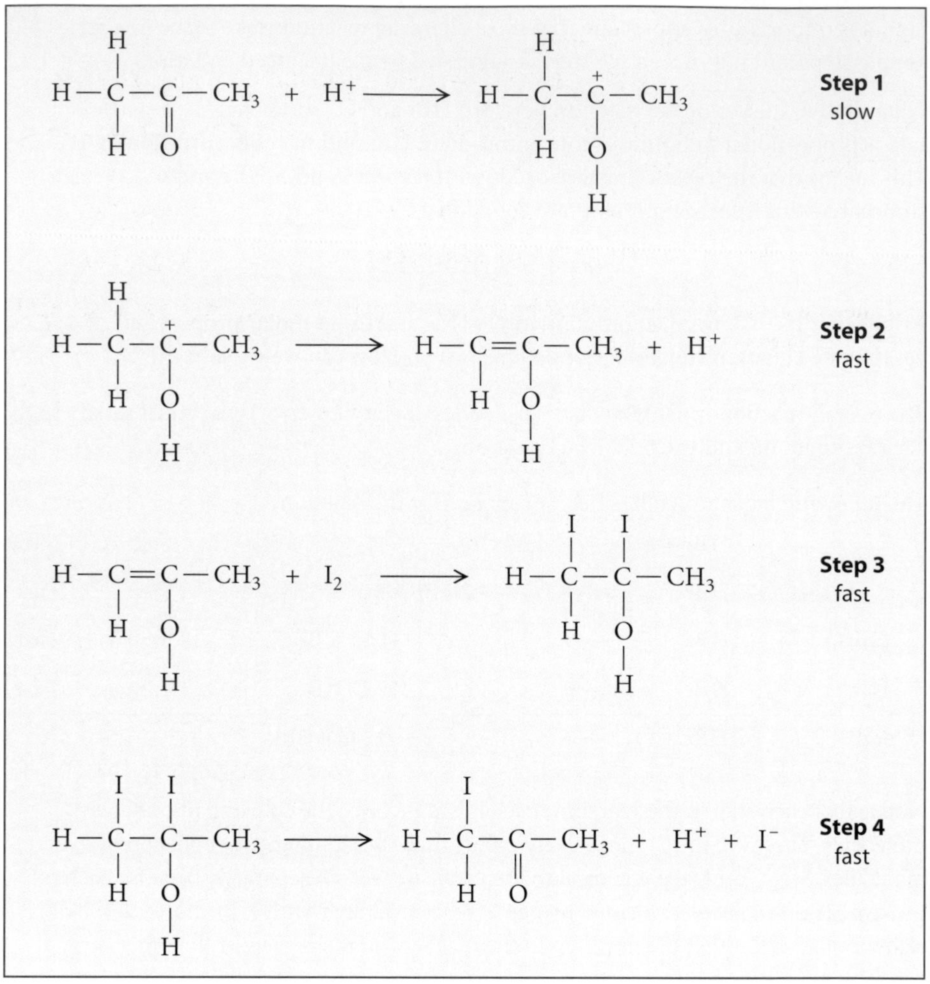

Figure 22.18
The mechanism for the reaction of iodine with propanone in acid solution

21 a In the reaction of iodine with propanone, H^+ ions do not appear as reactants in the balanced equation. Explain why.
  b Chemists predicted and then verified by experiment that bromine reacts with propanone in acid solution at the same rate as iodine. Why is this?
22 H_2O_2 oxidises I^- ions to I_2 in acid solution. The reaction is first order with respect to H_2O_2, first order with respect to I^- and zero order with respect to H^+ ions.
  a What is the overall order of reaction?
  b Write a balanced equation for the reaction.
  c Write a rate equation for the reaction.
    A proposed mechanism for the reaction is:
    Step 1
    $H_2O_2 + I^- \longrightarrow H_2O + IO^-$
    Step 2
    $H^+ + IO^- \longrightarrow HIO$
    Step 3
    $HIO + H^+ + I^- \longrightarrow I_2 + H_2O$
  d Identify the rate-determining step and explain your choice.

The reaction rate is dictated by the first, slow step which involves only the participation of CH_3COCH_3 and H^+. Once this first step is completed, the remaining steps take place rapidly. Consequently, the reaction rate is independent of the concentration of iodine, so $[I_2]$ does not feature in the rate equation.

The reaction of bromide and bromate(v) ions in acid solution

Both reaction mechanisms that we have considered so far involve an initial slow step. This is not always the case. In order to illustrate this point, consider the reaction between bromide and bromate(v) ions in acid solution:

$$5Br^-(aq) + BrO_3^-(aq) + 6H^+(aq) \longrightarrow 3Br_2(aq) + 3H_2O(l)$$

Kinetic studies show that the reaction is fourth order overall; first order with respect to bromide, first order with respect to bromate(v) and second order with respect to H^+ ions:

$$\text{Rate} = k[Br^-][BrO_3^-][H^+]^2$$

The immediate deduction from this is that the rate-determining step involves one Br^-, one BrO_3^- and two H^+ ions. But, the simultaneous collision of four ions is most unlikely. A more likely explanation is that the slow rate-determining step is preceded by faster reactions. Hence, the suggested mechanism involves the initial formation of HBr and $HBrO_3$ in two fast reactions, followed by a reaction between these two substances in the slow rate-determining step.

$$H^+ + Br^- \longrightarrow HBr \qquad \text{Step 1 Fast}$$

$$H^+ + BrO_3^- \longrightarrow HBrO_3 \qquad \text{Step 2 Fast}$$

Then, the slow rate-determining step:

$$HBr + HBrO_3 \longrightarrow HBrO_2 + HBrO \qquad \text{Step 3 Slow}$$

The $HBrO_2$ and HBrO, produced in the slow step, then react rapidly with more HBr, eventually forming bromine and water:

$$HBrO_2 + HBr \longrightarrow 2HBrO \qquad \text{Step 4 Fast}$$

$$HBrO + HBr \longrightarrow H_2O + Br_2 \qquad \text{Step 5 Fast}$$

So, the rate equation is determined by the slow step 3, which depends on the two fast steps leading up to it.

Review questions

1 For the gaseous reaction,

$A(g) + B(g) \longrightarrow C(g) + D(g)$ it is found that,

Reaction rate $= k\,[A]^2[B]$

How many times does the rate increase or decrease if:

a the partial pressures of both A and B are doubled,

b the partial pressure of A doubles, but that of B remains constant,

c the volume of the reacting vessel is doubled,

d an inert gas is added, which doubles the overall pressure whilst the partial pressures of A and B remain constant.

2 a Draw a sketch graph of the percentage reactant remaining against time, for a zero-order and a first-order reaction. (Assume, in each case, that it takes 10 minutes for the amount of reactant to fall from 100% to 50%.)

b Rate constants (k) for the decomposition of hydrogen iodide at different temperatures are given in the table below.

   i Plot a graph of ln k against $1/T$.

   ii Use this graph to obtain a value for the activation energy for the decomposition of hydrogen iodide. (The gas constant, $R = 8.3\,J\,K^{-1}\,mol^{-1}$.)

Rate constant, k /mol^{-1}dm^{3}s^{-1}	Temperature/K
3.75×10^{-9}	500
6.65×10^{-6}	600
1.15×10^{-3}	700
7.75×10^{-2}	800

3 The oxidation of nitrogen monoxide occurs readily according to the following equation.

$$NO(g) + \tfrac{1}{2}O_2(g) \longrightarrow NO_2(g)$$

The following table shows how the initial rate of this reaction depends on the concentrations of the two reactants.

[NO] /mol dm^{-3}	[O$_2$] /mol dm^{-3}	Initial rate /mol dm^{-3} s^{-1}
0.0050	0.0050	0.02
0.0050	0.0075	0.03
0.0100	0.0075	0.12

a i Use the data to determine the order of reaction with respect to each of the reactants.

   ii Write the rate equation for the reaction, and use it to calculate a value for the rate constant, k, stating its units.

   iii Use your rate equation in ii to calculate the rate of reaction when $[NO] = [O_2] = 0.0025\,mol\,dm^{-3}$.

b Nitrogen monoxide plays an important catalytic role in the oxidation of atmospheric SO_2 in the formation of acid rain.

   i State the type of catalysis shown in this process.

   ii Explain the steps involved in this process by writing equations for the reactions that occur.

Cambridge Paper 4 Q1 June 2006

4 Hydrogen peroxide reacts with iodide ions in acid solution according to the following equation:

$$H_2O_2(aq) + 2I^-(aq) + 2H^+(aq) \longrightarrow I_2(aq) + 2H_2O(l)$$

The rate of the reaction can be calculated by measuring the time for the first appearance of I_2 in the solution. When iodine first appears the concentration of I_2 is $10^{-5}\,mol\,dm^{-3}$.

a For a particular experiment, the initial concentrations are $[H_2O_2] = 0.010\,mol\,dm^{-3}$, $[I^-] = 0.010\,mol\,dm^{-3}$ and $[H^+] = 0.10\,mol\,dm^{-3}$.

   Calculate the reaction rate if I_2 first appears after 6 seconds.

b In a second experiment, the initial concentrations are $[H_2O_2] = 0.005\,mol\,dm^{-3}$, $[I^-] = 0.010\,mol\,dm^{-3}$ and $[H^+] = 0.10\,mol\,dm^{-3}$.

   Calculate the reaction rate if I_2 first appears after 12 seconds.

c From these calculations explain why the reaction is first order with respect to H_2O_2.

d Given the further information that the rate law is, Reaction rate $= k[H_2O_2][H^+][I^-]$, calculate the rate constant, k.

e What are the units of k?

f Predict the rate of reaction when $[H_2O_2] = 0.05\,mol\,dm^{-3}$, $[H^+] = 0.10\,mol\,dm^{-3}$ and $[I^-] = 0.02\,mol\,dm^{-3}$.

5 Two gases, X and Y, react according to the equation

$$X(g) + 2Y(g) \longrightarrow XY_2(g)$$

Experiments were performed at 400 K in order to determine the order of this reaction and the following results were obtained.

Experiment number	Initial concentration of X/mol dm^{-3}	Initial concentration of Y/mol dm^{-3}	Initial rate of formation of XY$_2$ /mol dm^{-3} s^{-1}
1	0.10	0.10	0.0001
2	0.10	0.20	0.0004
3	0.10	0.30	0.0009
4	0.20	0.10	0.0001
5	0.30	0.10	0.0001

a What is the order of this reaction with respect to

　i X, 　　**ii** Y?

b Write a rate equation for the reaction of X with Y.

c Using the rate equation, predict a possible mechanism for this reaction.

d Using the results from experiment 1, calculate the value of the rate constant, k and state its units.

e What further experiments would you carry out to find the activation energy of the reaction between X and Y?

f Why are chemists interested in obtaining orders of reaction and rate equations?

6 a Draw an energy profile curve for the reaction:

$$H_2(g) + I_2(g) \longrightarrow 2HI(g) \qquad \Delta H^\ominus = -10\,kJ\,mol^{-1}$$

Put labelled arrows on your diagram to indicate the activation energy of the reaction and the enthalpy change of the reaction.

b The rate of this reaction is given by:

Rate = $k[H_2][I_2]$

　i What is the order of the reaction with respect to iodine?

　ii What is the overall order of the reaction?

c When 0.1 mol of H_2 and 0.2 mol of I_2 were mixed at 400 °C in a 1 dm^3 vessel, the initial rate of formation of hydrogen iodide was $2.3 \times 10^{-5}\,mol\,dm^{-3}\,s^{-1}$. What is the value of k at 400 °C?

7 The following results were obtained from a study of the isomerisation of cyclopropane to propene in the gas phase at 433 °C.

Time/hours	0	2	5	10	20	30
% of cyclopropane remaining	100	91	79	63	40	25

a Write an equation for the reaction involved.

b Show that the reaction is first order with respect to cyclopropane.

8 In an experiment to study the acid-catalysed reaction of propanone (CH$_3$COCH$_3$) with iodine, 50 cm^3 of 0.02 mol dm^{-3} I$_2$ were mixed with 50 cm^3 of acidified 0.25 mol dm^{-3} propanone solution. 10 cm^3 portions of the reaction mixture were removed at 5 minute intervals and added rapidly to excess NaHCO$_3$(aq). The remaining iodine was then titrated against Na$_2$S$_2$O$_3$(aq). The graph in Figure 22.19 shows the volume of Na$_2$S$_2$O$_3$(aq) required to react with the remaining iodine at different times from the start of the reaction.

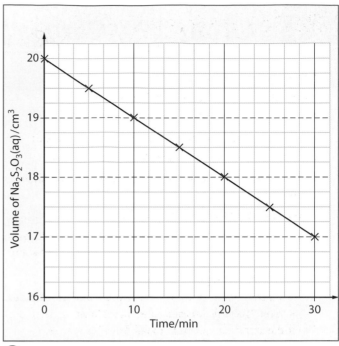

Figure 22.19
Volume of Na$_2$S$_2$O$_3$(aq) required to react with the remaining iodine at different times during the reaction of iodine with propanone.

a Why are the 10 cm^3 portions of the reaction mixture added rapidly to excess NaHCO$_3$(aq) before titration with Na$_2$S$_2$O$_3$(aq)?

b What is the rate of reaction in terms of cm^3 of Na$_2$S$_2$O$_3$(aq) min^{-1}?

c How does the *rate* of change of iodine concentration vary during the experiment?

d Is the reaction rate dependent on the concentration of iodine?

e What is the order of reaction with respect to iodine?

f Write an equation for the reaction between S$_2$O$_3$$^{2-}$ and I$_2$.

g What is the concentration of I$_2$ in the 100 cm^3 of reaction mixture at time = 0 min?

h Use the graph to predict the volume of Na$_2$S$_2$O$_3$(aq) which reacts with 10 cm^3 of the reaction mixture at time = 0 min.

i What is the concentration of the Na$_2$S$_2$O$_3$(aq) used in the titrations?

j Suppose the reaction is first order with respect to propanone. What would be the rate of reaction (in cm^3 Na$_2$S$_2$O$_3$(aq) min^{-1}) if 0.50 mol dm^{-3} propanone were used in place of 0.25 mol dm^{-3}?

23 Group IV – Carbon to lead, non-metal to metal

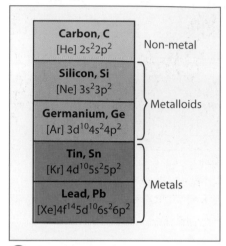

Figure 23.1
The elements in Group IV and their electron configurations

Carbon, C [He] $2s^2 2p^2$ — Non-metal

Silicon, Si [Ne] $3s^2 3p^2$

Germanium, Ge [Ar] $3d^{10} 4s^2 4p^2$ — Metalloids

Tin, Sn [Kr] $4d^{10} 5s^2 5p^2$

Lead, Pb [Xe] $4f^{14} 5d^{10} 6s^2 6p^2$ — Metals

Key Point

As proton number increases, the elements in Group IV change from non-metal, through metalloids to metals.

Note

Silicon(IV) oxide, SiO_2, is also named silicon dioxide and it is commonly called silica. Don't confuse silica, SiO_2, with silicon, Si.

23.1 Introducing Group IV

The similarity between elements in the same family which is so obvious in Groups I, II and VII is much less apparent in Group IV. Here the elements change from non-metal to metal, and there is a greater variation in their chemistry than in any other group of the periodic table.

Carbon is usually regarded as a non-metal. This is certainly true of diamond, but graphite shows some metalloid character. Silicon and germanium are metalloids. Tin and lead show typical metallic properties (Figure 23.1).

23.2 The occurrence, extraction and uses of Group IV elements

Although carbon occurs naturally as diamonds (in Brazil and South Africa) and as graphite (in Sri Lanka, Germany and the USA), it occurs much more abundantly as carbon compounds. These are present in coal, oil and natural gas, in carbonates such as limestone and as fats, proteins and carbohydrates in living things.

Figure 23.2
A rough diamond being polished

Figure 23.3
Large diamond-studded drills being used to bore holes through concrete in a sea wall to improve drainage

The structure, properties and uses of diamond and graphite are discussed in Section 4.13 and fullerenes in Section 4.16. The high refractive index and dispersive power of diamonds led to their use as jewellery in the very earliest civilisations. The modern industrial uses of diamond almost all result from its hardness – drilling (Figure 23.3), cutting, grinding and for bearings in precision instruments such as watches.

The uses of graphite are related to its conducting and lubricating qualities. It is used in 'lead' pencils, as a high-temperature lubricant and as inert electrodes in various electrolytic processes. It is also used as a moderator for slowing down neutrons in nuclear reactors.

After oxygen, silicon is the most abundant element on the Earth. It occurs as silicon(IV) oxide (SiO_2) in sand and sandstone and as many forms of silicate in rocks and clays. The structure and properties of silicon(IV) oxide as the basic component of ceramics is covered in Section 4.14.

In comparison with carbon and silicon, germanium, tin and lead are rare elements. Traces of germanium are present in coal and accumulate in flue dust as germanium(IV) oxide, GeO_2. Silicon and germanium are used as semi-conductors in electronic components and microprocessors. These two elements are obtained from their oxides, SiO_2 and GeO_2, by very similar processes. Silicon is the basis of the modern computer, whether large or small. It is also the basis of photovoltaic cells for generating electricity from sunlight.

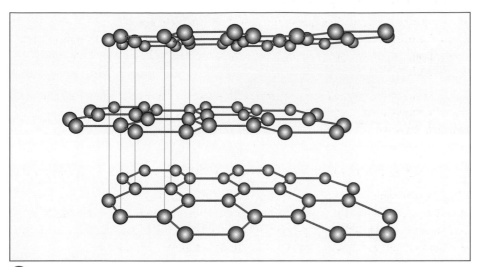

Figure 23.5
A view of the planes of atoms in the structure of graphite. Within the horizontal planes, the carbon atoms are arranged in interconnected hexagons.

The silicon for integrated circuits must be very pure. The flow scheme in Figure 23.6 shows how pure silicon is obtained industrially. The final zone-refining process allows the production of ultra-pure silicon. This is necessary for use in the integrated circuits of microprocessors. The impure silicon is packed in a small cylindrical tube and suspended vertically (Figure 23.7). At the top of the tube, a short length of the cylinder is surrounded by an electrical heating coil. This melts a narrow band of material within the tube. During the zone-refining operation, the tube is slowly raised through the heating element and the zone of molten material moves down the tube. Once the sample gets above the heating element, it begins to recrystallise and impurities are left in the molten zone. In this way, the impurities collect in the molten zone and end up concentrated at the bottom of the tube.

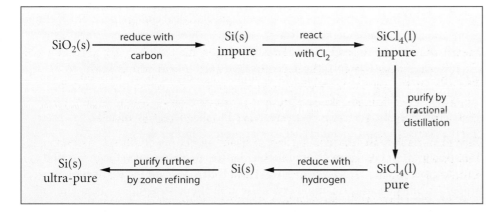

Figure 23.6
A flow scheme showing the industrial manufacture of pure silicon

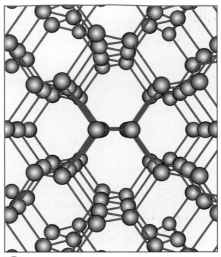

Figure 23.4
A view through the structure of diamond. Notice the tetrahedral arrangement of carbon atoms.

Quick Questions

1 a Write equations for:
   i the reduction of SiO_2 with carbon,
   ii the reaction of Cl_2 with Si.
  b Why is the impure silicon converted to $SiCl_4$?
  c Why can the impure silicon not be zone-refined?
2 Why is graphite sometimes described as a metalloid?

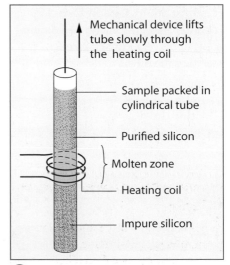

Figure 23.7
The production of ultra-pure silicon by zone refining

> Figure 23.8
Graphite can be made into extremely thin fibres which have an exceptional strength-to-mass ratio. This speed boat has a plastic hull reinforced with graphite fibres.

> Figure 23.9
Woven carbon-fibre cloth contains fibres of graphite which can absorb large molecules between the layers of carbon atoms. Carbon-fibre cloth is hung between cooking and dining areas to prevent kitchen smells reaching the diners.

Key Point

The structures of Group IV elements range from giant covalent structures (C, Si, Ge) to giant metallic (Sn, Pb).

Tin occurs as tin(IV) oxide (SnO_2) in cassiterite (tinstone). Lead occurs as the sulfide ore, galena (PbS). Lead is obtained by roasting the sulfide in air to obtain PbO, which is then reduced with carbon:

$$2PbS(s) + 3O_2(g) \longrightarrow 2PbO(s) + 2SO_2(g)$$

$$PbO(s) + C(s) \longrightarrow Pb(l) + CO(g)$$

Tin can be obtained by direct reduction of SnO_2 with carbon:

$$SnO_2(s) + 2C(s) \longrightarrow Sn(l) + 2CO(g)$$

Tin and lead have important uses as relatively inert metals. Tin is used to tin-plate steel which is then used to make the so called 'tins' for canning meats, soup, fruit, etc. In the past, lead was used as an inert material for gas and water pipes, for cable sheathing and for chemical vessels. Nowadays, it is used for the plates of lead–acid cells (batteries), as an inert and malleable sheeting for traditional roofs and as a screen from radioactivity.

Another important use of both tin and lead is in alloying. The most important alloys containing both of these metals are solder (50% Sn, 50% Pb) and pewter (80% Sn, 20% Pb).

23.3 Variation in the physical properties of the elements

Table 23.1 shows some selected atomic and physical properties for the elements in Group IV. Notice how these properties show greater variation from one element to the next than in Group II or Group VII.

Look closely at the variations in melting point and electrical conductivity. The melting points change from an exceptionally high value in carbon to high values for silicon and germanium and then relatively low values for tin and lead.

In contrast, electrical conductivity rises from nil in diamond to semi-conduction in silicon and germanium and finally good conductivity for tin and lead.

These changes in property are, of course, related to the increasing metallic character and the decreasing non-metallic character as proton number rises.

The structural changes from giant molecular lattices in carbon and silicon (Section 4.13) to giant metallic structures in tin and lead help to explain the changes in physical properties. Silicon and germanium crystallise in the same structure as diamond. Tin and lead have distorted close-packed metal structures. As the atoms get larger and the atomic radius increases, the interatomic bonding becomes weaker. As a result the attraction of neighbouring nuclei for intervening electrons gets less.

Table 23.1

Selected atomic and physical properties of the elements in Group IV

Element	C	Si	Ge	Sn	Pb
Proton (atomic) number	6	14	32	50	82
Atomic radius/nm	0.077	0.117	0.122	0.141	0.154
Melting point/°C	3730$^{d.}$	1410	937	232	327
Boiling point/°C	4830$^{d.}$	2680	2830	2270	1730
Density/g cm^{-3}	2.26$^{gr.}$ 3.51$^{d.}$	2.33	5.32	7.3	11.44
Electrical conductivity	fairly good$^{gr.}$ non-conductor^d	semi-conductor	semi-conductor	good	good
Enthalpy change of atomisation/kJ mol^{-1}	+716$^{gr.}$	+456	+376	+302	+195
First ionisation energy/kJ mol^{-1}	1086	787	760	707	715
Type of structure	giant molecular	giant molecular similar to diamond	giant molecular similar to diamond	giant metallic	giant metallic

gr. = graphite   d. = diamond

The weaker interatomic forces result in a change in bonding from covalent to metallic down the Group. Hence there is a decrease in melting point, boiling point, enthalpy change of atomisation and first ionisation energy. At the same time, the increasing metallic character causes a general increase in electrical conductivity.

The first ionisation energy decreases considerably from carbon to silicon. After that, it falls relatively little. The reason for this is that, after silicon, there is a larger increase in nuclear charge which counterbalances the increase in atomic radius.

23.4 Variation in the chemical properties of the elements

Information related to the chemical properties of the Group IV elements is given in Table 23.2. The group trends further emphasise the increase in metallic character down the group. Notice first how carbon, the first member of the group, is much more electronegative than the others.

In general, chemical reactivity increases from carbon to lead. Electrode potentials show that only tin and lead are strong enough reducing agents to liberate hydrogen from dilute acids.

$$Pb(s) \longrightarrow Pb^{2+}(aq) + 2e^- \qquad E^{\ominus} = +0.13 \text{ volts}$$

Lead will react very slowly with soft water containing dissolved oxygen to form $Pb(OH)_2$. This reaction has resulted in cases of lead poisoning in areas where householders have used hot water directly from lead pipes to make hot drinks. The same problem does not arise in hard-water areas. Here the lead piping develops a protective layer of insoluble lead sulfate or lead carbonate.

As the exothermic nature of the enthalpy changes of formation suggest, all Group IV elements react with oxygen on heating to form the dioxide.

$$C(s) + O_2(g) \longrightarrow CO_2(g)$$

$$Sn(s) + O_2(g) \longrightarrow SnO_2(s)$$

In the case of lead, however, PbO_2 decomposes to form PbO.

The enthalpy changes of formation of hydrides, XH_4, suggest that only carbon could react directly with hydrogen. In practice, carbon does not react with hydrogen, even at very high temperatures, because of the large activation energy needed to break up its giant covalent structure.

Quick Questions

3 Why do you think the density of Group IV elements increases with proton number?

Figure 23.10
This leaded window from a church in Toga in the South Pacific shows Jonah and the Whale. Lead is an ideal material for holding the different-coloured pieces of glass in place on account of its malleability and inertness.

Quick Questions

4 **a** Use the values of $\Delta H_f^\ominus$ in Table 23.2 to calculate $\Delta H^\ominus$ for the reaction:
$$PbO_2(s) + Pb(s) \longrightarrow 2PbO(s)$$
b Suggest why lead reacts with oxygen to form PbO and not PbO_2.

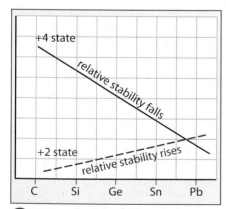

Figure 23.11
This photograph shows the extremely small size of some silicon chips. The ant is carrying a very complex electrical circuit with many components. Silicon chips such as this are used extensively as microprocessors in today's electrical equipment.

Figure 23.12
Relative stabilities of the +4 and +2 oxidation states for the elements in Group IV

Table 23.2
Selected chemical properties of the elements in Group IV

Element	C	Si	Ge	Sn	Pb
Electronegativity (Pauling scale)	2.5	1.8	1.8	1.8	1.8
Electrode potential $M^{2+}(aq) + 2e^- \longrightarrow M(s)/V$			0.23	−0.14	−0.13
$\Delta H_f^\ominus(XO_2)/kJ\,mol^{-1}$	−394	−910	−551	−581	−277
$\Delta H_f^\ominus(XO)/kJ\,mol^{-1}$	−111		−212	−286	−217
$\Delta H_f^\ominus(XH_4)/kJ\,mol^{-1}$	−75	+34	+90	+163	
Bond energy, $D(X—H)/kJ\,mol^{-1}$	435	318	285	251	
$\Delta H_f^\ominus(XCl_4)/kJ\,mol^{-1}$	−136	−640	−544	−511	−320
Bond energy, $D(X—Cl)/kJ\,mol^{-1}$	327	402	339	314	235

All the elements in Group IV react directly on heating with chlorine. They form the tetrachloride, except lead, which forms the dichloride (see $\Delta H_f^\ominus(XCl_4)$ values in Table 23.2).

23.5 General features of the compounds

The most striking feature of the compounds of Group IV elements is the existence of two oxidation states, +2 and +4. The existence of compounds in which the elements show different oxidation numbers is typical of the p-block elements.

It is interesting to consider the relative stabilities of the +2 and +4 oxidation states for elements in Group IV. In carbon and silicon compounds, the +4 state is very stable relative to +2. The +2 state is rare and easily oxidised to +4. Thus, CO reacts very exothermically to form CO_2 and SiO is too unstable to exist under normal conditions.

Germanium forms oxides in both +4 and +2 states. However, GeO_2 is rather more stable than GeO. GeO_2 does not act as an oxidising agent and GeO is readily converted to GeO_2.

In tin compounds, the +4 state is only slightly more stable than the +2 state. Thus, aqueous tin(II) ions are mild reducing agents. They will convert mercury(II) ions to mercury and iodine to iodide:

$$Sn^{2+}(aq) + Hg^{2+}(aq) \longrightarrow Sn^{4+}(aq) + Hg(l)$$

$$Sn^{2+}(aq) + I_2(aq) \longrightarrow Sn^{4+}(aq) + 2I^-(aq)$$

In lead compounds, however, the +2 state is unquestionably more stable. PbO_2 is a strong oxidising agent, whilst PbO is relatively stable. Thus, PbO_2 can oxidise hydrochloric acid to chlorine and hydrogen sulfide to sulfur.

$$PbO_2(s) + 4HCl(aq) \longrightarrow PbCl_2(s) + Cl_2(g) + 2H_2O(l)$$

Notice the steady increase in the stability of the lower oxidation state relative to the higher oxidation state on moving down the group from carbon to lead (Figure 23.12).

The increasing stability of the +2 oxidation state with respect to the +4 state as proton number rises is well illustrated by the standard electrode potentials of the $M^{4+}(aq)/M^{2+}(aq)$ systems for germanium, tin and lead.

$$Ge^{4+}(aq) + 2e^- \longrightarrow Ge^{2+}(aq) \qquad E^\ominus = -1.6\,V$$

$$Sn^{4+}(aq) + 2e^- \longrightarrow Sn^{2+}(aq) \qquad E^\ominus = +0.15\,V$$

$$Pb^{4+}(aq) + 2e^- \longrightarrow Pb^{2+}(aq) \qquad E^\ominus = +1.8\,V$$

As the electrode potentials get more positive from Ge^{4+} to Pb^{4+}, the oxidised form is more readily reduced to the +2 state.

All the Group IV elements have four electrons in their outermost shell (ns^2np^2), so it is not surprising that they show a well-defined oxidation state of +4 (–4 in the hydrides). However, none of the elements forms an M^{4+} cation in its solid compounds. This is due to the high ionisation energies needed to remove four successive electrons from an atom. Consequently, the bonding in the +4 oxidation state is predominantly covalent.

Compounds of tin and lead in which the Group IV element has an oxidation number of +2 (e.g. PbF_2, $PbCl_2$, PbO) are normally regarded as ionic. In these compounds, the Sn^{2+} and Pb^{2+} ions are believed to form by loss of the two 'p' electrons in the outer shell. The two 's' electrons remain relatively stable and unreactive in the filled sub-shell. This is sometimes referred to as the **'inert pair' effect**.

One of the most curious features of Group IV is the unique ability of carbon to form stable compounds containing long chains and rings of carbon atoms. This property is called **catenation**. It results in an enormous range of compounds for carbon. Indeed, there are many thousands of compounds containing only carbon and hydrogen of which methane (CH_4), ethane (C_2H_6) and butane (C_4H_{10}) are three of the simplest.

This ability of carbon to catenate results from the fact that the C—C bond is almost as strong as the C—O bond (Table 23.3). This makes the oxidation of carbon compounds to such products as carbon dioxide less energetically favourable than in the case of silicon. With silicon, the Si—Si bond is much weaker than the Si—O bond. Consequently, catenated compounds of silicon are energetically unstable with respect to SiO_2 and therefore do not occur naturally. Nevertheless, chemists have succeeded in synthesising a series of silicon hydrides, called silanes. Some of these have as many as 11 silicon atoms linked together. In a similar fashion, three hydrides have been synthesised for germanium, but, tin and lead form only one hydride each, SnH_4 and PbH_4. The ability of carbon to catenate is the basis of organic chemistry, and this is considered further in Section 13.1.

Key Point

Moving down Group IV from carbon to lead:

▶ The +2 oxidation state becomes more stable relative to the +4 state.
▶ The bonding in compounds changes from covalent to ionic.

Definition

Catenation is the ability of an element to form chains or rings of its atoms bonded to one another.

 Table 23.3
Average energies for corresponding carbon and silicon bonds

Bond	Average bond energy /kJ mol^{-1}
C—C	346
C—O	360
Si—Si	226
Si—O	464

23.6 Bonding and properties of the tetrachlorides

Table 23.4 shows the structure and selected chemical properties of the tetrachlorides of Group IV elements. All these compounds from tetrachloromethane, CCl_4, to lead tetrachloride, $PbCl_4$, are volatile, simple molecular substances.

Table 23.4
The structure and selected chemical properties of Group IV tetrachlorides

Formula	CCl_4	$SiCl_4$	$GeCl_4$	$SnCl_4$	$PbCl_4$
Structure and bonding	Simple molecules with a tetrahedral shape. Covalent bonds link the central atom to four chlorine atoms				
Volatility	Low melting points and boiling points. All are volatile liquids at room temperature				
Thermal stability	Stable even at high temperatures			Decomposes on heating to form $SnCl_2$ + Cl_2	Decomposes at room temperature to form $PbCl_2$ + Cl_2
Reaction with water	No reaction	Readily hydrolysed to form a hydroxy compound + HCl			

Quick Questions

5 **a** Draw a 'dot-and-cross' diagram (outermost shell electrons only) for $GeCl_4$.
   b Why is the $GeCl_4$ molecule tetrahedral in shape?
   c Molecules of $GeCl_4$ are non-polar even though the Ge—Cl bonds are all polar. Why is this?
   d How would you expect the boiling points of Group IV tetrachlorides to change from CCl_4 to $PbCl_4$?
   e How would you expect the stability of tetrachlorides and dichlorides to change as relative molecular mass increases?

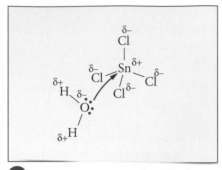

Figure 23.13
The first stage in the hydrolysis of Group IV tetrachlorides

As the tetrachlorides increase in relative molecular mass, their electrons become more readily polarised. This results in larger induced dipoles and a general decrease in volatility. At the same time, X—Cl bonds become longer and weaker causing the tetrachlorides to become less stable.

All the tetrachlorides (except CCl_4) are readily hydrolysed to a hydroxy compound and hydrogen chloride.

For example,

$$SnCl_4(l) + 4H_2O(l) \longrightarrow Sn(OH)_4(aq) + 4HCl(aq)$$

The first stage in the hydrolysis of these chlorides involves nucleophilic attack of the Group IV element by a lone pair of electrons on the oxygen atom in water (Figure 23.13). The Group IV elements carry a partial positive charge owing to the electron-withdrawing effect of four chlorine atoms and this attracts the oxygen's lone pair.

During the reaction, the Group IV atom is briefly surrounded by five pairs of electrons. In silicon and the elements below it, the extra pair of electrons can be accommodated using unoccupied d orbitals in their outer shells. But carbon has no d orbitals in its outer shell and so CCl_4 cannot be hydrolysed.

23.7 Bonding and properties of the oxides

The information in Table 23.5 shows the structure, bonding and selected chemical properties of the oxides of Group IV elements in the +4 oxidation state (dioxides). Look carefully at Table 23.5 and answer Quick questions 6 and 7.

Table 23.5
The structure and chemical properties of the oxides of Group IV elements in the +4 oxidation state

Formula	CO_2	SiO_2	GeO_2	SnO_2	PbO_2
Structure	Simple molecular	Giant molecular	Intermediate between giant molecular and ionic		
Reaction with acids	No reaction		React with conc. acids to give salts $SnO_2 + 4HCl \longrightarrow SnCl_4 + 2H_2O$		
Reaction with alkalis	React to form $XO_3{}^{2-}$ $SiO_2 + 2OH^- \longrightarrow SiO_3{}^{2-} + H_2O$ silicate		React with conc. alkalis to form $X(OH)_6{}^{2-}$ $SnO_2 + 2OH^- + 2H_2O \longrightarrow Sn(OH)_6{}^{2-}$ tetrahydroxostannate(IV)		
Thermal stability	Stable even at high temperatures				Decomposes to PbO on warming

Note

The naming of complex ions such as $Sn(OH)_6{}^{2-}$ is explained in Section 24.8.

Quick Questions

6   How does the structure and bonding in the dioxides change from CO_2 to PbO_2?

7 **a** Write equations for the reactions of:
    i CO_2 with dilute sodium hydroxide solution,
    ii GeO_2 with concentrated hydrochloric acid,
    iii PbO_2 with concentrated sodium hydroxide.
  b Are the following oxides acidic, amphoteric or basic?
    i SiO_2, **ii** GeO_2, **iii** PbO_2.

Figure 23.14
Dry ice (solid CO_2) being used to create mist in a production at the Edinburgh International Festival in Scotland

Table 23.6 shows the structure, acid–base nature and thermal stability of the oxides of Group IV elements in the +2 oxidation state (monoxides).

 Table 23.6
The structure, acid–base nature and thermal stability of the oxides of Group IV elements in the +2 oxidation state

Formula	CO	SiO	GeO	SnO	PbO
Structure	Simple molecular		Intermediate but increasingly ionic towards PbO		
Acid–base nature	Neutral oxides • react with neither acids nor alkalis		Amphoteric oxides • react with acids to form salts, e.g. $SnO + 2H^+ \longrightarrow Sn^{2+} + H_2O$ • react with alkalis to form $X(OH)_3^-$ ions, e.g. $SnO + OH^- + H_2O \longrightarrow Sn(OH)_3^-$ trihydroxostannate(II)		
Thermal stability	Readily oxidised to dioxide	SiO, GeO and SnO form dioxide on standing in air			Stable

Notice in Table 23.6 that the structures of the monoxides, like the dioxides, change from simple molecular through intermediate (giant molecular/ionic) to predominantly ionic. The nature of the monoxides changes from neutral in CO and SiO to amphoteric in GeO, SnO and PbO, whereas the corresponding dioxides change from acidic to amphoteric.

The thermal stabilities of the oxides in both the +4 and +2 oxidation states illustrate the increasing stability from carbon to lead of the +2 state relative to the +4 state.

Definition

A **neutral oxide** is an oxide which does not react with acids or alkalis.

Key Points

Moving down Group IV from carbon to lead:

▶ oxides change from simple molecular to predominantly ionic,
▶ oxides of oxidation state +4 change from acidic to amphoteric,
▶ oxides of oxidation state +2 change from neutral to amphoteric.

Figure 23.15
Watercolour paintings containing lead(II) compounds in their pigments darken over many decades as traces of hydrogen sulfide in the air react with lead(II) salts. 'White lead' pigment containing lead(II) carbonate becomes darker due to the formation of black lead(II) sulfide. The original colour of the pigment can be restored by oxidation of the black lead(II) sulfide to white lead(II) sulfate using hydrogen peroxide. These photos show a watercolour painted in 1674 by Matthew Snelling of Susanna, Lady Dormer before and after its restoration. Notice how black patches which had developed on her face and shoulder were removed.

Review questions

1 The principal ore of tin is tinstone, SnO_2.

  a Explain briefly, with an equation, how tin is obtained from tinstone.

  b State two main uses of tin or its alloys.

  c What are the relative advantages and disadvantages of tin-plating and galvanising (zinc-plating) iron in order to prevent corrosion?

2 a By choosing the chlorides of two of the Group IV elements as examples, describe the trend in the reactions of these chlorides with water. Suggest an explanation for any differences, and write equations for any reactions that occur.

  b The standard enthalpy changes of formation of lead(II) chloride and lead(IV) chloride are given in the following table:

Compound	$\Delta H_f^{\ominus}/\,kJ\,mol^{-1}$
$PbCl_2(s)$	−359
$PbCl_4(l)$	−329

  Use these data, and also bond energy data from Table 5.2, to calculate the enthalpy changes for the following two reactions:

  i $CCl_2(g) + Cl_2(g) \longrightarrow CCl_4(g)$

  ii $PbCl_2(s) + Cl_2(g) \longrightarrow PbCl_4(l)$

  iii Make use of your answers in parts i and ii to suggest how the relative stabilities of the two oxidation states vary down the group.

  Cambridge Paper 4 Q4 November 2007

3 Carbon forms two stable oxides, CO and CO_2. Lead forms three oxides: yellow PbO, black PbO_2 and red Pb_3O_4.

  a Carbon monoxide burns readily in air. Heating black lead oxide produces oxygen gas, leaving a yellow residue.

  i Suggest a balanced equation for each reaction.

  ii Explain how these two reactions illustrate the relative stabilities of the +2 and +4 oxidation states down Group IV.

  b Red lead oxide contains lead atoms in two different oxidation states.

  i Suggest what these oxidation states are, and calculate the ratio in which they occur in red lead oxide.

  ii Predict the equation for the action of heat on red lead oxide.

When red lead oxide is heated with dilute nitric acid, HNO_3, a solution of lead(II) nitrate is formed and a black solid is left.

  iii Suggest an equation for this reaction.

  iv Explain how this reaction illustrates the relative basicities of the two oxidation states of lead.

  c Both tin(II) oxide and tin(IV) oxide are amphoteric.

  Write a balanced equation for the reaction between tin(II) oxide and aqueous sodium hydroxide.

  Cambridge Paper 4 Q3 June 2007

4 The following passage describes the preparation of tin(IV) iodide.

  Add 4.0 g of powdered tin to a solution of 12.7 g of iodine (harmful) in 100 cm^3 of tetrachloromethane (toxic). Reflux the mixture gently until the reaction is complete. Now filter the mixture through a pre-heated funnel. Wash the residue with hot tetrachloromethane, adding the washings to the filtrate.
  Cool the filtrate in ice until orange crystals of tin(IV) iodide form. Filter and dry the crystals.

  a Write an equation for the formation of tin(IV) iodide in the above preparation.

  b Give two reasons for using CCl_4 as the solvent in this preparation.

  c Calculate the maximum possible yield of tin(IV) iodide. Explain your calculation.

  d How would you know when the reaction was complete?

  e Why was the mixture filtered through a pre-heated funnel?

  f Why was the residue washed with hot CCl_4?

  g What would you predict for:

  i the structure of tin(IV) iodide,

  ii the thermal stability of tin(IV) iodide?

5 a Draw electron 'dot-and-cross' diagrams for carbon monoxide and carbon dioxide. (You need only show the electrons in the outer shells of the constituent atoms.)

  b 'Carbon monoxide is iso-electronic with nitrogen.' Explain what is meant by this statement.

  c Use the standard enthalpy changes of formation of CO_2 and CO in Table 23.2 to find the enthalpy changes of the following reactions.

  $$CO_2(g) + C(s) \longrightarrow 2CO(g)$$
  $$CO(g) + \tfrac{1}{2}O_2(g) \longrightarrow CO_2(g)$$

  Comment on the relative stabilities of CO_2 and CO.

24 The transition elements

24.1 Introduction

The elements from scandium (proton number 21) to zinc (proton number 30) in the periodic table form what is usually regarded as the first row of transition elements. But, what exactly is a transition element? And why is it that these elements have similarities to each other across their period, unlike elements in other parts of the periodic table?

The answer to these questions concerning the transition metals lies in their electronic structures shown in Table 24.1.

In fact, most of the properties of transition elements are related to their electronic structures and the relative energy levels of the orbitals available for their electrons.

 Table 24.1

The electronic structures of the atoms and common ions of the elements K to Zn in the periodic table ([Ar] ≡ electron structure of argon)

Element	Symbol	Electronic structure of atom	Common ion	Electronic structure of ion
potassium	K	$[Ar]4s^1$	K^+	$[Ar]$
calcium	Ca	$[Ar]4s^2$	Ca^{2+}	$[Ar]$
scandium	Sc	$[Ar]3d^14s^2$	Sc^{3+}	$[Ar]$
titanium	Ti	$[Ar]3d^24s^2$	Ti^{4+}	$[Ar]$
vanadium	V	$[Ar]3d^34s^2$	V^{3+}	$[Ar]3d^2$
chromium	Cr	$[Ar]3d^54s^1$	Cr^{3+}	$[Ar]3d^3$
manganese	Mn	$[Ar]3d^54s^2$	Mn^{2+}	$[Ar]3d^5$
iron	Fe	$[Ar]3d^64s^2$	Fe^{2+} Fe^{3+}	$[Ar]3d^6$ $[Ar]3d^5$
cobalt	Co	$[Ar]3d^74s^2$	Co^{2+}	$[Ar]3d^7$
nickel	Ni	$[Ar]3d^84s^2$	Ni^{2+}	$[Ar]3d^8$
copper	Cu	$[Ar]3d^{10}4s^1$	Cu^+ Cu^{2+}	$[Ar]3d^{10}$ $[Ar]3d^9$
zinc	Zn	$[Ar]3d^{10}4s^2$	Zn^{2+}	$[Ar]3d^{10}$

 Figure 24.1
Coloured pigments used in the make-up of this tribal dancer include several compounds of transition elements: chromium(III) oxide – green, iron oxides – red and yellow, manganese salts – violet.

Key Point

Unlike elements in other rows of the periodic table, elements in a transition series have similar properties.

Note

Electronic structures are sometimes called electronic configurations.

Key Point

Most of the properties of transition elements can be attributed to their electronic structures and the energy levels of the orbitals available for their electrons.

As the shells of electrons get further and further from the nucleus, successive shells become closer in *energy*. Thus, the difference in *energy* between the second and third shells is less than that between the first and second. By the time the fourth shell is reached, there is, in fact, an overlap between the third and fourth shells. In other words, from potassium onwards, the orbitals of highest energy in the third shell (i.e. the 3d orbitals) have higher energy than that of lowest energy in the fourth shell (the 4s orbital) (Figure 24.2).

The 3d sub-shell is 'on average' nearer the nucleus than the 4s sub-shell, but at a higher energy level. So, once the 3s and 3p sub-shells are filled at argon, further electrons go in the 4s sub-shell because it has a lower energy level than the 3d sub-shell. Hence, potassium and calcium have, respectively, one and two electrons in the 4s sub-shell (Table 24.1). Once the 4s sub-shell is filled at calcium, electrons enter the 3d level. Therefore, scandium has the electron structure $[Ar]3d^14s^2$, titanium has the electron structure $[Ar]3d^24s^2$ and so on (Table 24.1 and Figure 24.3).

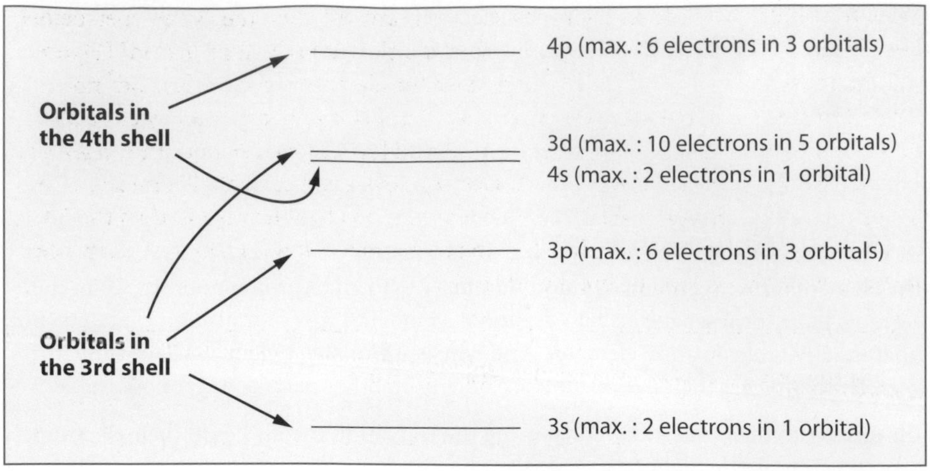

Figure 24.2
Relative energy levels of the 3s, 3p, 3d, 4s and 4p orbitals before they contain electrons

Notice the unexpected electron configurations of chromium and copper. The arrangement of electrons in chromium is $[Ar]3d^5 4s^1$ (Figure 24.3) not $[Ar]3d^4 4s^2$ which we might have predicted. The explanation of this anomaly is that the electron structure $[Ar]3d^5 4s^1$ with half-filled 3d and 4s sub-shells is more stable than $[Ar]3d^4 4s^2$. The extra stability of a half-filled sub-shell results from the occupation of each orbital by one electron which produces an equal distribution of charge around an atom.

Copper atoms have an electron structure $[Ar]3d^{10} 4s^1$ (Figure 24.3) rather than $[Ar]3d^9 4s^2$. In this case, it appears that $[Ar]3d^{10} 4s^1$ with a filled 3d sub-shell and a half-filled 4s sub-shell is more stable than $[Ar]3d^9 4s^2$.

Element		Electron structure 3d					4s
Scandium	[Ar]	↑					↑↓
Titanium	[Ar]	↑	↑				↑↓
Chromium	[Ar]	↑	↑	↑	↑	↑	↑
Iron	[Ar]	↑↓	↑	↑	↑	↑	↑↓
Copper	[Ar]	↑↓	↑↓	↑↓	↑↓	↑↓	↑

Figure 24.3
The 'electrons-in-boxes' representation of the electronic configurations of certain transition metals

24.2 Ions of the transition metals

Quick Questions

1 Look closely at Table 24.1.
- **a** Which electrons does a calcium atom lose in forming a Ca^{2+} ion?
- **b i** Which electrons would you expect an iron atom to lose in forming an Fe^{2+} ion?
  - **ii** Which electrons are lost in practice by the Fe atom?
- **c** Why do you think Fe^{3+} ions are more stable than Fe^{2+} ions?

When transition metals form their ions, electrons are lost from the 4s sub-shell before those in the 3d sub-shell. Thus, Fe^{2+} ions have the electron structure $[Ar]3d^6$ (Figure 24.5) rather than $[Ar]3d^44s^2$ and V^{3+} ions have the electron structure $[Ar]3d^2$ not $[Ar]4s^2$. At first sight, this seems very odd because, prior to occupation by electrons, the 4s level is energetically more stable than the 3d level. However, once the 3d level is occupied by electrons, these repel the 4s electrons even further from the nucleus. The 4s electrons are therefore pushed to a higher energy level, higher in fact than the 3d level now occupied. Consequently, when transition metal atoms form ions, they lose electrons from the 4s level before the 3d level.

Atom/ion		Electron structure						4s
				3d				
Fe	[Ar]	↑↓	↑	↑	↑	↑		↑↓
Fe^{2+}	[Ar]	↑↓	↑	↑	↑	↑		
Fe^{3+}	[Ar]	↑	↑	↑	↑	↑		

This means that all the first row transition metals will have similar chemical properties because these are dictated by the behaviour of the 4s electrons in the outermost shell.

This horizontal similarity contrasts sharply with the marked changes across a period of the s- and p-block elements, such as that from sodium to argon.

24.3 What is a transition element?

The simplest answer to this question is to say that *transition elements are those in the 'd-block' of the periodic table*. The neatness of this definition lies in the fact that it emphasises the four blocks of elements in the periodic table. This division is useful for the s-block elements in Groups I and II which are so alike in their properties, but it is much less satisfactory for the elements in the p-block which include metals such as aluminium, reactive non-metals within Group VII and the noble gases. Likewise, the simple definition of transition metals as 'd-block elements' is also unsatisfactory because it leads to the inclusion of scandium and zinc as transition elements. As you may have realised already, scandium and zinc show some obvious differences to the other metals in the first row of d-block elements.

▸ They have only one oxidation state in their compounds (scandium +3, zinc +2), whereas the others have two or more.
▸ Their compounds are usually white, unlike the compounds of transition metals which are generally coloured.
▸ They show little catalytic activity.

As scandium and zinc do not show typical transition metal properties, we need to find a more satisfactory definition for transition metals. This definition should exclude these two elements, but include all the other elements in the first transition series from titanium to copper.

In order to achieve this, we can define a transition metal as *an element which forms at least one ion with a partially filled d sub-shell*.

Look closely at Table 24.1 and you will see that neither Sc^{3+} nor Zn^{2+} has a partly filled d sub-shell.

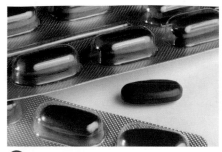

Figure 24.4
Iron is an important constituent of haemoglobin and patients suffering from anaemia may have a diet which is deficient in iron. A supplement of 'iron' tablets containing iron(II) sulfate can help to cure the anaemia.

Figure 24.5
The 'electrons-in-boxes' representation of the electronic structures of Fe, Fe^{2+} and Fe^{3+}

Key Point

Before occupation by electrons, the 4s level is energetically more stable than the 3d level. But, once the 3d level is occupied, its electrons repel the 4s electrons to an energy level higher than the 3d electrons.

Therefore, transition metals lose electrons from the 4s level before the 3d level when they form ions.

Quick Questions

2 Use Table 24.1 to answer the following.
   a Why does titanium form Ti^{4+} ions?
   b Why is the commonest ion of manganese Mn^{2+}?
   c Why might you expect chromium to form compounds containing Cr^+ ions?
   d Would you expect Cu^+ or Cu^{2+} ions to be more stable? Explain your answer.

A **transition element** is an element that forms one or more stable ions with a partially filled d sub-shell.

24.4 Trends in atomic properties across the transition metals

In building up the elements across Period 3 from sodium to argon, electrons are being added to the outer shell, while the nuclear charge is increasing by the addition of protons. The added electrons, entering the same shell, shield each other only weakly from the extra nuclear charge. So, atomic radii decrease sharply from sodium to argon (Figure 9.12, Section 9.7). At the same time, electronegativities and ionisation energies steadily rise. The increasing number of outer-shell electrons from sodium to argon also results in major differences in structure and chemical properties from one element to the next.

In moving across the first row of transition elements from titanium to copper, the nuclear charge is also increasing, but electrons are being added to an *inner* d sub-shell. These inner d electrons shield the outer 4s electrons from the increasing nuclear charge much more effectively than outer shell electrons can shield each other. Consequently, the atomic radii decrease much less rapidly from titanium to copper (Table 24.2) than from sodium to argon.

▼ Table 24.2

Atomic radii, electronegativities, ionisation energies and electrode potentials for the elements Sc to Zn

	Sc	Ti	V	Cr	Mn	Fe	Co	Ni	Cu	Zn
Atomic radius/nm	0.16	0.15	0.14	0.13	0.14	0.13	0.13	0.13	0.13	0.13
Electronegativity	1.2	1.3	1.45	1.55	1.6	1.65	1.7	1.75	1.75	1.6
First ionisation energy/kJ mol^{-1}	+630	+660	+650	+650	+720	+760	+760	+740	+750	+910
Second ionisation energy/kJ mol^{-1}	+1240	+1310	+1410	+1590	+1510	+1560	+1640	+1750	+1960	+1700
Third ionisation energy/kJ mol^{-1}	+2390	+2650	+2870	+2990	+3260	+2960	+3230	+3390	+3560	+3800
Standard electrode potential for $M^{2+}(aq) + 2e^- \longrightarrow M(s)$/V			−1.20	−0.91	−1.19	−0.44	−0.28	−0.25	+0.34	−0.76
Standard electrode potential for $M^{3+}(aq) + 3e^- \longrightarrow M(s)$/V	−2.1	−1.2	−0.86	−0.74	−0.28	−0.04	+0.40			

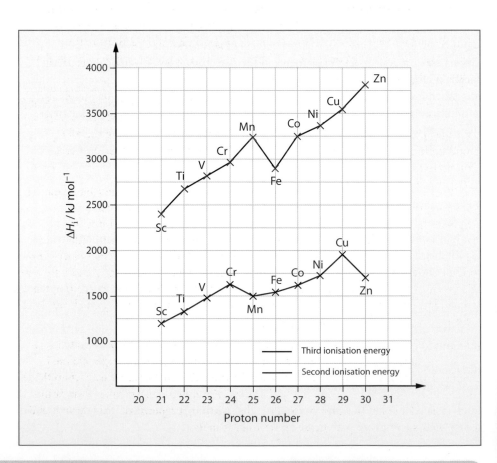

▶ Figure 24.6

Graphs of the second and third ionisation energies of the elements from scandium to zinc

Now look at Table 24.2 again and notice that electronegativities and first ionisation energies increase only marginally from titanium to copper, compared with the large increase across Period 3 from sodium to argon (Figures 9.15 and 9.16, Section 9.7).

The increasing electronegativity from titanium to copper means that the elements become slightly less metallic. This is reflected in the increasingly positive electrode potentials (Section 20.4) of their M^{2+} and M^{3+} ions.

Quick Questions

3 Look closely at Figure 24.6 which shows graphs of the second and the third ionisation energies of the elements from scandium to zinc.

Remember that the second ionisation energy refers to the process
$$M^+(g) \longrightarrow M^{2+}(g) + e^-$$
and the third ionisation to
$$M^{2+}(g) \longrightarrow M^{3+}(g) + e^-$$
Write the electron structures of the following:
a Cu **b** Cu^+ **c** Cu^{2+} **d** Cr **e** Cr^+ **f** Mn^{2+} **g** Zn^{2+}
(Use [Ar] to represent the electron structure of argon as we have done before.)

4 By referring to electron structures explain why:
a The second ionisation energies of both Cr and Cu are higher than those of the next element.
b The third ionisation energies of both Mn and Zn are higher than those of the next element.

24.5 General properties of the first transition series (Ti to Cu)

Most of the transition metals have a close-packed structure in which each atom has twelve nearest neighbours. In addition, transition metals also have a relatively small atomic radius compared to s-block metals in the same period because the electrons being added to the d sub-shell are nearer the nucleus than the electrons in the outermost s orbital. The double effect of close packing and small atomic radii results in strong metallic bonds between atoms.

Hence, *the transition metals in Period 4 have higher melting points, higher boiling points, higher densities and higher enthalpy changes of fusion and vaporisation than potassium and calcium, s-block metals in the same period* (Table 24.3). The strong metallic bonding in transition metals is also reflected in high tensile strengths, electrical conductivity and good mechanical properties.

The transition metals are less electropositive (weaker metals) than the s-block metals. However, their electrode potentials (Table 24.2) indicate that all of them (except copper) should react with dilute solutions of strong acids, such as $1.0\,mol\,dm^{-3}$ HCl(aq). In practice, however, many of the metals react only slowly with dilute acids. This is due to protection of the metal from chemical attack by a thin impervious and unreactive layer of oxide. Chromium provides an excellent example of this. Despite its electrode potential, it can be used as a protective, non-oxidising, non-corroding metal owing to the presence of an unreactive layer of Cr_2O_3 similar to Al_2O_3 on aluminium. This makes it useful as chromium plating.

The ions of transition metals are smaller than those of the s-block metals in the same period (Table 24.3). Owing to their smaller ionic radii and also to a larger charge in many of their ions, the charge/radius ratios for transition metals are greater than those of s-block metals. Their small, highly charged cations also cause greater polarisation of associated anions. These factors result in the following properties of transition metal compounds compared with those of the s-block metals.

- Their oxides and hydroxides in oxidation states +2 and +3 are less basic and less soluble.
- Their salts are less ionic and less thermally stable.
- Their salts and aqueous ions are more hydrated and more readily hydrolysed forming acidic solutions.
- Their ions are more easily reduced to the metal.

There are relatively small changes in ionic radii from titanium to copper. Because of this, compounds of the simple hydrated +2 and +3 ions have very similar crystalline structures, hydration and solubility. Thus, all the M^{3+} ions form compounds, known as alums, with the formula $K_2SO_4 \cdot M_2(SO_4)_3 \cdot 24H_2O$ and all the M^{2+} ions form double sulfates of formula $(NH_4)_2SO_4 \cdot MSO_4 \cdot 6H_2O$, all of which have the same crystalline structure.

Table 24.3
Physical properties of the elements K to Zn

| Element | s-block metals | | transition metals | | | | | | | | | |
	K	Ca	Sc	Ti	V	Cr	Mn	Fe	Co	Ni	Cu	Zn
Atomic radius/nm	0.24	0.20	0.16	0.15	0.14	0.13	0.14	0.13	0.13	0.13	0.13	0.13
Melting point/°C	64	850	1540	1680	1900	1890	1240	1540	1500	1450	1080	420
Boiling point/°C	770	1490	2730	3260	3400	2480	2100	3000	2900	2730	2600	910
Density/g cm^{-3}	0.86	1.54	3.0	4.5	6.1	7.2	7.4	7.9	8.9	8.9	8.9	7.1
Ionic radius/nm												
M$^+$	0.130											
M^{2+}		0.094		0.090	0.088	0.084	0.080	0.076	0.074	0.072	0.070	0.074
M^{3+}			0.081	0.076	0.074	0.069	0.066	0.064	0.063	0.062		

Key Points

Transition metals show the following characteristic properties:

- variable oxidation states,
- formation of complex ions,
- coloured compounds,
- catalytic properties.

In addition to theses general properties of the elements from titanium to copper, transition metals share some important characteristic properties:

- They show variable oxidation states in their compounds (Section 24.6).
- They form a wide range of complex ions with various ligands (Section 24.7).
- They form coloured compounds and coloured complex ions (Section 24.8).
- They have catalytic properties as elements and compounds (Section 24.13).

24.6 Variable oxidation states

Transition metals from Ti to Cu have electrons of similar energy in both the 3d and 4s levels. This means that one particular element can form ions of roughly the same stability by losing different numbers of electrons. Thus, all the transition metals from titanium to copper exhibit two or more oxidation states in their compounds.

The formulae of the common oxides and chlorides of the elements from Sc to Zn are shown in Figure 24.7. All the oxidation numbers of the elements in their compounds are also shown below these formulae. The more important oxidation states are emphasised by bold print. Notice that both scandium and zinc have only one common oxide, one common chloride, and only one oxidation state in their compounds. Notice also, how closely the oxidation states of all the elements in their common oxides and chlorides compare with the more important oxidation states of the elements.

We can make the following generalisations about these oxidation states.

- *The common oxidation states for each element include +2 or +3 or both.* +3 states are relatively more common at the beginning of the series. +2 states are more common towards the end.
- The highest oxidation states from titanium to manganese correspond to the involvement of all the electrons outside the argon core (4 for Ti, 5 for V, 6 for Cr and 7 for Mn). After this, the increasing nuclear charge binds the d electrons more strongly. So, one of the more important oxidation states is that which involves the weakly held electrons in the outer 4s shell only (2 for Fe, 2 for Co, 2 for Ni and 1 for Cu).

	Sc	Ti	V	Cr	Mn	Fe	Co	Ni	Cu	Zn
Common oxides	Sc_2O_3	Ti_2O_3 TiO_2	V_2O_3 V_2O_5	Cr_2O_3 CrO_3	MnO MnO_2 Mn_2O_7	FeO Fe_2O_3	CoO Co_2O_3	NiO	Cu_2O CuO	ZnO
Common chlorides	$ScCl_3$	$TiCl_3$ $TiCl_4$	VCl_3	$CrCl_2$ $CrCl_3$	$MnCl_2$ $MnCl_3$	$FeCl_2$ Fe_2Cl_6	$CoCl_2$	$NiCl_2$	$CuCl$ $CuCl_2$	$ZnCl_2$
Oxidation numbers that occur in compounds	3	4 3 2 1	5 4 3 2 1	6 5 4 3 2 1	7 6 5 4 3 2 1	6 5 4 3 2 1	5 4 3 2 1	4 3 2 1	3 2 1	2

Figure 24.7
Oxidation states of the elements Sc to Zn. (Common oxidation numbers are in bold print.)

► The transition metals usually exhibit their highest oxidation states in compounds with oxygen or fluorine. These are the two most electronegative elements.

► Ti, V, Cr and Mn never form simple ions in their highest oxidation state as this would result in ions of extremely high charge density. Hence the compounds of these elements in their highest oxidation state are either covalently bonded (e.g. TiO_2, V_2O_5, CrO_3, Mn_2O_7) or contain complex ions (e.g. VO_3^-, CrO_4^{2-}, MnO_4^-).

One of the most beautiful and effective demonstrations of the range of oxidation states of a transition metal can be shown by shaking a solution containing a compound of vanadium(v) with zinc and dilute acid. The solution of vanadium(v) can be made by dissolving ammonium vanadate(v) (NH_4VO_3) in dilute sodium hydroxide solution and then adding excess dilute sulfuric acid. This mixture contains yellow dioxovanadium(v) ions, VO_2^+, formed by a reaction of VO_3^- ions with H^+ ions.

$$VO_3^-(aq) + 2H^+(aq) \longrightarrow VO_2^+(aq) + H_2O(l)$$

When the yellow solution is shaken with granulated zinc or zinc amalgam, it changes gradually through green to blue oxovanadium(iv) ions, $VO^{2+}(aq)$. It then changes to green vanadium(iii) ions, $V^{3+}(aq)$, and finally to violet vanadium(ii) ions, $V^{2+}(aq)$ (Table 24.4).

Figure 24.8
This photograph shows the common oxidation states in chromium compounds very colourfully. From left to right, the solutions contain violet chromium(iii) chloride, green chromium(iii) nitrate, yellow potassium chromate(vi) and orange potassium dichromate(vi).

Oxidation state	+5	+4	+3	+2
Colour in aqueous solution	yellow	blue	green	violet
Ion	$VO_2^+(aq)$	$VO^{2+}(aq)$	$V^{3+}(aq)$	$V^{2+}(aq)$
Name	dioxovanadium(v) ion	oxovanadium(iv) ion	vanadium(iii) ion	vanadium(ii) ion

Table 24.4
The oxidation states of vanadium

All the transition metals from titanium to copper exhibit oxidation numbers of +3 and +2 in their compounds. But what are the relative stabilities of the +3 and +2 states for the different elements? Why, for example, is manganese more stable in the +2 state than the +3 state, whilst the reverse is true for iron? Look closely at Figure 24.9 which shows the electrode potentials of the $M^{3+}(aq)/M^{2+}(aq)$ systems for the elements from titanium to cobalt. Notice that the steady rise in $E^{\ominus}$ values across the series is interrupted by what appears to be an abnormally high value for manganese and a low value for iron.

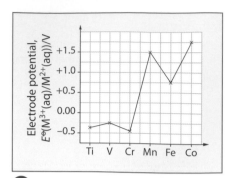

Figure 24.9
Electrode potentials of the M^{3+}/M^{2+} systems for the transition metals Ti to Co

The more positive the value for $E^{\ominus}$, the more readily will $M^{3+}(aq)$ be reduced to $M^{2+}(aq)$.

$$M^{3+}(aq) + e^- \longrightarrow M^{2+}(aq)$$

The negative values of $E^{\ominus}$ for Ti, V, and Cr and the exceptionally lower value for Fe indicate that the 3+ oxidation state is relatively more stable in these elements than in Mn and Co which have much higher values of $E^{\ominus}$.

The relative stabilities of the +3 and +2 states in manganese and iron can be interpreted using an 'electrons-in-boxes' representation of their electronic structures (Figure 24.10).

Atom/ion		Electron structure					
		3d					4s
Mn	[Ar]	↑	↑	↑	↑	↑	↑↓
Mn²⁺	[Ar]	↑	↑	↑	↑	↑	
Mn³⁺	[Ar]	↑	↑	↑	↑		
Fe	[Ar]	↑↓	↑	↑	↑	↑	↑↓
Fe²⁺	[Ar]	↑↓	↑	↑	↑	↑	
Fe³⁺	[Ar]	↑	↑	↑	↑	↑	

Figure 24.10
The electron structures of manganese, iron and their respective 2+ and 3+ ions

Mn^{2+} and Fe^{3+} each have half-filled 3d orbitals. This makes them more stable than Mn^{3+} and Fe^{2+}, respectively. Hence, in manganese the +2 state is more stable than +3, whereas in iron the +3 state is more stable than +2.

Key Point

The important oxidation states for each transition metal include +2 or +3 or both.

The relative stabilities of oxidation states in transition metals can be interpreted in terms of the stability of their electronic structures.

24.7 The formation of complex ions

From our studies of enthalpy changes in solution, we know that metal ions exist in aqueous solution as hydrated ions. So, $Cu^{2+}(aq)$, $Al^{3+}(aq)$ and $Ag^+(aq)$ can be represented more accurately as $[Cu(H_2O)_6]^{2+}(aq)$, $[Al(H_2O)_4]^{3+}(aq)$ and $[Ag(H_2O)_2]^+(aq)$ respectively. In these ions, the water molecules are bound to the central cation by co-ordinate (dative covalent) bonds (Figure 24.11).

Other polar molecules, besides water, can also co-ordinate to metal ions. In excess ammonia solution, Cu^{2+} exists as $[Cu(NH_3)_4(H_2O)_2]^{2+}(aq)$ and Ag^+ exists as $[Ag(NH_3)_2]^+(aq)$.

In addition to polar molecules, anions can also form co-ordinate bonds to metal ions. For example, when anhydrous copper(II) sulfate is added to concentrated hydrochloric acid, the solution becomes yellow-green due to the formation of $[CuCl_4]^{2-}$ ions.

5 a Write the electron structures of Cu, Cu^+ and Cu^{2+} using the 'electrons-in-boxes' notation.

   b In which oxidation state would you expect copper to be more stable, +1 or +2?

   c From experience you will know that copper compounds usually exist in the +2 state. What factors might increase the relative stability of the +2 state over the +1 state for copper?

6 In section 20.7, we found that $E^{\ominus}$ values could be used to predict the likelihood of redox reactions.

   a Which of the $M^{3+}(aq)/M^{2+}(aq)$ systems in Figure 24.9 will oxidise the system:
   $2Br^-(aq) \longrightarrow Br_2(aq) + 2e^-$
   $E^{\ominus} = -1.09\,V$

   b Deduce $E^{\ominus}$ values of $M^{2+}(aq) \longrightarrow M^{3+}(aq) + e^-$ for the systems in Figure 24.9. Then decide which of these systems could be oxidised by:
   $Cr_2O_7^{2-}(aq) + 14H^+(aq) + 6e^-$
   $\longrightarrow 2Cr^{3+}(aq) + 7H_2O(l)$
   for which $E^{\ominus} = +1.33\,V$.

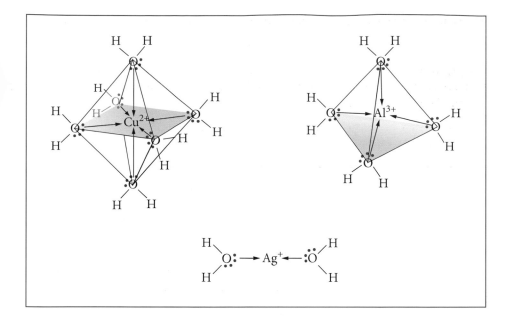

Figure 24.11
Co-ordinate bonding in hydrated cations

Ions such as $[Cu(H_2O)_6]^{2+}$, $[Cu(NH_3)_4(H_2O)_2]^{2+}$ and $[CuCl_4]^{2-}$, in which a metal ion is associated with a number of molecules or anions are known as **complex ions**. The molecules and anions bonded to the central metal ion are called **ligands**. Each ligand contains at least one atom with a lone pair of electrons. These can be donated to the central metal ion forming a co-ordinate bond. The ligands are said to be co-ordinated to the central metal ion and the number of ligands in a complex is usually two, four or six. The number of co-ordinate bonds from ligands to the central metal ion is described as its **co-ordination number**.

Water is by far the most common ligand in complex ions. The strengths of the co-ordinate bonds between water molecules and different metal ions can be compared using their enthalpy changes of hydration (hydration energies) which we studied in Section 5.15. Hydration energies relate to the process:

$$M^{x+}(g) + nH_2O(l) \longrightarrow [M(H_2O)_n]^{x+}(aq)$$

The hydration energies of some cations are listed in Table 24.5. Use this table to answer Quick question 7.

Definitions

A **complex ion** is an ion in which a number of molecules or anions are bound to a central metal ion by co-ordinate (dative covalent) bonds.

A **ligand** is a molecule or anion bound to the central metal ion in a complex ion by co-ordinate bonding.

The **co-ordination number** of a metal ion in a complex ion is the number of co-ordinate bonds to it from surrounding ligands.

Quick Questions

7 Look closely at Table 24.5.
  a How does the hydration energy change as:
    i   ionic radius increases,
    ii the charge on the metal ion increases?
  b In general, how will the strength of co-ordinate bonds from ligands to a central metal ion vary with:
    i   the size of the ion,
    ii the charge on the ion?

 Table 24.5
The hydration energies of some metal ions

Ion	Ionic radius /nm	Hydration energy /kJ mol⁻¹
Li⁺	0.068	−499
Na⁺	0.098	−390
K⁺	0.133	−305
Rb⁺	0.148	−281
Mg²⁺	0.065	−1891
Ca²⁺	0.094	−1562
Sr²⁺	0.110	−1413
Al³⁺	0.045	−4613
Ga³⁺	0.062	−4650

24.8 Naming complex ions

The systematic naming of complex ions is based on four simple rules.

1 *State the number of ligands around the central metal ion* using Greek prefixes: mono-, di-, tri-, etc.
2 *Name the ligand* using names ending in -o for anions, e.g. F⁻ is fluoro; Cl⁻ is chloro; CN⁻ is cyano; OH⁻ is hydroxo. Use aqua- for H_2O and ammine for NH_3.

Note

In complex ions, ammonia, NH_3 is named 'ammine', whereas the $-NH_2$ group in organic compounds such as $CH_3CH_2NH_2$ is named 'amine'.

Quick Questions

8 What is the co-ordination number of the central metal ion in each of the complex ions in Table 24.6?

9 Name the following complex ions:
  a $[Al(OH)_4]^-$
  b $[Zn(NH_3)_4]^{2+}$
  c $[Fe(CN)_6]^{4-}$
  d $[CrCl_2(H_2O)_4]^+$

3 *Name the central metal ion* using the English name for the ion in a positively charged complex, but the Latinised name ending in –ate for the ion in a negative complex ion, e.g. aluminate for Al, cuprate for Cu, ferrate for Fe, plumbate for Pb, zincate for Zn, etc.

4 *Add the oxidation number of the central metal ion* (I, II, III, etc.).

The following examples in Table 24.6 show how these rules are applied.

▼ Table 24.6
Naming some complex ions

Rule	1	2	3	4
Formula of complex ion	**State the number of ligands**	**Name the ligand**	**Name the central metal ion (suffix –ate for anions)**	**Add oxidation number of central metal ion**
$[Cu(NH_3)_4]^{2+}$	tetra…	ammine…	copper…	(II)
$[CuCl_4]^{2-}$	tetra…	chloro…	cuprate…	(II)
$[Fe(CN)_6]^{3-}$	hexa…	cyano…	ferrate…	(III)
$[Ag(NH_3)_2]^+$	di…	ammine…	silver…	(I)

24.9 The structure of complex ions

The formula of a complex ion tells us the number and type of ligands, but there are no definitive rules relating the formula of a complex ion to its structure, i.e. the arrangement of ligands around the central metal ion. However:

▶ complex ions with a co-ordination number of *six* usually have ligands in *octahedral* positions so that the six electron pairs around the central ion are repelled as far apart as possible (Figure 24.12),

▶ complex ions with a co-ordination number of *four* usually have ligands in *tetrahedral* positions, although a few four co-ordinated complexes, such as $[PtCl_4]^{2-}$ have a *square planar* structure (Figure 24.12),

▶ complexes with a co-ordination number of *two* usually have a *linear* arrangement of ligands (Figure 24.12).

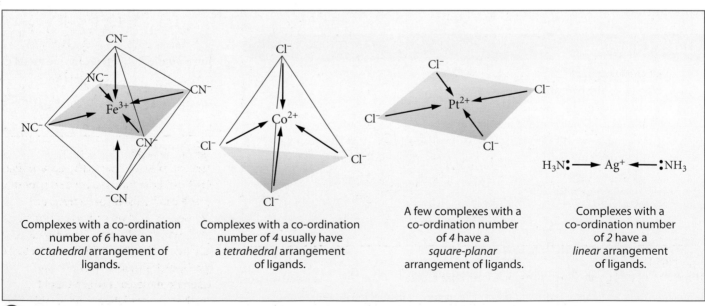

Complexes with a co-ordination number of *6* have an *octahedral* arrangement of ligands.

Complexes with a co-ordination number of *4* usually have a *tetrahedral* arrangement of ligands.

A few complexes with a co-ordination number of *4* have a *square-planar* arrangement of ligands.

Complexes with a co-ordination number of *2* have a *linear* arrangement of ligands.

▲ Figure 24.12
The structure of complex ions

Aqueous complex ions

In aqueous solution, transition metal ions generally exist as octahedral complex ions with the formula $[M(H_2O)_6]^{2+}(aq)$ or $[M(H_2O)_6]^{3+}(aq)$. However, the high charge density on the central cation draws electrons in the O–H bonds of the water molecules towards itself. This polarising effect causes some of the complexed water molecules to dissociate producing H^+ ions.

$$[Fe(H_2O)_6]^{3+}(aq) + H_2O(l) \rightleftharpoons [Fe(H_2O)_5OH]^{2+}(aq) + H_3O^+(aq)$$

This explains why the aqueous solutions of most transition metal compounds such as $CuSO_4$, $FeCl_3$, $Co(NO_3)_2$ and $CrCl_3$ are moderately acidic.

In oxidation states higher than 3+, the polarising power of the central metal ion is even greater. This causes a release of even more H^+ ions and the loss of water molecules resulting in oxyanions. For example, neither $[Cr(H_2O)_6]^{6+}$ nor $[Mn(H_2O)_6]^{7+}$ exist in aqueous solution. Instead, the loss of two water molecules and eight H^+ ions from each of these ions results in the formation of CrO_4^{2-} and MnO_4^{-} respectively.

$$[Cr(H_2O)_6]^{6+}(aq) \longrightarrow 2H_2O(l) + 8H^+(aq) + CrO_4^{2-}(aq)$$
$$[Mn(H_2O)_6]^{7+}(aq) \longrightarrow 2H_2O(l) + 8H^+(aq) + MnO_4^{-}(aq)$$

Types of ligand

Most ligands form only one co-ordinate bond with the central metal ion in a complex. These ligands are described as monodentate (or unidentate) because they have only 'one tooth' with which to bind with the central cation. (The word *dens* in Latin means 'tooth'.)

In some cases, ligands can form more than one bond with a metal ion and these are described as **polydentate** (or multidentate). Examples of bidentate ligands which form two co-ordinate bonds with the central cation in complexes are ethanedioate, $C_2O_4^{2-}$, 1,2-diaminoethane, $H_2NCH_2CH_2NH_2$ and amino acids such as glycine (Figure 24.13).

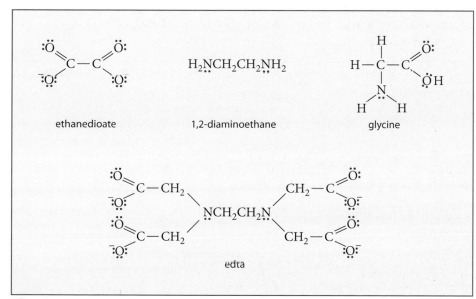

Figure 24.13
Three bidentate ligands and the hexadentate ligand edta

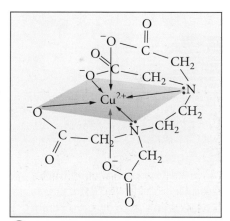

Figure 24.14
The complex ion formed by edta with a Cu^{2+} ion. Edta binds cations so strongly that it can be used sparingly as an antidote to lead poisoning by trapping Pb^{2+} ions. It is also used as a means of extending the shelf life of manufactured foods, such as salad dressings, by trapping metal ions which would catalyse the oxidation of vegetable oils.

A few ligands, such as edta (**ethanediaminetetraethanoate**), are particularly strong because they can form six co-ordinate bonds with the central metal ion in complexes. Edta binds so strongly with metal ions that it makes them chemically inactive (Figure 24.14).

Definition

A **chelate** is a complex ion in which each ligand forms two or more co-ordinate bonds with the central metal ion.

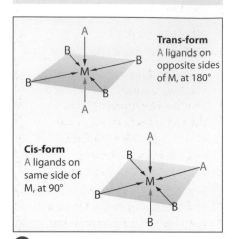

Trans-form
A ligands on opposite sides of M, at 180°

Cis-form
A ligands on same side of M, at 90°

Figure 24.15
Isomers of complexes of the type MA_2B_4

Key Point

Square planar complexes with two or more different ligands can show cis-trans isomerism.

Figure 24.16
Lance Amstrong was diagnosed with testicular cancer in 1995. After surgery, he was cured using a mixture of drugs including *cis*-platin. He began cycling again in 1998 and won the Tour de France seven times in a row from 1999 to 2005.

Note

The reaction forming the pale blue precipitate of copper(II) hydroxide is usually simplified to:

$Cu^{2+}(aq) + 2OH^-(aq)$
$\longrightarrow Cu(OH)_2(s)$

The complex ions which form between polydentate ligands and cations are known as **chelates** (pronounced 'keelates'). This name comes from the Greek word *chelos* meaning 'crab's claw'. In these complexes the ligand forms a clawlike pincer-grip on the central metal ion.

In general, polydentate ligands are more powerful than simple monodentate ligands. The stability of a complex ion is enhanced by chelation as the pincer-grip of the polydentate ligand can hold the central metal ion more securely. Even if one of the bonds is broken, the others still hold the metal ion tightly.

The stereochemistry and isomerism of complex ions

Owing to the three-dimensional shape of complex ions and the positions of ligands around the central metal ion, isomers can occur in four co-ordinated and six co-ordinated complexes. Isomers are compounds with the same formula, but different arrangements of their constituent atoms (Section 13.6).

Octahedral complexes of the type MA_2B_4 exhibit a kind of *cis–trans* (geometric) isomerism as shown in Figure 24.15. One of the most striking examples of this type of isomerism occurs with tetraamminedichlorocobalt(III) chloride, $[Co(NH_3)_4Cl_2]^+Cl^-$. The two isomers with this formula have different-coloured crystals. The *cis*-form is blue-violet, but the *trans*-form is green.

Quick Questions

10 Predict the shape of:
   a $[Cu(CN)_2]^-$,
   b $[Al(OH)_4]^-$,
   c $[Ni(H_2NCH_2CH_2NH_2)_3]^{2+}$,
   d the complex ion in Figure 24.14.
11 Explain the following statements:
   a Ethanedioate, $^-OOC—COO^-$, is a bidentate ligand.
   b Aqueous copper(II) sulfate turns blue litmus red.
12 Write the following ions in order of increasing stability and explain your order: $[Cu(NH_3)_4(H_2O)_2]^{2+}$, $[Cu(H_2NCH_2CH_2NH_2)_3]^{2+}$, $[Cu(edta)]^{2-}$
13 The neutral complex, $PtCl_2(NH_3)_2$, in which Cl^- ions and NH_3 molecules act as ligands to the platinum ion, Pt^{2+}, has two isomers.
   a Explain why a square-planar structure can have isomers, but a tetrahedral structure cannot.
   b The isomers of $PtCl_2(NH_3)_2$ are called *cis*- and *trans*-platin. *Cis*-platin can be used in chemotherapy for the treatment of certain cancers, but the *trans*-isomer has no effect. Draw the structure of *cis*-platin and write its correct chemical name.

24.10 Ligand exchange reactions

In Section 7.16 we noticed that acid–base reactions involve competitions between different bases for H^+ ions. In a similar fashion, complexing reactions involve competitions between different ligands for metal ions. This results in reactions in which metal ions exchange one ligand for another.

For example, when sodium hydroxide solution is added to aqueous copper(II) sulfate, hydroxide ions displace water molecules from hydrated copper(II) ions forming a pale blue precipitate of copper(II) hydroxide. In effect, this is an uncharged insoluble complex.

$[Cu(H_2O)_6]^{2+}(aq) + 2OH^-(aq) \longrightarrow [Cu(H_2O)_4(OH)_2](s) + 2H_2O(l)$
pale blue soln   pale blue ppt

If excess ammonia solution is now added, the pale blue precipitate dissolves as water molecules and hydroxide ions around the Cu^{2+} ion are exchanged for ammonia molecules. A deep blue solution of tetraamminecopper(II) ions is produced:

$$[Cu(H_2O)_4(OH)_2](s) + 4NH_3(aq) \longrightarrow [Cu(NH_3)_4(H_2O)_2]^{2+}(aq) + 2OH^-(aq) + 2H_2O(l)$$

pale blue ppte   deep blue soln

An exchange of ligands also occurs when concentrated hydrochloric acid reacts with copper(II) sulfate solution. In this case, the colour changes from pale blue to green and eventually yellow, as chloride ions replace water molecules in pale blue $[Cu(H_2O)_6]^{2+}(aq)$ ions to form yellow $[CuCl_4]^{2-}(aq)$ ions.

$$[Cu(H_2O)_6]^{2+}(aq) + 4Cl^-(aq) \longrightarrow [CuCl_4]^{2-}(aq) + 6H_2O(l)$$

pale blue soln   yellow soln

Quick Questions

14 Write equations for the ligand exchange reactions which occur when:
   a hexaaquanickel(II) ions react with ammonia molecules to form hexaamminenickel(II) ions,
   b hexaamminenickel(II) ions react with chloride ions to form tetrachloronickelate(II) ions,
   c hexaaquanickel(II) ions react with edta^{4-} ions to form edtanickelate(II) ions.

15 Write equations for the ligand exchange reactions which occur when:
   a colourless diaminoethane solution is added to copper(II) sulfate solution and the colour changes from pale blue to violet,
   b silver nitrate solution is added to sodium chloride solution, forming a white precipitate which dissolves in ammonia to give a colourless solution.

24.11 The importance and use of complex ions

Complex ions are important in biology and in industry.

Complex ions of biological importance

Two important biological macromolecules containing complex ions are **chlorophyll** and **haemoglobin**. Chlorophyll is the green pigment in plant cells. It is responsible for absorbing the radiant energy of sunlight and converting it into chemical energy in the bonds of carbohydrate molecules synthesised by the plant.

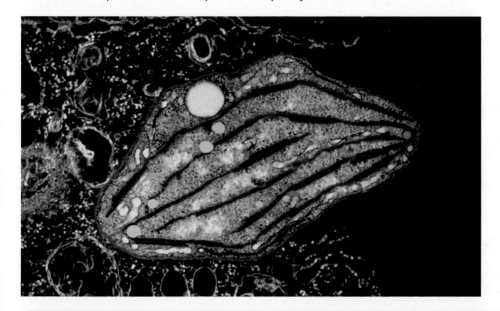

Note

As a result of the reaction of ammonia with water, ammonia solution contains $NH_4^+(aq)$ and $OH^-(aq)$ in addition to $NH_3(aq)$ (Section 12.5):

$$NH_3(aq) + H_2O(l) \rightleftharpoons NH_4^+(aq) + OH^-(aq)$$

So, when dilute ammonia solution is added to an aqueous solution of copper(II) ions, the OH^- ions first react with $[Cu(H_2O)_6]^{2+}$ ions to form a pale blue precipitate of $[Cu(H_2O)_4(OH)_2](s)$. But, when excess ammonia is added, this dissolves to form a deep blue solution containing $[Cu(NH_3)_4(H_2O)_2]^{2+}(aq)$.

Key Point

Ligand exchange reactions can occur when one ligand in a complex is exchanged for another. This often results in a colour change.

◀ Figure 24.17
An electron microscope photograph of a chloroplast. The green material inside the chloroplast is chlorophyll. The white and yellow circles are the sites of carbohydrate (starch) produced during photosynthesis.

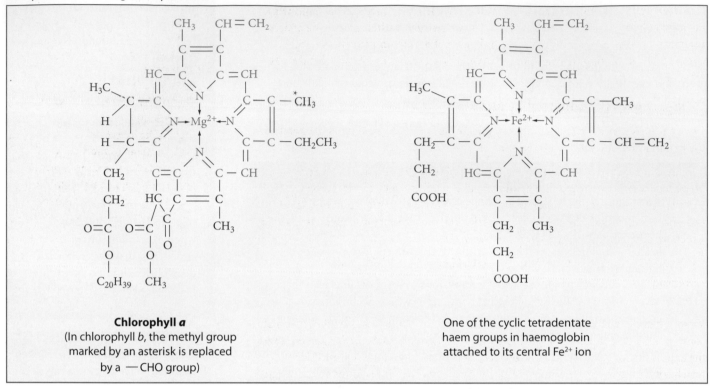

Figure 24.18
Complex ions of biological importance

The chlorophyll molecule is composed of a complicated cyclic tetradentate ligand. This contains four nitrogen atoms surrounding an Mg^{2+} ion in a square planar arrangement (Figure 24.18).

Chlorophyll _a_
(In chlorophyll _b_, the methyl group marked by an asterisk is replaced by a —CHO group)

One of the cyclic tetradentate haem groups in haemoglobin attached to its central Fe^{2+} ion

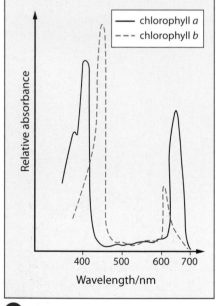

Figure 24.19
The absorption spectrum of chlorophyll

Sunlight reaching the Earth's surface has a maximum intensity in the blue–green region of the spectrum in the wavelength range 450–550 nm. Curiously, the chlorophyll molecule has its weakest absorption in this portion of the visible spectrum.

The maximum absorption peaks for chlorophyll are, in fact, at 680 nm in the red region of the spectrum and at 440 nm in the violet (Figure 24.19). In spite of this inefficiency in absorbing the radiation of greatest intensity, chlorophyll has another property that makes it particularly suitable as a photosynthetic pigment. It can receive energy both directly from light and indirectly from other pigments in plants such as carotenoids.

Figure 24.18 also shows the structure of one of the four cyclic tetradentate haem groups in haemoglobin with its central Fe^{2+} ion. Notice the similarity of the haem group with the chlorophyll molecule. This suggests that the two may have evolved from the same original molecule during the course of evolution. Haemoglobin is the red pigment present in red blood cells. It acts as the oxygen-carrying constituent in the blood. Each of the four haem groups in haemoglobin is bound to a protein molecule and these four components make up haemoglobin.

The active parts of the haemoglobin complex are the four Fe^{2+} ions. Each of these Fe^{2+} ions acts as the central ion in an octahedral complex co-ordinately bonded to four nitrogen atoms in the haem group and to two nitrogen atoms in its protein.

One of the two ligand positions occupied by each protein can be exchanged weakly and reversibly with oxygen. When this happens, a lone pair of electrons on one atom in an oxygen molecule forms a co-ordinate bond to an Fe^{2+} ion displacing one of the bonds to a nitrogen atom in its protein. Each haem group can bind one oxygen atom and all four haem groups bind oxygen simultaneously. So, the equation for this is:

$$Hb(aq) + 4O_2(aq) \rightleftharpoons Hb(O_2)_4(aq)$$

deoxygenated haemoglobin
(bluish- red)

oxygenated haemoglobin
(red)

This weak reversible binding enables oxygen to be delivered from the lungs to other parts of the body. Oxygenated haemoglobin is blood-red. Deoxygenated haemoglobin is bluish-red, the colour of your lips when you are very cold.

Unfortunately, other ligands, such as cyanide and carbon monoxide, can bind with haemoglobin more strongly than oxygen. Unlike oxygen, carbon monoxide attaches itself irreversibly to Fe^{2+} ions in haemoglobin. So, it acts as an acute poison by completely eliminating the transport of oxygen by any haemoglobin molecule with which it combines.

Complex ions of industrial importance

Complex ions play an important part in the methods used to soften hard water and in the extraction of metals from their ores.

Soap contains the sodium salts of long-chain carboxylic acids such as sodium hexadecanoate (sodium palmitate) and sodium octadecanoate (sodium stearate) (Section 19.2). When hard water, containing calcium and magnesium ions, is mixed with soap, it reacts with the anions in these salts. The products are insoluble compounds which we see as 'scum' on the surface of the water.

$$Ca^{2+}(aq) + 2CH_3(CH_2)_{16}COO^-(aq) \longrightarrow Ca(CH_3(CH_2)_{16}COO)_2(s)$$

ions in hard water	octadecanoate (stearate) ions in soap	insoluble ppte of calcium octadecanoate (stearate) in 'scum'

Various ligands will react with calcium and magnesium ions to form soluble, very stable complex ions. As a result of this, the calcium and magnesium ions no longer react with the anions in soap and the water is softened. Thus, sodium polyphosphate ($Na_6P_6O_{18}$), which contains the powerful polyphosphate ligand ($P_6O_{18}^{6-}$), is sold for domestic and industrial water softening under the trade name Calgon. The name Calgon is derived from the expression 'calcium gone'. Edta and its sodium salt are also used in water softening. However, these two substances would be poisonous in the quantities required so they cannot be used to soften drinking water. Consequently, their use is restricted to the softening of water required for industrial processes such as the dyeing of textiles.

Complex formation sometimes plays an essential part in the purification of metal ores and the subsequent extraction of the metal. For example the extraction of gold and silver involves the formation of complex cyanides. Impure silver ores such as argentite (impure Ag_2S) and horn silver (impure $AgCl$) are first mixed with a solution of sodium cyanide. This forms a soluble complex cyanide $[Ag(CN)_2]^-$.

$$Ag_2S(s) + 4CN^-(aq) \longrightarrow 2[Ag(CN)_2]^-(aq) + S^{2-}(aq)$$

$$AgCl(s) + 2CN^-(aq) \longrightarrow [Ag(CN)_2]^-(aq) + Cl^-(aq)$$

Silver is then precipitated by adding zinc dust to the solution, which reduces Ag^+ to Ag. Any excess zinc is removed by adding dilute acid.

$$2[Ag(CN)_2]^-(aq) + Zn(s) \longrightarrow [Zn(CN)_4]^{2-}(aq) + 2Ag(s)$$

24.12 Coloured compounds and coloured ions

Most of the compounds of transition elements are coloured. The colour of these compounds can be related to incompletely filled d orbitals in the transition metal ion.

When light hits a substance, part is absorbed, part is transmitted (if the substance is transparent) and part may be reflected. If all the incident radiation is absorbed, then the substance looks black. If all the incident radiation is reflected, the substance looks white or silvery. If only a small proportion of the incident light is absorbed and all the radiations in the visible region of the spectrum are transmitted equally, then the substance will appear colourless, like water.

Quick Questions

16 a How many haem groups are present in one molecule of haemoglobin?

b Why is the haem group described as a tetradentate ligand?

c Explain how an O_2 molecule attaches itself to haemoglobin.

d Write an equation to summarise the reaction of carbon monoxide with oxygenated haemoglobin.

e Why is carbon monoxide a fatal poison?

Figure 24.20
A stained glass window in the cathedral at Florence in Italy. The coloured constituents in stained glass are compounds of transition elements.

However, many compounds of the transition metals only absorb light in certain areas of the visible spectrum. In this case, the compound takes on the colour of the light which it transmits or reflects. If, for example, a substance absorbs all radiations in the red–orange–yellow region of the spectrum of white light, it will appear blue.

When radiation is absorbed by substances, electrons are promoted from lower to higher energy levels. But, in most compounds, the 'electron jumps' between one sub-shell and the next are so large that the radiation absorbed must be in the ultraviolet region of the electromagnetic spectrum. So, these compounds are white or colourless because they are not absorbing any radiation from the visible region of the spectrum. The colour of transition metal ions arises, however, from the promotion of electrons between orbitals *within* the d sub-shell. These energy transitions are small enough to correspond to the energy of visible light.

In an isolated gaseous atom or ion, the five 3d orbitals are all at the same energy level and are described as **degenerate** (Figure 24.21(a)). However, when the ion of a transition metal is surrounded by ligands in a complex ion, the electrostatic interactions raise all five d orbitals to a higher energy level. If these ligands provided a uniform electric field around the central cation, all five d orbitals would be raised to the same level (Figure 24.21(b)). But, because the ligands are arranged around the central metal ion in specific directions, the electrostatic interactions are not uniform and they split the five orbitals into two groups (Figure 24.21(c)). In octahedral complexes such as $[Cu(H_2O)_6]^{2+}$ and $[Cu(NH_3)_4(H_2O)_2]^{2+}$, two of the five 3d orbitals move to a slightly higher energy level than the other three.

▶ Figure 24.21
The effect of ligands on the energy levels of the five 3d orbitals of the central metal cation

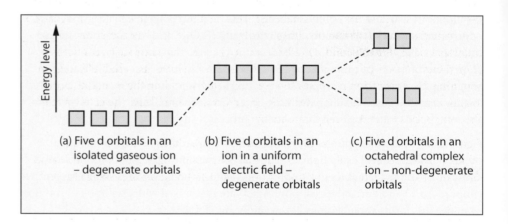

(a) Five d orbitals in an isolated gaseous ion – degenerate orbitals

(b) Five d orbitals in an ion in a uniform electric field – degenerate orbitals

(c) Five d orbitals in an octahedral complex ion – non-degenerate orbitals

This means that complex ions of transition metals such as Cu^{2+}, Fe^{3+} and Ti^{3+} are coloured because radiation of a particular frequency and colour can be absorbed from visible light as electrons are promoted from a lower to a higher 3d orbital (Figure 24.22).

The energy gap between the two sets of orbitals is known as the **crystal field splitting energy**. This is given the symbol Δ. The size of Δ depends on the metal ion and the surrounding ligands. Because of this, the colour of the compounds of one particular metal may vary depending on the ligands associated with the metal ion. So, for example, $[Cu(H_2O)_6]^{2+}(aq)$ and $[Cu(H_2O)_4(OH)_2](s)$ are pale blue. $[Cu(NH_3)_4(H_2O)_2]^{2+}(aq)$ is dark blue and $[Cu(H_2NCH_2CH_2NH_2)_3]^{2+}(aq)$ is violet. All these complexes contain Cu^{2+}, but the different ligands cause different values of Δ, so they have slightly different colours.

The blue colour observed for $[Cu(H_2O)_6]^{2+}(aq)$ is the colour of the light transmitted through the solution. This is white light minus the light that has been absorbed and caused the electron 'jump' in Figure 24.22. The observed blue colour is the complementary colour of that absorbed. The colour wheel in Figure 24.23 shows the relationship between absorbed and complementary colours.

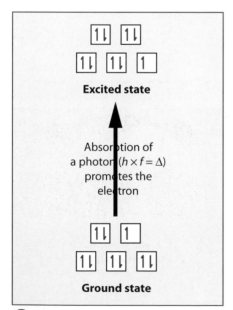

Excited state

Absorption of a photon ($h \times f = \Delta$) promotes the electron

Ground state

▲ Figure 24.22
Promotion of an electron from a lower to a higher 3d orbital in the Cu^{2+} ion of the octahedral complex, $[Cu(H_2O)_6]^{2+}(aq)$

Key Point

Colour in transition metal complexes is caused by electrons jumping between different energy levels caused by crystal field splitting.

Quick Questions

17 a Write the electron structure of:
    i a Ti atom,
    ii a Ti^{3+} ion,
    iii a Ti^{4+} ion.
  b Explain why you would expect $[Ti(H_2O)_6]^{4+}(aq)$ to be colourless.
  c Draw a diagram similar to Figure 24.22 showing how an electron is promoted from the ground state to an excited state in $[Ti(H_2O)_6]^{3+}(aq)$.
  d $[Ti(H_2O)_6]^{3+}(aq)$ is violet in colour. What colour of visible light does it absorb?

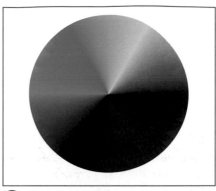

Figure 24.23
A colour wheel. The observed colour of a complex is opposite the colour of the light it absorbs on the colour wheel. So, complexes, such as $[Cu(H_2O)_6]^{2+}(aq)$ which are blue, absorb red light.

24.13 Catalytic properties

The importance of transition metals and their compounds as catalysts and enzyme co-factors has already been discussed in Section 8.10.

When transition metals acts as catalysts, they usually do so as heterogeneous catalysts (Section 22.8), such as iron in the Haber process, nickel in the catalytic hydrogenation of unsaturated fats and platinum alloys in catalytic converters. The transition metal catalysts work by loosely binding (adsorbing) reactants onto their surface. Because of this, they are usually given a very large surface area to maximise their effect on the reaction rate. Adsorption increases the reaction rate by holding the reactants close together on the catalyst surface and making a reaction more likely. The catalyst may also help bonds to weaken and begin to break.

Figure 24.24 shows how the platinum alloy catalyst in the catalytic converter of a vehicle's exhaust system enables poisonous gases such as nitrogen monoxide and carbon monoxide to be converted to non-toxic nitrogen and carbon dioxide (Section 14.8).

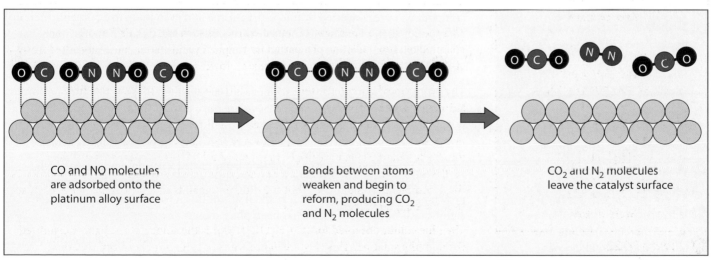

CO and NO molecules are adsorbed onto the platinum alloy surface

Bonds between atoms weaken and begin to reform, producing CO_2 and N_2 molecules

CO_2 and N_2 molecules leave the catalyst surface

Figure 24.24
The stages in heterogeneous catalysis in a catalytic converter

In some cases, impurities can get adsorbed on the catalyst surface. This hinders adsorption of the reactant molecules and reduces the efficiency of the catalyst, which is then said to be poisoned. Lead compounds bind strongly to the catalyst in catalytic converters, so lead-free petrol must always be used in modern vehicles with catalytic converters.

When transition metal compounds act as catalysts in aqueous solution, they usually do so as homogeneous catalysts (Section 22.8). The effective catalyst is the transition metal ion which acts by changing from one oxidation state to another.

For example, the reaction between iodide ions and peroxodisulfate(VI) ions occurs very slowly without a catalyst:

$$2I^-(aq) + S_2O_8^{2-}(aq) \longrightarrow I_2(aq) + 2SO_4^{2-}(aq)$$
peroxodisulfate(VI)

However, in the presence of Fe^{3+} ions, the reaction occurs much faster. Fe^{3+} ions catalyse the reaction by oxidising I^- ions to I_2 as they get reduced to Fe^{2+}.

$$2I^-(aq) \longrightarrow I_2(aq) + 2e^-$$
$$2Fe^{3+}(aq) + 2e^- \longrightarrow 2Fe^{2+}(aq)$$

Then, the Fe^{2+} ions reduce $S_2O_8^{2-}$ ions to SO_4^{2-} ions as they get oxidised back to Fe^{3+} ready to oxidise more I^- ions.

$$2Fe^{2+}(aq) \longrightarrow 2Fe^{3+}(aq) + 2e^-$$
$$S_2O_8^{2-} + 2e^- \longrightarrow 2SO_4^{2-}(aq)$$

Quick Questions

18 Which one of the following ions will not catalyse the reaction between $I^-(aq)$ and $S_2O_8^{2-}(aq)$? Explain your answer.
  a $Fe^{2+}(aq)$
  b $Co^{2+}(aq)$
  c $Zn^{2+}(aq)$

24. 14 Toxic trace metals in the environment

Some metal ions, particularly those of lead and mercury, can have completely the opposite effect of catalysts and enzyme cofactors by disrupting the action of enzymes. Lead compounds can cause mental health problems, particularly in young children, while mercury compounds cause a loss of muscle co-ordination and paralysis. The key problem with lead and mercury compounds is that trace quantities in the environment can accumulate along a food chain and build up to toxic levels.

Mercury contamination at Minamata in Japan during the 1950s is probably the most notorious case of metal poisoning. Mercury(II) compounds from electrolytic cells, which at the time were used in the production of chlorine and sodium hydroxide, leaked into Minamata Bay in the waste water from nearby factories. These mercury(II) compounds were converted by micro-organisms into even more toxic organic mercury compounds such as methymercury ethanoate, $CH_3—Hg^+CH_3COO^-$. Inevitably, the Hg^{2+} and CH_3Hg^+ ions were ingested by aquatic organisms, accumulated and passed along the food chain via fish to humans.

The problems of lead poisoning may be less dramatic than those of mercury poisoning, but they are just as insidious. Lead pipes were once used in water supply systems. As a result of this, minute quantities of lead compounds were ingested over many years. The problems became more acute in areas where the water supply was slightly acidic.

In recent times, lead poisoning has arisen from the lead(II) compounds in vehicle exhaust fumes. These compounds were detectable in the air and on the crops grown near busy roads. Fortunately, the levels of this form of pollution have fallen significantly as leaded gasoline has been phased out.

The mechanism by which Pb^{2+}, Hg^{2+} and $CH_3—Hg^+$ ions disrupt metabolism is by interfering with the catalytic activity of enzymes. Hg^{2+} and $CH_3—Hg^+$ ions are able to form covalent bonds with the sulfur atoms in the cysteine units in enzymes (Section 28.4). These ions react with one or more —SH groups, replacing the hydrogen atoms with heavy metal atoms or methylmercury groups (Figure 24.26).

A Mad Tea-party

Figure 24.25
The use of mercury(II) nitrate in treating the felt used in hat-making was probably responsible for the expression 'mad as a hatter' and for the Mad Hatter's strange behaviour in 'Alice in Wonderland'.

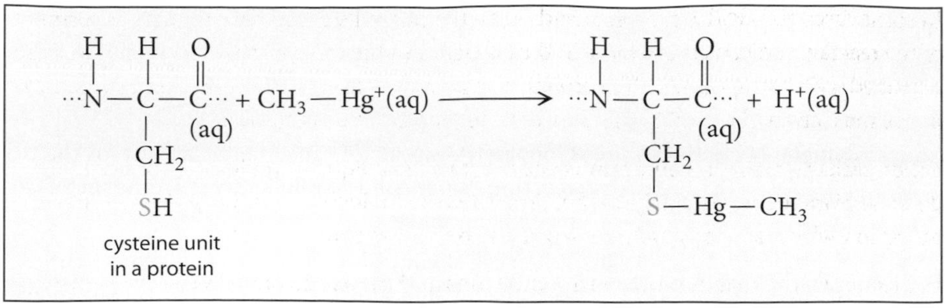

Figure 24.26
The reaction of a methylmercury ion, $CH_3 — Hg^+$, with the cysteine unit in an enzyme

Hg^{2+} and Pb^{2+} ions also interfere with the van der Waals interactions which operate between the chains and side-chains in the protein structure of enzymes. Shape is crucially important in the structure of an enzyme. Therefore, any disruption in the forces holding its structure in place will seriously affect the catalytic activity of an enzyme.

24.15 Fundamental reactions in inorganic chemistry

In studying the formation of complex ions and the competition between ligands for metal ions, we have completed the study of all the fundamental types of reaction involving inorganic chemicals. This is therefore an appropriate point to summarise these key reaction types.

The reactions of inorganic compounds can be divided into four major classes: redox, acid–base, precipitation and complex formation. Other processes such as decomposition and synthesis usually fit into one of these four categories, each of which involves some form of competition.

Redox reactions have already been discussed in Chapter 6. They involve competition for electrons. For example, when sodium reacts with chlorine, sodium atoms lose electrons to chlorine atoms forming sodium chloride, Na^+Cl^-.

$$2Na \longrightarrow 2Na^+ + 2e^-$$

$$2e^- + Cl_2 \longrightarrow 2Cl^-$$

Acid–base reactions, which we discussed in Sections 7.14 to 7.16, involve competition for H^+ ions. For example, when copper oxide reacts with hydrochloric acid, H_3O^+ ions in the acid lose H^+ ions to oxide ions forming water.

$$Cu^{2+}O^{2-}(s) + 2H_3O^+(aq) \longrightarrow Cu^{2+}(aq) + 3H_2O(l)$$

Precipitation occurs when aqueous solutions are mixed and an insoluble substance forms. For example, 'scum' is precipitated when calcium ions in hard water react with stearate ions in soap (Section 24.11):

$$Ca^{2+}(aq) + 2CH_3(CH_2)_{16}COO^-(aq) \longrightarrow Ca^{2+}(CH_3(CH_2)_{16}COO^-)_2(s)$$

calcium stearate (scum)

Essentially, precipitation involves reactions in which cations and anions compete with each other and with water molecules. In solution, polar water molecules attract the ions, but as precipitation occurs, water molecules lose out as the attraction between cations and anions results in an insoluble ionic solid.

Complex formation was studied earlier in this Chapter (Section 24.7). This involves competition between ligands for metal ions. For example, when copper(II) sulfate solution reacts with ammonia to form a deep blue solution of tetraamminecopper(II) ions, ammonia molecules have out-competed water molecules for the Cu^{2+} ions:

$$[Cu(H_2O)_6]^{2+}(aq) + 4NH_3(aq) \longrightarrow [Cu(NH_3)_4(H_2O)_2]^{2+}(aq) + 4H_2O(l)$$

Key Point

The reactions of inorganic compounds can be divided into four major processes each of which involves competition:

1 Redox
2 Acid–base
3 Precipitation
4 Complex formation

Figure 24.27
'Scum' is a white precipitate which forms when soap is used with hard water.

Definition

Precipitation is a reaction which occurs when aqueous solutions are mixed and an insoluble solid forms.

Review questions

1. a. What do you understand by the term 'transition metal'?

   b. Which of the following do you regard as transition metals? Explain your answer.

      i scandium,   ii iron,   iii zinc.

   c. Although the salts of transition elements are usually coloured, there are several copper(I) compounds which are white. Suggest an explanation for this.

   d. The densities of transition elements in the same period gradually increase with relative atomic mass. Why is this?

2. a. Give the name and formula of:

      i one complex cation containing a transition metal,

      ii one complex anion containing a transition metal.

   b. Describe the overall shape (octahedral, tetrahedral, etc.) of the two ions in part a.

   c. Outline the electron structures of the ions that you chose in a, restricting yourself to a consideration of the outer shell of each atom.

3. a. How do the electron structures of transition elements differ from those of the elements across the main groups of the periodic table?

   b. Describe the electron structure of either manganese or chromium. How are the important oxidation states of the element determined by this electron structure?

   c. Give three characteristic features of transition metals or their compounds (other than variable oxidation states). Illustrate your answer with reference to manganese or chromium.

4. Complex ions can sometimes exhibit isomerism.

   a. The compound $[Co(NH_3)_5Br]^{2+}SO_4^{2-}$ is isomeric with the compound $[Co(NH_3)_5SO_4]^+Br^-$.

      i What ions will these two isomers yield in solution?

      ii How would you confirm which isomer was which?

      iii What is the oxidation state and the co-ordination number of cobalt in each complex ion?

      iv Draw the structure of the $[Co(NH_3)_5Br]^{2+}$ ion indicating its shape and the co-ordinate bonds involved.

   b. The compound $NiCl_2(NH_3)_2$ has cis–trans isomers.

      i Does $NiCl_2(NH_3)_2$ have a tetrahedral or a square-planar structure? Explain your answer.

      ii Draw the cis- and trans-isomers for $NiCl_2(NH_3)_2$.

5. Ascorbate oxidase is a metallo-protein enzyme in plants. The enzyme consists of a protein associated with a copper(II) ion. The following observations were made:

      i The enzyme protein alone has no catalytic activity.

      ii Copper(II) ions alone can act as a catalyst for the reaction, but much less efficiently than the metal–protein combination.

      iii When egg albumen is added to aqueous copper(II) ions, the mixture shows greater catalytic activity than copper(II) ions alone. But, this activity is not so good as with the specific metal–protein in ascorbate oxidase.

   a. Write the electron structure of copper(II) ions.

   b. Explain how copper(II) ions might act as catalysts.

   c. What explanation can you offer for observations i, ii and iii above?

6. Early in the 20th century, the German scientist Werner succeeded in clarifying the structure of the five compounds of $PtCl_4$ and ammonia. The properties of these compounds are listed in the table below.

Compound	Formula	Total no. of free ions in the formula	No. of free Cl⁻ ions in the formula
A	$PtCl_4 \cdot 6NH_3$	5	4
B	$PtCl_4 \cdot 5NH_3$	4	3
C	$PtCl_4 \cdot 4NH_3$	3	2
D	$PtCl_4 \cdot 3NH_3$	2	1
E	$PtCl_4 \cdot 2NH_3$	0	0

   a. What is the oxidation state of Pt in each of the compounds, A to E?

   b. The co-ordination number of Pt in each compound is six. Write a formula for each of the five compounds. Show the complex ion and the other ions and/or molecules present.

   c. Each of the compounds forms an octahedral complex ion. Draw structures for the complex ions in A, B, C and D.

   d. Which of the complex ions in c have isomers?

   e. Draw structures to show the various structural isomers of A, B, C and D.

7. a. A transition element X has the electronic configuration $[Ar]3d^34s^2$.

      i Predict its likely oxidation states.

      ii State the electronic configuration of the ion X^{3+}.

   b. Potassium manganate(VII), $KMnO_4$, is a useful oxidising agent in titrimetric analysis.

      i Describe how you could use a $0.020\,mol\,dm^{-3}$ solution of $KMnO_4$ to determine accurately the $[Fe^{2+}]$ in a solution. Include in your description how you would recognise the end-point in the titration and write an equation for the titration reaction.

ii A 2.00 g sample of iron ore was dissolved in dilute H_2SO_4 and all the iron in the salts produced was reduced to $Fe^{2+}(aq)$.

The solution was made up to a total volume of 100 cm^3. A 25.0 cm^3 portion of the solution required 14.0 cm^3 of 0.020 mol dm^{-3} KMnO$_4$ to reach the end-point. Calculate the percentage of iron in the ore.

c High-strength low-alloy (HSLA) steels are used to fabricate TV masts and long span bridges. They contain very low amounts of phosphorus and sulfur, but about 1% copper, to improve resistance to atmospheric corrosion. When dissolved in nitric acid, a sample of this steel gives a pale blue solution.

i What species is responsible for the pale blue colour?

ii Describe and explain what you would see when dilute aqueous ammonia is added to this solution.

Cambridge Paper 4 Q3 June 2006

25 Aromatic hydrocarbons and phenol

Figure 25.1
F.A. Kekulé (1829–96), who proposed the ring structure of benzene

25.1 Aromatic hydrocarbons

In Chapter 15 we studied the alkanes – unsaturated compounds with double bonds. There is another important class of unsaturated compounds that we need to look at separately because their properties are so different from the alkenes.

This class of compounds is the **aromatic hydrocarbons** or **arenes**, and its simplest and most important member is **benzene, C_6H_6**. The name 'aromatic' was originally used because some derivatives of these hydrocarbons have pleasant smells. We now know that just as many of them smell unpleasant, and in any case many of the aromatic vapours are toxic, so it is unwise to smell them. Benzene itself is particularly toxic, and long-term exposure to benzene may cause certain kinds of cancer. The name 'aromatic' has been retained to indicate certain chemical characteristics rather than smell.

Detergents, polystyrene, nylon and insecticides can all be made from benzene, which is industrially the most important arene. Most benzene is normally manufactured from oil by **catalytic reforming**. In the presence of a catalyst, C_6–C_8 hydrocarbons from the gasoline fraction rearrange their molecules, producing a variety of aromatic hydrocarbons, including benzene.

25.2 The structure of benzene

Since 1834 the molecular formula of benzene has been known to be C_6H_6. The exact structural formula, however, posed a problem for many years. With such a high C : H ratio, benzene must be highly unsaturated. A possible structure might be hexatetraene, e.g. $CH_2{=}C{=}CH{-}CH{=}C{=}CH_2$. Such a structure would be expected to have two isomeric forms of C_6H_5Cl:

$$ClCH{=}C{=}CH{-}CH{=}C{=}CH_2 \quad \text{and} \quad CH_2{=}C{=}CCl{-}CH{=}C{=}CH_2$$

Only one form of chlorobenzene has ever been isolated. Therefore all six hydrogen atoms in benzene occupy equivalent positions.

The problem was solved by Friedrich August Kekulé in 1865. He proposed a ring structure in which alternate carbon atoms were joined by double bonds:

This structure, called the Kekulé structure, explains many of the properties of benzene and was accepted for a long time. But even this structure leaves some problems, particularly concerning bond length and thermochemistry.

Bond lengths in benzene

X-ray diffraction studies show that benzene contains planar (flat) molecules in which all the C—C bonds in benzene are the same length:

carbon–carbon bond length in all bonds in benzene = 0.139 nm

Compare this with:

carbon–carbon single bond length in cyclohexane = 0.154 nm

carbon–carbon double bond length in cyclohexene = 0.133 nm

The Kekulé model would suggest unequal carbon–carbon bond lengths, alternating between double and single bond values. In fact X-ray studies show a constant bond length, somewhere between the value for a single and a double bond.

Thermochemistry of benzene

We can work out a theoretical value for the enthalpy change of formation of benzene on the basis of the Kekulé model, and compare this with the experimental value obtained from the enthalpy change of combustion.

The enthalpy change of formation of gaseous benzene is the enthalpy change when a mole of gaseous benzene is formed from its elements:

$$6C(s) + 3H_2(g) \longrightarrow C_6H_6(g)$$

Relevant data are:

enthalpy change of atomisation of C(s)   = +715 kJ (mol of C atoms)$^{-1}$
enthalpy change of atomisation of H$_2$(g) = +218 kJ (mol of H atoms)$^{-1}$
bond energy of C$=$C (average)   = 610 kJ mol^{-1}
bond energy of C$-$C (average)   = 346 kJ mol^{-1}
bond energy of C$-$H (average)   = 413 kJ mol^{-1}

Quick Questions

1 Work out the enthalpy change of formation of benzene by the following stages.
  a Calculate the energy needed to produce:
    i   six moles of gaseous carbon atoms from C(s),
    ii six moles of gaseous hydrogen atoms from H$_2$(g).
  b Calculate the energy released when:
    i   three moles of C$-$C bonds are formed from gaseous atoms,
    ii   three moles of C$=$C bonds are formed from gaseous atoms,
    iii six moles of C$-$H bonds are formed from gaseous atoms.
  c Use your answers to parts a and b to calculate the total energy change when a mole of gaseous benzene is formed from its elements.
  d Compare your answer with the experimental value of +82 kJ mol^{-1}.
  e Do your results suggest that 'real' benzene is more or less stable than the Kekulé structure?

From the calculation in Quick question 1, we find that the theoretical enthalpy change of formation of gaseous benzene based on the Kekulé structure is +252 kJ mol^{-1}. This is 170 kJ mol^{-1} greater (i.e. more endothermic) than the experimental value of +82 kJ mol^{-1}. This means that the actual structure of benzene is considerably more stable than the Kekulé structure. The 170 kJ mol^{-1} of extra stability is called the **delocalisation energy**. The result agrees reasonably closely with the delocalisation energy of benzene obtained using enthalpy changes of hydrogenation. (See Review question 3 at the end of this chapter.)

Electron delocalisation in benzene

The extra stability of benzene and the fact that its C$-$C bonds are all of equal length can be explained using the following model.

The carbon atoms in the benzene ring are bonded to one another and to their hydrogen atoms by σ-bonds. This leaves one unused p orbital on each carbon, each

Key Point

Benzene has a flat, symmetrical ring structure, with all carbon–carbon bonds of the same length.

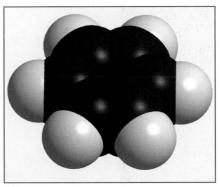

▲ Figure 25.2
A space-filling representation of the benzene molecule, C$_6$H$_6$

Key Point

The delocalisation of electrons around the benzene ring gives benzene extra stability.

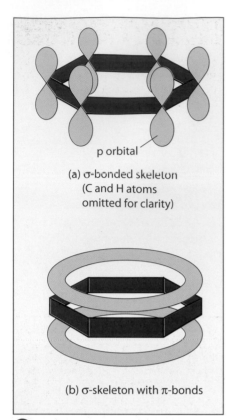

(a) σ-bonded skeleton (C and H atoms omitted for clarity)

(b) σ-skeleton with π-bonds

Figure 25.3
Bonding in benzene. Note that the plane of the molecule is perpendicular to the paper.

containing a single electron. These p orbitals are perpendicular to the plane of the ring, with one lobe above and one below this plane (Figure 25.3 (a)). Each p orbital overlaps sideways with the two neighbouring orbitals to form a single π-bond extending as a ring of charge above and below the plane of the molecule (Figure 25.3 (b)).

The electrons in the π-bond do not 'belong to' any particular carbon atom. Each electron is free to move throughout the entire π system, so the electrons are said to be **delocalised.** It is this delocalisation that gives benzene its extra stability: any system in which electron delocalisation occurs is stabilised. The reason for this is not hard to see. Electrons tend to repel one another, so a system in which they are as far apart from one another as possible will involve minimum repulsion and will therefore be stabilised.

To conform with this model, the structural formula of benzene is often written as

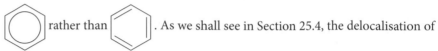

π-electrons has a profound effect on the chemical properties of benzene.

The word **aromatic** is used to describe any system like that in benzene which is stabilised by a ring of delocalised π-electrons. Non-aromatic compounds (such as alkanes and alkenes) are called **aliphatic.**

The structure described above means that the benzene molecule is planar, symmetrical and non-polar. Its lack of polarity results in benzene being a liquid at room temperature and makes it immiscible with water. Its boiling point is 80 °C and its melting point 6 °C. The surprisingly high melting point is due to the ease with which highly symmetrical benzene rings can pack into a crystal lattice. Compare this with that of the structurally similar but less symmetrical compound methylbenzene $C_6H_5CH_3$, which melts at –95 °C.

25.3 Naming aromatic compounds

Benzene is not the only aromatic hydrocarbon. Many others exist: they are either substituted forms of benzene or compounds containing different ring systems. Some other aromatic hydrocarbons are discussed in Section 25.8.

The name 'benzene' comes from *gum benzoin*, a natural product containing benzene derivatives. The name *phene*, derived from the Greek *pheno* 'I bear light', was at one time suggested as an alternative to 'benzene' (benzene was originally isolated from illuminating gas by Michael Faraday). It was not adopted, but it survives in the word **phenyl** used for the C_6H_5— group. The phenyl group is an example of an **aryl** group.

The hydrogen atoms in benzene can be substituted by other atoms and groups, as we shall see in Section 25.4. Some examples are given in Table 25.1.

Table 25.1
Some derivatives of benzene

Substituent group	Benezene derivative	Systematic name	Other name
methyl — CH_3	C_6H_5 — CH_3	methylbenzene	toluene
chloro — Cl	C_6H_5 — Cl	chlorobenzene	—
nitro — NO_2	C_6H_5 — NO_2	nitrobenzene	—
hydroxy — OH	C_6H_5 — OH	phenol	—
amino — NH_2	C_6H_5 — NH_2	phenylamine	aniline
carboxylic acid — COOH	C_6H_5 — COOH	benzenecarboxylic acid	benzoic acid

Where more than one hydrogen atom is substituted, numbers are used to indicate which of the six possible ring positions are occupied. So these three possible methylchlorobenzenes are named:

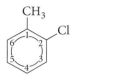

2-chloromethylbenzene   3-chloromethylbenzene   4-chloromethylbenzene

The methyl group is regarded as occupying the 1 position and the ring is numbered clockwise as shown in the first formula. Two more examples are:

3-nitrophenylamine   1,2-dimethylbenzene

Note that other numbers could be used (e.g. 3,4-dimethylbenzene instead of 1,2-dimethylbenzene), but the numbers actually used are the lowest ones possible.

Quick Questions

2 Use these rules and Table 25.1 to write the structural formulae of the following compounds:
  a ethylbenzene,
  b 2-methylphenol,
  c 1,3-dinitrobenzene.
3 a Name the following:

i

ii

iii

b Why are there no compounds named 5-chloromethylbenzene or 6-chloromethylbenzene?

You may come across the terms *ortho*, *meta* and *para* in the names of di-substituted benzene derivatives, particularly in older books. Under this system, 1,2 derivatives are given the prefix *ortho-*, 1,3 derivatives are *meta-* and 1,4 derivatives are *para-*. Thus the three chloromethylbenzenes above become *ortho*-chloromethylbenzene, *meta*-chloromethylbenzene and *para*-chloromethylbenzene, respectively.

Key Point

Benzene derivatives are named using numbers to indicate the position on the ring of the second substituent.

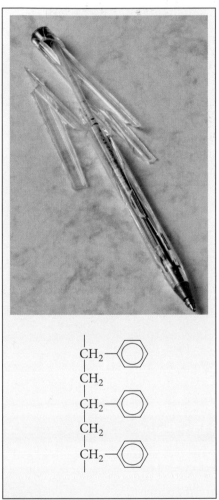

▲ Figure 25.4
Polystyrene contains alkane chains with benzene rings as side groups. The bulky benzene rings make it difficult for the polymer chains to slide over each other so polystyrene is brittle, not flexible like polythene.

25.4 Chemical characteristics of benzene

Table 25.2 compares some reactions of cyclohexane, cyclohexene and benzene.

Table 25.2
Some reactions of cyclohexane, cyclohexene and benzene

Reagent	cyclohexane 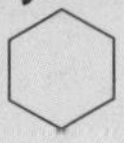	cyclohexene 	benzene 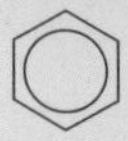
bromine (in the dark)	no reaction	bromine is decolorised, no HBr evolved	no reaction with bromine alone. In presence of iron filings, bromine is decolorised and HBr fumes are evolved
acidified potassium manganate(VII) (potassium permanganate)	no reaction	manganate(VII) is decolorised	no reaction
hydrogen over very finely divided nickel catalyst	no reaction	one mole absorbs one mole of hydrogen at room temperature	one mole absorbs three moles of hydrogen at 150 °C
mixture of concentrated nitric and concentrated sulfuric acids	no reaction	oxidised	substitution reaction: a yellow oil is formed

Quick Questions

4   Look at Table 25.2 and answer these questions.
  a   Does the evidence suggest that
    i   cyclohexene,
    ii   benzene undergo addition with bromine in the dark?
  Explain your answer.
  b   Does benzene undergo
    i   addition,
    ii   substitution with bromine in the presence of iron filings?
  Explain your answer.
  c   Does benzene undergo catalytic hydrogenation as readily as cyclohexene?

As Table 25.2 shows, benzene undergoes addition reactions far less readily than we might expect for such an unsaturated compound. In fact, *substitution* reactions are more characteristic of benzene than addition reactions. You can see this by looking at the reaction of benzene with bromine in the presence of iron filings. HBr is evolved, which implies a substitution reaction in which H is replaced by Br.

If we remember the delocalisation of π-electrons in benzene, it is quite easy to see why addition reactions are difficult. For example, if a molecule of benzene carried out an addition reaction with a molecule of bromine, the ring of delocalised π-electrons would be broken:

Key Point

The typical reactions of benzene and its derivatives involve electrophilic substitution.

In order for this to happen, the delocalisation energy would need to be supplied which is considerably more energy than is needed to break the one double bond in cyclohexene. Therefore, benzene tends to maintain its π-electron ring intact and to undergo substitution reactions rather than addition (see Review question 10 at the end of this chapter).

25.5 The mechanism of substitution reactions of benzene

The substitution reactions of benzene involve **electrophilic substitution.** The high density of negative charge in the delocalised electron system of the benzene ring tends to attract electrophiles.

Consider as an example the nitration of benzene. Benzene reacts with a mixture of concentrated nitric acid and concentrated sulfuric acid (called a **nitrating mixture**) at 50 °C. The product is nitrobenzene:

$$\text{benzene} + HNO_3 \xrightarrow[\text{50 °C}]{\text{conc. } H_2SO_4} \text{nitrobenzene} + H_2O$$

nitrobenzene
(a yellow oil)

Key Point

The nitration of benzene to form nitrobenzene involves electrophilic attack by NO_2^+.

This is a substitution reaction. A hydrogen atom on the benzene ring has been substituted by a nitro group, $-NO_2$.

The reaction of benzene with concentrated nitric acid alone is slow, whilst pure sulfuric acid at 50 °C has practically no effect on benzene. This suggests that sulfuric acid must somehow react with nitric acid, producing a species that then reacts with benzene. There is good evidence that this species is NO_2^+, the **nitronium ion**, also called the **nitryl cation**. This is formed by the removal of OH^- from HNO_3 by sulfuric acid:

$$HNO_3 + 2H_2SO_4 \longrightarrow NO_2^+ + 2HSO_4^- + H_3O^+$$

nitryl cation

In this reaction, nitric acid is acting as a base in the presence of the stronger sulfuric acid. The NO_2^+ ion is a strong electrophile and is attracted to the negative π-electron system in benzene. First a loose association is formed:

$$\text{benzene} + NO_2^+ \longrightarrow \text{benzene} \cdots NO_2^+$$

Quick Questions

5 The typical reactions of benzene are electrophilic substitution. Explain why they are:
  a electrophilic not nucleophilic,
  b substitution, not addition.

The NO_2^+ then attacks one of the carbon atoms of the ring, forming a bond to it and disrupting the delocalised π-electron system:

$$\text{benzene} \cdots NO_2^+ \longrightarrow \text{intermediate cation}$$

The formation of this intermediate requires the input of considerable energy to break the delocalised π-electron system. Because of this, the reaction has a fairly high activation energy, which is why it needs heating.

Finally, the intermediate cation breaks down, either reforming benzene or producing nitrobenzene. As this happens, the delocalised π system re-forms, releasing energy:

$$\text{benzene } H + NO_2^+ \longleftarrow \text{intermediate cation} \longrightarrow \text{nitrobenzene } NO_2 + H^+$$

Notice that the first two stages of this mechanism, up to the formation of the intermediate cation, are similar to the stages in addition to an alkene double bond. The difference is that with aromatic compounds, the intermediate cation loses its charge by loss of H^+, thus regaining delocalisation energy. With alkenes there is no delocalisation energy to be lost or gained, so the intermediate cation combines with an anion to form an addition product.

Figure 25.5
Artificial food colourings contain dyes that are synthesised from aromatic hydrocarbons. These dyes are rigorously checked for safety before they can be used in foods.

25.6 Important electrophilic substitution reactions of benzene

Nitration

The detailed mechanism of this reaction has already been considered in the last section. It is used industrially to manufacture nitrobenzene, from which phenylamine (aniline), $C_6H_5NH_2$, is produced by reduction. Phenylamine is used to manufacture dyes (see Chapter 26).

Halogenation

Benzene does not react with chlorine, bromine or iodine on their own in the dark. This is because the non-polar halogen molecule has no centre of positive charge to initiate electrophilic attack on the benzene ring. However, in the presence of a catalyst such as iron filings, iron(III) bromide or aluminium chloride, benzene is substituted by chlorine or bromine:

chlorobenzene

When iron filings act as the catalyst, they first react with the halogen, forming iron(III) chloride or iron(III) bromide.

The catalyst (called a 'halogen carrier') accepts a lone pair from one of the halogen atoms. This induces polarisation in the halogen molecule:

$$\underset{\underset{\delta+}{} \quad \underset{\delta-}{}}{Cl - Cl:} \longrightarrow AlCl_3$$

The positively charged Cl atom in the halogen molecule is now electrophilic and attacks the benzene ring.

Nucleophilic substitution of chlorobenzene

The electron-withdrawing effect of the Cl atom in chlorobenzene makes the carbon atom to which it is attached attractive to *nucleophiles*. (This contrasts strongly with the typical reactions of benzene which involve *electrophilic* substitution.) This allows both chlorobenzene and bromobenzene to undergo nucleophilic substitution with nucleophiles such as ^-OH.

However, these reactions do not go as easily as with halogenoalkanes (Chapter 16). The lone pairs of electrons on the halogen atom interact with the delocalised π-electron system of the benzene ring, strengthening the C–Hal bond. Furthermore, the high electron density on the benzene ring tends to repel the approaching negative ^-OH ion. As a result, chlorobenzene requires the extreme conditions of 300 °C and a pressure of 200 atmospheres to react with sodium hydroxide. Compare this with the behaviour of chlorobutane, which is readily converted to butanol by gently refluxing with aqueous sodium hydroxide at atmospheric pressure.

The behaviour of the —Cl group in chlorobenzene and chlorobutane illustrates the fact that functional groups behave differently depending on whether they are attached to an alkyl or an aryl group. The high electron density on the benzene ring has a modifying effect on any group that is attached to it.

The reaction of chlorobenzene with sodium hydroxide described above provides one way of manufacturing phenol (Section 25.10).

Quick Questions

6 What would you expect to be the product of a substitution reaction between benzene and bromine(I) chloride, BrCl? (Hint: Think about the way this molecule is polarised.)

7 Iodine is too unreactive to substitute benzene even in the presence of a halogen carrier. Quite good yields of iodobenzene can however be obtained by reacting benzene with iodine(I) chloride, ICl. Explain why.

Key Point

The high electron density on the benzene ring has a modifying effect on any group that is attached to it. Functional groups behave differently depending on whether they are attached to an alkyl or an aryl group.

Alkylation: Friedel–Crafts reaction

Aluminium chloride is used as a catalyst to polarise halogen molecules and cause them to substitute a benzene ring. The same catalyst can be used to bring about the substitution of benzene by an alkyl group, R. For example, if benzene is warmed with chloromethane and aluminium chloride under anhydrous conditions, a substitution reaction occurs and methylbenzene is formed:

$$
\text{benzene} + CH_3Cl \xrightarrow[\text{heat}]{AlCl_3} \text{methylbenzene} + HCl
$$

methylbenzene

As in the reaction with halogens, the aluminium chloride accepts an electron pair from the chlorine atom, polarising the chloromethane molecule:

$$
\overset{\delta+}{CH_3} - \overset{\delta-}{Cl:} \longrightarrow AlCl_3
$$

The positively charged methyl group attacks the benzene ring and electrophilic substitution occurs. This is an example of a **Friedel–Crafts** reaction. Such reactions occur between aromatic hydrocarbons and any combination of reagents that give rise to a positively charged carbon atom. These include alkenes and alcohols as well as halogenoalkanes. For example, ethene and benzene undergo a Friedel–Crafts reaction in the presence of hydrogen chloride and aluminium chloride, to form ethylbenzene:

$$
\text{benzene} + CH_2{=}CH_2 \xrightarrow[\text{HCl, 95\,°C}]{AlCl_3} \text{ethylbenzene } (CH_2CH_3)
$$

ethylbenzene

This reaction is used industrially to manufacture ethylbenzene, from which phenylethene (styrene) is made by catalytic dehydrogenation:

$$
\text{ethylbenzene } (CH_2CH_3) \xrightarrow[\text{600\,°C}]{Zn} \text{phenylethene } (CH{=}CH_2) + H_2
$$

ethylbenzene   phenylethene

Dodecylbenzene, important in the manufacture of detergents (Figure 25.6) is made by a Friedel–Crafts reaction between benzene and dodecene similar to that just described between benzene and ethene:

$$
\text{benzene} + CH_2(CH_2)_9CH{=}CH_2 \xrightarrow[\text{HCl}]{AlCl_3} \text{dodecylbenzene } (CH_2CH_2(CH_2)_9CH_3)
$$

dodecene   dodecylbenzene

Figure 25.6
Many solid detergents contain dodecylbenzenesulfonate. This is manufactured by reacting dodecylbenzene with concentrated sulfuric acid.

$$
CH_3CH_2CH_2CH_2CH_2CH_2CH_2CH_2CH_2CH_2CH_2CH_2 -\!\!\bigcirc\!\!- SO_3{}^-Na^+
$$

A synthetic detergent

10 If one mole of benzene and one mole of hydrogen are reacted, one-third of the benzene is converted into cyclohexane and the remainder is left unreacted. We might expect some cyclohexadiene,

and cyclohexene,

to be formed, but this does not in fact happen. Can you suggest a reason why?

Key Point

▶ Benzene does not undergo addition reactions as readily as alkenes.

Figure 25.7
Benzene is added to gasoline to increase the octane number. However, its use is limited to less than 1% of the mixture because of its toxicity.

25.7 Other important reactions of benzene

The characteristic reactions of benzene and other aromatic hydrocarbons involve electrophilic substitution, because this type of reaction keeps the delocalised π-electron system intact. Addition reactions, which disrupt the π-electron system, are less characteristic of benzene but some do occur.

Addition with hydrogen

Alkenes undergo addition with hydrogen in the presence of a nickel catalyst. Benzene also gives this reaction, but considerably higher temperatures are required. The higher temperature is needed because extra energy must be supplied to break up the delocalised π-electron system.

$$3H_2 \ + \ \bigcirc \xrightarrow[\text{150°C}]{\text{Raney nickel}} \ \bigcirc$$

cyclohexane

The catalyst, Raney nickel, is a very active form of nickel with an extremely high surface area.

The catalytic hydrogenation of benzene is important industrially in the manufacture of cyclohexane, from which nylon is made (Section 27.4).

Burning

Benzene burns in air with a sooty, smoky flame. This sort of flame is characteristic of all hydrocarbons containing a high percentage of carbon.

Benzene has a high octane rating, and is used as a component of gasoline. However, its use is strictly limited because of its toxicity.

Figure 25.8 summarises some important reactions of benzene.

⌄ **Figure 25.8**
Some important reactions of benzene

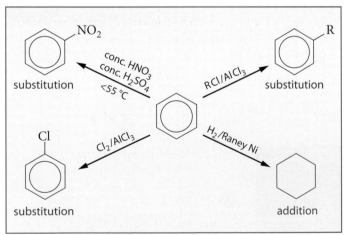

25.8 Other arenes

There are many other arenes apart from benzene. Some of these arenes are shown in Table 25.3. Naphthalene and anthracene are examples of fused ring arenes, in which two or more rings are joined.

Systematic name	Other name	Molecular formula	Structural formula
methylbenzene	toluene	C_7H_8	
1,4-dimethylbenzene	*para*-xylene	C_8H_{10}	
naphthalene	—	$C_{10}H_8$	
anthracene	—	$C_{14}H_{10}$	

◀ **Table 25.3**
Some arenes other than benzene

> **Key Point**
>
> There are many arenes apart from benzene.
>
> These include:
>
> ▶ benzene rings with alkyl side groups,
> ▶ fused benzene rings

Methylbenzene

The reforming reactions used to manufacture benzene from crude oil (see Section 25.3) also produce methylbenzene (commonly called toluene). This substance is used in the manufacture of plastics and explosives (see below), but much more is produced than can be used in this way. The majority of the methylbenzene is added to gasoline to increase its octane rating: it is much less toxic than benzene.

Reactions of methylbenzene

The properties of aromatic compounds are very different from those of aliphatic ones. Methylbenzene's molecule has an aromatic portion (the benzene ring) and an aliphatic portion (the —CH_3 group). These two portions make different contributions to the properties of methylbenzene and have a modifying effect on each other.

i The — CH₃ group

The —CH_3 group shows some reactions we would expect of an alkyl group. For example, its H atoms can be substituted by chlorine when chlorine is bubbled into boiling methylbenzene in sunlight:

The reaction has a free-radical mechanism similar to the reaction of methane with chlorine described in Section 14.6.

Oxidation of the side group in methylbenzene

The —CH_3 group in methylbenzene does not behave as a typical alkyl group in all its reactions. The benzene ring with its regions of high electron density has a modifying effect on any group that is attached to it.

One example of the way in which the ring modifies the properties of the —CH_3 group in methylbenzene is the reaction with potassium manganate(VII). Alkanes are inert to oxidation, but the alkyl group in methylbenzene can be oxidised by alkaline manganate(VII) to give benzoic acid. The purple manganate(VII) is reduced to green manganate(VI).

> **Quick Questions**
>
> 11 Write the structure of two other compounds that might form when the —CH_3 group is substituted by chlorine.
>
> 12 Under certain conditions chlorine will substitute H atoms in the *ring* of methylbenzene instead of those in the —CH_3 group. What will these conditions be?

> **Key Point**
>
> Methylbenzene has two different substitution reactions with chlorine, depending on the conditions.
>
> ▶ When chlorine is bubbled into boiling methylbenzene in sunlight, the —CH_3 *side group* is substituted.
> ▶ When chlorine is bubbled into methylbenzene in the presence of $AlCl_3$ as a catalyst, the *ring* is substituted.

13 Starting with ethylbenzene, how would you produce the following? Give the reagents and conditions.

a

b

Br

Figure 25.9

Manufacturing high explosive shells in 1940. TNT was an important explosive in both world wars.

Figure 25.10

One of the earliest high explosives was nitroglycerine. It is made by heating glycerol with a nitrating mixture, but the reaction must be watched carefully to make sure it does not get out of control. In this print of early nitroglycerine manufacture, the operator minding the reaction is sitting on a one-legged stool to prevent him nodding off to sleep.

Note that the ring is not affected in the reaction, which is another indication of its stability.

ii The aromatic ring

If chlorine is bubbled through methylbenzene in the absence of sunlight and in the presence of a halogen carrier such as $AlCl_3$, the *ring* is substituted instead of the CH_3 side group. This reaction proceeds by the electrophilic substitution mechanism described in Section 25.6. A mixture of two isomers is obtained:

2-chloromethylbenzene
58%

4-chloromethylbenzene
42%

Virtually none of the other possible isomer, 3-chloromethylbenzene, is produced. Furthermore, the reaction proceeds at a considerably higher rate than the corresponding reaction of chlorine with benzene. The $—CH_3$ group influences the aromatic ring, making it more susceptible to electrophilic substitution and dictating the positions in which it is substituted. This effect is discussed further in the next section.

An important substitution reaction of methylbenzene is nitration. When methylbenzene is heated with a nitrating mixture of concentrated nitric acid and concentrated sulfuric acid, it is substituted by one, two or three $—NO_2$ groups, depending on the conditions. The main products are:

2-nitromethylbenzene

4-nitromethylbenzene

2,4-dinitromethylbenzene

2,4,6-trinitromethylbenzene

Note the effect of the $—CH_3$ group in determining that substitution in positions 2, 4 and 6 rather than 3 or 5.

2,4,6-Trinitromethylbenzene (trinitrotoluene or TNT) is an important high explosive. It is fairly resistant to shock and so can be used in shells without risk of explosion under the shock of firing a gun. When detonated it decomposes forming large volumes of CO, H_2O and N_2 at high temperature. The sudden expansion of these gases gives TNT its explosive force. TNT has the additional advantage that it is a solid which melts below $100\,°C$, so it can be melted with steam and poured into its container. See Section 12.9 for more about high explosives such as TNT and nitroglycerine.

25.9 The position of substitution in benzene derivatives

The methyl group in methylbenzene activates the ring towards electrophilic substitution. It also favours substitution in positions 2, 4 and 6 rather than 3 or 5. In fact, *any substituent group* attached to a benzene ring affects the rate and the position at which further substitution occurs. Table 25.4 shows the main products of mononitration (i.e. substitution by one nitro group) of different benzene derivatives, and whether they are nitrated faster or slower than benzene.

⌄ Table 25.4
Mononitration products of benzene derivatives

Compound	Main products of mononitration	Rate of nitration relative to benzene
methylbenzene 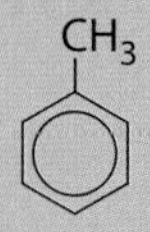	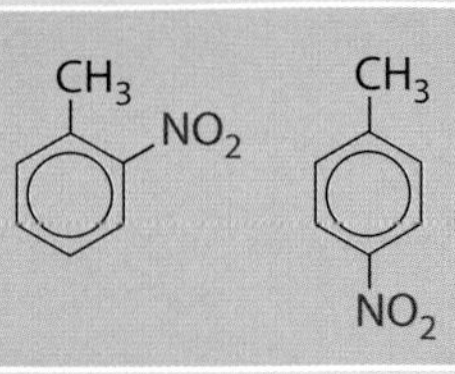	Faster
phenol 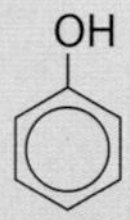	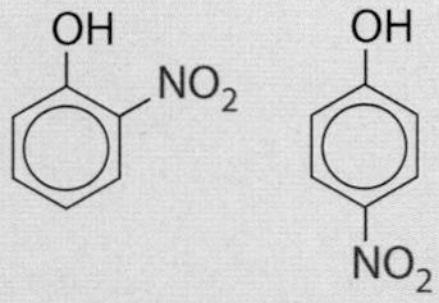	Faster
nitrobenzene 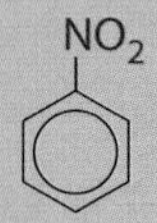	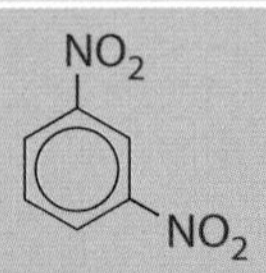	Slower
phenylamine 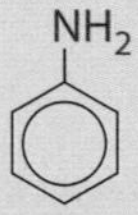	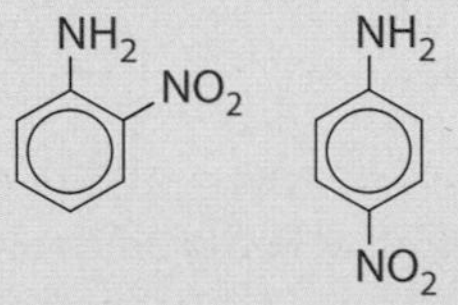	Faster
benzoic acid 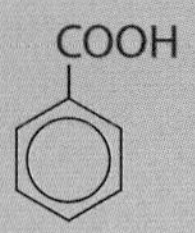	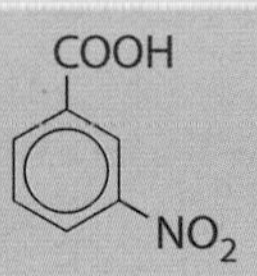	Slower

You can see from Table 25.4 that benzene derivatives fall into two classes as far as further substitution is concerned.

▶ Those which substitute faster than benzene and in which the new substituent is directed to the 2 and 4 position, a mixture of the two isomers being obtained. Functional groups causing this behaviour include:
  — CH_3 and all alkyl groups, — OH, — NH_2, — OCH_3.
▶ Those which substitute more slowly than benzene and in which the new substituent is directed to the 3 position. Functional groups causing this behaviour include:
  — NO_2, — COOH, — SO_3H.

In practice, a mixture of all possible isomers is obtained, but these rules give the *main* products. Note that the rules apply whatever the nature of the new substituent, not just to nitration. Note too, that as far as monosubstitution is concerned, the 6 position is equivalent to the 2 position and the 5 position is equivalent to the 3 position.

Quick Questions

14 Look at Table 25.4 and answer these questions:
  a Which groups tend to direct substitution to the 2 or 4 position?
  b Which groups tend to direct substitution to the 3 position?
  c Is there any correlation between the *position* to which a group directs substitution and the *rate* at which it causes the ring to substitute?

Key Point

A substituent that is already present on a benzene ring directs further substituents to either:

▶ the 2 and 4 position, or
▶ the 3 position.

Quick Questions

15 Write structural formulae for the main products you would expect from the following substitution reactions. Assume monosubstitution occurs in each case.
  a 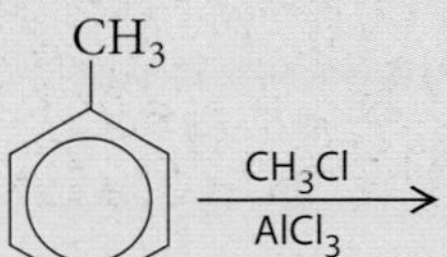
  b 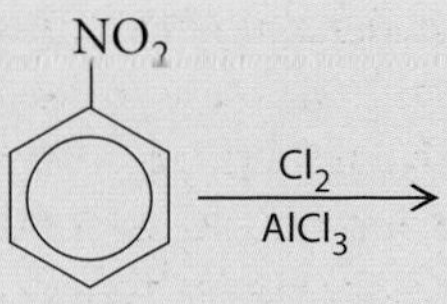
  c

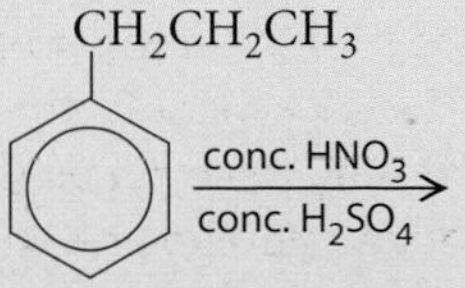

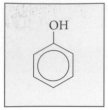

(a) Skeletal formula

(b) Space-filling model

Figure 25.11
Phenol, C_6H_5OH

Key Point

Phenols are a family of compounds with an —OH group attached directly to an arene ring. The simplest member of the family is phenol itself, C_6H_5OH. Phenols have rather different properties from aliphatic alcohols like ethanol.

Table 25.5

The reaction of a small piece of sodium with ethanol and phenol in ethanol

Ethanol	Solution of phenol in ethanol
Sodium sinks, evolves hydrogen steadily	Sodium sinks, evolves hydrogen rapidly

25.10 Phenol – an aromatic hydroxyl compound

Phenol (Figure 25.11) might also be called hydroxybenzene. Like the aliphatic alcohols in Chapter 17, phenol has an —OH group in its molecule, but the benzene ring has a significant effect on the behaviour of the —OH group. So phenol has rather different properties from aliphatic alcohols such as ethanol. In fact, aromatic hydroxyl compounds with the —OH attached directly to the arene ring are treated as a group of compounds in their own right, and are given the family name of **phenols.** So phenol is both a specific compound, C_6H_5OH, and the name given to a family of compounds.

If an —OH group is attached to the alkyl side group of an arene, but not directly attached to the ring, the compound is called an aromatic alcohol. Aromatic alcohols behave more like alcohols than phenols (see Quick question 16).

Quick Questions

16 Which of the following compounds is/are:
   a an alcohol,
   b a phenol,
   c an aromatic alcohol?

| A | B | C | D |

The non-polar benzene ring affects the physical properties of phenol as well as its chemical properties. So, while short-chain alcohols like ethanol and hexanol are liquids, phenol is a crystalline solid, melting point 43 °C. Phenol is only partially soluble in water (9.3 g phenol dissolves in 100 g of water at 20 °C).

Phenol as an acid

Table 25.5 shows how sodium reacts with ethanol and with a solution of phenol in ethanol.

Quick Questions

17 Look at Table 25.5 and answer the following questions:
   a The reaction in each case involves the reduction of H^+ ions by sodium. Write an ionic equation for this reaction.
   b Judging from these experimental results, which contains the higher concentration of H^+ ions, ethanol or phenol?
   c Which is the stronger acid, ethanol or phenol?

The reactions with sodium suggest that phenol is a stronger acid than ethanol. In fact, it is also a stronger acid than water (K_a for phenol = 1.3×10^{-10} mol dm^{-3}; K_a for water = 5.6×10^{-16} mol dm^{-3}; K_a for ethanol = 1×10^{-18} mol dm^{-3} at 25 °C).

If we consider the equilibrium:

$$H_2O \; + \; \text{phenol} \; \rightleftharpoons \; \text{phenoxide ion} \; + \; H_3O^+$$

we can see why phenol is a stronger acid than aliphatic alcohols such as ethanol. In the phenoxide ion, the negative charge on the O atom can be partly delocalised around the ring. This reduces the tendency of the phenoxide ion to attract H^+. In other words it reduces its strength as a base. This increases the strength of its conjugate acid, phenol. So phenol is a stronger acid than aliphatic alcohols, though it is still a weaker acid than carboxylic acids such as ethanoic acid.

Being an acid, phenol reacts with alkalis such as sodium hydroxide. Hydroxide ions from sodium hydroxide remove H^+ (H_3O^+) ions, displacing the above equilibrium to the right and forming a solution of sodium phenoxide, $C_6H_5O^-Na^+$. Because of this reaction, phenol is much more soluble in aqueous sodium hydroxide than it is in water.

$$NaOH + \text{(phenol)} \longrightarrow \text{(sodium phenoxide)} + H_2O$$

sodium phenoxide

Phenol was once known as 'carbolic acid', and it was used as one of the earliest antiseptics. The Edinburgh doctor, Joseph Lister first used it in the 1860s to prevent wounds going septic after surgery. Unintentionally, it had been used as an antiseptic even before this, because the old-fashioned way of treating an amputation wound was to cover it with coal tar, which contains phenol (Figure 25.12). Phenol is effective in killing bacteria, but it is also very corrosive to the skin, so it must be used in low concentrations. Some modern antiseptics still contain phenol or phenol derivatives (Figures 25.13 and 25.14).

The major industrial use of phenol is in making plastics such as nylon (Chapter 27). There are several methods for manufacturing phenol: the most economical is the cumene process (see Review question 7 at the end of this chapter).

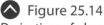
2,4,6-trichlorophenol: 23 times as effective a germicide as phenol in a given concentration

4-chloro-3,5-dimethylphenol: 280 times as effective as phenol

Figure 25.14
Derivatives of phenol used in antiseptics

Esterification of phenol

Like aliphatic alcohols, phenol forms esters. However, unlike alcohols, phenol does not react directly with carboxylic acids. Phenyl esters are normally made by reacting phenol with acyl halides (Section 19.5). For example, phenyl benzoate is made by reacting phenol with benzoyl chloride in the presence of aqueous alkali:

$$\text{(benzoyl chloride)} + \text{(phenol)} \longrightarrow \text{(phenyl benzoate)} + HCl$$

benzoyl chloride   phenol   phenyl benzoate

Key Point

Phenol is a weak acid. It reacts with alkalis to form salts called phenoxides.

Figure 25.12
The antiseptic action of 'Coal Tar Soap' depended on the phenol and related compounds that it contained. Phenol and its derivatives are still used as antiseptics (Figure 25.13).

Figure 25.13
TCP is a common household antiseptic which contains phenol and phenol derivatives.

Quick Questions

18 **a** How would you convert phenol to phenyl ethanoate?
     b Why do you think aqueous alkali is added to the mixture of phenol and acyl halide in the preparation of phenyl esters?

25.11 Substitution reactions of the aromatic ring in phenol

When aqueous bromine is added to a solution of phenol, the bromine is immediately decolorised and a white precipitate is formed. This is a substitution reaction and the white precipitate is 2, 4, 6-tribromophenol.

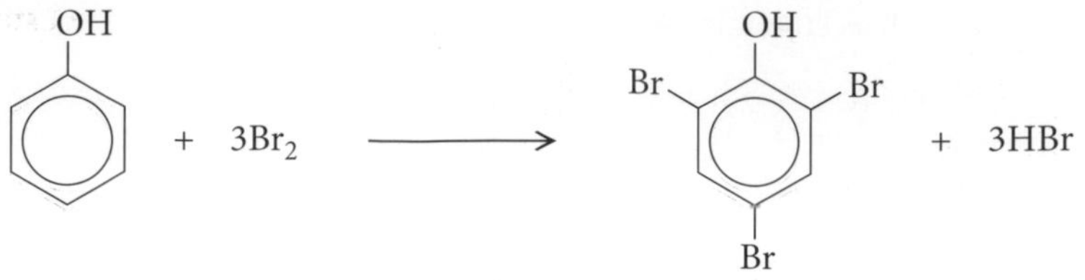

2,4,6-tribromophenol

Benzene does not react with bromine except in the presence of a halogen carrier catalyst (Section 25. 6). In general, phenol is more readily attacked than benzene by electrophiles. This is because the lone pairs of electrons on the oxygen of the —OH group in phenol become partially delocalised around the ring. This increases the electron density on the ring and makes it more attractive to electrophiles. The 2, 4 and 6 positions on the ring are preferentially substituted (Section 25.9).

Key Point

The aromatic ring in phenol is readily substituted by electrophiles. The 2, 4 and 6 positions are preferentially substituted.

Phenol can also be nitrated far more readily than benzene. Dilute nitric acid alone is sufficient, and the products are 2- and 4-nitrophenol. If you use a 'nitrating mixture' of concentrated nitric and sulfuric acids, 2,4,6-trinitrophenol, commonly known as picric acid, is formed. Picric acid, like other highly nitrated molecules, is a high explosive.

Testing for phenols

Phenols, unlike alcohols, give a violet colour when added to neutral iron(III) chloride solution. The colour is probably due to a complex with Fe^{3+}, with phenol as the ligand.

Figure 25.15 summarises some of the important reactions of phenol.

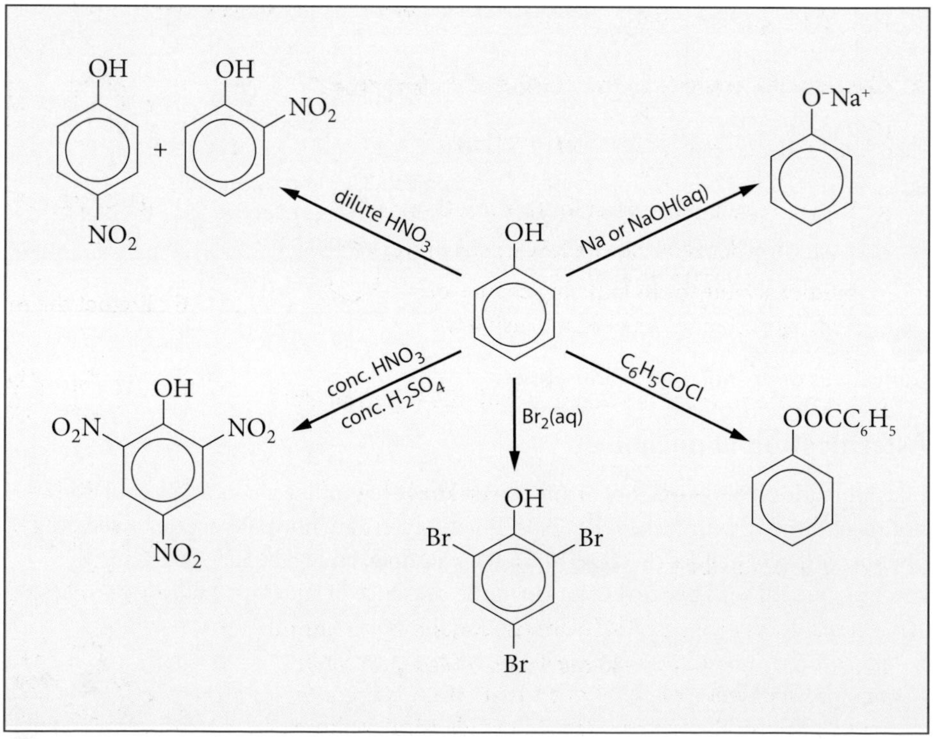

Figure 25.15
Some important reactions of phenol

Review questions

1 Name the following benzene derivatives:

a (benzene ring with Br at position 1 and Br at position 3)

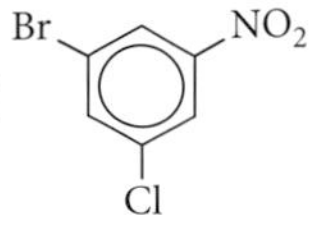

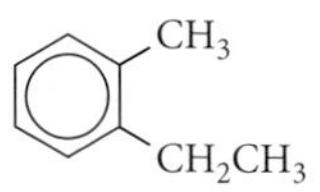

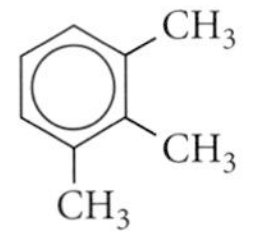

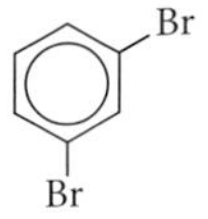

d (benzene ring with Br, NO₂, and Cl)

b (benzene ring with CH₃ and CH₂CH₃)

e (benzene ring with COOH and NO₂)

c (benzene ring with three CH₃ groups)

f (benzene ring with OH and Cl)

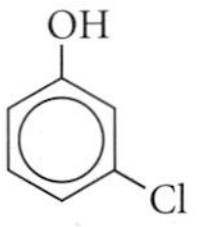

2 Write structural formulae for the following benzene derivatives:

a 2,4-dinitrophenol,

b 1,4-dichlorobenzene,

c 4-nitrophenylamine,

d 2-hydroxybenzoic acid,

e 2-chlorophenylamine.

3 Consider the catalytic hydrogenation of cyclohexene:

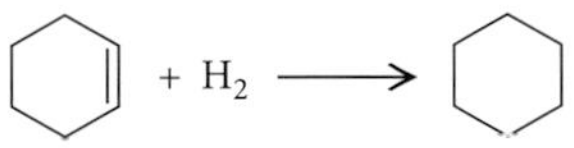

⬡ + H₂ ⟶ ⬡ $\Delta H = -120\,\text{kJ}\,\text{mol}^{-1}$

a Suggest a suitable catalyst for this reaction.

b Assuming benzene has the Kekulé structure,  predict a value for its enthalpy change of hydrogenation to cyclohexane, using the data above.

c Compare your answer in part b with the experimental value:

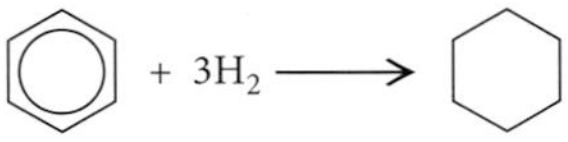

⬡ + 3H₂ ⟶ ⬡ $\Delta H = -208\,\text{kJ}\,\text{mol}^{-1}$

d Suggest why the two values differ.

4 Deuterium is an isotope of hydrogen containing a neutron in its nucleus in addition to the single proton present in normal hydrogen. The symbol for deuterium is $_{1}^{2}\text{H}$ or D, and its chemical properties are almost identical to those of normal hydrogen, $_{1}^{1}\text{H}$.

If deuterium chloride, DCl, is dissolved in methylbenzene, no reaction occurs. However, if anhydrous aluminium chloride is added to the solution, hydrogen atoms on the aromatic ring are rapidly substituted by deuterium atoms. If excess DCl is present, all five ring hydrogens are substituted, forming:

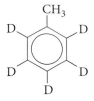

a Write the structure of the complex formed between AlCl₃ and DCl.

b What is the effect of the AlCl₃ on the polarisation of the DCl molecule?

c Why is AlCl₃ effective in causing DCl to substitute H atoms in the aromatic ring?

d Why are none of the hydrogen atoms on the CH₃ side chain substituted, even in the presence of excess DCl?

5 How would you distinguish between the members of the following pairs of compounds, using simple chemical tests?

a ⬡(benzene) and ⬡(cyclohexene)

b ⬡(methylbenzene, CH₃) and ⬡(benzene)

c ⬡(benzene) and ⬡(cyclohexane)

d CH₃CH₂CH₂CH₂CH₂CH₂OH and ⬡(phenol, OH)

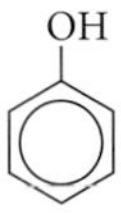

6 Predict the major products of the following reactions.

a (methylbenzene, CH₃) $\xrightarrow[300\,°C]{H_2/Ni}$

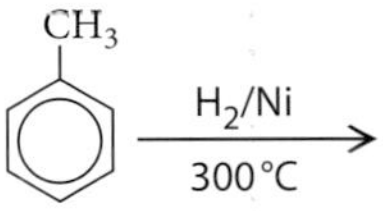

b (phenol, OH) $\xrightarrow{Cl_2(aq)}$

c (dimethylbenzene, CH₃ and CH₃) $\xrightarrow[\text{warm}]{\text{alkaline KMnO}_4(aq)}$

d (benzene with NO₂ and NO₂) $\xrightarrow[120\,°C]{\text{conc.H}_2SO_4 \ \text{conc.HNO}_3}$

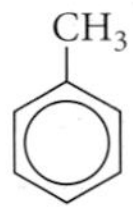

e (benzene) $\xrightarrow[AlCl_3, \text{ warm}]{CH_3CHClCH_3}$ (mono-substituted product only)

f (methylbenzene, CH₃) $\xrightarrow[\text{cold, in dark}]{Br_2/FeBr_3}$

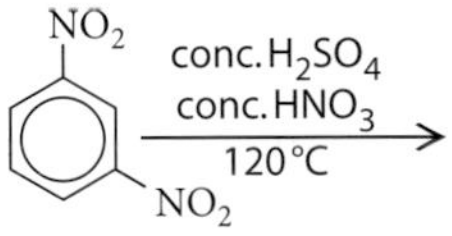

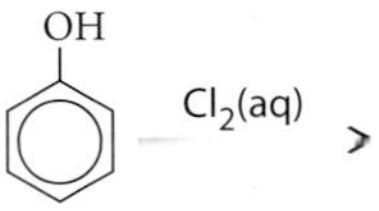

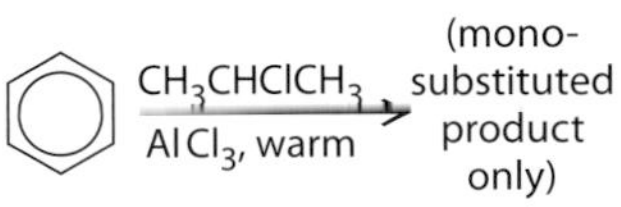

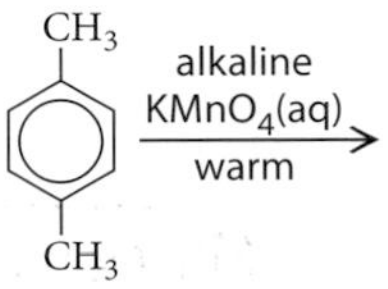

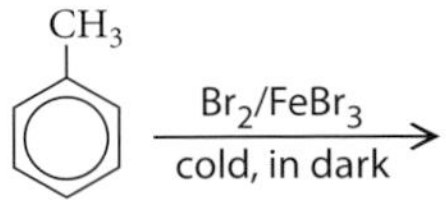

7 Ethyl-1-methylethylbenzene, commonly called cumene, has the structure:

$$CH_3-CH-CH_3$$

It is an important intermediate in a process used to manufacture phenol and propanone (acetone).

Ethyl-1-methylethylbenzene is manufactured by a Friedel–Crafts-type reaction between propene and benzene in the presence of an acid catalyst:

(benzene) + $CH_3-CH=CH_2$ $\xrightarrow{H^+}$ (product with $CH_3-CH-CH_3$ group)

The reaction is believed to proceed via a carbocation intermediate.

a Write the structures of the two possible carbocations formed by the attack of H^+ on the double bond of propene. (Sections 15.4 and 15.5 may help you.)

b Which is the more stable of these two carbocations?

c Show how the carbocation you identified in b can attack and substitute the benzene ring. Explain why ethyl-1-methylethylbenzene is virtually the only product of this reaction, hardly any propylbenzene being produced.

Treatment of the product with air followed by dilute acid gives propanone and phenol:

(cumene) $+ O_2 \longrightarrow$ (phenol, OH) $+ CH_3-C-CH_3$ (with =O)

d Suggest a reason why this is the most economic of the several methods available for manufacturing phenol.

8 a Write structural formulae for all the compounds of molecular formula C_8H_{10} containing one benzene ring.

b For each of these compounds, write the formulae of all the possible mononitration products (not just the ones you would expect as the major product).

c For one of the compounds, state which of the mononitration products you would expect to be produced in the majority.

9 Consider the following compounds:

i (methylbenzene, CH_3)   **ii** (benzene)

iii (cyclohexane)   **iv** (cyclohexene)

v $CH_3CH_2CH=CHCH_2CH_3$

a Which are aromatic hydrocarbons?

b Which are cyclic compounds?

c Which are unsaturated hydrocarbons?

d Which have a planar ring in their molecule?

e Which would decolorise aqueous bromine in the dark?

f Which would evolve fumes of HBr when treated with bromine and iron filings in the dark?

g Which would react with alkaline potassium manganate(VII) solution?

10 Consider two possible reactions of chlorine with benzene in the gas phase:

A (benzene)(g) $+ 2Cl_2$(g) $\longrightarrow$ (1,2-dichlorobenzene)(g) $+ 2HCl$(g)

B (benzene)(g) $+ Cl_2$(g) $\longrightarrow$ (dichloro addition product)(g)

Some relevant bond energies (in kJ mol^{-1}) are: Cl—Cl 242; C—Cl (general value) 339; C—H (in benzene) 430; H—Cl 431.

a Which of the two reactions is addition and which is substitution?

b Calculate the energy change in reaction **A** by the following stages.

  i How much energy is needed to split two moles of chlorine molecules into atoms?

  ii How much energy is needed to break two moles of C—H bonds in benzene?

  iii How much energy is released when two moles of new C—Cl bonds form?

  iv How much energy is released when two moles of HCl molecules are formed?

  v What is the total energy change for reaction **A**?

c Calculate the energy change in reaction **B** by the following stages, assuming that the energy needed to break one C—C bond in benzene, i.e. convert

(benzene) to (cyclohexadiene-type) is 434 kJ mol^{-1}

  i How much energy is needed to split one mole of chlorine molecules into atoms?

  ii How much energy is released when two moles of C—Cl bonds form?

  iii What is the total energy change in reaction **B**?

d Which reaction is more likely to occur between chlorine and benzene, **A** or **B**? Give reasons for your answer.

26 Organic nitrogen compounds

26.1 Dyes and the development of the organic chemical industry

The worldwide chemical industry manufactures the basic chemicals that are used to make the materials we need: fertilisers, plastics, medicines, detergents and many other essential materials.

Inorganic chemicals, such as sodium hydroxide and sulfuric acid, are often needed in large quantities and have been manufactured for hundreds of years. Developments in the inorganic chemical industry are often concerned with finding 'greener' manufacturing processes and better catalysts which create less pollution.

The organic chemical industry is rather different. The number of organic compounds, and the number of routes by which they can be made, is very large. As a result, the organic chemical industry is very dynamic, constantly discovering new compounds for a particular job, and new ways of making them. Pharmaceuticals and polymers are just two areas of the organic chemical industry that show this rapid evolution. With new compounds and new processes to be investigated, research chemists are closely involved.

How it all began

The organic chemical industry really started to develop in the second half of the nineteenth century with the discovery of synthetic dyes. Until the 1850s, dyes were produced from living materials, usually plants. They were expensive, limited in colour range and tended to fade. Most people's clothes were drab and dull in colour.

In 1856 in London, William Perkin, an 18-year-old student of the German chemist August Hofmann, was trying to prepare the drug quinine from aniline. Although he failed, he managed, quite by accident, to produce a brilliant purple dye that was named *mauveine*. This became particularly popular when Queen Victoria wore a silk gown dyed with mauveine at the 1862 Royal Exhibition. Mauveine was the first synthetic dye and the first of many produced from aniline, or phenylamine as it is now called. The presence of an $-NH_2$ group attached to a benzene ring makes phenylamine particularly suited to the preparation of coloured materials, particularly azo dyes (Section 26.7).

The synthetic dye industry developed quickly in Britain, and even more quickly in Germany. Brightly coloured clothing became affordable to ordinary people. The experience gained by the German chemical industry in the development of dyes enabled its industrial chemists to discover ways of synthesising new organic chemicals in other industrial sectors, such as pharmaceuticals (aspirin was first used in Germany in 1898), synthetic polymers and photography.

Today, dyestuffs are a small but important part of the worldwide organic chemical industry whose research chemists are constantly finding new and valuable compounds to improve the quality of our lives.

Figure 26.1
Sir William Perkin discovered mauveine, the first synthetic dye, in 1856 when he was 18.

Definitions

The functional groups defining the key families of nitrogen compounds are listed below.

Family	Functional group
Amines	$-NH_2$
Amides	$-CNH_2$ $\parallel$ O
Nitro-compounds	$-NO_2$
Nitriles	$-CN$
Amino acids	$-CHCOOH$ $\mid$ NH_2

26.2 Important organic nitrogen compounds

In this chapter we shall be mainly concerned with compounds containing the $-NH_2$ group, but we will briefly meet other nitrogen compounds.

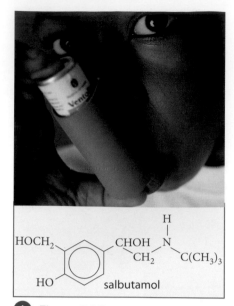

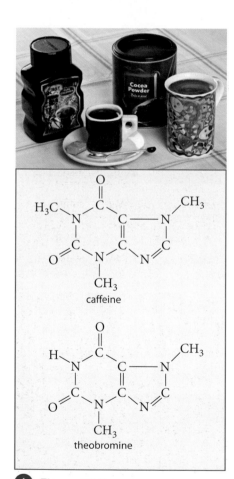

HOCH₂—[benzene ring]—CHOH—CH₂—N(H)—C(CH₃)₃

HO— salbutamol

The structure shown:

$$HOCH_2 \text{—} \bigcirc \text{—} CHOH\text{—}CH_2\text{—}NH\text{—}C(CH_3)_3$$
$$HO\text{—}$$

Figure 26.2
Salbutamol is a very effective drug for treating asthma attacks. (See Review question 10.)

Figure 26.3
Caffeine is a stimulant found in coffee; theobromine is the stimulant in chocolate. Note the similarity of the two molecules – can you spot the difference?

Amines

Amines can be considered as compounds formed by substituting the hydrogen atoms in ammonia, NH_3, with alkyl or aryl groups. If one of the hydrogen atoms in NH_3 is substituted, we get a compound of the form RNH_2, called a **primary amine**. Such compounds are named using the suffix **-ylamine** after a prefix indicating the number of carbon atoms in the molecule. Thus CH_3NH_2 is methylamine.

Substitution of two of the hydrogens of NH_3 gives compounds of the form R^1R^2NH, called **secondary amines**, named in a similar way to primary amines. Thus $(CH_3)_2NH$ is dimethylamine. You should be able to work out the general structure and nomenclature of **tertiary amines**. **Quaternary ammonium salts** have four alkyl groups substituted on a nitrogen atom. This gives the nitrogen a positive charge, which is balanced by a negative ion such as Cl^-. Thus, $(CH_3)_4 N^+Br^-$ is tetramethylammonium bromide. Table 26.1 gives some examples of amines and other nitrogen compounds together with their systematic and common names.

The prefix **amino-** can be used to indicate the presence of an —NH_2 group in molecules containing more than one functional group, e.g. aminoethanoic acid.

Table 26.1
Some important organic nitrogen compounds

Formula	Type of compound	Name	Other name
$CH_3CH_2NH_2$	primary amine	ethylamine	aminoethane
[benzene ring]—NH_2	primary amine	phenylamine	aniline
$(CH_3)_3N$	tertiary amine	trimethylamine	—
CH_3CONH_2	amide	ethanamide	acetamide
[benzene ring]—NO_2	nitro-compound	nitrobenzene	—
CH_3CN	nitrile	ethanenitrile	acetonitrile
$CH_3CHCOOH$ $\mid$ NH_2	amino acid	2-aminopropanoic acid	alanine

Amides

Not to be confused with amines, **amides** have the general structure $R\text{—}\underset{\underset{O}{\|}}{C}\text{—}NH_2$.

The presence of a —C=O group 'next door' has a big effect on the properties of an —NH_2 group, so amides behave very differently from amines (Section 26.8). Amides are named using the suffix **-amide** after a stem indicating the number of carbon atoms in the molecule, including that in the —C=O group. Thus CH_3CONH_2 is ethanamide, *not* methanamide.

Nitro-compounds

Compounds containing the —NO_2 group are called **nitro-compounds**. Aromatic nitro compounds are particularly important. They are named using the prefix **nitro-**. Thus $C_6H_5NO_2$ is called nitrobenzene.

Nitriles (cyano-compounds)

Compounds containing the cyano group $-C \equiv N$ are called **nitriles** or **cyano-compounds**. They are named using the suffix **-nitrile** after a stem indicating the number of carbon atoms, including that in the $-CN$ group. Thus CH_3CN is ethanenitrile.

Amino acids

As their name implies, **amino acids** contain both the amino group, $-NH_2$ and the carboxylic acid group, $-COOH$. By far the most important are those in which the $-NH_2$ and the $-COOH$ are both attached to the same carbon, thus:

$$R-CH-COOH$$
$$|$$
$$NH_2$$

Most of the amino acids found in nature are of this form. They are called 2-amino acids or α-amino acids. Amino acids are named as amino derivatives of carboxylic acids. Thus

$$CH_3CHCOOH$$
$$|$$
$$NH_2$$

is called 2-aminopropanoic acid. Since many natural amino acids have quite complex structures, they are often referred to by common names, such as alanine in this case. There is more about amino acids in Chapter 28.

Quick Questions

1 Name the following compounds:
 a $CH_3CH_2CH_2CONH_2$
 b CH_3CH_2CN
 c $(CH_3CH_2)_2NH$
2 Give the formulae of the following compounds:
 a 2-nitrophenylamine
 b aminoethanoic acid
 c propylamine
 d propanamide

Key Point

Organic nitrogen compounds are named as follows:

Family	Named using	
Amines	**-ylamine** amino-	suffix *or* prefix
Amides	**-amide**	suffix
Nitro-compounds	**nitro-**	prefix
Nitriles	**-nitrile**	suffix

26.3 The nature and occurrence of amines

The $-NH_2$ group is very widespread in biological molecules, especially proteins. Normally the group is associated with other functional groups. Free amines are relatively rare in nature, though they do occur in decomposing protein materials (Figure 26.4). Amines are formed by the action of bacteria on amino acids. For example, $NH_2(CH_2)_4NH_2$ and $NH_2(CH_2)_5NH_2$ are found in decaying animal flesh, as their common names, putrescine and cadaverine, suggest. Dimethylamine and trimethylamine are found in rotting fish and are partly responsible for its peculiar smell.

The physical and chemical properties of the early members of the amine series resemble those of ammonia. Their smell is very similar to that of ammonia, though with a slightly fishy character. Like ammonia, the early amines are gaseous and very soluble in water. Their high solubility is due to hydrogen bonding between the $-NH_2$ group and water molecules. As the R group in RNH_2 gets larger, amines get less like ammonia.

 Figure 26.4
The peculiar smell of fish is mainly due to amines.

Table 26.2 gives some physical properties of ammonia and some amines.

 Table 26.2
The physical properties of ammonia and some amines

Name	Formula	State at 25 °C	Boiling point/°C
ammonia	NH_3	g	−33
methylamine	CH_3NH_2	g	−6
ethylamine	$CH_3CH_2NH_2$	g	17
propylamine	$CH_3CH_2CH_2NH_2$	l	49
butylamine	$CH_3CH_2CH_2CH_2NH_2$	l	78
phenylamine	$C_6H_5NH_2$	l	184
dimethylamine	$(CH_3)_2NH$	g	7
trimethylamine	$(CH_3)_3N$	g	3
triethylamine	$(CH_3CH_2)_3N$	l	90

Phenylamine, with its large hydrocarbon portion, is only sparingly soluble in water, but it dissolves well in organic solvents. Its ability to dissolve in fats means that it is readily absorbed through the skin. It is also toxic, so great care must be taken when using phenylamine.

26.4 Making amines

There are many ways of making amines. Three of the most important methods are listed here.

By reduction of nitriles

Primary amines can be produced by the reduction of nitriles using powerful reducing agents such as lithium tetrahydridoaluminate(III) (LiAlH$_4$). For example:

$$CH_3 - C \equiv N \xrightarrow[\text{dry ether}]{\text{LiAlH}_4} CH_3CH_2NH_2$$
$$\text{ethylamine}$$

From ammonia and halogenoalkanes

Halogenoalkanes undergo substitution reactions with ammonia to form a mixture of primary, secondary and tertiary amines plus a quaternary ammonium salt (Section 16.5):

$$RCl + NH_3 \longrightarrow RNH_2 + HCl$$
$$RCl + RNH_2 \longrightarrow R_2NH + HCl$$
$$RCl + R_2NH \longrightarrow R_3N + HCl$$
$$RCl + R_3N \longrightarrow R_4N^+Cl^-$$

In practice the products are the salts of the amines, formed by combination of the HCl with the free amines (Section 26.5). The problem with this method is that it produces a mixture of products which must be separated by distillation. Nevertheless, the method is widely used both in the laboratory and industrially, because the starting materials are readily available.

By reduction of nitro-compounds

This method is used only for aromatic amines, especially phenylamine.

Quick Questions

3 Suggest a reason why trimethylamine has a lower boiling point than dimethylamine, even though its relative molecular mass is higher.

Key Point

The physical and chemical properties of the early members of the amine family resemble those of ammonia.

Key Point

Aliphatic amines can be made:

▶ by reducing nitriles,
▶ by reacting halogenoalkanes with ammonia.

Aromatic amines can be made:

▶ by reducing nitro-compounds.

Aromatic nitro-compounds are readily prepared by the nitration of aromatic hydrocarbons (Section 25.5). Reduction of nitro-compounds in acid solution produces amines:

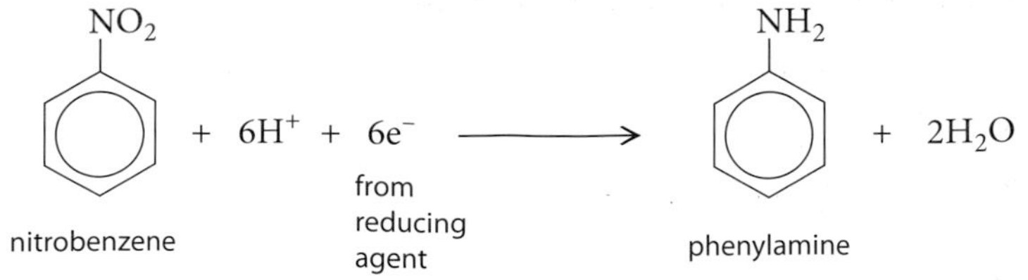

NO$_2$ (nitrobenzene) + 6H$^+$ + 6e$^-$ (from reducing agent) ⟶ NH$_2$ (phenylamine) + 2H$_2$O

In the laboratory the reducing agent used can be lithium tetrahydridoaluminate(III), LiAlH$_4$ or tin in concentrated hydrochloric acid:

$$Sn \longrightarrow Sn^{4+} + 4e^-$$

Overall:

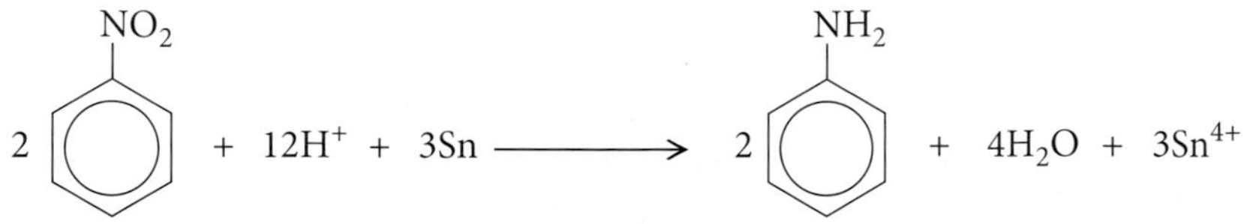

2 NO$_2$ + 12H$^+$ + 3Sn ⟶ 2 NH$_2$ + 4H$_2$O + 3Sn^{4+}

When this method is used industrially to manufacture phenylamine, the reducing agent is iron in concentrated hydrochloric acid.

26.5 Amines as bases

Figure 26.5 illustrates an experiment in which acid is added to a solution of ethylamine.

Quick Questions

4 How would you prepare:
  a ethylamine from chloroethane,
  b propylamine from chloroethane (2 stages),
  c 2-methylphenylamine from methylbenzene (2 stages)?

Quick Questions

5 Look at Figure 26.5 and answer these questions.
  a What evidence is there for a chemical reaction between ethylamine and hydrochloric acid?
  b Why does the smell of ethylamine disappear when hydrochloric acid is added?
  c Why does the smell reappear when sodium hydroxide is added?

⌄ Figure 26.5
The effect of acid on aqueous ethylamine

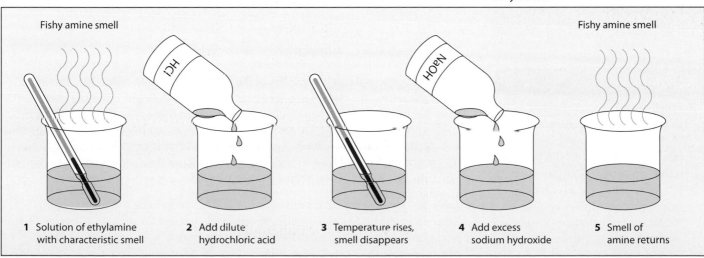

1 Solution of ethylamine with characteristic smell
2 Add dilute hydrochloric acid
3 Temperature rises, smell disappears
4 Add excess sodium hydroxide
5 Smell of amine returns

Key Points

Primary amines react with acids to form salts with the general formula $RNH_3^+X^-$ (from the acid HX).

The reaction is reversed by adding an alkali.

Similar reactions occur with secondary and tertiary amines.

Like ammonia and all primary amines, ethylamine carries a lone pair of electrons on its nitrogen atom. This enables it to bond to a hydrogen ion:

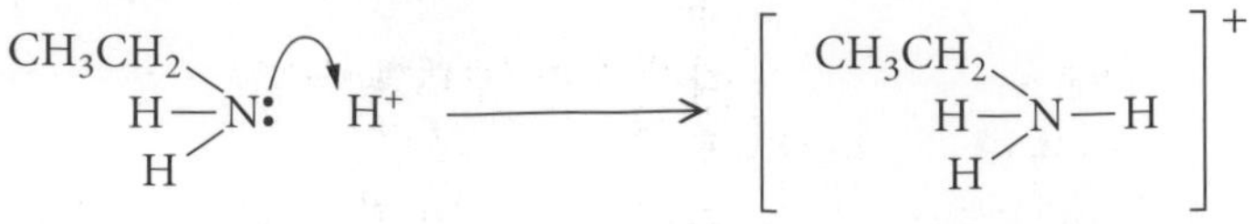

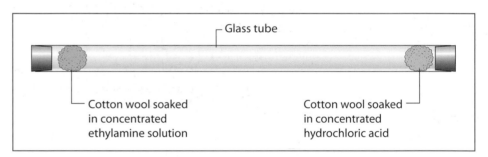

ethylammonium ion

Ethylamine is therefore a *base*, like ammonia.

When an acid is added to a solution of ethylamine, a salt is formed – in this case the salt is **ethylammonium chloride**, $CH_3CH_2NH_3^+Cl^-$. Like all salts, it is involatile and therefore has no smell. The reaction is reversible, and when a strong alkali such as sodium hydroxide is added to this salt, $H^+(aq)$ ions are removed from it. This reforms the free amine.

$$CH_3CH_2NH_3^+(aq) + OH(aq) \longrightarrow CH_3CH_2NH_2(aq) + H_2O(l)$$

ethylammonium ion   ethylamine

Compare these two reactions with the corresponding reactions of ammonia:

$$NH_3(aq) + H^+(aq) \longrightarrow NH_4^+(aq)$$

$$NH_4^+(aq) + {}^-OH(aq) \longrightarrow NH_3(aq) + H_2O(l)$$

Salts of the early members of the amine series, such as ethylammonium chloride, are soluble white crystalline solids, similar to ammonium compounds.

Quick Questions

6 Figure 26.6 shows another experiment that can be carried out with ethylamine and HCl.
  a What would you expect to happen?
  b Why?
7 By considering the electron-pushing effect of alkyl groups and the electron-withdrawing effect of the $-C{=}O$ group, try to arrange the following in order of basic strength, starting with the strongest:
  a NH_3
  b CH_3CNH_2
  $\quad\quad \overset{\|}{O}$
  c $(CH_3CH_2)_2NH$
  d $CH_3CH_2NH_2$

```
┌──────────────────────────── Glass tube
│
│   ▒▒                                                    ▒▒
│
└── Cotton wool soaked          Cotton wool soaked ──
    in concentrated             in concentrated
    ethylamine solution         hydrochloric acid
```

Is ethylamine a stronger or weaker base than ammonia?

To answer this question, look at the following equilibria:

$$CH_3CH_2NH_2(aq) + H^+(aq) \rightleftharpoons CH_3CH_2NH_3^+(aq)$$

$$NH_3(aq) + H^+(aq) \rightleftharpoons NH_4^+(aq)$$

The stronger the base, the further the equilibrium is to the right. The position of equilibrium depends on the stability of the cation on the right-hand side. The more stable the cation, the further the equilibrium will be to the right – and the stronger the base. Remember that alkyl groups have a positive inductive effect (Section 15.5): the CH_3CH_2- group in $CH_3CH_2-NH_3^+$ pushes electrons onto the $-NH_3^+$ group. This reduces the positive charge and stabilises the ion, so $CH_3CH_2NH_2$ is a stronger base than NH_3. We can use similar ideas to predict the basic strength of other compounds containing the $-NH_2$ group (Quick question 7).

Table 26.3 gives K_b values for different compounds containing the $-NH_2$ group. Remember, the higher the K_b value, the stronger the base (Section 21.10). Note that phenylamine is a much weaker base than aliphatic amines. This is because the lone pair of electrons on the nitrogen atom is partially delocalised around the benzene ring, so it is less available to bond to an H^+ ion.

Key Points

▶ Ethylamine is a stronger base than ammonia. This is because the CH_3CH_2- group pushes electrons onto the $-NH_3^+$ group, stablising the $CH_3CH_2NH_3^+$ ion.
▶ Phenylamine is a weaker base than ammonia. This is because the lone pair of electrons on the N atom is partially delocalised around the benzene ring.

Table 26.3
K_b values of ammonia, amines and ethanamide

Formula	Name	K_b at 25 °C/mol dm^{-3}
NH_3	ammonia	1.8×10^{-5}
$CH_3CH_2NH_2$	ethylamine	5.4×10^{-4}
$(CH_3CH_2)_2NH$	diethylamine	1.3×10^{-3}
CH_3CONH_2	ethanamide	10^{-15}
$C_6H_5NH_2$	phenylamine	5×10^{-10}

26.6 Other reactions of phenylamine

Reactions of the aromatic ring in phenylamine

The lone pair of electrons on the nitrogen atom in phenylamine tends to get partly delocalised round the ring. We have already seen how this reduces the basic strength of phenylamine relative to aliphatic amines. Another result is that the electron density round the ring in phenylamine is considerably increased. This allows it to undergo electrophilic substitution much more readily than benzene. For example, phenylamine reacts with bromine water even in the absence of a halogen carrier catalyst:

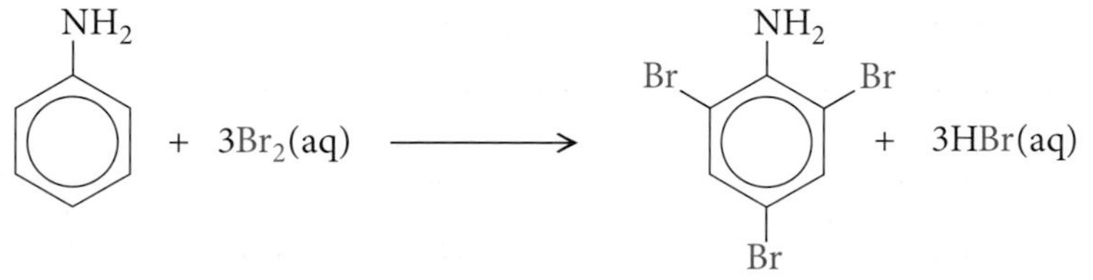

2,4,6-tribromophenylamine

Notice that the 2, 4 and 6 positions are preferentially substituted.

The high electron density in the phenylamine molecule also makes it very easy to oxidise. Pure phenylamine is colourless, but it quickly darkens owing to atmospheric oxidation. Chemical oxidising agents can convert it to a host of different products, including the pigment Aniline Black (Figure 26.7).

Reaction with nitrous acid

Nitrous acid (HNO_2) is rather unstable, so when it is used as a chemical reagent it is usually generated on the spot from sodium nitrite and hydrochloric acid.

When phenylamine is reacted with nitrous acid at low temperatures (below 10 °C), a clear solution is formed. When this solution is warmed, nitrogen is evolved and phenol is formed.

These observations are explained as follows. When amines react with nitrous acid, **diazonium compounds** are formed:

$$RNH_2 + HNO_2 \longrightarrow R-\overset{+}{N}{\equiv}N + {}^-OH + H_2O$$
diazonium ion

The diazonium group, $-\overset{+}{N}{\equiv}N$, is rather unstable. If it is attached to an alkyl group, it decomposes at once, producing $N_2(g)$. But, if it is attached to an aromatic ring, the ion is stabilised to some extent by the delocalised electrons of the ring. Even so, the benzenediazonium ion decomposes at temperatures above 10 °C, giving phenol and nitrogen:

$$C_6H_5-\overset{+}{N}{\equiv}N(aq) + H_2O(l) \longrightarrow C_6H_5OH(aq) + N_2(g) + H^+(aq)$$
benzenediazonium ion

Aromatic diazonium ions are important in the manufacture of dyes (Section 26.7).

Key Point

Phenylamine undergoes electrophilic substitution of the aromatic ring much more readily than benzene. The 2, 4 and 6 positions are preferentially substituted.

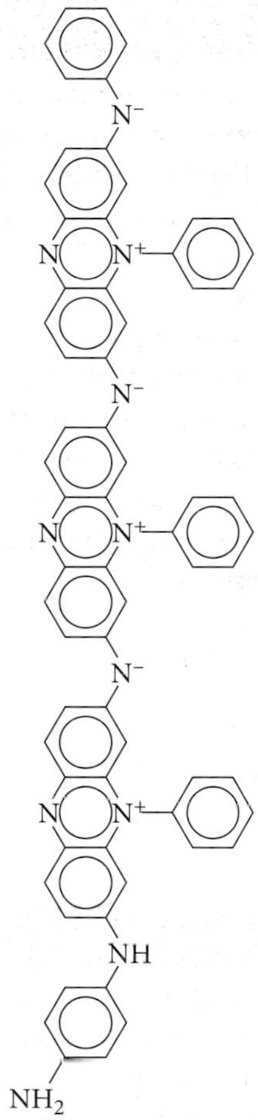

Figure 26.7
The pigment Aniline Black is a mixture of several substances. The structure of one of them is believed to be that shown here (the + and – charges are effectively delocalised over the molecule, which is electrically neutral overall).

8 Predict the structures of the
   products when phenylamine
   reacts with:
   a dilute nitric acid, to give two
      monosubstituted products,
   b concentrated nitric acid, to
      give a tri-substituted product.

▶ Figure 26.8
Some important reactions of
phenylamine

Figure 26.8 summarises some important reactions of phenylamine.

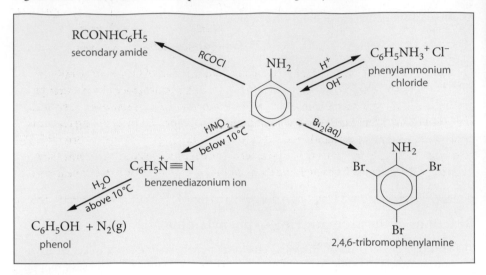

26.7 Diazonium salts

The only stable diazonium salts are aromatic ones (Section 26.6), and even these are
not particularly stable. Benzenediazonium chloride decomposes in aqueous solution
above about 10°C, and the compound is explosive when solid. However, the fact that
it is reactive makes it useful in synthesis.

Benzenediazonium chloride is prepared by adding a cold solution of sodium nitrite to
a solution of phenylamine in concentrated hydrochloric acid below 5°C:

$$C_6H_5NH_2(aq) + HNO_2(aq) + HCl\,(aq) \longrightarrow C_6H_5\overset{+}{N}_2Cl^-(aq) + 2H_2O\,(l)$$

phenylamine   benzenediazonium
chloride

Owing to the explosive nature of the solid, the compound is always used in solution.

The benzenediazonium ion reacts readily with nucleophiles. With a general
nucleophile X^-:

The introduction of the diazonium group is therefore a way of making the aromatic
ring susceptible to *nucleophilic* substitution. (Remember the characteristic reaction
of an aromatic ring is normally *electrophilic* substitution.) This makes diazonium
compounds useful in synthesis. For example:

$$C_6H_5N_2^+ + I^- \longrightarrow C_6H_5I + N_2$$

iodobenzene   (warm benzenediazonium
chloride with KI solution)

$$C_6H_6N_2^+ + H_2O \longrightarrow C_6H_5OH + N_2 + H^+$$

phenol   (warm the aqueous
solution)

$$C_6H_5N_2^+ + Cl^- \longrightarrow C_6H_5Cl + N_2$$

chlorobenzene   (warm benzenediazonium
chloride with CuCl catalyst)

Coupling reactions

The positive charge on the $-\overset{+}{N}\equiv N$ group of the benzenediazonium ion means that
this group is itself a strong electrophile. Thus we might expect it to attack another benzene
ring, particularly one that has an electron-donating group such as —OH attached to it:

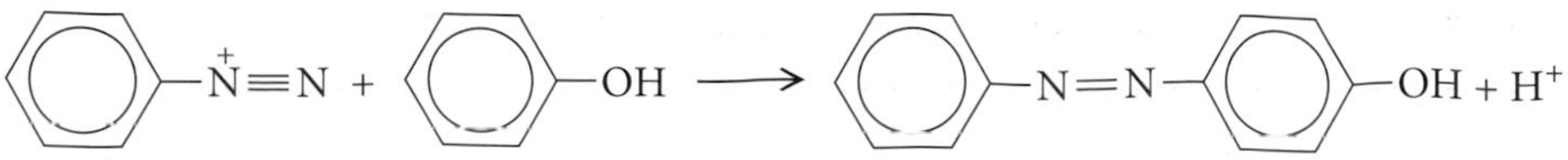
OH
N=N
OH + H+

(4-hydroxyphenyl)azobenzene

This is an example of a **coupling reaction**. If a cold solution of benzenediazonium chloride is added to a cold solution of phenol in sodium hydroxide, a bright orange precipitate is immediately formed. This is (4-hydroxyphenyl)azobenzene – a typical **azo compound**. Many different azo compounds can be formed by coupling reactions between diazonium compounds and activated aromatic rings. They are all brightly coloured. Their colour results from the extensive delocalised electron systems they possess, which extends from one ring through the —N=N— group to the next ring.

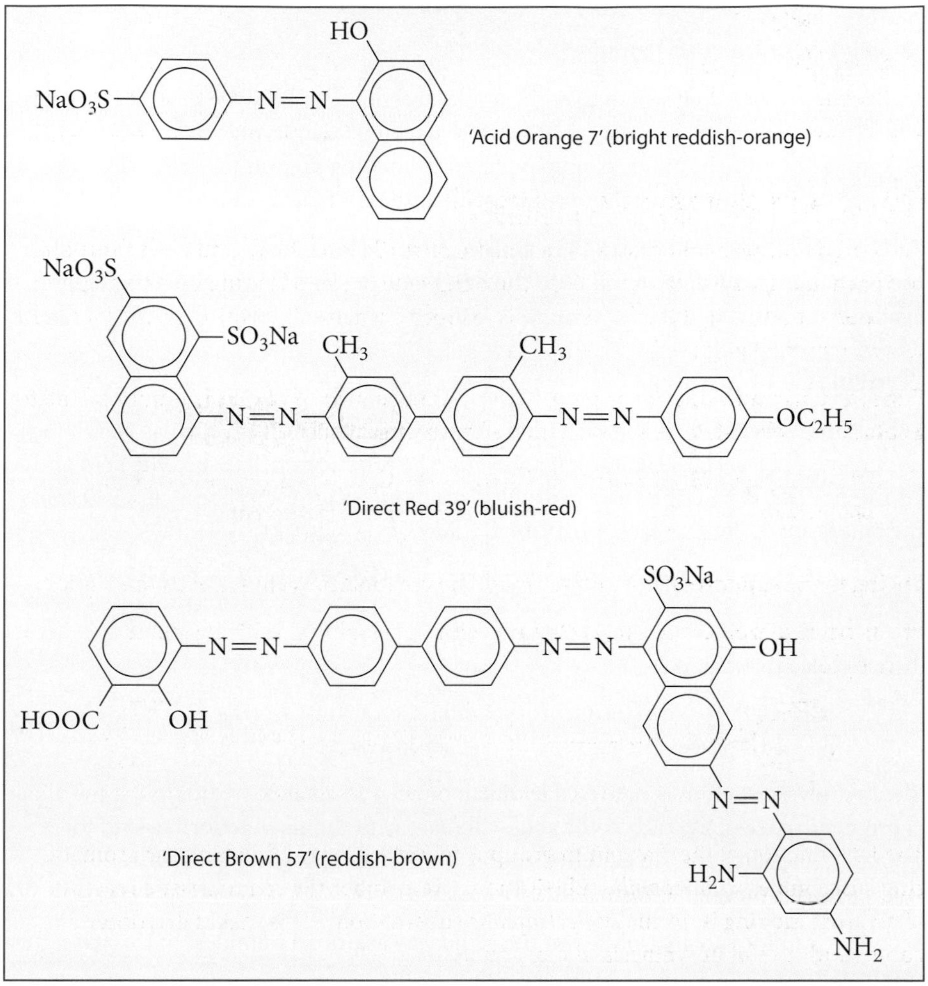
'Acid Orange 7' (bright reddish-orange)
'Direct Red 39' (bluish-red)
'Direct Brown 57' (reddish-brown)

Figure 26.9
Many of the dyes used for clothes and fabrics are azo dyes.

Figure 26.10
Structures of some azo dyes

Coupling reactions are important in the dyestuffs industry for making azo dyes. Many different colours can be obtained by adjusting the structure of the azo compound. These can be quite complex as Figure 26.10 shows.

Unlike diazonium compounds, azo compounds are quite stable. Azo dyes do not fade or lose their colour, unlike many natural dyes.

Quick Questions

9 Write the structures of the products you would expect to be formed at each stage when:

a phenylamine is dissolved in excess concentrated hydrochloric acid,

b sodium nitrite solution is added to the cooled solution from part a,

c the product from part b is added to a fresh solution of phenylamine.

26.8 Amides

Primary amides have the general formula $R{-}\underset{\underset{\displaystyle O}{\|}}{C}{-}NH_2$, often written as $RCONH_2$.

The simplest primary amide is ethanamide, CH_3CONH_2.

Primary amides are made by reacting acyl halides with ammonia (Section 19.5):

$$RCOCl(l) + NH_3(aq) \longrightarrow RCONH_2(aq) + HCl(aq)$$

A similar reaction occurs between acyl halides and primary amines. In this case, a **secondary amide** is formed. For example:

phenylamine   ethanoyl chloride   *N*–phenylethanamide (m.pt. 114 °C)

The general reaction can be written as:

$$RCOCl + R'NH_2 \longrightarrow RCONHR' + HCl$$

(R and R' are alkyl or aryl groups)

Unlike the amines from which they are made, secondary amides are crystalline solids with sharp melting points. They are therefore useful in identifying (*characterising*) unknown amines. The amide is prepared and its melting point is taken. This is then checked against the melting points of known amides in published tables.

Amides are much weaker bases than amines, because the lone pair of electrons on the N atom is partly delocalised onto the neighbouring $C{=}O$ group, making it less available for bonding to H^+.

Hydrolysis of amides

When a primary amide is heated with a solution of aqueous alkali, such as NaOH(aq), or aqueous acid, such as HCl(aq), it splits up. The products are a carboxylic acid and ammonia. This is described as a hydrolysis reaction, because a molecule of H_2O splits the amide apart. For example, with ethanamide:

This can also be written as:

$$CH_3CONH_2(aq) + H_2O(l) \longrightarrow CH_3COOH(aq) + NH_3(aq)$$
ethanamide   ethanoic acid

The hydrolysis reaction is catalysed by alkali or acid. In alkaline hydrolysis, if the alkali is present in excess, the carboxylic acid will react with the alkali to form a salt, for example $CH_3COO^-Na^+$. In acid hydrolysis, if the acid is present in excess, the NH_3 will react with the acid to form a salt, for example $NH_4^+Cl^-$.

Secondary amides are hydrolysed in the same way as primary amides. In this case, the products are a carboxylic acid and a primary amine.

The $-\underset{\underset{O}{\|}}{C}-NH_2$ group is present in all amides. This is also the group that links together amino acids in *proteins* (Chapter 28). In proteins, the amide link is also called a *peptide* link. Like amides, proteins are hydrolysed by heating with aqueous acid or alkali (Section 28.2)

26.9 Amino acids, proteins and polymers

Amino acids contain both the $-NH_2$ group and the $-COOH$ group. They are the building blocks for proteins, which are covered in detail in Chapter 28.

The $-NH_2$ group is important in a number of polymers, including nylon and aramids such as Kevlar. These are covered in detail in Chapter 27.

Key Points

Primary amides have the formula $RCONH_2$.
Secondary amides have the formula RCONHR'.
Primary amides are made by reacting RCOCl with NH_3.
Secondary amides are made by reacting RCOCl with $R'NH_2$.

Amides are hydrolysed (split) by heating with aqueous acid or alkali. With acid, the products are a carboxylic acid and an ammonium salt. With alkali, the products are a carboxylic acid salt and ammonia.

Quick Questions

10 What products would you expect to be formed from the hydrolysis of:
  a $CH_3CH_2CONHCH_3$ in excess HCl(aq),
  b $CH_3CH_2CH_2CONH_2$ in excess NaOH(aq)?

Review questions

1 Consider the following compounds:

A CH_3CHCH_3
       |
       CN

B $CH_3CH_2CH_2NH_2$

C O_2N ⬡ NO_2

D CH_3
      \\
        NH
   /
CH_3CH_2

E $CH_3CH_2CONH_2$

F $(CH_3CH_2CH_2)_3N$

a Which is a primary amine?

b Which is a nitrile?

c Which is an amide?

d Which is a tertiary amine?

e Name each compound.

2 a Place the following in order of increasing basic strength. Give reasons for your answer.

  i ⬡NH_2 with CH_3
  ii ⬡CH_2NH
  iii ⬡$NHCH_3$

b Place the following in order of increasing boiling point. Give reasons for your answer.

  i $CH_3CH_2CH_2NH_2$
  ii $CH_3CH_2CH_2CH_3$
  iii $CH_3CH_2CH_2OH$

3 Draw up a table comparing the main physical and chemical properties of methylamine and ammonia. Do the two compounds in general behave similarly?

4 Consider the following compounds:

A ⬡ with CH_3 and NH_2

B $CH_3CH_2NH_2$

C $(CH_3CH_2)_2NH$

D $H_2N(CH_2)_5NH_2$

a Which would be among the products of the reaction of chloroethane with ammonia?

b Which would be converted to a diazonium compound by the action of nitrous acid below 10 °C?

c Which could be made by reduction of ethanenitrile?

d Which would react with hydrochloric acid in the ratio one mole of the compound to two moles of hydrochloric acid?

e Which is the weakest base?

5 Suggest explanations for the following.

a Ethylamine can be readily made from chloroethane and ammonia, but it is difficult to make phenylamine from chlorobenzene and ammonia.

b Phenylamine is much more soluble in dilute hydrochloric acid than in water.

c When butylamine is added to a solution of copper(II) sulfate, a deep blue colour is formed.

6 In the following, explain how you would convert the first compound to the second. Each conversion may involve one or more steps.

a $CH_2{=}CH_2$ to $H_2NCH_2CH_2NH_2$

b ⬡ to (benzene ring with NH_2, Br, Br, and Br substituents)

c $CH_2{=}CH_2$ to $CH_3CH_2CH_2NH_2$

d $CH_3CH_2NH_2$ to $CH_3CH_2NH-C(=O)-$⬡

e ⬡NO_2 to ⬡OH

7 A compound **X** containing carbon, hydrogen and nitrogen only has relative molecular mass 88. When reacted with nitrous acid, 0.1 g of **X** released 50.9 cm^3 of nitrogen gas, measured at 0 °C and 1 atm pressure (1 mole of a gas occupies 22 400 cm^3).

a How many moles of nitrogen gas are produced by 1 mole of **X** when it reacts with nitrous acid?

b How many —NH_2 groups are there in one molecule of **X**?

c Write three possible structural formulae for **X**.

d What volume of 0.1 mol dm^{-3} hydrochloric acid would be needed to neutralise 50 cm^3 of a 0.2 mol dm^{-3} solution of **X**?

8 Copy out the following reaction sequences, inserting the formulae of the products formed in the blank spaces.

a

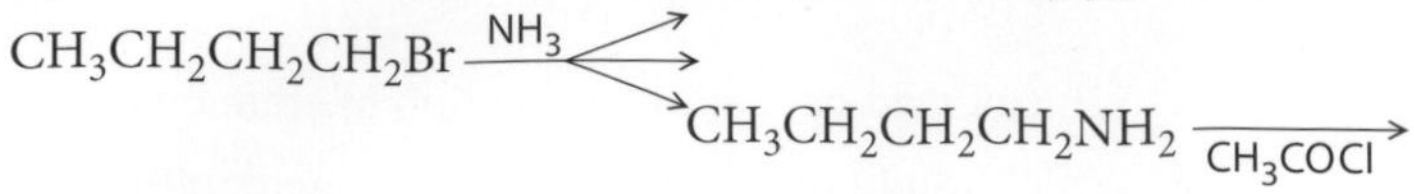

$$CH_3CH_2CH_2CH_2Br \xrightarrow{NH_3} $$

$$CH_3CH_2CH_2CH_2NH_2 \xrightarrow{CH_3COCl}$$

b

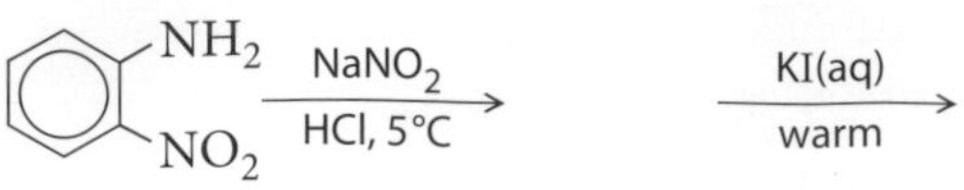

$$\xrightarrow[\text{HCl, 5°C}]{\text{NaNO}_2} \xrightarrow[\text{warm}]{\text{KI(aq)}}$$

c

$$\xrightarrow[\text{HCl, <10°C}]{\text{NaNO}_2} \longrightarrow$$

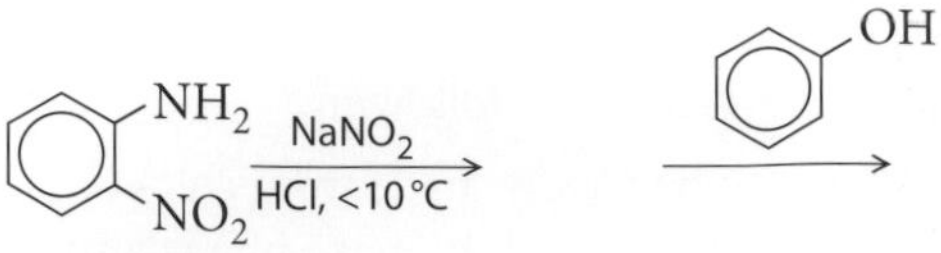

9 A compound **A** of molecular formula C_7H_8 was treated with a mixture of concentrated nitric acid and concentrated sulfuric acid to form two isomeric compounds. One of these isomers was **B**, $C_7H_7O_2N$. **B** could be converted to **C**, C_7H_9N by reduction. **C** reacted with dilute hydrochloric acid to form **D**, $C_7H_{10}NCl$. When cold sodium nitrite solution was added to a cold solution of **D** in hydrochloric acid, a solution of **E**, $C_7H_7N_2Cl$, was produced. When a cold solution of **E** was added to a cold solution of phenol, a brightly coloured substance **F** was produced. When the solution **E** was warmed alone an unreactive gas was evolved and **G**, C_7H_8O, was formed. Give possible structural formulae for **A** to **G**.

10 Salbutamol is a very effective treatment for asthma. It is the active ingredient in asthma inhalers.

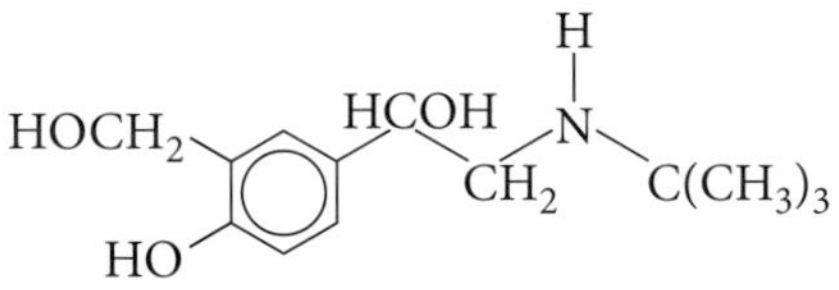

salbutamol

  a Is the amine group in salbutamol primary, secondary or tertiary?

  b Draw the structure of the product of the reaction of salbutamol with:

   i HCl(aq)

   ii CH$_3$COCl

27 Polymers

27.1 The importance of polymers

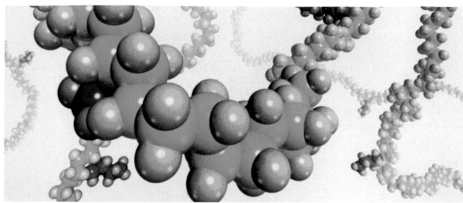

Figure 27.1
A computer-generated image of low-density poly(ethene)

About 80% of the world's output of organic chemicals is used to make polymers. The use of polymers by humans is not new. Wood, cotton, wool and rubber are all naturally occurring polymeric materials, two of which have been used by humans for centuries. However, the manufacture and use of *synthetic* polymers has only really developed since the 1940s, stimulated by the demand for new materials during and after the Second World War. Since then, their use has increased very rapidly and they have steadily replaced traditional, natural materials. Synthetic polymers are often better suited to their particular use, because chemists have been able to produce polymers to suit most purposes.

Polymers are long-chain molecules made by joining together many small molecules, called **monomers,** in a repeating pattern. Some polymers have a single monomer (such as poly(ethene), whose monomer is ethene) while others, such as polyesters, may have two or more monomers. Proteins have over 20 different monomers, called amino acids. Polymers are examples of **macromolecules** – very large molecules. They may be natural or synthetic, and Table 27.1 gives examples of both types.

Table 27.1
Examples of natural and synthetic polymers

	Polymer	**Monomer**	**Where you find it**
Natural	protein	amino acids	wool, silk, muscle, etc.
	starch	glucose	potato, wheat, etc.
	cellulose	glucose	paper, wood, dietary fibre
	DNA	nucleotides	chromosomes, genes
Synthetic	poly(ethene)	ethene	bags, washing-up bowls, etc.
	poly(chloroethene) (PVC)	chloroethene	fabric coatings, electrical insulation, etc.
	poly(phenylethene) (polystyrene)	phenylethene	toys, expanded polystyrene
	polyester	ethane-1,2-diol and benzene-1,4-dicarboxylic acid	skirts, shirts, trousers

Definition

Polymers are macromolecules made by joining together many small **monomers** in a repeating pattern.

Figure 27.2
Polymer chains can be compared to cooked spaghetti

Natural polymers include many materials in common use. Many biological chemicals are polymers, including the proteins in your hair, the DNA in your genes and the starch in your food. The structures of proteins and DNA are studied in Chapter 28.

Polymers vary widely in properties such as tensile strength, flexibility and softening temperature. Some, such as polyester and nylon, are strong and not easily stretched, and are therefore suitable for use as **fibres**. Others, such as polythene and PVC, are more easily shaped and are classified as **plastics**.

27.2 Polymer properties

Have you ever cooked spaghetti? You cook it by boiling in water, then straining it off from the hot water. The cooked spaghetti is a flexible, slithery mass that takes the shape of the bowl you put it in. But when it's cold, the strands stick to each other and it sets into a solid lump (Figure 27.2).

Most polymers behave like this. The long, flexible chains slide over one another when the temperature is high, but they bond to one another when the polymer cools. Polymers that behave in this way are called **thermoplastics**. They are flexible, yet strong, and can be moulded into whatever shape you need. Polythene and PVC are thermoplastics.

Some polymers set into a rigid solid as soon as they are made, and cannot be moulded. They are called **thermosetting** polymers. The hard melamine resin used to make kitchen worktops is a thermosetting ploymer.

Different polymers have different characteristics, depending on the properties of the molecular chains. The five most important properties are listed here:

1 **Chain length** The tensile strength of a polymer increases with chain length. Intermolecular forces attract the chains to each other all along their length, and the longer the chain, the more intermolecular forces there are. Typical polymer properties of flexibility and strength begin to be shown for chains with over 50 monomer units. The strength goes on increasing up to about 500 monomer units, after which it changes very little. Chemists talk of the *average* length of a polymer chain, because individual chain lengths vary, even in a pure sample of the polymer.

2 **Chain flexibility** Poly(ethene) contains alkane chains which are very flexible, with 360 degrees rotation around every carbon–carbon bond. So, poly(ethene) is a flexible polymer. Some polymer chains, such as those in Kevlar (Section 27.4) are much more rigid, and this results in a strong, tough rigid polymer.

3 **Side groups** Different polymers have different side-groups on their chains. For example, poly(chloroethene), commonly called PVC, has Cl atoms as side groups (Figure 27.3). Polar side groups, such as Cl, increase the intermolecular forces between the chains, so they attract each other more strongly than in polythene. As a result, PVC is stronger and harder to melt.

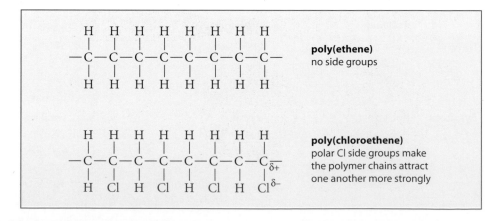

Figure 27.3
Poly(ethene) (polythene) and poly(chloroethene) (PVC)

4 Branching Highly branched polymer chains cannot pack together as tightly and neatly as unbranched chains. Highly branched polymers therefore tend to have lower tensile strength and to melt more easily than unbranched polymers (Figure 27.4(a)) and (b)). This is the reason for the different properties of low-density and high-density polythene (Section 15.6). In most polymers, there are regions in which the chains are neatly and regularly packed. These are called *crystalline* regions, and they make the polymer stronger than in the non-crystalline, or *amorphous* regions. Stretching or *cold-drawing* a polymer makes the polymer chains line up to form more crystalline regions (Figure 27.6), and this is used to turn polymers like nylon into tough fibres.

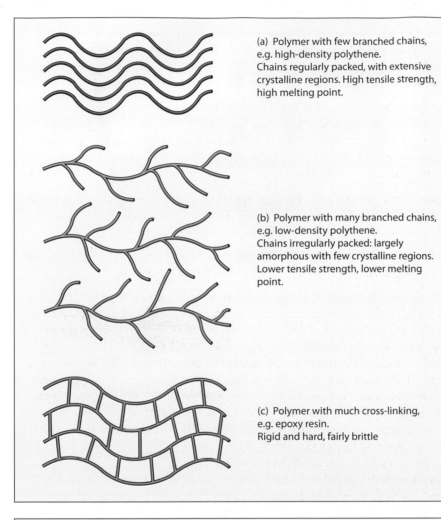

(a) Polymer with few branched chains, e.g. high-density polythene.
Chains regularly packed, with extensive crystalline regions. High tensile strength, high melting point.

(b) Polymer with many branched chains, e.g. low-density polythene.
Chains irregularly packed: largely amorphous with few crystalline regions. Lower tensile strength, lower melting point.

(c) Polymer with much cross-linking, e.g. epoxy resin.
Rigid and hard, fairly brittle

Figure 27.4
Unbranched, branched and cross-linked polymers

Figure 27.5
Spiders' webs and silkworm silk are made from a strong protein called fibroin. Strong intermolecular forces between the side groups make fibroin stronger than steel with the same diameter.

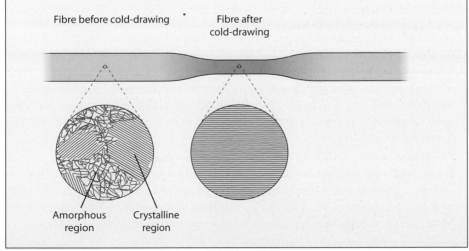

Fibre before cold-drawing   Fibre after cold-drawing

Amorphous region   Crystalline region

Figure 27.6
Cold-drawing increases the crystallinity of a fibre

5 Cross-linking Thermosetting polymers have extensive cross-linking between chains (Figure 27.4(c)). This forms a rigid network, making a hard polymer. The more cross-linking, the more rigid the polymer. When an epoxy resin glue sets, the monomers in the liquid glue form a strong cross-linked network to one another, giving a hard, rigid join between the surfaces that are being glued together.

27.3 Types of polymerisation: addition and condensation

When a polymer forms, the monomer molecules join together. This can happen in two ways: by addition or by condensation.

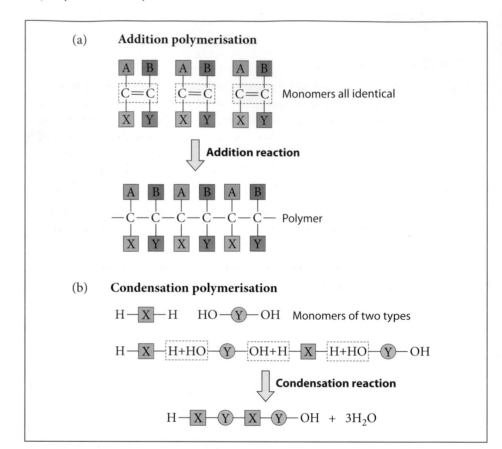

▶ **Figure 27.7**
(a) Addition polymerisation
(b) Condensation polymerisation

▶ **Addition polymerisation** involves addition reactions between monomer molecules. Usually, only a single type of monomer is involved, and the monomers usually contain a double bond. When polymerisation occurs the double bonds break, allowing the monomers to join up (Figure 27.7(a)). Most addition polymers are synthetic: poly(ethene) and poly(chloroethene) are examples (Section 15.6). Rubber is an unusual example of a natural addition polymer (Section 15.7).

▶ **Condensation polymerisation** involves a condensation reaction between two types of monomer. A small molecule, usually H_2O or HCl, is eliminated from between two monomer molecules. In the process, a bond forms between the two monomers. Each monomer has a reactive group at both ends of its molecule, so a chain can form (Figure 27.7(b)).

Condensation polymerisation can sometimes be reversed by hydrolysis. The condensation polymer is heated for a long time with aqueous acid or alkali. This reverses the condensation reaction, putting H_2O molecules back into the chain and re-forming the monomers. Addition polymerisation cannot be reversed in this way. Most natural polymers are condensation polymers, and so are some synthetics. Table 27.2 gives examples of both types.

Note

A condensation reaction is a reaction in which a small molecule, usually H_2O or HCl, is eliminated from between two larger molecules, resulting in a bond between the larger molecules.

Table 27.2
Addition and condensation polymers

Addition polymers	Condensation polymers
Rubber (natural)	Cellulose (natural)
Polythene (synthetic)	Starch (natural)
Polystyrene (synthetic)	Proteins (natural)
PVC (synthetic)	DNA and RNA (natural)
PTFE (synthetic)	Polyester (synthetic)
Acrylic (synthetic)	Nylon (synthetic)

Quick Questions

1 For the polymer named poly(propene):
   a What is the monomer?
   b Is this an addition or a condensation polymer?
   c Draw a section of the polymer chain containing three monomers.
   d What is the repeating unit?

Naming polymers

The systematic names for polymers are formed from the prefix 'poly', followed by the name of the monomer in brackets. For example, the polymer made from ethene is called poly(ethene), though this is often shortened to 'polythene'. Where there are two or more monomers, both names are given in brackets. However, this often leads to long, unwieldy names, so shorter common names are normally used. For example, the commonest polyester is systematically named poly(ethane-1,2-diyl benzene-1,4-dicarboxylate), but it is usually called by its common name, polyethylene terephthalate, shortened to PETE.

Every polymer has a characteristic **repeating unit** which is repeated over and over along the polymer chain. For example, the repeating unit in poly(chloroethene) is $-CH_2CHCl-$ (Figure 27.8).

$$CH_2=CHCl \qquad CH_2=CHCl \qquad CH_2=CHCl$$

$$\downarrow$$

$$-CH_2=CHCl-CH_2=CHCl-CH_2=CHCl-$$

repeating unit

Figure 27.8
The repeating unit in poly(chloroethene)

Predicting polymers

From the structure of a monomer, it is possible to predict the type of polymerisation that will occur. If the monomer has a double bond, addition polymerisation will occur. If there are two monomers, carrying groups that can combine together to eliminate a small molecule such as H_2O or HCl, condensation polymerisation will occur.

Addition polymerisation is covered in detail in Section 15.6.

27.4 Important condensation polymers: polyesters and polyamides

Polyesters

As their name suggests, polyesters are polymers joined by an ester linkage between a carboxylic acid group on one monomer and a hydroxy group on another monomer. One of the monomers has two $-COOH$ groups, and the other has two $-OH$ groups. By far the commonest polyester is poly(ethane-1,2-diylbenzene-1,4-dicarboxylate), also known as 'Terylene' or PETE. It is made by esterifying ethane-1,2-diol with benzene-1,4-dicarboxylic acid (terephthalic acid) as shown in Figure 27.9. Notice how each link in the polyester chain results from the elimination of a molecule of water – a condensation reaction.

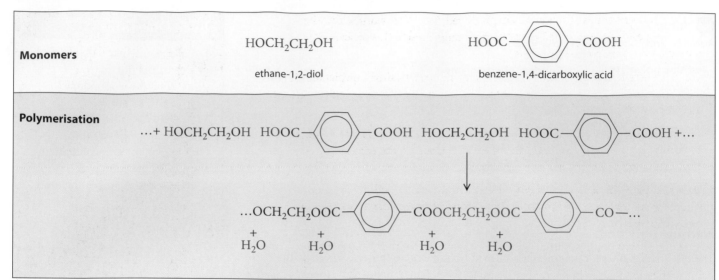

Monomers	HOCH$_2$CH$_2$OH ethane-1,2-diol	HOOC—〈 〉—COOH benzene-1,4-dicarboxylic acid
Polymerisation		

Figure 27.9
The formation of polyester. The repeating unit is shown in green.

Polyesters have high tensile strength, especially when cold-drawn to form fibres. Polyester fibre is often used as a substitute for cotton: it is cheaper than cotton and does not crease so easily. Polyester is also used as a plastic, for example in PETE plastic bottles. Some types of polyester form cross-links, which makes them very hard. The bonding resin in glass fibre reinforced plastic is a cross-linked polyester.

Polyamides: nylons and aramids

Polyamides are polymers joined by an amide linkage, —C—NH— between a carboxylic acid group on one monomer and an amino group on another monomer.

Proteins are natural polyamides. In proteins the monomers are amino acids and the polyamide link in proteins is usually called a **peptide link**. There is more detail about proteins in Section 28.2. The commonest synthetic polyamides are the **nylons**. There are several different types of nylon, formed from different monomers.

Nylon 66 is made by a condensation reaction between 1,6-diaminohexane and hexanedioic acid (Figure 27.10). The product is named nylon 66 because both the monomers contain 6 carbon atoms.

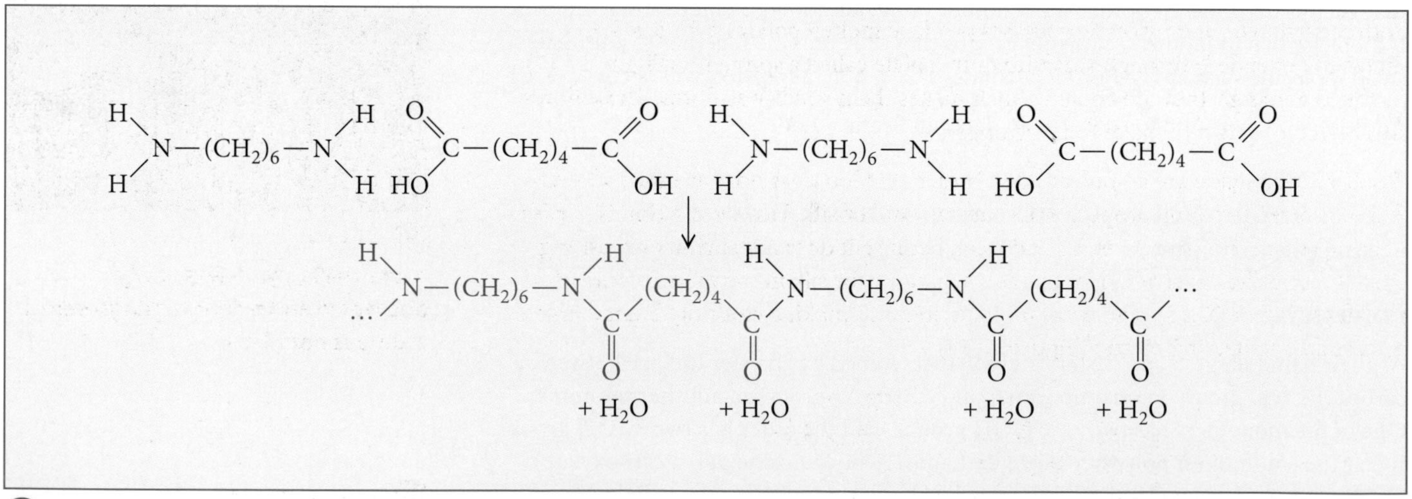

Figure 27.10
The formation of nylon 66. The repeating unit is shown in green.

A condensation reaction occurs when the monomers are heated together in aqueous solution at about 500 K. By controlling the temperature, the pressure and the time of heating, the chain length of the polymer, which determines its strength, can be controlled.

About 3.4 million tonnes of nylon 66 are produced every year worldwide. The 1,6-diaminohexane and hexanedioic acid monomers used in this process are manufactured from benzene or phenol.

Nylon 6 is made by a reaction involving a single monomer called caprolactam (Figure 27.12). The product is named nylon 6 because it is produced from a single monomer containing 6 carbon atoms.

Caprolactam has a ring structure which breaks open when polymerisation occurs. The —NH— group from one ring joins to the —C— group on the next ring.
$$\begin{array}{c} \| \\ O \end{array}$$

Strictly speaking, this is not a condensation reaction because neither H_2O nor any other small molecule is eliminated, but nylon 6 is nevertheless classed as a condensation polymer.

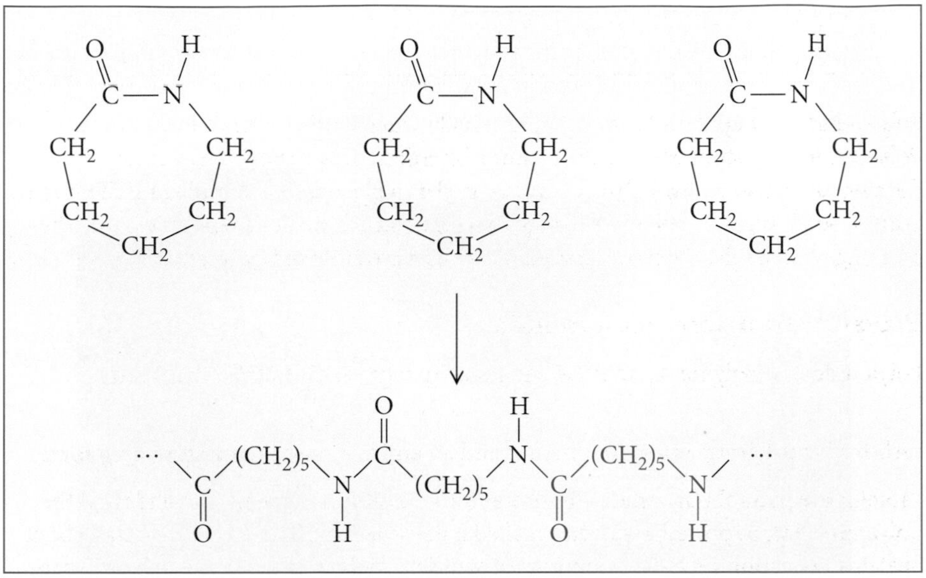

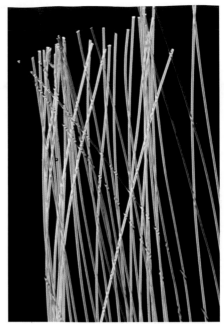

Figure 27.11
Nylon fibres. These fibres have been photographed using polarised light and are magnified 12 times.

Figure 27.12
The formation of nylon 6. The repeating unit is shown in green.

The condensation reaction occurs when caprolactam is heated with water at about 500 K. As with nylon 66, the chain length can be varied by varying the conditions.

About 4.3 million tonnes of nylon 6 are produced every year worldwide. The caprolactam monomer used in this process is manufactured from benzene or phenol.

Both the nylon 66 and nylon 6 processes produce molten polymer. This is then squeezed (extruded) through small holes in a plate called a spinneret (Figure 27.11). As the jets emerge, they are cooled which makes them solidify as fibres. These fibres are then cold-drawn to make them stronger.

Wool, silk and nylon are all polyamide fibres, but nylon does not have the softness and moisture-absorbing properties of natural wool or silk. However, nylon is harder wearing and much easier to wash and wear, because it does not shrink on washing. One of the earliest uses of nylon was as a substitute for silk in the manufacture of stockings (Figure 27.13). The main use of nylon is in making clothing, but it is also used as an engineering plastic (Figure 27.14).

Figure 27.13
In the 1940s, nylon was newly discovered and nylon stockings were a luxurious novelty.

Figure 27.14
Nylon is tough and strong enough to make engineering parts such as skateboard wheels.

3 Nylon 6,10 is used as an engineering plastic. The repeating unit in nylon 6,10 is shown in Figure 27.15(a). Draw the structures of the monomers of nylon 6,10.

4 The monomers of nylon 6,4 are shown in Figure 27.15(b). Draw the structure of the repeating unit in nylon 6,4.

(a) The repeating unit of nylon 6,10

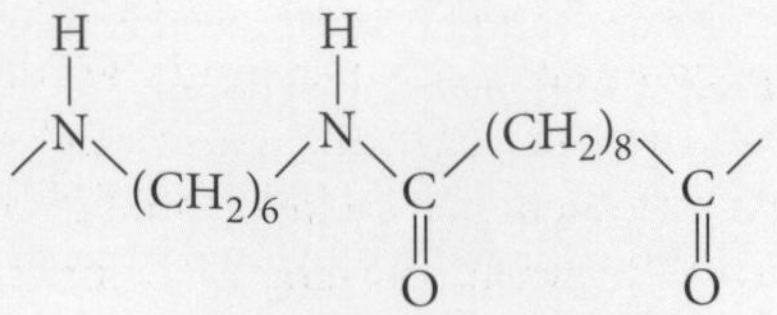

(b) The monomers of nylon 6,4

$$H_2N(CH_2)_6NH_2 \qquad HOOC(CH_2)_2COOH$$

▲ Figure 27.15

Aramids are aromatic polyamides. They are structurally similar to nylon, but they have benzene rings instead of alkane chains between the amide groups. **Kevlar** is an aramid, with the structure shown in Figure 27.16.

The flat, rigid benzene rings make the polymer chains in Kevlar very rigid. Because of this, it is a very strong material; weight for weight five times stronger than steel (Figure 27.17). Kevlar is used in ropes, tyres, sports equipment and bullet-proof vests.

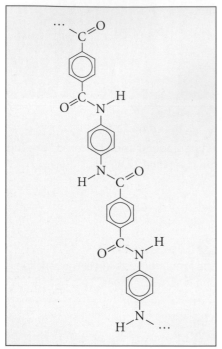

▲ Figure 27.16
The polymer chain in Kevlar

Quick Questions

5 Look at the structure of Kevlar in Figure 27.16.
   a Draw the repeating unit in Kevlar.
   b The monomers of Kevlar are shown in Figure 27.18. Draw a diagram to show the condensation reaction that occurs when Kevlar is manufactured.

▲ Figure 27.17
Kevlar is used to make bullet-proof vests, for dogs as well as humans

▶ Figure 27.18
The monomers of Kevlar

$$H_2N-\langle\!\bigcirc\!\rangle-NH_2 \qquad ClOC-\langle\!\bigcirc\!\rangle-COCl$$

1,4-diaminobenzene   1,4-benzenedicarbonyl chloride

27.5 Recycling polymers

Synthetic polymers like polythene and PVC degrade very slowly, so they create a serious waste disposal problem. Plastic articles like bottles and packaging tend to be bulky so burying them takes a lot of landfill space. Disposing of plastics by burning may produce toxic waste. Burning PVC, for example, can produce hydrogen chloride gas and toxic dioxins.

Recycling polymeric materials makes sense because:

▸ It solves the problem of waste disposal.
▸ It conserves resources by saving the crude oil that is needed to make new polymer.
▸ It can save energy and therefore save carbon emissions.
▸ It can save money.

The easy way to recycle plastics would be to melt down all plastic waste and then allow it to solidify as a mixed polymer. But, when a molten mixture of different polymeric materials solidifies, the chains tend to pack with chains of their own type rather than with those of different types. This means that the different polymers do not mix well together, and the result is a weak and low-grade plastic.

To recycle plastics effectively, they need to be sorted into the different polymer types. This can be difficult, because different plastics often look similar. This is very different to metal recycling, where different metals are often easily identified. It is easy to separate iron from copper, but PVC and polythene are much more difficult to distinguish and separate.

One way round this is to identify the type of polymer used to make a plastic article, by means of a special symbol and number (Figure 27.19). This makes it possible to separate the plastics into similar groups. Once this has been done, the plastic is shredded and can be melted down to make new articles. Sometimes, the polymer is depolymerised to reform the monomer, which is then purified and re-polymerised.

Recycled polymer is never as pure as the new material, so recycled plastics tend to be lower grade than the newly made polymer. Nevertheless, recycled plastics are useful for making all sorts of articles, including plastic bags and plastic dustbins, where the appearance does not matter too much.

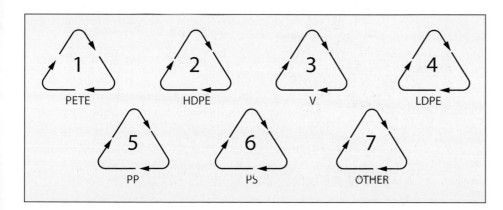

◀ Figure 27.19
These recycling symbols are used on plastic articles to identify the polymer from which they are made, and aid recycling.

Key Point

Polymeric materials present a waste disposal problem because they are bulky to bury and they produce toxic substances when burned. Recycling is desirable, but they need to be separated into different polymer types first.

Quick Questions

6   Look at Figure 27.19. HDPE stands for high-density poly(ethene) and V stands for vinyl chloride. What do you think the following stand for?
   a PETE   **b** LDPE   **c** PP   **d** PS

Review questions

1 Answer the questions that follow for each of the following monomers or pairs of monomers:

a 1,1-dichloroethene, $Cl_2C=CH_2$ (shown as structure with Cl, Cl on one carbon and H, H on the other)

b $H_2N(CH_2)_5NH_2$ and $HOOC(CH_2)_5COOH$,

c 3-aminobenzoic acid, $HOOC-\langle benzene\rangle-NH_2$

d ethene $H_2C=CH_2$ and propene $CH_3CH=CH_2$.

   i Say whether you would expect an addition polymer or a condensation polymer.

   ii Draw the structure of a section of the polymer chain showing three monomers or pairs of monomers.

   iii Draw the structure of the repeating unit.

2 The viscosity of engine oil can be improved by the addition of certain medium chain-length polymers.

A portion of the chain of one such polymer is shown below:

$$-CH_2CH(CH_2CH_2CH_3)CH_2CH(CH_2CH_2CH_3)CH_2-$$

On average, the molecules of the medium-chain polymer contain 40 carbon atoms.

a Suggest the structure of the monomer.

b How many monomer units are incorporated into the average molecule of the polymer?

Cambridge Paper 4 Q4a November 2008

3 Silk from silkworms and spiders is an example of a condensation polymer:

a Explain what is meant by a condensation polymer.

b Another type of polymer is called an addition polymer. Name an example of an addition polymer.

c Suggest why condensation polymers, such as proteins, show a wider range of properties than addition polymers.

Cambridge Paper 4 Q10b November 2008

4 Short sections of different polymer chains are shown below.:

a $\cdots-O-CH_2-CH_2-O-\underset{\underset{O}{\|}}{C}-CH_2-CH_2-\underset{\underset{O}{\|}}{C}-\cdots$

b $\cdots-CF_2-CF_2-CF_2-CF_2-CF_2-CF_2-\cdots$

c $\cdots-HN-(CH_2)_9-NHCO-(CH_2)_7-CO\cdots$

d $\cdots-\underset{\underset{CH_3}{|}}{CH}-\underset{\underset{CH_3}{|}}{CH}-\underset{\underset{CH_3}{|}}{CH}-\underset{\underset{CH_3}{|}}{CH}-\underset{\underset{CH_3}{|}}{CH}-\underset{\underset{CH_3}{|}}{CH}\cdots$

   i In each case, say whether it is an addition or a condensation polymer.

   ii Draw the structures of the monomer(s) from which each polymer is made.

5 a What are the advantages of recycling plastics compared with disposing of them by burning?

b Explain why plastics must be separated into different polymer types before they can be recycled.

c Explain why it is harder to separate plastics into different polymer types than to separate different metals.

d One method for separating polymers uses their density differences. Suggest a system for separating the three polymers below by this method.

	Density/ $g\,cm^{-3}$
Low-density poly(ethene), LDPE	0.92
High-density poly(ethene), HDPE	0.96
Poly(chloroethene), PVC	1.40

6 Consider the following:

a increasing polymer chain length,

b increasing cross-linking between polymer chains,

c the presence of polar side-groups.

In each case, describe the effect on the strength of the polymer, and explain why it has this effect.

7 A section of the polymer chain in poly(ethenol) is shown below.

$$\cdots-CH_2-\underset{\underset{OH}{|}}{CH}-CH_2-\underset{\underset{OH}{|}}{CH}-CH_2-\underset{\underset{OH}{|}}{CH}-CH_2-\underset{\underset{OH}{|}}{CH}-\cdots$$

a Is poly(ethenol) an addition polymer or a condensation polymer?

b What kind of intermolecular forces exist between poly(ethenol) molecules?

c Pure poly(ethenol) is insoluble in water. However, a soluble form of the polymer can be made by replacing some of the $-OH$ groups with CH_3COO- groups.

   i Draw a section of the poly(ethenol) chain in which half the $-OH$ groups have been replaced by CH_3COO- groups.

   ii Explain why replacing some of the $-OH$ groups helps to make the polymer soluble in water.

d If 2% of the $-OH$ groups are replaced, the polymer becomes soluble in hot water, but insoluble in cold water. What practical uses can you suggest for such a polymer?

28 Proteins and DNA

28.1 Sickle cell disease

Red blood cells are the body's oxygen-carriers, transporting oxygen into the organs where it is needed. Red blood cells are filled with haemoglobin, a protein with a high oxygen-carrying capacity. Sickle cell disease is a blood disease in which the normally disc-shaped red blood cells are mis-shapen and have a bent 'sickle' shape (Figure 28.1). These sickle cells are not flexible like normal red cells, so they cannot make their way easily through narrow blood capillaries. This means they cannot reach into the organs where oxygen is needed. The result can be acute pain in the affected organs. Sickle cell disease also leads to anaemia, because the body tries to destroy the defective blood cells, so the blood has less oxygen-carrying capacity.

Sickle cell disease is common in sub-Saharan Africa. It is a genetic disease which is inherited from parents who carry a defective gene for making haemoglobin. If a child inherits a defective gene from both parents, he or she will have sickle cell disease throughout their lives because all the haemoglobin they produce will have the sickle defect. Males with sickle cell disease have an average life expectancy of only 42 years.

Genes are made from DNA, and the defective haemoglobin-making gene in people with sickle cell disease has just one mistake in over 10 000 letters that make up the gene's DNA code. This single mistake is converted into a single error in the amino acid sequence of haemoglobin. One of the 287 amino acid residues in the haemoglobin chain is the wrong one (valine instead of glutamic acid). This single misplaced amino acid causes the haemoglobin protein to form fibres, rather than folding up normally (Figure 28.2) and this is why the red blood cells are mis-shapen.

So a single error in the DNA molecule causes a single error in the haemoglobin protein chain, which results in a very serious blood disease. This illustrates just how sensitive proteins are to tiny changes, and why controlling their shape is critical to life.

28.2 Amino acids, the building blocks for proteins

All proteins are chemically similar, but they perform many different functions in living things. They make structural materials, such as skin and fingernails, they make the fibre in muscles and they are the essential material of which enzymes are made. Protein molecules are chemically very similar, yet their diverse functions mean there must be many different ways of putting their molecules together.

Proteins are condensation polymers made by linking amino acids. There are 20 different naturally-occurring amino acids, and many proteins contain several thousand amino acid units. For a protein containing 5000 amino acid units, the number of possible arrangements, using 20 different amino acids, is 20^{5000}, so it isn't really surprising that proteins come in so many different forms. Indeed, the surprising thing is that an organism can control precisely the form of protein it produces.

Amino acids

Amino acids were introduced in Section 26.2. The amino acids that make up proteins are all 2-amino acids, also called α-amino acids, with the general form:

$$R-CH-COOH$$
$$|$$
$$NH_2$$

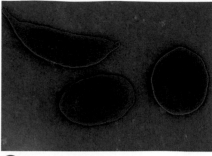

Figure 28.1
In sickle cell disease, the patient's red blood cells are bent and sickle-shaped, instead of disc-shaped. They are less flexible and cannot move easily through narrow blood capillaries.

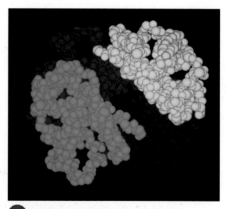

Figure 28.2
Normal haemoglobin naturally folds itself into the shape shown in this computer-generated image. In sickle cell disease, a single error in the protein chain stops the haemoglobin folding like this, and instead it forms fibres. Haemoglobin has four identical protein units, shown yellow, red, purple and green in this picture.

Each of the 20 naturally occurring amino acids found in proteins has a different R group, as shown in Table 28.1. Amino acids are usually called by their common names, rather than their systematic names, and that is what we will do in this chapter. Each amino acid also has an abbreviated three letter code – for example, glycine is **Gly**.

Amino acid	Abbreviated code	R group	Amino acid	Abbreviated code	R group
glycine	Gly	—H	cysteine	Cys	—CH_2—SH
alanine	Ala	—CH_3	methionine	Met	—CH_2—CH_2—S—CH_3
valine	Val	—CH(CH_3)(CH_3)	aspartic acid	Asp	—CH_2—C(=O)OH
leucine	Leu	—CH_2—CH(CH_3)(CH_3)	glutamic acid	Glu	—CH_2—CH_2—C(=O)OH
isoleucine	Ile	—CH(CH_2CH_3)(CH_3)	asparagine	Asn	—CH_2—C(=O)NH_2
phenylalanine	Phe	—CH_2—C₆H₅	glutamine	Gln	—CH_2—CH_2—C(=O)NH_2
proline	Pro	HN⟩COOH (ring)	tyrosine	Tyr	—CH_2—C₆H₄—OH
tryptophan	Trp	—CH_2 (indole)	histidine	His	—CH_2 (imidazole, HN—N)
serine	Ser	—CH_2—OH	lysine	Lys	—CH_2—CH_2—CH_2—CH_2—NH_2
threonine	Thr	—CH(CH_3)(OH)	arginine	Arg	—CH_2—CH_2—CH_2—NH—C(=NH)(NH_2)

Table 28.1

The 20 amino acids found in proteins. The structure of the R group is shown in each case, except for proline, where the whole molecule has been drawn.

All amino acids, except glycine, have four different groups attached to the central carbon atom (Figure 28.3). Because of this, they exhibit optical isomerism (Section 13.6).

Each amino acid carries at least two functional groups, —NH_2 and —COOH. They therefore show the properties of both amines and carboxylic acids, and they can act as both a base and an acid.

Consider the simplest amino acid, glycine, H_2NCH_2COOH. It can react with an acid:

$$H_2NCH_2COOH(aq) + H^+(aq) \longrightarrow H_3\overset{+}{N}CH_2COOH(aq)$$

It can also react with a base:

$$H_2NCH_2COOH(aq) + {}^-OH(aq) \longrightarrow H_2NCH_2COO^-(aq) + H_2O(l)$$

Having both an acidic and a basic group, glycine can also react with itself, in an *internal* acid–base reaction in which the COOH group donates H^+ to the —NH_2 group:

$$H_2NCH_2COOH(aq) \longrightarrow H_3\overset{+}{N}CH_2COO^-(aq)$$

In neutral solution and in the solid state, glycine exists as this dipolar ion:

$$H_3\overset{+}{N}CH_2COO^-$$

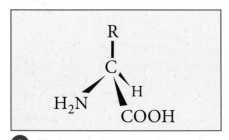

Figure 28.3

The asymmetry of the amino acid molecule

This kind of ion is called a **zwitterion**, from the German *zwei* meaning 'two'. Because of their ionic character, glycine and other amino acids are soluble in water and have high melting points. The melting point of glycine is 234 °C.

Glycine can therefore exist in three different forms, depending on the pH (Figure 28.4).

$H_3\overset{+}{N}-CH_2-COOH$	$H_3\overset{+}{N}-CH_2-COO^-$	$H_3N-CH_2-COO^-$
In acidic conditions	In neutral conditions	In basic conditions

Other amino acids behave in a similar way to glycine, although the pH ranges in which the three forms exist differ according to the amino acid. Some amino acids have extra —NH_2 and —COOH groups in their molecule, and this significantly affects their behaviour towards acids and bases.

Quick Questions

1 Look at the amino acid structures shown in Table 28.1. Remember this table shows the R groups, not the full amino acid structure, except for proline. Which amino acids:
   a have an extra carboxylic group, and will therefore be acidic overall,
   b have an extra amino group (primary, secondary or tertiary) and will therefore be basic overall?

Joining amino acids: the peptide link

The amino acids in a protein chain are linked together by condensation reactions between the —NH_2 group of one amino acid and the —COOH group of the next. Such a link is called a **peptide link** (Figure 28.5).

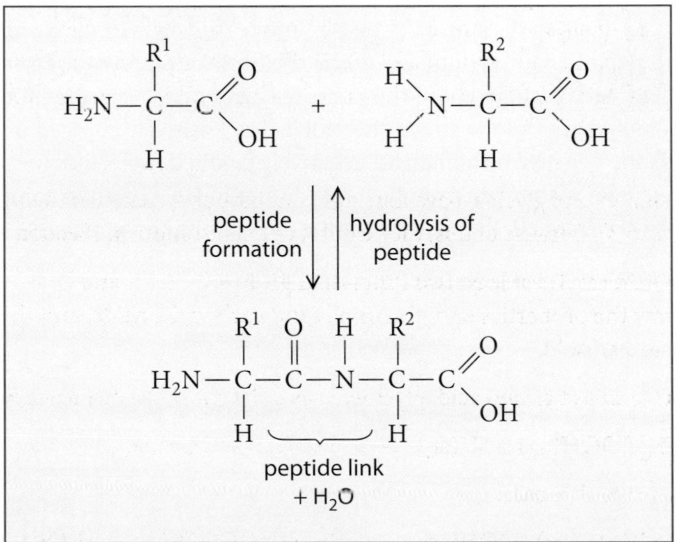

The peptide link is not easily formed under laboratory conditions, but in living systems it forms readily because the reaction is catalysed by enzymes. Chains of amino acids, called **polypeptides,** are formed in this way. Once formed, and under the right conditions of temperature and pH, polypeptide chains can take up precise three-dimensional shapes, leading to the formation of proteins. A **protein** is a polypeptide with a specific biological function.

Definition

A **zwitterion** is a dipolar ion of the form:

$$H_3\overset{+}{N}-\underset{\underset{R}{|}}{CH}-COO^-$$

◀ Figure 28.4
Glycine exists in three different forms, depending on the pH.

Key Points

▶ The natural amino acids which make up proteins are all 2-amino acids with the form:

$$R-\underset{\underset{NH_2}{|}}{CH}-COOH$$

▶ All natural amino acids except glycine show optical isomerism.
▶ Amino acids can behave as both acids and bases.

◀ Figure 28.5
A peptide link

Definitions

A **peptide link** is formed between two amino acids by the condensation reaction between an —NH_2 group on one amino acid and a —COOH group on another.

A **polypeptide** is a chain of amino acids joined by peptide links.

A **protein** is a polypeptide with a specific biological function.

Rather than writing out their full structural formulae, polypeptides are often written using the abbreviated three-letter amino acid codes (Figure 28.6). One end of the polypeptide always has a free $-NH_2$ group, called the N-terminal and the other end has a free $-COOH$ group, called the C-terminal. By convention the amino acid at the N-terminal is always shown on the left.

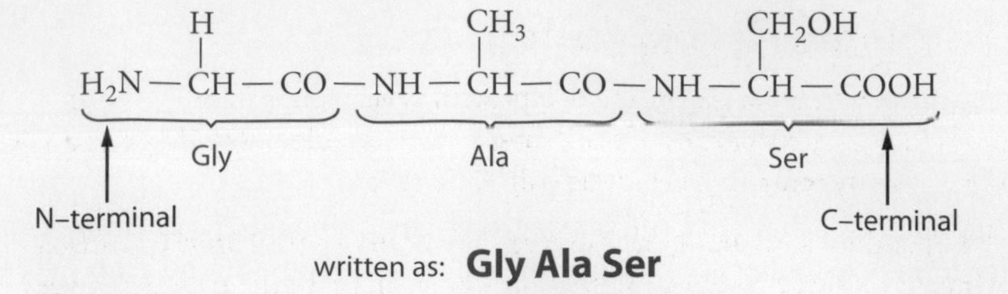

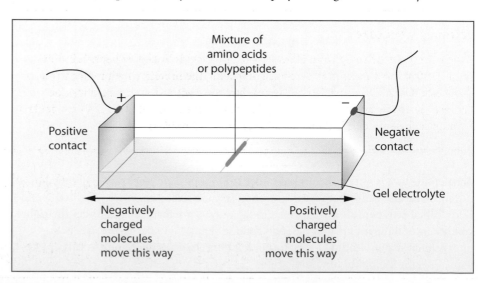

Figure 28.6
Polypeptides are written using the three-letter amino acid codes. The N-terminal is always on the left.

Hydrolysis of polypeptides

The formation of a peptide link between amino acids is reversible. So, polypeptides can be hydrolysed to re-form amino acids. Hydrolysis is carried out by boiling with aqueous acid, but it is performed more effectively by digestive enzymes like pepsin and trypsin.

Hydrolysis is used to separate the amino acids present in a protein, so they can be identified. The protein is hydrolysed, and then the amino acids are identified by chromatography (Section 13.9) or by electrophoresis.

Electrophoresis

Amino acids carry electrical charges arising from their amino and carboxylic acid groups. The overall electrical charge depends on the pH, as shown in Figure 28.4. Polypeptides and proteins also carry charges, because of the amino and carboxylic acid units on some R groups.

These electrical charges provide a way of separating mixtures of amino acids or polypeptides in aqueous solution. In **electrophoresis**, the mixture is placed in an electric field, as shown in Figure 28.7. The electrolyte is usually in the form of a gel, which slows down the natural diffusion of the mixture and improves separation. In the electric field, each component in the mixture is separated according to the overall charge on its molecules and the size of the molecules. Positively charged molecules move towards the negative terminal and negatively charged molecules to the positive terminal. Molecules without an overall charge do not move. Once the components of the mixture have separated, they can be shown up by staining them with a dye.

Definition

Electrophoresis is the separation of charged molecules by placing them in an electric field.

Figure 28.7
The principle of electrophoresis

Electrophoresis can be used to separate amino acids, following the hydrolysis of a protein. It can also be used to separate proteins themselves. In this case the separation depends on the fact that the larger the protein molecule, the slower it moves through the gel. Electrophoresis is also used to separate other biological macromolecules such as RNA and DNA – it is used in DNA fingerprinting (Section 28.7).

28.3 The structure of proteins

The precise three-dimensional shape of a protein is very important, because it determines the exact properties of the protein. For example, if the shape of an enzyme molecule is slightly wrong, the enzyme will not work.

The order of amino acids along the polypeptide chain of a protein is described as its **primary structure** (Figure 28.8). Once the primary structure has formed, under normal physiological conditions (that is, the conditions of temperature and pH that exist within the living organism), the protein chain *automatically* forms a precise shape. First, the chain folds or twists itself into the **secondary structure**, which is usually either a helix (coil) or a sheet. The secondary structure then folds further to form a **tertiary structure**. This folding happens spontaneously because of intermolecular forces between different parts of the chain. Small changes to the primary structure can lead to big changes in the tertiary structure, by changing the intermolecular forces. For example, changing Glu to Val in haemoglobin's primary structure makes the haemoglobin fibrous and causes sickle cell disease (Section 28.1).

Quick Questions

2 Look back at your answer to Quick question 1. If you used electrophoresis to separate the 20 amino acids, which amino acids would you expect to move towards the *negative* terminal, if the pH of the solution were:
a 2,
b 12?

Definitions

Primary structure is the order of amino acids in a protein chain.

Secondary structure is the coiling or folding of chains into a helix or sheet.

Tertiary structure is the folding of helices and sheets into the final shape.

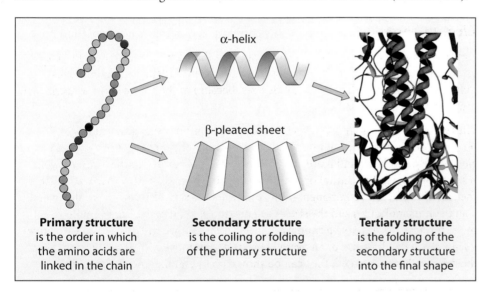

α-helix

β-pleated sheet

Primary structure is the order in which the amino acids are linked in the chain

Secondary structure is the coiling or folding of the primary structure

Tertiary structure is the folding of the secondary structure into the final shape

◀ Figure 28.8
The primary, secondary and tertiary structure of a protein

Primary structure

Proteins contain anything from 50 to 10 000 amino acids linked by peptide bonds. Figure 28.9 shows the primary structure of insulin, the protein which regulates sugars in the blood. Insulin has a relatively simple structure with 51 amino acid residues. Because the amino acids in the polypeptide chain have eliminated an H_2O molecule in forming each peptide link, they are called amino acid *residues*.

▼ Figure 28.9
The primary structure of insulin, which has two polypeptide chains. The —S—S— links between Cys residues are called disulfide bridges.

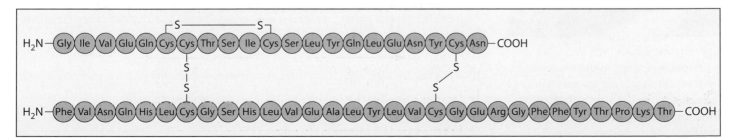

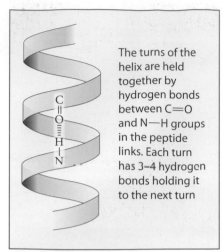

The turns of the helix are held together by hydrogen bonds between C=O and N—H groups in the peptide links. Each turn has 3–4 hydrogen bonds holding it to the next turn

C=O
H—N

Figure 28.10
The α-helix. α means that the helix turns to the right.

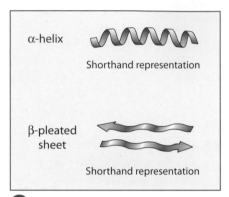

α-helix

Shorthand representation

β-pleated sheet

Shorthand representation

Figure 28.11
Shorthand representations used for the α-helix and β-pleated sheet

Figure 28.13
A hedgehog's sharp spines are made from keratin, a protein whose secondary structure is made almost entirely from α-helices. Human hair is also made from keratin.

Secondary structure

Primary structures usually fold themselves into one of two secondary structures.

The **α-helix** (Figure 28.10) is a tight coil with the R groups sticking out from the coiled polypeptide chain. The coil is held together by hydrogen bonding between NH and CO groups along the polypeptide chain.

The **β-pleated sheet** is a flat sheet made from polypeptide chains lying alongside each other and held together by hydrogen bonds.

To simplify things when structures are drawn, chemists often use the simplified shorthand representations in Figure 28.11 to represent the α-helix and the β-sheet.

Tertiary structure

Most proteins contain a combination of helixes and sheets, folded together into a tertiary structure with the exact final shape. Figure 28.12 shows the tertiary structure of insulin.

The tertiary structure of a protein is held together in the correct shape by a combination of:

▶ hydrogen bonds,
▶ permanent dipole and induced dipole attractions,
▶ ionic attractions between COO^- and NH_3^+ on the R side groups,
▶ disulfide (—S—S—) bridges between neighbouring Cys residues.

These bonds work together to stabilise the protein in exactly the right shape. It is a delicate balance which is easily upset, particularly by changes in pH and temperature. When this happens, the protein loses its correct shape, becomes *denatured* and no longer works as it should.

Figure 28.12
A computer-generated model of a protein from the Spanish Flu virus, the virus responsible for the disastrous flu pandemic in 1918. This model uses the shorthand notation shown in Figure 28.11.

Sometimes, several protein molecules cluster together to form a **quaternary structure**. For example, under physiological conditions, insulin molecules usually cluster together in groups of six, called insulin hexamers.

Proteins fall into two main classes as far as their overall shape and properties are concerned.

▶ **Fibrous proteins** such as keratin in hair and collagen in muscle.
▶ **Globular proteins,** which are folded compactly and often dissolve in water. Enzymes are globular proteins.

Understanding protein structure

Understanding the three-dimensional structure of a protein molecule is very important, because it helps scientists to explain how the protein functions and how it is valuable in the treatment of diseases. For example, careful analysis of the structure of haemoglobin in people with sickle cell disease revealed that it had a Val where a Glu should be. This led to an understanding of the mistake in the DNA coding in the gene that codes for haemoglobin.

Insulin was the first protein to have its primary structure determined, by Frederick Sanger in 1958. But it was another 11 years before Dorothy Hodgkin worked out the tertiary structure of insulin using X-ray crystallography (Section 4.8). Once the primary and tertiary structures of insulin were known, it became possible to develop new treatments for diabetes using modified forms of insulin.

Today, the tertiary structures of proteins can be worked out much more quickly, using automated X-ray crystallography to map the density of electrons in a crystal of the protein. Nuclear magnetic resonance, NMR (Section 13.8) is used to find the positions of C and H atoms in the structure. Because proteins are so complex, the X-ray and NMR measurements give enormous amounts of data. These measurements are processed by powerful computers which then draw a three-dimensional representation of the structure, like the one shown in Figure 28.12. Protein structures can also be *predicted* from the primary structure, using computer programs that work out which groups on the protein chain will interact with one another.

Quick Questions

3 Suggest reasons for the following:
 a Most enzymes stop working above about 50 °C.
 b Fibrous proteins like keratin are made almost completely of α-helices (Figure 28.13).
 c Albumen, a globular protein found in egg white, sets into an insoluble white solid when the egg white is heated.

28.4 Enzymes

Enzymes are biological catalysts which are far more efficient than most inorganic catalysts. Enzymes are also highly specific, meaning that they catalyse only one type of reaction with a specific type of reactant, called the **substrate**. Catalase, for example, which is an enzyme that destroys toxic hydrogen peroxide, can decompose 100 000 molecules of hydrogen peroxide every second, but it has no effect on other compounds.

Enzymes are usually named using the suffix *-ase*, and the name also often refers to the substrate – for example *glucose oxidase* catalyses the oxidation of glucose.

As well as being the key to all life processes, enzymes are increasingly used in industry. For example, papain, an enzyme extracted from papaya fruit which breaks down proteins, is used to tenderise meat. Amylases, enzymes which break down starch, are used to turn corn starch into a sugary syrup.

The **lock and key model** is a simple explanation of enzyme action. Like all proteins, enzymes have a precise molecular shape. The enzyme has a particular location on its surface, called the **active site**, into which molecules of substrate fit exactly (Figure 28.14). This is an example of *molecular recognition*, where one molecule recognises another by its shape. The active site is shaped exactly to fit the particular shape of the substrate molecule, but no other. An enzyme may have more than one active site.

◯ Figure 28.14
The lock and key model of enzyme action. The exact shape of the substrate and the active site depend on the enzyme and substrate involved.

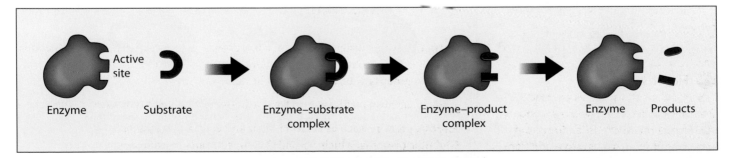

Enzyme   Substrate   Enzyme–substrate complex   Enzyme–product complex   Enzyme   Products

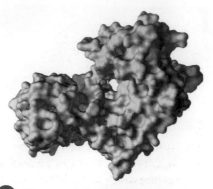

Figure 28.15
A computer-generated space-filling model of the enzyme hexokinase, which catalyses the conversion of glucose to glucose-6-phosphate. The glucose substrate (yellow) is shown fitting neatly into the active site.

Figure 28.16
Competitive enzyme inhibition. The inhibitor is similar in shape to the substrate and binds to the active site, preventing the substrate from entering.

Figure 28.17
Non-competitive inhibition

The active site is a cleft in the enzyme molecule created by folding of the protein chain (Figure 28.15). The substrate binds to the active site by intermolecular attractions such as hydrogen bonding. This weakens bonds in the substrate so they break more easily, or it may bring atoms in the substrate into the correct configuration for reaction. The substrate is converted to products at the active site, which then leave the enzyme, so it is free to accept another substrate molecule. Usually the substrate concentration is in great excess; typically there might be 10^5 substrate molecules for every enzyme molecule. All this can be summarised as:

$$\text{enzyme} + \text{substrate} \rightleftharpoons \text{enzyme–substrate complex} \rightleftharpoons \text{enzyme–product complex} \rightleftharpoons \text{enzyme} + \text{product}$$

Enzyme inhibition

Enzymes are delicate molecules whose activity depends on having exactly the right shape. They are inactivated by the wrong conditions of temperature and pH, and inhibited from working properly by certain molecules or ions. There are two types of inhibition.

In **competitive inhibition**, a molecule of inhibitor that is similar in shape to the substrate enters and binds to the active site, preventing the substrate molecule from entering (Figure 28.16). The inhibitor and the substrate are *competing* for the enzyme's active site. The more inhibitor there is, the fewer active sites there are for the substrate, and the less efficient the enzyme becomes.

An example of competitive inhibition occurs in the enzyme succinate dehydrogenase, which catalyses the oxidation of succinate ions (butanedioate ions). This enzyme is competitively inhibited by malonate ions (propanedioate ions) which have one CH_2 group less.

$$^-OOC—CH_2—CH_2—COO^-$$
succinate ion

$$^-OOC—CH_2—COO^-$$
malonate ion

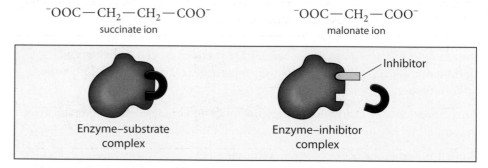

Enzyme–substrate complex   Enzyme–inhibitor complex   Inhibitor

In **non-competitive inhibition**, a smaller molecule or metal ion binds onto the enzyme and changes the shape of its active site, so that it no longer fits the substrate (Figure 28.17). Many toxins are non-competitive enzyme inhibitors. For example, nerve gases contain molecules that bind onto an enzyme called acetylcholinesterase, which plays an important role in transmitting nerve impulses. This changes the shape of the enzyme's active site, so it no longer works properly, leading to paralysis and death.

Many drugs are non-competitive inhibitors which correct an imbalance in enzyme activity in the patient.

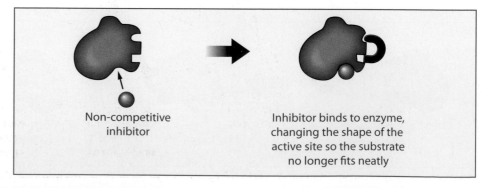

Non-competitive inhibitor   Inhibitor binds to enzyme, changing the shape of the active site so the substrate no longer fits neatly

In contrast to enzyme inhibitors, **co-enzymes** are molecules that *help* enzymes to do their job efficiently. Co-enzymes sometimes bind onto the substrate and help it fit into the active site. Many vitamins, such as the B vitamins, are needed by the body as starting materials for making co-enzymes, without which vital enzymes could not work.

Figure 28.18
Soft-centred chocolates start off with hard sugar centres. An enzyme is added to the centre which slowly breaks down the sugar, making it soft.

Quick Questions

4 In what ways do enzymes differ from inorganic catalysts?
5 Why are enzymes made inactive by small changes of temperature or pH?
6 Pineapples contain a protease enzyme that breaks down proteins. If you try to make a jelly with chunks of fresh pineapple in it, the jelly won't set – but it does set if you use canned pineapple. Explain.
7 At low substrate concentrations, the rate of an enzyme-catalysed reaction is proportional to the concentration of the substrate. But at high concentrations, the rate does not change with increasing substrate concentration. Suggest an explanation in terms of the lock and key model.

28.5 Nucleic acids: DNA and RNA

Proteins are the building blocks for making living organisms, and nucleic acids carry the code for making proteins. So nucleic acids are sometimes called the 'blueprint of life'.

Nucleic acids are polymers, consisting of a chain of alternating sugar molecules and phosphate groups, with different organic bases attached to the sugar molecules and sticking out from the chain (Figure 28.19). In **DNA** (deoxyribonucleic acid), the sugar is deoxyribose, which has 5 carbon atoms in each sugar molecule. In **RNA** (ribonucleic acid), the sugar is ribose, which also has 5 carbon atoms, but with one extra oxygen atom.

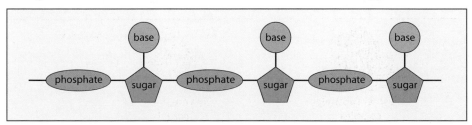

Figure 28.19
Nucleic acids (DNA and RNA) have a chain of alternating phosphate and sugar groups, with different organic bases sticking out from the chain.

DNA is found in the nuclei of cells, and is the permanent coded plan (or 'blueprint') for making proteins. Each specific protein is made by a section of DNA called a **gene** (Figure 28.20). Genes are joined together into structures called chromosomes: humans have 23 pairs of chromosomes which we inherit from our parents. The nucleus of every cell in your body (except sperm cells and ova) contain all 23 pairs of chromosomes, but each cell only makes the proteins it actually needs. For example, a blood cell expresses ('turns on') the gene for haemoglobin, but it does not express the gene for keratin, which would be expressed by a hair-producing cell.

Figure 28.20
Chromosomes, genes, DNA and codons

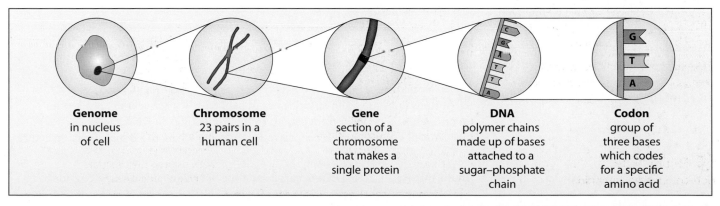

Genome	Chromosome	Gene	DNA	Codon
in nucleus of cell	23 pairs in a human cell	section of a chromosome that makes a single protein	polymer chains made up of bases attached to a sugar–phosphate chain	group of three bases which codes for a specific amino acid

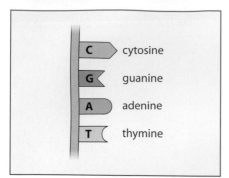

Figure 28.21
The four bases in DNA

RNA is found in the main body of the cell, and is the go-between or link between DNA and the protein production process (Section 28.6).

The structure and function of DNA

How does the DNA code work?

There are four different bases that can be attached to the sugar–phosphate backbone in DNA. They are called adenine (A for short), guanine (G), thymine (T) and cytosine (C). We will give their detailed structures later – for now, we will just use the letters A, G, T and C. The DNA chain looks like Figure 28.21, with the bases A, G, T and C occurring over and over again.

These four letters provide the code for different amino acids in the protein chain. Each different amino acid has a triplet code, called a **codon**, made up of three bases. There are 64 (4 × 4 × 4) ways that the four letters A, G, T and C can form a triplet, and these are given in Table 28.2, together with the amino acids that they code for. Notice that most amino acids have more than one triplet code, but that each triplet code is for only one amino acid. Notice also that some triplets which show 'Stop' in Table 28.2 mean 'Stop the chain growing'. The way these triplets are decoded in protein synthesis is explained in Section 28.6.

Table 28.2
The triplet codes in DNA, together with the amino acids that they code for. For example, TTT codes for phenylalanine (Phe). 'Stop' means the end of the chain.

First base	Second base				Third base
	T	**C**	**A**	**G**	
T	TTT Phe	TCT Ser	TAT Tyr	TGT Cys	T
	TTC Phe	TCC Ser	TAC Tyr	TGC Cys	C
	TTA Leu	TCA Ser	TAA Stop	TGA Stop	A
	TTG Leu	TCG Ser	TAG Stop	TGG Trp	G
C	CTT Leu	CCT Pro	CAT His	CGT Arg	T
	CTC Leu	CCC Pro	CAC His	CGC Arg	C
	CTA Leu	CCA Pro	CAA Gln	CGA Arg	A
	CTG Leu	CCG Pro	CAG Gln	CGG Arg	G
A	ATT Ile	ACT Thr	AAT Asn	AGT Ser	T
	ATC Ile	ACC Thr	AAC Asn	AGC Ser	C
	ATA Ile	ACA Thr	AAA Lys	AGA Arg	A
	ATG Met	ACG Thr	AAG Lys	AGG Arg	G
G	GTT Val	GCT Ala	GAT Asp	GGT Gly	T
	GTC Val	GCC Ala	GAC Asp	GGC Gly	C
	GTA Val	GCA Ala	GAA Glu	GGA Gly	A
	GTG Val	GCG Ala	GAG Glu	GGG Gly	G

How does DNA copy itself?

When a cell divides to form two new cells, it needs to pass on an exact copy of the DNA in its nucleus. A similar thing has to happen when sex cells are produced, in order to pass on DNA from parents to their offspring. So DNA must have a mechanism for making exact copies of itself. This is where the 'double helix' structure of DNA comes in.

In the nucleus of a cell, DNA occurs as not one but *two* chains coiled alongside each other in a double helix (Figure 28.22). The two helices are held together by hydrogen bonds between pairs of neighbouring bases. It's like a spiral staircase in which the sugar–phosphate chains make up the two sides of the staircase, with the base pairs making up the steps. Notice that two pairs of bases match each other in shape, so they fit together in the double helix. They are called **complementary** pairs: **A** matches **T** and **G** matches **C**. Later we will show how this matching occurs because of hydrogen bonds between exactly-matching N, H and O atoms on the base pairs.

As a cell divides into two, the two coils of DNA in each chromosome separate, and the two halves quickly build a new matching coil alongside each separated coil. In this way, two new double helices are produced – one for each new cell (Figure 28.22).

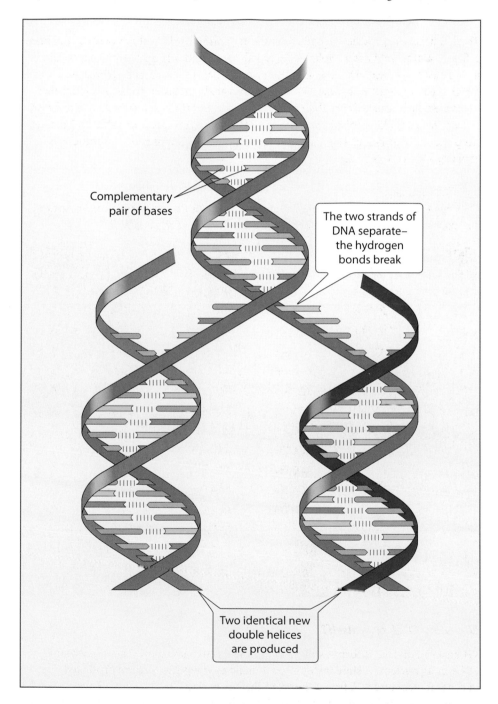

Complementary pair of bases

The two strands of DNA separate– the hydrogen bonds break

Two identical new double helices are produced

◀ Figure 28.22
The double helix structure of DNA is held together by hydrogen bonding between complementary base pairs. When the cell divides, two new copies of the DNA are made.

So DNA not only has a coding system built into its structure, it also has a built-in way of copying itself. It is the molecule of life, and it is used by practically every living organism, from viruses and bacteria to plants and humans.

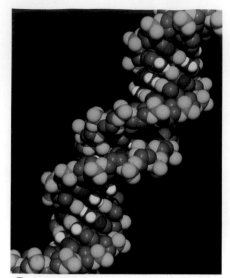

Figure 28.23 shows a computer-generated space-filling model of DNA. Although this is more complex than the simplified models we have been using, you can clearly see the double-helix shape.

Figure 28.23
A computer-generated space-filling model of DNA

Key Points

DNA carries four different bases, A, G, T and C, on its chain which provide the code for the sequence of amino acids in a protein. Each different amino acid code consists of one or more triplet codons made up of combinations of A, G, T and C.

DNA forms a double helix in which the bases on one strand are hydrogen-bonded to complementary bases on the other strand. A pairs with T and G pairs with C.

When a cell divides, the two coils of DNA separate, and the cell quickly builds a new matching coil alongside each separate coil, making two new double helices.

Matching base pairs

Figure 28.24(a) shows the structural formulae of the bases adenine, guanine, thymine and cytosine. Notice the following points about these structures.

1 All four are cyclic compounds, containing rings made up of carbon and nitrogen atoms.
2 Because of their N atoms, all four are amines, which is why they are described as *bases*.
3 A and G each have two rings fused together. T and C each have a single ring.

Figure 28.24
(a) The four bases in DNA
(b) Complementary base-pairings by hydrogen bonding in DNA

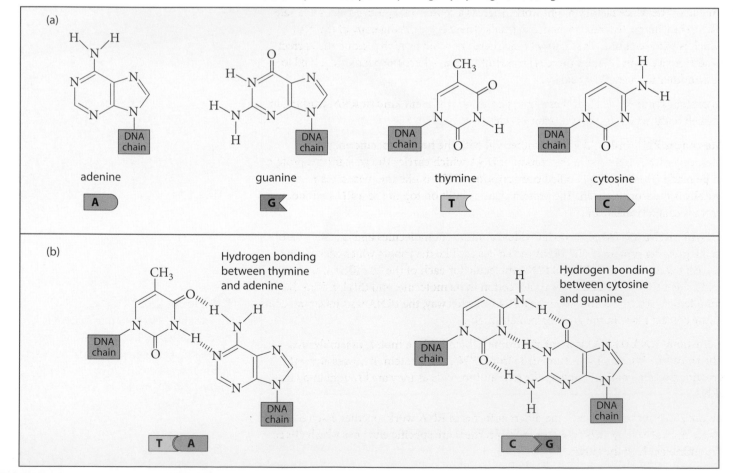

Figure 28.24(b) shows how the complementary bases pair up, A with T and G with C. The pairs are held together by hydrogen bonding involving N, O and H atoms on the bases. Certain N, O and H atoms are in exactly the right position to bond the base-pairs together. It is an example of molecular recognition.

So here is the reason why DNA is the molecule of life. Its bases are able to pair up exactly using hydrogen bonds. The hydrogen bonds are strong enough to hold the two strands of the double helix together, but weak enough to break when the time comes for the two strands to separate and make copies. Normally the copying is very accurate, but occasionally mistakes are made, and tiny mistakes can lead to serious consequences, including genetic conditions like sickle cell disease.

The structure of RNA

RNA has a similar structure to DNA, with the following differences.

1 Its backbone contains the sugar ribose instead of deoxyribose.
2 It has the same bases as DNA, except that thymine, T, is replaced by uracil, U. Uracil is like thymine but without the CH_3 group on the ring. When RNA copies DNA, it puts a U in the RNA chain where there would be a T in DNA.
3 It does not form a double helix, because it does a different job to DNA in making proteins (Section 28.6).

28.6 From DNA to proteins

DNA in the cell nucleus carries the permanent record for making proteins, but it is RNA in the body of the cell that does most of the work.

Making a protein chain from amino acids can be compared to making a long necklace from beads with 20 different colours arranged in an exact sequence. Here is how an imaginary necklace factory might work. There is a master plan, locked away in a safe place, showing the correct sequence of beads. There is a working copy of the plan which is taken out onto the factory floor. There are robots which pick up the correct bead from a supply. Finally, there is an assembly area, where the beads are joined to one another to make the necklace.

In protein synthesis, each of these jobs is done by a different kind of RNA, working in the cell body, which is like the factory floor.

Messenger RNA (mRNA) is first synthesised with the help of specific enzymes. Messenger RNA is a copy of the section of DNA which carries the gene for the protein to be made. This copying is called **transcription**. DNA is like the master plan, and mRNA is the working plan. The protein chain is built on top of the mRNA strand, in a process called **translation**.

Transfer RNA (tRNA) picks up individual amino acid molecules and transfers them to the growing protein chain. tRNA molecules are like the robots which pick up the correct bead. There is a different tRNA molecule for each of the 20 different amino acids. Each tRNA has a different triplet codon in its molecule, and this binds to the complementary triplet codon on the mRNA. In this way, the tRNA puts its amino acid in the correct place in the growing protein chain.

Ribosomal RNA (rRNA) is in the **ribosome**, which is like a mobile assembly area. The ribosome is a small structure made from rRNA and protein. It moves along the growing protein chain, joining up the new amino acids as they are brought in by the tRNA.

Figure 28.25 summarises how the different forms of RNA work together. Each of the stages is catalysed by enzymes – for example, there are specific enzymes which attach the amino acids to the tRNA.

Quick Questions

10 The double helix structure of DNA is easily destroyed by changes in temperature or pH. Explain why.
11 Suppose the bonds holding the DNA strands of the double helix together were **a** covalent bonds, **b** induced dipole attractions. What would be the problem in each case?

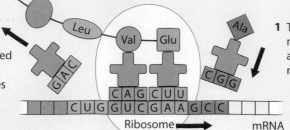

mRNA is copied from DNA.
It contains U where T would be in DNA

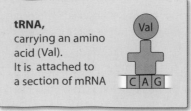

tRNA, carrying an amino acid (Val).
It is attached to a section of mRNA

2 Amino acids are assembled into the growing protein chain

Leu | Val | Glu | Ala

1 Transfer RNA (tRNA) molecules bring amino acids to the mRNA in the ribosome

3 Having delivered its amino acid, the tRNA leaves the ribosome

GAC

CGG

CAGCUU

CUGGUCGAAGCC

Ribosome

mRNA

 Figure 28.25
RNA and protein synthesis

Key Points

In protein synthesis:

DNA is transcribed onto mRNA, which carries the code which is translated into the correct amino acid sequence in the protein. tRNA carries amino acids to the mRNA, and ribosomes provide the assembly area in which the amino acids are joined onto the growing protein chain.

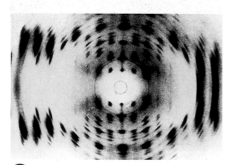

Figure 28.26
An X-ray diffraction photograph of DNA. The repeated pattern provides evidence for the structure of DNA as a double helix.

28.7 Using DNA

The structure of DNA was discovered in 1953 by James Watson and Francis Crick, two scientists working at Cambridge University in the UK. They used X-ray diffraction data to work out the arrangement of the groups in DNA, and deduced it must be a double-helix (Figure 28.26). Until that time, scientists had no idea how genes could code for proteins, nor how genes could copy themselves.

Once the structure of DNA was known, scientists began to realise the great things that could be done with it.

Genetic modification

Traditionally, new varieties of plants are produced by selective breeding. Genetic modification (GM) can be used to speed up selective breeding by inserting new genes directly into the DNA chain, using *restriction enzymes* to snip open DNA chains and then insert new sections.

For example, a gene that produces a natural insecticide can be inserted into the genome of the cotton plant, to protect the growing plant from insects that attack it (Figure 28.27). Today, nearly half of the cotton grown in the world is genetically modified in this way.

GM is controversial. Environmentalists are worried about the unexpected results that might come from the release of a genetically modified organism into the environment. On the other hand, GM offers great potential for producing more food for the world's growing population, and for producing new sources of energy. The important thing is to carry out carefully controlled trials before going ahead with the full-scale use of GM technologies.

DNA fingerprinting

Only about 2% of the DNA in your cells actually codes for proteins. The remaining 98% is often called 'junk DNA', but it is better described as 'non-coding DNA' because it actually has useful functions in regulating protein translation.

Non-coding DNA is also useful because it varies slightly between individual people. 99.9% of your DNA is identical to everyone else's, but the remaining 0.1% varies between individuals, and some non-coding DNA is particularly variable. No two people have exactly the same DNA, unless they are identical twins, but the more closely related they are, the more similar is their DNA.

This is the basis of genetic fingerprinting, which was invented in 1985 by the British scientist Alec Jeffries. Figure 28.28 outlines the method, which involves breaking the DNA into specific fragments, separating the fragments by electrophoresis and then making them visible by staining. The resulting pattern, or 'DNA fingerprint' (Figure 28.29), looking rather like a barcode, is characteristic of the individual person who provided the DNA.

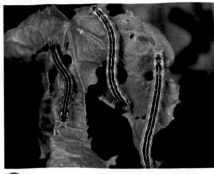

Figure 28.27
Cotton plants can be protected from insects by inserting into the plant's genome a gene that produces a natural insecticide.

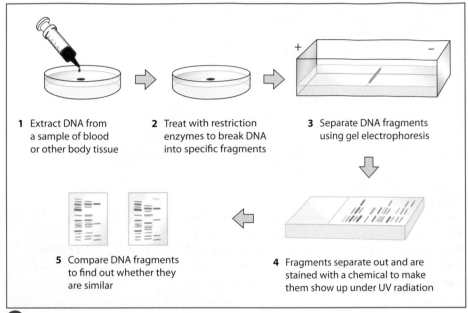

1 Extract DNA from a sample of blood or other body tissue

2 Treat with restriction enzymes to break DNA into specific fragments

3 Separate DNA fragments using gel electrophoresis

5 Compare DNA fragments to find out whether they are similar

4 Fragments separate out and are stained with a chemical to make them show up under UV radiation

Figure 28.28
Outline of the method for DNA fingerprinting

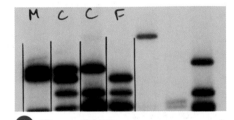

Figure 28.29
A DNA fingerprint. The pattern of the bands is unique to each individual, but some bands are shared by related people, such as a parent and a child. The bands in these DNA fingerprints are marked M for mother, C for child, F for father. Both children share some bands with each parent, proving that they are indeed related.

The first use of DNA fingerprinting in a legal case was in an immigration case in the UK in 1985. A young Ghanaian boy was saved from deportation when DNA fingerprinting proved who his mother was. Since then, it has been used in thousands of forensic science cases to solve crimes, and it has proved valuable in archaeology (Figure 28.30) and to help people trace their ancestors.

DNA and the future

In 1990, scientists around the world set out to map the **human genome** – that is, to find the entire sequence of DNA in all 23 chromosomes of the human cell, consisting of about 3.3 billion base-pairs. The complete genome was published in 2003 and the whole project cost several billion dollars. Today, automated methods have been developed that make it possible to map the genome of an individual person for $50 in a few days. It will soon become easy for people to have a map their own genome, or those of their children, if they want it.

But there is still an enormous amount of work to do before scientists understand exactly what all this genetic information actually does: which genes translate into which proteins, and what the non-coding DNA does. The time is still a long way off when doctors will be able to forecast a patient's medical future from their genome, but the potential of genomic science is very great.

Figure 28.30
Anna Anderson died in 1984 in the USA. During her life, Anna had always claimed she was the Grand Duchess Anastasia, daughter of the last Tsar of Russia whose family were shot in 1918. In 2008, the bodies of the Tsar and his family were exhumed and DNA fingerprinting was carried out on them. A DNA fingerprint was also made from a lock of Anna Anderson's hair, which showed conclusively that she was *not* the Grand Duchess.

Review questions

1 a i In a protein, amino acids are joined together by a process called *condensation polymerisation. Addition polymerisation* is used in some synthetic polymers, such as poly(propene).

State *two* important differences between condensation polymerisation and addition polymerisation.

ii Using the amino acids glycine and alanine below, draw the displayed formula of the dipeptide Ala–Gly, clearly labelling the peptide link.

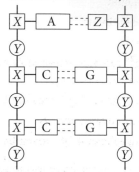

glycine   alanine

b The diagram below shows a section of DNA. Identify the blocks labelled *X*, *Y* and *Z*.

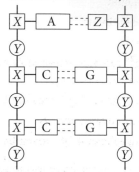

c The table below shows the three base codes used by RNA.

UUU	Phe	UCU	Ser	UAU	Tyr	UGU	Cys
UUC	Phe	UCC	Ser	UAC	Tyr	UGC	Cys
UUA	Leu	UCA	Ser	UAA	stop	UGA	stop
UUG	Leu	UCG	Ser	UAG	stop	UGG	Trp
CUU	Leu	CCU	Pro	CAU	His	CGU	Arg
CUC	Leu	CCC	Pro	CAC	His	CGC	Arg
CUA	Leu	CCA	Pro	CAA	Gln	CGA	Arg
CUG	Leu	CCG	Pro	CAG	Gln	CGG	Arg
AUU	Ile	ACU	Thr	AAU	Asn	AGU	Ser
AUC	Ile	ACC	Thr	AAC	Asn	AGC	Ser
AUA	Ile	ACA	Thr	AAA	Lys	AGA	Arg
AUG	Met/ start	ACG	Thr	AAG	Lys	AGG	Arg
GUU	Val	GCU	Ala	GAU	Asp	GGU	Gly
GUC	Val	GCC	Ala	GAC	Asp	GGC	Gly
GUA	Val	GCA	Ala	GAA	Glu	GGA	Gly
GUG	Val	GCG	Ala	GAG	Glu	GGG	Gly

i What amino acid sequence would the following base code produce? (You may use abbreviations in your answer.)

— AUG UCU AGA GAC GGG UAA —

ii What would be the effect in the amino acid sequence if a mutation caused the base G at position 13 in the sequence to be replaced by U?

d i Name a disease which results from a genetic defect.

ii Explain how the genetic defect can bring about the disease you named.

Cambridge Paper 4 Q7 June 2007

2 a Electrophoresis can be used to separate amino acids which are produced by the hydrolysis of a polypeptide.

Using glycine as an example, explain why the result of electrophoresis depends on pH.

b The diagram below shows the results of electrophoresis in neutral solution. At the start of the experiment a spot of a solution containing a mixture of amino acids P, Q, R and S was placed in the middle of the plate. Following electrophoresis the amino acids had moved to the positions shown in the lower diagram.

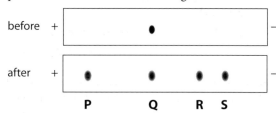

i Which amino acid existed mainly as zwitterions in the buffer solution? Explain your answer.

ii Assuming amino acids R and S carry the same charge when in this buffer solution, which is likely to be the larger molecule? Explain your answer.

Cambridge Paper 4 Q8a–b June 2007

3 a DNA carries the genetic code in living organisms and consists of a double helix.

i Describe what is meant by a *double helix*.

ii How are the strands of the double helix held together?

b In replicating the genetic code two RNA molecules, mRNA and tRNA, are used to perform functions called *transcription* and *translation*. Describe the role of the RNA molecules in these two functions.

c When an egg is boiled, the protein changes from a viscous liquid to a solid.

   i Suggest what causes this change as the protein is heated.

   ii Why is there no change to the primary structure of the protein under these conditions?

   Cambridge paper 4 Q8 a–c November 2007

4 a Explain briefly what is meant by the word *protein*.

   b Describe how peptide bonds are formed between amino acids during the formation of a tripeptide. Include diagrams and displayed formulae in your answer.

   c Describe how proteins can be broken down into amino acids in the laboratory *without* the aid of enzymes.

   d When a small polypeptide, S, was broken down in this way, three different amino acids were produced according to the following reaction.

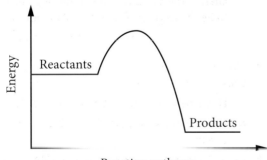

$S \rightarrow 3NH_2CH_2CO_2H + 2NH_2CHCO_2H + 2NH_2CHCO_2H$

$M_r = 7 \qquad M_r = 89 \qquad M_r = 165$

   i How many peptide bonds were broken during this reaction?

   ii Calculate the M_r of the polypeptide S.

   Cambridge paper 4 Q7 November 2008

5 a Enzymes play a vital role in all living organisms, helping chemical reactions to take place at body temperature.

   i The diagram below shows the reaction pathway of an enzyme-catalysed reaction without an enzyme present. Redraw this and on the diagram sketch the pathway if the enzyme was present.

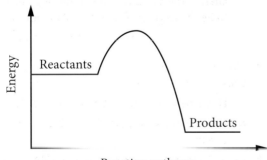

Reaction pathway

   ii What type of molecule are most enzymes?

   iii Why do many enzymes lose their catalytic effectiveness above 40 °C?

b i Explain the difference between competitive and non-competitive inhibition of an enzyme.

   ii The graph below shows how the rate of an enzyme-catalysed reaction varies with substrate concentration in the absence of an inhibitor.

   For a given amount of enzyme, V_{max} represents the rate when all the active sites on the enzyme are being used.

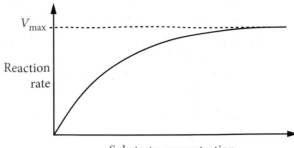

Substrate concentration

   Redraw this graph and sketch on the diagram curves to show the effect on the rate of reaction of:

   I a competitive inhibitor,

   II a non-competitive inhibitor.

   Clearly label your curves.

c Heavy metal ions like Hg^{2+} can bind irreversibly to enzymes and this can result in poisoning.

   i Suggest to what atom or group Hg^{2+} ions bind.

   ii Explain how this affects enzyme activity.

   Cambridge Paper 4 Q8 November 2008

6 The technology of DNA fingerprinting has enormously advanced scientific identification techniques in medicine, crime detection and archaeology in recent years.

   a i In order to prepare a DNA sample for analysis, the DNA is treated with restriction enzymes. What do restriction enzymes do?

   ii What is the next stage in DNA analysis, after the treatment with restriction enzymes?

   iii How are the DNA fragments made visible?

   b NMR and X-ray crystallography have made significant contributions to our knowledge of the structure of proteins and, in the pharmaceutical industry, how drugs react with target proteins.

   i Suggest an advantage of *each* technique in helping to determine protein structure.

   ii MRI scanning is a medical technique based on NMR spectroscopy. It is particularly useful for looking for tumours in healthy tissue. Suggest how this technique can distinguish tumour tissue from healthy tissue?

   Cambridge Paper 4 Q9a–b November 2008

Table of relative atomic masses

The table gives the relative atomic masses of elements correct to one decimal place.

Element	Symbol	A_r	Element	Symbol	A_r
Aluminium	Al	27.0	Molybdenum	Mo	95.9
Antimony	Sb	121.8	Neodymium	Nd	144.2
Argon	Ar	39.9	Neon	Ne	20.2
Arsenic	As	74.9	Nickel	Ni	58.7
Barium	Ba	137.3	Niobium	Nb	92.9
Beryllium	Be	9.0	Nitrogen	N	14.0
Bismuth	Bi	209.0	Osmium	Os	190.2
Boron	B	10.8	Oxygen	O	16.0
Bromine	Br	79.9	Palladium	Pd	106.4
Cadmium	Cd	112.4	Phosphorus	P	31.0
Caesium	Cs	132.9	Platinum	Pt	195.1
Calcium	Ca	40.1	Potassium	K	39.1
Carbon	C	12.0	Praseodymium	Pr	140.9
Cerium	Ce	140.1	Rhenium	Re	186.2
Chlorine	Cl	35.5	Rhodium	Rh	102.9
Chromium	Cr	52.0	Rubidium	Rb	85.5
Cobalt	Co	58.9	Ruthenium	Ru	101.1
Copper	Cu	63.5	Samarium	Sm	150.4
Dysprosium	Dy	162.5	Scandium	Sc	45.0
Erbium	Er	167.3	Selenium	Se	79.0
Europium	Eu	152.0	Silicon	Si	28.1
Fluorine	F	19.0	Silver	Ag	107.9
Gadolinium	Gd	157.3	Sodium	Na	23.0
Gallium	Ga	69.7	Strontium	Sr	87.6
Germanium	Ge	72.6	Sulfur	S	32.1
Gold	Au	197.0	Tantalum	Ta	180.9
Hafnium	Hf	178.5	Tellurium	Te	127.6
Helium	He	4.0	Terbium	Tb	158.9
Holmium	Ho	164.9	Thallium	Tl	204.4
Hydrogen	H	1.0	Thorium	Th	232.0
Indium	In	114.8	Thulium	Tm	169.9
Iodine	I	126.9	Tin	Sn	118.7
Iridium	Ir	192.2	Titanium	Ti	47.9
Iron	Fe	55.8	Tungsten	W	183.9
Krypton	Kr	83.8	Uranium	U	238.0
Lanthanum	La	138.9	Vanadium	V	50.9
Lead	Pb	207.2	Xenon	Xe	131.3
Lithium	Li	6.9	Ytterbium	Yb	173.0
Lutetium	Lu	175.0	Yttrium	Y	88.9
Magnesium	Mg	24.3	Zinc	Zn	65.4
Manganese	Mn	54.9	Zirconium	Zr	91.2
Mercury	Hg	200.6			

Index

First published in 1978 by:
Thomas Nelson & Sons Ltd

This edition published in 2011 by:
Nelson Thornes Ltd
Delta Place
27 Bath Road
CHELTENHAM
GL53 7TH
United Kingdom

11 12 13 14 15 / 10 9 8 7 6 5 4 3 2 1

A catalogue record for this book is available from the British Library

ISBN 978 1 4085 1496 2

Cover photograph: iStockphoto

Illustrations by Russell Parry and GreenGate Publishing Services; illustrations from 5th edition by IFA Design Ltd

Page make-up by GreenGate Publishing Services, Tonbridge, Kent

Printed in China by 1010 Printing International Ltd

Acknowledgments

Advertising Archives: 15.4; Alamy: /Anglia Images 17.7, /Ashley Cooper pics 10.4, /David J. Green – environment 14.17, /Jack Sullivan 20.12, /Jerome Yeats 10.6, /Mike Kipling Photography 23.3, /PCL 26.9, / PHOTOTAKE Inc 4.34, /sciencephotos 11.3, /Studioshots 22.16, / Trinity Mirror /Mirrorpix 15.9b; Courtesy Argonne National Laboratory 9.8; Banana Rite 15.1; BASF Corporate Archives, Ludwigshafen/ Rhine, Germany 12.10, 12.11; Camra 17.1; Castrol (UK) Ltd 14.6; Cephas/Mick Rock 17.11; Dalton Collecting Marsh Fire Gas, 1879-93 (mural painting), Brown, Ford Madox (1821-93) /Manchester Town Hall, Manchester, UK /The Bridgeman Art Library 1.1; Corbis 24.25, / Adam Woolfitt 1.13c, /altrendo travel 3.37, /Bettmann 27.13, /Chaiwat Subprasom/Heuters 17.12, /Charles E. Rotkin 20.11, /Charles O'Rear 23.2, /Construction Photography 15.9a, /David Madison 25.7, /Dean Conger 1.14, /Jonathan Blair 2.10, /Lester V. Bergman 2.11, /Owen Franken 8.2, /Richard T Nowitz 11.4, /Robbie Jack 23.14, /STRINGER/ TURKEY/Reuters 1.13r; FLPA/David Hosking 4.8, /Ingo Arndt/Minden Pictures 27.5, /M. Szadzuik , R. Zinck 19.11, /Nigel Catlin 28.27, /Piotr Naskrecki/Minden Pictures 19.1, /Terry Whittaker 22.13; Fotolia.com/ joerg kemmler 28.13, /Jose Manuel Gelpi 13.5, /Maceo 21.4, /My 3 kids 11.2, /ste72 3.10, /wuttichok 15.12, /XtravaganT 3.3; Getty Images 23.8, 24.16, 28.1, /AFP 1.10, 14.4, /Greg Pease 4.25, /Hulton Archive 12.9, 25.9, 28.30, /Laurie Noble 26.4, /Michael Melford 6.4, /Oxford Scientific /Photolibrary 21.14, /Thomas Marent 4.2, /Time & Life Pictures 13.11, / Tooga 5.17; IBM 1.3; Imagestate/The Print Collector 25.12; iStockphoto 1.12, 4.5, 4.22, 7.19. 8.3, 10.10, 11.17, 12.4, 12.5, 13.2, 13.3, 13.9, 13.27, 14.1, 15.8, 16.4, 18.1, 19.5, 19.10, 20.13, 21.6, 21.15, 24.1, 24.20, 25.5, 27.2, 27.14, 28.18; J.V.Barrett & Co Ltd 17.2; Lafarge 7.10; Martyn Chillmaid 1.7, 1.8, 6.5, 11.9, 15.6, 16.9, 19.2, 19.7, 24.27, 25.4, 26.3; Mary Evans Picture Library 4.1; NASA 5.13, 7.20; PA Photos 27.17; Palm Press Inc 8.16; Perrier 7.16; Photolibrary.com: 2.14, 4.7, 4.10, 5.25, 6.9, 6.15, 7.1, 7.7, 8.4, 10.1, 10.5, 10.9, 11.3, 11.15, 13.4, 13.17, 18.6, 19.4, 21.1, 21.2, 23.10; Rex Features 15.10, /Andy Hooper /Daily Mail 4.36, / Brian Harris 6.16, /Chris Gascoigne /View Pictures 2.24, /Dave Penman 17.10, /KeystoneUSA-ZUMA 14.8, /Newspix 16.2, /Nick Randall 6.13, / OJO Images 14.10, /Sipa Press 10.7, /Stewart Cook 20.15; Science and Society Picture Library 1.5, 2.1; Science Photo Library 1.2, 2.4, 6.2, 9.1, 12.15, 25.1, 25.10, 26.1, 28.2, /Adam Hart-Davis 8.1, /Alex Bartel 12.18, /Alexis Rosenfeld 11.2, /Andrew Lambert Photography 5.5, 18.7, 18.9, 21.8, 21.9, /Andrew McClenaghan 7.13, /Astrid & Hanns-Frieder Michler 8.17, /Bruce Frisch 23.9, /CCI Archives 11.11, /Charles D. Winters 21.5, 24.8, /Chris R. Sharp 6.12, /Claude Nuridsany & Marie Perennou 3.36, /Cordelia Molloy 24.4, /Courtesy Of Crown Copyright FERA 13.22, / David Aubrey 7.8, /David Campione 21.3, /David Nunuk 12.1, /David Parker 6.8, 28.29, /David R. Frazier 1.13l, /Dept. of Physics, Imperial College 2.12, /Dr Jeremy Burgess 4.26, 14.13, /Dr Mark J. Winter 28.12, /Dr Patricia J. Shulz, Peter Arnold Inc. 24.17, /Edward Kinsman 5.1, / Geoff Tompkinson 1.11, 21.16, /George Steinmetz 16.5, /Giphotostock 20.17, /Harvey Pincis 18.4, /Health Protection Agency 4.39, /HR Bramaz, ISM 16.8, /Ian Boddy 26.2, /Jim Varney 8.6, /Kenneth Eward /Biografx 27.1, 28.15, /Malcolm Fielding, Johnson Matthey PLC 12.16, /Martyn F. Chillmaid 22.17, /Mauro Fermariello 12.3, /Maximilian Stock Ltd 22.1, / NASA 5.16, 10.3, /Oscar Burriel 18.11, /Patrice Latron /Look at Sciences 12.19, /Paul RaPSON 14.15, /Peticolas /Megna /Fundamental Photos 6.3, /Philippe Psaila 4.33, /Power And Syred 23.11, /Prof. K.Seddon & Dr T.Evans, Queen's University, Belfast 28.23, /Richard Megna / Fundamental Photos 18.12, /Ria Novosti 6.11, /Royal Institution of Great Britain 20.19, /Science Source 28.26, /Sidney Moulds 27.11, /Sheila Terry 3.32, /Takeshi Takahara 4.16, /TEK Image 4.17; Still Pictures/ sinopictures/dinodia 12.21; TCP 25.13; TopFoto 6.1, 9.5, /PA 4.18; University of California, Berkley 2.21; V&A Images 23.15; Waters 13.35; Wellcome Photo Library 4.21.

Picture Research for this edition by Sue Sharp.

Review questions credited to 'Cambridge' are reproduced by kind permission of the University of Cambridge Local Examinations Syndicate and are taken from past examination papers from the University of Cambridge International Examinations: General Certificate of Education Advanced Subsidiary Level and Advanced Level Chemistry (9701).